MATHEMATICAL AND COMPUTATIONAL METHODS IN BIOMECHANICS OF HUMAN SKELETAL SYSTEMS

Wiley Series on

Bioinformatics: Computational Techniques and Engineering

Bioinformatics and computational biology involve the comprehensive application of mathematics, statistics, science, and computer science to the understanding of living systems. Research and development in these areas require cooperation among specialists from the fields of biology, computer science, mathematics, statistics, physics, and related sciences. The objective of this book series is to provide timely treatments of the different aspects of bioinformatics spanning theory, new and established techniques, technologies and tools, and application domains. This series emphasizes algorithmic, mathematical, statistical, and computational methods that are central in bioinformatics and computational biology.

Series Editors: **Professor Yi Pan** and **Professor Albert Y. Zomaya**
pan@cs.gsu.edu zomaya@it.usyd.edu.au

Knowledge Discovery in Bioinformatics: Techniques, Methods, and Applications
Xiaohua Hu and Yi Pan

Grid Computing for Bioinformatics and Computational Biology
Edited by El-Ghazali Talbi and Albert Y. Zomaya

Bioinformatics Algorithms: Techniques and Applications
Ion Mandiou and Alexander Zelikovsky

Analysis of Biological Networks
Edited by Björn H. Junker and Falk Schreiber

Computational Intelligence and Pattern Analysis in Biological Informatics
Edited by Ujjwal Maulik, Sanghamitra Bandyopadhyay, and Jason T. L. Wang

Mathematics of Bioinformatics: Theory, Practice, and Applications
Matthew He and Sergey Petoukhov

Algorithms in Computational Molecular Biology: Techniques, Approaches and Applications
Edited by Mourad Elloumi and Albert Y. Zomaya

Mathematical and Computational Methods in Biomechanics of Human Skeletal Systems: An Introduction
Jiří Nedoma, Jiří Stehlík, Ivan Hlaváček, Josef Daněk, Taťjana Dostálová, and Petra Přečková

MATHEMATICAL AND COMPUTATIONAL METHODS IN BIOMECHANICS OF HUMAN SKELETAL SYSTEMS

An Introduction

JIŘÍ NEDOMA
Academy of Sciences of the Czech Republic

JIŘÍ STEHLÍK
The Hospital České Budějovice

IVAN HLAVÁČEK
Academy of Sciences of the Czech Republic

JOSEF DANĚK
University of West Bohemia

TATJANA DOSTÁLOVÁ
Charles University Prague

PETRA PŘEČKOVÁ
Academy of Sciences of the Czech Republic

WILEY

A JOHN WILEY & SONS, INC., PUBLICATION

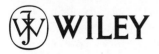

Published by John Wiley & Sons, Inc., Hoboken, New Jersey
Published simultaneously in Canada

Library of Congress Cataloging-in-Publication Data:

Mathematical and computational methods in biomechanics of human skeletal systems : an introduction / Jiří Nedoma . . . [et al.].
p. ; cm.
Includes bibliographical references and index.
ISBN 978-0-470-40824-7 (cloth)
1. Biomechanics. 2. Human mechanics—Mathematics. 3. Musculoskeletal system—Mathematical models. 4. Artificial joints—Mathematical models. I. Nedoma, Jiří.
[DNLM: 1. Biomechanics. 2. Arthroplasty, Replacement. 3. Joints–surgery. 4. Models, Theoretical. WE 103]
QP303.M346 2011
612′.014410151—dc22

2011012258

Printed in Singapore

oBook ISBN: 978-1-118-00647-4
ePDF ISBN: 978-1-118-00642-9
ePub ISBN: 978-1-118-00646-7

10 9 8 7 6 5 4 3 2 1

Dedicated to our families

CONTENTS

PREFACE

At present, alloarthoplastics of human body joints is a dominant method in orthopedics, and it could be considered as one of the most significant innovations in medicine in the twentieth century. Despite the successful spreading of the method and the achievement of good joint implantation results, it is essential to further optimize the shapes of implants enabling them to function under extreme long-term mechanical demands. Long-term success is becoming more important because more and more young patients (less than 50 years old) require artificial joints and, therefore, the requirement to extend the implant's life is fundamental. In addition to the obvious need for new materials and correct surgical techniques, the fundamental requirement is full knowledge of the biomechanical conditions of each joint and the maximum approximation to its physiological conditions.

It is well known that without a compliance with the correct position of implants toward the surrounding skeleton, even the best implants are damaged due to mechanical overloading. The required improvements cannot be achieved without detailed mathematical modeling and its verification by biomechanical experimental measurements under loading conditions, without reviewing the measured values in three areas: in the implant itself, in the implant–bone contact area, and in the surrounding bone. Only after a review of values gained from different models can a suitable implant be developed together with recommendations for its optimal position. At present, biomedical informatics is a field of great interest. From the global point of view, it also covers computational biomechanics as well as computer simulation methods in biomechanics.

Therefore, computer simulations, in close connection with information technology, can be used in clinical practice in orthopedics and also for controlled operational techniques, the so-called navigated surgery. Due to the multidisciplinary character of the problem, and its rigorous biomechanical, medical, technological, and mathematical characteristics, this book must take into account novel trends in all these branches as well as in computational science. With the advent of personal computers (PCs), graphical techniques, and information technology, their applications are accepted by healthcare professionals as well as by specialists in the industry.

Computer simulations are cognitive methods. Computer models are also mostly used to analyze dynamics of biological, physiological, and biomechanical systems.

At present, the term *biomechanics* includes a great number of different areas of the biological sciences. In this book, we focus on mathematical (numerical) modeling of the human skeletal system and the artificial replacements of its parts. The aim of the book is to give biomechanical and above all mathematical bases for numerical modeling of total human joint replacements and simulations of their functions, together with rigorous biomechanics of human joints and other skeletal parts, as well as a basis for mathematical simulations of navigated surgical techniques that lead to a construction of biomechanical and mathematical (numerical) models. While writing this book the authors had to take into account the fact that at present mathematical modeling and mathematical simulations play an important role in biological sciences, and they are essential for a better understanding of biological processes in the human body.

Therefore, the book is written from an interdisciplinary point of view, but above all with respect to the specialists working in the fields of mathematical modeling in biomechanics of the skeletal system, in the fields of bioengineering and medical biomechanics having a connection with human joint replacement and navigated surgical techniques as it is necessary to find common language between specialists working in different sciences.

The main goal of this book is to give solutions to specific current problems in orthopedics and to present up-to-date mathematical modeling and numerical analyses as well as to stimulate growth and development in the existing field of mathematical modeling in biomechanics. The book is divided into three parts. Part I, Anatomy, Biomechanics, and Alloarthroplasty of Human Joints, deals with a definition of biomechanics, with an introduction to the anatomy and biomechanics of the human skeleton, biomaterials, problems of alloarthroplasty, and with a discussion of a definition of geometry of the skeletal system and its parts by using magnetic resonance imaging (MRI) and computed tomography (CT) techniques. The definition of selected orthopedic problems to be simulated is also presented. This part includes Chapters 1–3.

Part II, Mathematical Models in Biomechanics, deals with the introduction of mathematical models to different problems in orthopedics, constructions of mathematical model problems of biomechanics of the human skeleton and its parts, as well as of replacements of different parts of the human skeleton and corresponding mathematical model problems. Foundations of continuum mechanics are also presented. Mathematical analyses of model problems investigate mathematical models in different rheologies and with numerical methods of their solutions. This part represents the main section of the book. This part includes Chapters 4, 5, and 6. Chapters 5 and 6 deal with detailed mathematical analyses of mathematical models based on the functional analysis and finite element methods. These models are based on several rheologies such as linear and nonlinear elasticity, thermoelasticity, thermoviscoelasticity, and thermoviscoplasticity. The human skeleton is characterized by the joint connections between two neighboring bones. Therefore, the phenomena of contacts between deformable bones occur in biomechanics. In the first approximation the bones can be assumed to be linearly elastic, and the processes can be assumed to be static with frictionless and/or with the given frictional contact conditions. However, at present, there is a need to consider unilateral contact models involving elastic, thermoelastic, thermoviscoelastic, and thermoviscoplastic biomaterials as well as a number of contact and frictional boundary conditions, like contact conditions with the Tresca and Coulombian or with viscoelastic and viscoplastic frictions, respectively, that can predict the evolution of the contact processes in different situations and different conditions in the human body more reliably. Mathematical problems of solidification and recrystallization and their analyses concern metallic alloys and ceramics. Chapter 6 deals with mathematical models, mathematical formulations, and variational and numerical analyses of the model problems investigated. The aim is to give complete solutions and analyses of the model problems.

Part III, Biomechanical Analyses of Particular Parts of the Human Skeleton, Joints, and Their Replacements, contains Chapter 7. In this chapter the theory presented in the previous chapters is applied to several orthopedic problems and numerical results are discussed. Determination of the geometry of the human skeleton and its parts is the crucial problem in mathematical modeling in biomechanics. Therefore, problems of data processing from nuclear magnetic resonance imaging and computed tomography are briefly discussed.

This book assumes that readers have a basic knowledge of mechanics and finite element techniques. Nevertheless, a short introduction to functional and convex analyses and the backgrounds of numerical solutions of variational equations and variational inequalities is also given. The aim of this book is to give orthopedic, biomechanical, and mathematical basics for the mathematical simulation of surgical techniques in orthopedics as well as of numerical modeling of total human joint replacements and of other parts of the human skeleton, and, moreover, of surgical methods in orthopedics by using results of methods of numerical modeling and of mathematical simulation.

This book is very timely and contains a wealth of information on current research in the broad subject of biomechanical and numerical concepts and their applications in orthopedics. Moreover, the book presents rigorous phenomenological biomechanics of human joints and spine processes that lead to the construction of biomechanical and mathematical (numerical) models, which can be stated as a boundary value, an initial boundary value, and contact problems in linear and nonlinear elastic, thermoelastic, thermoviscoelastic, and viscoplastic rheologies. The emphasis is placed on the variational approach to the investigated model problems, which involves numerical analyses while preserving the orthopedic nature of the investigated problems. A study of algorithmic procedures based on these simulation models is also presented. The theories presented in the book will be applied to specific problems of orthopedics, and numerical results will be presented and discussed from the biomechanical and orthopedic points of view. Treatment methods are also briefly discussed. The major objective of this book is to stimulate new multidisciplinary research and to give rigorous mathematical, biomechanical, and orthopedic analyses.

The book also aims to fill the gap in applied and computational biomechanics and to create the bridge between the specialists in orthopedics, biomechanics, and applied mathematics. Its aim is to give the solution of specific current problems in orthopedics and to present up-to-date mathematical modeling and numerical methods. New problems concerning globalization problems in medicine as the role of medical informatics, research diagnostic criteria, the international classification systems and the standardized clinical terminology, for simplicity, applied to the temporomandibular joint analyses, are briefly presented. It is evident that in all disciplines in orthopedics the methods of medical informatics are fundamental, but these problems are, to a greater extent, far beyond the scope of this book.

This book is intended for the designers and manufacturers of joint implants, who require the results of suggested experiments to improve existing shapes or to design new shapes. Chapters 2 and 7 on the anatomy and human skeleton biomechanics, modified for technicians' requirements, enable good-quality understanding between these particular fields. This monograph should serve orthopedists and traumatologists involved in implanting by providing information both for planning and performing the operation. The adherence to the basic procedure scheme with placing the component into the optimum position, considering also the anatomic situation of the whole limb, enables long-term trouble-free functioning of the implanted joint.

This book will have the positive benefit for students of graduate and postgraduate studies in medical, technical, and applied mathematical fields, who have to get familiar with kinetic apparatus and alloplastics. This book represents the result of a cooperation between the first two authors, who started to investigate and mathematically simulate functions of human joints and their replacements in the years 1986–1987, and later they started to cooperate with the future co-authors of this book. In the book we have used also the results of our joint collaboration with numerous colleagues and our students. The list of the references concerning the problems solved in the book is not fully exhaustive; it only contains the books and papers that were used for the solution of problems discussed in the book or that are closely connected with these subjects.

ACKNOWLEDGMENTS

The authors are most grateful to our institutes and corporations that facilitated our research on these problems and enabled us to work on this book. Our thanks are directed also to the EuroMISE Centre and, namely, to Prof. Jana Zvárová, who helped to publish some of our first results, which are also partly presented in this book, in the *Series of Biomedical Informatics*. We are honored and grateful for the contributions of everyone who has shared in the preparation of this book, namely Ms. Kateřina Jermanová, who translated the first three chapters of this book, Ms. Hana Bílková, who greatly helped with the technical preparation of the book including the great number of figures presented in the book, and Dr. Petra Přečková, who made the language correction of the book. We would like to thank all our students and co-workers who participated in some research problems, namely Drs. Z. Milka, A. Dusová, Z. Kestřánek, A. Baštová, R. Krejčí, and J. Dvořák, students of the first author, Dr. M. Lanzendorfer, and, moreover, to all who helped with the preparation of this book. Special thanks to Profs. Yi Pan and A.Y. Zomaya, the Editors of the Wiley Series on Bioinformatics: Computational Techniques and Engineering, for initiating this project; to the copyeditor Ernestine Franco; to Nalini Satish from MPS Ltd.; and Mr. Paul Petralia and Michael Christian as well as Kristen Parrish, Simone Taylor, Diana Gialo, and others from John Wiley & Sons for helping us to bring it to realization.

Jiří Nedoma
Jiří Stehlík

Prague
March, 2010

PART I

ANATOMY, BIOMECHANICS, AND ALLOARTHROPLASTY OF HUMAN JOINTS

CHAPTER 1

BIOMECHANICS OF THE HUMAN SKELETON AND THE PROBLEM OF ALLOARTHROPLASTY

1.1 INTRODUCTION TO HISTORY OF BIOMECHANICS AND ALLOARTHROPLASTY

Scientific branches were historically divided into five separate branches—physics, mechanics, chemistry, geology, and biology. Until the first half of the nineteenth century these branches were mutually isolated and independent. From the second half of the nineteenth century, and mainly during the twentieth century, these scientific branches started to cooperate. As a result new scientific branches, such as biophysics, biomechanics, biochemistry, and geophysics, have developed.

Modern physics creates theoretical background for our understanding of nature and its regularities. The principles and laws were found in physics, namely in classical mechanics, relativity, and quantum physics. Quantum mechanics facilitates an understanding of the structure of matter and the chemical coupling as well as further properties of mass. Statistical mechanics facilitates analyzing separate chemical reactions, therefore, facilitating the study of the fundamental basis of chemical reaction theory. Knowledge of the physical principles and regularities of the physics of nonliving mass was then applied to the physics of living mass, which creates a new branch—biomechanics. Understanding a living organism, its functions, its reproduction, as well as its further evolution on the higher qualitative level, yet realizing its dependence on previous evolutionary states, represents the assignment of biology. Biology as a scientific branch is divided into such fields as biophysics, biomechanics, and biochemistry.

Mechanics is the part of physics that is concerned with motion and deformation of bodies, which are or are not in mutual contact and which are loaded by external

Mathematical and Computational Methods in Biomechanics of Human Skeletal Systems: An Introduction, First Edition. Jiří Nedoma, Jiří Stehlík, Ivan Hlaváček, Josef Daněk, Taťjana Dostálová, and Petra Přečková.

and internal forces. Mechanics is a very old branch of physics, dating back to the time of Aristotle (384–322 BC) and Archimedes (287–212 BC) and later to the time of Galileo Galilei (1564–1642) and Isaac Newton (1642–1727). The word "mechanics" was first used by Galileo in his book *Discorsi e dimostrazioni matematiche intorno a due nuove scienze, attenenti alla mecanica & i movimenti locali* (1638), where he describes force, motion, and strength of materials. Galileo made the first fundamental analyses in dynamics and mechanical experiments, while Newton formulated the laws of motion and gravity. As the study of mechanics evolved, it was shown that with greater experimental experience in the field of elastic mass properties our conceptions changed, but the fundamental idea that mass is continuously spread and that this mass spreading is due to the force actions, including thermal loading, was changed. Then we spoke of the continuum and about a branch called continuum mechanics.

Classical continuum mechanics of nonliving nature studies mass properties by considering the thermodynamics of closed systems. This idea was based on the fact that these systems do not need any further energy or any change of the mass with their neighborhood. Therefore, a change of entropy with the neighborhood is always positive in the closed system. Owing to dissipative processes the degree of the system organization decreases continuously.

Systems having properties of regularization, reproduction, and possibility of conservation of information represent open systems, for example, systems having an ability to change mass as well as energy with the surroundings. Such systems have dynamic equilibrium and their invariability is ensured by means of a mass exchange. Such systems describe the biological systems. The living systems receive mass by means of meals. The most important assignment of the control system is the conservation of its energetic foundation with the aim to conserve a certain level and stability of the function of the biosystem, also under the changed conditions of an outer medium. It is evident that the law of conservation of the living mass is a part of other conservation laws, known from classical and relativistic mechanics. The processes of acquisition, cumulation, transfer, and employment of energy in biosystems ensure both the growth of a living mass and conservation of the structure and realization of the function of the biological system, which is realized in the cooperation with receiving, processing, protection, and employment of the information.

Biomechanics is a scientific branch that combines the field of applied (e.g., engineering) mechanics with the fields of biology and physiology and it is concerned with a human body. In biomechanics, the principles of mechanics are applied to the conception, design, development, and analysis of the equipment and systems in biology and medicine. One of the main goals of biomechanics is to study responses of living tissues on an external energetic function from the physiological point of view, where we assume that the living tissue is a composite material with controlled properties. A biological material is a strongly organized material with the ability of self-evolution, reproduction, and a possibility of conforming to the surroundings. The modern development of biomechanics started in the fifteenth century, based on the studies of Leonardo da Vinci (1452–1519), though the concepts of biomechanics were probably given in Greek (Aristotle 384–322 BC) and Chinese writings. Later the contributors to biomechanics were Galileo Galilei (1564–1642), René Descartes

(1596–1650), Giovanni Alfonso Borelli (1608–1679), Robert Boyle (1627–1691), Robert Hooke (1635–1703), Isaac Newton (1642–1727), Leonard Euler (1707–1783), Thomas Young (1773–1829), Jean Poiseuille (1797–1869), Hermann von Helmholtz (1821–1894), and the others. The development of the biomechanics as a separate branch has improved our understanding of the mechanics of human joints. In the last several decades this understanding has grown to include the total replacements of joints as well as implants, the mechanics of a blood flow, the mechanics of an airflow in the lungs, the mechanics of soft tissues, and the mechanics of the growth and a form of joints. Moreover, biomechanics has contributed to the development of medical diagnostic and treatment procedures as well as to the development of designing and manufacturing medical instruments and devices for the handicapped. It has also contributed to the development of sport and forensic medicine.

Materials of nonliving nature, because of their low organization, are only slightly accepted or not accepted by highly organized living systems because they cannot be quickly regenerated and renovated. Another disadvantage is the insufficient adaptability of contacts between living and nonliving systems. This contributes to the problems that artificial replacements for biological organs and their parts experience. One of the main aims of biomechanics is the detailed research of composite materials suitable for the development of artificial replacements for human organs and joints. To apply materials of nonliving nature for the development of artificial replacements for human organs the following fundamental criteria and properties must be satisfied:

1. Unconditional adaptability to the surrounding materials of the living systems
2. Sufficient range of elastic deformations with reasonable nonlinearity and with useful elastic modules
3. Useful orientation of deformable properties with regard to the type and direction of force actions
4. Useful nonreturned deformations rendering an adaptation possible without great time and space changes of properties and evoked damages, but rendering a stress relaxation and precluding microdamage origins possible
5. Limitation of a total deformation during the growth of stresses with reinsurance of an elastic behavior with high local solidity and with a minimal requirement of a further delivery of energy
6. High biocorrosion resistance
7. High solidity against cyclical loading with high initial damping
8. High quality of the surface design, ensuring biocompatibility and a decrease in the possibility of biocorrosion
9. Having an ability of certain regeneration in connection with a neighboring living mass–tissue, as a higher form of biocompatibility

At present a most vigorous development of biomechanics is associated with orthopedics because the most frequent use of surgical intervention is with patients with musculoskeletal problems. In orthopedics, results of biomechanics have become everyday clinical tools. Therefore, the most urgent problems in biomechanics are problems connected with static and dynamic loading of human joint systems and their artificial joint replacements. Then fundamental research has included not only surgery, prostheses, implantable materials, and artificial limbs but also cellular and molecular aspects of healing in relation to stress and strain and, moreover, tissue engineering of a cartilage, tendon, muscle, and bone. Thus, rheology of biological tissues, transfer of a synovial fluid between both parts of joint systems, mass transfer through membranes, microcirculation, and interfacial phenomena must be investigated. We see that biomechanics represents a strongly interdisciplinary branch.

The first replacement of a human joint was made by Carnochan in 1840 in New York. The first replaced human joint was the temporomandibular joint. This attempt was unsuccessful because the implanted material was a wood. The first successful attempts of the hip prosthesis were made at the turn of the twentieth century. A successful replacement was made by Jones. His attempt was based on a gold plate used as an insert into the hip joint. This replacement functioned for 21 years, that is, until the patient's death. In the twentieth century Smith-Petersen accomplished marked successes by putting a cap on the femur head. Next attempts are connected with Delbet, Hey-Growes, and Judet. Delbet and Hey-Groves were the first to implant an artificial replacement of the whole head of the femur (in the twentieth century). Later, Judet used a type of replacement that was further developed by Thompson, Zanoli, Townley, Movin, Güntz, Merle d'Aubigue, Lange, Neff Marine-Zuco, and Gosett. Moore implanted a new type of a whole-metal femoral component—an artificial head of the femur with the stem fixed in the medullar cavity. This type was modified by Thompson, Eichler, Lipmann, McKee, Reiley, and others. In 1951 Haboush introduced a bone cement for fixation of the stem into the marrow channel. This technique was also applied by McKee and Charnley. In 1950 Urist and in 1960 McBride implanted in addition to the stem also the acetabular cup. In this way they were able to do a total hip joint replacements. From this time on we speak about total hip replacements. The modern arthroplasty of the hip joint and then of the knee and other joint replacements started with the Charnley shape of the hip joint replacement and mainly with his "low-friction arthroplasty" (Charnley, 1979). After a transient failure in which Teflon was used like a material for an artificial acetabulum, Charnley established the ultra-high molecular polyethylene—UHMWPE—into the construction. His replacement was solved as a metal femoral component with the metal head, whose stem was fixed by a bone cement. The artificial acetabular cup made of UHMWPE was also fixed with bone cement, and a mutual motion between the head of the femur and the cup of the acetabulum was realized by pairing metal–polyethylene. This "low-friction arthroplasty" was a model for many other authors and it has been used in most up to the present time. A change in the shape of the femoral stem, made by Müller in the late 1960s, was important and it enabled a component implantation without the necessity to apply a large femur trochanter. Because of this all new types of artificial joints were introduced. Ceramics were the preferred material for the production of the femur head (pairing ceramics–polyethylene), although some replacements were made without contact with polyethylene (pairing metal–metal or ceramics–ceramics).

The problem of fixation of individual components in a bone without using bone cement was solved by developing cementless types of joints.

Subsequently primal and revised types of implants were developed. Both types of implants were available as cemented or cementless. At present, replacements for temporomandibular joints have been developed and are available for use.

1.2 BIOMECHANICS OF HUMAN JOINTS AND TISSUES

Biomechanics as a modern science combines the developments in engineering mechanics with developments in biology and physiology. Biomechanics is concerned with the human body; thus it is a natural science concerned with living systems mechanics. It is the study of mechanical movements, their function in the whole biosystem as well as its individual parts. Modeling biosystems and the simulation of their functions developed into a significant understanding of the construction and function of living mass (Valenta, 1985, 1993). In biomechanics, the principles of classical mechanics are applied to the conception, design, development, and an analysis of equipment and systems in medicine.

Modeling of biosystems can be divided into two categories of models. The first one introduces mathematical models that model structures, functions, and quantities of investigated biosystems. Into this group we can add mathematical-mechanical models, which model specific biomechanical systems and simulate their functions. Models of physical-mechanical properties, thermodynamic models, structural models, and function models also belong here. Theoretical biomechanics is concerned with these problems. Real experimental models, modeling of specified biomechanical actions, or biomechanical objects belong in the second category. These models' task is to verify theoretical possibilities concerning their structure and function or why a biomechanical object such as a unit or its parts can verify the rightness, exactness, and accuracy of a solution of abstract mathematical models, to investigate biomechanical problems such as stress states, deformation ability, and the like. Experimental biomechanics is concerned with this topic. For that reason when artificial joint replacements have to be reliable and function long term, their construction must follow the principles of biomechanics and biomechanical relationships in an appropriate joint sector of the movement system. The study of the forces that have an effect on the joint system, natural or artificial, concerns fields such as biostatics, biodynamics, biokinematics, biokinetics, and tribology. In biomechanics we are concerned with the study of external and internal forces, which are summarized and transferred by joints. As a consequence of external and internal forces, we study the distributions of deformation and stresses in the movement apparatus, their character and relationship in the course of movement through time.

Arrangements and shapes of human joints establish their kinematic and dynamic characteristics. A kinematic characteristic is given by a joint geometry, by a shape of contact areas, and by their cartilaginous surface. Ligaments satisfy a function of mechanical stops or leading and stabilizing elements. From analyses of movements of other living creatures it is evident that their movement organs are constructed on the principle of the lever system with an alternating movement. Shapes and forms of individual structural elements of joint connection are so various that it is not possible to construct a universal artificial replacement for whatever human joint to fully comprehend its real functional properties in a human organism. It is the main goal of orthopedic specialists and design engineers to approach this ideal state. In that case mathematical modeling and mathematical simulation of the function of a joint and of its optimal replacements can be also helpful.

CHAPTER 2

INTRODUCTION TO THE ANATOMY OF THE SKELETAL SYSTEM

2.1 ANATOMY OF THE SKELETAL SYSTEM

From a biomechanical point of view in the human locomotor apparatus we can speak about tissues, biological structures, and passive and active formations. **Passive elements** are tissues that transfer originated and acting forces. Simultaneously, they must resist acting forces and they must satisfy certain conditions of strength and elasticity. These tissues would orient acting forces in different ways, regulate them, and change their ordering. Bones and their parts, including joint surfaces covered by a hyaline cartilage and further ligaments, tendons, and fasciae, belong among passive elements. Generally, bones are nonhomogenous and anisotropic materials.

Among **active elements** we can include muscles, which are able to change the energy of biomechanical reactions into work and develop some power for achieving movement. Mechanical qualities of active and passive elements of a human locomotive organ have been changing during its life.

A bone tissue is a connective type of tissue whose solid composition enhances its supportive and protective roles. It consists of cells and an organic extracellular matrix of fibers and a ground substance produced by the cells. Moreover, bones contain a high content of inorganic materials in the form of mineral salts, which are combined intimately with the organic matrix. These inorganic components make bone tissues hard and rigid; organic components give a bone its flexibility and elasticity. These inorganic components consist of calcium and phosphate in the form of small crystals of $Ca_{10}(PO_4)_6(OH)_2$. Bone minerals are embedded in variously oriented fibers of the protein collagen, in the inorganic matrix. Water is fairly abundant in a living bone, $\sim$ up to 25% of its total weight. From a macroscopic point of view bones are divided into two types of osseous tissue—that is, **cortical** or **compact** and **cancellous** or **trabecular** or **spongy** bones. Cortical bones form the outer cover of

Mathematical and Computational Methods in Biomechanics of Human Skeletal Systems: An Introduction, First Edition. Jiří Nedoma, Jiří Stehlík, Ivan Hlaváček, Josef Daněk, Taťjana Dostálová, and Petra Přečková.

the bone and they have dense structures, while a cancellous bone within the cortical shell is composed of thin plates, or trabeculae, in loose mesh structures, where the interstices between the trabeculae are filled with a red marrow. The cancellous bone tissue is arranged in concentric lacunae-containing lamellae. Bone tissues behave like composite (bio)materials.

The role of the ligaments and joint capsules, which connect a bone with a bone, is to augment the mechanical stability of the joints, to guide joint motion, and to prevent excessive motion. Ligaments and joint capsules act as static restraints. The muscle–tendon units are formed by tendons and muscles and they act as dynamic restraints. Moreover, the tendons also enable an optimal distance between a muscle and a joint. Tendons and ligaments are dense connective tissues, that is, parallel-fibered collagenous tissues. The great mechanical stability of collagen gives the tendons and ligaments their characteristic strength and flexibility. Like other connective tissues tendons and ligaments also consist of relatively few cells, the so-called fibroblasts, and an abundant extracellular matrix.

The human body consists of three muscle types—the cardiac muscles (which compose the heart), smooth muscles (which line hollow internal organs), and skeletal (striated or voluntary) muscles (which are attached to the skeleton via tendons). Skeletal muscles create ~40–45% of the total weight of the human body. In the human body there can be found more than 430 skeletal muscles; they are in pairs on the right and left sides of the body. The most vigorous movements are produced by fewer than 80 pairs. The muscles provide strength and protection to the skeleton by distributing loads and absorbing shock. Moreover, they enable the bones to move at the joints. The skeletal muscles perform both static and dynamic work, which permits locomotion.

The human body has three types of joints—fibrous, cartilaginous, and synovial. The synovial joints are the main joints of the skeletal system as they allow large degrees of motions. The specific property of synovial joints, with which the articulating bone ends, are covered by a thin (1–6 mm), dense, translucent, white connective tissue layer, known as the articular cartilage. This cartilage is a highly specialized tissue that enables one to transform and transmit the highly loaded joint environment without any failure during a relatively long lifetime. Its cellular density is less than that of any other tissues. Its function is to distribute joint loads over a relatively wide area and to allow a relative movement of the opposing joint surfaces with minimal values of friction and wear. Its properties are incompressibility, immiscibility, and distinct phases—an interstitial fluid phase and a porous-permeable solid phase. The water contributes to its mechanical properties, so that the articular cartilage can be considered as a fluid-filled porous-permeable biphasic medium. Due to the articulation cartilage the coefficient of friction is very low. From the point of view of its nature the articular cartilage can be assumed as a bioviscoelastic material. Like bone, the articular cartilage is also anisotropic and its material properties depend on the direction of loading and it allows also a creep. For more on the compositions and properties of bones, muscles, cartilages, tendons, and ligaments see Fung (1981), Nordin and Frankel (2001), and Covin (2001).

We can summarize components of the locomotive apparatus from a biomechanical point of view into the following systems:

1. *System of Skeletal Muscles* Its main function is the production of active mechanical movements.
2. *System of Body Segments and Skeleton Elements* Its function is the passive transmission and facilitation of active forces onto the surroundings; it creates movable and carrying bases for muscle fixing, ligaments, and fasciae.
3. *System of Intermediate Elements* It connects body segments by ligaments, cartilage, bone, joints; it connects muscles with tendon segments; its main function is a mechanical connection between the first and second systems and among themselves.
4. *Informative System* It is created by receptive elements in muscles, joints, tendons, and proprioceptors; then ear, eye, dermatic detectors, the exteroceptors; visceroceptors and afferent neural tracks. Its function is mechanical reception and transfer of information.
5. *Innervation System* It is formed by motor neurons, myoneural junctions, and efferent neuron tracks, synaptic transfers, and the like; its function is activation of motor muscle units.
6. *Central Neural System* It is created by the spinal cord, inter- and intrasegmental connections, subcortical and cortical mobile subsystems; its function is a collection, selection, saving, and an information analysis, representation of mechanical properties of an environment, a reflective activity, determination, and the like.

The human skeleton bones are mutually connected. These connections can be continuous and movable. The movable connection is a joint. We can distinguish the following formations:

1. Articular contact surfaces.
2. Articular cartilages, which cover joint contact surfaces.
3. An articular capsule, anticular ligaments. The capsule is created by two layers—a surface external layer and a synovial internal layer secreting into an articular cavity a viscous synovial liquid, which nourishes a cartilage and increases articular surface adhesion.
4. An articular cavity, which is a slot space between small-sized contact surfaces and it is closed by an articular capsule and filled by a microscopic amount of the viscous liquid, the so-called synovia.

Mobility of the joints is a combination of movements around three axes:

1. Around a horizontal axis in the sagittal plane, the flexion and extension
2. Around a horizontal axis in the frontal plane, the adduction and abduction
3. Around a longitudinal axis of the bone, the rotation

With regard to the human skeleton construction and a way of human walking in the upright position, these are the most exercised loading joints of lower limbs, that is, the hip and knee joints as well as the ankle joint.

2.2 HUMAN JOINTS AND THEIR FUNCTIONS

The synovial joints are subjected to an enormous range of loading conditions. Moreover, because of the very good properties of the articular cartilage, the cartilage surface sustains little wear. The minimal wear of the cartilage, associated with loads indicates the high quality of cartilage materials as well as the possibility of lubrication based on the high quality of the synovial liquid. There are two types of lubrication. The first is boundary lubrication, which involves a single monolayer of lubricant molecules adsorbed on each bearing surface. The second is fluid-film lubrication in which a thin fluid-film layer provides surface-to-surface separation. The coefficient of friction has an extremely low value of 0.02 and, therefore, the low rates of wear as well.

Developing artificial joint replacements must be based on knowledge of basic functions of natural joints. We must try to simulate them mathematically because only comparing mathematical simulation of basic functions of natural joints and the results of simulation of natural joint functions replaced by total replacements allow us to decide which surgical method to use for joint implantation. Osteointegration of the artificial joint represents, similar to a natural joint, a balance system, where a shape corresponds to a function. Violation of this balance in the natural joint enables arthritic changes with an ensuing destruction of the bone joint structures. The artificial joint leads to infringement of this balance and to the wear and tear of individual components of the replacement joint and afterward to their mechanical breakdown and untimely loosening. Elimination of consequences of violation of this balance is a goal of the mathematical simulation of the function of natural and artificial joints.

Investigating human joints leads to studies from a kinematic, static, dynamic, and tribilogy points of view. Results of an analysis of articular kinematics of the locomotive apparatus lead to the explanation of relative movements in the joint, not only from a quantitative but also from a qualitative point of view. Kinematic articular connection characteristics, for example, degrees of freedom and transference ratio, are determined by contiguous shapes of articular surfaces, a bone epiphysis shape and a cartilage surface shape. Roentgen kinematography or stroboscopic photography are technical devices that help us to obtain necessary information. A reliable human joint function is influenced not only by its skeleton shape and its cartilage covering but also by the effectiveness of different muscle groups, tissues, and viscous elastic characteristics of the synovial liquid. Individual articular connections are strained during ordinary limb movements in different ways. Basically, we can characterize forces affecting an articular connection of the upper limb, such as a tension force, and a force affecting the upper limb joints, such as a compressive force.

During a tension force, articular contact surfaces are pulled away from each other, and the effect of operating force is absorbed by a tissue apparatus of an individual joint. In this case a joint functions as a movable connection, allowing relative movements of skeleton-connected parts.

By a pressure character force contact surfaces are pressed against one another; thus, the relative movements in connected skeleton parts occur mainly in conditions of a close contact with sliding articular surfaces under pressure. In this case the joint represents a flexible pressure connection. These joints are strained much more than upper limb joints, and that is also the reason why degenerative changes happen more often. A pressure force value, acting, for example, on the femur head, can be determined by a force parallelogram, that is, from a body weight and a resultant of muscle forces retaining the balance. During a static situation, forces acting in the hip joint, the knee joint, and the ankle joint become an anatomical structure whose influence is bigger than the body's weight.

Dynamics in the sense of a human movement puts high demands on the locomotive apparatus, then specially on joints and their contact surfaces. During walking, lower limb joints are subjected to dynamic forces, which with regard to their long-term cyclic operation, leads often to the destruction of contact cartilage surfaces.

The hip, knee, and ankle joints have contact surfaces adapted in such a way as to absorb dynamical forces corresponding to the human weight. For example, the diameter of the adult human femur head is 38–56 mm. When walking, the pelvic acetabulum is in contact with the femur head spreading on an area of about 80% of a half-circle, which is the same as the head diameter. Thanks to the elasticity of the cartilage covering the articular surfaces, this contact surface, which takes in the hip joint a value of $\sim$430–920 mm^2, increases during overloading a value depending on the cartilage size, its elasticity, and an operating pressure force. For example, in the knee joint case it was shown that this contact surface enlargement can be up to 50%, during flexion the contact surface size is decreasing proportionately with the knee flexion increasing. The knee joint adapts to the strain to which it is exposed most of the time.

In consequence of straining lower limb joints by considerable forces and pressure tension concentration on a relatively small area, lubrication has a substantial significance for a right joint function; in human joints this function is represented by a synovial liquid. The synovial liquid is secreted into the articular cavity by an internal layer of the articular capsule. It gives elasticity to cartilage coverings of articular surfaces and, because of its viscoelastic properties, it is able to absorb a certain value of pressures. Human joints are constituted by components from living tissues, which have a regeneration ability. It means that microcracks created as a consequence of overloading do not lead by a consequence of tissue regeneration straight to the joint destruction.

According to shapes of articular contact surfaces, human anatomy displays the following types of joints:

1. Spherically shaped (ball-and-socket shaped)—contact surfaces are created from parts of spherical areas.

2. Bounded shaped (walnut shaped, enarthrosis)—articular surfaces are larger than half of the spherical surfaces or of rotational ellipsoidal surfaces, for example, the hip joint.
3. Free shaped (arthrodis)—articular surfaces are smaller than half of spherical surfaces, for example, the shoulder joint.
4. Cylindrically shaped—articulatio radiohumeralis.
5. Trochlear shaped (gynglimus)—articular surfaces have a leading edge and groove.
6. Pivot shaped (wheel shaped)—a movement around an axis parallel with a longitudinal bone axis.
7. Egg shaped—contact surfaces are parts of an ellipsoid.
8. Saddle shaped—contact surfaces have a horse saddle shape.
9. Flat shaped—contact surfaces are of a plane type, for example, joints of the cervical spine.
10. Amphiarthrosis—irregular contact surfaces; they have minimal movements.
11. Combinated shaped—an anatomically separated joint, it is functionally connected with another joint and movements act simultaneously.

2.3 TRIBOLOGY OF HUMAN JOINTS

Tribology is a theory concerned with the friction mechanism and wear of rigid bodies. **Friction** is a term expressing the resistance against a movement that originates between two bodies being in mutual contacts in the area of their surface contact in a transverse direction toward them. Between bodies, which are in contact, a certain medium can be present—the so-called frictional interlayer, which can be liquid, solid, or gaseous. **Wear** is an unfavorable change of a surface or surfaces of rigid bodies, which are in mutual close contact, caused by a mutual activity of functional surfaces or of a functional surface and a medium, respectively, that evokes the wear. Wear is not a material property but a property of a body system, including an interlayer. Between the friction and the wear there is no straight connection. Friction in the Coulombian sense depends on the absolute value of a normal component of an acting force, with an increasing normal force the friction and wear increase, too.

We speak about friction and sliding pairs. Friction pairs are two bodies in mutual contact, inside which friction appears like a demanding effect (e.g., brakes), sliding pairs are two bodies in mutual contact, inside which friction is an unfavorable effect (e.g., arthroplasty). From tribology of human joints the viscoelastic properties of the synovial liquid are sufficiently known. It seems that according to experimental results of many authors (Walker, Erkman, Weightman, Duff-Barclay, and Spillman), the synovial liquid after an artificial joint application is secreted into the joint in the same chemical constitution as in the natural joint.

A mathematical simulation of the joint function based on a mathematical theory of contact problems of the Signorini type allows us to study stress and force conditions on the contact surfaces, magnitude of deformation of contact surfaces as well as the transfer of loading forces, transferred from the acetabulum onto the head of the femur. A comparison of the results of a mathematical simulation of natural and artificial joints allows one to judge whether the function of the hip joint arthroplasty will be able to satisfy fully functional demands, and whether it can provide determination about what kind of clearance in the joint can be optimal for the full functional ability of the total hip arthroplasty (THA).

2.4 BIOMECHANICS OF THE SKELETAL SYSTEM

Generalized continuum mechanics of nonliving nature is used when studing mass properties of the thermodynamics of closed systems, and these systems do not need further energy or metabolism from their surroundings to maintain a balanced state. That is why entropy change in the surroundings of a closed system has been always positive. As a result of a dissipative process the degree of system organization decreases all the time. Systems that have a regulation property, a reproduction property, and possibilities to keep the information are open systems, that is, systems, that are able to change the mass and energy with the surroundings. For systems with a dynamic balance constancy is supplied by metabolism. These systems describe biological systems. Living systems increase their mass by an intake of energy, which enables them to make new structures, that is, to eliminate and correct mistakes in their organization structure. Thus biosystems are autoregulating and directing systems at the same time. The most important assignment of the control system is to conserve its energetic background with the aim to conserve the level and constancy of biosystem functioning, mainly by changing the conditions of the exterior surroundings. It is evident that the law of conservation of living mass is used here as a part of other conservation laws known from classical and relativistic mechanics. Processes of obtaining, accumulating, transforming, and fully utilizing the energy in biosystems ensure that a living mass grows as a conservation of the structure and realization of a biological systems function as well. What is realized in cooperation is the receiving, elaborating, protection, and a full use of the information. Informative control mechanisms influence the quality energetic processes of biosystems and the speed of all structural and functional changes.

Biosystems are characterized by (i) a high level of complexity, characterized by control systems that provide an information exchange, their processing for a need to provide the function and structure, (ii) a collective production principle, (iii) mechanisms of selection, (iv) significant stochasticity of structures, providing biosystem evolution, and (v) stability and lability, where high stability provides vitality and liveness and lability provides further evolution.

Biosystems are studied by biological, biochemical, biophysical, biomechanical, and medical sciences as well as combinations of these fields because living tissues must be studied as the coupled problems. Hereafter we will concentrate on biomechanical problems of living tissues only.

Biomechanics is based on applications of mechanical and biological laws to biological and medical sciences. One of the main goals of biomechanics is to study living tissue responses to exterior energetic actions from a physiological point of view, whereas we suppose that the living tissue is basically a composite material with controlling properties. A nonliving natural material is relatively little organized (e.g., a structure and a construction of crystals), and it is not able to do self-evolution—sense and function is added by humans only. On the contrary, biological materials are highly organized with a self-evolution possibility, reproduction, and adaptation possibility. The system function then provides its existence. Nonliving materials, thanks to their low organization, are very neither accepted nor nonaccepted by highly organized living systems because it is not possible to regenerate and recover them sufficiently fast. Their disadvantage is also their insufficient adaptation to the dividing line between living and nonliving systems. This is why there are problems with artificial replacements of biological organs and their parts. One goal of biomechanics is the detailed research of composite materials suitable for use in the development of artificial replacements of human organs. In order to use these materials in artificial human organs, they must have certain fundamental properties:

1. Unconditional adaptation to the surrounding materials of the living systems
2. Sufficient range of elastic deformation with a reasonable nonlinearity and a needed elastic modulus
3. Suitable orientation of deforming properties with regard to a type and direction of a force operation
4. Suitable irreversible deformations providing adaptation without greater time and space property changes and evoked damaging, which allow relaxation of stresses and prevention of microfailure formations
5. Limitation of deformation because of increasing stresses while providing an elastic behavior of the biomaterials with high strength and a minimum need of more energy delivery
6. High immunity against biocorrosion
7. High strength against cyclic loading with high beginning damping
8. High-quality surface treatment that will provide biocompatibility and not allow biocorrosion
9. Ability of certain regeneration in relation to surrounding living mass—a tissue—as a higher level of biocompatibility.

An increase of further biochemical knowledge will bring new possibilities of tissue and organ replacements in connection with new methods of living tissue cloning. For that reason future research is going to be focused on mechanical properties research of the muscle–skeletal system based on macro- and microscopic, that is, cellular, living tissue structures. This research will be focused not only on an experimental study of living tissue structure but also on mathematical modeling and mathematical simulation of processes in living tissues and organisms. Presently, the most urgent and most accessible seems to be a study of biomechanical and biodynamical processes of the locomotive human system and its relation to surrounding tissues and its influences on the human locomotive apparatus. This research will be the focus of research in fields such as bioengineering, medical engineering, and in the medicine disciplines and applied mathematics. In biomechanics the main focus in medicine will be the skeletal and muscular apparatus with consideration of the analysis and synthesis of stiffness and dynamics of the human skeleton together with considering the influences of applied muscle forces, rheological properties of synovial liquids, research of muscle utilitization and temperature regulation as well as their influences on human skeleton pathological appearances such as osteophytes and other pathological deformations of the human skeleton and its parts. Another task of biomechanics is its utilization in the elimination of congenital and obtained defects of the locomotive apparatus.

2.4.1 The Hip Joint

The hip joint (see Fig. 2.1) has a unique importance in the human body. It is one of the largest and most stable joints. As a consequence of bipedal walking, from the mechanical point of view, it becomes the most exposed place, where the carrying free limb is connected with the solid pelvic girdle on which the spine system is connected.

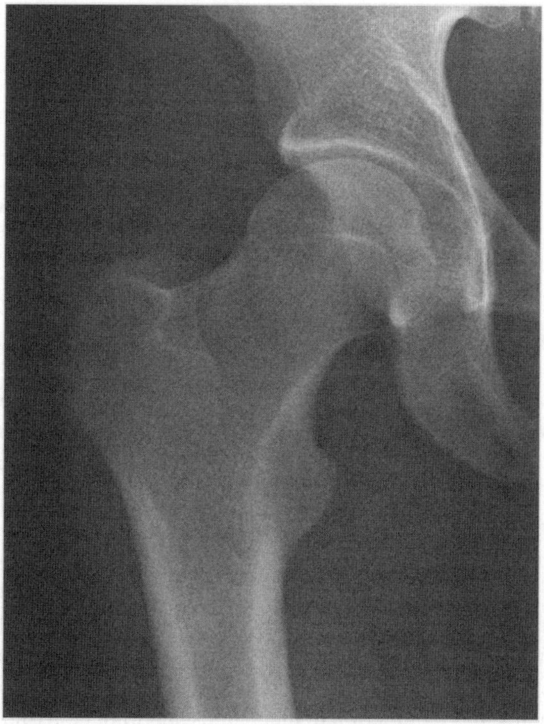

FIG. 2.1 Hip joint (X ray).

A right bipedal system is observed only in humans and in birds. The difference is in the location of a central point. Birds (and in the past dinosaurs too) have the central point bellow the hip joint level, so the bird body resembles a hanged pendulum. This system is very stable. It demands a minimal force for keeping an erect position, which was very important for dinosaurs with their enormous weight. Humans are a different example. Their central point is over the hip joint in the area of the Th 10–11 disks. Therefore, to keep the erect position the muscle apparatus keeps the balance still functional. The hip joint has an intrinsic stability provided by its relatively rigid ball-and-socket configuration. It has also a great deal of mobility, which allows normal locomotion in the performance of daily activities. For this posture the whole human skeleton is adapted, but especially the skeleton of the pelvis, which is flat with an open pelvis muscle apparatus. By this the origin of abductors shifted laterally, far from the center of hip joint rotation, which causes extension and enables these muscles to act as levers. Loading relations during standing and moving are very complicated. The hip joint has four degrees of freedom, which are rotation movements along the x, y, z axes and one displacement movement in the direction of the y axis (this movement occurs only during luxation). Thus, from the mechanical point of view, 6 muscles would be enough to move the hip. In the hip joint area there are 20 different muscles, whose combinative operation causes different phases of individual movements of the joint itself.

The hip joint is made up of the acetabulum of the pelvis and the head of the femur. This articulation has a loose joint capsule surrounded by large and strong muscles. The acetabulum is a concave component of a ball-and-socket configuration of the hip joint. The acetabular surface is covered with the articular cartilage. The cavity of the acetabulum faces obliquely forward, outward, and downward. Osseous acetabulum in the hip joint is deep and provides a substantial static stability to the hip (Nordin and Frankel, 2001). The femoral head is the convex component of the ball-and-socket configuration of the hip joint. The hip joint is created by approximately two-thirds of a sphere. The articular cartilage covers the femoral head; the variations in its thickness result in a different strength and stiffness in various regions of the femoral head.

Moreover, the hip joint is basically a spherically shaped joint with bounded dynamic mobility, with the shape of a rotational ellipsoid, whose lengths of axes are not that different. Joint cartilage covering the femur head and the acetabulum is very flexible hyaline cartilage, which because of its deformations allows movements corresponding to the ball-shaped joint, which is its main function from a biomechanical point of view. The femoral head is covered with hyaline cartilage from 2.2 to 3.7 mm thickness. Hyaline cartilage of the acetabulum covers the facies lunata acetabuli, which has a horseshoe shape and is of about 0.5–0.9 mm thick and places the horseshoe of ~0.8–3.0 mm thick approximately in the convex part of the horseshoe and on the periphery of the facies lunatum acetabuli. A part of the facies lunatum acetabuli is created from cartilage, that is, the fossa acetabuli, which is filled with pulvinar acetabuli with the insertion of ligamentum capitis femoris. The synovial membrane covering the pulvinar acetabuli, whose background is a ligamentous stroma with an amount of fat cellular elements, has its function in mechanics and joint nourishment (an analogy with Hoff's knee joint corpus). The hip joint has a unique status in the body because it connects a limb carrying the whole body weight with a relatively nonmovable pelvic round. It is substantially different, for example, from the shoulder joint, inside which the humerus is connected with a trunk by an interlink—scapula and the heterogeneous muscles. Moreover, the proximal limb does not have any carrying function. The hip joint skeleton architecture is therefore determined by the combination of statical loading and a dynamical component determined by the muscle traction and a tonus ligament. Both of the components get the same denominator factor in a period, when the hip joint begins to realize the biomechanical requirements that walking imposes. The joint capsule of the hip joint is very thick and together with ligaments, which strengthen it, creates an anatomical and functional unit. The joint capsule is strengthened on the front side by the strongest ligament in the body, the so-called ligamentum iliofemorale, which is stretched during standing with the hyperextended hip joint as the arresting mechanism. In order to reach an elastic bounded extension of the hip joint its normal structure is necessary. Two other ligaments, which strengthen the articular capsula, are the ligamenta ischiocapsulare and pubocapsulare, circulary rounded by sheafs connecting all the ligaments and unifying them functionally (zone orbicularis).

From the value of the compression force operating on the head of the femur due to classical biomechanics it is possible to obtain the force paralelogram, that is, from body weight and the resultant muscle forces that maintain balance. Movement of this muscle force comes from a great trochanter, where all the mentioned abductors are fixed, in the direction of the upper pelvic edge. The muscle moment is determined by many factors, for example, the length of the neck of the femur, the size of the trochanter, the size of the angle that contains the neck with the femur axis, and by the femur location against the pelvis. The total limb length of the lower limb has a certain influence on the absolute value of this muscle moment.

To determine the value of pressure force functioning on the femur head we will assume that individual skeleton segments are perfectly solid and mutually connected bodies. Let point C be the center of the femoral head (Fig. 2.5). Then from a moment balance condition to this point we determine a reactive force F_s as a resultant of muscle forces keeping the balance. We have

$$F_s = G'\frac{a}{b} \quad [\mathrm{N}] \tag{2.1}$$

where $G' = \frac{5}{6}G$ is the weight of the body G lowered by the weight of the supporting lower limb, a is an arm of the acting weight force G with respect to the point C, and b is an arm of the acting force F_s with respect to the point C.

For a rate of the moment arm from the radiographs we find $2 < a/b < 3.5$. Hence and from (2.1) it follows that $1.7G < F_s < 2.9G$.

The force of the abductors, F_s, acts on the angle $\beta = 22°$. For $G = 800$ N components F_{sv} and F_{sh} are equal to $F_{sv} = F_s \cos\beta = 1237$–$2110$ N, $F_{sh} = F_s \sin\beta = 500$–$853$ N (see Fig. 2.3). From the balance of forces acting on the pelvis it follows that $P = [F_{sh}^2 + (F_{sv} + G')^2]^{1/2} = 1956$–$2893$ N, $\mathrm{tg}\,\alpha = F_{sh}/(F_{sv} + G')$, then $\alpha = 15$–$17°$ (see Fig. 2.3).

It means that when standing on one leg a force acts on the femur head that is 3.7 times greater than the weight of the human body. This force acts on the femur head in a direction diverted medially from 16° from the vertical. The resultant muscle forces F_s and the resulting pressure force P in this static situation act in one plane. This plane goes in a direction of an acting pressure force of the human body weight and the center of rotation of the femur head (Beznoska et al., 1987). During walking this tension plane is turned probably to a small angle against the frontal plane going through both centers of femur heads, so that the mathematical factor for a muscle force resultant will be $\approx$0.94. Probably the direction of the resulting pressure force P acting in the hip joint in the relation with the vertical is medially diverted in different phases of gait in the range of 2°–20°. The pressure force resultant acting on the femur head is possible with an advantage and with a better exactness to get an analysis of the biomechanical model problem based on the contact theory, which are going to be the main parts of this study concerning the hip joint. As a result we will not get only the value of the resulting force at the point C but also the distribution of the normal and tangential components of stresses on a contact surface between the femur and the acetabulum and also the distribution of the stress and strain fields in the whole articular system. Figure 2.2 shows the schematic cross section of the hip joint. Figures 2.3 and 2.4 present the hip joint models according to Pauwels and Nedoma. The Nedoma model is based on the contact theory in linear (visco-)elasticity and thermoelasticity, and it will be analyzed in detail from the mathematical point of view in Chapter 6.

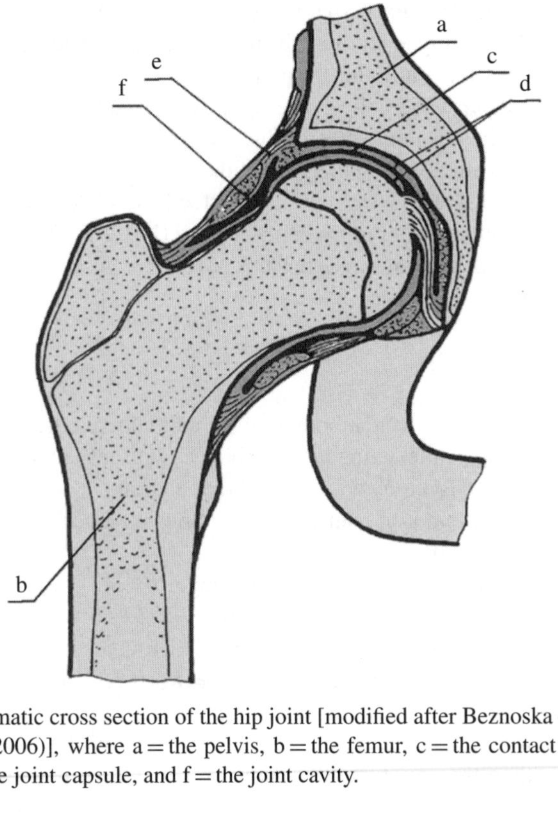

FIG. 2.2 Schematic cross section of the hip joint [modified after Beznoska et al. (1987) and Nedoma et al. (2006)], where a = the pelvis, b = the femur, c = the contact surface, d = the cartilages, e = the joint capsule, and f = the joint cavity.

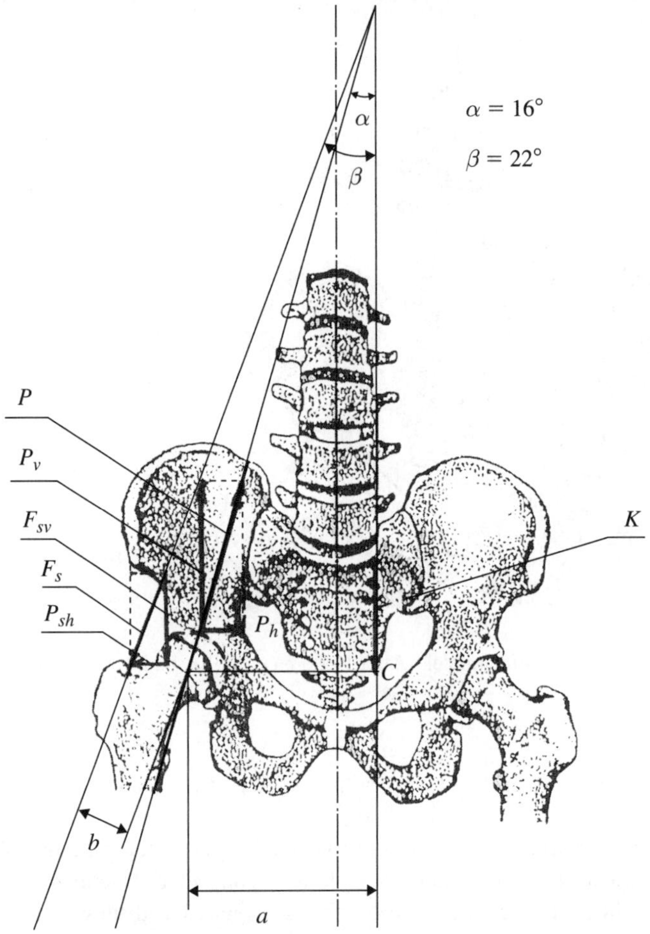

FIG. 2.3 The Pauwels model. [Modified after Bartoš (1998) and Nedoma et al. (2006).]

The hip joint is stabilized by the muscle system, which is important for biomechanical relationships in a joint, because it affects force relationships in a limb and its movements. Movements in the hip joint are controlled by muscles, which do not lay directly on the hip joint. Where muscles grip and the direction of their operation are very important from a biomechanical point of views. They are muscles, which provide flexion, extension, abduction, adduction, and internal and external rotation.

The distribution of muscles and their positions surrounding the hip and the forces they are developing will be necessary to derive the boundary conditions at the simulation of the hip joint function and tension and deformation relations inside the articular system.

For the hip joint arthroplasty following morphological data are important: (i) a colodiaphysaire (CD) angle about 126° with a range of 115°–140°. This angle changes

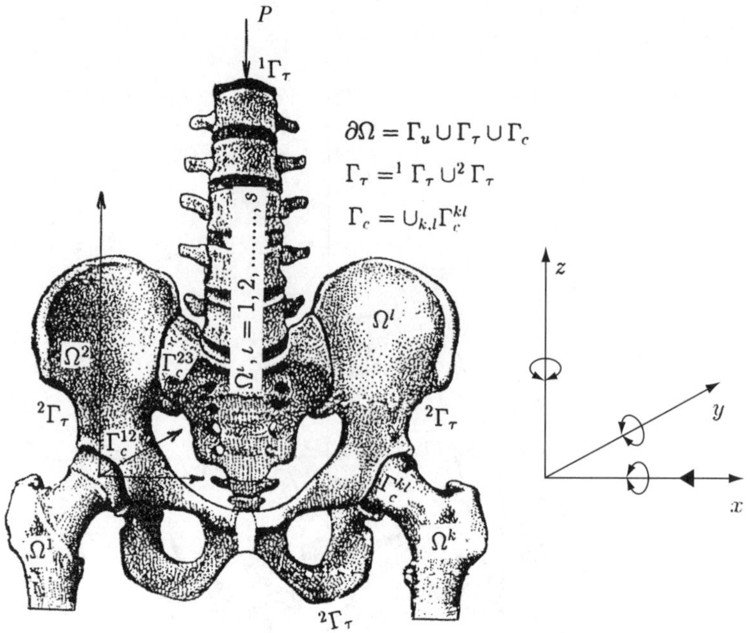

FIG. 2.4 The Nedoma model. [After Nedoma (1993b) and Nedoma et al. (2006).]

with age. (ii) An angle neck anteversion is quoted as having an average of 12° with a range of 4°–20°.

Fundamental morphological constitution of the upper femur end establishes biomechanical characteristics on which musculatures and other structures also establish. The bone architecture depends also on changes related with flows of forces in the neighborhood of the upper end of the femur. Two basic trajectory systems—tension and pressure, as it follows from experimental and theoretical biomechanics—are a morphological substrate of tensile and compression forces during loading of the end of the upper femur. Biomechanically less active, less exposed places do not have such a structure, and they are labeled like the so-called Ward triangle, seen on the frontal cut of the end of the upper femur. These anatomical experiences and basic knowledge are necessary to the mathematical simulation function of the hip joint and during the mathematical simulation as well as design of total replacement, but also during its construction and own implantation of an artificial joint (THA).

Many models try to explain the biomechanics of the hip joint. The basic ordinary model is the static Fischer hip joint model; see Fig 2.5 (Pauwels, 1973; Bombelli, 1983; Bartoš, 1998). The center of gravity is situated in the middle sagittal plane at the Th 10–11 disk area. Both joints carry the weight of the body, which is distributed between both limbs and the median halves vertically cut the line connecting the centers of both hip joints. The head is in contact with the acetabular surface and we refer to it as the load zone. The resultant of post-operating forces act on the head of the hip joint in a vertical direction. The Pauwels model (see Fig. 2.3) developed from the previous

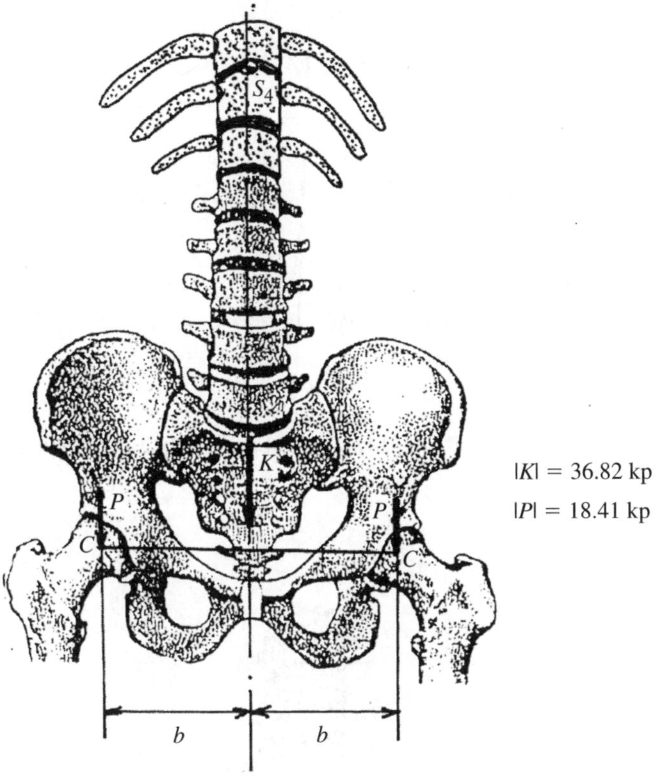

FIG. 2.5 The Fischer model. [Modified after Bartoš (1998) and Nedoma et al. (2006).]

model of the so-called sixteenth period of the gait, where the gravity center S is situated in the vertical frontal plane led by the joint center O. Biomechanical relations in a joint correspond to a double-reversible lever with a center of rotation O in the center of the joint of the carrying limb, where on a longer arm of a lever system OC the resulting force corresponds to the body weight without a carrying limb, and this moment is eliminated by a force of muscles acting on a lever arm, corresponding to a distance from a joint center to where muscles are inserted in the great trochanter. According to Pauwels, the direction of adductoral muscle movement is of about 22°, the resulting force acting on the hip joint during normal relations corresponds to 2.5 times the weight, and it is deviated laterally at 16° from the vertical. The Pauwels model does not correspond to the requirement of statics to conserve balance, the median does not come through a supporting plane of the limb, and the model of the stand is unstable. In spite of this the model is often used (Huggler and Schreiber, 1978; Rodin, 1980; Bombelli, 1983).

Others models were developed by Inman, Denham, and also by Rydell (1966), Bombelli (1983), [see Fig. 2.6, modified by Bartoš (1998) and Nedoma et al. (2006)], Debrunner (1975) [see Fig. 2.7, modified by Bartoš (1998) and Nedoma

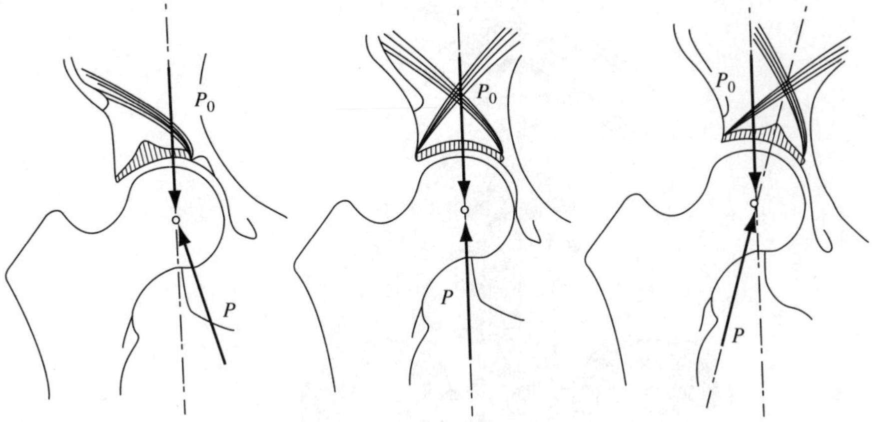

FIG. 2.6 The Bombelli model. [Modified after Bombelli (1983), Bartoš (1998), and Nedoma et al. (2006).]

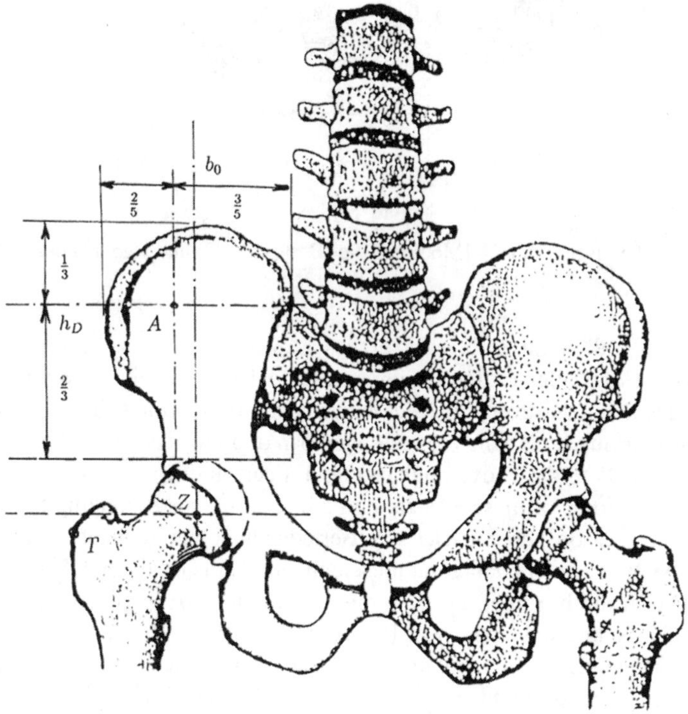

FIG. 2.7 The Debrunner model. [Modified after Bartoš (1998) and Nedoma et al. (2006).]

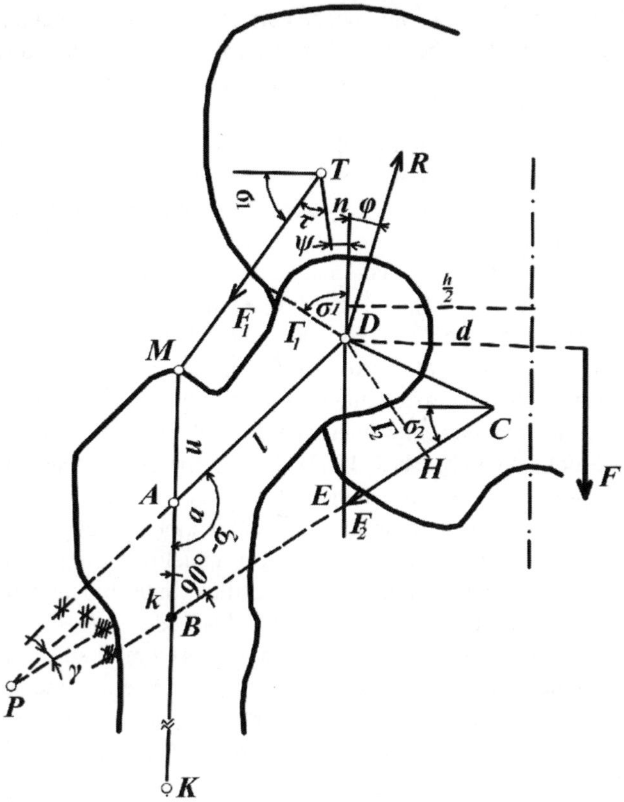

FIG. 2.8 The Karas model. [Modified after Bartoš (1998) and Nedoma et al. (2006).]

et al. (2006)], Jensen (1978), and Karas et al. (1985, 1986) [see Fig. 2.8, modified by Bartoš (1998) and Nedoma et al. (2006)]. Jensen (1978) studied the biomechanical problems of the hip joint with the help of photoelasticimetry. He created a plane model of a proximal femur affected by different loading forces. Photoelasticimetry results of osseous trabeculae of the neck were studied. From the results Jensen derived the direction of the resulting forces acting on the hip joint and the direction acting on the abductors. Bombelli (1983) showed that the forces acting on the hip joint were vertical, going through the center of the loading surface, the so-called weight-bearing surface (WBS). Under normal anatomical relations this force goes through the top of the Gothic arch. During the inclination of the loading plane from a horizontal line, damage to the biomechanical balance is observed. Debrunner's model is based on the Pauwels principles, but in addition it also considers the influence of the geometrical relations of the pelvis and an area of the hip on the loading of this joint. Bombelli points out the relationship between the beginning of the abductors and the insertion of the great trochanter into the joint center (see Fig. 2.7). His model was modified by Karas et al. (1985, 1986), who in addition to body weight considers also the dynamical overloading caused by inertial body effects during the gait (see Fig. 2.8). Despite a complicated model, it does not comprehend correctly the loading relations in the hip area because it does not describe tensile forces of individual muscles correctly. These forces are reduced to two groups of abductors and adductors (Karas et al., 1985, 1986).

With the development of computer technology mathematical modeling of individual parts of the skeleton and biological processes ongoing in the human body and mathematical simulation of functions of these parts of the skeleton start to be developed. The first period is characterized by a study of the geometrical character of the problems (Johnson and Larson, 1964; Johnson, 1979; Johnson et al., 1986). In 1973 Brekelmans applied the finite element method on biomechanical problems to orthopedics (Brekelmans et al., 1973), and later this technique was developed by Huiskes and his collaborators (Huiskes, 1987; Kuiper, 1993; Weinaus, 1991) who compared some surgical achievements in the proximal femur area (Huiskes and Chao, 1983). The improvement of the process facilitates also a study of acetabular articular parts and compares reconstruction achievements on the skeleton against own implantation (Huiskes and Sloof, 1986; Huiskes, 1987). During the years 1986–1989 Nedoma formulated the problem of the distribution of stresses in the hip joint based on the theory of contact problems in linear elasticity and thermoelasticity, and on this theory he presented the fundamental biomechanical model of the hip joint (Fig. 2.4). This model allows one to study biomechanical relations in the hip joint as a system of viscoelastic bodies, which are in mutual contact. If this model is investigated as a three-dimensional model, then it describes the hip joint and its biomechanical behavior very exactly. The exactness of this model depends on the exactness of the determination of the geometry of the investigated skeleton part and a location and a size of acting muscular forces. Contemporary computed tomography (CT) and magnetic resonance imaging (MRI) techniques allow one to obtain a real geometry of the investigated part of the skeleton with sufficiently high accuracy. The fundamental theory of static and dynamic loading of the hip joint was elaborated in Nedoma (1983, 1987, 1991, 1993b, 1994b, 1997a), Nedoma and Stehlík (1995), Nedoma et al. (1999a), Nedoma and Hlaváček (2002), and Stehlík and Nedoma (1989). These models will be discussed in this book. In cases that we do not know the initial data with sufficient accuracy, then it is possible to use the worst-scenario method with an advantage (Hlaváček, 2003, Hlaváček and Nedoma, 2002b, 2004; Hlaváček et al. 2004; Nedoma et al. 2006).

2.4.2 The Knee Joint

The knee joint is created by the contact of three bones: femur, tibia, and patella. The joint contact surfaces on the femur condyles are egg shaped, whereas the medial condyle is more convex than the lateral one. On the front side between the femur condyles there is a groove-shaped area leading to the patella. The joint contact surfaces on the tibia condyles are slightly concave and do not correspond to curved surfaces of the femur condyles. This lack of harmony is balanced by ligament cartilages—menisci that are situated medially and laterally. The knee joint is basically a combination of the troclear-shaped and wheel-shaped connections [Fig. 2.9 (a) and (b)].

In comparison with the hip joint the mobility of the knee joint is much more complicated. Nonconcentric curvatures of the femur condyles and shallow hollows

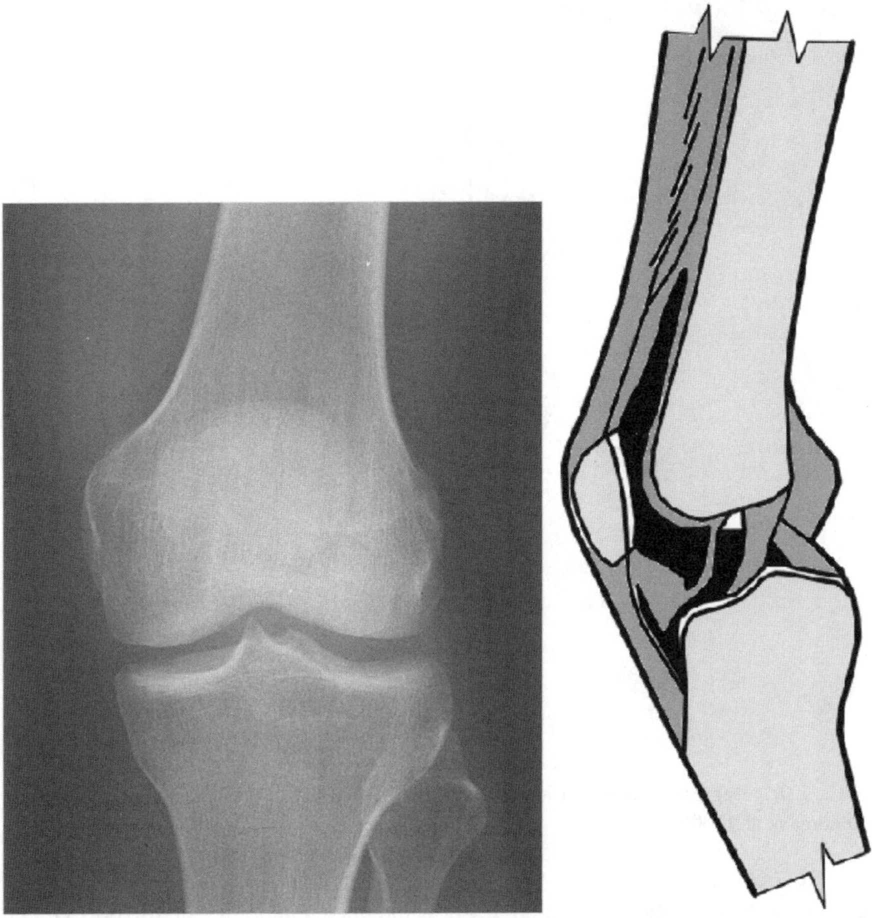

(a) (b)

FIG. 2.9 Knee joint: (a) the X-ray image and (b) the schematic design. [Modified after Beznoska et al. (1987) and Nedoma et al. (2006).]

of the tibia condyles together with a functional influence of the ligament apparatus give the knee joint from 5° to 6° of freedom. There are 2° of freedom in rotation and 2° of freedom in displacement (Fig. 2.10). The elastic connection and stability of the femoral and tibial joint parts are established mainly by the cruciate ligaments, which are situated among condyles and collateral ligaments (medial and lateral).

The rotary motion along the y axis is created during normal knee bending. But the center of bending is moved along the surfaces that form a circle of 20–25 mm in diameter. It is established by the curvature of the femoral condyles. Rotation along the z axis occurs when the knee is influenced by different femur condyle shapes, mainly before reflexion ends. Limited rotation is allowed also along the x axis. Displacement between the femur condyle and the tibia occurs along x and y axes and during flexion along the y axis. It is also established by the condyle femur shape. The curvature of

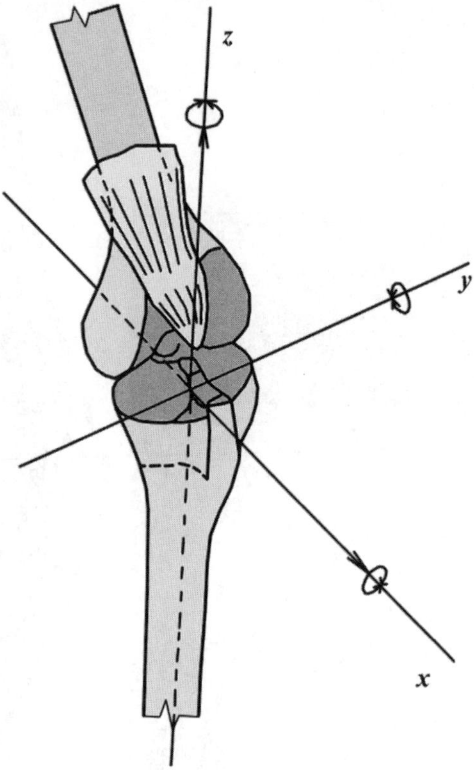

FIG. 2.10 Degrees of freedom in the knee joint. [Modified after Rybka et al. (1993) and Nedoma et al. (2006).]

contact surfaces of the femur condyle allow the knee joint to flex up to 160°. On the longitudinal – sagittal plane the femur condyle has the shape of a spiral. The curvature gets smaller toward the back. The result of this shaping is the continual transfer of the center of rotation, which is dependent on the angle of the knee bending. As a consequence of this shaping of femur condyles and the difference between the size of the medial and lateral condyles, there is a combination of their generating and sliding along surfaces of the tibia condyles during joint flexion. This finding is important in determining the reasons for the degenerative changes in the knee joint contact surfaces. Force relations in the knee joint are shown in Fig. 2.11.

2.4.3 The Shoulder Joint

The shoulder connects the upper limb to the trunk, it functions together with the elbow to position the hand in space, and allows for its effective function. The shoulder consists of the glenohumeral, acronioclavicular, sternoclavicular, and scapulothoracic joints and the musculature structures that support these joints. Shoulder motions include flection and extension, abduction, and internal-external rotation [for more

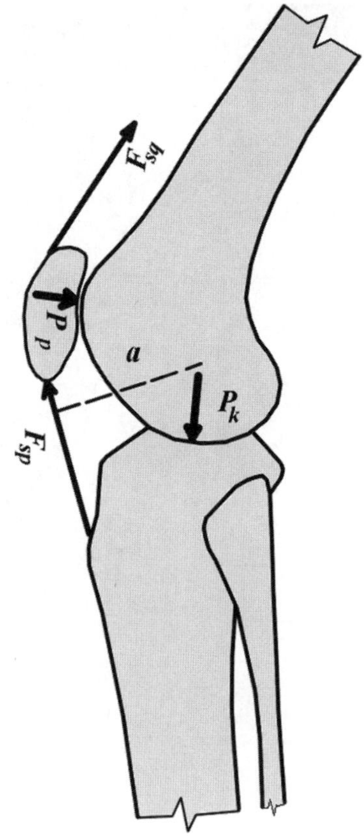

FIG. 2.11 Force relations in the knee joint; P_k = a compression force, F_{sp} = a tensile (traction) force, F_{sq} = a quadriceps force, P_p = a force acting in the patello-femoral connection, and a = an arm of the patellar tendom force.

details see Agur et al. (1991) and Valle et al. (2001)]. The joint pit is situated on a shoulder scapula and the head is situated on the humerus. The contact joint surfaces are covered by cartilage. On the periphery the joint pit is enlarged about the cartilage border. According to the shape of the contact surface, the shoulder joint is a ball-and-socket joint (Fig. 2.12). It has from 4° to 6° degrees of freedom, that is, three in the rotation around the x, y, and z axes and one during the shifting process, which occurs only in luxation. Theoretically, the forward elevation is possible at about 180°, the average value in men is 167° and in women it is 171° extension or posterior elevation averages. These values are limited by capsular torsion. Abduction in the coronal plane is limited by bony tuberosity on the acromion. Forward elevation in the scapula plane is more functional because in this plane the inferior portion of the capsule is not twisted and the musculature of the shoulder is optimally aligned for elevation of the arm. The synovial fluid acts by cohesion and adhesion to stabilize the glenohumeral joint, and it adheres to the articular cartilage overlying the glenoid and

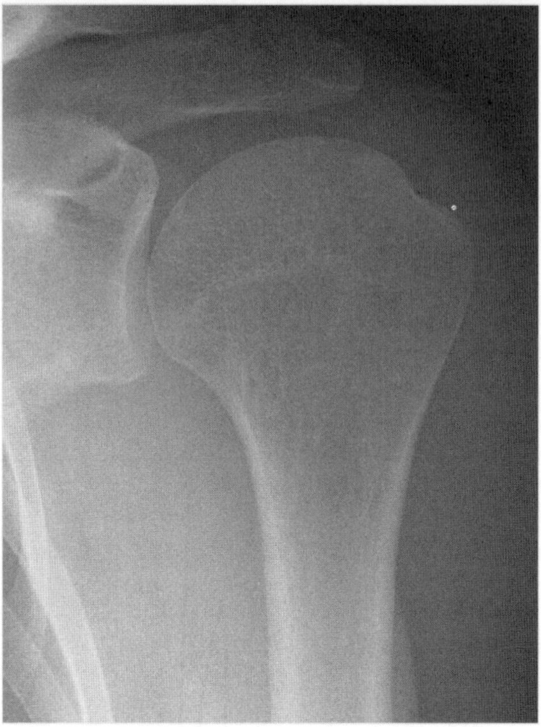

FIG. 2.12 Shoulder joint (X ray).

proximal humerus. Numerous muscles act on various components in the shoulder area to provide mobility and dynamic stability. To understand muscle functions and force acting, one must consider a given orientation, size, and activity of the acting muscles. For more details see Boone and Azen (1979), Moore (1999), and Valle et al. (2001).

2.4.4 The Elbow Joint

The elbow joint (Fig. 2.13) is a compound joint that allows two types of motion, that is, flexion–extension and pronatation–supination. Elbow flexion and extension take place at the humeroulnar and humeroradial articulations. The flexion–extension is in the range of $\sim$0°–146° with a functional range of $\sim$30°–130°. The forearm pronatation–supination is in the range from 71° of pronatation to 81° of supination. According to several specialists, the axis of rotation for flexion–extension is situated at the center of the trochlea. In the elbow joint the joint surfaces of the humerus, the ulna and the radius, are contiguous on the head of radius. The contact surfaces are covered by cartilage. The joint consists of three parts. The first is of the trochlea-shaped type, and it is situated between the humerus and the ulna. The second is the spherically shaped type situated between the humerus and the radius, and the third one is the cylindrically shaped type situated between the ulna and the radius. The bending of

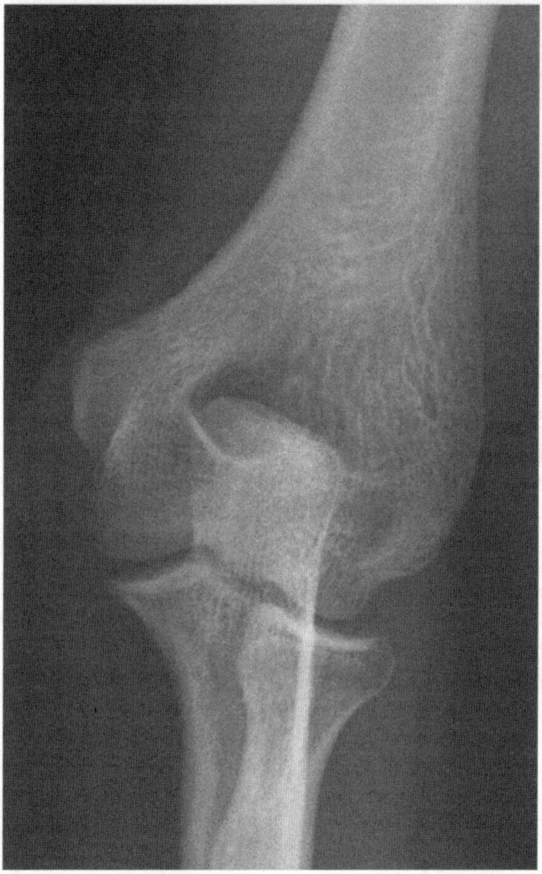

FIG. 2.13 Elbow joint (X ray).

the elbow occurs in the first and the second parts and the rotation of the radius occurs in the third part. The center of rotation coincides with the center of bending. For more details see Morray et al. (1981), Morray (1993), and Jazrawi et al. (2001).

2.4.5 The Ankle Joint

The function of the lower limb is adapted to a supporting function. This is why the skeleton of the leg and the foot are fundamentally different types of the limb construction, dependent on posture and gait. The ankle consists of three bones forming the ankle mortise. This joint system consists of the tibiotalar, the fibullotalar, and the tibiofibullar joints. The ankle joint is of a hinge-type joint. Its stability depends on joint congruency and the medial, lateral, and syndesmotic ligaments. Ankle stability increases and depends more on the articular surface congruency during weight-bearing activities.

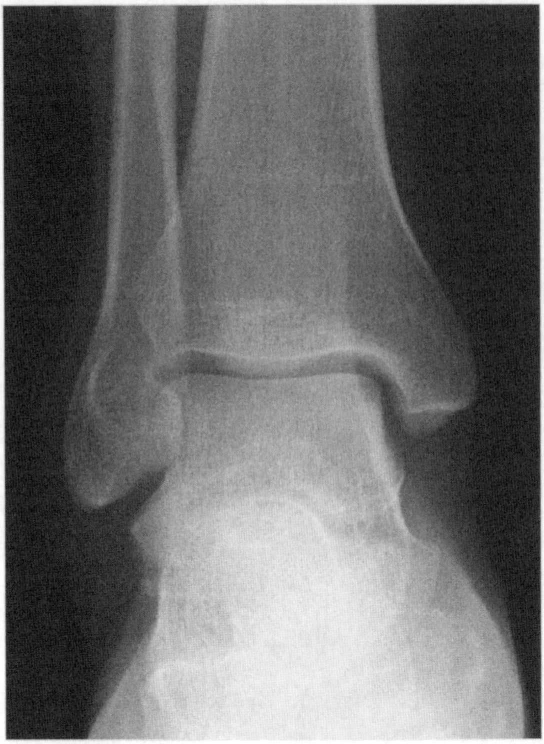

(a)

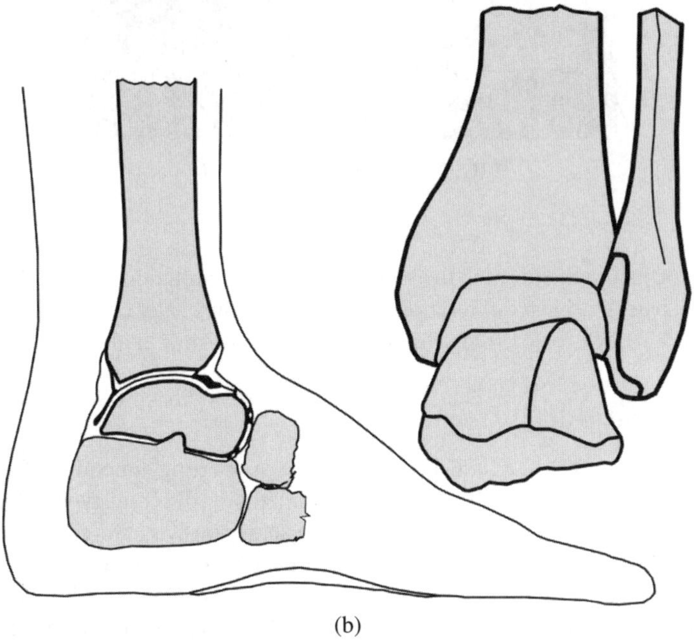

(b)

FIG. 2.14 Ankle joint. (a) The X-ray image and (b) the schematic image. [Modified after Beznoska et al. (1987), Sammarco and Hockenburg (2001), and Nedoma (2006)].

The skeleton of the foot is similar to that of the hand. It consists of 28 bones, of which 7 bones are ossa tarsi (talus, os naviculare, os cuneiforme I–III, that is, mediale, intermedium, laterale, calcareus, and os cuboideum), 5 ossa metatarsalia, and digiti pedis—created by the phalanges.

The greatest mobility is in the ankle joint. The motion of the foot occurs around three axes (i.e., abduction–adduction, dorsiflexion–plantarflexion, and eversion–inversion) and on three planes. Flexion–extension acts in the sagittal plane, abduction–adduction acts in the horizontal or tranverse plane, and inversion–eversion acts in the coronal or frontal plane.

The function of the ankle joint is rendered possible because the joint surfaces are kept in contact by the lateral ligaments. The ankle joint is a special kind of the trochlea-shaped joint with screw-shaped surfaces (helicoid surfaces and helical surfaces) [Fig. 2.14(a) and (b)].

For more details see Inman (1976), Proctor and Paul (1982), Mann (1993), Sammarco and Hockenburg (2001).

CHAPTER 3

TOTAL REPLACEMENT OF HUMAN JOINTS

3.1 VIEW OF ARTHROPLASTY DEVELOPMENTS

We can divide joint replacements into a few groups according to fixation or type of operation. According to fixation, we can divide joint replacements into cemented and cementless, and according to the type of operation, they are divided into primary and revised implants. Then we have:

1. Primary implants, which are of the following types:

    a. Cemented and polished femoral stem with or without a collar.
    b. Cementless femoral stem without a collar, the surface of which is rough by jetting corundum powder or a porous surface possibly completed by spraying with hydroxyapatite $Ca_{10}(PO_4)_6(OH)_2$.
    c. Cemented acetabular cup of the hip joint, which is made from polyethylene.
    d. Cementless acetabular cups of the hip joint are of the following types: the metal fixing casing with a polyethylene insert of a self-tapping type; the press fit type without spraying; the press fit type with a porous surface and with spraying of hydroxyapatite; the type with an anchoring tip or with the holes for spongy screws; and a Spotorno expansive type.

2. Revised, which are of the following types:

    1. Cemented femoral stem
    2. Cementless femoral stem with a surface arrangement

Mathematical and Computational Methods in Biomechanics of Human Skeletal Systems: An Introduction, First Edition. Jiří Nedoma, Jiří Stehlík, Ivan Hlaváček, Josef Daněk, Taťjana Dostálová, and Petra Přečková.

3. Cementless acetabular cup with an rough surface and with anchoring holes for screws
4. Cementless acetabular cup with a porous surface and with spraying of hydroxyapatite and anchoring holes for screws

For the manufacture of cemented femoral stems the anticorrosive steels are used mainly, and possibly cobalt alloys are used as well. Cementless hip stems are conversely manufactured mainly from titanium alloys, possibly from cobalt alloys, and generally they are provided by with porous surface in the whole length or in the proximal part only. Cemented acetabular cups and contact inserts of the cementless acetabulae are manufactured from low-pressure high-molecular polyethylene (UHMWPE), and for the metal casing of cementless artificial acetabulae the titanium alloys with porous surfaces are used almost solely. Over the last several decades the acetabular cups with the metal contact insert (the so-called metal–metal pairing) and the acetabular cups manufactured from the cobalt alloy with the ceramic insert (the so-called ceramics–ceramics pairing) have started to be used as well. Individual components of the hip replacements are created from the hip heads, which are manufactured mainly with diameters of 28, 32, or 36 mm and some of them with a diameter of 22 mm. Materials used are mainly ceramics, and cobalt alloys.

Treatment methods for destroyed knee joints are dependent on the development of surgical techniques, on our knowledge of properties of the used materials, and biomechanical studies that have been developing for more than 130 years. First works began in the second half of the nineteenth century, when Verneuil in 1863 and Ollier in 1886 were trying an interposition of soft tissues among damaged knee joint surfaces. In 1918 Baer described 23 successful attempts of the knee arthroplasty by using a chromized mucous membrane of pig urinary bladder. In 1940 Cambel applied a metal plate as an interposition of the knee joint. His method was developed by McKeever in the late 1950s and by McIntosh in the beginning of the 1960s. They used chromium–cobalt plates. Further developments led to improvements in implants—the so-called hemiarthroplastics, which closly copyied the surface of the femoral and the opposite tibial condyle of the knee joint. Later a combination of the metal (alloy) and polyethylene started to be used. Their advantage was in the limitation of surgical performances. Their disadvantages complicated the surgical technique, insufficient treatment of the femoro-patellar joint, continuing destruction of an opposite part of the joint, a small resistance at overloading, and, therefore, also their short service life.

In 1967 the Judet brothers manufactured a model of the hinge knee joint arthroplasty from acrylic resins. Seedhorm implanted his knee joint type in 1952. Modification of the previous type manufactured from the CoCrMo alloy was implanted by Jackson Burrows (in the years 1954–1958). Next variations of this type are known as Staumov's, and then Waldius' (in 1951) and Schiers' (in 1953) joint prostheses. The Waldius hinge knee arthroplasty (from the year 1957) represents the first real artificial knee joint alloarthroplasty. It was constructed as the vitalium implant, anchored into the femur and the tibia by massive stems. Both components—femoral and tibial—were connected by a joint, which in contrast to a triplanar movement at the healthy knee joint allows only movement in the sagittal plane. This led to transmission of unfavorable forces on the bone, to fatigue fractures of diaphysis, to the release of implants, and occasionally to mechanical failure. In the 1970s Matthews (in 1973) and Attenborough (in 1978) improved this type so that the implant allowed for partial rotation inside the joint. Because of its complexity, short service life, and very difficult revised operation it has not been successful. Today hinge implants are used only in the context of oncology indications. In 1968 Guston introduced a hingeless polycentric model, which was then further developed by Sheehan (in 1971), Gschwend (in 1972), and Attenborough (in 1973), who introduced the so-called spherocentric type in 1973. The first types of geometrically shaped implants respected an anatomical shape and a physiological movement in the knee joint [Gunston (in 1971) and Freeman and Swanson (in 1973)]. Therefore, Insall and Townley introduced condylar joint replacements, which were later modified by alternative stabilized elements for the surgical treatment of bruised cruciate ligaments and as a patella replacement [Insall, Lachiewicz and Burnstein (in 1979)]. At the end of the 1970s rotating meniscular elements were developed [New Jersey Knee, Pappas and Buechel (in 1979)], which enabled rotations as well as shifts in the knee joint that were similar to a healthy knee joint.

At present the following artificial knee joints are manufactured:

1. With or without rotating elements
2. Cemented or cementless types or hybrid ones
3. Modular and nonmodular types
4. Primary or revised types
5. Hinge types

The first attempt at a shoulder replacement, made from platinum fixated with india rubber, was accomplished by Péan in Paris in 1892. The next attempt was not until 1950. It was done by Judet and the replacement joint was manufactured from Plexiglass. The first modern shoulder joint prosthesis manufactured from metal was used by Krüger in 1951 and by Neer in 1953.

Artificial elbow joint and ankle joint replacements were only marginally successful. There are very unfavorable conditions in the elbow joint for anchoring an artificial joint to the skeleton. Nevertheless they have already been implanted using, for example, the St. Georg's model. In the case of the ankle joint, which is a very complicated joint, the first attempts were made during the 1970s. Much progress in hand joint replacements has been achieved in the last decade. At present artificial joints of the shoulder, elbow, radio-carpal, ankle, and finger joint arthroplasty can be manufactured using silicon. Until now all human joint replacements were merely a compromise among the anatomical-physiological reality and the current technological and economical manufacturing possibilities. A reader can find details in the monographs of Huggler and Schreiber (1978), Charnley (1979), and in many others.

3.2 STATIC AND DYNAMIC LOADING OF HUMAN JOINT REPLACEMENTS

To simulate correctly the function of tension in a total joint replacement, we must know the physiological distribution of forces acting on the joint, which are transferred on the joint from the skeleton. We must also know the physiologically correct direction of acting forces. Thus, we present an introduction to the anatomy of human joints and muscles as well as mathematical descriptions of their functions.

Most authors start from Pauwels' considerations, which solved only the one-dimensional problem that considered the weight of a body and the muscle force of the abductors. These considerations were extended into the frontal plane for the static stand on one leg and under simplified muscular functioning. From this consideration we can see that the force **P**, the result of the body weight force and of the function of muscle groups, is crucial for the determination of compression loading of the head and the neck of the femur, the stress of the upper femur bone end and, therefore, also of the artificial hip joint replacement. It will have a bending character.

An analysis of the individual muscle group's functions in the hip joint sector has not been done by anyone yet. It is evident that the acting horizontal muscle forces influence a direction of the force **P**.

Loading of an implanted joint replacement is influenced by the following:

1. Walking—a crucial factor influencing the character of loading of lower limb joints. Walking is a forward motion in a vertical direction toward the frontal plane. That is why we analyze a function of the force in a plane of movement direction, that is, in the sagittal plane.
2. The method and direction of loading—Artificial acetabular cups of the hip joint, which were implanted a long time ago and which were functionally loaded, give us information about the direction of the force on the head of the artificial hip joint replacement during walking.

The body movement during walking is assumed to be uniform. From experimental measurements it follows that a reactive force predominates the acting force, evoked by the patient's weight only slightly. An increase is a reaction of force impulses. The second is concerned with an ordered increased force acting on the hip joint head. The characteristic course of this force in a step cycle corresponds to a force registered on the contact desk, but it is of a higher order.

The direction of the force acting on all joints of the lower limbs when one walks without support will act in the direction of walking, that is, in the sagittal plane. After an artificial joint replacement biomechanical relations will change in the articular sector, above all dependent on the intensity of patient's movements. From an activity point of view, it follows that it is not only the magnitude of the resulting force acting on the hip joint head but also the pelvis sway in the frontal plane. With high patient activity, the resulting force direction in relation to a vertical pelvis axis declines laterally. With low patient activity the wear will be also lower and the reaction force acting on the acetabulum leans toward an ideal vertical axis of the pelvis.

Artificial hip arthroplasty is performed on the most stressed joint suffering from the effects of bending. Anchoring the stem of the joint arthroplasty in the medullar channel plays an important role. The stem has a wedge-shaped design that narrows toward its lower end. This shape is chosen because during the stem implementation into the medullar channel of the femoral bone, which is filled with a poly(methyl methacrylate) (PMMA) bone cement, the stem penetrates through the PMMA mass and at the same time it presses the PMMA into the spongy bone on the internal channel surface. Since the bone cement (PMMA) fills in all irregularities in the released channel of the upper end of the femur, it creates an ideal bed for the stem of the joint arthroplasty. As for its strength, a wedge-shaped stem has a critical section. In case the stem is closely fixed and its collar leans on the plane of the resected femoral bone neck, then the prosthesis acts as a rigid two-arm lever and stress created by a bending moment is observed only on the neck of the prosthesis. On the stem a bending moment manifests as a compressive force, which acts on the opposite end of the lever, that is, on the end of the stem, by a compression on the internal surface of the bone channel. In case the stem is partly released, then the prosthesis acts as an anchored thin prismatic bar. Based on our numerical results, presented in Chapter 7, the problem must be studied as the contact problem with the Coulombian friction, where the contact surface is observed between the loosened stem and the femoral bone. An extreme loading by a bending moment of the artificial hip joint after implantation sometimes evokes failures in PMMA anchoring. It relates with a different elasticity of the femur, PMMA, and the THA stem. That is why new design solutions of the stems as well as their anchoring are being researched.

3.3 MECHANICAL DESTRUCTION OF IMPLANTS AND DEMANDS ON HUMAN JOINT ARTHROPLASTY

3.3.1 Fatigue Fractures of Human Joint Replacements, Corrosion of Metal Implants

A fatigue strength or a material fatigue limit is an expression of an indirect dependence between a loading force amplitude and a joint replacement service life. It is given by a tension value and a number of cycles that lead to a fracture of the stem. As a fatigue limit of the material we understand an oscillating tension in bending, which until it reaches 10^7 cycles did not lead to a fracture of a tested sample. The fatigue fracture is characterized by the fact that it is not accompanied by permanent material deformation as well as no crystalline structure changes.

From the fatigue strength point of view the anchoring stem is the weak point in the artificial replacement of the hip joint. The main reason is that a fatigue fracture can be assumed when the stem is cyclically stressed in bending in a cross section, whose shape fatigue strength is lower than the tension evoked by a bending moment. There exist many other factors that shift a critical section closer to a stem collar of the joint arthroplasty or decrease a value of tolerated tension. These factors are: the construction, material used, design of the surface, and the surgical treatment selected. Therefore, the stem must be constructionally adapted to a channel size of the femoral bone. That is why a dimension array of joint prostheses with differently shaped modifying stems is made. The aim is that during an implantation the surgeon can use a minimal quantity of bone cement (~2–7 mm of PMMA) for the stem fixation.

Fatigue fractures are also caused by corrosion. They occur during cyclic loading in a corrosive environment during overfulfilment of the fatigue strength. When there is a concentration of tension, for example, in a fatigue crack where an anode occurs, corrosion occurs. Other effects include:

1. An intercrystalline corrosion caused by the prescence of chromium carbide granules.
2. A crack-stressed corrosion, originating at a parallel function of the corrosion environment and the straining. It leads to a creation of cracks that decrease the supporting section up to a static fracture.

In Chapter 6 we will present methods and algorithms that study the accumulation of deformation energy and where the stresses are concentrated.

3.3.2 Requirements Imposed on Human Joint Arthroplasty

The fundamental presumption for a long-term satisfactory result of the surgical treatment is the long-term function of the joint prosthesis. Therefore, we must impose stringent requirements on the material and the construction of artificial human joint prostheses. These requirements can be summarized in the following points:

1. Sufficient static and dynamic strengths
2. Contact surfaces that not to an excessive wear, that is, characterized by a very low friction
3. A modulus of elasticity that is close to the modulus of elasticity of human bone
4. Good ability to absorb energy and also to absorb shocks
5. Possibility of sterilization without influencing material properties
6. Good corrosion resistance in a living tissue environment
7. Biocompatibility of implants
8. Relatively easy construction from the point of view of implantation and reimplantation.

3.4 BIOMATERIALS IN OSTHEOSYNTHESIS AND ALLOARTHROPLASTY

3.4.1 Introduction

The long bones of the human skeletal system are inclined to injury and internal and external fixations are used in their treatments. It seems that human joints are in very bad states, making their replacements necessary. At present about 500,000 hip joint replacements are done in the United States, and the same number of hip joint replacements are done in Europe. About 250,000 knee joint replacements are done in the United States with the same number in Europe. And these numbers increase all the time. The response of a bone to the biomaterial used affects the regeneration process. These biomaterials implanted into the bone will cause local and systemic biological responses even if they are known to be inert. A host response with joint replacement and fixation materials will initiate adaptive and reactive processes.

The development of contemporary osteosynthesis and alloarthroplasty during the last 50 years has been influenced significantly by two special scientific disciplines, that is, biomechanics and material sciences. Based on these disciplines, the development and manufacture of joint arthroplasty are major contemporary problems in an orthopedic practice. The first experiments to replace parts of human joints and of the skeleton by artificial implants were made from nonmetallic materials, that is, from synthetic elastic resins. After failures caused by excessive wear of the contact surfaces of the implanted components of the joints, nonmetallic materials started to be replaced by metallic materials and later also by ceramic materials. During the 1970s, nonmetallic biomaterials began to be applied again. This trend continues and at present joint arthroplasty manufactured entirely without the use of a metal has been developed. The composite materials are denoted as future materials.

The year 1958 represents the year of new trends in osteosynthesis. In that year the Swiss company Association Orthopedic AO-ASIF in Bern introduced sets of instruments and implants into modern osteosynthesis together with well-developed application techniques. From 1958 to 1986 more than 150,000 successfully cured fractures were recorded by this AO system. In 1984 the AO classification of fractures was elaborated based on these cures. At present it is one of the most used methods in orthopedics. In 1995 the software Classification of Fractures, the computer version of the book *The Comprehensive Classification of Fractures of Long Bones* [see Müller et al. (1990)] was put on the market. This time the material for implants is stainless steel CrNiMo. In that year Charnley introduced his new type of the total hip arthroplasty (THA), where the femoral part was made from stainless steel and the acetabulum was made from polytetrafluorethylene (PTFE), which was replaced by ultra-high-molecular-weight polyethylene (UHMWPE) during the following two years. Materials used in human joint arthroplasty must satisfy certain properties, such as useful chemical compositions, desirable mechanical properties, as well as their biological adaptability in the human body. The corresponding quality criteria were formulated in international standards. The materials satisfying such international standards are denoted by the symbol ISO (International Standards Organization).

For decades metals have been used successfully in fracture fixation techniques as well as in human joint arthroplasty and in spondylosurgical fixations. Fixation devices and the artificial joint arthroplasty represent about 44% of all medical devices. The demands for such orthopedic device implants are expected to increase in future years. At the present time metal implants are required to be inert if implanted into human bone. They are supposed to be bioactive as their surfaces are porous and coated. Metallic fixators are used alone or as a system of several metal components. The joints are comprised from several different parts of metal, polymer (PMMA–UHMWPE), and ceramics (Al_2O_3, ZrO_2). The joint prosthesis are either fully metal or as metals combined with polymers and ceramics. Cemented types are popular owing to the fourth-generation cementing techniques, that is, medullary plugs, viscous cements, vacuum mixing, pressuring, and centralizers. It is evident that the rate of complications will increase as the number of materials used for applications in orthopedic surgery increases. The types of metals, their manufacture, their standards, alloys, composition, processing conditions, and mechanical properties influence the interactions between metals and bones. Stainless steel, cobalt, titanium, and their alloys are widely used in the production of artificial joints and fixators. The advantages of titanium over cobalt alloys are a lower modulus of elasticity and a higher biocompatibility (Santavirta et al., 1992; Head et al., 1995; Jinno et al., 1998). Long-term stability of implants is required for bone–implant integrations. Both titanium and cobalt alloys demonstrate good biocompatibility.

3.4.2 Metal, Plastic, Ceramic, Composite Materials, and Bone Cement

Metals During the last 50 years metals have been used successfully in fracture fixations and joint replacements. The first use of metallic materials for implants probably dates back to the year 1565, when Petronious used a golden plate to repair a cleft of the hard palate. In the seventeenth century Hieronymus Fabricius applied iron, golden wires, and bronze in his medical practice. In 1829 Levart presents the first work on reactions of living tissues to metals as metal toxicity was already known. Between 1875 and 1883 Lister introduced an antiseptic surgical technique and in 1895 Röentgen demonstrated that bones and joints can be seen by X rays. In 1880 von Langenbeck and Hansman introduced the metal bone splints. In 1912 the steel alloys with vanadium were introduced by O'Neil. During the 1920s the first experiences with metallic implants made from various materials were applied. In 1936, a new alloy, the so-called Vitalium based on a cobalt alloy, was introduced and applied by Venable and Stuck. Pure titanium for clinical applications was used in 1951 (Semlitsch, 1986). In the 1970s two companies, the Down Bros. and Zimmer, introduced commercial implants made from a titanium alloy, Ti6Al4V, and in a modified (by Zimmer) version Ti6Al4V/ELI. Zwicker in 1980 developed the titanium alloy Ti5Al2.5Fe and the titanium alloy Ti6Al7Nb was developed by Semlitsch.

Metals in nature contain a large group of materials with different physical properties. The most outstanding property of metals is their ability to conduct electricity and heat. Their physical properties are characterized by their atomic structure (bond). One of the fundamental properties of this bond is the lack of an orientation of the bonding forces, making possible the easy reformation of the ions into the most suitable configuration. For example, in 1 cm^3 there are 10^{23} ions. Metals have a crystalline structure in the solid state, that is, the spatial configuration of their atoms is symmetrical. Metals have three spatial configurations: the spatially centered cubic lattice (β-Ti), the plane centered cubic lattice (γ-Fe), and the hexagonal lattice (α-Ti, ε-Co). Many of the properties of metals and their alloys are related directly to the configuration of the atomic lattice. The useful properties of metals or their alloys are obtained because the alloys are based on a combination of different metallic components. An alloy is homogeneous if all its ions are located in a uniform lattice. If, for a given temperature, the concentration of one of the alloy components involved reaches the limit of its solubility, a new phase is formed, consisting of mixed crystals of a type different from the initial alloy. Such alloys are of two-phase or polyphase origins and we speak about heterogeneous alloys. All types of alloys used in joint replacements are required to have a stable temperature that correspondes to the temperature of the human body. For more details see Friedel (1965), Parsons and Ruff (1973), Beznoska et al. (1987), Black (1988), Valenta (1993), and Nedoma et al. (2006).

The metal alloys are made from Fe, Co, Ti, Ni, Cr, Mo, C, Si, Mg, P, S, Al, H, O, V, Cr, and N components. At present the following types of materials are used: (i) stainless steel of the CrNiMo type, (ii) an alloy based on Co, Cr, and Mo, (iii) an alloy based on Ti, and (iv) partially also alloys based on noble metals. The Young's modulus E (MPa) for corrosian-resistant CrNiMo steel is in the range of 2–2.1 $\times$ 10^5 MPa, for a CoCrMo alloy (e.g., Protasul 2) it is in the range of 2–2.2 $\times$ 10^5 MPa, and for an MoNiCrTi alloy (e.g., Protasul 10) it is in the range of 2.2–2.4 $\times$ 10^5 MPa and the Poisson's constant ν is in the range of 0.25–0.33 and their densities ρ are in the range of 7800–8100 kg m^{-3}.

The noble metals and their alloys in the physiological environment are thermodynamically stable and resistant to corrosion and they are biocompatible with human tissues. Their disadvantage is their lower strength, brittleness, and fretting, that is, a corrosion owing to the friction. Suitable materials for implants are metals with resistance, that is, materials with a thin nonporous layer on their surface that shelters a material against the neighboring environment. This thin (microscopic) layer is created from oxides of fundamental materials.

Next, the following symbols will be used: R_m is the ultimate tensile strength (MPa), R_p is the yield point (MPa), R_f is the fatigue strength (MPa), E is Young's modulus (MPa), ν is Poisson's constant, HV is hardness, and A is the ultimate of elongation (%). The introduced data are based on the results of Friedel (1965), Parsons and Ruff (1973), Stevens (1986), Beznoska et al. (1987), Black (1988), Valenta (1993), Rabinowitz (1995), Helsen and Breme (1998), Yaszemski et al. (2004), and Nedoma et al. (2006).

A. Stainless Steels of the Cr–Mo–Ni type At present, from stainless steels the high alloyed austenic Cr–Mo–Ni type steels with carbon contents below 0.08 percent are used in orthopedics. Their mechanical properties can be increased by applying 0.1–0.2% of nitrogen. The chemical composition of this steel ensures high resistance against corrosion beyond fretting. The following stainless steels are used: (i) stainless steel ISO 5832-1-alloy FeCr18Ni14Mo3. Because a large amount of nickel is problematic in certain cases in orthopedic practice (possibility of an allergy), it is used for hip joint stems; (ii) Stainless steel ISO 5832-9-alloy FeCr22Ni10Mn4Mo2NNb, the fine-grain austenite structure with a high mechanical strength and high resistance against wear, possibility of an allergy. (See Tables 3.1 and 3.2.)

B. Cobalt Alloys of the CoCrMo or CoCrMoNi Type These types of alloys are used in orthopedic surgery. Solid cobalt is of two crystallographic types: (i) the so-called α phase with a plane-centered cubic lattice, which is stable between 420 and

TABLE 3.1 Mechanical Properties under ISO 5832-9 (Annealing)

State	$R_{p0.2}$ (MPa)	R_m (MPa)	A (%)
After dissolving annealing	Min. 430	Min. 740	Min. 35
Intermediate hard	Min. 700	Min. 1000	Min. 20
Hard	Min. 1000	Min. 1100	Min. 10

TABLE 3.2 Chemical Composition under ISO 5832-9

Mass (%)	Fe	Cr	Ni	Mn	Mo	Nb	N	C	Si	Cr	S	P
Min.	Rest	19.5	9	2	2	0.25	0.25					
Max.	Rest	22	11	4.25	3	0.8	0.5	0.08	0.75	0.25	0.01	0.025

1495°C, the temperature of the melting point; and (ii) bellow 420°C, the modification with a hexagonal lattice of a close atomic configuration known as the ε phase, is also stable. At a temperature of 1125°C phase-change transitions, that is, transitions between ferromagnetic and paramagnetic states, are observed.

The cobalt alloys are divided into two types—with a large content of carbon and with a small content of carbon (~0.35%). The implants are made from the latter.

The phase change between modifications α and ε is significantly influenced by the change in grain size of the material. For fine-grained cobalt the α phase remains stable even at room temperature. For large grains, the phase-change transition—the so-called recrystallization—takes place during cooling and the ε-phase becomes stable in this temperature range. In the case of cobalt with different grain sizes a stable mixture of both phases, that is, of α and ε phases, is observed at room temperature.

Cobalt alloys have good strength properties and they have very good biocompatibility with bone and other tissues. The price of these materials is very high. Examples of these materials are Stellit 21 and Vitallium. In Switzerland they are named Protasul 2 and Protasul 10 (with nickel). The CoCrNiW alloy is called Stellit 25. This material is of a high quality, with a high strength and a high resistance to corrosion. The heads of the prostheses are made from the classical Vitallia (Protasul 2) and the stems of prostheses are made from Vitallium (Protasul 10). The distinguishing quality of these types of alloys can be provided by carbides. Therefore, the Co–C alloys are the basis for all cobalt alloys.

The following Co alloys are used:

1. The cobalt alloy under ISO 5832-4—the alloy Co28Cr6Mo; it has a high resistance to wear, good articulate partners with UHMWPE and with the ISO 5832-4 alloy, that is, with the same type of an alloy. It is used for knee joint replacements. (See Tables 3.3 and 3.4.)
2. The cobalt alloy under ISO 5832-6—the alloy CoNi35Cr20Mo10. It is not useful as an articular partner with UHMWPE.
3. The cobalt alloy under ISO 5832-12—the alloy Co28Cr6Mo with 0.05% of carbon. (See Tables 3.5 and 3.6.)

TABLE 3.3 Mechanical Properties under ISO 5832-4

State	$R_{p0.2}$ (MPa)	R_m (MPa)	A (%)
Cast	Min. 450	Min. 665	Min. 8

TABLE 3.4 Chemical composition under ISO 5832-4

Mass (%)	Co	Cr	Mo	Ni	Fe	Mn	Si	C
Min.	Rest	26.5	4.5					
Max.	Rest	30.0	7.0	1.0	1.0	1.0	1.0	0.35

TABLE 3.5 Mechanical Properties under ISO 5832-12

State	$R_{p0.2}$ (MPa)	R_m (MPa)	A (%)
After dissolving annealing	Min. 550	Min. 750	Min. 16
After hot forming	Min. 700	Min. 1000	Min. 12
After cold forming	Min. 827	Min. 1172	Min. 12

TABLE 3.6 Chemical Composition under ISO 5832-12

Mass (%)	Co	Cr	Mo	Ni	Fe	Mn	C	Si	N
Min.	Rest	26.0	5.0						
Max.	Rest	30.0	7.0	1.0	0.75	1.0	0.35	1.0	0.25

Its structure is fine-grain austenitic, with high mechanical strength values and with a high resistance to wear. It is a good articulatory partner of UHMWPE.

4. The cobalt alloy under ISO 5832-12—the alloy Co28Cr6Mo with 0.2% of carbon. (See Tables 3.7 and 3.8.)

Its structure is fine-grain austenitic with high mechanical strength values and with a high wear resistance. It is a good articulatory partner with UHMWPE or with itself (e.g., Metasul). Typically, it is used for the Metasul hip head and the Metasul acetabulum in the metal–metal application.

C. Titanium and Its Alloys Titanium (Ti) is one of the most inert metals. Its strength is lower than the strength of steel. The asset of titanium and its alloys is high corrosion resistance, both in chemically aggressive environments and in aggressive human tissues. The density of titanium and its alloys is about 4.5 g cm^{-3}. The technology of titanium metallurgy is a very complicated process. The melting point of titanium is relatively high, ~1725°C, which is about 100–200°C higher than the melting point of iron. Its thermal conductivity is about 5 times lower than that of iron and about 15 times lower than that of aluminum. Titanium has a high resistance to corrosion in air at low as well as high temperatures. Pure titanium is ductile, its

TABLE 3.7 Mechanical Properties under ISO 5832-12 with 0.2% Carbon

State	$R_{p0.2}$ (MPa)	R_m (MPa)	A (%)
After dissolving annealing	Min. 550	Min. 750	Min. 16
After hot forming	Min. 700	Min. 1000	Min. 12
After cold forming	Min. 827	Min. 1172	Min. 12

TABLE 3.8 Chemical Composition under ISO 5832-12 with 0.2% Carbon

Mass (%)	Co	Cr	Mo	Ni	Fe	Mn	C	Si	N
Min.	Rest	26.0	5.0				0.2		
Max.	Rest	30.0	7.0	1.0	0.75	1.0	0.2	1.0	0.25

tensile strength is ~220–260 MPa at an elongation of 60%. Even the most negligible amount of additives increases rapidly its strength and decreases its ductility. Titanium can be cold formed. Oxygen, nitrogen, and hydrogen have a tendency to change their fatiguing strength. Therefore, their metallurgy is made in a vacuum with helium.

Aluminum is the fundamental alloying element of titanium alloys. Vanadium increases the strength of titanium alloys and at the same time it diminishes its ductility. Molybdenum stabilizes the strength of titanium alloys by lowering its ductility. Manganese increases the hardness and the strength of titanium alloys. Chromium also increases the strength of titanium alloys. Analogously with manganese, the chromium content decreases the transition temperature. Carbon of about 0.2 weight percent increases the strength of titanium alloys without a reduction in ductility. It is evident that the alloying application leads to the discovery of new types of alloys with high strength and acceptable ductility.

The following titanium materials are used in orthopedic surgery: pure titanium, Ti6Al4V, Ti6Al7Nb, and Ti5Al2.5Fe. All these alloys are of the two-phase type, that is, $\alpha + \beta$, where the β phase is of the martensitic type.

Titanium itself exists in two phases corresponding to two crystalline modifications. The first has a plane centered hexagonal crystalline lattice at room temperature, which is the α phase. The second one is heated to above 882°C. This lattice changes to a spatially centered cubic latice, called the β phase. The O, H, and N have a tendency to extend the region of the α phase, with Al stabilizing the α phase and Cr, Fe, Mg, Nb, Mo, V, and H stabilizing the β phase. Zirconium and its alloys have similar properties to titanium and its alloys but their applications can be realized after detailed analyses of their properties.

The used titanium materials are the following:

1. *Pure Titanium under ISO 5832-2Ti* Its structure is a one-phase structure, and its properties include high ductility. Osteointegration is ensured by means of direct ingrowth of the bone into the rough surface or into the porous surface

TABLE 3.9 Mechanical Properties under ISO 5832-2 with 0.2% of carbon

State	$R_{p0.2}$ (MPa)	R_m (MPa)	A (%)
Class 1	Min. 170	Min. 240	Min. 24
Class 4A	Min. 440	Min. 550	Min. 15
Class 4B	Min. 520	Min. 680	Min. 10

TABLE 3.10 Chemical Composition under ISO 5832-2

	Mass (%)	Ti	C	H	O	N	Fe
Class 1	Min.	Rest	0.1	0.0125	0.18	0.03	0.15
Class 4	Max.	Rest	0.1	0.0125	0.45	0.05	0.3

of the stems. Noncemented acetabulum and inner fixations implants are typical applications of pure titanium. (See Tables 3.9 and 3.10.)

2. *Titanium Alloy under ISO 5832-3, Alloy Ti6Al4V* Its structure is of a fine-grained globular α/β structure, and its properties include high values of mechanical strength, the osteointegration by means of direct ingrowth of a bone into the rough surface, low resistance toward wear but unsuitable as the articular partner of UHMWPE. (See Tables 3.11 and 3.12.)
3. *Titanium Alloy under ISO 5832-11, alloy Ti6Al7Nb* Its structure is a fine-grained globular α/β structure. Its properties include high mechanical strength values, the osteointegration by means of direct ingrowth of a bone into the rough surface, and low resistance to wear. It is suitable as the articular partner of UHMWPE. (See Tables 3.13 and 3.14.)

The titanium alloys are used for manufacturing hip joint stems and acetabulae. The acetabular components are manufactured from two parts—from the titanium cup and the polyethylene insert.

For more details see Valenta (1993), Yaszemski et al. (2004), and Nedoma et al. (2006). The problem with phase-change transitions will be mathematically discussed in Chapter 5 and in the literature introduced here.

D. Materials with a Shape Memory Materials (alloys) with a shape memory are equiatomic alloys of nickel and titanium. The most outstanding property of this alloy is its ability to regain, after heating to a certain temperature, its shape prior to plastic deformation. This transformation of a shape is realized suddenly; the temperature of this shape transformation process depends on the composition of the alloy. Owing to these properties, this alloy, though developed for different purposes, is suitable for use in orthopedics, traumatology, and stomatology. The composition of such an alloy is 55% nickel and 45% titanium. The melting point is 1240°C, density is 6.6 g cm^{-3}, and $R_m = 850$ MPa.

TABLE 3.11 Mechanical Properties under ISO 5832-3

State	$R_{p0.2}$ (MPa)	R_m (MPa)	A (%)
After dissolving annealing	Min. 780	Min. 860	Min. 10

TABLE 3.12 Chemical Composition under ISO 5832-3

Mass (%)	Ti	Al	V	Fe	H	N	O	C
Min.	Rest	5.50	3.50					
Max.	Rest	6.75	4.50	0.30	0.015	0.05	0.20	0.08

TABLE 3.13 Mechanical Properties under ISO 5832-11

State	$R_{p0.2}$ (MPa)	R_m (MPa)	A (%)
After dissolving annealing	Min. 800	Min. 900	Min. 10

TABLE 3.14 Chemical Composition under ISO 5832-11

Mass (%)	Ti	Al	Nb	Fe	Ta	N	O	C	H
Min.	Rest	5.50	6.50						
Max.	Rest	6.50	7.5	0.25	0.50	0.05	0.20	0.08	0.009

Nonmetallic Materials

A. Plastic Biomaterials Plastic biomaterials are macromolecular media known as polymers. Polymers are mainly used in bone replacements, fixations of ligaments, and drug delivery. Polyclides (PLA), polyglycolides (PGA), and polyhydroxybutyrates (PHBV) are polymers used in hard tissue engineering. Their structures are characterized by long molecules with repetitive basic structures. These materials are predominantly of an organic origin, and besides carbon they also contain oxygen, hydrogen, nitrogen, fluorine, and chlorine. The structure of monomers is based on saturated hydrocarbon ethylene CH_2–CH_2. The polyethylenes are used for orthopedic objects. Plastics are of low molecular weight, that is, those of low relative molecular weight, and high molecular weight, that is, those of a high relative molecular weight. Their physical, chemical, and mechanical properties are different. The higher molecular weight of a polymer tends to increase its toughness, ductility, and resistance to stress corrosion and decreases the tendency of the material to creep at low temperatures. High-molecular-weight polyethylenes have been used in the manufacture of implants due to their good mechanical properties and due to a low coefficient of friction as well as a relatively high resistance to wear.

Table 3.15 shows the material properties of UHMWPE, the ultra-high-molecular weight polyethylene. UHMWPE was developed during the 1950s by Philips–Standard Oil in the United States and by Zeigler in Germany. We classify these materials

TABLE 3.15 Properties of UHMWPE Used in Alloarthroplasty

Parameter	UHMWPE
Tensile strength (MPa)	43
Bending strength (MPa)	40
Strength in compression (MPa)	16
Density ($g\,cm^{-3}$)	0.9–0.96
Ductility (%)	600
Absorbability (%)	0.01
Melting point (°C)	130–135

as low-molecular, middle-molecular, and high-molecular materials. Polymerization defines the property of polyethylene. The low-pressure polyethylene has a long linear molecular structure. This biomaterial can be processed easily, its biocompatibility is very good, and its absorptivity is minimal (very low).

B. Ceramic Materials–Oxide Ceramics Ceramic materials used in alloarthroplasty are inorganic nonmetallic materials produced by sintering from powdery raw materials and strengthened by firing, during which manifold increases in strength occur. During this process the material from one phase state changes into the second one. The process is described by the phase-change problems, which mathematically lead to solve free boundary value problems for parabolic differential equations with initial boundary conditions, where a part of the boundary is free and where additional conditions are given at this free boundary (see Chapter 5 on the mathematical model of heat generation). In 1960 and 1961 Charnley opened the period of low-friction alloarthroplasty with the introduction of polyethylene sockets in an articulation with metal balls. In 1970 Bountin [see Bountin (1971)] initiated the age of the low-wear alloarthroplasty by the application of alumina ceramics to the hip surgery, and later he was the first to replace polyethylene sockets and metals heads by Al_2O_3 ceramics, the so-called corundum ceramics, and they were applied in alloarthroplasty. The other types of ceramics such as silicone ceramics SiO_2, Si_3N_4, SiC (silicon carbide), or ZrO_2 (stabilized by calcium oxide) and compound ceramics SiO_2, Al_2O_3, and Si_3N_4, were developed and tested. Alumina Al_2O_3 and zirconias ZrO_2 ceramics belong to this group of oxide ceramics.

1. *Alumina Al_2O_3 Ceramics* The crystal lattice of aluminum oxide is hexagonal. Al_2O_3 consists solely of densely packed tiny corrundum crystals. This crystallographic phase is called α alumina. Corundum is crystallographically identical with sapphire and ruby, but it does not have traces of metal ions, which originate their colors. The melting point of alumina is 2050°C. Corundum is one of the hardest materials. Its level of free enthalpy is one of the lowest among ceramics. The distance between atoms is the smallest, therefore, its bonding energy is the longest. From this very good mechanical, chemical, and electrical properties follow, but three basis criteria—high purity, high density, and fine grain—must be satisfied. High purity is necessary because Al_2O_3 has a tendency to lose its strength if impurities are present. High density reduces the incidence of pores,

TABLE 3.16 Properties of Ceramics

Property	Values
Density (g cm^{-3})	≥ 3.9
Grain size (μm) ($<7\ \mu$m—mean grain size)	0.003–0.01
Thermal conductivity (W mK^{-1})	30
Coefficient of thermal elongation α (K^{-1})	$6–8.1 \times 10^{-6}$
Hardness [HV]	20,000–30,000
Compression strength R_m (MPa)	$2.96–3.8 \times 10^5$
Bending strength (MPa)	500–600
Purity—content of Al_2O_3 (%)	≥ 99.5
Impurities (sum of SiO_2+CaO+Na_2) (%)	≤ 0.1
Additivities (%)	≤ 0.3
Porosity (%)	0.2
Flexural strength (MPa)	>250
Impact strength (N/mm^2)	4
Resistance against wear (ring-on-disk test) (mm^3/h)	~0.01
Resistance against corrosion (in Ringer's solution) (mg/m^2 per day)	~0.1
Modulus of elasticity in tension E (MPa)	$2.96–3.8 \times 10^5$
Poisson's constant ν	0.28–0.30

which diminishes the mechanical strength of the ceramic material and increases the surface roughness. The sintering temperatures are in the range from 1300 to 1600°C, depending on the grain sizes and activities of the raw material used and the desired properties. The tendency of the mechanical strength and resistance to wear to decrease with an increasing grain size is known even for other materials. The properties of ceramic materials are given in Table 3.16.

For more details see Stevens (1986), Yaszemski et al. (2004), and Nedoma et al. (2006).

Some of the physical properties of corundum ceramics make these materials more suitable for the manufacture of replacement joint components than the metals in current use. Hardness of corundum ceramics is from 5 to 10 times higher than that of known metals and their alloys. They are hard and brittle materials, but with very good tribological properties as well as very good biocompatible properties. It is a typical bioinert material.

2. *Zirconium Ceramics ZrO_2—zirconia* As a result of the complicated phase relations of zirconia, pure zirconia (ZrO_2) powders cannot be transformed into polycrystalline solids by sintering because of several phase changes occurring during cooling from the molten state to room temperature. In passing down through the melting temperature at 2706°C a cubic phase with density ~6.09 g cm^{-3} appears. During further cooling a monoclinic structure appears at 1170–950°C with its density of 5.83 g cm^{-3}.

TABLE 3.17 Comparison of Properties of Al_2O_3 and ZrO_2 Ceramics

Property	Units	Al_2O_3 Ceramics	ZrO_2 Ceramics Mg Stabilizer	ZrO_2 Ceramics Yttria Stabilizer (Y)
Composition	%	>99.97	91 ZrO_2 + 9 Mg + 9 Mn	97 ZrO_2 + 3Y
Density	$g\,cm^{-3}$	>3.98	5.75	>6
Young's modulus *E*	GPa	380	200	210
Bending strength	MPa	580	450–700	900–1200
Impact strength	$MPa\,m^{-1}$	4	7–15	7–10
Thermal conductivity	$W\,mK^{-1}$	30	2	2
Hardness	HV	23,000	12,000	12,000

Table 3.17 gives a comparison of both ceramics. For more details see Stevens (1986), Yaszemski et al. (2004), and Nedoma et al. (2006).

Advantages of ceramics and polymers are evident. Ceramic polymer composites are used as bone graft substitutes and also in glue therapy approach to bone renegeneration. The future hard tissue engineering lies between the approximate composition of a facilitating matrix, mediators, and osteogenic cells. The need to create tissues close to original tissues is essential.

C. Composites Composites are materials with a high modulus of elasticity and strength, and at the same time they are strong and tough, with a relatively low density, and with a number of other good properties. They are composed of two or more components, the so-called phases, whose chemical and physical compositions are different. In the constituent phase of the composite system each separate component is chemically and physically defined and is separated from the other components by a distinct phase boundary. Present knowledge of composite materials, such as bones, has been used for the design of artificial composite materials. The properties of components of composite materials are almost isotropic, made from polycrystalline metal or nonmetal materials, the resulting properties of composite materials are anisotropic properties. Todays' composite materials are composed of reinforming fibers. The role of the fibers is to stiffen the basic matrix of a material, which has relatively low strength and toughness. The ratio between the modulus of elasticity or the strength of the materials available for use as reinforming fibers and their specific weight represents the criterion of their suitability for specific purposes. The composite materials use relatively short fibers of an extreme strength for their reinforcing structure—the so-called whiskers, that is, short monocrystalline fibers of metallic, carbide, nitride, or silicon origin. These fibers are of lengths ~1 mm and of diameters <10 μm. Composite materials based on carbon are of the following types: polycrystalline isotropic carbon, carbon reinforced by silicon carbide, and carbide reinforced by carbon fibers. All these materials are biologically acceptable, inert in a physiological environment, and can be easily sterilized. Moreover, they have good mechanical and tribological properties, and, therefore, they satisfy demands required for friction components of joint replacements. Their rigidity is similar to that of bone tissues and their fatigue properties are good. Their coefficient of friction is 0.06. Their resistance to wear is extremely high even under the extreme friction conditions in the knee and hip joints, with their high pressures, low friction velocities, and cyclic movements. One of these materials is BIOCARB, the other one are SEPCARB and CERASEP, which were developed in France and Germany. These materials are transparent on X rays due to their low density of $\sim 2\,g\,cm^{-3}$. Table 3.18 gives a comparison of composites with other biomaterials used in orthopedics [after Rabinowitz (1995), Valenta (1993), and Nedoma et al. (2006)].

The material simulation of these type of biomaterials is very complicated. Since the bone structures represent basically a porous structure, they can be assumed to be composite materials with various (periodic or stochastic) solid and fluid phases. Chapter 5 will discuss how to study such a type of composite material. The method is known as the homogenization method.

Porous Surfaces

A. Porous Metal Surfaces Porous metal surfaces applied on to joint replacements are of the same metal materials as the original materials of the joint replacements. Due to their excellent biocompatibility, titanium (Ti) and cobalt (Co) permit good tissue integration. Therefore, titanium and cobalt alloys are used for these technologies. These technologies are based on three types:

1. Technology of sintering in which spherical fractions are under very high temperatures connected with the original surface of the joint replacement. The spherical fractions are mutually connected and they create a porous structure defined by the sizes of used fragments.
2. Technology of diffusive coupling is based on metal fibers of small thickness, which are adapted under high pressures into the needed design and then are

TABLE 3.18 Comparison of Properties of Biomaterials

Material	Density ($g\,cm^{-3}$)	R_m (MPa)	R_f (MPa)	E (10^{11} Pa)	BF 10^{-3}	A (%)	Hardness (kg/mm^2)
CrCoMo alloy	8.3	1050	550–560	2.08	1.5	8–16	290
CoNiCr alloy	7.8	800–1200	550–560	2.2	2.3	8–40	900–1250
Naturally pure Ti	4.51	390–540	150–200	1.0	1.8	22–30	65
Ti alloy	4.4	930–1140	350–650	1.15	5.2	8–15	2450 (TiC)
Al_2O_3	3.9	350	400	3.8	1.05	<1	1860–2110
ZrO_2	5.5–6.05	350–400	450	1.7	2.6	>1	1150–1400
Composite	~2.0	500		0.45			900–1250
UHMWPE	0.9	37–46	16	0.12	13.3	600	
Cortical bone	1.0	80–150	30	1.71	1.8		0.85–1.45
Bone cement PMMA		24–48	30	0.25	1.2	<1	

connected with the implant surface by the same process. Porosity of the surface is parametrically defined by the diameters of metal fibers, by the highest used temperature and pressure.

3. Technology of plasma spreading is one of the methods for producing TPS coating (Ti plasma spraying). This technology can be applied in plasma spreading of hydroxyapatite (HA). The principle of plasma spraying is the following: a direct current (dc) electric arc is struck between two electrodes while a stream of gases passes through this arc. This results in an ionized high-temperature gas ~30,000°C. A large gaseous expansion occurs as a result of the increase in gaseous temperature, thereby causing the carrier gas stream to pass through the arc at a speed approaching the speed of sound. A coating powder is suspended in the carrier gas stream, which has fallen into the plasma flame. Properties of plasma-sprayed coating depend on the relation between the particle size of a powder, the gas used, the speed of the plasma, the distance between the plasma nozzle and the basic foundation, and the cooling process and crystallinity. The optimal size of porosity is ~150–350 μm. The metal fibers are created from metal fibers of the length of ~4–6 mm and the diameter of ~50–100 μm. The fibers are welded on the stem surface at a temperature of 1250°C and a pressure of ~100–400 MPa for 1–4 h inside the hydrogen atmosphere or in vacuum. The tensile strength is ~200 MPa, that is, ~35–40% of the strengths of the initial materials. The modulus of elasticity in compression is ~10 times lower than the modulus of elasticity in tension.

B. Porous Ceramic Surfaces Porous ceramic surfaces are developed for better coupling between an artificial joint replacement and a bone tissue. The mineral substances are applied onto the implant surface, and its chemical compositions are biologically closed to the bone tissue. Porous ceramic surfaces are made from the plasma-spraying of hydroxyapatite (HA). The coating can be of the following types: (i) HA coating on Ti foundation; (ii) TPS coating on Ti foundation; (iii) ZrO_2 coating on Ti foundation; (iv) HA coating on ZrO_2-coated Ti foundation; and (v) HA coating on TPS-coated Ti foundation. The adhesive strengths of HA coating on ZrO_2 coating Ti foundation and HA on TPS-coated Ti foundation were determined as the lowest at less than 30 MPa (Yang et al., 2004). Table 3.19 presents the adhesive strength of different coatings on Ti foundation. For more details see Yang et al. (2004), Nedoma et al. (2006), and the references presented here.

Bone Cement Materials used as a filling material for the space between the total joint replacement and the bone are known as bone cements, and they are used in orthopedic and dental specializations. These materials must satisfy the properties of biocompatibility, chemical and physical stabilities, sufficient strength, and other mechanical properties as well as their case of preparations.

Acrylic bone cement is made up of powder and liquid components. The powder component is based on the polymer—PMMA in the form of tiny grains and an initiator such as benzol peroxide. A commercial powder component includes copolymers of methylmethacrylate with methacrylate (MA), styrene (S), butyl metacrylate

(BMA), ethyl acrylate (EA), or ethyl methacrylate (EMA) (Kühn, 2000). Radio-opacity is provided by barium sulfate ($BaSO_4$) or zirconium dioxide (ZrO_2), and they may be added to the powder component. The liquid part consists substantially of MA monomer. Some commercial products contain also butyl methacrylate, *N*-decyl methacrylate (DMA), or isobornyl methacrylate (IBMA). Bone cement has a polymer–monomer ratio of 2 : 1. The initiator—benzyl peroxide—is the most common initiator used in bone cement formulations and it is added to the powder component. It is made of tri-*n*-butylborane and $B(C_4H_9)_3$. After mixing the liquid and the powder parts, the initiator produces radicals at room temperature and starts the polymerization reaction, causing an increase in temperature (Serbetci and Hasirci, 2004). Moreover, tertiary aryl-amine, N, and *N*-dimethyl-*p*-toluidine (DMPT) as an accelerator for polymerization reaction are applied as the liquid components. Liquid components of bone cements contain also inhibitors as radical scavengers—less toxic materials such as food-grade di-*tert*-butyl-*p*-cresol [see Brauer et al. (1986)]. Some additives such as radio-opacity antibiotics and fibers are added. Bone cements have viscoelastic properties. A few minutes after mixing both solid and liquid parts start to have a liquidlike consistency and change to a hard solid viscoelastic form (Dunne and Orr, 1998). The polymerization reaction process will be discussed in Chapter 5. For more details see also Kühn (2000) and Serbetci and Hasirci (2004), where some commercially available bone cement prescriptions are presented. Table 3.20 presents the main properties of bone cements.

Bone cement was first applied by Haboush in 1953 and by Sir John Charnley in 1958. Charnley succeeded with the bone cement in anchoring femoral head prostheses in the femur with in-site autopolymerization of PMMA. His studies represent new surgical techniques in orthopedic surgery (Charnley, 1960, 1970). For more details

TABLE 3.19 Adhesive Strength of Different Coating to Ti Foundation

Surface	Adhesive Strength [MPa]
HA coating on Ti foundation	32,50 ± 3,56
Ti coating on Ti foundation	54,10 ± 1,33
ZrO_2 coating on Ti foundation	32,30 ± 1,80

Source: Modified Yang et al. (2004).

TABLE 3.20 Main Properties of Bone Cements

Materials	Temperature (°C)	Strength in Bending (kp/cm^2)	Impact Strength (kp/cm^2)	Temperature Duration above 50°C
Palacros	98	650	7.6	7 min 24 sec
CMW bone cement (UK)	104	570	7.2	7 min 6 sec
Sufix (Switzerland)	83	550	5.5	6 min 9 sec
Akrylon (CR)	99	782	5.1	8 min 30 sec

see Charnley (1970), Brauer et al. (1986), Valenta (1993), Kühn (2000), Serbetci and Hasirci (2004), and Nedoma et al. (2006).

3.5 ARTIFICIAL JOINT REPLACEMENTS

3.5.1 Hip Arthroplasty

Indications for Hip Arthroplasty Total hip arthroplasty (THA) is indicated at joints that are so devastated by degenerative processes that they cannot function properly, and the patient will experience pain during any conservative approach to healing. The most frequent reason leading to a THA indication is primary coxarthritis, postdysplastic coxarthritis, posttraumatic coxarthritis, and degenerative changes as a consequence of rheumatic arthritis as well as other rheumatic diseases, specifically in some fractures of the upper part of the femur in older people. **Coxarthritis** is an osteoarthrosis of the hip joint. The cause of primary coxarthritis is unknown, but there exist some factors, such as heredity or chronical overloading, respectively, that influence its development. Secondary coxarthritis develops as a consequence of joint incongruity, originated at the base of hip joint dysplasia as well as of other disorders, traumatic changes, and inflammations. Arthroplasty is the surgical technique used when a damaged joint is replaced by an implant. It can be used almost for all human joints. THA is indicated at primary coxarthritis when all possibilities of the conservative methods have been exhausted and the progression of difficulties is evident also from an X-ray film.

During THA application, it is necessary to take into consideration the following:

1. Whether there is possibility for good anchoring of the artificial acetabulum and which type of the artificial acetabulum can be used
2. Whether it is necessary to apply the THA with shorter stems or thin stems because a marrow channel can often be narrow
3. Whether the short muscles permit reconstruction of the hip joint
4. Whether changes and complications of previous surgical treatments can endanger the success of the THA implantation.

At present in Europe approximately 450,000 THA implants are done each year and a similar situation exists in the United States.

Construction of the THA Alloarthroplastics is the surgical technique used when the destroyed acetabulum and the head of femur are replaced by artificial replacements, that is, by the THA. Total hip replacements are divided as follows:

1. Surface type, where the contact surface of the femur head is replaced only by an implant.
2. Cervicocapital type, when the implant replaces the whole head of the femur. These types are manufactured as (i) all metal, when the stem and the head

FIG. 3.1 Cervicocapital all-metal type.

create one body (see Fig. 3.1); (ii) combined, when the stem is metallic and the head is made from another material and is situated on the neck; and (iii) as a composite type.

3. Total type, when the implant substitutes both components of the hip joint, that is, the head as well as the acetabulum. They are manufactured (i) from one kind of the material as the all-metal type (this type has a very high coefficient of friction or it can be allergic to the cobalt) or as a ceramic type; (ii) as a combined type, when the head and the stem are metallic and the acetabulum is made from UHMWPE—the classical type, or as a postclassical type, where the ceramic head is situated on the metallic stem; and (iii) another types of the THA, but they are not often used.
4. Anatomical or tumorous types, when the implant replaces the acetabulum, the head, and also a part of the femur.

According to their use the THA can be divided as primary, revised, and tumorous, as follows:

A. Primary

   a. Cemented (see Fig. 3.2)

      i. With or without collars
      ii. Depending on the surface cutting the THA, they are divided into high polished and unpolished

   b. Uncemented (see Fig. 3.3)

      i. With a proximal fixation
      ii. Combined (proximal and distal)

B. Revised (see Fig. 3.4)

   a. Uncemented with a distal fixation (porocoated)
   b. Combined fixation
   c. Cemented with a long stem

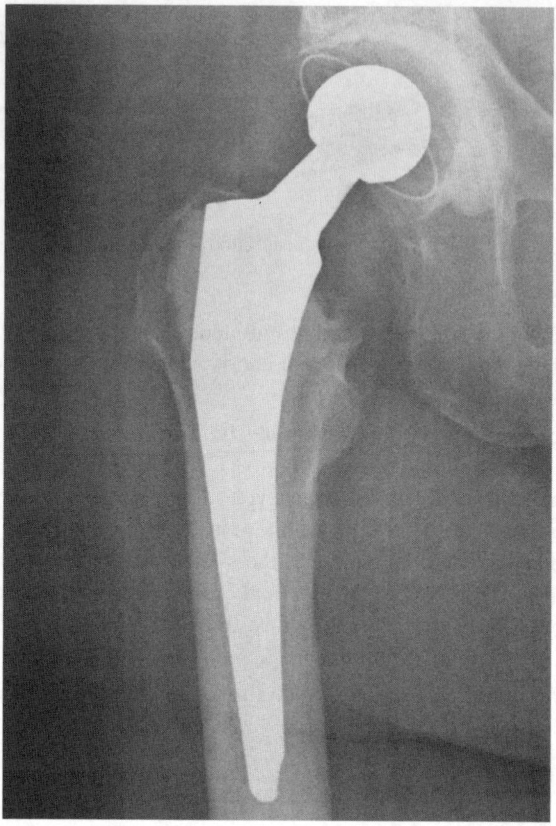

FIG. 3.2 Cemented THA (X ray).

C. Tumorous, used in the case of a bigger resection of the proximal femur (see Fig. 3.5)

The artificial acetabulae are divided as follows:

A. Primary

   a. Cemented
   b. Uncemented, which are divided into press-fit, expansional, and threaded

B. Revised, which are of special shapes solving defects of the acetabulum and they are of spherical-shaped or oval-shaped types

Stems According to the kind of fixation needed, the stems are divided into cemented and uncemented and then according to stabilizations into primary and revised.

A. Cemented Stems The cemented THA prostheses are made from two components—the artificial acetabulum manufactured from UHMWPE and from the

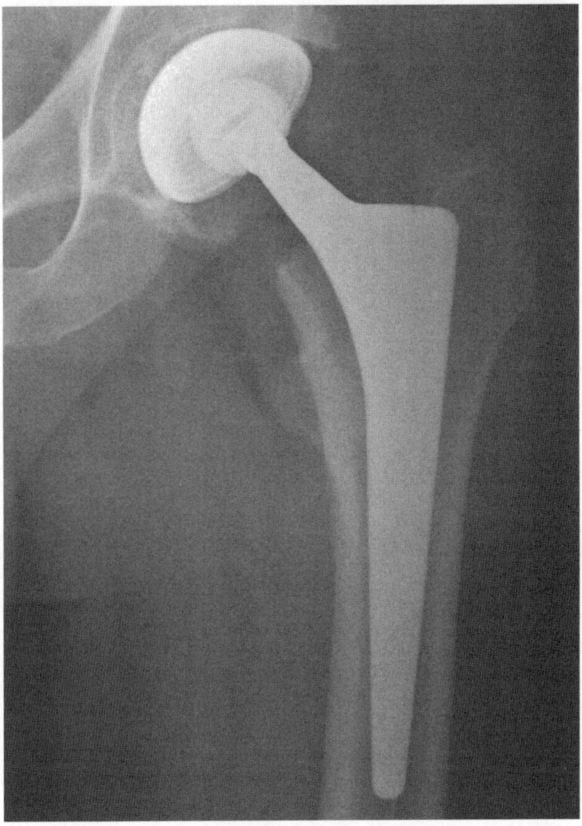

FIG. 3.3 Uncemented THA, pressfit cup (X ray).

metallic stem with the metallic or ceramic heads. The femoral stems are fixed with bone cement injected into a cavity of the proximal femur where has momentarily deteriorated contact with spongy bone. The stem end is made from a cone on which a metal head (from a cobalt–chrome–molybdenum alloy, special anticorrosive and nonmagnetic steels) or ceramic head in harmony with the need for stability and the length of the lower limb are applied. Therefore, the stem must be centered as exactly as possible such that between the wall of the bone and the stem surface a uniform space will be of about 2–7 mm thick, and this space will be later filled with bone cement. The margins of the stem must be rounded due to a minimalization of tension concentrations. The stem surfaces are finely roughened for a better stem fixation. The roughness of the stem is made by a bombardment of a biocompatible crushed material of a powdered corundum. Other ways of fixing the stems produce highly polished stems. The highly polished stem surface prevents wear between the bone cement and the implant. Existing small friction of the Coulomb type between the polished stem surface and the bone cement creates minimal wear.

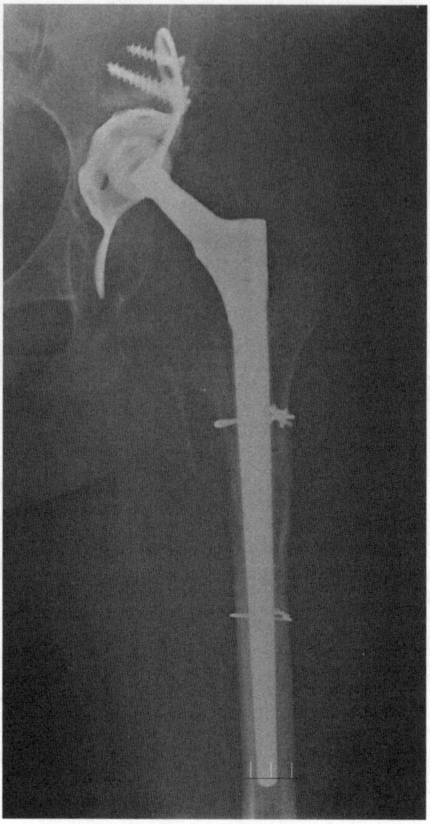

FIG. 3.4 Revisional THA (X ray).

Generally, from the mechanical point of view, the following minimal demands are imposed:

Tensile strength R_m	min. 650–1000 MPa
Yield point $R_{p0.2}$	min. 500–650 MPa
Ductility *A*	min. 10%
Corrosion fatigue limit	min. 400–450 MPa.

B. Uncemented Stems The optimal stem shape of uncemented THA is important to attain perfect primary THA stability. A coarsened stem surface or its further treatment by hydroxyapatite, thanks to an osteointegration, guarantees long-term secondary stability. The systems with proximal anchoring are the most suitable systems because physiological force transfers to the proximal part of the femur. By using the **press-fit technique**, that is, by a fixed pressing into an accurate cutting bone bed, stems of different shapes are fixed in the proximal end of the femur. Stems are

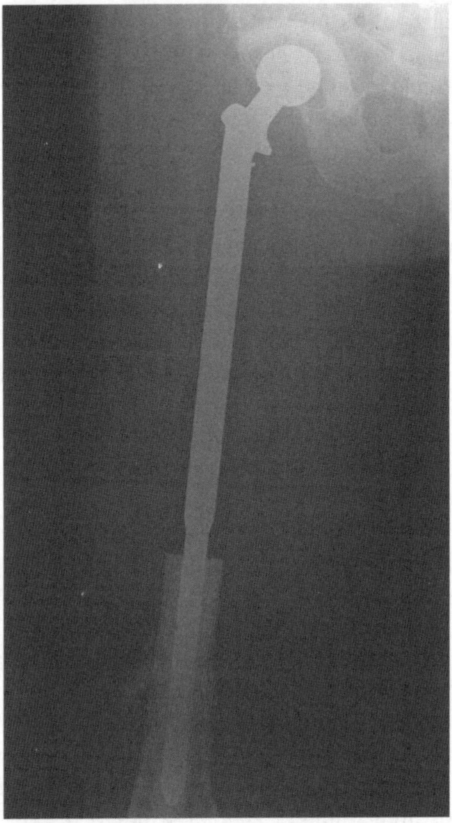

FIG. 3.5 Tumorous hip arthroplasty (X ray).

provided with a porous surface, the **porous coating**, which is made by a special technology called **sintering** or by plasma spraying of a porous titanium alloy of the same chemical composition as an initial forged implant material or a fibrous structure. Over this underlying layer a spray of hydroxyapatite [$Ca_{10}(PO_4)_6(OH)_2$] is often applied. The average surface roughness of the stem is ~3–5 μm. The aim of this process is to increase the contact surface with the bone wall and, therefore, increase significantly the secondary stability. In the case of **hydroxyapatite spraying**, the goal is to activate osteoblasts to the structural (coupling) osteogenesis, because lamellae of newly created bone tissues coalesce together with hydroxyapatite spraying.

Acetabular Cups

A. Cemented Acetabular Cups Optimal fixations of cemented cups of the acetabulum in the pelvis can be used after the space has been milled out and the damaged cartilage can be reached by a suitable polyethylene surface design either with the aid of a stiffened ring or by hybrid constructions. The layer of bone cement must be inserted in the milled-out space of the acetabulum, disposed of the damaged cartilage, and distributed uniformly with a minimal thickness of 3 mm. The cup must be in an exact position in the pelvis because after THA the joint is stabilized against a luxation by optimal centration both components and the muscles around the joint.

The artificial acetabulum is manufactured from UHMWPE, which has a good biological tolerance with the organism, a very small friction coefficient, and small abrasive wear. The manufacture of these artificial joints is based on a cutting technology or on a power pressing into a precision mold. The artificial acetabulum has a classical shape and is fixed in the pelvis with bone cement. The concentric slots on the external side of the cup are situated here for a better connection between the bone cement and the artificial acetabulum. The cross slots facilitate the elimination of any excessive bone cement. The layer of bone cement cannot be very thin (minimal 3 mm), the cups are manufactured in several sizes and then applied according to the acetabulum size.

B. Uncemented Acetabular Cups Uncemented acetabular cups use a precise Beaming of the bone bed for fixation into which the cup is inserted. The anchoring parts of the cup are manufactured of titanium and its alloys, eventually of CoCrMo alloys of spherical, conical, and biconical shapes and an articular part is manufactured from UHMWPE or ceramics (Al_2O_3, ZrO_2), rarely are they manufactured from highly smooth metals—the special alloys. Primary stability is reached by press-fitting, that is, by pressing the cup into a bone bed. Primary stability is also ensured by a polar flattening of the cup and by an angular anchoring that prevents luxation and ensures a rotating stability. Secondary stability is ensured by roughening the surface of the cup casing, eventually creating a porous macrocoersening and hydroxyapatite spraying as well as by an osteointegration. In special cases a titanium wire netting is used. For a higher rotational and tilting stability angular wide-open screw sleeves, pyramids, anchoring wedges, one or two bits, stabilizing fins, and helical cones are applied, and alternatively in addition the casing is secured by special screws, which increase the primary stability or expand the fixation principle. (See Fig. 3.6.)

Head of THA The heads are manufactured from metal alloys and from alumina (Al_2O_3) ceramics and zirkonium (ZrO_2) ceramics (Zirconia), alternatively from perfectly polished and surface-finished metallic alloys.

3.5.2 Knee Joint Arthroplasty

Gonarthritis is osteoarthritis of knee joints that afflicts the femoro-tibial and femoro-patellar joints. Gonarthritis is primarily a hereditary disorder in which local mechanical and chronicle overloading are exerted. Second, it ensues after traumas such as fractures and injuries of the cartilage or after joint knee arthritis.

In advanced stages of the condition, if a conservative treatment is not effective, a surgical therapy may be needed. In its early phases arthroscopy is used. Later, when only one compartment of the knee joint is handicapped, having a connection with an axis deformity in the sense of varosity and valgosity, the osteotomy approach is used. In case that the damage creates large axis changes, then hemiarthroplasty, that is, the

FIG. 3.6 Uncemented expansion cup.

special monocondylar knee replacement, is applied. If the destruction of the knee joint is greater, then **total knee joint arthroplasty (TKA)** is applied. (See Fig. 3.7.)

The great number of different types of implants facilitate the treatment of individual parts of the knee joint—the tibia, femur, and patella surfaces—and make possible the partial substitution joint ligaments (i.e., the **cruciate ligaments**) by using the ultracongruent insert. Total knee arthroplasty is manufactured as primary objects (primary implants, stabilized joint, etc.) or as building blocks for modular systems and certain problems, that is they must allow for revised surgical treatments. (See Fig 3.8.)

The artificial knee joint replacements can be divided using many factors:

A. According to construction—systems with or without movable parts or hinge type
B. According to kind of fixation—cemented, uncemented, and hybrid. Furthermore, TKA can be divided as follows:
   i. With a conservation of the posterior cruciate ligament (PCL), which can be used as standard types for primary knee joint arthroplasty
   ii. Without conservation of the PCL, which can be used as the rotating primary and revised arthroplasty
C. According to an indication—primary or revisional
D. According to a type of a tibial plateau—a fixed plateau (fixed bearing) and mobile plateau (mobile bearing)

At present new ways to improve functionality and to increase mobility of the total knee joint arthroplasty are being researched. An effort is directed at building compatible systems that allow alternative solutions depending on amount of joint destruction and secure revised operations and higher resistance against infections.

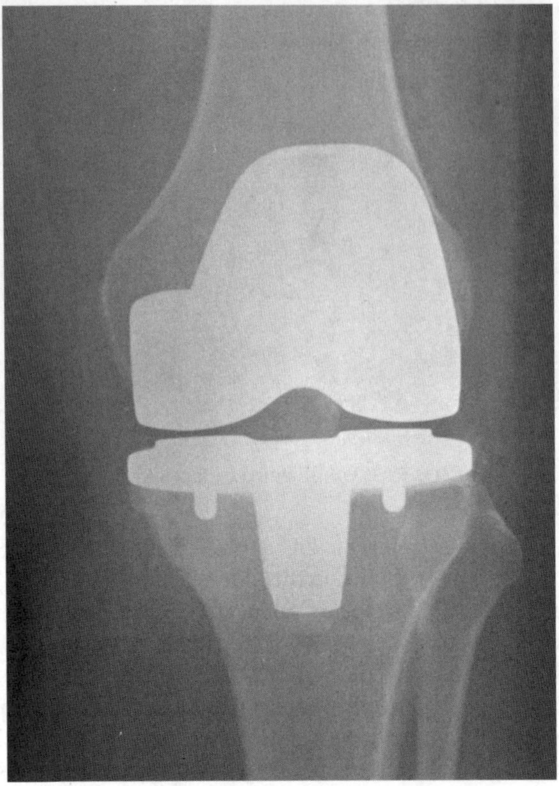

FIG. 3.7 Total knee arthroplasty (A-P X ray).

3.5.3 Replacements of Other Joints of the Upper and Lower Limbs

Artifical Shoulder Arthroplasty The first attempt at shoulder joint replacement was done by Péan in Paris 1892 when he manufactured an implant from platinum and rubber. The attempt was not successful because the patient contracted tuberculosis. In 1950 Judet accomplished a humerus head replacement, whose construction (based on Plexiglass–Perspex) was a modification of the hip arthroplasty. In 1951 Krüger applied a metal shoulder joint replacement. During the 1970s several new types were developed.

The Shoulder pain is caused by the destruction of the joint surfaces with consequences of restriction of motion. The artificial shoulder joint arthroplasty is divided as follows:

A. Partial shoulder joint arthroplasty, where the head of the joint or the proximal part of the humerus is replaced (see Fig. 3.9)

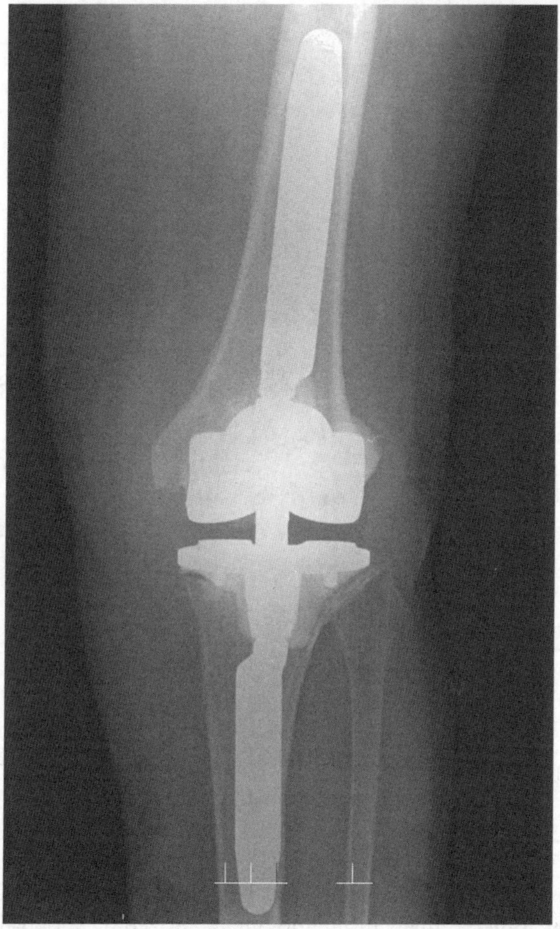

FIG. 3.8 Revisional knee arthroplasty (A-P X ray).

B. Total shoulder joint arthroplasty (see Fig. 3.10), which is divided into:

   i. Nonhinge system if the head and the pit are replaced, that is, a total shoulder joint arthroplasty
   ii. Reverse type into which a pit is implanted and the spherically shaped head and end of humerus are replaced by a cup fixed on the stem

Another division of the shoulder joint arthroplasty is (i) a primary type and (ii) a revised type. For these types of shoulder joint arthroplasty the principle of low friction, introduced by Charnley, can also be applied.

Artificial Elbow Arthroplasty The development of the artificial elbow arthroplasty (see Fig. 3.11), similar to other smaller joints starts at the end of the 1970s.

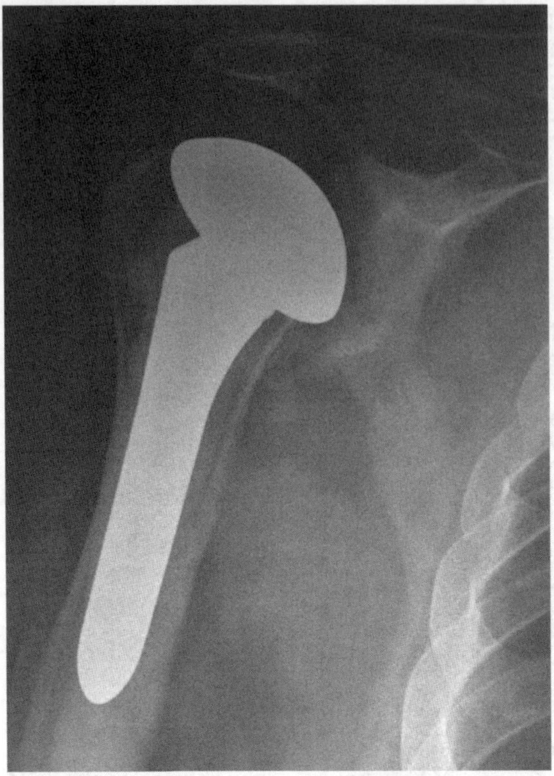

FIG. 3.9 Cervicocapital shoulder arthroplasty (X ray).

During the last decade greater attention has been paid to its development. Unfortunately, the elbow joint has many difficulties characterized by very bad conditions for the fixation or for anchoring the implant in the skeleton.

Artificial Replacements of the Ankle, Wrist, and Joints of the Hand

Ankle joint replacements (see Fig. 3.12) are not used on a large scale because of their relatively short service life. Therefore, in most cases an arthrodesis the joint is strengthened without the possibility of motion.

Wrist replacement (see Fig. 3.13) has not been used often because its development has not receive much attention. Since the biomechanics of the wrist and the small joints of the hand and of their functions is very complicated, their development needs different solutions in comparison with the arthroplasty of large joints.

Replacements of the small joints of the hand (see Fig. 3.14) are often used for the metacarpophalanged (MCP) joint and on the proximal interphalangeal (PIP) joints. In view of the fact that the MCP joint is important for finger function, its destruction leads to a malfunction of the hand. Surgical results have been satisfactory, but success depends on a well-timed indication, on the state of surrounding articular tissues and

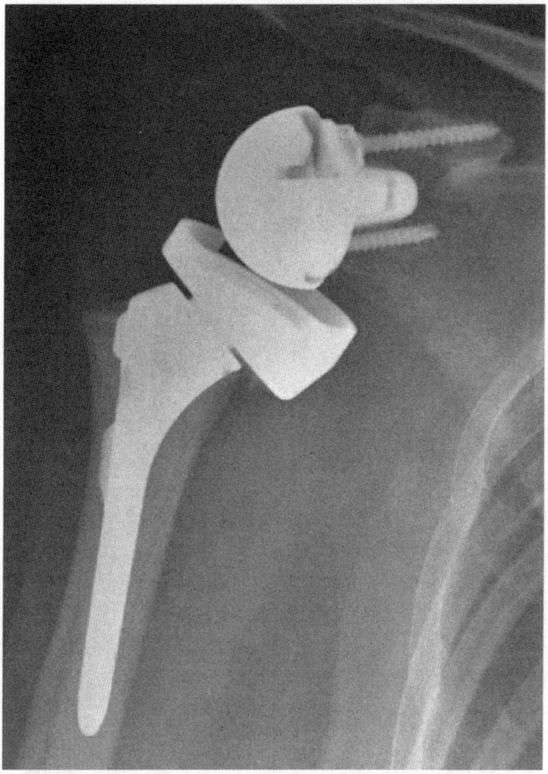

FIG. 3.10 Reverse total shoulder arthroplasty (X ray).

on the loadings and working of the hand. The service life of these prostheses is about 10 years.

3.5.4 Treatment of Toracolumbar Fractures

The human spine is a complex structure whose function is to transfer loads from the head and trunk to the pelvis and lower limbs and to protect the spinal cord. The spine ensures stability from the intervertebral disks and from the surrounding ligaments and muscles. The term spinal stability has different meanings and depends on the setting in which it has been used [see White and Panjabi 1978]. At spine injuries, namely the thoracic (Th) and lumbar (L) spine, many different types of fractures occur. Spine fractures can occur because of axial compression, axial distraction, axial rotation, and shear. The mechanism of the fracture origin has been explained by many classifications, which are discussed by Nicoll (in 1949), Holdsworth (1963, 1970), Whitesides (in 1977), Denis (1983, 1984), McAfee et al. (1983), Magerl (1984a,b), Ferguson (in 1984), and others. According to Magerl's classification, two fundamental mechanisms of the fracture origin are created from a flexion in the sagittal plane and a translation, that is, a shear, and a rotation in the transversal plane. Both mechanisms

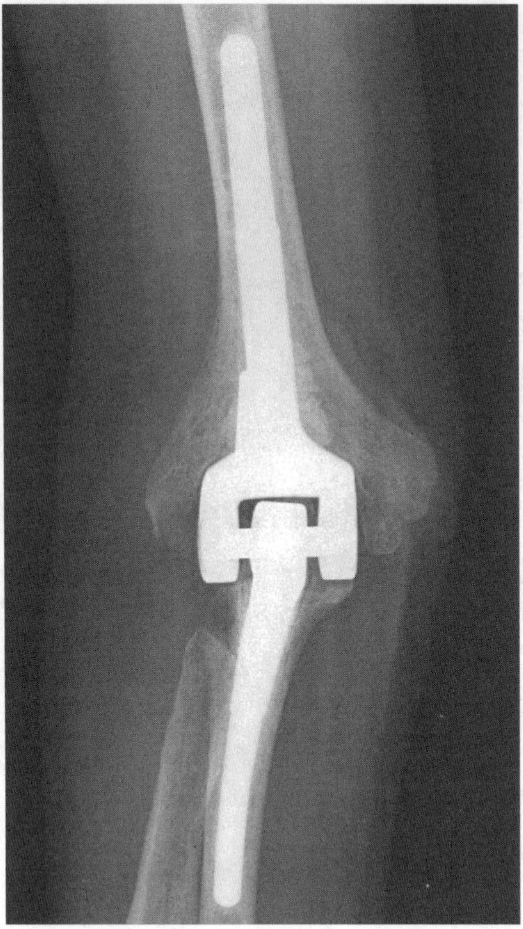

FIG. 3.11 Total elbow arthroplasty (A-P X ray).

are often combined. By a **reposition** we understand a renovation of the shape of a vertebra or a renovation of a mutual position of individual vertebrae (see Fig. 3.15).

By a **reconstruction** we understand a composition of the vertebral body. The replacement of the vertebral body is made by using a bone fragment fussion, which is then overbridged by an implant that stops the bone fragment in the correct position, where the implant is fixed to the neighbouring vertebrae. The bone fragment immediately carries part of the loading. The osteosynthesis in the case of a spine injury is always the overbridging osteosynthesis because no osteosynthesis of individual bone fragments can be possible (see Fig. 3.16). According to the classification of the origin of the fracture the fractures are divided as follows:

1. Fractures as a consequence of a flexion in the sagittal plane, where by a flexion we understand a rotating motion around the x axis. The structures situated in

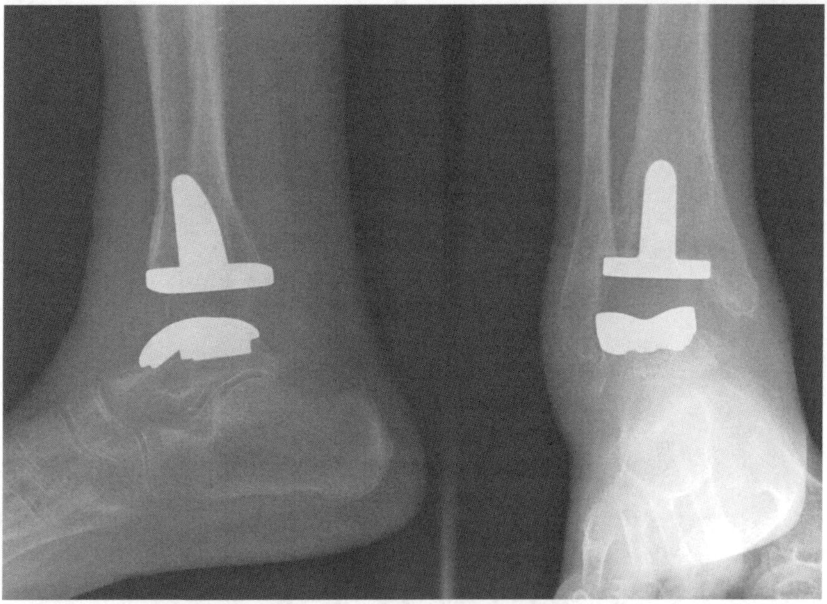

FIG. 3.12 Total ankle arthroplasty (X ray).

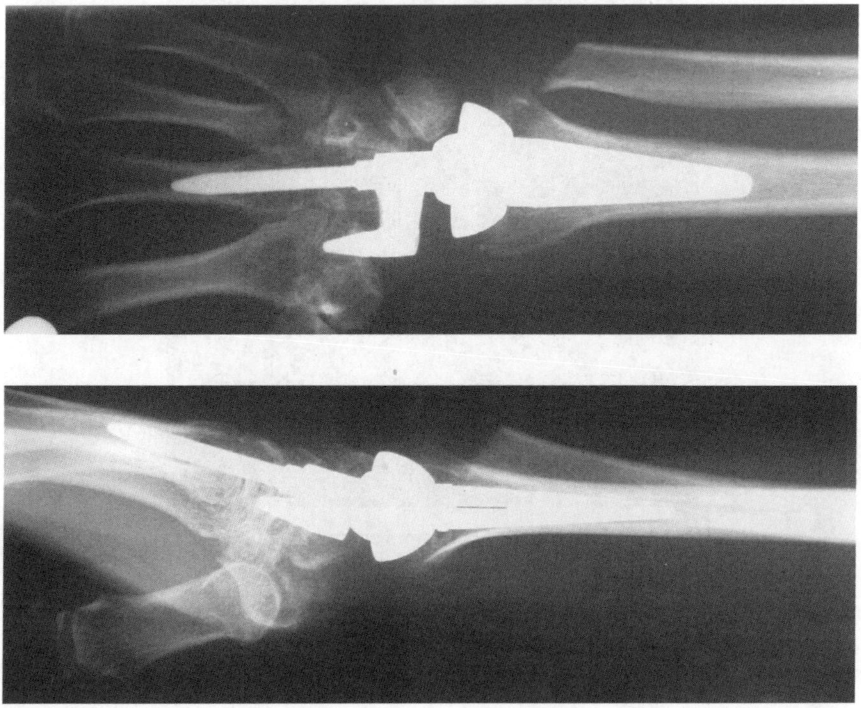

FIG. 3.13 Total wrist arthroplasty (X ray).

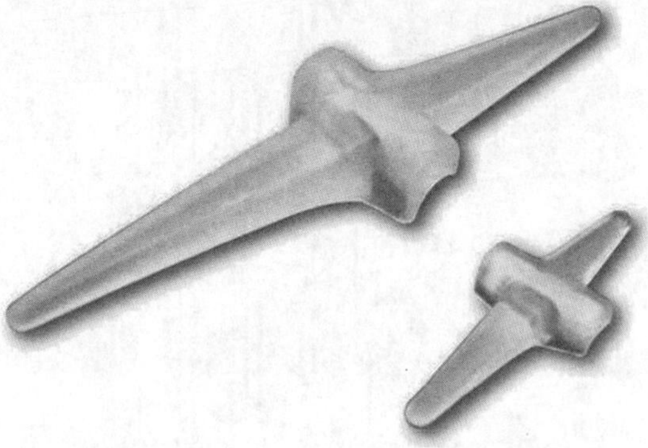

FIG. 3.14 Metacarpophalangeal joint implants.

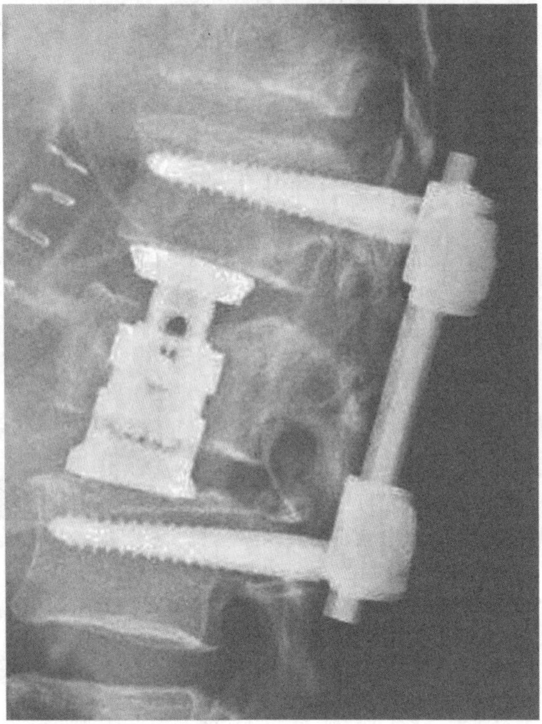

FIG. 3.15 Intervertebral cage from anterior approach (X ray).

front of this axis are exposed to a compression, and structures situated dorsally are exposed to a distraction. There are Chance fracture types, wedge fractures, stable and unstable burst fractures, and flexion–distraction fractures.

2. Fractures as a consequence of a translation in the transversal plane—these fractures originated due to a shear acting perpendicular to the spinal column or

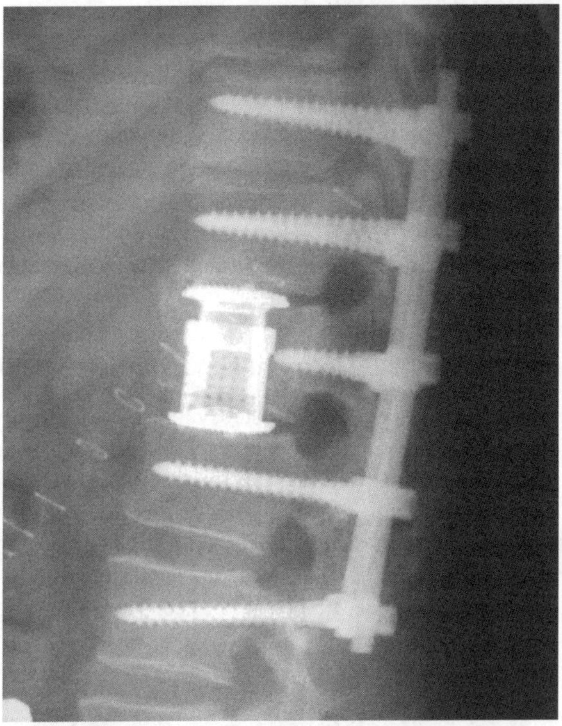

FIG. 3.16 Stabilization of thoracolumbar spine with combination of approaches (X ray).

due to a rotation around it. After a reposition the stability is ensured only in a compression and in other directions (i.e., in a shear, rotation, and distraction) there is instability.

3. Fractures as a consequence of a combination of both previous types—by this type of the mechanism the most complicated and unstable fractures originate. There are the fractures of dislocation types, of a flexion rotation type, burst fractures with a rotational component, and pure dislocation fractures.

The aim of the surgical treatment of spinal fractures is a renovation of neurological functions, retaining well-preserved neurological functions, and preserving and regeneration of the spinal stability. A surgical execution on the injured spine involves mostly decompression (release of the spinal cord and neural structures), reposition, and stabilization.

To ensure a reposition and a reconstruction overbridging with the help of implants has been used. Implants used to enable to a spinal stabilization must fulfill the following requirements:

1. The possibility to apply the implant over a small number of segments, where the segment is created by adjacent vertebrae, a vertical disk and ligamental structures

2. To ensure stability of the used assembly
3. To allow a correction in all directions and planes, that is, to allow a maximal number of degrees of freedom
4. To allow its biocompatibility and universality for applying in all ranges of the toracolumbar spine
5. To allow for the possibility of easy application and then for its easy extraction

For more details see Weber and Magerl (1984), and White and Panjabi (1978).

3.5.5 Fracture Treatments by Internal and External Fixations

In the 1970s in clinical practice external osteosynthesis started to be significantly promoted as a very effective method in open fractures, infected fractures, and bone deformities. Fracture treatments used internal fixations (screws, splints, nails, etc.) and external fixations (stabilizing screws with a different length of threads, swing balts, fixing bolts, telescopic stiffeners, telescopic arches of different length, etc.).

Internal fixation (see Fig. 3.17) is used in fractures of the bone body. It uses fixation splints situated directly on the bone. In the case of neck fractures, surgical treatments

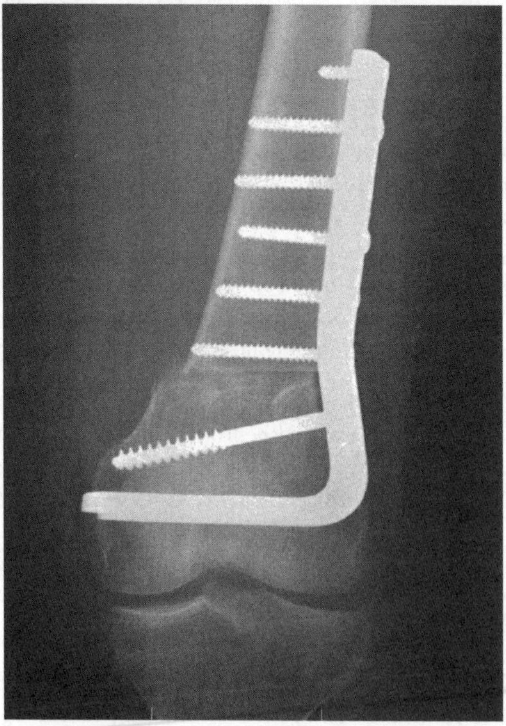

FIG. 3.17 Internal fixation—condylar plate.

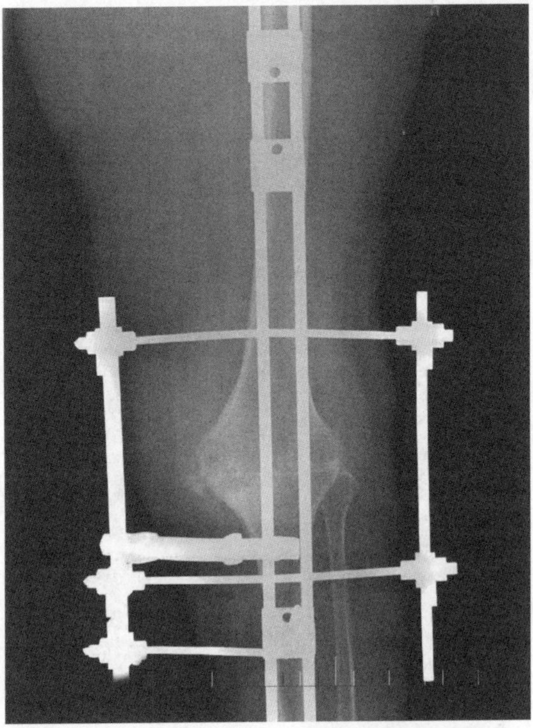

FIG. 3.18 Knee arthrodesis with external fixation.

of different types of osteotomy and types of arthrodesis, are used with internal fixation by using special fixation splints, nails, and screws, for example, the gamma nail.

External fixators (see Figs 3.18 and 3.19) are used in different kinds of fractures, from the simple type of fractures to multifragmental fractures and defective fractures. The surgical goal is as follows:

1. To reach maximal stability of the fitting using a minimal number of constructive elements. The stiffness of the fitting is determined by constructive elements of the devices, by a configuration of the fitting, as well as by the technique of application.
2. During the treatment to allow changes in stiffness of the fitting, by the biomechanical relations in the areas of the fracture would conform with biological needs, that is, a positive influence on the process of fracture healing.

To choose the right type (i.e., configuration) of the external fixation, it is necessary to take into consideration the following factors:

1. Type of fracture (i.e., stable or unstable fractures)
2. Localization of the fracture (i.e., the center of diaphysis, epiphysis, etc.)

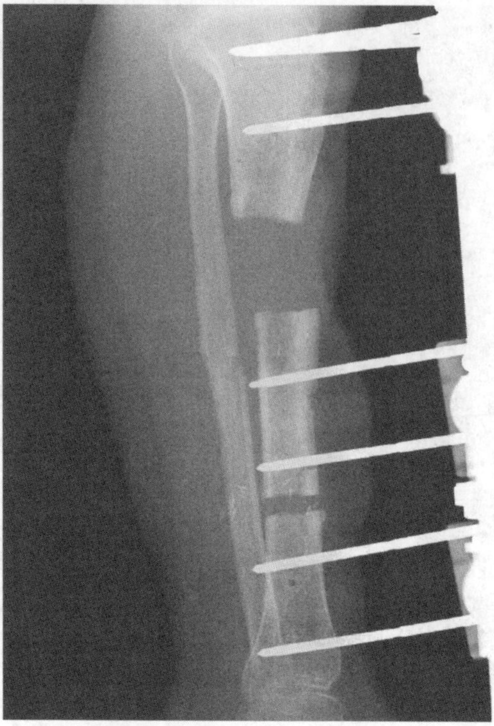

FIG. 3.19 External fixation for tibial nonunion (resection–distraction).

3. Anatomical localization (i.e., if the fracture is situated in the upper or lower limbs)
4. Anatomical and biomechanical relations in the given area (i.e., in the middle leg, thigh, etc.)
5. Height and weight of the patient

These factors cannot be ignored, but we can influence the choice of fixators and the way it is applied, that is, the configuration of devices and the surgical techniques used.

PART II

MATHEMATICAL MODELS OF BIOMECHANICS

CHAPTER 4

BACKGROUND OF BIOMECHANICS

4.1 INTRODUCTION

Biomechanics is concerned with the living materials that make up the human body. Therefore, biomechanics combines applied mechanics with biology and physiology and takes into account aspects of medicine, such as orthopedia. Applied mechanics is then divided into three main parts: (i) rigid body mechanics, which is divided into statics and dynamics, which is divided into kinematics and kinetics, (ii) deformable body mechanics, which is divided into elasticity, plasticity, viscoelasticity, and viscoplasticity, and (iii) fluid mechanics, which is divided into mechanics of liquids and gases. Solid materials can be rigid or they can be deformable. A rigid body is one that cannot be deformable. Statics deals with the study of forces on rigid bodies at rest or moving with a constant velocity. Therefore, statics represents an area of mechanics that is concerned with the analysis of rigid bodies in equilibrium. The term equilibrium implies that the body is either at rest or moving with a constant velocity. The body is rigid and it can maintain a posture and flexible in the sense that it can change its posture and movement. Dynamics deals with bodies in motion. One of its part is kinematics, which deals with geometry and time-dependent aspects of motion without considering the forces causing the motion. Kinetics includes the effects of forces and masses on mechanics analysis and, therefore, it depends on kinematics. Then statics and dynamics deal with the study of the external effects of forces on rigid bodies; the mechanics of deformable bodies deals with the relations between externally applied loads and temperature and their internal effects on bodies, that is, it is concerned with the deformability of objects. Therefore, we can assume elastic, viscoelastic, and plastic bodies. Elastic bodies are bodies in which all deformations are recoverable upon removal of the forces acting on them. Plastic bodies are bodies in which unrecoverable (permanent) deformations remain even when the forces acting on them are removed. In biology there exist materials that have properties of

Mathematical and Computational Methods in Biomechanics of Human Skeletal Systems: An Introduction, First Edition. Jiří Nedoma, Jiří Stehlík, Ivan Hlaváček, Josef Daněk, Taťjana Dostálová, and Petra Přečková.

solid materials and at the same time properties of a fluid. The viscosity is defined as a quantitative measure of resistance of flow. Many biological materials exhibit viscoelastic properties. Since in biomechanics the results of applied mechanics, biology, and physiology are combined, therefore, in biomechanics the kinematics, kinetics, deformable human body mechanics, and fluid mechanics are also investigated.

The development of the special disciplines in biomechanics has improved our understanding of many aspects in medicine, including normal and pathological situations, mechanics of growth, and different forms, and it has contributed to the development of medical diagnostic and treatment procedures as well as to the development of designing and manufacturing medical instruments, artificial replacements, and implants of different parts of the human body as well as devices for the handicapped patients. A great portion of such developments is associated with orthopedics. The most frequently preformed surgery techniques in the world are used on patients with musculoskeletal problems. Therefore, biomechanics has become an everyday part of clinical practice. Present medical and orthopedic research includes partly surgery, artificial prosthesis of different parts of the human skeleton, implantable materials, artificial limbs but partly also cellular and molecular aspects of healing in relation to stress and strain as well as tissue engineering of bone, cartilage, and tendon. Moreover, the biomechanical research of trauma, injury, and rehabilitation is more important in orthopedical therapy. The most important contribution of modern biomechanics to medicine represents its promotion of a better understanding of physiology.

Therefore, in biomechanics, the principles of mechanics are applied to the conception, design, development, and analysis of equipment and systems in medicine, especially in orthopedics and in biology. Thus our approach to the study of biomechanical problems in orthopedia will be based on the following steps: (1) First, the morphology of the organism, anatomy of single parts of the human body, histology of the tissues, the structure of the biomaterials, and the geometry of the investigated part of the human skeleton are studied in detail. (2) Second, mechanical properties of the tissues and biomaterials of the investigated object are determined. Due to nonlinearity many biological tissues are often subjected to large deformations; and, moreover, in many cases they are historically dependent on the previous deformations. (3) Third, on the basis of the fundamental laws of physics such as the conservation laws of mass, momentum, and energy, in their integral or differential forms, and the constitutive equations of biomaterials, the governing integral or differential equations are derived. (4) Then on the basis of the function of the studied object and about the environment in which an object works we find the meaningful boundary and contact as well as the initial conditions. (5) Then we analyze the model problem mathematically, that is, we prove the existence and uniqueness of the solution in the continuous domain. Also, after the approximation of the model problem by an optimal numerical method, we analyze the numerical model problem, that is, we prove the existence of the numerical solution, its uniqueness, as well as its convergence to the solution of the initial continuous model problem. Moreover, we analyze all convergences of the used algorithms. (6) Then the obtained results must be compared with the reality and experimental data. By means of this comparison, we determine whether the hypotheses applied in our theory and the data used are correct or not and whether the choice of the numerical approximation is usable or not.

Therefore, determination of the constitutive relations starts to be very difficult. We must accept that from the biological reality to the biomechanical analyses of model problems there are several steps that cannot be proved in detail because they are based on intuition, knowledge of orthopedics, and knowledge of biomechanics and mathematics. Only if we finished the construction of the final mathematical model, that is, if we define equations of equilibrium or of motion, the constitutive relations, boundary, contact, and initial conditions, can we then analyze mathematically the model problem and its numerical approximation. All previous steps are only intuitive steps without any possibility to prove their validities. We verify their validities on our previous knowledge about the biomechanical problem studied and about the behavior of our investigated object in reality.

The most weighty problems for biomechanics analysts represent the decision about acceptability of the constitutive equations of living tissues because without the constitutive laws, no analysis can be done. On the other hand, without the solution of (initial) boundary value and/or contact problems, the constitutive laws cannot be determined.

4.2 FUNDAMENTALS OF CONTINUUM MECHANICS

4.2.1 Introduction

Continuum mechanics is a branch of physics that is concerned with the motion and deformations of bodies that are acted on by forces. In continuum mechanics, kinematics refers to the mathematical description of the deformation and motion of a material particle (we speak also about a material point), as opposed to dynamics, that is, the study of the influence of dynamic forces on motion, and to statics, that is, the study of the influence of static forces on deformation of a continuum (which are independent of time). The objective of continuum mechanics is to provide models for the macroscopic behavior of solids, fluids, and structures, where inhomogeneities such as molecular, grain, or crystal structures are ignored, but they are introduced through the constitutive equations, and, moreover, where the response and the properties are assumed to be smooth with a finite number of discontinuities. For more details see Duvaut and Lions (1976), Segal (1977), Fung (1977), Nečas and Hlaváček (1981), Ciarlet (1988), Marsden and Hughes (1983, 1994), Zhong (1993), Bonet and Wood (1997), Simo and Hughes (1997), Nedoma (1998a), Belytschko et al. (2000), and Zienkiewicz and Taylor (2000).

4.2.2 Deformation, Motion, Stresses, Strains, and Conservation Equations

Deformation and Motion of a Body We assume that an origin O and an orthogonal basis $\{\mathbf{e}_1, \ldots, \mathbf{e}_N\}$, $N = 2, 3$, have been chosen in the N-dimensional Euclidean space, which will be then identified with the space $\mathbb{R}^N$. Let Ω_0^ι be a domain in $\mathbb{R}^N$,

$N=2,3$, $\iota=1,2,\ldots,s$ and let us assume that Ω_0^ι is occupied by a continuum, that is, bodies being or not being in contact, in an initial state at time $t=t_0(\equiv 0)$ and by $\Omega^\iota(t)$, $\iota=1,\ldots,s$, the subset of $\mathbb{R}^N$ occupied by a continuum at time $t>0$. We assume that Ω_0^ι are bounded, open, and connected subsets of $\mathbb{R}^N$. Then we speak about the initial or undeformed configuration (Fig. 4.1). The boundaries of Ω_0^ι will be denoted as $\partial\Omega_0^\iota$, $\iota=1,2,\ldots,s$, $\partial\Omega_0^\iota=\overline{\Gamma}_{0\tau}^\iota\cup\overline{\Gamma}_{0c}^\iota\cup\overline{\Gamma}_{0u}^\iota$, where $\overline{\Gamma}_{0\tau}^\iota$ denotes part of the boundary where the body is loaded, $\overline{\Gamma}_{0c}^\iota$ is the contact part of the boundary, and $\overline{\Gamma}_{0u}^\iota$ denotes the part of the boundary where the displacements are or are not prescribed, all in the initial configuration. By the reference configuration we mean the configuration to which various equations are referred. As the reference configuration the initial configuration can be used. The domain corresponding to the deformed configuration (or current configuration) is denoted by $\Omega^\iota(t)$, $\iota=1,2,\ldots,s$, and the boundary by $\partial\Omega(t)=\overline{\Gamma}_\tau(t)\cup\overline{\Gamma}_c(t)\cup\overline{\Gamma}_u(t)$, where $\Gamma_\tau(t)$ denotes the part of the boundary where the continuum is loaded, $\Gamma_c(t)$ is the contact part of the boundary, and $\Gamma_u(t)$ denotes the part of the boundary where displacements are prescribed, all in the deformable configuration.

As a continuum (bodies) moves, it is natural to determine the position of each point as a function of time $t\in\overline{I}\equiv[t_0,t_1]$, where t_0 is an initial and t_1 a final time of the time interval I. The position vector of a material point in the reference (initial) configuration is given by $\mathbf{x}_0$, where $\mathbf{x}_0=x_{0i}\mathbf{e}_i\equiv\sum_{i=1}^N x_{0i}\mathbf{e}_i$, where x_{0i} are the components of the position vector in the reference configuration and $\mathbf{e}_i$, $i=1,\ldots,N$, are unit vectors of a rectangular Cartesian coordinate system. The vector variable $\mathbf{x}_0$ for

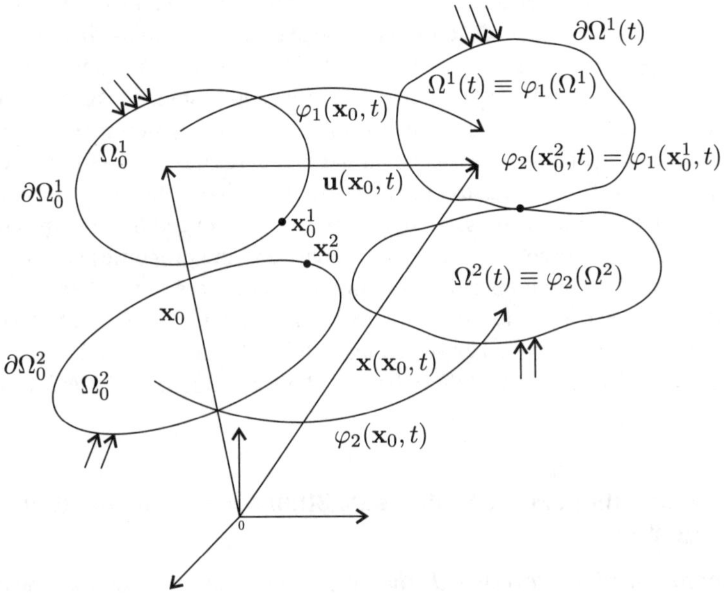

FIG. 4.1 Nondeformed (initial) and deformed (current) configurations of the system of bodies and the finite deformation contact.

a given material point does not change with time t, and the variable $\mathbf{x}_0$ is called the material or the Lagrangian coordinate. The position of a material point in the deformed (current) configuration is given by $\mathbf{x} = x_i \mathbf{e}_i \equiv \sum_{i=1}^{N} x_i \mathbf{e}_i$, where x_i are the components of the position in the deformed configuration, Eulerian coordinates.

Remark 4.1 Next, we will use the following notation for the functions in Eulerian and Lagrangian coordinate systems: the function, which in the Eulerian description is $f(\mathbf{x}, t)$, is denoted by $F(\mathbf{x}_0, t)$ in the Lagrangian description and it holds

$$F(\mathbf{x}_0, t) = f(\varphi(\mathbf{x}_0, t), t) \quad \text{or} \quad F = f \circ \varphi.$$

The motion of the continuum is described by

$$\mathbf{x} = \varphi(\mathbf{x}_0, t), \qquad [\text{or } \mathbf{x} = (x_i), \quad x_i = \varphi_i(\mathbf{x}_0, t)], \tag{4.1}$$

where $\mathbf{x}$ is the position of the material point $\mathbf{x}_0$ at the time $t \in \bar{I}$. The coordinates x_i give the spatial position of a particle of the continuum and they are called the **spatial or the Eulerian coordinates**. The function $\varphi(\mathbf{x}_0, t)$ maps the initial configuration into the deformed configuration at the time $t \in \bar{I}$. The word "particle" is used to denote the point that starts at a fixed initial position $\mathbf{x}_0$ and moves in accordance with Eq. (4.1). For fixed $\mathbf{x}_0$ the curve $\mathbf{x} = \mathbf{x}(\mathbf{x}_0, t) \equiv \varphi(\mathbf{x}_0, t), t \geq t_0$, is called a particle path. When the reference configuration is identical to the initial configuration, the position vector $\mathbf{x}$ of any point at the time $t = t_0$ coincides with the material coordinates, so that

$$\mathbf{x}_0 = \mathbf{x}_0(\mathbf{x}_0, t_0) \equiv \varphi(\mathbf{x}_0, t_0) \qquad [\text{or } x_{0i} = x_{0i}(\mathbf{x}_0, t_0) \equiv \varphi_i(\mathbf{x}_0, t_0)] \tag{4.2}$$

and the mapping $\varphi(\mathbf{x}_0, t_0)$ is the identity mapping. We speak about a **material description** or a **Lagrangian description** if the independent variables are the material coordinates x_{0i} and the time t, and about a **spatial** or **Eulerian description** if the independent variables are the spatial coordinates $\mathbf{x}$ and the time t.

In solids, since the stress generally depends on the deformation and its history, an undeformed configuration must be specified. Therefore, Lagrangian descriptions are mostly used. In fluid mechanics, since it is impossible and unnecessary to describe the motion with respect to a reference (initial) configuration, the Eulerian descriptions are used. For example, in Newtonian fluids stresses and other properties are independent of their histories. The displacement of a material point $\mathbf{u}(\mathbf{x}_0, t)$ is given by the difference between the deformable (current) position and the original position of a material point (see Fig. 4.1), so that

$$\mathbf{u}(\mathbf{x}_0, t) = \varphi(\mathbf{x}_0, t) - \varphi(\mathbf{x}_0, t_0) = \varphi(\mathbf{x}_0, t) - \mathbf{x}_0, \tag{4.3}$$

where $\mathbf{u}(\mathbf{x}_0, t) = u_i \mathbf{e}_i$ or in an equivalent form

$$\mathbf{u} = \mathbf{x} - \mathbf{x}_0, \qquad u_i = x_i - x_{0i}. \tag{4.4}$$

Definition 4.1 Let Φ be a scalar or vector function defined on $\Omega \times \mathbb{R}_+$ in the Lagrangian frame, where $\mathbb{R}_+$ is the set $[0, +\infty)$. The derivative of Φ with respect to time with the $\mathbf{x}_0$ constant is called the material or the Lagrangian derivative (or the total derivative) of Φ and it is denoted by $d\Phi/dt \equiv \Phi'$ $(\equiv \dot{\Phi})$, that is, $\Phi'(\mathbf{x}_0, t) = (d\Phi/dt)(\mathbf{x}_0, t)$ $\forall \mathbf{x}_0 \in \Omega_0$, $t > 0$. Now let Φ be a scalar or a vector function defined in the Eulerian frame, that is, Φ depends on $\mathbf{x} \in \boldsymbol{\Omega}(t)$, $t > 0$. The material derivative of Φ is the derivative of the function $t \to \Phi(\varphi(\mathbf{x}_0, t), t)$, for $\mathbf{x}_0 \in \Omega_0$ fixed, and generally, it is different from the time derivative of the function $t \to \Phi(\mathbf{x}, t)$ for a fixed $\mathbf{x} \in \Omega(t)$, and it is called the Eulerian or the spatial derivative of Φ and it is denoted by $\partial\Phi/\partial t$. Then

$$\Phi'(\mathbf{x}, t) = \frac{d}{dt}\Phi(\mathbf{x}, t) = \frac{d}{dt}(\Phi(\varphi(\mathbf{x}_0, t), t)) = \frac{\partial \Phi}{\partial t}(\mathbf{x}, t) + \frac{\partial \Phi}{\partial x_i}(\mathbf{x}, t)\frac{\partial \varphi_i}{\partial t}(\mathbf{x}, t).$$

The first term is known as the local derivative; the second one is known as the convective part of the time derivative.

The velocity $\mathbf{v}(\mathbf{x}_0, t)$ is defined as the rate of the change of the position vector for a material point. Then the velocity can be written as

$$\mathbf{v}(\mathbf{x}_0, t) = \frac{\partial \varphi(\mathbf{x}_0, t)}{\partial t} = \frac{\partial \mathbf{u}(\mathbf{x}_0, t)}{\partial t} \equiv \mathbf{u}'(\mathbf{x}_0, t) \quad (\equiv \dot{\mathbf{u}}). \tag{4.5}$$

The acceleration $\mathbf{a}(\mathbf{x}_0, t)$ is the rate of change of the velocity in a material point, that is, the material time derivative of the velocity, and then it can be written as

$$\mathbf{a}(\mathbf{x}_0, t) = \frac{\partial \mathbf{v}(\mathbf{x}_0, t)}{\partial t} = \frac{\partial^2 \mathbf{u}(\mathbf{x}_0, t)}{\partial t^2} \equiv \mathbf{v}'(\mathbf{x}_0, t) \quad (\equiv \dot{\mathbf{v}}). \tag{4.6}$$

When the velocity is expressed in an Eulerian description, that is, in terms of the spatial coordinates and the time t, then

$$\frac{dv_i(\mathbf{x}, t)}{dt} = \frac{\partial v_i(\mathbf{x}, t)}{\partial t} + \frac{\partial v_i(\mathbf{x}, t)}{\partial x_j}\frac{\partial \varphi_j(\mathbf{x}_0, t)}{\partial t} = \frac{\partial v_i}{\partial t} + \frac{\partial v_i}{\partial x_j}v_j, \tag{4.7}$$

where the second term on the right-hand side is the convective term or the transport term, $\partial v_i/\partial t$ is the spatial time derivative. Equation (4.7) can be written in an equivalent tensor notation as

$$\frac{d\mathbf{v}(\mathbf{x}, t)}{dt} = \frac{\partial \mathbf{v}(\mathbf{x}, t)}{\partial t} + \mathbf{v} \cdot \nabla \mathbf{v} = \frac{\partial \mathbf{v}}{\partial t} + \mathbf{v} \cdot \operatorname{grad} \mathbf{v}, \tag{4.8}$$

where $\nabla \mathbf{v}$ and $\operatorname{grad} \mathbf{v}$ are the left gradients of a vector field (another notation is $\partial_j v_i$) and $\partial \mathbf{v}/\partial t$ is the so-called local acceleration and the term $\mathbf{v} \cdot \nabla \mathbf{v}$ is the convective acceleration.

The material time derivative for a tensor function is

$$\frac{d\tau_{ij}(\mathbf{x}, t)}{dt} = \frac{\partial \tau_{ij}(\mathbf{x}, t)}{\partial t} + v_k \frac{\partial \tau_{ij}}{\partial x_k}; \qquad \frac{d\boldsymbol{\tau}}{dt} = \frac{\partial \boldsymbol{\tau}}{\partial t} + \mathbf{v} \cdot \nabla \boldsymbol{\tau} = \frac{\partial \boldsymbol{\tau}}{\partial t} + \mathbf{v} \cdot \operatorname{grad} \boldsymbol{\tau},$$

where the first term in each equality on the right-hand side is the spatial time derivative and the second one is the convective term.

Let $\Omega(t)$ denote the domain occupied by the moving continuum at time $t > 0$ and let $\Omega(t_0)$ denote the domain occupied by the continuum at the initial time t_0. Then the domain $\Omega(t)$ occupied by the designated continuum at time t is the image of $\Omega(t_0)$ at time t under the mapping $\mathbf{x} = \varphi(\mathbf{x}_0, t)$. Let $V(t_0)$ denote the volume of $\Omega(t_0)$, then

$$V(t_0) = \int_{\Omega(t_0)} d\mathbf{x}_0 \quad \text{and} \quad V(t) = \int_{\Omega(t)} d\mathbf{x} = \int_{\Omega(t_0)} J(\mathbf{x}_0, t)\, d\mathbf{x}_0, \tag{4.9}$$

where $J(\mathbf{x}_0, t) = \partial(x_i)/\partial(x_{0i})$, $i = 1, \ldots, N$, is the Jacobian of the transformation of variables. It is evident that $V(t)/V(t_0) \to J$ as $V(t_0) \to 0$. The Jacobian at point $\mathbf{x}_0$ and time t represents the dilatation of an infinitesimal volume initially at $\mathbf{x}_0$, where dilatation is the ratio of a volume occupied by the infinitesimal material region at the time t to its initial volume. For the time derivative of the Jacobian we have

$$\frac{dJ}{dt} = J(\nabla \cdot \mathbf{v}) = J \operatorname{div} \mathbf{v} \equiv J\frac{\partial v_i}{\partial x_i} \qquad \text{in the spatial variables,}$$

$$\frac{\partial J}{\partial t} = J[(\nabla \cdot \mathbf{v})|_{\mathbf{x}=\mathbf{x}(\mathbf{x}_0,t)}] \qquad \text{in the material variables.}$$

Stress Vector The stress measure defined in total Lagrangian formulations does not correspond to the physical stress. The physical stress, the so-called **Cauchy stress**, defined as the force per unit area, that is, $\tau = F/|\Omega|$, is referred to as the current area $|\Omega|$, where F is a total force across a given cross-sectional area. In the total Lagrangian formulation, we introduce the **nominal stress** $P = F/|\Omega_0|$. It differs from the physical stress in that the force is divided by the initial, undeformed, area $|\Omega_0|$. It can be seen that the physical and nominal stresses are related by $\tau = (|\Omega_0|/|\Omega|)P$ and $P = (|\Omega|/|\Omega_0|)\tau$. Next, we first introduce the Cauchy stress.

Remark 4.2 The basic unit of the force is the newton (N), thus the basic unit of stress is the newton per square meter (N/m^2) or pascal (Pa); 1 Pa = 1 N/m^2. A force of 1 N can accelerate a body of mass 1 kg to 1 m/sec^2. For the viscosity the unit 1 poice = 0.1 N sec/m^2 was introduced.

Consider a domain $\Omega \subset \mathbb{R}^N$, $N = 2, 3$, with the boundary $\partial\Omega$. Let $\mathbf{f}(\mathbf{x}) = (f_i(\mathbf{x})) \in [C(\Omega)]^N$ be a density of forces acting on each volume element G, such that $\overline{G} \subset \Omega$. We will assume that G has a Lipschitz boundary ∂G. Let $\mathbf{x} \in \partial G$ and let $\mathbf{n}$ be the unit outward normal to $\partial\Omega$ at point $\mathbf{x}$. Then, $\mathbf{t}(\mathbf{x}, \mathbf{n})$ is the force per unit area exerted at the point $\mathbf{x} \in \partial G$ by the material exterior to G on the material interior to G (Fig. 4.2), and it is called the **Cauchy stress vector**. It depends on the point $\mathbf{x}$ and on the direction of the normal $\mathbf{n}$.

In view of Newton's third law, that is, the principle of action and reaction

$$\mathbf{t}(\mathbf{x}, \mathbf{n}) = -\mathbf{t}(\mathbf{x}, -\mathbf{n}).$$

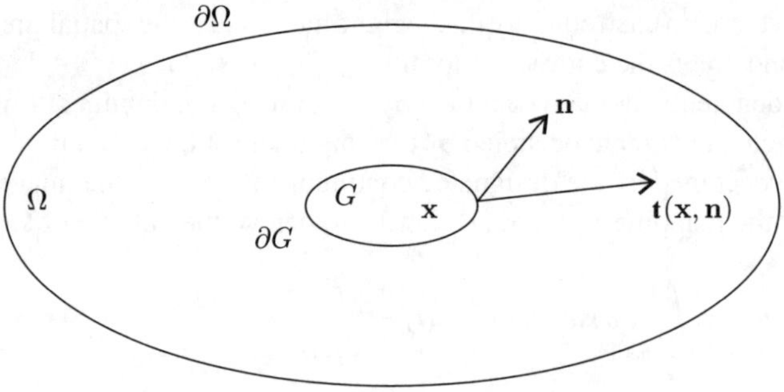

FIG. 4.2 Derivation of the Cauchy stress vector.

The projection of $\mathbf{t}(\mathbf{x},\mathbf{n})$ onto the normal $\mathbf{n}$ we denote as $N(\mathbf{x},\mathbf{n})$, and it is called the **normal stress** and it holds $N(\mathbf{x},\mathbf{n})=\boldsymbol{\tau}(\mathbf{x},\mathbf{n})\mathbf{n}$. The projection onto the tangential plane to ∂G we denote as $\boldsymbol{\Sigma}(\mathbf{x},\mathbf{n})$ and is called the **shear stress** or the **tangential stress**, and

$$\boldsymbol{\Sigma}(\mathbf{x},\mathbf{n}) = \mathbf{t}(\mathbf{x},\mathbf{n}) - N(\mathbf{x},\mathbf{n})\mathbf{n}$$

holds. If we choose for the particular directions of the normal those parallel to the coordinate axes for which $\mathbf{e}_1=(1,0,0)$, $\mathbf{e}_2=(0,1,0)$, and $\mathbf{e}_3=(0,0,1)$, then $\tau_{ij}=\tau_i(\mathbf{x},\mathbf{e}_j)$, where τ_{ii}, $i=1,\ldots,N$, are normal stress components or normal stresses and τ_{ij}, $i\neq j$, $i,j=1,\ldots,N$, are tangential (or shear) stress components or shear (tangential) stresses. Normal stresses are positive in the case of material elongation (i.e., by applying the traction) or in the case of dilatation, respectively, and negative in the case of material compression (i.e., by applying pressure).

Let us consider a tetrahedron $G\subset\Omega$ defined by $OABC$ (Fig. 4.3), three edges of which are parallel to the coordinate system, defined by the orthogonal basis vector $\mathbf{e}_i$, $i=1,\ldots,3$. The domain $OABC$ has a Lipschitz boundary and the outward normal $\mathbf{n}$ to ABC exists. The mutually perpendicular faces of the area S_i are normal to the orthonormal basis vector $\mathbf{e}_i$, and they meet at point $O\equiv\mathbf{x}$. The surface ABC has the area S. Then $S_i=S\cdot\mathbf{n}_i=S\cdot[\mathbf{n}\cdot\mathbf{e}_i]$. Then h is the distance of point $O\equiv\mathbf{x}$ from the plane ABC and the volume $|G|=\frac{1}{3}hS$. Then we investigate the equilibrium state applied to the tetrahedron [see Nečas and Hlaváček (1981)]. After some modification we find that

$$t_i(\mathbf{x},\mathbf{n}) = \tau_{ij}(\mathbf{x})n_j, \qquad i,j=1,2,3,$$

where the matrix $\tau_{ij}(\mathbf{x})$ is called the **Cauchy stress tensor** at point $\mathbf{x}$. It can be shown that the stress tensor has the character of a tensor of order 2. Denote by $\boldsymbol{\tau}=(\tau_{ij})$ the Cauchy stress tensor.

To define the **principal stresses** and **principal directions of stress**, we find a direction $\mathbf{n}$ for which the tangential stress vanishes. For such a direction $\mathbf{n}$ it holds

$$\tau_{ij}(\mathbf{x})n_j = \tau(\mathbf{x})n_i, \qquad i,j=1,2,3,$$

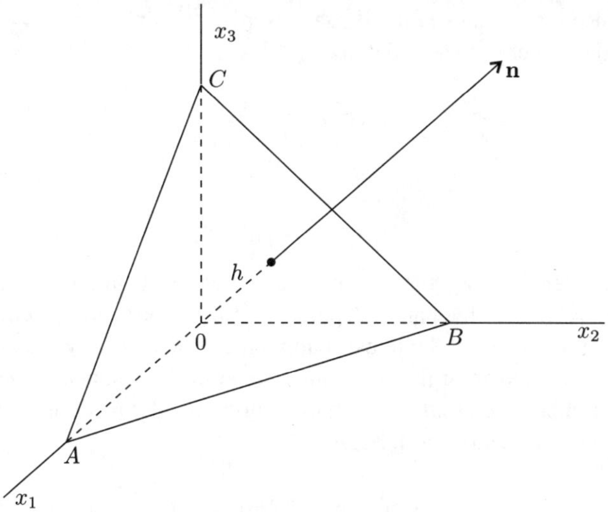

FIG. 4.3 Derivation of the Cauchy stress tensor.

representing a system of three homogeneous linear equations for normal components n_i, where τ is a parameter of proportionality. The necessary condition for the existence of a nonzero solution is

$$\det(\tau_{ij} - \tau\delta_{ij}) = 0 \tag{4.10}$$

or in an equivalent form

$$-\tau^3 + \Theta_1\tau^2 - \Theta_2\tau + \Theta_3 = 0 \tag{4.11}$$

having three real roots τ_1, τ_2, τ_3 in general, where

$$\Theta_1 = \tau_{11} + \tau_{22} + \tau_{33},$$

$$\Theta_2 = \begin{vmatrix} \tau_{11} & \tau_{12} \\ \tau_{12} & \tau_{22} \end{vmatrix} + \begin{vmatrix} \tau_{11} & \tau_{13} \\ \tau_{13} & \tau_{33} \end{vmatrix} + \begin{vmatrix} \tau_{22} & \tau_{23} \\ \tau_{23} & \tau_{33} \end{vmatrix},$$

$$\Theta_3 = \det(\tau_{ij}) = \frac{1}{3!}\varepsilon_{ijk}\varepsilon_{lmn}\tau_{il}\tau_{jm}\tau_{kn},$$

where the permutation symbol or the alternator or the Levi–Civita tensor ε_{ijk}, $i, j, k = 1, 2, 3$, which is the tensor of the third-order asymmetric in all indices and as such has 27 values and is defined as follows:

$\varepsilon_{ijk} = 0$ if any two of the subscripts i, j, k are equal or if all three are equal.
$\varepsilon_{ijk} = 1$ if (i, j, k) is an even permutation of $(1, 2, 3)$.
$\varepsilon_{ijk} = -1$ if (i, j, k) is an odd permutation of $(1, 2, 3)$.

We introduce the deviator of the stress tensor by $\tau_{ij}^D = \tau_{ij} - \frac{1}{3}\tau_{ii}\delta_{ij} = \tau_{ij} - \frac{1}{3}\Theta_1\delta_{ij}$. The trace of the Cauchy stress is defined as

$$\text{trace}(\boldsymbol{\tau}) \equiv \text{tr}(\boldsymbol{\tau}) = \tau_{ii}$$

and

$$\tfrac{1}{3}\text{tr}(\boldsymbol{\tau}) = \tfrac{1}{3}\tau_{ii}.$$

As the stress tensor is symmetric (see later), then the solution of (4.11) has three real roots τ_i, $i = 1, 2, 3$, which are independent of the coordinate system because the characteristic polynomial of the linear mapping, characterized by the matrix (τ_{ij}), is independent of the choice of the coordinate system. The numbers τ_i, $i = 1, 2, 3$, are called the **principal stresses**, for which three orthogonal directions $\mathbf{n}^i$ exist. We see that for a principal directions $\mathbf{n}^i$ it holds

$$|\mathbf{t}(\mathbf{x}, \mathbf{n}^i)| = |\tau_i| \cdot |\mathbf{n}^i| = |\tau_i|.$$

In a new coordinate system, whose axes coincide with the principal directions, we have $\tau'_{ii} = \tau_i$ (i is not summed), $\tau'_{ij} = 0$ ($i \neq j$). Then we can define the Cauchy quadric of stress.

Definition 4.2 Let $k \neq 0$ be an arbitrary real number. The quadratic surface defined by

$$\tau_{ij}\xi_i\xi_j = \pm k^2,$$

where $k \neq 0$, is called the **Cauchy quadric of stress**.

Under the transformation into the principal directions, the Cauchy quadric is of the form

$$\tau_1\xi_1^2 + \tau_2\xi_2^2 + \tau_3\xi_3^2 = \pm k^2$$

and the principal directions thus coincide with the axes of the quadric.

If $\tau_1 = \tau_2$, there is a symmetry with respect to the axis ξ_3. If $\tau_1 = \tau_2 = \tau_3$, then the quadric is the sphere. If $\tau_1 \geq \tau_2 \geq \tau_3 > 0$, then for $+k^2$ we have a general ellipsoid, and for $-k^2$ the empty set. In this case there is tension in each direction. If $0 > \tau_1 \geq \tau_2 \geq \tau_3$, then for $+k^2$ we have the empty set and for $-k^2$ an ellipsoid, and there is a pressure in each direction. If $\tau_1 \geq \tau_2 > 0 > \tau_3$, then for $+k^2$ we have a hyperboloid of one sheet, and for $-k^2$ we have a hyperboloid of two sheets.

The expression

$$\hat{S} = \left\{\tfrac{1}{6}[(\tau_1 - \tau_2)^2 + (\tau_2 - \tau_3)^2 + (\tau_1 - \tau_3)^2]\right\}^{1/2}$$

is called the **intensity of shear stress** and represents an average of the maximum shear stresses. Moreover, $\hat{S}$ is an invariant. Its importance is in inelastic mechanics.

Determination of the distribution of stresses in the body from the conditions of equilibrium represents the problem that is statically undetermined because the six components of the stress tensor will be determined from the three equilibrium equations. From this point of view we must investigate the geometry of the deformed body. We will show that the finite strain tensor will characterize the change of distance between two points of the body. While the stress tensor is related to the body after deformation, the finite strain tensor will be related to the body before deformation.

Strain and Strain Rate Tensors The displacement vector $\mathbf{u}(\mathbf{x}_0, t)$ is defined as the difference between a particle present position $\mathbf{x}(\mathbf{x}_0, t)$ and its initial position (see Fig. 4.1). That is

$$\mathbf{u}(\mathbf{x}_0, t) = \mathbf{x}(\mathbf{x}_0, t) - \mathbf{x}_0 \quad \text{or} \quad u_i(\mathbf{x}_0, t) = x_i(\mathbf{x}_0, t) - x_{0i}.$$

Let us consider the position of two near points after displacement. Then we have

$$x_i(\mathbf{x}_2, t) - x_i(\mathbf{x}_1, t) = x_{2i} - x_{1i} + \frac{\partial u_i}{\partial x_j}(x_{2i} - x_{1i}) + \cdots,$$

where $\partial u_i/\partial x_j \equiv u_{i,j}$ denotes the displacement gradient tensor.

The displacement field does not give complete information about the deformation of the continuum. We need to be able to distinguish between a rigid continuum motion (in which the continuum is translated and rotated to a new position without any deformation) and a situation in which the continuum changes its shape.

Let P, Q be two points in a nondeformed elastic continuum such that $\overline{PQ} \equiv d\mathbf{x}_0$, and such that their coordinates are x_{0i} and $x_{0i} + dx_{0i}$, respectively. After deformation let the continuum points P, Q be transformed into points P', Q' and let $\overline{P'Q'} \equiv d\mathbf{y}$ and let their coordinates be y_i and $y_i + dy_i$ (Fig. 4.4). Then from the Taylor expansion of the function $u_i(x_{0j} + dx_{0j})$, where we restrict ourselves to the first degrees only, that is,

$$dy_i = dx_{0i} + \frac{\partial u_i}{\partial x_{0j}} dx_{0j} = (\delta_{ij} + u_{i,j})\, dx_{0j},$$

where δ_{ij} is the Kronecker symbol. Since $|d\mathbf{x}_0|^2 = dx_{0i}\, dx_{0i} = \delta_{jk}\, dx_{0j}\, dx_{0k}$, and $|d\mathbf{y}|^2 = dy_i\, dy_i = \delta_{jk}\, dy_j\, dy_k$, then (Fig. 4.2)

$$|d\mathbf{y}|^2 - |d\mathbf{x}_0|^2 = ((\delta_{ij} + u_{i,j})(\delta_{ik} + u_{i,k}) - \delta_{jk})\, dx_{0j}\, dx_{0k} = 2\varepsilon_{jk}\, dx_{0j}\, dx_{0k},$$

where $F_{ij} = \partial y_i/\partial x_{0j}$, $F_{ij} = \delta_{ij} + u_{i,j}$ or $\mathbb{F} = \mathbb{I} + \nabla_0 \mathbf{u}$ is the **position gradient tensor** (or the **deformation gradient**) and

$$\varepsilon_{jk}(\mathbf{u}) = \frac{1}{2}\left(\frac{\partial u_j}{\partial x_{0k}} + \frac{\partial u_k}{\partial x_{0j}} + \frac{\partial u_i}{\partial x_{0j}}\frac{\partial u_i}{\partial x_{0k}}\right), \qquad \varepsilon_{jk}(\mathbf{u}) = \varepsilon_{kj}(\mathbf{u}),$$

$$i, j, k, = 1, \ldots, N,$$

or $$\epsilon = \tfrac{1}{2}((\nabla_0 \mathbf{u})^{\mathrm{T}} + \nabla_0 \mathbf{u} + \nabla_0 \mathbf{u} \cdot (\nabla_0 \mathbf{u})^{\mathrm{T}}) = \tfrac{1}{2}(\mathbb{F}^T \cdot \mathbb{F} - \mathbb{I}) = \tfrac{1}{2}(\mathbb{C} - \mathbb{I}),$$

where $$\mathbb{C} = \mathbb{I} + (\nabla_0 \mathbf{u})^{\mathrm{T}} + \nabla_0 \mathbf{u} + \nabla_0 \mathbf{u} \cdot (\nabla_0 \mathbf{u})^{\mathrm{T}}, \tag{4.12}$$

is the **finite strain tensor** or the **Green strain tensor** and it is a symmetric tensor.

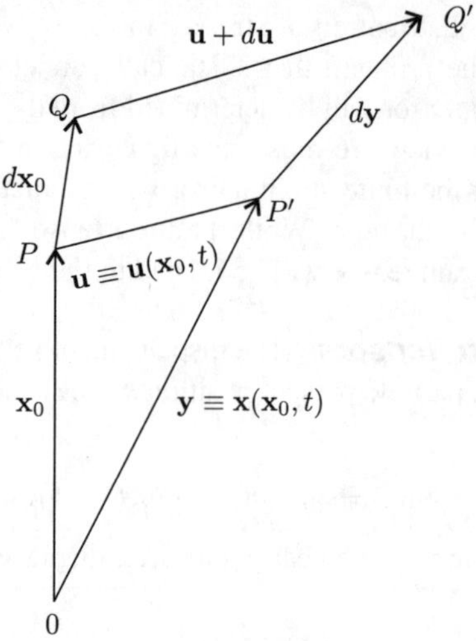

FIG. 4.4 Elastic displacements in nondeformed and deformed configurations.

A rigid body motion consisting of a translation $\mathbf{x}_T(t)$ and a rotation about the origin is written as

$$\mathbf{x}(\mathbf{x}_0, t) = \mathbf{R}(t) \cdot \mathbf{x}_0 + \mathbf{x}_T(t); \qquad x_i(\mathbf{x}_0, t) = R_{ij}(t)x_{oj} + x_{Ti}(t),$$

where $\mathbf{R}(t)$ is the rotation tensor, that is, a rotation matrix, which is an orthogonal matrix for which $\mathbf{R}^{-1} = \mathbf{R}^{\mathrm{T}}$, $R_{ij}^{-1} = R_{ij}^{\mathrm{T}} = R_{ij}$.

The deformation gradient $\mathbb{F}$ can be uniquely decomposed as

$$\mathbb{F} = \mathbf{RU} = \mathbf{VR},$$

where $\mathbf{R}$ is a proper orthogonal matrix, the **rotation matrix**, and $\mathbf{U}$ and $\mathbf{V}$ are the **right and left stretch tensors**, which are positive definite and symmetric, and moreover, $\mathbf{U} = (\mathbb{F}^{\mathrm{T}}\mathbb{F})^{1/2}$, $\mathbf{V} = (\mathbb{F}\mathbb{F}^{\mathrm{T}})^{1/2}$. Furthermore, $\mathbf{C} = \mathbb{F}^{\mathrm{T}}\mathbb{F}$ is the **right Cauchy–Green tensor** and $\mathbf{B} = \mathbb{F}\mathbb{F}^{\mathrm{T}}$ is the **left Cauchy–Green tensor**.

The linear part of the finite strain tensor ε_{ij} defines the **small strain tensor** e_{ij}, that is,

$$e_{ij}(\mathbf{u}) = \frac{1}{2}\left(\frac{\partial u_i}{\partial x_{0j}} + \frac{\partial u_j}{\partial x_{0i}}\right), \qquad e_{ij}(\mathbf{u}) = e_{ji}(\mathbf{u}), \qquad i, j = 1, \ldots, N,$$

or

$$\mathbb{E} = \tfrac{1}{2}((\nabla_0 \mathbf{u})^{\mathrm{T}} + \nabla_0 \mathbf{u}). \tag{4.13}$$

Since $dy_i - dx_{0i} = u_{i,j}\, dx_{0j} \equiv \delta(du_i)$ and since

$$\frac{\partial u_i}{\partial x_{0j}} = \frac{1}{2}\left(\frac{\partial u_i}{\partial x_{0j}} + \frac{\partial u_j}{\partial x_{0i}}\right) + \frac{1}{2}\left(\frac{\partial u_i}{\partial x_{0j}} - \frac{\partial u_j}{\partial x_{0i}}\right) = e_{ij}(\mathbf{u}) + \omega_{ij}(\mathbf{u}),$$

then

$$\delta(du_i) = e_{ij}(\mathbf{u})dx_{0j} + \omega_{ij}(\mathbf{u})dx_{0j},$$

where $\omega_{ij}(\mathbf{u})$ is the antisymmetric (skew-symmetric) tensor of the second order connected with a small rotation of the considered rigid particle and where

$$\omega_{ij}(\mathbf{u}) = \frac{1}{2}\left(\frac{\partial u_i}{\partial x_j} - \frac{\partial u_j}{\partial x_i}\right), \qquad \omega_{ij}(\mathbf{u}) = -\omega_{ji}(\mathbf{u}), \qquad i,j = 1,\ldots,N. \tag{4.14}$$

To derive the **rate of deformation** (or the velocity strain) $\mathbb{D} = (D_{ij})$, we first define the velocity gradient $\mathbb{L} = (L_{ij})$ by $\mathbb{L}\mathbf{v} = \partial\mathbf{v}/\partial\mathbf{x} = (\nabla\mathbf{v})^{\mathrm{T}} = (\mathrm{grad}\,\mathbf{v})^{\mathrm{T}}$, $L_{ij} = \partial v_i/\partial x_j$, where the derivatives are taken with respect to the Eulerian (i.e., spatial) coordinates. The symbol ∇ denotes the spatial gradient, the symbol ∇_0 denotes the material gradient. The velocity gradient tensor can be decomposed into a symmetric and the skew-symmetric parts as any second-order tensor can be expressed as the sum of its symmetric and skew-symmetric parts, that is, $\mathbb{L} = \frac{1}{2}(\mathbb{L} + \mathbb{L}^{\mathrm{T}}) + \frac{1}{2}(\mathbb{L} - \mathbb{L}^{\mathrm{T}}) = \mathbb{D} + \mathbb{W}$ or $L_{ij} = \frac{1}{2}(L_{ij} + L_{ji}) + \frac{1}{2}(L_{ij} - L_{ji}) = D_{ij} + W_{ij}$, where $\mathbb{D}$ is defined as the symmetric part of $\mathbb{L} = \partial\mathbf{v}/\partial\mathbf{x} \equiv (\nabla\mathbf{v})^{\mathrm{T}}$ and $\mathbb{W}$—the spin—is defined as the skew-symmetric part of $\mathbb{L}$, that is,

$$D_{ij}(\mathbf{v}) = \frac{1}{2}\left(\frac{\partial v_i}{\partial x_j} + \frac{\partial v_j}{\partial x_i}\right), \qquad D_{ij}(\mathbf{u}) = D_{ji}(\mathbf{u}), \qquad i,j = 1,\ldots,N,$$

or

$$\mathbb{D} = \frac{1}{2}((\nabla\mathbf{v})^{\mathrm{T}} + \nabla\mathbf{v}) = \frac{1}{2}(\mathbb{L}^{\mathrm{T}} + \mathbb{L}), \qquad \mathbb{L} = \frac{\partial\mathbf{v}}{\partial\mathbf{x}} \equiv (\nabla\mathbf{v})^{\mathrm{T}},$$

$$W_{ij}(\mathbf{v}) = \frac{1}{2}\left(\frac{\partial v_i}{\partial x_j} - \frac{\partial v_j}{\partial x_i}\right), \qquad W_{ij}(\mathbf{u}) = -W_{ji}(\mathbf{u}), \qquad i,j = 1,\ldots,N,$$

or

$$\mathbb{W}\mathbf{v} = \tfrac{1}{2}((\nabla\mathbf{v})^{\mathrm{T}} - \nabla\mathbf{v}) = \tfrac{1}{2}(\mathbb{L}^{\mathrm{T}} - \mathbb{L})\mathbf{v}. \tag{4.15}$$

For more details see Marsden and Hughes (1983), Belytschko et al. (2000), and Segel (1977).

Similar to the previous stress case we can construct the quadratic form $e_{ij}n_in_j$, and the principal directions of the small strain, the invariants as well as the eigenvalues e_i, $i = 1,2,3$, the **principal strains (principal extensions)**. Consider the quadratic form $e_{ij}n_in_j$ and for a coordinate system such that $e'_{ij} = 0$, $i \neq j$, we seek numbers e and a vector $\mathbf{n}$ such that

$$e_{ij}n_j = en_i, \qquad i = 1,2,3.$$

Then, there are three eigenvalues e_i, $i = 1,2,3$, and three corresponding orthonormal vectors (eigenvectors) $\mathbf{n}^i$, $i = 1,2,3$. Then

$$e_{ij}n_in_j = \sum_{k=1}^{3} e_k(n'_k)^2,$$

where n'_k are the components of the vector $\mathbf{n}$ in the new Cartesian coordinate system. Then $e'_{ij}=0$ for $i \neq j$, $e'_{ii}=e_i$, where i is not summed. The eigenvectors $\mathbf{n}^i$, $i=1,2,3$, are the **principal directions of small strain** at point $\mathbf{x}$. The characteristic equation is

$$\det(e_{ij} - e\delta_{ij}) = 0 \quad \text{or} \quad -e^3 + \vartheta_1 e^2 - \vartheta_2 e + \vartheta_3 = 0.$$

Since the principal directions and eigenvalues do not depend on the coordinate system, then the numbers ϑ_i, $i=1,2,3$, are invariants and

$$\vartheta_1 = e_{11} + e_{22} + e_{33},$$

$$\vartheta_2 = \begin{vmatrix} e_{11} & e_{12} \\ e_{12} & e_{22} \end{vmatrix} + \begin{vmatrix} e_{11} & e_{13} \\ e_{13} & e_{33} \end{vmatrix} + \begin{vmatrix} e_{22} & e_{23} \\ e_{23} & e_{33} \end{vmatrix},$$

$$\vartheta_3 = \det(e_{ij}) = \frac{1}{3!}\varepsilon_{ijk}\varepsilon_{lmn}e_{il}e_{jm}e_{kn}.$$

For the small strain the invariant $\vartheta_1 = e_{11} + e_{22} + e_{33}$ represents the principal linear part of the relative volume increment, the **cubical dilatation**.

Now, we introduce the **intensity of shear strain** $\bar{\varepsilon}$ by

$$\bar{\varepsilon} = \left\{\tfrac{1}{6}[(e_{22} - e_{33})^2 + (e_{33} - e_{11})^2 + (e_{11} - e_{22})^2] + e_{12}^2 + e_{13}^2 + e_{23}^2\right\}^{1/2}.$$

Solving the system of partial differential equations

$$\frac{\partial u_i}{\partial x_j} + \frac{\partial u_j}{\partial x_i} = 2e_{ij}(\mathbf{u})$$

for unknown functions u_i, $i=1,\ldots,N$, then certain conditions, the necessary conditions, must be satisfied. The necessary condition for $e_{ij} \in C^2(\Omega)$, $i,j=1,2,3$, to be components of a small strain tensor are the **Saint-Venant equations of compatibility**, that is,

$$\varepsilon_{ikl}\varepsilon_{jmn}\frac{\partial^2 e_{km}}{\partial x_l \partial x_n} = 0, \qquad i,j = 1,2,3. \tag{4.16}$$

4.3 BACKGROUND OF THE STATIC AND DYNAMIC CONTINUUM MECHANICS IN DIFFERENT RHEOLOGIES

4.3.1 Conservation Laws in Their Integral and Differential Forms

Above we gave a characterization of a specified portion of a continuum occupying a material domain $\Omega(t)$ at time $t \in \bar{I}$. Mass conservation requires that the mass of any material domain will be constant. Since no materials can flow over the boundary $\partial\Omega$ and since we do not consider mass-to-energy conversion, then

$$\frac{dm(\Omega)}{dt} = \frac{d}{dt}\int_{\Omega}\rho(\mathbf{x}_0, t)\, d\mathbf{x} = 0. \tag{4.17}$$

Applying the Reynold's transport theorem [see the Appendix, Belytschko et al. (2000), and Nedoma (1998a)] we have

$$\int_{\Omega}\left(\frac{d\rho}{dt}+\rho\,\mathrm{div}\,(\mathbf{v})\right)\,d\mathbf{x}=0. \tag{4.18}$$

Hence, since the above holds for any subdomain Ω, we obtain

$$\frac{d\rho}{dt}+\rho\,\mathrm{div}(\mathbf{v})=0 \quad \text{or} \quad \frac{d\rho}{dt}+\rho\frac{\partial v_i}{\partial x_i}=0, \tag{4.19}$$

which is the **equation of mass conservation** or the **continuity equation**. If the material of the continuum is incompressible, then the material time derivative of the density vanishes and then

$$\mathrm{div}(\mathbf{v})=0 \quad \text{or} \quad \frac{\partial v_i}{\partial x_i}=0. \tag{4.20}$$

After linearization (4.19) yields

$$\frac{\partial\rho(\mathbf{x},t)}{\partial t}+\rho(\mathbf{x},t)\frac{\partial v_i(\mathbf{x},t)}{\partial x_i}=0, \tag{4.21}$$

then for the constant $\mathbf{x}$ and after integration over $\overline{I}=[t_0,t_1]$ we have

$$\int_{t_0}^{t_1}\left(\frac{1}{\rho}\frac{\partial\rho}{\partial t}+\frac{\partial v_i}{\partial x_i}\right)dt=0 \Longrightarrow \ln\left(\frac{\rho(\mathbf{x},t_1)}{\rho(\mathbf{x},t_0)}\right)+\int_{t_0}^{t_1}\frac{\partial v_i(\mathbf{x},\tau)}{\partial x_i}d\tau=0.$$

Since $u_i(\mathbf{x},t)$ is small, then $\partial v_i/\partial x_i=\partial^2 u_i/\partial x_i\,\partial t$ is also small and then

$$\rho(\mathbf{x},t)=\rho_0(\mathbf{x})\left[1-\int_{t_0}^{t_1}\frac{\partial v_i(\mathbf{x},\tau)}{\partial x_i}d\tau\right], \quad \text{where } \rho_0(\mathbf{x})=\rho(\mathbf{x},t_0). \tag{4.22}$$

To derive the laws of balance of linear and angular momentum or the laws of conservation of linear momentum and conservation of angular momentum, we will use the ideas of the particle mechanics momentum [see Nedoma (1998a) and Belytschko et al. (2000)]. **Linear momentum conservation** or the conservation principle or the balance of momentum principle is equivalent to Newton's second law of motion, relating the forces acting on a body to its acceleration. Then the rate of the change of linear momentum in a material domain Ω is equal to the sum of body and surface forces, that is,

$$\frac{d}{dt}\int_{\Omega(t)}\rho\mathbf{v}\,d\mathbf{x}=\int_{\Omega(t)}\rho\mathbf{f}\,d\mathbf{x}+\int_{\partial\Omega(t)}\mathbf{t}\,ds, \tag{4.23}$$

where ρ is density, $\mathbf{v}$ is velocity, $\mathbf{f}$ is body forces, $\mathbf{t}$ is surface forces, all assumed per unit area exerted at point $\mathbf{x}$ of $\partial\Omega$, $\mathbf{x}$ is a position vector, and $\mathbf{n}$ is the unit outward normal to $\partial\Omega$. The surface forces have the meaning of surface stress vectors. Then using the conservation of mass, the conservation of linear momentum (4.23) in the integral form can be written as

$$\int_{\Omega(t)} \rho\left(\frac{\partial \mathbf{v}}{\partial t} + (\mathbf{v}\cdot\nabla)\mathbf{v}\right) d\mathbf{x} - \int_{\Omega(t)} \rho\mathbf{f}\, d\mathbf{x} = \int_{\partial\Omega(t)} \mathbf{t}\, ds. \tag{4.24}$$

Since $\mathbf{t} = \mathbf{t}(\mathbf{x}, t, \mathbf{n}) = \mathbf{n}\cdot\mathbf{t}(\mathbf{x}, t)$ is the stress vector with components $t_i(\mathbf{x}, t, \mathbf{n}) = \tau_{ij}(\mathbf{x}, t)n_j$, where τ_{ij} is the stress tensor, using the divergence theorem (4.24) yields

$$\int_{\Omega(t)} \rho\left(\frac{\partial \mathbf{v}}{\partial t} + (\mathbf{v}\cdot\nabla)\mathbf{v}\right) d\mathbf{x} - \int_{\Omega(t)} \rho\mathbf{f}\, d\mathbf{x} = \int_{\Omega(t)} \nabla\cdot\boldsymbol{\tau}\, ds. \tag{4.25}$$

Since $\Omega(t)$ is arbitrary and if we assume that the integrand is continuous, then

$$\rho\left(\frac{\partial \mathbf{v}}{\partial t} + (\mathbf{v}\cdot\nabla)\mathbf{v}\right) - \rho\mathbf{f} = \nabla\cdot\boldsymbol{\tau} \equiv \operatorname{div}\boldsymbol{\tau} \tag{4.26}$$

or in the form

$$\rho\left(\frac{\partial v_i}{\partial t} + v_j\frac{\partial v_i}{\partial x_j}\right) - \rho f_i = \frac{\partial \tau_{ij}}{\partial x_j}, \tag{4.27}$$

where the body forces and surface stresses are defined as follows: (i) **body forces**: at time t, a force per unit mass $\mathbf{f}(\mathbf{x}, t)$ is assumed to act at each point $\mathbf{x}$ of Ω; (ii) surface forces at time t, consider a point $\mathbf{x} \in \partial\Omega$ and let $\mathbf{n}$ be the unit exterior normal to Ω at $\mathbf{x} \in \partial\Omega$. Then $\mathbf{t}(\mathbf{x}, t, \mathbf{n})$ is a force per unit area exerted at point $\mathbf{x} \in \partial\Omega$ by the material exterior to Ω on the material interior to Ω.

In many problems of biomechanics, the loads are applied statically or applied very slowly, and, therefore, the inertia forces are equal to zero in the statically loaded joints or they are very small, and, therefore, they can be neglected in the second case. Then (4.26) and (4.27) leads to

$$\nabla\cdot\boldsymbol{\tau} + \rho\mathbf{f} = \mathbf{0} \quad \text{or} \quad \frac{\partial \tau_{ij}}{\partial x_j} + \rho f_i = 0, \tag{4.28}$$

and we speak about the **equilibrium equation.**

To derive the **law of angular momentum** (i.e., momentum of momentum) or conservation of angular momentum, we retain the idea from particle mechanics. If a mass point at point $\mathbf{x}$ possesses momentum M, then its moment of momentum about O is defined as $\mathbf{x} \times M \equiv [\mathbf{x}, M]$. From the condition of equilibrium of moments for a volume and surface forces the conservation of angular momentum is obtained by taking the cross-product applying the position vector $\mathbf{x}$ on each term in (4.23), that is,

$$\frac{d}{dt}\int_{\Omega(t)} (\mathbf{x} \times \rho\mathbf{v})\, d\mathbf{x} = \int_{\Omega(t)} (\mathbf{x} \times \rho\mathbf{f})\, d\mathbf{x} + \int_{\partial\Omega(t)} (\mathbf{x} \times \mathbf{t})\, ds,$$

or

$$\begin{aligned}\frac{d}{dt}\int_{\Omega(t)} \varepsilon_{ijk} x_j \rho v_k\, d\mathbf{x} &= \int_{\Omega(t)} \varepsilon_{jk} x_j \rho f_k\, d\mathbf{x} + \int_{\partial\Omega(t)} \varepsilon_{ijk} x_j t_k\, ds \\ &= \int_{\Omega(t)} \varepsilon_{jk} x_j \rho f_k\, d\mathbf{x} + \int_{\partial\Omega(t)} \varepsilon_{ijk} x_j \tau_{lk} n_l\, ds. \end{aligned} \tag{4.29}$$

Hence, using the equilibrium conditions, continuity of the integrand and since the element $d\mathbf{x}$ is arbitrary, we obtain

$$\varepsilon_{ijk}\tau_{jk} = 0.$$

The indices j, k are the summing indices. From the definition of the Levi–Civita ε_{ijk}, which is the tensor of the third-order asymmetric in all indices and defined above, then from (4.29) it follows that the Cauchy stress is a symmetric tensor, that is,

$$\boldsymbol{\tau} = \boldsymbol{\tau}^{\mathrm{T}} \quad \text{or} \quad \tau_{jk} = \tau_{kj}. \tag{4.30}$$

Equations (4.26) and (4.30) are referred to as the current configuration.

The conservation of energy then requires that the rate of the change of the total energy in the body, which includes both the internal energy and the kinetic energy, equal the power of the applied forces and the energy added to the body by heat conduction and any heat sources.

To derive the **law of energy conservation** from thermodynamics, it is known that the rate of the total energy of the system (i.e., the sum of kinetic and internal energy) is the sum of the effect of external forces acting on the system and of the energy flux per unit time. Thus

$$\begin{aligned}\frac{d}{dt}\int_{\Omega(t)} \rho\left(\frac{1}{2}\mathbf{v}^2 + e\right) d\mathbf{x} &= \int_{\Omega(t)} \rho\mathbf{f}\cdot\mathbf{v}\, d\mathbf{x} + \int_{\partial\Omega(t)} \mathbf{t}\cdot\mathbf{v}\, ds + \int_{\Omega(t)} \rho w\, d\mathbf{x} \\ &\quad - \int_{\partial\Omega(t)} \mathbf{q}\cdot\mathbf{n}\, ds, \end{aligned} \tag{4.31}$$

or

$$\begin{aligned}\frac{d}{dt}\int_{\Omega(t)} \rho\left(\frac{1}{2}\mathbf{v}^2 + e\right) d\mathbf{x} &= \int_{\Omega(t)} \rho f_i v_i\, d\mathbf{x} + \int_{\partial\Omega(t)} \tau_{ij} n_j v_i\, ds + \int_{\Omega(t)} \rho w\, d\mathbf{x} \\ &\quad - \int_{\partial\Omega(t)} q_i n_i\, ds, \end{aligned} \tag{4.32}$$

where the scalar e represents an internal energy of the medium, w represents an influx of an energy, and the vector $\mathbf{q}$ represents a transport of an energy. Hence applying (4.19), (4.26), or (4.27), using (4.30) and putting

$$\mathbb{D}\mathbf{v} = \frac{1}{2}(\mathbb{L} + \mathbb{L}^{\mathrm{T}})\mathbf{v} \quad \text{or} \quad D_{ij}v = \frac{1}{2}\left(\frac{\partial v_i}{\partial x_j} + \frac{\partial v_j}{\partial x_i}\right),$$
$$\mathbb{L}\mathbf{v} = \frac{\partial \mathbf{v}}{\partial \mathbf{x}} = (\nabla \mathbf{v})^{\mathrm{T}} = (\mathrm{grad}\, \mathbf{v})^{\mathrm{T}} \quad \text{or} \quad L_{ij} = \frac{\partial v_i}{\partial x_j}, \tag{4.33}$$

where D_{ij} is the strain rate tensor, we obtain

$$\int_{\Omega(t)} \rho \left(\frac{de}{dt}\right) d\mathbf{x} = \int_{\Omega(t)} \tau_{ij} D_{ij}\, d\mathbf{x} + \int_{\Omega(t)} \left(\rho w - \frac{\partial q_i}{\partial x_i}\right) d\mathbf{x}. \tag{4.34}$$

Since $\Omega(t)$ is arbitrary and if we assume that the integrand is continuous, then

$$\rho \frac{de}{dt} = \tau_{ij} D_{ij} + \rho w - \frac{\partial q_i}{\partial x_i} \quad \text{or} \quad \rho \frac{de}{dt} = \mathbb{D} : \boldsymbol{\tau} + \rho w - \mathrm{div}\, \mathbf{q}, \tag{4.35}$$

representing the **differential form of the law of energy conservation**.

Notice that $\boldsymbol{\tau}(\mathbf{x}, t, \mathbf{n})$, the Cauchy stress vector, represents physically the force per unit area exerted on a surface element oriented with normal $\mathbf{n}$, and it is referred to the surfaces after deformation.

We saw that Eqs. (4.26) and (4.30) were referred to the current configuration. But often we need a formulation of these equations in quantities that are related to the initial (reference) configuration, that is, to use material quantities rather than spatial ones. If this idea is applied to the Cauchy stress, we get the Piola–Kirchhoff stress. The first Piola–Kirchhoff stress vector is the vector $\mathbf{t}_0(\mathbf{x}_0, t, \mathbf{n}_0)$, which is parallel to the Cauchy stress $\mathbf{t}(\mathbf{x}, t, \mathbf{n})$ but measures the force per unit undeformed area, where $\mathbf{x} = \varphi(\mathbf{x}_0, t)$ and $\mathbf{n}_0, \mathbf{n}$ are unit normal vectors to the corresponding undeformed and deformed area elements.

Then the stresses are defined by the Cauchy law

$$\mathbf{n} \cdot \boldsymbol{\tau} ds = \mathbf{t}\, ds, \tag{4.36}$$

where $\mathbf{t}$ is the traction. In the initial (reference) configuration we have

$$\mathbf{n}_0 \cdot \mathbb{P}\, ds_0 = \mathbf{t}_0\, ds_0, \tag{4.37}$$

where $\mathbb{P}$ is the nominal stress, the **first Piola–Kirchhoff stress tensor**. The second Piola–Kirhoff stress tensor is defined by

$$\mathbf{n}_0 \cdot \mathbb{S}\, ds_0 = \mathbb{F}^{-1} \mathbf{t}_0\, ds_0, \tag{4.38}$$

where $\mathbb{S}$ is the **second Piola–Kirchhoff stress**. Then

$$
\begin{aligned}
&\mathbb{P} = J\mathbb{F}^{-1}\cdot\boldsymbol{\tau} &&\text{or}\quad P_{ij} = JF^{-1}_{ik}\tau_{kj} = J\frac{\partial x_{0i}}{\partial x_i}\tau_{kj},\\
&\boldsymbol{\tau} = J^{-1}\mathbb{F}\cdot\mathbb{P} &&\text{or}\quad \tau_{ij} = J^{-1}F_{ik}P_{kj},\\
&\mathbb{P} = \mathbb{S}\cdot\mathbb{F}^{\mathrm{T}} &&\text{or}\quad P_{ij} = S_{ik}F^{\mathrm{T}}_{kj} = S_{ik}F_{jk},\\
&\boldsymbol{\tau} = J^{-1}\mathbb{F}\cdot\mathbb{S}\cdot\mathbb{F}^{\mathrm{T}} &&\text{or}\quad \tau_{ij} = J^{-1}F_{ik}S_{kl}F^{\mathrm{T}}_{lj},\\
&\mathbb{S} = J\mathbb{F}^{-1}\cdot\boldsymbol{\tau}\cdot\mathbb{F}^{-\mathrm{T}} &&\text{or}\quad S_{ij} = J^{-1}F^{-1}_{ik}\tau_{kl}F^{-\mathrm{T}}_{lj}.
\end{aligned}
\tag{4.39}
$$

Notice that the traces of stress measures $\mathbb{P}$ and $\mathbb{S}$ do not give the true pressure because they are referred to the undeformed area.

In a Lagrangian description, the rate of the change of linear momentum in a material domain Ω_0 is equal to the sum of body and surface forces in terms of an integral over the initial (reference) configuration, that is,

$$
\frac{d}{dt}\int_{\Omega_0}\rho_0\mathbf{v}\,d\mathbf{x}_0 = \int_{\Omega_0}\rho_0\mathbf{f}\,d\mathbf{x}_0 + \int_{\partial\Omega_0}\mathbf{t}_0\,ds_0. \tag{4.40}
$$

In (4.40) in the term on the left-hand side, the material derivative can be taken inside the integral because the initial domain is constant in time, thus

$$
\frac{d}{dt}\int_{\Omega_0}\rho_0\mathbf{v}\,d\mathbf{x}_0 = \int_{\Omega_0}\rho_0\frac{\partial\mathbf{v}(\mathbf{x}_0,t)}{\partial t}d\mathbf{x}_0.
$$

Using the Cauchy law, thus

$$
\int_{\partial\Omega_0}\mathbf{t}_0\,ds_0 = \int_{\partial\Omega_0}\mathbf{n}_0\cdot\mathbb{P}\,ds_0 \quad\text{or}\quad \int_{\partial\Omega_0}t_{0i}\,ds_0 = \int_{\partial\Omega_0}n_{0j}P_{ji}\,ds_0 = \int_{\Omega_0}\frac{\partial P_{ji}}{\partial x_{0j}}d\mathbf{x}_0
$$

and the Gauss theorem [see Belytschko et al. (2000)] then is

$$
\int_{\Omega_0}\left(\rho_0\frac{\partial\mathbf{v}(\mathbf{x}_0,t)}{\partial t} - \rho_0\mathbf{f} - \nabla_0\cdot\mathbb{P}\right)d\mathbf{x}_0 = 0. \tag{4.41}
$$

Hence, because the domain Ω_0 is arbitrary and the integrand is assumed to be continuous, we have

$$
\rho_0\frac{\partial\mathbf{v}(\mathbf{x}_0,t)}{\partial t} = \nabla_0\cdot\mathbb{P} + \rho_0\mathbf{f} \quad\text{or}\quad \rho_0\frac{\partial v_i(\mathbf{x}_0,t)}{\partial t} = \frac{\partial P_{ji}}{\partial x_{0j}} + \rho_0 f_i, \tag{4.42}
$$

representing the **Lagrangian form of the momentum equation** or the **Lagrangian form of the conservation linear momentum**.

The conservation of energy in the Lagrangian description can be written as

$$\frac{d}{dt}\int_{\Omega_0}\rho_0\left(\frac{1}{2}\mathbf{v}^2+e\right)d\mathbf{x}_0=\int_{\Omega_0}\rho_0\mathbf{f}\cdot\mathbf{v}\,d\mathbf{x}_0+\int_{\partial\Omega_0}\mathbf{t}_0\cdot\mathbf{v}\,ds_0+\int_{\Omega_0}\rho_0 w\,d\mathbf{x}_0$$
$$-\int_{\partial\Omega_0}\mathbf{q}_0\cdot\mathbf{n}_0\,ds_0, \tag{4.43}$$

where $\mathbf{q}_0 = J^{-1}\mathbb{F}^T \cdot q$ or $q_{0i} = J^{-1}F^T_{ij}q_j$ and

$$\int_{\Omega_0}\rho_0\left(\frac{\partial e(\mathbf{x}_0,t)}{\partial t}+v_j\frac{\partial v_j(\mathbf{x}_0,t)}{\partial t}\right)d\mathbf{x}_0$$
$$=\int_{\Omega_0}\rho_0 f_i v_i\,d\mathbf{x}+\int_{\Omega_0}\left(\frac{\partial F_{ij}}{\partial t}P_{ij}+\frac{\partial P_{ij}}{\partial x_{oi}}v_j\right)d\mathbf{x}_0+\int_{\Omega_0}\left(\rho_0 w-\frac{\partial q_{0i}}{\partial x_{0i}}\right)d\mathbf{x}_0. \tag{4.44}$$

4.3.2 Constitutive Laws

Generalized Hooke's Law in Elasticity The well-known experiment of loading a bar of the length l and constant cross section q gives us the relation between the relative extension $e = \Delta l/l$ and the stress $\tau = P/q$, where P is a loading force (see Fig. 4.5). The part of the diagram between points O and A is linear and corresponds practically to the elastic behavior of the bone tissue. The stress τ_A corresponding to point A is the **proportional limit**. The part between points A and B is curved and the extension increases suddenly at point B without any change in the stress. Point B is the **yield point**. The value τ_B is the **yield strength of the material.** If the stress further increases, the extension also increases up to point C. Point C is the highest stress point, and the corresponding value τ_C is the ultimate strength of the material or the **strength limit.** Further stretching leads to breaking of the bar. Point R represents the **rupture or the failure point** and the corresponding stress at which the rupture occurs is the **rupture strength of the material.** Now let us stop the loading of the bar at a certain load P and make the loading force decrease to zero. If we come back to the origin along the same curve, then the maximum of the corresponding stresses is called the **elastic limit** τ_E and it satisfies $\tau_E \geq \tau_A$. The deformation corresponding to stresses $\tau \leq \tau_E$ is called the **elastic deformation**. If the loading force was removed after a stress greater than τ_E, a part of the extension vanishes and a part remains permanently. This part of the deformation is the **plastic deformation**. The magnitude of the permanent deformation depends on the physical properties of a bone, such as viscosity, temperature, pressure, and the like. Elastic deformations are studied in the theories of (visco)elasticity and thermoviscoelasticity; the plastic deformations are studied in theory of (visco)plasticity and thermoviscoplasticity. In the case of the linear theory of elasticity, $\tau < \tau_A$ and the relative extension e is proportional to the stress τ as $\tau = Ee$, where E is the **Young's modulus** of elasticity. The relation $\tau = Ec$ is known as the **Hooke law of elasticity**.

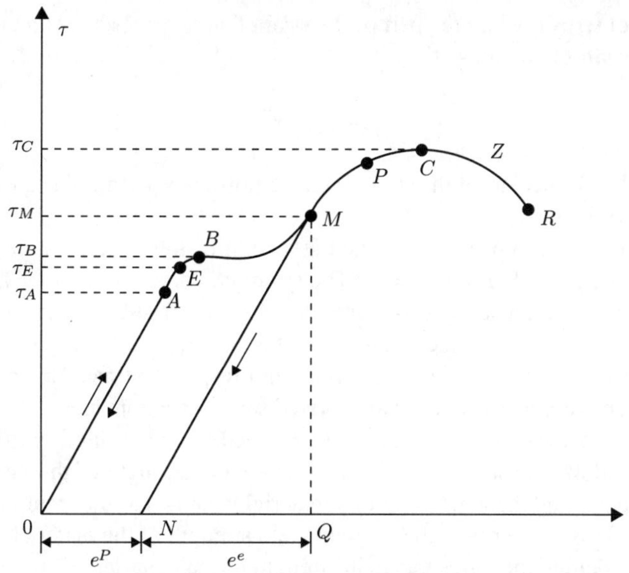

FIG. 4.5 Stress–strain diagram for axial loading.

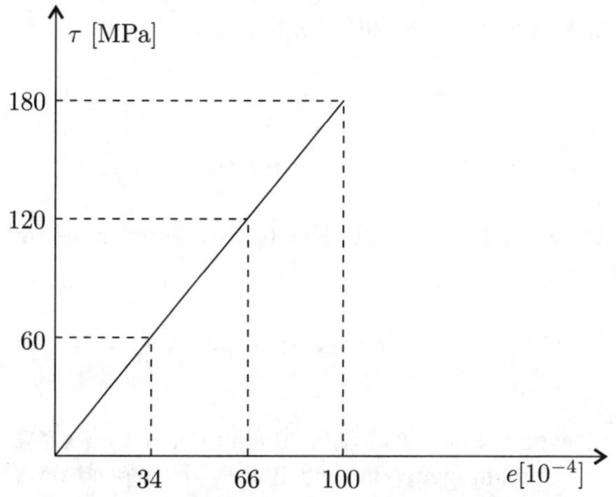

FIG. 4.6 Stress–strain diagram for the bone.

The stress–strain diagram for the bone is given in Fig. 4.6 and the average mechanical properties of selected material are given in Table 4.1.

Let us suppose that after the loading reaches point M such that τ_M exceeds the yield limit τ_B, an unloading process is started. We see (Fig. 4.5) that after the unloading process, the part NQ of the strain disappears, although the part ON remains permanently. The part of the strain that remains permanently is called the **plastic**

(irreversible) strain, while the part of the strain that vanishes after unloading is the **elastic (reversible) strain**, so that

$$e = e^e + e^p,$$

where e^e is the elastic part of the strain (i.e., the portion NQ) and e^p is the plastic part of the strain (i.e., the portion ON).

Now let a loading process be stopped at a certain stress and a strain state corresponding to the point P at time $t = t_p$. The strain $e(t)$ at any instant $t \geq t_p$ coincides with the strain $e(t_p)$ for a purely elastic material; for an inelastic material, an additional strain can be observed with increasing time, that is, $e(t) > e(t_p)$, $e'(t) > 0$ for $t > t_p$, where the prime denotes the differentiation with respect to time. This phenomenon is called **creep**. Biomaterials are characterized by this phenomenon.

But, if the material after unloading is reloaded, it will exhibit elastic behavior between N and M, the stress at M being a new yield strength of the material. This behavior of changing the yield point of a material is known as the **strain hardening**. The stress at M is greater than the original yield strength of the material, and we see that strain hardening increases the yield strength of the material.

The generalized Hooke law is as follows: Let us assume that at point $\mathbf{x} \in \Omega$, $t \in I$, the relations between stress and strain tensors are defined by

$$\tau_{ij}(\mathbf{x}, t) = c_{ijkl}(\mathbf{x}) e_{kl}(\mathbf{u}(\mathbf{x}, t)), \qquad i, j, k, l = 1, \ldots, N,$$

for the anisotropic materials or

$$\tau_{ij}(\mathbf{x}, t) = \lambda(\mathbf{x}) \vartheta(\mathbf{u}(\mathbf{x}, t)) \delta_{ij} + 2\mu(\mathbf{x}) e_{ij}(\mathbf{u}(\mathbf{x}, t)), \qquad i, j = 1, \ldots, N, \tag{4.45}$$

for the isotropic materials. The coefficients c_{ijkl} are symmetric and Lipschitz continuous, so that

$$c_{ijkl} = c_{jikl} = c_{ijlk} = c_{klij}, \qquad c_{ijkl} e_{ij} e_{kl} \geq c_0 e_{ij} e_{ij}, \quad c_0 > 0 \quad \text{for all } e_{ij} = e_{ji}. \tag{4.46}$$

Assume that the elastic coefficients $c_{ijkl}(\mathbf{x})$ do not depend on $\mathbf{x} \in \Omega$; then the bone material is said to be **homogeneous**; and if they do depend on $\mathbf{x} \in \Omega$, then it is **nonhomogeneous**. If the coefficients $c_{ijkl}(\mathbf{x})$ do not depend on the choice of the coordinate system, then the bone material is said to be **isotropic** at $\mathbf{x} \in \Omega$, while if the coefficients $c_{ijkl}(\mathbf{x})$ do not depend on the choice of the coordinate system, it is **anisotropic** at $\mathbf{x} \in \Omega$. For an isotropic bone material

$$c_{ijkl} = \lambda \delta_{ij} \delta_{kl} + \mu(\delta_{ik} \delta_{jl} + \delta_{il} \delta_{jk}), \tag{4.47}$$

where $\lambda = c_{1122}$ and $\mu = c_{1212}$ are the **Lamé coefficients**. Note that

$$\tau_{kk} = (3\lambda + 2\mu) e_{kk}; \tag{4.48}$$

then (4.45) and (4.47) yield

$$e_{ij} = (2\mu)^{-1}\tau_{ij} - [2\mu(3\lambda + 2\mu)]^{-1}\lambda\tau_{kk}\delta_{ij}, \qquad (3\lambda + 2\mu) \neq 0, \tag{4.49}$$

for isotropic bone materials or in general

$$e_{ij} = s_{ijkl}\tau_{kl} \tag{4.50}$$

for anisotropic bone materials, where

$$s_{ijkl} = s_{jikl} = s_{ijlk} = s_{klij}, \qquad s_{ijkl}\tau_{ij}\tau_{kl} \geq s_0\tau_{ij}\tau_{ij}, \qquad s_0 > 0 \tag{4.51}$$

are the **inverse Hooke laws** for isotropic and anisotropic bone materials, respectively.

In the tension test under consideration, the ratio of the axial stress $\mathbf{P}/q$ to the axial strain e_{11} is called the **Young modulus** E; the ratio of lateral contraction to longitudinal extension is the **Poisson ratio** $\nu = |e_{22}/e_{11}| = |e_{33}/e_{11}|$. Moreover, it holds that

$$\begin{aligned} E &= \mu(3\lambda + 2\mu)(\lambda + \mu)^{-1}, \qquad & \nu &= \tfrac{1}{2}\lambda(\lambda + \mu)^{-1}, \\ \lambda &= E\nu[(1 + \nu)(1 - 2\nu)]^{-1}, \qquad & \mu &= \tfrac{1}{2}E(1 + \nu)^{-1}. \end{aligned} \tag{4.52}$$

Thus

$$e_{ij} = \frac{1 + \nu}{E}\tau_{ij} - \frac{\nu}{e}\theta\delta_{ij}, \qquad \theta = \tau_{ii}.$$

Let us assume the case if $\tau_{ii} = -p$, $\tau_{ij} = 0$ for $i \neq j$, where $p > 0$ is a pressure. As $\tau_{11} + \tau_{22} + \tau_{33} = -3p$, then for the volume dilatation we have

$$\vartheta = -\frac{3p}{3\lambda + 2\mu} = -\frac{p}{\lambda + \frac{2}{3}\mu} = \frac{p}{k}, \tag{4.53}$$

characterizing the relative reduction of the volume, where $k = \lambda + \frac{2}{3}\mu > 0$ is the **bulk modulus**.

Next Table 4.1 gives the average mechanical properties of biomaterials.

Constitutive Laws of Linear Viscoelasticity There exists a group of materials such as almost all biomaterials, metals at high temperatures, and polymer plastics that exhibits gradual deformation and recovery when they are subject to loading and unloading. The response of these materials depends upon how quickly the load was applied or removed. Such materials exhibit time-dependent material behaviors that are known as viscoelastic properties, where elastic properties are represented by solid material properties and viscous behaviors are represented by fluid properties, and they are characterized by measures of resistance to flow.

The relationship between the stress and the strain can be written as

$$\tau = \tau(e, e'), \qquad \text{where } e' \equiv \frac{de}{dt}, \ t \text{ is time.}$$

TABLE 4.1 Average Mechanical Properties of Biomaterials

Material	Yield Strength R_e(MPa)	Strength Limit R_m (MPa)	Young Modulus E (Pa)	Shear Modulus G (MPa)	Poisson's Ratio	Density (g cm^{-3})
Cortical bone	80–121	120–193	$0.96–1.74 \times 10^{10}$	3.3–5	0.2–0.32	1.0–1.94
Spongy bone	48–59	78	$0.445–1.4 \times 10^{10}$	3–3.5	0.3	0.5–1.875
Muscle		0.1–0.2	$0.1–2.8 \times 10^{6}$		0.4–0.49	
Tendon		35–70	$0.4–0.7 \times 10^{5}$		0.4	
Cartilage		68	$0.42–0.43 \times 10^{6}$		0.347	
Steel	200–700	400–850	$2.0–2.08 \times 10^{11}$	73–80	0.21–0.3	7.8–8.3
Titanium	280–1100	500–900	$1.0–1.3 \times 10^{11}$	41–45	0.3–0.34	4.4
Composite	200	500	$1.5–4.5 \times 10^{10}$			1.57–2.0
Ceramics	2440–4000	350–5000	$2.1–4.2 \times 10^{11}$		0.2–0.25	3.1–6.0
UHMWPE	23–33	16–43	$2.45–3.4 \times 10^{8}$	7	0.4–0.46	0.9–1.18
Bone Cement		550–780	$0.239–0.265 \times 10^{11}$		0.4	1.57–2.0

Source: After Brown and Ferguson (1980), Hvid et al. (1983), Valenta et al. (1985), (1993), Beznoska et al. (1987), Covin (2001), Feldmann (2005), Nedoma et al. (2006), Jarm et al. (2007), and after some technical data of JJOsly, Ticona, SCI, etc.

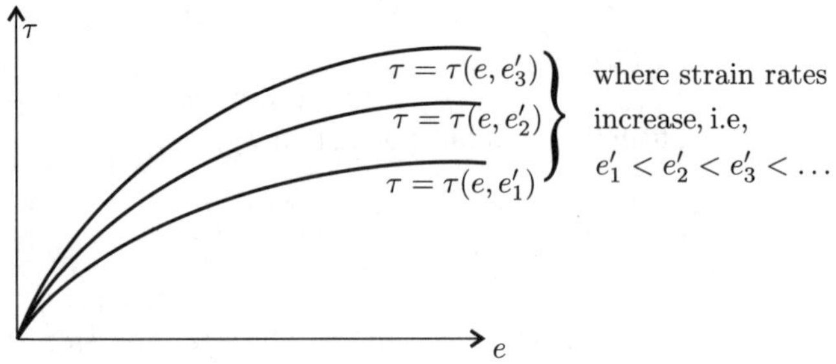

FIG. 4.7 Stress–strain diagram for viscoelastic materials.

The stress–strain diagram of such a material depends upon the rate at which the strain originates in the material (see Fig. 4.7).

The creep and recovery are eliminated by observing the response of a material to a constant stress τ_0 applied at the time t_0 and removed at a later time t_1. The typical diagram for a creep and recovery for a viscoelastic material is given in Fig. 4.8.

Let us consider the stress–strain diagram (Fig. 4.9) of the hysteresis loop. The tensile force is applied to the material between O and A. The material is deformed beyond its elastic limit. Then the material is unloaded and the line $\overline{AB}$ represents the unloading path. Then the material is reloaded by a compressive load. At point C the compressive load is removed and then the material is unloaded. Finally, the material has its original shape. During this process of loading and unloading, the total strain energy is

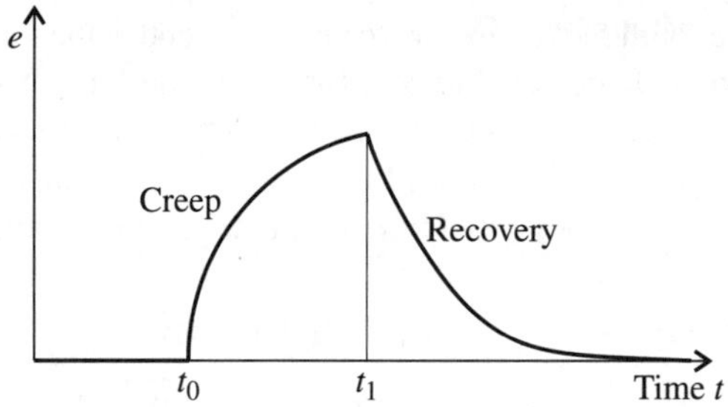

FIG. 4.8 Diagram of creep and recovery for a viscoelastic material.

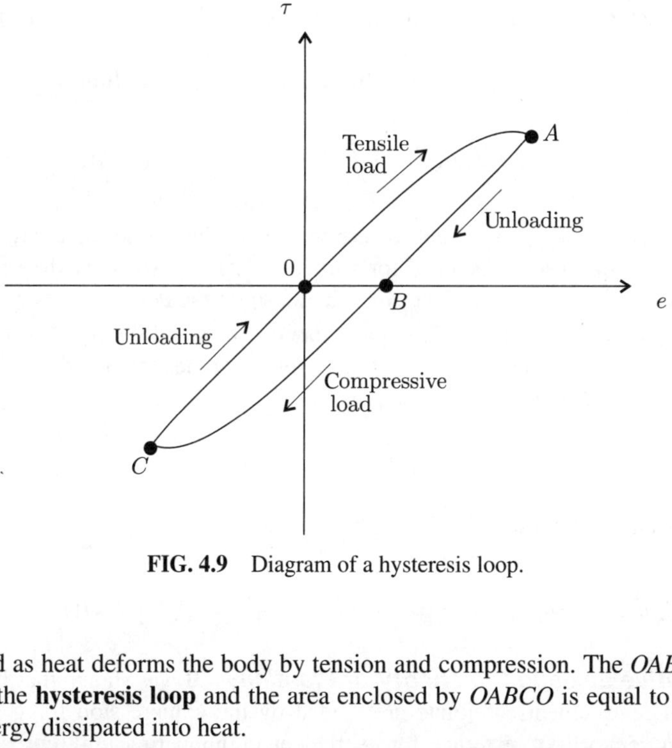

FIG. 4.9 Diagram of a hysteresis loop.

dissipated as heat deforms the body by tension and compression. The *OABCO* loop is called the **hysteresis loop** and the area enclosed by *OABCO* is equal to this total strain energy dissipated into heat.

Viscoelastic properties of a bone tissue characterize materials with a memory in the sense that the state of the stress at the instant t depends on the deformations undergone by the materials in the previous time period. The bone tissues are linear if the constitutive law establishes a linear relation between the stress tensor and the strain tensor. In biomechanics it is known that the bone tissues are endowed with a memory. One type of bone tissue is endowed with a "short memory," while others are endowed with a "long memory." For bone materials with a short memory the constitutive law is

$$\tau_{ij}(\mathbf{x}, t) = c_{ijkl}^{(0)}(\mathbf{x})e_{kl}(\mathbf{u}(\mathbf{x}, t)) + c_{ijkl}^{(1)}(\mathbf{x})e_{kl}(\mathbf{u}'(\mathbf{x}, t)), \tag{4.54}$$

where the derivatives with respect to t are to be understood in the sense of distribution, and the tensors τ_{ij} and e_{ij} are equal to zero for $t < 0$. The state of stress at moment t depends only on the deformation at time t and the immediately preceding time period. The coefficients $c_{ijkl}^{(0)}(\mathbf{x})$ and $c_{ijkl}^{(1)}(\mathbf{x})$ have a meaning of coefficients of elasticity and of viscosity, respectively, and they satisfy the symmetry and Lipschitz conditions

$$c_{ijkl}^{(n)} = c_{jikl}^{(n)} = c_{ijlk}^{(n)} = c_{klij}^{rn}, \qquad c_{ijkl}^{(n)} e_{ij} e_{kl} \geq c_0^{(n)} e_{ij} e_{ij},$$
$$\forall e_{ij} = e_{ji}, \; c_0^{(n)} > 0, \qquad n = 0, 1. \tag{4.55}$$

The scalar $c_{ijkl}^{(n)} e_{ij} e_{kl}$ has a meaning of dissipation density of energy for $n = 1$ and it is positive, and it has a meaning of a deformation energy density for $n = 0$ and it is also positive.

For materials with a long memory, the constitutive law is as follows:

$$\tau_{ij}(\mathbf{x}, t) = c_{ijkl}^{(0)}(\mathbf{x}) e_{kl}(\mathbf{u}(\mathbf{x}, t)) + \int_0^t b_{ijkl}(\mathbf{x}, t - s) e_{kl}(\mathbf{u}(\mathbf{x}, s))\, ds, \tag{4.56}$$

where tensors τ_{ij} and e_{ij} are equal to zero for $t < 0$. This constitutive law describes the materials with a long memory for which the state of stress at time t depends on the "history of" deformation at time t and also on the deformations at the times proceeding t. The coefficients $c_{ijkl}^{(0)}(\mathbf{x})$ are bounded in $\mathbf{x}$ and satisfy (4.55) for $n = 0$ and the coefficients $b_{ijkl}(\mathbf{x}, t)$, representing the effect of the memory, depend on $\mathbf{x}$ and t, are bounded in $\mathbf{x}$ and t and satisfy

$$b_{ijkl}(\mathbf{x}, t) = b_{jikl}(\mathbf{x}, t) \tag{4.57}$$

and the regularity conditions

$$b_{ijkl}, \qquad \frac{\partial}{\partial t} b_{ijkl}, \qquad \frac{\partial^2}{\partial t^2} b_{ijkl} \in L^\infty(\Omega \times I), \qquad I = (0, t_1). \tag{4.58}$$

Constitutive Laws in the Theory of Plasticity It was shown that if the artificial total replacements of joints are loaded by the compression force $\sim 2500N$, then upon removal of external forces they undergo permanent (unrecoverable) deformations. These experiments were made on the total replacement of the hip joint.

To formulate the constitutive law of plasticity (Fig. 4.5), we see that the segment $\overline{OA}$ characterizes the elastic property of the bone material. Now let us place ourselves at point M and let e decrease. Then we observe that the representative point (e, τ) describes a line starting from M, parallel to $\overline{OA}$, and crossing the e axis at point N.

At point M, the behavior of a material is then no longer reversible and therefore represents a plastic region as the rest of the deformation remains. Moreover, $\overline{MN} > \overline{OA}$ as long as the arc $\vec{Az}$ is not a half-line parallel to $\overline{Oe}$; this effect represents the phenomenon of work hardening. If the segment $\vec{Az}$ is parallel to $\overline{Oe}$, then we speak of the **perfectly plastic materials**. For the perfectly plastic material for each point (e, τ) of the region between the Oe axis and the graph $\vec{O}Az$, it is impossible to find a constituent law in the form e being a function of τ, or τ being a function of e. If starting from a state (e, τ), $\tau \leq g$, we increase e by a small value de, then τ changes by a small value $d\tau$ such that

$$\begin{aligned} &de = c\, d\tau + \tilde{\lambda}, \\ &\text{where } c = \text{const and} \\ &\tilde{\lambda} = 0 \quad \text{when} \quad \tau < g \quad \text{or when} \quad \tau = g,\ d\tau < 0, \\ &\tilde{\lambda} \geq 0 \quad \text{when} \quad \tau = g \quad \text{and,} \quad d\tau = 0. \end{aligned} \tag{4.59}$$

If increase de takes place during time dt, then (4.59) can be rewritten as follows:

$$\begin{aligned} &\frac{de}{dt} \equiv e' = c\, d\tau' + \lambda, \quad \text{where } e' = e(\mathbf{u}') \equiv \mathbf{D}, \\ &\text{where} \\ &\lambda = 0 \quad \text{when} \quad \tau < g \quad \text{or when} \quad \tau = g, \tau' < 0, \\ &\lambda \leq 0 \quad \text{when} \quad \tau = g \quad \text{and,} \quad \tau' = 0, \end{aligned} \tag{4.60}$$

Now, we will generalize the constituent law to three dimensions perfectly plastic isotropic continuum. We will assume that the elastic and the plastic regions are defined by

$$\mathcal{F}(\tau_{ij}) < 0 \quad \text{or} \quad \mathcal{F}(\tau_{ij}) = 0, \text{ respectively,} \tag{4.61}$$

where τ_{ij} are components of the stress tensor and $\tau_{ij} = \tau_{ji}$ holds. The function $\mathcal{F}$ is convex and continuous with respect to τ_{ij}, so that the domain of $\mathbb{R}^6$, where $\mathcal{F}(\tau_{ij}) \leq 0$, is a closed convex set. The function $\mathcal{F}(\tau_{ij})$ involves only the deviator τ_{ij}^D of the tensor τ and it can be defined as (i) the von Mises model

$$\mathcal{F}(\tau_{ij}) = \tfrac{1}{2}\tau_{ij}^D \tau_{ij}^D - k^2, \tag{4.62}$$

where k is a given constant, or (ii) as the Tresca model

$$\mathcal{F}(\tau_{ij}) = \sup_{I,J;I \neq J} |\tau_I - \tau_J| - k^2, \tag{4.63}$$

or (iii) as the Yang model

$$\mathcal{F}(\tau_{ij}) = \max\left(|2\tau_1 - \tau_2 - \tau_3|, |2\tau_2 - \tau_3 - \tau_1|, |2\tau_3 - \tau_1 - \tau_2|\right) - k^2, \tag{4.64}$$

where the τ_I are the eigenvalues of the stress tensor $\boldsymbol{\tau}$ and where k is a positive constant. Since the function $\mathcal{F}(\tau_{ij})$ is independent in both cases of the deformation, then both cases describe perfectly plastic isotropic materials, the Prandtl–Reuss materials.

The constitutive law (4.60) can be generalized. We saw that the plastic deformations originate if stresses reach a certain value. In the six-dimensional space of stress tensor components τ_{ij} we can define the region G_e such that for τ_{ij}, which is situated inside this region, the material is in an elastic state, and in the opposite case the material is manifested by permanent plastic deformation. The boundary ∂G_e represents the set of yield points for different stresses of a material, and it is denoted as the **surface of plasticity** or the **condition of plasticity**. At plastic deformation or at continuous transition from elastic to plastic states, the resulting stress τ_{ij} is situated on the surface ∂G_e. In the case of ideally plastic materials this surface at the isothermal process is constant and a point τ_{ij} moves along it. For materials with hardening the surface changes with changes of τ_{ij}, and, moreover, the shape of the surface also changes. The initial surface of plasticity is known as the primary (original) surface of plasticity, the ensuing surfaces of plasticity ∂G_e are denoted as ensuing or instantaneous surfaces of plasticity. The surface of plasticity can be described by the so-called **function of plasticity** $\mathcal{F}$, about which we assume that it is twice differentiable and such that $\mathcal{F}(0)=0$, $\mathcal{F}(\boldsymbol{\tau})>0$ for $\boldsymbol{\tau}\neq 0$ and $\partial^2\mathcal{F}/\partial\tau_{ij}\,\partial\tau_{kl}\zeta_{ij}\zeta_{kl}\geq 0$. Let us put

$$e_{ij} = e_{ij}^e + e_{ij}^p,$$

where e_{ij} is the **tensor of total deformations** that can be expressed as the sum of elastic and plastic components. Let us put $(e_{ij}^e)'=s_{ijkl}\tau_{kl}'$, $(e_{ij}^p)'=\lambda_{ij}$, where $(e_{ij}^e)'$ is the strain rate tensor of an elastic component and $(e_{ij}^p)'$ of a plastic component (the velocity of plastic deformations) and where s_{ijkl} $[s_{ijkl}\in L^\infty(\Omega)]$ are the coefficients of elasticity. Then (4.60) yields

$$e_{ij}(\mathbf{u}') = s_{ijkl}\tau_{kl}' + \lambda_{ij}, \tag{4.65a}$$

$$\lambda_{ij}(\sigma_{ij} - \tau_{ij}) \leq 0 \quad \text{for all } \boldsymbol{\sigma} \text{ with } \mathcal{F}(\sigma_{ij}) \leq 0, \tag{4.65b}$$

$$\lambda_{ij}\tau_{ij}' = 0, \tag{4.65c}$$

where s_{ijkl} are the coefficients of elasticity, satisfying (4.51) and where (4.65b,c) describe the rate of plastic deformation. In every case $\lambda_{ij}\geq 0$, that is, the plastic change of the strain tensor has the direction of the outward normal to the surface of plasticity $\mathcal{F}(\tau_{ij})=0$.

This law can be generalized for elastoviscoplastic materials. For this reason we introduce the closed convex set

$$K = \{\boldsymbol{\tau}|\boldsymbol{\tau} = (\tau_{ij}) \in \mathbb{R}^6, \qquad \tau_{ij} = \tau_{ji}, \qquad \mathcal{F}(\tau_{ij}) \leq 0\},$$

where $\mathcal{F}$ is a convex continuous function of the six scalar variables τ_{ij}, and the mapping $\boldsymbol{\tau}\to P_K(\overline{\boldsymbol{\tau}})$ is an orthogonal projection in the usual Euclidean structure of $\boldsymbol{\tau}\in\mathbb{R}^6$ on the convex set K. Then the constitutive law can be written as

$$e_{ij}(\mathbf{u}') = s_{ijkl}\tau_{kl}' + \lambda_{ij}, \tag{4.66a}$$

$$\{\lambda_{ij}\} = 0 \quad \text{if } \mathcal{F}(\tau_{ij}) < 0, \tag{4.66b}$$

$$\lambda_{ij} = (2\mu)^{-1}[\tau_{ij} - (P_K\boldsymbol{\tau})_{ij}] \quad \text{if } \mathcal{F}(\tau_{ij}) \geq 0, \tag{4.66c}$$

where μ is a positive scalar, physically having the meaning of a viscosity coefficient. If $\tau'_{ij}=0$, then (4.66) leads to

$$\tau_{ij} = (P_K\boldsymbol{\tau})_{ij} + 2\mu e_{ij}(\mathbf{u}'). \tag{4.67}$$

In the special case if $(P_K\boldsymbol{\tau})_{ij} = -p\delta_{ij}$, we have the **constitutive law for the Newtonian liquid**. In the other special case if the function $\mathcal{F}(\tau_{ij})$ will be defined by the von Mises model $\mathcal{F}(\tau_{ij})=\frac{1}{2}\tau_{ij}^D\tau_{ij}^D - k^2$, where k is a given constant, then (4.66) leads to

$$e_{ij}(\mathbf{u}') = s_{ijkl}\tau'_{kl} + \lambda_{ij}, \tag{4.68a}$$

$$\lambda_{ij} = 0 \quad \text{if } \mathcal{F}(\tau_{ij}) < 0, \tag{4.68b}$$

$$\lambda_{ij} = (2\mu)^{-1}[(\tau_{II}^{1/2} - k)\tau_{II}^{-1/2}]\tau_{ij}^D \quad \text{if } \mathcal{F}(\tau_{ij}) \geq 0, \tag{4.68c}$$

where $\tau_{II} = \frac{1}{2}\tau_{ij}^D\tau_{ij}^D$ and τ_{ij}^D is the deviator of the stress tensor.

The special case of the constitutive law, if

$$e_{ij}(\mathbf{u}) = s_{ijkl}\tau_{kl} + \lambda_{ij}, \quad \text{with } \mathcal{F}(\tau_{ij}) \leq 0,$$

and

$$\lambda_{ij}(\sigma_{ij} - \tau_{ij}) \leq 0, \quad \text{for all } \sigma_{ij}, \qquad \mathcal{F}(\sigma_{ij}) \leq 0, \tag{4.69}$$

is known as the **Hencky constitutive law**.

Moreover, a special case of the nonlinear Hooke law is the stress–strain relation for elastoplastic materials in the deformation theory of plasticity, in which the Lamé coefficients depend on the first two invariants ϑ_1 and ϑ_2 of the small strain tensor, and the relation between the first invariants of stress and strain is linear. Then

$$\tau_{ij}(\mathbf{u}) = \left[k(\mathbf{x}) - \tfrac{2}{3}\mu(\mathbf{x},\gamma)\right]\vartheta_1\delta_{ij} + 2\mu(\mathbf{x},\gamma)e_{ij}(\mathbf{u}), \tag{4.70}$$

where $k(\mathbf{x})$ is the bulk modulus and γ is defined by

$$\begin{aligned}\gamma &= \left\{\tfrac{1}{3}[(e_{11}-e_{22})^2 + (e_{22}-e_{33})^2 + (e_{11}-e_{33})^2 + 6(e_{12}^2 + e_{23}^2 + e_{13}^2)]\right\}^{1/2} \\ &= \left(-\tfrac{1}{3}\vartheta_1^2 + e_{ij}e_{ij}\right)^{1/2} = \left(\tfrac{2}{3}(\vartheta_1^2 - 3\vartheta_2)\right)^{1/2}.\end{aligned}$$

More details for these models of plasticity used in biomechanics can be found in Washizu (1968), Duvaut and Lions (1976), Nečas and Hlaváček (1981), Marsden and Hughes (1983), Ionescu and Sofonea (1993), Khan and Huang (1995), and Belytschko et al. (2000), and for plastic models with hardening in Hlaváček et al. (1988) and Han and Reddy (1999) and for hypoelastic, hyperelastic, and hypoelastic-plastic models in Belytschko et al. (2000).

4.3.3 Basic Boundary Value Problems of the Theory of Linear Elasticity

Let the domain Ω be occupied by a deformed body, and let $\partial\Omega$ be its Lipschitz boundary. Let $\partial\Omega = \overline{\Gamma}_u \cup \overline{\Gamma}_\tau \cup R$, where Γ_u and Γ_τ are two disjoint sets, and where Γ_u represents a part of the boundary $\partial\Omega$ where a displacement is prescribed, Γ_τ represents a part of the boundary $\partial\Omega$ where loading is prescribed, and R is the set of zero surface measure. Let us assume that the generalized Hooke law with $c_{ijkl} \in C^1(\overline{\Omega})$ is given and let $\mathbf{u} \in C^{2,N}(\Omega)$. Consider the equilibrium equation

$$\frac{\partial \tau_{ij}}{\partial x_j} + f_i = 0, \tag{4.71}$$

with $f_i(\mathbf{x}) \in C^1(\overline{\Omega})$ in Ω and look for the displacement $\mathbf{u}$ satisfying (4.71) and the boundary conditions.

If $\Gamma_u = \emptyset$, then we speak about the first basic boundary value problem of elasticity, if $\Gamma_\tau = \emptyset$, then we speak about the second basic boundary value problem of elasticity, and if $\Gamma_u \neq \emptyset$, $\Gamma_\tau \neq \emptyset$, we speak about a mixed boundary value problem of elasticity.

First Basic Boundary Value Problem of Linear Elasticity We seek the displacement $\mathbf{u} \in C^{1,N}(\overline{\Omega}) \cap C^{2,N}(\overline{\Omega})$, which satisfies the equilibrium equation (4.71) and

$$\tau_{ij}(\mathbf{x}) n_j(\mathbf{x}) = P_i(\mathbf{x}), \qquad \mathbf{x} \in \partial\Omega, \tag{4.72}$$

where $P_i(\mathbf{x}) \in C(\partial\Omega)$ is prescribed on the boundary $\partial\Omega$ and where $\mathbf{n}$ denotes the outward normal to $\partial\Omega$.

Second Basic Boundary Value Problem of Linear Elasticity We seek the displacement $\mathbf{u} \in [C(\overline{\Omega})]^N \cap C^{2,N}(\overline{\Omega})$, which satisfies the equilibrium equation (4.71) and

$$\mathbf{u}(\mathbf{x}) = \mathbf{u}_0(\mathbf{x}), \qquad \mathbf{x} \in \partial\Omega, \tag{4.73}$$

where $\mathbf{u}_0(\mathbf{x}) \in [C(\partial\Omega)]^N$ is prescribed on the boundary $\partial\Omega$.

Mixed Basic Boundary Value Problem of Linear Elasticity We seek the displacement $\mathbf{u} \in C^{1,N}(\Omega \cup \Gamma_\tau) \cap [C(\Omega \cup \Gamma_u)]^N \cap C^{2,N}(\Omega)$, which satisfies the equilibrium equation (4.71) and the boundary conditions (4.72) and (4.73), where $P_i(\mathbf{x}) \in C(\Gamma_\tau)$ is prescribed on the boundary Γ_τ and $\mathbf{u}_0(\mathbf{x}) \in [C(\Gamma_u)]^N$ is prescribed on the boundary Γ_u.

4.3.4 Energetic Considerations

Let the outer forces on the considered part of the human skeleton be operated. These outer forces evoke its deformation, and, therefore, they perform the work. The inner energy of the body is changed, and, therefore, its temperature, its thermal capacity, as well as its entropy are changed. One part of the deformation energy is changed into heat, and on the contrary, in the heated elastic body the heat evokes thermal stresses. Therefore, between mechanical and thermodynamical quantities one can define the energy of the deformed body.

Let dQ be an added heat, dE a total elementary work of volume and surface forces acting on the nondeformed body, dE_{kin} a change of kinetic energy, and dU a change of inner energy of the body. Then the first thermodynamic theorem yields

$$dE + dQ = dE_{\text{kin}} + dU. \tag{4.74}$$

The processes acting in the elastic body are reversible, that is, the unloaded elastic body returns to the initial state. Then from the second thermodynamic theorem we have

$$dQ - T\,dS = 0, \tag{4.75}$$

where T is an absolute temperature and S is an entropy of the body. From (4.74) and (4.75) it follows that

$$dE - dE_{\text{kin}} = dU - T\,dS = d(T - TS) + S\,dT, \tag{4.76}$$

where U is the potential energy.

The processes acting in the body will be (i) adiabatic, when the considered system is isolated, and, therefore, no changes of heat with the neighborhood of the body can be exchanged, and (ii) isothermic, if the temperature is constant and because the process acts very slowly, then the heat can be exchanged with the neighborhood of the body, and thus the temperature of the body and of its neighborhood will be the same. The isothermic process occurs during a slow change of the elastic body from its initial nondeformed state into its final deformed state. In an adiabatic state $dS = 0$ holds, and then from (4.76) we obtain

$$dE = dE_{\text{kin}} + dU, \tag{4.77}$$

where U represents the potential energy of the body. In the case of the isothermic process $dT = 0$ and then from (4.76) we obtain

$$dE = dE_{\text{kin}} + d(U - TS), \tag{4.78}$$

where $U - TS$ represents a potential energy and it is known as free energy. For the adiabatic process

$$U = \int_V F\,dV, \tag{4.79}$$

and for the isothermic process

$$U - TS = \int_V \Phi\,dV, \tag{4.80}$$

where V is the volume of the body in its nondeformed state, F and Φ denote the density of the deformation energy, for which we introduce only one notation W. From (4.77) to (4.80) we have

$$\frac{dE}{dt} = \frac{dE_{\mathrm{kin}}}{dt} + \frac{d}{dt}\int_V W\, dV = \frac{dE_{\mathrm{kin}}}{dt} + \int_V \frac{\partial W}{\partial t} dV. \tag{4.81}$$

For a total work of volume and surface forces during a time unit we have

$$\frac{dE}{dt} = \int_V F_i \frac{\partial u_i}{\partial t} dV + \int_S P_i \frac{\partial u_i}{\partial t} ds, \tag{4.82}$$

where S represents the surface of the nondeformed body. Hence, applying the Green theorem and using the equation of motion, we obtain

$$\frac{dE}{dt} = \int_V \rho \frac{\partial^2 u_i}{\partial t^2}\frac{\partial u_i}{\partial t} dV + \int_V \tau_{ij} \frac{\partial}{\partial t}\left(\frac{\partial u_i}{\partial x_j}\right) dV = \frac{dE_{\mathrm{kin}}}{dt} + \int_V \tau_{ij} \frac{\partial}{\partial t}\left(\frac{\partial u_i}{\partial x_j}\right) dV, \tag{4.83}$$

as the kinetic energy of the whole body is

$$E_{\mathrm{kin}} = \frac{1}{2}\int_V \rho \frac{\partial u_i}{\partial t}\frac{\partial u_i}{\partial t} dV. \tag{4.84}$$

Because $\partial u_i/\partial x_j = e_{ij} + \omega_{ij}$, where e_{ij} is a symmetric tensor and ω_{ij} is a skew-symmetric tensor and thus $\tau_{ij}(\partial\omega_{ij}/\partial t) = 0$. Then

$$\frac{dE}{dt} = \frac{dE_{\mathrm{kin}}}{dt} + \int_V \tau_{ij}\frac{\partial e_{ij}}{\partial t} dV. \tag{4.85}$$

From (4.81) and (4.85) it follows that

$$\int_V \frac{\partial W}{\partial t} dV = \int_V \tau_{ij}\frac{\partial e_{ij}}{\partial t} dV, \tag{4.86}$$

and because the volume V is arbitrary and the integrands are assumed to be continuous, then

$$\frac{\partial W}{\partial t} = \tau_{ij}\frac{\partial e_{ij}}{\partial t}. \tag{4.87}$$

At the fixed chosen values of coordinates x_i for a change of the density deformation energy (the work of deformation) W in the course of time dt, we obtain

$$dW = \tau_{ij}\, de_{ij} = \tau_{ij}\frac{\partial e_{ij}}{\partial t} dt. \tag{4.88}$$

Considering the stress τ_{ij} to be a function of the small strain tensor e_{ij}, then dW in (4.88) is a total differential. Considering dW to be a function of differentials of independent variables, then

$$\tau_{ij} = \frac{\partial W}{\partial e_{ij}}. \tag{4.89}$$

Since the density of deformation energy W is a function of small strain components, then it can be expanded into a power series of variables e_{ij}, that is,

$$2W = c_0 + 2c_{ij}e_{ij} + c_{ijkl}e_{ij}e_{kl} + \cdots . \tag{4.90}$$

Hence, neglecting the constant c_0 and according to the initial state of the body, we can neglect also the linear term $2c_{ij}e_{ij}$, so that for the density of deformation energy we obtain

$$W = \tfrac{1}{2}c_{ijkl}e_{ij}e_{kl}. \tag{4.91}$$

Analyzing (4.90) we obtain the symmetry condition $c_{ijkl} = c_{klij}$ [for more details see, e.g., Brdička (1959, p. 194)]. As $e_{ij} = s_{ijkl}\tau_{kl}$, then

$$\begin{aligned} W &= \tfrac{1}{2}\tau_{ij}e_{ij} = \tfrac{1}{2}s_{ijkl}\tau_{ij}\tau_{kl} && \text{for anisotropic materials,} \\ W &= \tfrac{1}{2}\tau_{ij}e_{ij} = \tfrac{1}{2}\lambda\theta^2 + \mu e_{ij}e_{ij} && \text{for isotropic materials,} \end{aligned} \tag{4.92}$$

and we see that W is also a positive-definite quadratic form of τ_{ij}. Hence

$$e_{ij} = \frac{\partial W}{\partial \tau_{ij}}. \tag{4.93}$$

4.3.5 Variational Principles in Small Displacement Theory

Let us consider a domain Ω being occupied by a deformed body with the Lipschitzian boundary $\partial\Omega = \overline{\Gamma}_u \cup \overline{\Gamma}_\tau \cup R$, as defined above. Next, we will introduce a statically admissible stress field and a geometrically admissible displacement field and then variational principles in the theory of elasticity.

Definition 4.3 A **statically admissible stress field** is a tensor function $\boldsymbol{\tau}(\mathbf{x})$, being symmetric, that is, $\tau_{ij} = \tau_{ji}$, and its components τ_{ij} satisfy the equation of equilibrium

$$\frac{\partial \tau_{ij}}{\partial x_j} + f_i = 0 \quad \text{in } \Omega, \tag{4.94}$$

and the boundary conditions

$$\tau_{ij}n_j = P_i \quad \text{on } \Gamma_\tau. \tag{4.95}$$

A **geometrically admissible displacement field** is a vector function $\mathbf{u}(\mathbf{x})$ satisfying the boundary condition

$$\mathbf{u} = \mathbf{u}_0 \quad \text{on } \Gamma_u. \tag{4.96}$$

Theorem 4.1 (Principle of Virtual Work) Let $\boldsymbol{\tau}(\mathbf{x})$ be a statically admissible stress field, and let $\mathbf{u}(\mathbf{x})$ be a geometrically admissible displacement field. Let $e_{ij}(\mathbf{u}) = \frac{1}{2}(\partial u_i/\partial x_j + \partial u_j/\partial x_i)$ be a small strain tensor. Then we find

$$\int_{\Omega} \tau_{ij} e_{ij}(\mathbf{u})\, d\mathbf{x} = \int_{\Omega} f_i u_i\, d\mathbf{x} + \int_{\Gamma_\tau} P_i u_i\, ds + \int_{\Gamma_u} \tau_{ij} n_j u_{0i}\, ds. \tag{4.97}$$

The proof follows from the equation of equilibrium (4.94), the boundary conditions (4.95) and (4.96), and the Green theorem.

Remark 4.3 The virtual work of the internal forces is equal to the virtual work of the external forces. This principle holds for mutually independent stress and displacement fields, and, therefore, it holds for an arbitrary type of the generalized (linear and nonlinear) Hooke law.

Theorem 4.2 (Principle of Virtual Displacements) Let $\mathbf{u}^*$ and $\mathbf{u}^*(\mathbf{x}) + \delta\mathbf{u}$ be geometrically admissible displacement fields. Let the actual stress field $\boldsymbol{\tau}^*$ be statically admissible. Then

$$\int_{\Omega} \tau_{ij}^* \delta e_{ij}(\mathbf{u})\, d\mathbf{x} = \int_{\Omega} f_i \delta u_i\, d\mathbf{x} + \int_{\Gamma_\tau} P_i \delta u_i\, ds, \tag{4.98}$$

where $\delta e_{ij}(\mathbf{u}) = e_{ij}(\delta\mathbf{u})$, $\delta\mathbf{u}$ is the virtual displacement or the variation of $\mathbf{u}$.

Proof Let $\mathbf{u}^*$ be the classical solution of the mixed basic boundary value problem of linear elasticity. Hence, it is a geometrically admissible displacement field. Let τ^* be a statically admissible stress field. Into the principle of virtual work we substitute the actual stress field $\boldsymbol{\tau}^*$ and the actual displacement field $\mathbf{u}^*$ as well as an arbitrary geometrically admissible displacement field $\mathbf{u}^* + \delta\mathbf{u}^*$, where $\delta\mathbf{u}^*$ is the virtual displacements. Then subtracting the resultant equalities we find (4.98). □

Theorem 4.3 (Principle of Virtual Stresses) Let $\mathbf{u}^*$ be the actual displacement and let $\boldsymbol{\tau}^*$ be the actual stress field and $\boldsymbol{\tau}^* + \delta\boldsymbol{\tau}$ be an arbitrary stress field, which are both statically admissible stress fields. Then

$$\int_{\Omega} e_{ij}(\mathbf{u}^*) \delta\tau_{ij}\, d\mathbf{x} = \int_{\Gamma_u} \delta\tau_{ij} n_j u_{0i}\, ds, \tag{4.99}$$

where $\delta\tau_{ij}$ are components of the virtual stress or the variation of stress.

Proof We substitute the actual displacement field $\mathbf{u}^*$ into the principle of virtual work. Furthermore, we substitute the actual stress field $\boldsymbol{\tau}^*$ on the one hand and, on the other hand, an arbitrary field $\boldsymbol{\tau}^* + \delta\boldsymbol{\tau}$. Subtracting the obtained resultant equalities, we obtain (4.99). □

Remark 4.4 The principle of virtual displacements and the principle of virtual stresses can be applied for any type of generalized Hooke's stress–strain relations.

Theorem 4.4 (Principle of Minimum Potential Energy) Let us consider the generalized Hooke law (4.45). Let

$$c_{ijkl} = c_{jikl} = c_{ijlk} = c_{klij},$$
$$2W(e) = c_{ijkl}e_{ij}e_{kl} \geq c_0 e_{ij}e_{ij}, \qquad c_0 > 0, \tag{4.100}$$

be valid for all symmetric tensors e_{ij} and almost every $\mathbf{x} \in \Omega$. Let $\boldsymbol{\tau}^*$ be statically admissible. Then the solution $\mathbf{u}^*$ of the mixed problem (4.94)–(4.96) gives the functional of potential energy of an elastic body

$$J(\mathbf{u}) = \frac{1}{2}\int_\Omega c_{ijkl}(\mathbf{x})e_{ij}(\mathbf{u})e_{kl}(\mathbf{u})\,d\mathbf{x} - \int_\Omega f_i u_i\,d\mathbf{x} - \int_{\Gamma_\tau} P_i u_i\,ds \tag{4.101}$$

its minimum value over the set of geometrically admissible fields. Moreover, from the principle of minimum potential energy the principle of virtual displacements follows immediately.

For the proof see Nečas and Hlaváček (1981) and Nedoma (1998a).

Theorem 4.5 (Principle of Minimum Complementary Energy or Castigliano–Menabrea Principle) Let $\boldsymbol{\tau}^*$ be the actual stress field of the mixed problem of linear elasticity (4.94)–(4.96), and let it be statically admissible. Then it makes the functional of complementary energy

$$S(\boldsymbol{\tau}) = \frac{1}{2}\int_\Omega s_{ijkl}(\mathbf{x})\tau_{ij}\tau_{kl}\,d\mathbf{x} - \int_{\Gamma_u} u_{0i}\tau_{ij}n_j\,ds \tag{4.102}$$

a minimum over the set of all statically admissible stress fields.

For the proof see Nečas and Hlaváček (1981) and Nedoma (1998a).

Remark 4.5 In the theory of elasticity as well as in the practice the hybrid principle, the Hellinger–Reissner principle, and the other principles are also used. For more details see Brdička (1959), Washizu (1975), Nečas and Hlaváček (1981), Marsden and Hughes (1983), and Belytschko et al. (2000).

4.4 BACKGROUND OF THE QUASI-STATIC AND DYNAMIC CONTINUUM MECHANICS IN THERMO(VISCO)ELASTIC RHEOLOGY

4.4.1 Friction, Wear, and Lubrication in Contact Mechanics

The science and technology that study the interacting surfaces in contact and relative motion is **tribology**. Friction and wear represent two main parts of tribology. In addition to these two main parts of tribology, tribology also covers topics such as adhesion, lubrication, and thermal and electric contacts. Friction was first studied by Leonardo da Vinci, then by Amontons and Coulomb, but many frictional phenomena are currently completely understood. The term tribology is derived from the Greek word "tribos," which means rubbing. **Friction** is defined as the resistance (resting force) to motion that exists when a solid object is moved tangentially to the surface of another that it touches or is tended to move relatively to the surface of the other under the action of external forces. It is necessary to distinguish two situations in which the applied force is insufficient to cause motion and the other in which sliding is occurring (see Fig. 4.10). If an acting tangential force $\mathbf{P}_t$ is small, then sliding does not occur. It follows from the first of Newton's laws because the frictional force at the contact surface must be exactly equal and opposite to $\mathbf{P}_t$. Experiments show [see Rabinowicz (1995)] that in any situation where the resultant of the tangential forces is smaller than some force parameter specific to that particular situation, the friction force will be equal and opposite to the resultant of the applied forces and no tangential motion will occur.

The next situation is that in which the applied force $\mathbf{P}_t$ is sufficient to cause sliding. Experimentally, it was shown that the body moves in the direction $\mathbf{P}_t$ [Rabinowicz (1995)]. We see that when tangential motion occurs, the friction force always acts in a direction opposite to that of the relative velocity of the surfaces.

Friction depends on the nature of two sliding surfaces, and it is larger for materials that strongly interact. Moreover, friction depends on the quality of both surfaces and surface finish; a good surface finish (i.e., sufficiently smooth surfaces) reduces frictional effects. The frictional force does not depend on the total surface area of contact. In the tangential direction of the contact surface we have to distinguish two cases: (i) the first one is represented by a "stick state" in which points, which are in contact, are not allowed to move in a tangential direction; and (ii) the second one is "sliding," which means that a point moves in a tangential direction to the contact boundary.

There exist several types of friction. Consider the body resting on the floor (see Fig. 4.10). Its weight $\mathbf{W}$ acts on the rest and in return the rest evokes a normal force $\mathbf{N}$ on the body such that their magnitudes are equal, that is, $|\mathbf{N}| = |\mathbf{W}|$. Consider an acting horizontal force $\mathbf{P}_t$. This will cause a frictional force $\mathbf{f}_c$ between the body and the rest. Because the block is in equilibrium, the magnitude of the frictional force $\mathbf{f}_c$ would be equal to the magnitude of the acting force $\mathbf{P}_t$, and we denote it as the **static friction** ($\mathbf{f}_c^{\text{stat}}$). When the magnitude of the acting force $\mathbf{P}_t$ increases, the body will slip or begin sliding over the rest. When the body is in the situation before sliding, the magnitude of the static friction is equal to its maximal value $\mathbf{f}_c^{\max}$.

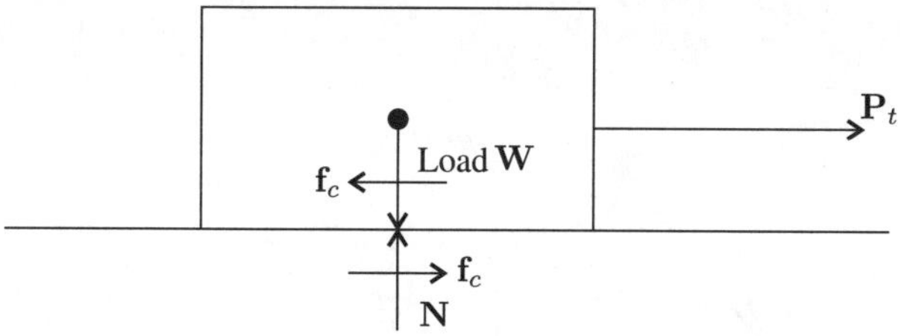

FIG. 4.10 Schematic view of a load on a horizontal contact surface.

When the magnitude of the acting force $\mathbf{P}_t$ exceeds $\mathbf{f}_c^{\max}$, the body starts to move. The resistance to its motion at the contact surface is the **kinetic or dynamic friction** ($\mathbf{f}_c^{\text{kin}}$ or $\mathbf{f}_c^{\text{dyn}}$). The difference between the magnitudes of the acting forces $\mathbf{P}_t$ and kinetic friction causes the acceleration of the body because the magnitude of kinetic friction is lower than $|\mathbf{f}_c^{\text{stat}}|$ and $|\mathbf{P}_t|$.

Experimentally, it was shown that the magnitudes of static and kinetic frictions are proportional to the normal force $\mathbf{N}$ with the coefficient of proportionality $\mathcal{F}_c$, the **coefficient of friction**, which depends on the material properties, the quality of the contact surface, the conditions of the surfaces in contact, and moreover, on whether the bodies in contact are stationary or move over each other. In the case of two or more bodies in contact, then two or more different friction coefficients are introduced [see Duvaut and Lions (1976)]. The coefficient of static friction $\mathcal{F}_c^{\text{stat}}$ is associated with static friction, and the coefficient of kinetic (dynamic) friction $\mathcal{F}_c^{\text{kin}}$ is associated with dynamic friction. At the moment before the body moves then the magnitude of the static friction force is $|\mathbf{f}_c^{\text{stat}}| = \mathcal{F}_c^{\text{stat}}|\mathbf{N}| \equiv f_c^{\max}$ and $|\mathbf{f}_c^{\text{stat}}| < \mathcal{F}_c^{\text{stat}}|\mathbf{N}|$ when the magnitude of the acting force $|\mathbf{P}_t|$ is less than the maximal value $f_c^{\max}$ of the frictional force, in which case the magnitude of the force of static friction is equal in magnitude to the acting force $\mathbf{P}_t$, that is, $|\mathbf{f}_c^{\text{stat}}| = |\mathbf{P}_t|$. The dependence of the frictional force as a function of the acting force $\mathbf{P}_t$ (see Fig. 4.11) and the average estimates of friction coefficients for different materials are given in Table 4.2.

In the case of kinetic friction $|\mathbf{f}_c^{\text{kin}}| = \mathcal{F}_c^{\text{kin}}|\mathbf{N}|$. The coefficient of kinetic friction is approximately constant at moderate sliding speeds, generally depending on the slip velocity, on the roughness of the contact surfaces, on the temperature, and in some cases on the tangential velocity. The temperature effect is very complicated to measure since frictional sliding causes considerable heat generation, which leads to uneven temperature rises.

Frictional forces always act in a direction tangent to the surfaces in contact. The component $\boldsymbol{\tau}_t = \boldsymbol{\tau} - \tau_n \mathbf{n}$ represents the friction force on the contact surface. To describe the contact of two bodies we need to consider the normal approach and the tangential process. By a contact condition we understand a condition that describes the normal approach, that is, a relation involving only the normal displacement, velocity, and stress components. The condition describing the tangential process is known as a friction condition, which is a relation involving the tangential stress $\boldsymbol{\tau}_t$ and the tangential velocity $\mathbf{u}_t'$.

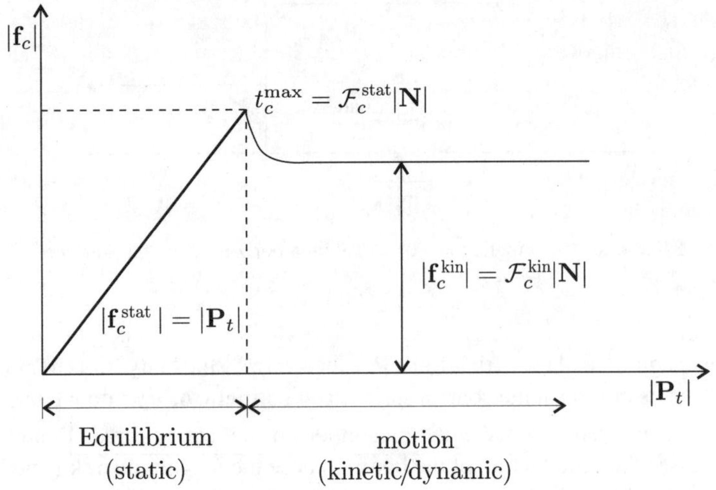

FIG. 4.11 Dependence of frictional force as a function of the acting force $\mathbf{P}_t$. [Modified after Özkaya and Nordin (1999) and Rabinowicz (1995).]

TABLE 4.2 Coefficient of Friction Compiled by Rankin (1926) and Rabinowitz (1995)

Type of Contact Surface	Friction Coefficient
Metal/metal	0.3–0.8
Plastic/plastic	0.1–0.3
Metal/plastic	0.1–0.2
Metal/ceramics	0.1
Cortical bone/metal	0.1–0.2
Cortical bone/ceramics	0.08
Ceramics/ceramics	0.1
Cartilage/cartilage	0.001–0.002
Composite materials/composite materials	0.1
Metal on metal, dry	0.15–0.20
Metal on metal, wet	0.3

Resulting **effects of friction** are (i) the **heat** to be generated between the surfaces in contact and (ii) **wear**.

Wear is defined as the removal of a material from solid surfaces as a result of a mechanical action. In orthopedics wear represents a very difficult problem connected with the application of total joint replacements. In general, wear is related to sliding contact. Present research has established that there are four main forms of wear mechanisms, that is, adhesive, abrasive, corrosive, and surface fatigue. Besides these types of wear there exist marginal processes that are often classified as forms of wear (like fretting wear and erosive wear, low-speed erosion, etc.) [see Rabinowicz (1995)].

Adhesive wear occurs when two smooth bodies slide over each other and fragments are pulled off one surface and adhere to the other. If these particles come off the surface, they can form loose wear particles.

Abrassive wear occurs when a rough hard surface and a soft surface containing hard particles slide over each other and the hard surface ploughs grooves in the soft one. The material from the grooves then forms the wear particles.

Corrosive wear occurs when sliding takes place in a corrosive environment. In the absence of sliding, the products of the corrosion form a film on the contact surfaces, which is removed by a sliding action, so that the corrosive attack can continue.

Surface fatigue wear is observed during repeated sliding or rolling over a track. The material undergoes many loading/unloading cycles that lead to cracks close to the contact surface, which might result in a breakup of the surface with the formation of large wear particles.

Fretting occurs when surfaces in contact undergo oscillatory tangential displacement of a small amplitude. As a result of the movement, adhering steel particles are produced (adhesive wear) and these oxidize (corrode) to Fe_2O_3. This substance is abrasive, and it causes abrasive wear of the surface. Fretting can be eliminated by eliminating slip at the contact surface.

From these definitions it is evident that wear is complicated, and it can also involve different mechanisms at different stages of the process. Wear depends on the properties of the material contact surfaces, their roughness, the sliding distance and sliding velocity, as well as the temperature. For more details see Rabinowicz (1995). Depending on the softness of the surfaces and on the material properties, we can consider the qualitative aspects of adhesive wear. To compute the volume of material lost by the wear process the Holm–Archard law in the form

$$V_{\text{wear}} = K_{\text{ad}} \frac{F_n g_t}{p},$$

can be used, where K_{ad} is the wear coefficient depending upon the materials in contact (see Table 4.3) and hence in relation to the friction coefficients, F_n is the normal force or normal loading, g_t is the relative sliding distance between materials, and p is the hardness of the surface that is worn away. This relation is linear and it is based on a simple energy relationship:

$$V_{\text{wear}} = k \frac{\text{energy expected}}{\text{hardness}}.$$

To reduce the frictional and wear effects in tribological joint systems, **lubrication** is applied. This phenomenon is produced by a thin layer of additional materials between the sliding surfaces. These materials can be solids (like grafit) or fluids, and they are known as **lubricants.**

Fluid lubrication occurs when a thick film of some liquid or a gas on the contact surface completely separates two solid bodies. Lubricants placed between the moving part of joint components reduce friction and wear.

In human joints such as the hip, knee, and elbow, the lubricants function as the **synovial fluid**. Synovial fluid is a viscous medium reducing the effects of friction and wear and tear of articulating surfaces by limiting direct contact between them.

TABLE 4.3 Wear Coefficient

Combination	Wear Coefficient K	References
Polyethylene on tool steel	0.0013	Hirst (1957)
Steel on steel	126×10^{-4}	Holm (1946)
Iron on iron (lubricated)	0.2	Holm (1946)
Mild steel on mild steel	150×10^{-4}	Archard (1953)
Stainless steel on stainless steel	70	Archard (1953)
Mild steel on cooper	1.7	Archard (1953)
Low-carbon steel on low-carbon steel	70×10^{-4}	Hirst (1957)
Tool steel on tool steel	1.3	Hirst (1957)
Stellite #1 on tool steel	0.55	Hirst (1957)
Ferritic stainless steel on tool steel	0.17	Hirst (1957)
Tungsten carbide on low-carbon steel	0.04	Hirst (1957)
Tungsten carbide on tungstan carbide	0.01	Hirst (1957)

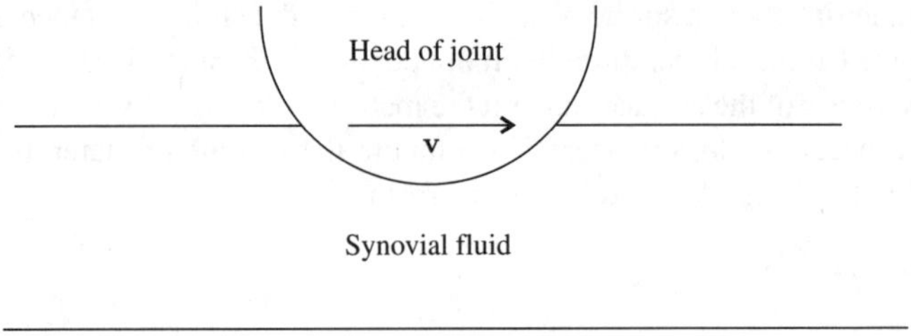

FIG. 4.12 Lubrication in the artificial human joint.

Moreover, synovial fluid nourishes joint cartilage and also plays a role in the hydrostatic transmission of loading. Lubrication can also appear when a chemical reaction in the contact zone leads to a layer of contaminants that reduce friction. Abrasive wear can also decrease the friction force in the contact zone. Lubrication only works when there are certain motion changes with relative velocity **v** between both solid joint components such that the fluid layer does not disappear (see Fig. 4.12). In this case we speak about **hydrodynamic lubrication.** If the velocity is too small, then the fluid can be squeezed out between both joint components, leading to the direct contact of both parts. Then we speak about **boundary lubrication** (see Fig. 4.13). But in the course of loading the human joint with different loads and under different motion conditions, a number of lubrication modes can be observed. These modes include hydrodynamic, boundary, squeeze-film, elastohydrodynamic, boosted, and weeping lubrications. For more details see Rabinowicz (1995). (See Fig. 4.13.)

Adhesion is the phenomenon that occurs when we press two surfaces together, either under a pure normal load τ_n or under combined normal and shear forces, and then find that a normal tension force must be exerted to separate the surfaces. Let F_n'

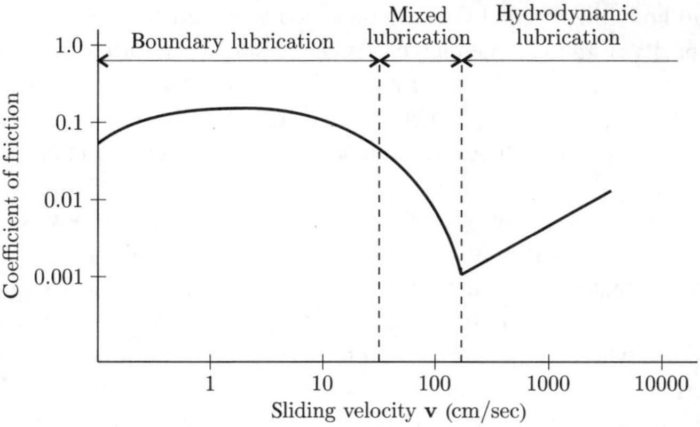

FIG. 4.13 Coefficient of friction versus sliding velocity **v** (cm/sec). [Modified after Rabinowicz (1995).]

be a normal tensile force required for separation and F_n a normal compressive force being initially applied. Then the ratio

$$f' = \frac{F'_n}{F_n}$$

is the **coefficient of adhesion.** Strong adhesion can be found in two-phase alloys in which adjoining grains differ in composition or in single-phase materials in which adjacent grains have the same composition. Other evidence for the strong tendencies of solids to adhere can be found in the process of adhesive wear, so that adhesive wear can be explained under the assumption that a strong adhesive force exists between solids that are in contact. Useful applications are in medical applications concerned with construction of prosthetic limbs, where the bonding (coupling) between the artificial limb and the tissue must be of a very high quality (Rojek and Telega, 2001). It is evident that adhesion changes the unilateral contact conditions on the contact surfaces. Frémond (1987) formulated adhesion by measuring the intensity of adhesion on the contact surface, which is similar to a description of damage mechanics by a damage function. Mutual relations between adhesion and friction were studied in Raous (1999) and Raous et al. (1999).

4.4.2 Equations of Equilibrium and Motion, Boundary, Contact, and Initial Conditions

The temperature of the human body is about 36–39°C. Therefore, for a better understanding of processes acting on the human skeleton before and after implantation of the artificial joint replacement these processes will be investigated as thermoelastic, thermoviscoelastic, and/or thermoviscoplastic, respectively. The implant, the bone tissue, and the bone cement behave like a rigid body.

In a rigid body the thermal field is connected with deformation. A change in the thermal capacity of the volume element evokes a state of deformation and a state of stress, and, on the contrary, the deformation field evokes the rise of a thermal field in the rigid body. Then part of the mechanical energy, evoked by a deformation of the body, is changed into heat. It attends to thermomechanical diffusion of the energy in the rigid body.

Let the elastic isotropic, in general anisotropic, homogeneous body of an absolute temperature T_0 be in a strainless and stressless state, that is, in the natural state. Due to the external load and heat the body becomes deformed. This generates a displacement $\mathbf{u}$, strain e_{ij}, and stress τ_{ij}, $i,j=1,2,3$, and its temperature will increase by $\Theta = T - T_0$, where T_0 is the initial (reference) temperature. Quantities $\mathbf{u}$, e_{ij}, τ_{ij}, T are functions of the coordinates $\mathbf{x}=(x_i)$, $i=1,2,3$, and of time t. To simplify our discussion we assume that the temperature change Θ is small compared to the temperature T_0, that is, $|\Theta/T_0| \ll 1$. Also we assume the increase in the temperature of about Θ has no effect on the elastic and thermal parameters of materials forming the human body system. Therefore, we are able to consider them independent of the temperature to the first approximation. We are thus restricting ourselves to the case of linear thermoelasticity. In the opposite case the problem will be investigated in the thermoviscoelasticity and thermoviscoplasticity.

From the theory of elasticity it is known that a small strain cannot be an arbitrary function, but it must satisfy the conditions of strain compatibility (the Saint-Venant equations)

$$e_{ij,kl} + e_{kl,ij} - e_{jl,ik} - e_{ik,jl} = 0, \qquad i,j,k,l = 1,2,3, \tag{4.103}$$

where $e_{ij,kl} = \partial^2 e_{ij}/\partial x_k\, \partial x_l$.

To derive the relations between components of stress and of strain, we will draw on the thermodynamics of a strain of an elastic body. We will start with thermodynamic identity, the Gibbs equation, for a strained body:

$$dU = \tau_{ij}\, de_{ij} + T\, dS, \tag{4.104}$$

where U is an internal energy, S an entropy, and $\tau_{ij}\, de_{ij}$ the strain work, where all quantities are related to the unit volume. If we introduce the Helmholtz's free energy $F = F(e_{ij}, T) = U - ST$, then

$$dF = \tau_{ij}\, de_{ij} - S\, dT, \tag{4.105}$$

with the independent variables e_{ij}, T. It holds

$$\tau_{ij} = \left(\frac{\partial U}{\partial e_{ij}}\right)_S = \left(\frac{\partial F}{\partial e_{ij}}\right)_T, \tag{4.106}$$

which is similar to the relation $\tau_{ij} = \partial W/\partial e_{ij}$, where W is the density of the strain energy from the classical linear theory of elasticity, which allows the stress τ_{ij} to be expressed in terms of the derivative of the internal energy with respect to e_{ij} at the constant entropy S and in terms of the derivative of the free energy F with respect to e_{ij} at the constant temperature T. If we develop the free energy $F(e_{ij}, T)$ in the neighborhood of the point $(0, T_0)$ into the Taylor series, we arrive at

$$F(e_{ij}, T) = F(0, T_0) + \frac{\partial F(0, T_0)}{\partial e_{ij}} e_{ij} + \frac{\partial F(0, T_0)}{\partial T}\Theta + \frac{1}{2}\frac{\partial^2 F(0, T_0)}{\partial e_{ij}\,\partial e_{kl}} e_{ij}e_{kl}$$
$$+ \frac{1}{2}\frac{\partial^2 F(0, T_0)}{\partial e_{ij}\,\partial T} e_{ij}\Theta + \frac{1}{2}\frac{\partial^2 F(0, T_0)}{\partial T^2}\Theta^2 + \cdots, \Theta = T - T_0. \quad (4.107)$$

Since $F(e_{ij}T)$ represents the free energy, then $F(0, T_0) = 0$. Due to (4.106), and since we assume that $|\Theta/T_0| \ll 1$ and assuming the linear theory and isothermal state,

$$\tau_{ij} = \left(\frac{\partial F}{\partial e_{ij}}\right)_T = a_{ij} + c_{ijkl}e_{kl} - \beta_{ij}\Theta, \quad (4.108)$$

where we denoted by

$$a_{ij} = \left(\frac{\partial F}{\partial e_{ij}}\right)_{(0,T_0)}, \qquad c_{ijkl} = \left(\frac{\partial^2 F}{\partial e_{ij}\,\partial e_{kl}}\right)_{(0,T_0)},$$
$$\beta_{ij} = -\left(\frac{\partial^2 F}{\partial e_{ij}\,\partial T}\right). \quad (4.109)$$

The natural state will be determined if for $T = T_0$, $e_{ij} = 0$ will be also $\tau_{ij} = 0$. Constants c_{ijkl}, β_{ij} are material constants, elastic constants (Lamé's constants in the case of an isotropic material), and the coefficient of thermal expansion. Then, from (4.107) and using (4.109), for the free energy we find

$$F(e_{ij}, T) = \tfrac{1}{2}c_{ijkl}e_{ij}e_{kl} - \beta_{ij}e_{ij}\Theta + \tfrac{1}{2}m\Theta^2, \qquad m = \left(\frac{\partial^2 F}{\partial T^2}\right)_{(0,T_0)}, \quad (4.110)$$

where we use $(\partial F/\partial T)_e = -S(0, T_0) = 0$, which follows from (4.105). Hence and from (4.106) and (4.109) it follows that

$$\tau_{ij} = c_{ijkl}e_{kl} - \beta_{ij}\Theta, \quad (4.111)$$

the **Duhamel–Neumann law** or **Hooke's law in linear thermoelasticity**.

Since

$$e_{ij} = \left(e_{ij} - \tfrac{1}{3}\delta_{ij}e_{kk}\right) + \tfrac{1}{3}\delta_{ij}e_{kk} = e_{ij}^D + \tfrac{1}{3}\delta_{ij}e \quad (4.112)$$

where $e = e_{kk}$, $e_{ij}^D = e_{ij} - \frac{1}{3}\delta_{ij}e_{kk}$ is the deviator of small strain tensor, then the term $\frac{1}{3}e\delta_{ij}$ is the spherical tensor, and it represents extension in all directions. After substitution into (4.110), then for the isotropic material we have

$$F = \mu e_{ij}e_{ij} + \frac{\lambda}{2}e_{kk}^2 - \gamma e_{kk}\Theta + \frac{1}{2}m\Theta^2 = \mu e_{ij}^D e_{ij}^D + \frac{1}{2}ke^2 - \gamma\Theta e + \frac{1}{2}m\Theta^2. \quad (4.113)$$

Here $k = \lambda + \frac{2}{3}\mu$ is the **modulus of compressibility** or **bulk modulus** and $\gamma = \text{const}$ will be defined later. From (4.106) and (4.113) we find

$$\tau_{ij} = 2\mu e_{ij} + (\lambda e_{kk} - \gamma\Theta)\delta_{ij} = 2\mu e_{ij}^D + (ke - \gamma\Theta)\delta_{ij}. \tag{4.114}$$

Hence, reducing the stress tensor, we have

$$\tau_{kk} = 3(ke - \gamma\Theta), \qquad e_{kk}^D = 0. \tag{4.115}$$

Now, let us investigate the free expansion of the body under which no internal stresses are generated, that is, $\tau_{ij} = 0$. From (4.114) and (4.115), which we set to be zero, we find

$$e^0 = \frac{\gamma\Theta}{k}, \qquad e_{ij}^0 = \frac{\gamma}{3k}\Theta\delta_{ij}, \tag{4.116}$$

where e^0 is the relative change of the volume, the **dilatation**. This dilatation depends on the change of temperature Θ, and the coefficient of proportionality is the volume extension $\alpha = \gamma/k$. Hence $\gamma = \alpha k$ or $\gamma = 3\alpha_t k$, where $\alpha_t = \alpha/3$ is the **coefficient of linear thermal expansion**. An elementary volume of the isotropic body under thermal changes, for the case if no stresses act onto its surface, changes only its volume but not its form, that is, $e_{ij}^0 = (\gamma/3k)\Theta\delta_{ij} = \alpha_t\Theta\delta_{ij}$. Then (4.114) can be written as

$$\tau_{ij} = 2\mu e_{ij} + (\lambda e_{kk} - \gamma\Theta)\delta_{ij} = 2\mu e_{ij}^D + k(e - 3\alpha_t\Theta)\delta_{ij} \quad \gamma = (3\lambda + 2\mu)\alpha_t, \tag{4.117}$$

representing the **Duhamel–Neumann law for isotropic materials.**

Since the processes in bone tissues and in artificial joint replacement are assumed to be elastically reversible (i.e., bone materials and materials for joint replacements after unloading and when $\Theta \to 0$ will be returned into their initial states), then the Lamé parameters will be related to the isothermic state, for which $e_{ij} = 0$ and $\Theta = 0$.

From (4.111) or (4.117), respectively, the inverse stress–strain relations

$$e_{ij} = s_{ijkl}\tau_{kl} + \alpha_{ij}\Theta \tag{4.118}$$

for anisotropic materials and

$$e_{ij} = \frac{1}{2\mu}\left(\tau_{ij} - \frac{\lambda}{2\mu + 3\lambda}\tau_{kk}\delta_{ij}\right) + \alpha_t\Theta\delta_{ij} \tag{4.119}$$

for isotropic materials are derived. Hence using (4.115) it follows that

$$\left(\frac{\partial e_{kk}}{\partial T}\right)_\tau = 3\alpha_t, \qquad \left(\frac{\partial e_{kk}}{\partial \tau_{kk}}\right)_T = \frac{1}{3k}. \tag{4.120}$$

Let us assume the case when all stress components are equal to zero. Then from (4.119) it follows that $e_{ij} = \alpha_t\Theta\delta_{ij} = e_{ij}^0$. The strain components must satisfy the compatibility conditions. The compatibility conditions will be satisfied, if all second derivatives of all components of e_{ij} will be equal to zero, that is, when the temperature distribution is linear ($\Theta = a_0 + a_i x_i$, a_0, $a_i = \text{const}$). Setting $e_{ij} = \alpha_t \Theta \delta_{ij}$ into the compatibility equations, then $\partial^2\Theta/\partial x_i\, \partial x_j = 0$, $i, j = 1, 2, 3$. The above-mentioned system of equations will be satisfied if the distribution of the temperature is linear.

Another model in thermo-(nonlinear)elasticity, which corresponds to the deformation theory of plasticity, can be found by the following procedure:

To develop the N-dimensional stress–strain relation ($N = 2, 3$) the strain energy density function $W(e_{ij})$ will be used. Let A be a scalar-valued function of strains e_{ij} and λ be a positive parameter, defined by [see Fung (1965, 1968)]

$$\begin{aligned} A &= A(e_{ij}) = c_{ijkl}(\mathbf{x})e_{ij}(\mathbf{u})e_{kl}(\mathbf{u}), \\ e_{ij}(\mathbf{u}) &= \frac{1}{2}\left(\frac{\partial u_i}{\partial x_j} + \frac{\partial u_j}{\partial x_i}\right), \qquad i, j, k, l = 1, \ldots, N, \end{aligned} \tag{4.121}$$

where $c_{ijkl}(\mathbf{x})$ are elastic coefficients and $e_{ij}(\mathbf{u})$ is the small strain tensor, $\mathbf{u} = (u_i)$ is the displacement vector, and N is the space dimension. For the tensor $c_{ijkl}(\mathbf{x})$ we assume

$$\begin{aligned} &c_{ijkl} \in L^{\infty}(\Omega), \\ &c_{ijkl} = c_{jikl} = c_{klij} = c_{ijlk}, \\ &c_{ijkl}e_{ij}e_{kl} \geq c_0 e_{ij}e_{ij} \quad \forall e_{ij}, e_{ij} = e_{ji} \quad \text{and} \quad \text{a.e.}; \mathbf{x} \in \Omega,\ c_0 > 0. \end{aligned} \tag{4.122}$$

A repeated index implies the summation from 1 to N.

The stress–strain relation can be derived by using the strain energy density function $W = A^{\lambda}(e_{ij}(\mathbf{u}))$, then

$$\tau^e_{ij}(\mathbf{u}) = \frac{\partial W(e_{ij}(\mathbf{u}))}{\partial e_{ij}} = \frac{\partial A^{\lambda}(e_{ij}(\mathbf{u}))}{\partial e_{ij}}. \tag{4.123}$$

Hence

$$\begin{aligned} \tau^e_{ij}(\mathbf{u}) &= \lambda[A(e_{ij}(\mathbf{u}))]^{\lambda-1}\frac{\partial A(e_{ij}(\mathbf{u}))}{\partial e_{ij}} = 2\lambda[A(e_{ij}(\mathbf{u}))]^{\lambda-1}c_{ijkl}(\mathbf{x})e_{kl}(\mathbf{u}) \\ &= c^*_{ijkl}(\mathbf{x},\mathbf{u})e_{kl}(\mathbf{u}), \quad c^*_{ijkl}(\mathbf{x},\mathbf{u}) = 2\lambda[A(e_{ij}(\mathbf{u}))]^{\lambda-1}c_{ijkl}(\mathbf{x}), \end{aligned} \tag{4.124}$$

where $c^*_{ijkl}(\mathbf{x},\mathbf{u})$ are strain-dependent nonlinear elastic coefficients. The coefficients c^*_{ijkl} satisfy the conditions

$$\begin{aligned} &c^*_{ijkl} = c^*_{jikl} = c^*_{klij} = c^*_{ijlk}, \\ &c^*_{ijkl}e_{ij}e_{kl} \geq c^*_0 e_{ij}e_{ij}, \qquad \forall e_{ij},\ e_{ij} = e_{ji} \quad \text{and} \quad \text{a.e. } \mathbf{x} \in \Omega,\ c^*_0 = 2\lambda c^{\lambda}_0 > 0. \end{aligned} \tag{4.125}$$

The scalar coefficient $2\lambda[A(e_{ij}(\mathbf{u}))]^{\lambda-1}$ depends on the state of a strain. The parameter λ simulates the degree of nonlinear properties of materials. The parameter λ

characterizes for $\lambda < 1$ softened materials, for $\lambda = 1$ elastic materials, and for $\lambda > 1$ hardened materials. Then the stress–strain relation in nonlinear thermoelasticity can be written as

$$\tau_{ij}(\mathbf{u}) = \tau_{ij}^{e}(\mathbf{u}) + \tau_{ij}^{T}(\mathbf{u}) = c_{ijkl}^{*}(\mathbf{x}, \mathbf{u})e_{kl}(\mathbf{u}) - \beta_{ij}\Theta. \tag{4.126}$$

To derive the equations of motion, equilibrium equations, and equation of thermal conduction we have the following:

Denote by G the region occupied by the investigated part of the loaded skeleton. Let G_1 be the subregion of G with the boundary S_1 and then we construct the equilibrium equation. Then

$$\int_{G_1} (F_i - \rho u_i'')\, dV + \int_{S_1} P_i\, ds = 0, \qquad i = 1, 2, 3, \tag{4.127}$$

where $\mathbf{F}$ is the volume forces, F_i their components, related to the unit volume, P_i are components of the surface forces $\mathbf{P}$ acting on the surface S_1 and $u_i'' = \partial^2 u_i/\partial t^2$. Since $P_i = \tau_{ij}n_j$, where n_j are components of the unit outward normal vector to the surface S_1, ρ is the density, then using the Green theorem ($\int_{S_1} \tau_{ij}n_j\, ds = \int_{G_1} \tau_{ij,j}\, dV$) we have

$$\int_{G_1} (F_i - \rho u_i'' + \tau_{ij,j})\, dV = 0.$$

Since the last relation is valid for every arbitrary domain $G_1 \subset G$, then the equilibrium equations are

$$\tau_{ij,j} + F_i = \rho u_i'', \qquad i,j = 1, 2, 3, \tag{4.128}$$

which are the **equations of motion in a thermoelastic medium**. If the inertial forces $\rho u_i''$ are zero, then we have the well-known **equations of equilibrium**

$$\tau_{ij,\,j} + F_i = 0, \qquad i,j, = 1, 2, 3. \tag{4.129}$$

In the bone tissue and in the materials used for joint replacements the heat transfer, similarly as in the rigid body, takes place by conduction, which represents a heat transfer from a place of higher temperature to a place of lower temperature. From thermodynamics it is known that processes of this kind are spontaneous and irreversible and that they are associated with an increase in entropy. From the second thermodynamic theorem, that is, from the law of energy conservation, expressed in the terms of the entropy flow

$$T\frac{\partial S}{\partial t} = -\text{div}\ \mathbf{q}, \tag{4.130}$$

where $\mathbf{q}$ is an energy flow, here the heat flow, and S is the entropy, it follows that

$$\frac{\partial S}{\partial t} = -\text{div}\left(\frac{\mathbf{q}}{T}\right) - \mathbf{q}\cdot\frac{\text{grad}\ T}{T^2}, \tag{4.131}$$

where $\mathbf{q}/T$ is the entropy and $-\mathbf{q} \cdot \operatorname{grad} T/T^2$ the source of entropy. Integration over $G_1 \subset G$ gives

$$\int_{G_1} \frac{\partial S}{\partial t}\, dV = -\int_{G_1} \operatorname{div}\left(\frac{\mathbf{q}}{T}\right) dV - \int_{G_1} \mathbf{q} \cdot \frac{\operatorname{grad} T}{T^2}\, dV$$
$$= -\int_{S_1} \frac{q_n}{T}\, ds - \int_{G_1} \mathbf{q} \cdot \frac{\operatorname{grad} T}{T^2}\, dV.$$

Here q_n/T represents the entropic flow through surface S_1 of G_1, and the integral $\int_{S_1} q_n/T\, ds$ represents the entropy exchange with the neighborhood, while the integral $\int_{G_1} \mathbf{q} \cdot \operatorname{grad} T/T^2\, dV$ represents the entropy increase (growth) in region G, as a consequence of an irreversible process of heat conduction. Since the entropy increases only, that is, $\partial S/\partial t > 0$, then the sum of both integrals is positive. Moreover, the second thermodynamic theorem states that every integrals is positive. Therefore, in every arbitrary small domain the entropy must increase, that is, $\sigma = -\mathbf{q} \cdot \operatorname{grad} T/T^2 > 0$, where σ denotes the source of entropy, $-\operatorname{grad} T/T^2$ is the thermodynamic impulse, and it represents the gradient of thermal intensity. But the energy flow is proportional to the thermodynamic impulse, that is,

$$\mathbf{q} = -\frac{L}{T^2} \cdot \operatorname{grad} T = -\kappa \cdot \operatorname{grad} T, \qquad \kappa = \frac{L}{T^2} > 0, \tag{4.132}$$

where κ is the **coefficient of heat conductivity**. Relation (4.132) is the **Fourier law**. The coefficient $\kappa > 0$, because the heat flow proceeds from a place with a higher temperature to a place with a lower temperature, the vectors $\mathbf{q}$ and grad T are of the opposite sense. Then from (4.130) and (4.132) it follows that

$$T\frac{\partial S}{\partial t} = \operatorname{div}(\kappa \operatorname{grad} T). \tag{4.133}$$

If in the unit volume of the body a quantity Q is created over a time unit, then (4.133) passes to

$$T\frac{\partial S}{\partial t} = \operatorname{div}(\kappa \operatorname{grad} T) + Q. \tag{4.134}$$

This equation couples the entropy increment in time onto the temperature T. But the entropy is a function of a strain because

$$T\, dS = dU - \tau_{ij}\, de_{ij} \quad \text{resp.} \quad dU = T\, dS + \tau_{ij}\, de_{ij}. \tag{4.135}$$

The increment of an internal energy and the increment of an entropy are the total differentials of variables T and de_{ij}, that is,

$$dU = \left(\frac{\partial U}{\partial T}\right)_e dT + \left(\frac{\partial U}{\partial e_{ij}}\right)_T de_{ij}, \qquad dS = \left(\frac{\partial S}{\partial T}\right)_e dT + \left(\frac{\partial S}{\partial e_{ij}}\right)_T de_{ij}. \tag{4.136}$$

Hence and from (4.135) we obtain

$$dU = T\left(\frac{\partial S}{\partial T}\right)_e dT + \left[T\left(\frac{\partial S}{\partial e_{ij}}\right)_T + \tau_{ij}\right] de_{ij},$$
$$dS = \frac{1}{T}\left(\frac{\partial U}{\partial T}\right)_e dT + \frac{1}{T}\left[\left(\frac{\partial U}{\partial e_{ij}}\right)_T - \tau_{ij}\right] de_{ij}. \tag{4.137}$$

From (4.136) and (4.137) it follows that

$$\left(\frac{\partial U}{\partial T}\right)_e = T\left(\frac{\partial S}{\partial T}\right)_e, \qquad \left(\frac{\partial U}{\partial e_{ij}}\right)_T = T\left(\frac{\partial S}{\partial e_{ij}}\right)_T + \tau_{ij}.$$

In order for dS to be a total differential with respect to T, as with respect to e_{ij}, then

$$\frac{\partial}{\partial e_{ij}}\left[\frac{1}{T}\left(\frac{\partial U}{\partial T}\right)_e\right] = \frac{\partial}{\partial T}\left[\frac{1}{T}\left(\left(\frac{\partial U}{\partial e_{ij}}\right)_T - \tau_{ij}\right)\right]$$

must be satisfied. Hence

$$\left(\frac{\partial U}{\partial e_{ij}}\right)_T = \tau_{ij} - T\left(\frac{\partial \tau_{ij}}{\partial T}\right)_e. \tag{4.138}$$

From the Duhamel–Neumann law it follows that

$$\left(\frac{\partial \tau_{ij}}{\partial T}\right)_e = -\gamma\delta_{ij} \qquad \text{for isotropic biomaterials,}$$
$$\left(\frac{\partial \tau_{ij}}{\partial T}\right)_e = -\beta_{ij} \qquad \text{for anisotropic biomaterials.} \tag{4.139}$$

Then

$$\left(\frac{\partial U}{\partial e_{ij}}\right)_T = \tau_{ij},+ T\gamma\delta_{ij} \quad \text{and} \quad \left(\frac{\partial S}{\partial e_{ij}}\right)_T = \gamma\delta_{ij}$$
$$\text{for isotropic biomaterials,}$$
$$\left(\frac{\partial U}{\partial e_{ij}}\right)_T = \tau_{ij} + T\beta_{ij} \quad \text{and} \quad \left(\frac{\partial S}{\partial e_{ij}}\right)_T = \beta_{ij}$$
$$\text{for anisotropic biomaterials.} \tag{4.140}$$

Thus (4.137) yields

$$dU = \tau_{ij}\, de_{ij} + \gamma T\, de_{kk} + c_e\, dT, \qquad dS = \gamma\, de_{kk} + c_e \frac{dT}{T}, \tag{4.141}$$

where $c_e = (\partial U/\partial T)_e$ is the **specific heat under constant strain**. If we express τ_{ij} as a function of e_{ij} and T from the Duhamel–Neumann law and integrating both equations, then under the assumption that $U=0$ for $e_{ij}=0, T=T_0$ and $S=0$ for $e_{kk}=0, T=T_0$, for the isotropic biomaterials we obtain

$$U = \tfrac{1}{2}\tau_{ij}e_{ij} + \tfrac{1}{2}\gamma e_{kk}(\Theta + 2T_0) + c_e\Theta, \tag{4.142}$$

$$S = \gamma e_{kk} + c_e \ln\left(1 + \frac{\Theta}{T_0}\right). \tag{4.143}$$

The term $\frac{1}{2}\tau_{ij}e_{ij}$ is the work of deformation, the second term in (4.142) arises from the mutual affection of the strain and the heat conduction, and the third term represents the amount of heat accumulated in a unit volume. In Eq. (4.143) the first term represents the relation between the thermal field and the strain field, and the second represents the growth of entropy evoked by thermal conduction. Since, due to our assumption, the elastic process is reversible, no entropy change exists, and, therefore, the elastic term is missing. Under the assumption that $|\Theta/T_0| \ll 1$, (4.143) leads to

$$S = \gamma e_{kk} + \frac{c_e\Theta}{T_0} \tag{4.144}$$

and hence

$$T_0\frac{\partial S}{\partial t} = \gamma T_0\frac{\partial e_{kk}}{\partial t} + c_e\frac{\partial \Theta}{\partial t}, \tag{4.145}$$

where $c_e\, \partial\Theta/\partial t$ is the heat, which was transferred into the investigated volume element by means of heat conduction, and the term $T_0\gamma\, \partial e_{kk}/\partial t$ represents the energy dissipated in the form of heat in the body and generated by its deformation. From (4.134) and (4.145)

$$\mathrm{div}(\kappa\, \mathrm{grad}\, \Theta) + Q = c_e\frac{\partial\Theta}{\partial t} + \gamma T_0\frac{\partial e_{kk}}{\partial t},$$

follows. Generally, for anisotropic biomaterials κ is a tensor, that is, $\kappa = (\kappa_{ij})$, then in a similar way we obtain

$$\mathrm{div}(\kappa\, \mathrm{grad}\, \Theta) + Q = c_e\frac{\partial\Theta}{\partial t} + T_0\beta_{ij}\frac{\partial e_{ij}}{\partial t}, \tag{4.146}$$

where β_{ij} is the coefficient of thermal expansion. In case the variation of a temperature with time is very slow, then the term $c_e\, \partial\Theta/\partial t$ can be neglected, and the dissipative term $T_0\beta_{ij}\, \partial e_{ij}/\partial t$ representing the dissipation of a deformation energy into the heat can be taken as a thermal source, and then the equation of heat conduction can be written in the form

$$-\mathrm{div}(\kappa\, \mathrm{grad}\, \Theta) = Q_0, \qquad Q_0 \equiv Q - T_0\beta_{ij}\frac{\partial e_{ij}}{\partial t}. \tag{4.147}$$

The coefficient of thermal conductivity $\kappa = (\kappa_{ij})$ does not depend or does depend on the temperature. Then Eqs (4.146) and (4.147) are linear or nonlinear equations of heat conduction.

Boundary, Contact, and Initial Conditions Muscular tissue evokes tendonous forces in insertions, which are across insertions and are transported onto bone tissue. To simulate the function of natural joints and the function of their arbitrary replacements, we must correctly know the physiological location of insertions on the bone tissue, absolute values of acting forces in these places, as well as the directions of their functions. Then these tendonous (muscular) forces represent the loading surface forces, and they can be written in the following form:

$$\tau_{ij}n_j = P_i, \quad \text{where } \tau_{ij} = 2\mu e_{ij} + (\lambda e_{kk} - \gamma\Theta)\delta_{ij} \quad \text{for an isotropic case,}$$
$$\tau_{ij} = c_{ijkl}e_{kl} - \beta_{ij}\Theta \quad \text{for an anisotropic case.} \tag{4.148}$$

If the bone is fixed, then the displacements are equal to zero, that is,

$$u_i = 0. \tag{4.149}$$

The displacements u_i in thermoelasticity can be determined as a solution of the model problem in which the body is loaded only by the tendonous (muscular) forces. From the equilibrium equations (equations of motion) and from (4.148), it follows that the effect of the temperature change is equivalent to the compensation of volume forces by forces $F_i - \gamma\ \partial\Theta/\partial x_i$ (or $F_i - \beta_{ij}\ \partial\Theta/\partial x_j$) and the surface forces P_i by surface forces $P_i + \gamma\,\Theta n_i$ (or $P_i + \beta_{ij}n_j\Theta$) in the boundary conditions. The term $\gamma\Theta$ (or $\beta_{ii}\Theta$) has the character of the hydrostatic pressure.

Contact Conditions There remains the question of which type of conditions will hold the contact boundary between components of the joint in a dynamically loaded (and moving) joint system or in a statically loaded immobile joint system or its artificial total joint replacement. For example, in classical biomechanics the loaded acetabulum and the head of the femur as well as their artificial replacements were computed separately since the loading of forces acting in the vertical direction was used, which in the case of the hip joint passes through the center of the head of the hip joint and through the center of the acetabulum. When the loaded surface from the horizontal direction is inclined, the biomechanical equilibrium is out of balance and in dependence on a surface inclination the head of the hip joints migrates laterally or medially, respectively. In addition, Debrunner (1975) also considers the influence of geometrical relationships in the area of the hip joint on loading, and he also suggests a relation between the abductor muscle origin and the hip joint. Investigations by other specialists introduce models that, besides the mass of the body, consider also the dynamic loading evoked by the inertia effects of the body when walking and the influence of the main muscles groups of abductors and adductors. The problem of simulation of the functions of natural and artificial joint is much more complicated, and it must be investigated from a complex point of view because both parts of single joints on the upper and lower limbs are or are not in contact in the joint box. These types of conditions are known as contact conditions.

To describe the contacts between separated parts of the human body we need to consider the normal approach and the tangential process. By **contact conditions** we mean conditions that describe the tangential process, which is the relation involving the tangential stress $\boldsymbol{\tau}_t$ and the tangential displacements $\mathbf{u}_t$ (or the tangential velocity $\mathbf{v}_t = \mathbf{u}'_t$, where we denote $\mathbf{u}' = \partial \mathbf{u}/\partial t$). The nature of friction forces acting on separated parts of human joints is extremely complex and depends on several factors such as the constitution of the interface, the time scales and frequencies of the contact during loading, the response of the interface to normal forces, the presence of inertia and thermal effects, the roughness of the contacting surfaces, and the history of loadings [see Wriggers (2002)]. These conditions can be formulated as nonlinear equalities or inequalities.

Let $u_n = \mathbf{u} \cdot \mathbf{n}$, $\mathbf{u}_t = \mathbf{u} - u_n \mathbf{n}$, $v_n = \mathbf{v} \cdot \mathbf{n}$, $\mathbf{v}_t = \mathbf{v} - v_n \mathbf{n}$, $\boldsymbol{\tau} = (\tau_{ij} n_j)$, $\tau_n = \tau_{ij} n_j n_i$, $\boldsymbol{\tau}_t = \boldsymbol{\tau} - \tau_n \mathbf{n}$ be the normal and tangential components of the displacement, velocity, and stress vectors, where $\mathbf{n}$ is the unit normal to the boundary $\partial\Omega$ outward with respect to Ω. The component $\boldsymbol{\tau}_t$ is connected with friction forces on the contact surface Γ_c. Generally, a mathematical theory of friction would be a generalization of experiments, and it must be in agreement with the knowledge of thermodynamics. In the case where there are no losses of the contacts during the process, the normal displacements or normal velocities vanish, that is, $u_n = 0$ or $v_n = 0$ and the tangential components of stresses $\boldsymbol{\tau}_t$ are equal to zero or are nonzero (i.e., satisfy the Coulomb friction law). Then we speak about **bilateral contact conditions**.

Other contact conditions must be used when the components of human joints being mutually in contact are more loaded or overloaded. These conditions are represented by the **unilateral contact conditions of the Signorini type (nonpenetration conditions)** with or without friction. In such cases when during the deformation process components of human joints are in mutual contact, the contact points will be displaced in different ways, but colliding components of human joints cannot penetrate, that is,

$$u_n^k - u_n^l \leq 0 \text{ in displacements} \quad \text{or} \quad v_n^k - v_n^l \leq 0 \text{ in velocities,} \tag{4.150a,b}$$

where the positive direction of the outward normal to the contact boundary $\mathbf{n}$ is related to the domain Ω^k occupying the kth part of the investigated skeleton and where $u_n^l = u_i^l n_i^k$ and $v_n^l = v_i^l n_i^k$.

For contact forces the law of action and reaction is satisfied so that $\tau_n^k = \tau_n^l \equiv \tau_n^{kl}$ and $\boldsymbol{\tau}_t^k = -\boldsymbol{\tau}_t^l$. Since normal components cannot be positive (tractions) if no adhesion is assumed, then

$$\tau_n^{kl} \leq 0. \tag{4.151}$$

During the deformation of both colliding components of human joints, the colliding components are in contact, that is, $u_n^k - u_n^l = 0$ in displacements or $v_n^k - v_n^l = 0$ in velocities or they are not in contact, that is, $u_n^k - u_n^l < 0$ in displacements or $v_n^k - v_n^l < 0$ in velocities. If they are not in contact $u_n^k - u_n^l < 0$, then the contact forces are equal to zero, that is, $\tau_n^k = \tau_n^l \equiv \tau_n^{kl} = 0$. If the colliding components of the joint are in contact $u_n^k - u_n^l = 0$, then there may exist nonzero contact forces $\tau_n^k = \tau_n^l \equiv \tau_n^{kl} \leq 0$. Thus

$$(u_n^k - u_n^l)\tau_n^{kl} = 0 \text{ in displacements} \quad \text{or} \quad (v_n^k - v_n^l)\tau_n^{kl} = 0 \text{ in velocities.} \quad (4.152a,b)$$

Equations (4.150)–(4.152) are known as the **nonpenetration conditions** [generalized Signorini conditions, proposed by Signorini (1933) for one body on the absolutely rigid foundation]. In some cases a gap d^{kl} between the colliding components of human joints measured along the outward normal can be also observed. Thus (4.150a) will be changed by $u_n^k - u_n^l \leq d^{kl}$ and (4.152a) by $(u_n^k - u_n^l - d^{kl})\tau_n^{kl} = 0$.

In general, frictional phenomena have been studied within the framework of the theory of thermoviscoelasticity and plasticity. At the contact boundary response in the tangential direction, it can be divided into two different actions—the stick and the slip. In the stick no tangential relative displacement occurs in the contact zone under the loading due to the acting force **F**. The stick represents the case in which the relative displacement or relative velocity are equal to zero. In the slip a relative tangential movement occurs on the contact interface. Whenever the tangential forces reach a certain limit (a critical value known as the friction bound), then the contact surfaces Γ_c no longer stick to each other but move relative to each other, and this relative tangential movement is known as sliding. Mathematically, it is described by the local frictional law of the form

$$|\boldsymbol{\tau}_t| \leq \mathcal{F}_c|\tau_n| \quad \left\{\begin{array}{lll} \text{if} & |\boldsymbol{\tau}_t| < \mathcal{F}_c|\tau_n| & \text{then } \mathbf{u}_t = 0 \\ \text{if} & |\boldsymbol{\tau}_t| = \mathcal{F}_c|\tau_n| & \text{then there exist } \lambda \geq 0 \\ & & \text{such that } \mathbf{u}_t = -\lambda\boldsymbol{\tau}_t \end{array}\right\} \quad \text{on } \Gamma_c \times I, \tag{4.153}$$

where $\mathcal{F}_c$ is the coefficient of friction in the Coulombian sense. If the strict inequality holds, the material contact point is in the stick zone, if the equality holds, the material contact point is in the slip zone. Then (4.153) is the **classical Coulombian law of friction**. Setting $g_c = \mathcal{F}_c|\tau_n|$, where $g_c \geq 0$ represents the friction bound or the slip limit (i.e., the limiting friction traction magnitude at which slip begins) and assuming that g_c is given a priori for $g_c = \text{const}$, we speak about the **Coulomb law for dry friction** or the **Tresca model of friction**.

In the quasi-static and dynamic cases frictional contacts are also modeled with the variant of the Coulomb law of dry friction. In this model the tangential traction $\boldsymbol{\tau}_t$ can reach the friction bound g_c, which is the maximal frictional resistance that the surface can generate and which is reached when a relative slip motion occurs. Thus it can be written

$$|\boldsymbol{\tau}_t| \leq g_c, \qquad \boldsymbol{\tau}_t = -g_c\frac{\mathbf{u}_t'}{|\mathbf{u}_t'|} \quad \text{if} \quad \mathbf{u}_t' \neq 0 \quad \text{on} \quad \Gamma_c \times I, \tag{4.154}$$

where $\mathbf{u}_t'$ is the relative tangential displacement velocity or slip rates, g_c is the friction bound, $g_c = \mathcal{F}_c|\tau_n|$, and here $\mathcal{F}_c \geq 0$ is the coefficient of friction and depends on the relative sliding speed, on the temperature, and it varies with the processes as the surface topography changes with respect to wear.

In the multibody quasi-static and dynamic (thermovisco)elastic cases, where we denote by Γ_c^{kl} the contact boundary between two neighboring bodies Ω^k and Ω^l, the contact conditions can be formulated in displacement velocities, that is,

$$\left.\begin{array}{l} u_n'^k - u_n'^l \le 0, \qquad \tau_n^{kl} \le 0, \qquad (u_n'^k - u_n'^l)\tau_n^{kl} = 0, \\ \text{if} \quad \mathbf{u}_t'^k - \mathbf{u}_t'^l = 0 \Rightarrow |\boldsymbol{\tau}_t^{kl}| \le \mathcal{F}_c^{kl}|\tau_n^{kl}| \\ \text{if} \quad \mathbf{u}_t'^k - \mathbf{u}_t'^l \ne 0 \Rightarrow |\boldsymbol{\tau}_t^{kl}| = -\mathcal{F}_c^{kl}|\tau_n^{kl}|\dfrac{\mathbf{u}_t'^k - \mathbf{u}_t'^l}{|\mathbf{u}_t'^k - \mathbf{u}_t'^l|} \end{array}\right\} \text{ on } \cup\Gamma_c^{kl} \tag{4.155}$$

and/or in displacements, that is,

$$\left.\begin{array}{l} u_n^k - u_n^l \le 0, \qquad \tau_n^{kl} \le 0, \qquad (u_n^k - u_n^l)\tau_n^{kl} = 0, \\ \text{if} \quad \mathbf{u}_t'^k - \mathbf{u}_t'^l = 0 \Rightarrow |\boldsymbol{\tau}_t^{kl}| \le \mathcal{F}_c^{kl}|\tau_n^{kl}| \\ \text{if} \quad \mathbf{u}_t'^k - \mathbf{u}_t'^l \ne 0 \Rightarrow |\boldsymbol{\tau}_t^{kl}| = -\mathcal{F}_c^{kl}|\tau_n^{kl}|\dfrac{\mathbf{u}_t'^k - \mathbf{u}_t'^l}{|\mathbf{u}_t'^k - \mathbf{u}_t'^l|} \end{array}\right\} \text{ on } \cup\Gamma_c^{kl} \tag{4.156}$$

But in some problems the coefficient of friction can be different on both components of the human joint as friction represents the resistance to motion when a solid object is moved tangentially with respect to the surface of another, and, therefore, it can be different on both surfaces of joint components. Next, we will introduce the Coulomb law for such cases for bilateral and unilateral contacts in the following form:

Let us consider s elastic bodies Ω^ι, $\iota = 1, \ldots, s$, being in contact at $\mathbf{x} \in \Gamma_c^{kl}$, $k, l \in \{1, \ldots, s\}$, where the exterior unit normal to Ω^ι is $\mathbf{n}^\iota$, $\iota = 1, \ldots, s$. Let $\boldsymbol{\tau}^\iota$ be the force exerted by Ω^l to Ω^k at Γ_c^{kl}. Let us decompose $\boldsymbol{\tau}^\iota$ as

$$\boldsymbol{\tau}^\iota = \tau_n^\iota \mathbf{n}^\iota + \boldsymbol{\tau}_t^\iota, \qquad \tau_n^\iota = \boldsymbol{\tau}^\iota \cdot \mathbf{n}^\iota, \qquad \boldsymbol{\tau}_t^\iota = \boldsymbol{\tau}^\iota - \tau_n^\iota \mathbf{n}^\iota.$$

Let $\mathcal{F}_c^{kl} = \mathcal{F}_c^{kl}(\mathbf{x})$ be a coefficient of friction at point $\mathbf{x} \in \Gamma_c^{kl}$. At instant t the Coulomb law is as follows:

$$\begin{array}{l} |\boldsymbol{\tau}_t^{kl}(\mathbf{x}, t)| < \mathcal{F}_c^{kl}(\mathbf{x})|\tau_n^{kl}(\mathbf{x}, t)| \Rightarrow \mathbf{u}_t'^k(\mathbf{x}, t) - \mathbf{u}_t'^l(\mathbf{x}, t) = 0, \\ |\boldsymbol{\tau}_t^{kl}(\mathbf{x}, t)| = \mathcal{F}_c^{kl}(\mathbf{x})|\tau_n^{kl}(\mathbf{x}, t)| \Rightarrow \text{there exists } \lambda \ge 0 \text{ such that} \\ \qquad\qquad \mathbf{u}_t'^k(\mathbf{x}, t) - \mathbf{u}_t'^l(\mathbf{x}, t) = -\lambda \boldsymbol{\tau}_t^{kl}(\mathbf{x}, t), \end{array} \tag{4.157}$$

or for the case in which we consider the situation where the friction coefficient $\mathcal{F}_c^{kl}$ is different with respect to $\tau_n^k < 0$ or $\tau_n^l > 0$ say $\mathcal{F}_c^k(\mathbf{x})$ and $\mathcal{F}_c^l(\mathbf{x})$:

$$\left.\begin{array}{l} |\boldsymbol{\tau}_t^k(\mathbf{x}, t)| < \mathcal{F}_c^k(\mathbf{x})|\tau_n^k(\mathbf{x}, t)| \\ |\boldsymbol{\tau}_t^l(\mathbf{x}, t)| < \mathcal{F}_c^l(\mathbf{x})|\tau_n^l(\mathbf{x}, t)| \end{array}\right\} \Rightarrow \mathbf{u}_t'^k(\mathbf{x}, t) - \mathbf{u}_t'^l(\mathbf{x}, t) = 0,$$

$$|\boldsymbol{\tau}_t^k(\mathbf{x}, t)| = \mathcal{F}_c^k(\mathbf{x})|\tau_n^k(\mathbf{x}, t)| \Rightarrow \text{there exists } \lambda_k \ge 0 \text{ such that}$$

$$\mathbf{u}_t'^k(\mathbf{x}, t) - \mathbf{u}_t'^l(\mathbf{x}, t) = -\lambda_k \boldsymbol{\tau}_t^k(\mathbf{x}, t),$$

$$|\boldsymbol{\tau}_t^l(\mathbf{x},t)| = \mathcal{F}_c^l(\mathbf{x})|\tau_n^l(\mathbf{x},t)| \Rightarrow \text{there exists } \lambda_l \geq 0 \text{ such that}$$
$$\mathbf{u}_t'^k(\mathbf{x},t) - \mathbf{u}_t'^l(\mathbf{x},t) = -\lambda_l \boldsymbol{\tau}_t^l(\mathbf{x},t). \qquad (4.158\text{a,b,c})$$

Due to contact conditions, that is, due to inequalities arising in the contact and friction laws, the variational formulations lead to variational inequalities that represent more difficulties for their solutions because the functionals that describe the friction effects turn out to be neither monotone nor compact. Since the solution method for variational inequalities cannot be applied directly, other contact and friction models were developed. According Duvaut (1980) a concept of **nonlocal friction** was introduced in which the normal component of contact traction in the Coulomb law was smoothed, and according Martins and Oden (1985) the concept of the normal compliance contact was introduced.

In this concept the foundation is deformable. Let the gap d between the potential contact surface Γ_c and the foundation, measured along the outward normal, be assumed. The exact nonpenetration condition is relaxed and replaced by a nonlinear contact law, where the normal contact traction is defined by a function of the magnitude of interpenetration of the bodies, that is,

$$\tau_n(\mathbf{u}) = -p_n(u_n - d). \qquad (4.159)$$

The function p_n is such that $p_n(\mathbf{u}) = 0$ if $\mathbf{u} \leq 0$. Here, $u_n - d$, when positive, represents the penetration of the surface asperities of the body into those of the foundation. Equation (4.159) is the **normal compliance contact condition.** This concept leads to a penalty approximation approach to the nonpenetration conditions. The idea can be extended for the multibody contact.

If wear of the surface is taken into account, the Coulomb law must be modified [see Strömberg et al. (1995, 1996) and Strömberg (1997)]. From the thermodynamic consideration it follows that

$$g_c = g_c(\tau_n) = \mathcal{F}_c|\tau_n|(1 - \delta|\tau_n|)_+, \qquad (4.160)$$

where δ is a very small positive parameter related to the wear constant of surfaces, $(r)_+ = \max(r, 0)$ denotes a positive part of r. Using the normal compliance condition (4.159), one gets

$$g_c = \mathcal{F}_c p_n (1 - \delta p_n)_+ (u_n - d). \qquad (4.161)$$

Hence using (4.153) the contact conditions can be written in the form

$$\tau_n = -p_n(u_n - d),$$
$$|\boldsymbol{\tau}_t| \leq \mathcal{F}_c p_n (1 - \delta p_n)_+ (u_n - d),$$
$$\boldsymbol{\tau}_t = -\mathcal{F}_c p_n (1 - \delta p_n)_+ \frac{\mathbf{u}_t'}{|\mathbf{u}_t'|} \quad \text{if} \quad \mathbf{u}_t' \neq 0. \qquad (4.162)$$

An extension for the multibody contact is possible.

To study processes with adhesion, which have great applications in the orthopedics of prosthetic limbs where the bounding (coupling) between the artificial limb and the tissue is of considerable importance, may be based on the normal compliance or the Signorini-type contact condition approaches.

The main idea is the introduction of a surface interval variable, the bounding field, describing the fractional density of active bounds on the contact surface (Raous et al., 1999; Rojek et al., 2001a,b), Shillor et al., 2004). With respect to the Frémond idea (Frémond, 1982, 1987, 2002), the bounding field β, defined on the contact surface and representing the intensity of adhesion, will be introduced as follows: As a fraction its values are restricted to $0 \leq \beta \leq 1$. If $\beta = 1$ at a point of contact surface, the adhesion is complete and all the bounds are active; if $\beta = 0$ all the bounds are inactive, points of contact are separated, and there is no adhesive; if $0 < \beta < 1$ there is partial adhesion. If the adhesion is described with a damage parameter D, we have $\beta = 1 - D$. The adhesive on the contact surface introduces tension that opposes the separation of the surfaces in the normal direction, and it opposes the relative motion in the tangential directions. The adhesive tensile traction is proportional to β^2 and to the displacements. The evolution of β depends on β and on the displacements (Wriggers, 2002). We do not discuss the problem of adhesive models in details, for more details see Frémond (1987, 2002), Raous (1999), Raous et al. (1999), Rojek et al. (2001a,b), Chau et al. (2003a,b, 2004), Shillor et al. (2004), and Sofonea et al. (2006).

In the case if the compressive part of the contact stress is expressed by the normal compliance, then the normal compliance contact condition with adhesion is as follows:

$$\tau_n = -p_n(u_n) + \gamma_n \beta^2 \overline{R}(u_n), \qquad \overline{R}(u_n) = (-R(u_n))_+, \tag{4.163}$$

where a gap is assumed to be zero, that is, $d \equiv 0$, p_n is the normal compliance function, γ_n is a positive adhesion coefficient, and R is the truncation operator defined by

$$R(s) = \begin{Bmatrix} -L & \text{if} & s < -L \\ s & \text{if} & -L \leq s \leq L \\ L & \text{if} & s > L \end{Bmatrix} \Rightarrow \overline{R}(s) = \begin{Bmatrix} L & \text{if} & s < -L \\ -s & \text{if} & -L \leq s \leq 0 \\ 0 & \text{if} & s \geq 0 \end{Bmatrix},$$

where $L > 0$ is the characteristic length of the bound [see Raous et al. (1999)]. When the unilateral conditions on the contact surface are assumed, then the contact condition with adhesion are as follows:

$$u_n \leq 0, \qquad \tau_n - \gamma_n \beta^2 \overline{R}(u_n) \leq 0, \qquad (\tau_n - \gamma_n \beta^2 \overline{R}(u_n))u_n = 0. \tag{4.164}$$

An extension for the multibody system is possible.

Moreover, in our models we can also assume that we have a certain information about a behavior of a thermal field in the neighborhood of the investigated human joint and that the initial temperature $T_0(\mathbf{x})$ is known.

The boundary condition for the heat problem can be of different types. Let us assume that on one part of the boundary of the investigated joint system, for example, on Γ_u, the temperature or the heat flow are prescribed, that is,

$$T = T_1 \quad \text{or} \quad \kappa_{ij} \partial T / \partial x_j n_i = q, \quad \text{resp.} \quad \text{on } \Gamma_u, \tag{4.165}$$

where T_1 is a given temperature, q a given heat flow, and on the other parts of the boundary of the investigated human joint system the following boundary condition

$$\kappa_{ij}\partial T/\partial x_j n_i = K_1(Y - T) \quad \text{on } \partial\Omega \setminus \Gamma_u$$

can be given, where K_1 is a constant, Y is the outer temperature of the investigated part of the surface of the studied system, and T is the actual temperature.

On the contact boundary the continuity of a temperature and of the heat flow can be assumed or the following condition

$$\kappa_{ij}\frac{\partial T}{\partial x_j}n_i = K_1(Y - T) + \mathcal{F}_c^{kl}|\tau_n^{kl}|\frac{\mathbf{u}'^k - \mathbf{u}'^l}{|\mathbf{u}'^k - \mathbf{u}'^l|} \quad \text{on } \Gamma_c,$$

where the second term on the right-hand side represents the thermal source evoked by friction, the frictional heat, can be used.

Moreover, the initial conditions

$$\mathbf{u}(0,\mathbf{x}) = \mathbf{u}_0(\mathbf{x}), \qquad \mathbf{u}'(0,\mathbf{x}) = \mathbf{u}_1(\mathbf{x}), \qquad T(0,\mathbf{x}) = T_0(\mathbf{x}),$$

are given.

Remark 4.6 The global models discussed above and based on different rheologies will be used for the simulation and modeling of biomechanical processes in different parts of the human skeleton. Model problems will be introduced in the next chapter.

CHAPTER 5

MATHEMATICAL MODELS OF PARTICULAR PARTS OF THE HUMAN SKELETON AND JOINTS AND THEIR REPLACEMENTS BASED ON BOUNDARY VALUE PROBLEM ANALYSES

5.1 INTRODUCTION

Modeling of stress distribution and functioning of human joints are related to the study of complicated mathematical model problems based on the boundary value problems or on the contact problems in different rheologies. The aim of this chapter and of the next chapters is further development of the study of different model problems describing different problems in orthopedics that can be modeled by the (initial-)boundary value and/or by the (initial-)boundary value and contact problems in different rheologies. While in this chapter we restrict ourselves to model problems based on the boundary value problems and their discrete approximations, in Chapters 6 and 7 we will discuss biomechanical problems based on the contact problems in different rheologies.

5.2 MATHEMATICAL MODELS OF HUMAN JOINTS AND OF THEIR TOTAL REPLACEMENTS AS WELL AS OF PARTS OF THE HUMAN BODY

Present knowledge of the human skeleton and its function as well as replacements of its parts depends on understanding biomechanical processes in the human skeleton as a whole as well as in its particular parts. Biomechanics as a branch of science is at the

Mathematical and Computational Methods in Biomechanics of Human Skeletal Systems: An Introduction, First Edition. Jiří Nedoma, Jiří Stehlík, Ivan Hlaváček, Josef Daněk, Taťjana Dostálová, and Petra Přečková.

beginning of its development but has nevertheless reached certain successes. The first experimental and mathematical models were based on the results of classical mechanics of continua. At present the models are based on complicated rheologies, namely on nonlinear rheologies, which need more complicated experimental approaches and devices and very complicated mathematical models. Experimental as well as mathematical models describe a reality only approximately, and therefore they permit one to understand, at least approximately, some biomechanical processes in the human skeleton. In the case of the mathematical models it will depend on the rheology we use to describe the studied part of the human skeleton. If linear elastic rheology, characterized by the linear Hooke law, is used to describe the parts of the human skeleton, then such models will be described using classical mechanics. Such models correspond to models that comply with behavior of the drain bone, which can be assumed to be isotropic or anisotropic. For example, the cancellous bone is characterized by beam structures, which are characterized by anisotropic behaviors of bone materials. Anisotropic elastic properties are characterized by a tensor of the fourth order. From experimental studies and analyses we can determine probable anisotropic material properties of the compact and cancellous bones. Both compact and cancellous bones are made up of composite biomaterials. In this case we can simulate their anisotropic properties by using the homogenization method, assuming their periodical or stochastic structures of materials, respectively, which build the compact and cancellous bones. But the reality is much more complicated, as bones are living tissues characterized by plastic appearances, by relaxations, creeps, as well as by a response of a living tissue onto external energetic actions, which from the physical point of view represents certain memory ability of a living tissue.

From the biomechanical point of view mathematical models represent certain approximations of the real living tissues. The main aim is to determine parameters of the geometry of investigated living tissues, their structures, and an understanding that their characters define the corresponding biomechanical model from the point of view of its geometry. The next problem is to determine an optimal rheology, that is, to determine the material properties of tissues that build the studied part of the human skeleton and then, with this, use rheology to determine constitutive material equations, which in biomechanics of a living tissue is a very complicated problem. It follows from the very complicated material properties of bone tissues and of soft tissues, characterized by their nonlinearities, great deformations, and their dependence on their previous states, that is, on the history of a loaded bone tissue. It is very difficult to determine material properties of a living bone, ligament, and cartilage tissues, which can be experimentally examined post morten, that is, after its extraction from the human body. Its properties will be different from the properties of living tissues under in vivo conditions that is, in a living human body. On the basis of the laws of biomechanics, on our knowledge about geometry of the investigated part of the skeleton, on the rheology used, and on the state of behavior of the investigated part of the skeleton in a given space and an initial time, we can derive then a mathematical model. Such a model will be based on several intuitive steps, which cannot ever be verified, unfortunately. Moreover, the obtained data can be determined with certain inaccuracies, given by the present possibilities of measurements of material parameters of bone, ligament, and cartilage tissues and their geometries. The derived mathematical model, including the boundary and initial conditions, characterizing the influence of the neighborhood in a space and time on the investigated part of the skeleton, can be analyzed in its continuous as well as discrete formulations and to prove existence and uniqueness (if possible) theorems of the continuous and discrete problems, and, moreover, convergences of the used numerical methods and algorithms. According to the inaccuracy (uncertain) of used input data, we must ask ourselves also about the accuracy of obtained results. This problem will be analyzed and discussed in the special section in Chapter 6. In the case of arbitrary human joint replacements the living tissues are replaced by nonliving biomaterials, which unfortunately have no ability of regeneration and, moreover, in many cases are not well accepted by the neighboring living tissues. The artificial replacements of skeletal parts in all cases are limited by small movements of the skeletal parts—joints, spine, and so on. Materials used for the construction of all types of artificial replacements of a human joint must satisfy the following criteria:

1. Unconditional adaptability to the neighboring materials of living tissues.
2. Sufficient extent of elastic deformation with small nonlinearity and the elastic modulus.
3. Suitable orientation of deformable properties with respect to a type and direction of a force function.
4. Suitable reversing deformation facilitating adaptation without unnecessary time and space changes of properties and evoked damages but rendering a relaxation of stress and prevention of a microviolation origin.
5. Limitation of total deformation with stress increasing together with the assurance of the elastic behavior of biomaterials with high local consistency and with minimal necessity of further energy delivery.
6. High resistance to biocorrosion.
7. High consistency against cyclical loading with high initial damping.
8. High quality of surface design, which make biocompatibility and biocorrosion impossible.
9. Ability of certain regeneration in relation to the neighboring living tissue as a higher form of biocompatibility.

Remark 5.1 Theoretical foundations concerning the functional analysis, Sobolev spaces, and the like used in this chapter are introduced in Chapter 6.

5.3 MATHEMATICAL MODELS OF HUMAN BODY PARTS AND HUMAN JOINTS AND THEIR TOTAL REPLACEMENTS BASED ON THE BOUNDARY VALUE PROBLEMS IN (THERMO)ELASTICITY

Let us assume that the region occupied by the acetabulum with the pelvis or the artificial acetabula replacement (see Figs. 5.1–5.3) and the head of the hip joint and

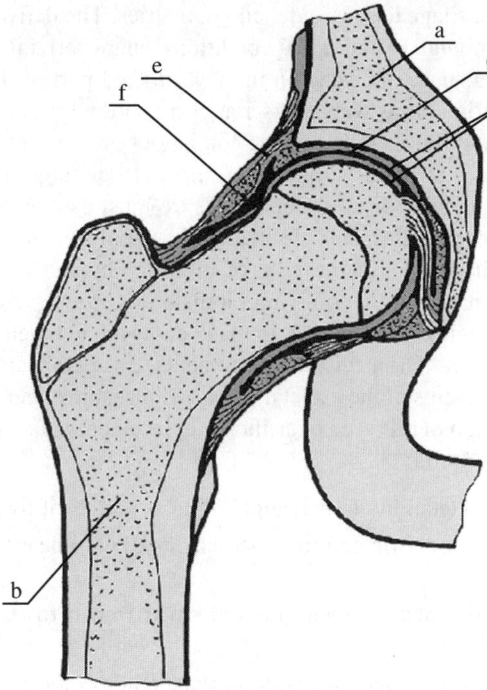

FIG. 5.1 Hip joint: a, pelvis; b, femur; c, contact surfaces; d, cartilage; e, joint capsule; and f, joint cavity. [Modified after Beznoska et al. (1987) and Nedoma et al. (2006).]

the femur or its artificial total hip arthroplasty (THA) fixed by the bone cement or the loaded long bone, respectively, occupy the N-dimensional region ($N = 2, 3$) denoted by Ω and its boundary by $\partial\Omega$. Next, for simplicity, we will study the models of their total arthroplasty. Distribution of fields of deformations and stresses in the artificial acetabulum with pelvis and/or in total hip replacement and the femur are described by the nonstationary heat equation and the equation of motion, that is, by

$$\begin{aligned} &\rho\frac{\partial T}{\partial t} - \frac{\partial}{\partial x_i}\left(\kappa_{ij}\frac{\partial T}{\partial x_j}\right) = W, \\ &\frac{\partial \tau_{ij}}{\partial x_j} + f_i = \rho\frac{\partial^2 u_i}{\partial t^2}, \end{aligned} \qquad \text{in } \Omega \times I, \tag{5.1}$$

where $I = (t_0, t_1)$ is a time interval. The repeated index denotes summation from 1 to N.

In the static case, if the displacement, deformation, and stresses are functions of coordinates, and thus the inertia forces are zero, the equation of motion is changed into

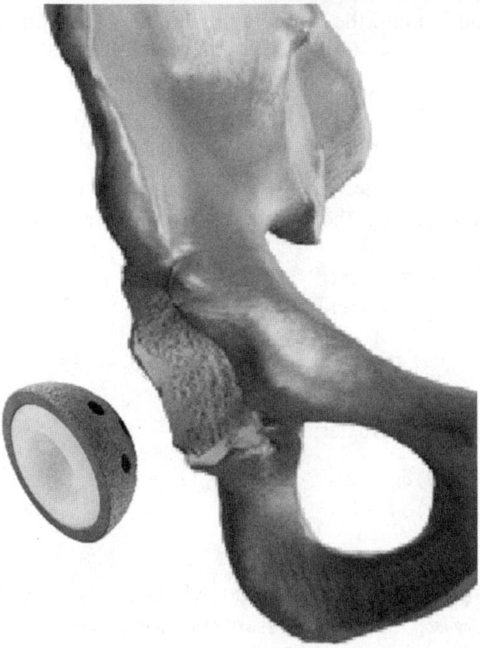

FIG. 5.2 Artificial cup and insert of THA with pelvis.

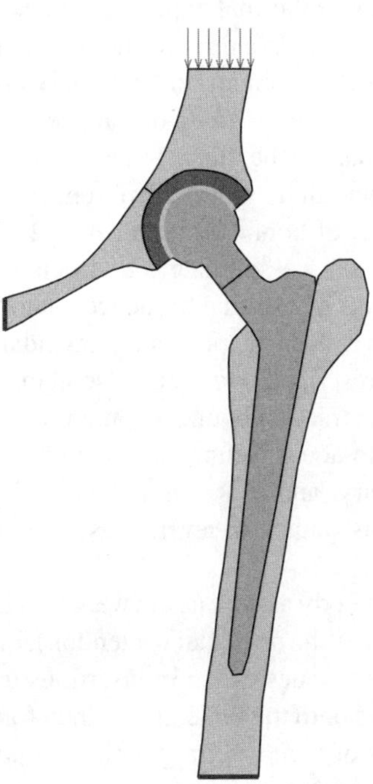

FIG. 5.3 Model of an artificial hip joint arthroplasty.

the equilibrium equation and the temperature satisfies the stationary heat equation, that is,

$$-\frac{\partial}{\partial x_i}\left(\kappa_{ij}\frac{\partial T}{\partial x_j}\right) = W, \qquad \frac{\partial \tau_{ij}}{\partial x_j} + f_i = 0 \quad \text{in } \Omega, \tag{5.2}$$

where the relation between stress and strain is defined by (i) the Duhamel–Neumann law in thermoelasticity and (ii) the generalized Hooke law in linear elasticity, that is,

$$\begin{aligned} &\text{(i)} && \tau_{ij} = c_{ijkl}e_{kl}(\mathbf{u}) - \beta_{ij}(T - T_0), \\ &\text{(ii)} && \tau_{ij} = c_{ijkl}e_{kl}(\mathbf{u}), \\ & && e_{ij} = \frac{1}{2}\left(\frac{\partial u_i}{\partial x_j} + \frac{\partial u_j}{\partial x_i}\right), \qquad i,j,k,l = 1,\ldots,N, \end{aligned} \tag{5.3}$$

in the case of anisotropic biomaterials and

$$\begin{aligned} &\text{(i)} && \tau_{ij} = \lambda e_{kk}(\mathbf{u})\,\delta_{ij} + 2\mu e_{kl}(\mathbf{u}) - \beta\delta_{ij}(T - T_0), \\ &\text{(ii)} && \tau_{ij} = \lambda e_{kk}(\mathbf{u})\,\delta_{ij} + 2\mu e_{kl}(\mathbf{u}), \end{aligned} \qquad i,j,k,l = 1,\ldots,N, \tag{5.4}$$

in the case of isotropic biomaterials, and where κ_{ij} is a coefficient of thermal conductivity and $\kappa_{ij} = \kappa\delta_{ij}$ in the case of isotropic biomaterials, W are thermal sources, τ_{ij} is the stress tensor, e_{ij} is the small strain tensor, f_i the components of body forces, c_{ijkl} the elastic coefficients, λ, μ are Lamé's elastic coefficients, δ_{ij} is the Kronecker delta, β_{ij} are the coefficients of thermal expansion, $\beta_{ij} = \gamma\delta_{ij}$ in isotropic cases, and T_0 is an initial temperature corresponding to the temperature of a human body. To complete the description of the problem, the boundary conditions and initial conditions in the case of time-dependent loading or a loaded moving human body, and/or only the boundary conditions in the static loaded human body and its parts must be given. The boundary conditions will be different in the different models of the problems. We denote the set of boundary points as $\partial\Omega$. The boundary $\partial\Omega$ can consist of several parts on which we simulate the effects of the neighboring parts of the human body or the effects of external influences onto the investigated part of the human body. A boundary is called a displacement boundary, and it will be denoted as Γ_u if the displacement is prescribed; we speak also about Dirichlet boundary conditions. A boundary is called a traction boundary, and it is denoted by Γ_τ if the traction is prescribed; we speak also about Neumann's boundary conditions. A boundary is called a symmetric boundary, and it is denoted as Γ_0 if normal displacement and tangential stress components vanish. In several cases the mixed boundary conditions are also used.

The weight of the human body as well as the weight of acting loads are transported into the bone tissue. Moreover, the muscular (or tendon) forces also must be taken into account. Since the muscular tissues evoke in insertions tendinous forces, which are across insertions transported onto the bone tissue, then for simulation of functions of natural joints and functions of their arbitrary replacements, we must correctly know the physiological location of insertions on the bone tissue, absolute values of acting forces in these places, as well as the directions of their functions. Then these tendinous (muscular) forces represent the loading surface forces and then they can be written in the following form:

$$\tau_{ij}n_j = P_i,$$

where n_j are components of the unit outward normal to Γ_τ, and where in the case of thermoelasticity

$$\tau_{ij} = 2\mu e_{ij} + (\lambda e_{kk} - \gamma\Theta)\delta_{ij}$$

for an isotropic case,

$$\tau_{ij} = c_{ijkl}e_{kl} - \beta_{ij}\Theta$$

for an anisotropic case, and in the case of linear elasticity

$$\tau_{ij} = c_{ijkl}e_{kl},$$

and where

$$\Theta \equiv T - T_0, \beta_{ij} = \gamma\delta_{ij} \tag{5.5}$$

for an isotropic case.

If the bone is fixed, then the displacements are equal to zero, that is,

$$u_i = 0 \quad \text{on} \quad \Gamma_u. \tag{5.6}$$

The displacements u_i in thermoelasticity can be determined as a solution of the model problem in which the body is loaded only by the tendinous (muscular) forces. From the equilibrium equations (or equations of motion) and from (5.5) it follows that the effect of the temperature change is equivalent with compensation of volume forces by forces $F_i - \gamma\,\partial\Theta/\partial x_i$ [or $F_i - \beta_{ij}\,(\partial\Theta/\partial x_j)$] and the surface forces P_i by surface forces $P_i + \gamma\Theta n_i$ (or $P_i + \beta_{ij}n_j$) in the boundary conditions. The term $\gamma\Theta$ (or $\beta_{ii}\Theta$) has a character of the hydrostatic pressure.

Since the heat equation and the equation of motion are of the first order (heat equation) and of the second order (equation of motion), two sets of initial conditions are needed. In the case of the heat equation the initial temperature T_0 must be prescribed. In the case of the equation of motion the initial conditions in terms of the displacements and velocities must be given, that is,

$$\begin{aligned} T(t_0, \mathbf{x}) &= T_0(\mathbf{x}), \\ \mathbf{u}(t_0, \mathbf{x}) &= \mathbf{u}_0(\mathbf{x}), \qquad \text{for } \mathbf{x} \in \Omega, \\ \mathbf{u}'(t_0, \mathbf{x}) &= \mathbf{v}_0(\mathbf{x}). \end{aligned} \tag{5.7}$$

If the human body is initially nondeformed and at rest, the initial conditions can be written as

$$\mathbf{u}(t_0, \mathbf{x}) = \mathbf{0}, \qquad \mathbf{u}'(t_0, \mathbf{x}) = \mathbf{0} \quad \text{for} \quad \mathbf{x} \in \Omega.$$

5.4 BIOMECHANICAL MODEL OF A LONG BONE

5.4.1 Introduction of Bone Biomechanics

This section describes the composition and structure of bone tissue, the mechanical properties of bone, and the behavior of bone under different loading conditions. Shortly, various factors that affect the mechanical behavior of bone in vitro and in vivo will be also discussed. A bone is the primarily structural element of the human body. Bones form the building blocks of the skeletal system. The function of the skeletal system is to protect the internal organs to provide kinematic links, to provide muscle attachment sites, and to facilitate muscle actions and body movements. A bone has unique structural and mechanical properties that allow it to carry out these functions. The bone tissue is a specialized connective tissue whose solid composition suits its supportive and protective roles. A bone is a composite material with various solid and fluid phases. The structure of a bone is different at many levels. A bone consists of cells and an organic universal matrix of fibers and a ground substance surrounding collagen fibers. Moreover, a bone contains inorganic substances in the form of mineral salts, which make it hard and relatively rigid. The organic components give a bone its flexibility and resilience. The inorganic, that is, mineral, portion of a bone consists of calcium and phosphate [$Ca_{10}(PO_4)_6(OH)_2$], which accounts for ~60–70% of its dry weight. Water accounts for ~5–8%, and the organic matrix makes up the remainder of the tissue.

From the macroscopic point of view, all bones consist of two types of tissues—the cortical or compact bone tissue and the cancelous, trabecular, or spongious bone tissue. The **compact bone tissue** is a dense material forming the outer shell of bones and the diaphysial part of the long bone. The **cancellous bone tissue** consists of thin plates, that is, trabeculae, in a loose mesh structure that is enclosed by the compact bone. The bone surface is covered by the periosteum, which covers the entire bone except for the joint surfaces, which are covered by an articular cartilage. Bones are surrounded by a dense fibrous membrane, the endoosteum.

In live bones water is ~25% of their total weight. Approximately 85% of water is in the organic matrix, around the collagen fibers and ground substances, and in the hydration shells surrounding the bone crystals. The other 15% of water is located in the cavities and channels in the bone. There are four main types of a bone (Dowson and Wright, 1981; Tortora and Anagnostakos, 1984; Nordin and Frankel, 2001):

1. The woven bone, which can be found only in a very young bone.
2. The primary lamellar bone, which has lamellae oriented parallel to the surface of the whole bone.
3. The cortical or compact bone, which is a bone of several types like in the Haversian system or the lamellae bone without any spaces except for blood channels and cells, it is formed from the cortex or an outer shell of the bone, and it has a dense structure similar to that of ivory.
4. The cancellous bone, which is formed from thin plates, or trabeculae, in a loose mesh structure, and it is made into slender ties and struts. Mechanically, it is

very different from the compact bone. Its structure is very complicated and it strongly depends on its loading, for example its structure in the upper part of the femur. The cancellous bone tissue is arranged in concentric lucunae-containing lamellae but does not contain Haversian channels.

Biomechanically, the bone tissue can be modeled as a biphasic composite material, with the mineral as one phase and the collagen and ground substance as the other. A bone is highly anisotropic (Dowson and Wright, 1981, Chapter 12). For bone to be anisotropic is to be expected from its structure. The apatite needles, collagen fibers, lamellae, laminae, and Haversian systems show a tendency to be oriented along the length of a long bone. Moreover, a bone is slightly viscoelastic. Long bones are **femur**, **tibia**, **ulna**, **radius**, and **humerus**; see Figs. 5.4 and Fig. 5.5, where a typical sectional view of a long bone is given.

5.4.2 Mathematical Model of Loaded Long Bones Based on Elastic Rheology in Two and Three Dimensions

Let us assume that the total artificial hip joint fixed in the femur occupies a bounded domain of an arbitrary shape Ω of $\mathbb{R}^N$, $N=2,3$. Let the boundary $\partial\Omega = \Gamma_\tau \cup \Gamma_u$, where Γ_u and Γ_τ are open subsets in $\partial\Omega$. Let **n** be the outward normal to the boundary $\partial\Omega$. The weight of the human body is transferred from the pelvis onto the head of the femur, this part of the boundary of the bone $\partial\Omega$ we denote by ${}^1\Gamma_\tau$ and the surface

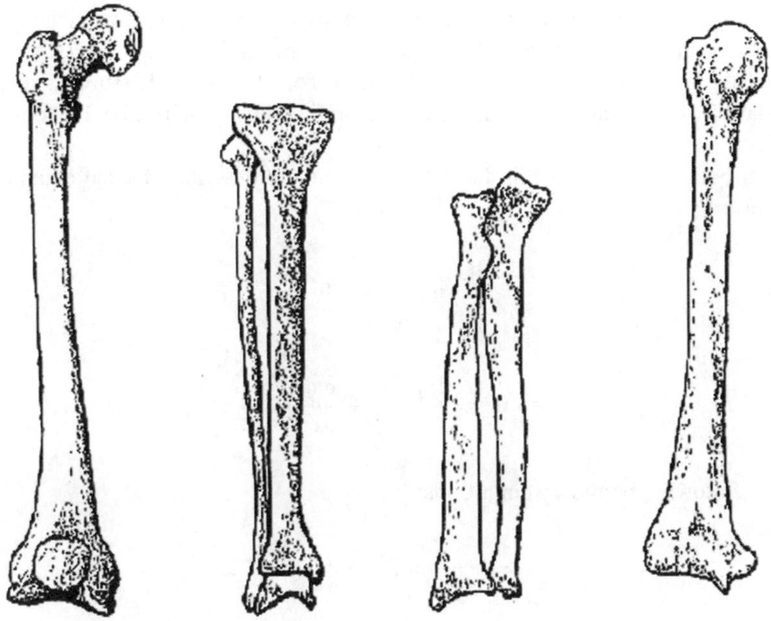

FIG. 5.4 Types of long bones: femur, fibula, tibia, ulna, radius, humerus.

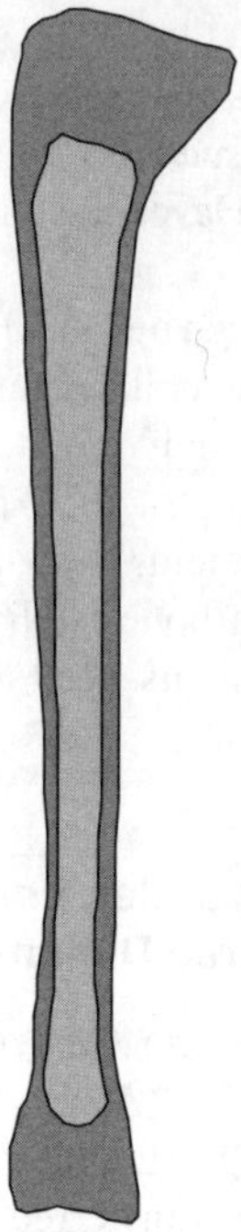

FIG. 5.5 Sectional view of a long bone.

force concerned with the weight of a human body functioning on it we denote as **P**. On the second part of the bone, where the force insertions do not exist, and, therefore, there also no muscle forces operate. Thus $\tau_{ij}n_j = 0$ on ${}^2\Gamma_\tau$. We will assume that on the boundary Γ_u the femur is fixed, that is, $u_i = 0$ on Γ_u. Thus we will solve the following problem:

Distribution of stress–strain fields in the bone are described by the equation of equilibrium:

$$\frac{\partial \tau_{ij}}{\partial x_j} + f_i = 0 \quad \text{in } \Omega \tag{5.8}$$

$$\tau_{ij} = c_{ijkl}e_{kl}(\mathbf{u}), \qquad e_{ij}(\mathbf{u}) = \frac{1}{2}\left(\frac{\partial u_i}{\partial x_j} + \frac{\partial u_j}{\partial x_i}\right), \qquad i,j,k,l = 1,\ldots,N,$$

and the following boundary conditions:

$$\tau_{ij}n_j = P_i \quad \text{on } \Gamma_\tau = {}^1\Gamma_\tau \cup {}^2\Gamma_\tau, \tag{5.9}$$

$$u_i = 0 \quad \text{on } \Gamma_u, \tag{5.10}$$

where $\mathbf{f} \in [L^2(\Omega)]^N$, $\mathbf{P} \in [L^2(\Gamma\tau)]^N$, where c_{ijkl} satisfy

$$\begin{aligned} c_{ijkl} = c_{jikl} = c_{klij} = c_{ijlk}, \qquad & i,j,k,l = 1,\dots N \\ 0 < \mu_0 \le c_{ijkl}(x)\xi_{ij}\xi_{kl}|\xi|^{-2} \le M_0 < \infty \quad & \text{for all } \mathbf{x} \in \Omega, \\ \xi = (\xi_{ij}) \in R^{N^2}, \qquad & \xi_{ij} = \xi_{ji}, i,j = 1,\dots,N, \end{aligned} \tag{5.11}$$

where μ_0, M_0 are positive constants independent of $\mathbf{x}$.

Let $H^k(\Omega) = W^{k,2}(\Omega)$, k integer, $k \ge 0$, be the Sobolev space in the usual sense and evidently $L^2(\Omega) = H^0(\Omega) \equiv W^{0,2}(\Omega)$ and $H^{k,N}(\Omega) = [H^k(\Omega)]^N$, and so forth.

Let

$$V = \{\mathbf{v} | \mathbf{v} \in H^{1,N}(\Omega), \quad \mathbf{v} = 0 \text{ on } \Gamma_u\}$$

be the space of virtual displacements. According to the scalar product $(\mathbf{u}, \mathbf{v}) = \int_\Omega (u_i v_i + u_{i,j} v_{i,j})\, d\mathbf{x}$ the space V is a Hilbert space. For $\mathbf{u}, \mathbf{v} \in H^{1,N}(\Omega)$ we define

$$\begin{aligned} a(\mathbf{u}, \mathbf{v}) &= \int_\Omega c_{ijkl} e_{ij}(\mathbf{u}) e_{kl}(\mathbf{v})\, d\mathbf{x}, \\ S(\mathbf{v}) &= \int_\Omega f_i v_i\, d\mathbf{x} + \int_{\Gamma_\tau} P_i v_i\, ds. \end{aligned} \tag{5.12}$$

Multiplying (5.8) by v_i and integrating over the domain Ω, using the boundary conditions and Green's theorem we obtain the Euler equations $\delta L = 0$, where

$$L(\mathbf{v}) = \tfrac{1}{2} a(\mathbf{v}, \mathbf{v}) - S(\mathbf{v}). \tag{5.13}$$

On the other hand we minimize on the space V the functional $L(\mathbf{v})$ or we seek the minimum of the functional $L(\mathbf{v})$.

Definition 5.1 By the variational or weak solution we understand the function $\mathbf{u} \in V$ such that

$$a(\mathbf{u}, \mathbf{v} - \mathbf{u}) \ge S(\mathbf{v} - \mathbf{u}) \quad \forall \mathbf{v} \in V,$$

or in the equivalent form (5.14)

$$a(\mathbf{u}, \mathbf{v}) = S(\mathbf{v}) \qquad \forall \mathbf{v} \in V.$$

It can be shown that every classical solution of the investigated problem is also its weak solution, and on the contrary if a weak solution is sufficiently smooth, then it is also a classical solution of the problem studied.

Theorem 5.1 Let coefficients satisfy symmetry and continuous conditions (5.11). Then for every $\mathbf{P} \in [L^2(\partial\Omega \backslash \Gamma_u)]^N$ and $\mathbf{f} \in [L^2(\Omega)]^N$ there exists a unique weak solution $\mathbf{u}$ of the problem.

Proof Bilinear form $a: V \times V \to \mathbb{R}$ is V-elliptic and bounded on V and $\mathbf{f} \in V'$. Then from the Lax–Milgram theorem the existence and uniqueness of the problem follows immediately. □

5.4.3 Numerical Solutions and Algorithms

A numerical solution will be based on the finite element method and the Ritz–Galerkin method. Let a region occupied by an investigated bone be approximated by an N-dimensional region Ω_h, the boundary of which, denoted by $\partial\Omega_h$, is a polygonal or polyhedral. Let the region $\overline{\Omega} = \Omega \cup \partial\Omega$ be divided into the finite number of triangles $\mathbb{T}_{hi}$, $i = 1, \ldots, M$, in the two-dimensional (2D) case and of tetrahedrons $\mathbb{T}_{hi}$, $i = 1, \ldots, M$, in the 3D case, such that $\overline{\Omega} = \cup_{i=1}^{M} \mathbb{T}_{hi}$ and such that two arbitrary neighboring triangles in the 2D case or tetrahedrons in the 3D case, respectively, have either no common point or have a common vertex, and a common edge in the 2D case such that two arbitrary tetrahedrons have either no common point or have a common vertex, a common edge, or a common face. Such a division of region Ω will be called the triangulation of the region Ω for the 2D case and the division of the region Ω for the 3D case and we denote it by $\mathcal{T}_h$, where $h = \max_{1 \le i \le M} (\text{diam } \mathbb{T}_{hi})$ is the largest edge of the division $\mathcal{T}_h$. Let ϑ_h denote the size of a minimal angle in the triangulation $\mathcal{T}_h$. In 3D division ϑ_h is defined as the minimal of all angles between the faces and between the edges of all tetrahedrons of $\mathcal{T}_h$. Then system $\{\mathcal{T}_h\}$, $h \to 0$, is called regular if there exists a positive number $\vartheta_0 > 0$ such that $\min_{h \to 0} \vartheta_h \ge \vartheta_0$.

We define the space V_h of all continuous functions that are linear on every triangle or on every tetrahedron, respectively, of the given division. We will assume that the given division is regular. Let V_h be the finite element approximation of V consisting of all functions from V, which on the triangles (tetrahedrons) $\mathbb{T}_{hi}$ are equal to interpolation polynomials of the first or the third degrees.

Let $R_j \in \Omega$ be an arbitrary vertex of the triangulation $\mathcal{T}_h$. Then a vector basis function $\mathbf{v}_h$ is defined as a continuous function, which is on triangles (tetrahedra) $\mathbb{T}_{hi} \in \mathcal{T}_h$ linear polynomial and in vertices of the triangulation (division) it has a value $\mathbf{v}_{hj}(R_k) = \delta_{jk}$, where δ_{jk} is the Kronecker symbol. The function $\mathbf{v}_h$ in the case of the polynomial of the first order represents the pyramid with height of 1 with vertex at point R_j and with the support supp $\mathbf{v}_h$, which consists of all triangles having the common point in R_j. The support of the basis functions $\mathbf{v}_h$ has diam(supp $v_{hj}) \le 2h$. The set of polynomials of the first order we denote by P_1. Let us define the finite dimensional space

$$V_h = \{\mathbf{v}_h | \mathbf{v}_h \in [C(\Omega)]^N \cap V;\ \mathbf{v}_h \in [P_1(\mathbb{T}_{hi})]^N, \ \text{for all } \mathbb{T}_{hi} \in \mathcal{T}_h,\ N = 2, 3\}.$$

Then this problem leads to the following problem:

Problem $(\mathcal{P})_h$ Find $\mathbf{u}_h \in V_h$ minimizing the functional

$$J(\mathbf{u}) = \frac{1}{2} a(\mathbf{u}, \mathbf{u}) - \int_\Omega f_i u_i \, d\mathbf{x} - \int_{\Gamma_\tau} P_i u_i \, ds \tag{5.15}$$

on the set V_h, where $a(\mathbf{u}, \mathbf{v})$ is the bilinear form defined by (5.12).

Let us denote by $\varepsilon(h) = \|\mathbf{u} - \mathbf{u}_h\|$ the error between the exact weak and the approximate solutions, where $\mathbf{u}_h \in V_h$. We will prove that $\varepsilon(h) \to 0$ as $h \to 0_+$.

Theorem 5.2 Let a: $V \times V \to \mathbb{R}$ be a symmetric, bounded, and V-elliptic bilinear form on V. Let V_h be the finite element approximation of V defined above with the following properties:

$$\forall \mathbf{v} \in V \quad \exists \mathbf{v}_h \in V_h : \mathbf{v}_h \to \mathbf{v}, \quad \text{as } h \to 0_+. \tag{5.16}$$

Then

$$\|\mathbf{u} - \mathbf{u}_h\| \to 0 \quad \text{as } h \to 0_+. \tag{5.17}$$

The proof is parallel to that of Axelsson and Barker (1984) and Nečas and Hlaváček (1981).

Problem $(\mathcal{P})_h$ leads to finding the minimum of the potential energy

$$J(\mathbf{v}_h) = \tfrac{1}{2}a(\mathbf{v}_h, \mathbf{v}_h) - S(\mathbf{v}_h) \quad \text{on } V_h.$$

As $\mathbf{v}_h(\mathbf{x}) = \sum_{k=1}^{M} v_k \varphi_k(\mathbf{x})$, then $a(\mathbf{v}_h, \mathbf{v}_h)$ is a quadratic form and $S(\mathbf{v}_h)$ is a linear form in the coordinates of the function $\mathbf{v}_h$ with respect to the basis $\{\varphi_k\}_{k=1}^{M}$ of V_h, that is,

$$J(\mathbf{v}_h) = \frac{1}{2}\sum_{k=1}^{M}\sum_{j=1}^{M} a(\varphi_k, \varphi_j) v_k v_j - \sum_{k=1}^{M} S(\varphi_k) v_k. \tag{5.18}$$

Defining the elements K_{ij} of the symmetric stiff matrix K by the relations

$$K_{ij} = a(\varphi_i, \varphi_j), \qquad i,j = 1,2,\ldots,M \tag{5.19}$$

and the components of the vector $\mathbf{F} = (F_i)$ by

$$F_i = S(\varphi_i), \qquad i = 1,2,\ldots,M, \tag{5.20}$$

we see that the matrix K is positive definite and therefore the function $\mathbf{u}_h \in V_h$ minimizes the functional $J(\mathbf{v}_h)$ on V_h if and only if the coordinates v_k, $k = 1,2,\ldots,M$, are the solution of the system of linear equations

$$\frac{\partial J(\mathbf{v}_h)}{\partial v_k} = 0, \tag{5.21}$$

that is,

$$K\Delta = \mathbf{F}, \tag{5.22}$$

where $\Delta = (v_1, \ldots, v_M)^{\mathrm{T}}$. The algorithm for computing K and $\mathbf{F}$ is a modification of the algorithm given in the next subsection for polynomials of the first order. To obtain a better approximation of our problem, polynomials of a higher order and adaptive finite element method (FEM) can be used (Ciarlet, 1978; Kolář et al., 1979; Axelsson and Barker, 1984; Babuška and Reinboldt, 1980; Nečas and Hlaváček, 1981; Nedoma, 1976, 1998a; Šolín, 2006; Šolín et al., 2004; Nedoma et al., 2006).

We assumed that the boundary $\partial\Omega$ was approximated by the polygonal (polyhedral) boundary. In many cases it is better to approximate the boundary $\partial\Omega$ by part by polynomials of a higher order $n > 1$, and in the triangulation (division) one side of it is curved. Such solutions will be of a higher order of accuracy [see, e.g., Babuška and Reinboldt (1980) and Šolín et al. (2004)].

Since the matrix K is positive definite, the system (5.22) has just one solution $\triangle$, which minimizes the functional $L(\mathbf{u}_h)$. Constructions of K and the vector $\mathbf{F}$ are built from two steps—at first we do it for every element.

The algorithm of computation starts from the following considerations: We assume that for every element $\mathbb{T}_{hk}$ we compute the matrix K_k and the vector $\mathbf{F}_k$, and then from these matrices and vectors we construct the resulting matrix K and the vector $\mathbf{F}$. Thus

$$\sum_{k=1}^{M} K_k \triangle = \sum_{k=1}^{M} \mathbf{F}_k. \tag{5.23}$$

A. Construction of Matrix K and Vector* F *for 3D Case and Linear (First Order) Finite Elements Let a 3D region Ω_h, the boundary of which is a polyhedron $\partial\Omega_h$, be divided into a finite number of tetrahedrons $\mathbb{T}_{hk}$, $k = 1, \ldots, M$. We restrict our considerations to the space V_h of all continuous functions that are linear on every tetrahedron of the given division. The polynomial of the first degree, denoted by $u_r^{(k)}$, $r = 1, 2, 3$, $k = 1, 2, \ldots, M$, is uniquely determined on the tetrahedron $\mathbb{T}_{hk}$ by the function values at the vertices of the tetrahedron. Let us denote the vertices of the tetrahedron by $P_i(x_i, y_i, z_i)$, $i = 1, \ldots, 4$ and let $\overline{x}_i = x_i - x_1$, $\overline{y}_i = y_i - y_1$, $\overline{z}_i = z_i - z_1$, $i = 2, 3, 4$.

Let $\mathbb{T}_{h0}$ be the tetrahedron in the Cartesian coordinate system (ξ, η, ζ) with vertices $R_1(0,0,0)$, $R_2(1,0,0)$, $R_3(0,1,0)$, and $R_4(0,0,1)$. It is well known that the transformation

$$\begin{aligned} x &= x(\xi,\eta,\zeta) = x_1 + \overline{x}_2\xi + \overline{x}_3\eta + \overline{x}_4\zeta, \\ y &= y(\xi,\eta,\zeta) = y_1 + \overline{y}_2\xi + \overline{y}_3\eta + \overline{y}_4\zeta, \\ z &= z(\xi,\eta,\zeta) = z_1 + \overline{z}_2\xi + \overline{z}_3\eta + \overline{z}_4\zeta, \end{aligned} \tag{5.24}$$

transforms vertices P_i onto R_i, $i = 1, \ldots, 4$. Then from the linearity of this transformation it follows that the function

$$\overline{\mathbf{u}}_r^{(k)}(\xi,\eta,\zeta) = \mathbf{u}_r^{(k)}(x(\xi,\eta,\zeta), y(\xi,\eta,\zeta), z(\xi,\eta,\zeta)) = \mathbf{G}\overline{\mathbf{a}}_r \tag{5.25}$$

is a polynomial of the first order in the variables ξ, η, ζ, where

$$\mathbf{G} = (1, \xi, \eta, \zeta)^{\mathrm{T}} \tag{5.26}$$

and

$$\overline{\mathbf{a}}_r = \left(a_1^{(r)}, a_2^{(r)}, a_3^{(r)}, a_4^{(r)}\right)^{\mathrm{T}} \tag{5.27}$$

denote the vector of the coefficients of the polynomial $\overline{\mathbf{u}}_r^{(k)}$. Moreover,

$$\overline{\mathbf{u}}_r^{(k)}(\xi(x,y,z), \eta(x,y,z), \zeta(x,y,z)) = \mathbf{u}_r^{(k)}(x,y,z) \tag{5.28}$$

and therefore

$$\frac{\partial \mathbf{u}_r^{(k)}}{\partial x} = \frac{\partial \mathbf{u}_r^{(k)}}{\partial \xi}\frac{\partial \xi}{\partial x} + \frac{\partial \mathbf{u}_r^{(k)}}{\partial \eta}\frac{\partial \eta}{\partial x} + \frac{\partial \mathbf{u}_r^{(k)}}{\partial \zeta}\frac{\partial \zeta}{\partial x}, \text{ etc.} \tag{5.29}$$

Let us denote

$$S = \begin{bmatrix} 1 & 0 & 0 & 1 \\ 0 & 1 & 0 & 1 \\ 0 & 0 & 1 & 1 \\ 0 & 0 & 0 & 1 \end{bmatrix}, \tag{5.30}$$

$$\triangle_r = \left(\overline{\mathbf{u}}_r^{(k)}(R_1), \overline{\mathbf{u}}_r^{(k)}(R_2), \overline{\mathbf{u}}_r^{(k)}(R_3)\right)^{\mathrm{T}}, \tag{5.31}$$

then

$$\triangle_r = S\overline{\mathbf{a}}_r. \tag{5.32}$$

Hence

$$\overline{\mathbf{a}}_r = S^{-1}\triangle_r. \tag{5.33}$$

Let us denote

$$T = \begin{bmatrix} \overline{x}_2 & \overline{x}_3 & \overline{x}_4 \\ \overline{y}_2 & \overline{y}_3 & \overline{y}_4 \\ \overline{z}_2 & \overline{z}_3 & \overline{z}_4 \end{bmatrix} \quad \text{and} \quad \mathbf{J} = \det(T). \tag{5.34}$$

According to the substitution theorem we have

$$\int_{\mathbb{T}_{hi}} f(x,y,z)\,dx\,dy\,dz = \int_{\mathbb{T}_{h0}} |\mathbf{J}|\overline{f}(\xi,\eta,\zeta)\,d\xi\,d\eta\,d\zeta.$$

Let us denote $\mathbf{s} = (x,y,z) = (s_1, s_2, s_3)$ and

$$\mathbf{t} = (\xi,\eta,\zeta) = (t_1,t_2,t_3).$$

Then

$$\begin{aligned}\int_{\mathbb{T}_{hi}} c_{ijgh}\varphi^m_{g,h}\varphi^n_{i,j}\,dx &= |\mathbf{J}|\,c_{ijgh}\int_{\mathbb{T}_{h0}} (S^{-1}\varphi^m_g)^{\mathrm{T}}\frac{\partial \mathbf{G}}{\partial t_k}\frac{\partial t_k}{\partial s_h}(S^{-1}\varphi^n_i)^{\mathrm{T}}\frac{\partial \mathbf{G}}{\partial t_l}\frac{\partial t_l}{\partial s_j}\,\mathbf{d}\mathbf{t} \\ &= |\mathbf{J}|c_{ijgh}(\varphi^m_g)^{\mathrm{T}}(S^{-1})^{\mathrm{T}}\int_{\mathbb{T}_{h0}}\frac{\partial \mathbf{G}}{\partial t_k}\frac{\partial \mathbf{G}^T}{\partial t_l}\frac{\partial t_k}{\partial s_h}\frac{\partial t_l}{\partial s_j}\,\mathbf{d}\mathbf{t}\,S^{-1}\varphi^n_i.\end{aligned}$$

Since

$$\frac{\partial t_i}{\partial s_j} = T_{ij}^{-1}, \qquad \frac{\partial \mathbf{G}}{\partial t_i} = (0, \ldots, 0, 1, 0, \ldots, 0),$$

where 1 is in the ith position, and since

$$\int_{\mathbb{T}_{h0}} \xi^a \eta^b \zeta^c \, d\mathbf{t} = \frac{a!b!c!}{(a+b+c+3)!}, \tag{5.35}$$

then

$$\int_{\mathbb{T}_{h0}} \frac{\partial \mathbf{G}}{\partial t_k} \frac{\partial \mathbf{G}^{\mathrm{T}}}{\partial t_l} \frac{\partial t_k}{\partial s_h} \frac{\partial t_l}{\partial s_j} \, d\mathbf{t} = \frac{1}{6} B_{hj},$$

where

$$B_{hj} = \begin{bmatrix} T_{1h}^{-1} T_{1j}^{-1} & T_{1h}^{-1} T_{2j}^{-1} & T_{1h}^{-1} T_{3j}^{-1} \\ T_{2h}^{-1} T_{1j}^{-1} & T_{2h}^{-1} T_{2j}^{-1} & T_{2h}^{-1} T_{3j}^{-1} \\ T_{3h}^{-1} T_{1j}^{-1} & T_{3h}^{-1} T_{2j}^{-1} & T_{3h}^{-1} T_{3j}^{-1} \\ 0 & 0 & 0 \end{bmatrix}.$$

Let us denote

$$\tilde{K}_{hj} = \frac{|\mathbf{J}|}{6} (S^{-1})^{\mathrm{T}} B_{hj} S^{-1}, \quad \varphi^m = ((\varphi_1^m)^{\mathrm{T}}, (\varphi_2^m)^{\mathrm{T}}, (\varphi_3^m)^{\mathrm{T}})^{\mathrm{T}},$$

then

$$(A)_{m,n} = \int_{\mathbb{T}_{h0}} c_{ijgh} \varphi_{g,h}^m \varphi_{i,j}^n \, d\mathbf{t} = (\varphi^m)^{\mathrm{T}} \tilde{K} \varphi^n,$$

where $\tilde{K}$ is the 12×12 dimensional matrix of the form

$$\tilde{K} = \begin{bmatrix} a_{1j1h} \tilde{K}_{hj} & a_{1j2h} \tilde{K}_{hj} & a_{1j3h} \tilde{K}_{hj} \\ a_{2j1h} \tilde{K}_{hj} & a_{2j2h} \tilde{K}_{hj} & a_{2j3h} \tilde{K}_{hj} \\ a_{3j1h} \tilde{K}_{hj} & a_{3j2h} \tilde{K}_{hj} & a_{3j3h} \tilde{K}_{hj} \end{bmatrix}. \tag{5.36}$$

If $\varphi^m = 0$ or $\varphi^n = 0$ in $\mathbb{T}_{hi}$, then $(A_{m,n}) = 0$. If $\varphi^m \neq 0$ or $\varphi^n \neq 0$ in $\mathbb{T}_{hi}$, then vectors φ^m, φ^n have on just one position number 1 and elsewhere 0. We denote these positions by i_m and i_n. Then $(A)_{m,n} = (\tilde{K})_{i_m, i_n}$. The Jacobian as well as the matrix $\tilde{K}$ are computed for every tetrahedron $\mathbb{T}_{hi}$.

To set up the matrix A we use the following algorithm: Let $(A_{ij}) = 0$, $i, j = 1, \ldots, M$. We compute the elementary matrix $\tilde{K}$ for all tetrahedrons $\mathbb{T}_{hi}$. We find all pairs of functions φ^m and φ^n, which are nonzero on the given tetrahedron. Then for such pairs of functions we find indices i_m and i_n and we put

$$(A)_{i_m, i_n} = (A)_{i_m, i_n} + (\tilde{K})_{m,n}.$$

To compute $(b^{kl})_n = \int_{\mathbb{T}_{hi}} c_{ijkl}\varphi^n_{i,j}\, d\mathbf{x}$ we use the technique that is a simple modification of that presented above. We have

$$\begin{aligned}(b^{kl})_n &= \int_{\mathbb{T}_{hi}} c_{ijkl}\varphi^n_{i,j}\, d\mathbf{x} = |\mathbf{J}| c_{ijkl} \int_{\mathbb{T}_{h0}} (S^{-1}\varphi^n_i)\frac{\partial \mathbf{G}}{\partial t_l}\frac{\partial t_l}{\partial s_j}\, d\mathbf{t} \\ &= |\mathbf{J}| c_{ijkl} \int_{\mathbb{T}_{h0}} (\varphi^n_i)^{\mathrm{T}}(S^{-1})^{\mathrm{T}}(T^{-1}_{1j}, T^{-1}_{2j}, T^{-1}_{3j}, 0)^{\mathrm{T}}\, d\mathbf{t} \\ &= (\varphi^n_i)^{\mathrm{T}}|\mathbf{J}| \int_{\mathbb{T}_{h0}} ((c_{1jkl}v_j)^{\mathrm{T}}, (c_{2jkl}v_j)^{\mathrm{T}}, (c_{3jkl}v_j)^{\mathrm{T}})^{\mathrm{T}}\, d\mathbf{t} = \frac{1}{6}(\varphi^n_i)^{\mathrm{T}}|\mathbf{J}|v^{kl}, \qquad (5.37)\end{aligned}$$

where $\mathbf{v}_j = (S^{-1})^{\mathrm{T}}(T^{-1}_{1j}, T^{-1}_{2j}, T^{-1}_{3j}, 0)^{\mathrm{T}}$, and where (5.35) was used. For $\varphi^n_i \neq 0$ in $\mathbb{T}_{hi}$ there exists just one number i such that $(\varphi^n)_i = 1$. Then

$$(b^{kl})_n = (b^{kl})_n + \tfrac{1}{6}|\mathbf{J}|(\mathbf{v}^{kl})_i.$$

B. Construction of Matrix K and Vector* F *the 3D Case and Third Order Finite Element Let a 3D region Ω_h, the boundary of which is a polyhedron $\partial\Omega_h$, be divided into a finite number of tetrahedrons $\mathbb{T}_{hk}, k = 1, \ldots, M$. The set of the tetrahedra originated in the used division of the region Ω_h we denote by $\{\mathbb{T}_{hk},\ k = 1, \ldots, M\}$. We will assume that coefficients c_{ijkl} on every tetrahedron are constant. We restrict our considerations to the space V_h of all continuous functions that are polynomials of the third order on every tetrahedron of the given division. The polynomial of the third degree will be denoted by $u^{(k)}_r$, $r = 1, 2, 3$, $k = 1, 2, \ldots, M$. Let us denote the vertices of an arbitrary tetrahedron by $P_i(x_i, y_i, z_i)$, $i = 1, \ldots, 4$, the centers of gravity of its faces are denoted as Q_i, $i = 1, \ldots, 4$, where Q_1 is the center of gravity of the face with vertices $P_2P_3P_4$, Q_2 is the center of gravity of the face with vertices $P_1P_3P_4$, Q_3 is the center of gravity of the face with vertices $P_1P_2P_4$, and Q_4 is the center of gravity of the face with vertices $P_1P_2P_3$.

Let $\mathbb{T}_{h0}$ be the tetrahedron in the Cartesian coordinate system (ξ, η, ζ) with vertices $R_1(0,0,0)$, $R_2(1,0,0)$, $R_3(0,1,0)$, and $R_4(0,0,1)$. Centers of gravity of faces we denote by S_i, $i = 1, \ldots, 4$, where S_1 is the center of gravity of the face with the vertices $R_2R_3R_4$, S_2 is the center of gravity of the face with the vertices $R_1R_3R_4$, S_3 is the center of the face with the vertices $R_1R_2R_4$, and S_4 is the center of gravity of the face with the vertices $R_1R_2R_3$. Let $\overline{x}_i = x_i - x_1$, $\overline{y}_i = y_i - y_1$, $\overline{z}_i = z_i - z_1$, $i = 2, 3, 4$.

It is well known that the transformation

$$\begin{aligned}x &= x(\xi, \eta, \zeta) = x_1 + \overline{x}_2\xi + \overline{x}_3\eta + \overline{x}_4\zeta, \\ y &= y(\xi, \eta, \zeta) = y_1 + \overline{y}_2\xi + \overline{y}_3\eta + \overline{y}_4\zeta, \qquad (5.38) \\ z &= z(\xi, \eta, \zeta) = z_1 + \overline{z}_2\xi + \overline{z}_3\eta + \overline{z}_4\zeta,\end{aligned}$$

transforms vertices P_i onto vertices R_i and the center of gravity onto S_i, $i = 1, 2, 3, 4$, where $s = (x, y, z) = (s_1, s_2, s_3)$ are the old coordinates and $\sigma = (\xi, \eta, \zeta) = (\sigma_1, \sigma_2, \sigma_3)$ are the new coordinates. Let us introduce the matrix of transformation $\mathcal{J}$ and its determinant J_p, the Jacobian, by

$$T_p = \begin{pmatrix} \overline{x}_2 & \overline{y}_2 & \overline{z}_2 \\ \overline{x}_3 & \overline{y}_3 & \overline{z}_3 \\ \overline{x}_4 & \overline{y}_4 & \overline{z}_4 \end{pmatrix}, \qquad J_p = \det(T_p). \tag{5.39}$$

The cubic polynomial (polynomial of the third order) can be written as

$$p = aG(x, y, z),$$

where $a = (a_1, \ldots, a_{20})$ is the 20 component vector of coefficients and

$$\begin{aligned} G(x, y, z) &= (1, x, y, z, x^2, xy, xz, y^2, yz, z^2, x^3, x^2y, \\ &\quad x^2z, xy^2, xyz, xz^2, y^3, y^2z, yz^2, z^3)^{\mathrm{T}}. \end{aligned}$$

Hence the polynomial p is uniquely determined by its 20 independent values. Therefore, for example, every component u_i, $i = 1, 2, 3$, of the displacement vector on $\mathbb{T}_{hk}$ is a polynomial of the third order uniquely determined by functional values in vertices of the tetrahedron, the centers of gravity of the faces, and by the first partial differences according to x, y, z in vertices of the tetrahedron. Let us denote by

$$\begin{aligned} \Delta_p^i = \bigg(& u_i(P_1), \frac{\partial u_i(P_1)}{\partial x}, \frac{\partial u_i(P_1)}{\partial y}, \frac{\partial u_i(P_1)}{\partial z}, u_i(P_2), \frac{\partial u_i(P_2)}{\partial x}, \frac{\partial u_i(P_2)}{\partial y}, \frac{\partial u_i(P_2)}{\partial z}, \\ & u_i(P_3), \frac{\partial u_i(P_3)}{\partial x}, \frac{\partial u_i(P_3)}{\partial y}, \frac{\partial u_i(P_3)}{\partial z}, u_i(P_4), \frac{\partial u_i(P_4)}{\partial x}, \frac{\partial u_i(P_4)}{\partial y}, \frac{\partial u_i(P_4)}{\partial z}, \\ & u_i(Q_1), u_i(Q_2), u_i(Q_3), u_i(Q_4) \bigg)^{\mathrm{T}}. \end{aligned} \tag{5.40}$$

Let us define the function

$$\tilde{\mathbf{u}}(\xi, \eta, \zeta) = \mathbf{u}(x, y, z),$$

where the old and the new coordinates are coupled by the transformation $\mathcal{J}$. From the linearity of the transformation $\mathcal{J}$ it follows that $\tilde{\mathbf{u}}$ are also polynomials of the third order and thus its components $\tilde{u}_i$ are on the fundamental tetrahedron $\mathbb{T}_{h0}$ uniquely determined by functional values in vertices of the tetrahedron and the centers of gravity of the faces and by the first partial derivatives according to ξ, η, ζ in the vertices, that is, by vectors of 20 components of the form

$$\begin{aligned} \tilde{\Delta}_p^i = \bigg(& \tilde{u}_i(R_1), \frac{\partial \tilde{u}_i(R_1)}{\partial \xi}, \frac{\partial \tilde{u}_i(R_1)}{\partial \eta}, \frac{\partial \tilde{u}_i(R_1)}{\partial \zeta}, \tilde{u}_i(R_2), \frac{\partial \tilde{u}_i(R_2)}{\partial \xi}, \frac{\partial \tilde{u}_i(R_2)}{\partial \eta}, \frac{\partial \tilde{u}_i(R_2)}{\partial \zeta}, \\ & \tilde{u}_i(R_3), \frac{\partial \tilde{u}_i(R_3)}{\partial \xi}, \frac{\partial \tilde{u}_i(R_3)}{\partial \eta}, \frac{\partial \tilde{u}_i(R_3)}{\partial \zeta}, \tilde{u}_i(R_4), \frac{\partial \tilde{u}_i(R_4)}{\partial \xi}, \frac{\partial \tilde{u}_i(R_4)}{\partial \eta}, \frac{\partial \tilde{u}_i(R_4)}{\partial \zeta}, \\ & \tilde{u}_i(S_1), \tilde{u}_i(S_2), \tilde{u}_i(S_3), \tilde{u}_i(S_4) \bigg)^{\mathrm{T}}. \end{aligned} \tag{5.41}$$

In the vertices of tetrahedrons it holds

$$\tilde{u}_i(R_j) = u_i(P_j), \qquad \tilde{u}_i(S_j) = u_i(S_j), \qquad i = 1,2,3, \quad j = 1,\ldots,4 \tag{5.42}$$

and for partial derivative of components of displacements we have

$$\begin{aligned}
\frac{\partial \tilde{u}_i(R_j)}{\partial \xi} &= \frac{\partial u_i(P_j)}{\partial x}\frac{\partial x}{\partial \xi} + \frac{\partial u_i(P_j)}{\partial y}\frac{\partial y}{\partial \xi} + \frac{\partial u_i(P_j)}{\partial z}\frac{\partial z}{\partial \xi} \\
&= \frac{\partial u_i(P_j)}{\partial x}\overline{x}_2 + \frac{\partial u_i(P_j)}{\partial y}\overline{y}_2 + \frac{\partial u_i(P_j)}{\partial z}\overline{z}_2, \\
\frac{\partial \tilde{u}_i(R_j)}{\partial \eta} &= \frac{\partial u_i(P_j)}{\partial x}\frac{\partial x}{\partial \eta} + \frac{\partial u_i(P_j)}{\partial y}\frac{\partial y}{\partial \eta} + \frac{\partial u_i(P_j)}{\partial z}\frac{\partial z}{\partial \eta} \\
&= \frac{\partial u_i(P_j)}{\partial x}\overline{x}_3 + \frac{\partial u_i(P_j)}{\partial y}\overline{y}_3 + \frac{\partial u_i(P_j)}{\partial z}\overline{z}_3 \\
\frac{\partial \tilde{u}_i(R_j)}{\partial \zeta} &= \frac{\partial u_i(P_j)}{\partial x}\frac{\partial x}{\partial \zeta} + \frac{\partial u_i(P_j)}{\partial y}\frac{\partial y}{\partial \zeta} + \frac{\partial u_i(P_j)}{\partial z}\frac{\partial z}{\partial \zeta} \\
&= \frac{\partial u_i(P_j)}{\partial x}\overline{x}_4 + \frac{\partial u_i(P_j)}{\partial y}\overline{y}_4 + \frac{\partial u_i(P_j)}{\partial z}\overline{z}_4.
\end{aligned} \tag{5.43}$$

Let us define the matrix of the dimension 3×3, the matrix H_p of the dimension 4×4, and the matrix of the dimension 20×20 by the following relations:

$$T_p = \begin{pmatrix} \overline{x}_2 & \overline{y}_2 & \overline{z}_2 \\ \overline{x}_3 & \overline{y}_3 & \overline{z}_3 \\ \overline{x}_4 & \overline{y}_4 & \overline{z}_4 \end{pmatrix}, \qquad H_p = \begin{pmatrix} 1 & \\ & T_p \end{pmatrix}, \qquad \Lambda_p = \begin{pmatrix} H_p & & & & \\ & H_p & & & \\ & & H_p & & \\ & & & H_p & \\ & & & & I \end{pmatrix} \tag{5.44}$$

where I is the unit matrix of the dimension 4×4. Then

$$\tilde{\Delta}_p^i = \Lambda_p \, \Delta_p^i \tag{5.45}$$

or

$$\begin{pmatrix} \tilde{u}_i(R_1) \\ \dfrac{\partial \tilde{u}_i(R_1)}{\partial \xi} \\ \dfrac{\partial \tilde{u}_i(R_1)}{\partial \eta} \\ \dfrac{\partial \tilde{u}_i(R_1)}{\partial \xi} \\ \cdots \\ \tilde{u}_i(S_2) \\ \tilde{u}_i(S_3) \\ \tilde{u}_i(S_4) \end{pmatrix} = \begin{pmatrix} 1 & 0 & 0 & 0 & & & & & \\ 0 & \overline{x}_2 & \overline{y}_2 & \overline{z}_2 & & & & & \\ 0 & \overline{x}_3 & \overline{y}_3 & \overline{z}_3 & & & & & \\ 0 & \overline{x}_4 & \overline{y}_4 & \overline{z}_4 & & & & & \\ \cdots & \cdots & \cdots & \cdots & \cdots & \cdots & \cdots & \cdots & \cdots \\ & & & & & 0 & 1 & 0 & 0 \\ & & & & & 0 & 0 & 1 & 0 \\ & & & & & 0 & 0 & 0 & 1 \end{pmatrix} \begin{pmatrix} u_i(P_1) \\ \dfrac{\partial u_i(P_1)}{\partial x} \\ \dfrac{\partial u_i(P_1)}{\partial y} \\ \dfrac{\partial u_i(P_1)}{\partial z} \\ \cdots \\ u_i(Q_2) \\ u_i(Q_3) \\ u_i(Q_4) \end{pmatrix}.$$

Then from (5.45) it follows that

$$\Delta_p^i = \Lambda_p^{-1} \tilde{\Delta}_p^i, \tag{5.46}$$

where

$$\Lambda_p^{-1} = \begin{pmatrix} H_p^{-1} & & & & \\ & H_p^{-1} & & & \\ & & H_p^{-1} & & \\ & & & H_p^{-1} & \\ & & & & I \end{pmatrix}, \qquad H_p^{-1} = \begin{pmatrix} 1 & \\ & T_p^{-1} \end{pmatrix}.$$

The function $\tilde{u}_i(\xi, \eta, \zeta)$ will be found in the form of the polynomial of the third order in components ξ, η, ζ of the form

$$\tilde{u}_i(\xi, \eta, \zeta) = \tilde{a}_i^T \tilde{G}(\xi, \eta, \zeta), \tag{5.47}$$

where $\tilde{a}_i(\tilde{a}_1^i, \tilde{a}_2^i, \ldots, \tilde{a}_{20}^i)^{\mathrm{T}}$, $i = 1, 2, 3$ are 20 component vectors of coefficients where

$$\tilde{G}(\xi, \eta, \zeta)$$
$$= (1, \xi, \eta, \zeta, \xi^2, \xi\eta, \xi\zeta, \eta^2, \eta\zeta, \zeta^2, \xi^3, \xi^2\eta, \xi^2\zeta, \xi\eta^2, \xi\eta\zeta, \xi\zeta^2, \eta^3, \eta^2\zeta, \eta\zeta^2, \zeta^3)^{\mathrm{T}}.$$

Let us define the matrix of the dimension 20×20 of the form

$$S = \Bigg(\tilde{G}(R_1), \frac{\partial \tilde{G}(R_1)}{\partial \xi}, \frac{\partial \tilde{G}(R_1)}{\partial \eta}, \frac{\partial \tilde{G}(R_1)}{\partial \zeta}, \tilde{G}(R_2), \ldots,$$
$$\frac{\partial \tilde{G}(R_4)}{\partial \zeta}, \tilde{G}(S_1), \tilde{G}(S_2), \tilde{G}(S_3), \tilde{G}(S_4)\Bigg).$$

Then it holds

$$(\tilde{\Delta}_p^i)^{\mathrm{T}} = \tilde{a}_i^{\mathrm{T}} S.$$

Hence

$$\tilde{a}_i^{\mathrm{T}} = (\tilde{\Delta}_p^i)^{\mathrm{T}} S^{-1}. \tag{5.48}$$

Let us define the nineth component vector

$$\tilde{\mathbf{B}} = \left(\frac{\partial \tilde{u}_1}{\partial \xi}, \frac{\partial \tilde{u}_1}{\partial \eta}, \frac{\partial \tilde{u}_1}{\partial \zeta}, \frac{\partial \tilde{u}_2}{\partial \xi}, \ldots, \frac{\partial \tilde{u}_3}{\partial \zeta} \right)^{\mathrm{T}},$$

and let us denote M_j the jth row of the matrix T_p^{-1}, $j = 1, 2, 3$. Let us define the 6×9 dimensional matrix F_p as

$$F_p = \begin{pmatrix} M_1 & \mathbf{o}_{13} & \mathbf{o}_{13} \\ \mathbf{o}_{13} & M_2 & \mathbf{o}_{13} \\ \mathbf{o}_{13} & \mathbf{o}_{13} & M_3 \\ M_2 & M_1 & \mathbf{o}_{13} \\ \mathbf{o}_{13} & M_3 & M_2 \\ M_3 & \mathbf{o}_{13} & M_2 \end{pmatrix}, \qquad \mathbf{o}_{13} = (0, 0, 0).$$

For the vector $\boldsymbol{\varepsilon}$ on the tetrahedron $\mathbb{T}_{hk}$ it holds that

$$\boldsymbol{\varepsilon} = F_p \tilde{\mathbf{B}}. \tag{5.49}$$

By the transformation of the Hooke coefficients c_{ijkl} on the tetrahedron $\mathbb{T}_{hk}$ we obtain the symmetric matrix

$$\tilde{D}_p = |J_p| F_p^{\mathrm{T}} D_p F_p = \{\tilde{d}^p_{mn}\}_{m,n=1}^{9}, \qquad \tilde{d}^p_{mn} = \tilde{d}^p_{nm}. \tag{5.50}$$

Computation of the integral $\int_{\mathbb{T}_{hk}} \varepsilon^{\mathrm{T}} D_p \varepsilon \, dx$ On the thetrahedron $\mathbb{T}_{hk}$ by step by step we obtain

$$\begin{aligned}\int_{\mathbb{T}_{hk}} \varepsilon^{\mathrm{T}} D_p \varepsilon \, d\mathbf{x} &= \int_{\mathbb{T}_{hk}} (F_p \tilde{\mathbf{B}})^{\mathrm{T}} D_p F_p \tilde{\mathbf{B}} \, dx\, dy\, dz \\ &= \int_{\mathbb{T}_{hk}} \tilde{\mathbf{B}}^{\mathrm{T}} F_p^{\mathrm{T}} D_p F_p \tilde{\mathbf{B}} \, dx\, dy\, dz. \end{aligned} \tag{5.51}$$

Relation (5.51) will be pretransformed on the fundamental tetrahedron $\mathbb{T}_{h0}$. Applying substitution theorem we obtain

$$\begin{aligned}\int_{\mathbb{T}_{hk}} \tilde{\mathbf{B}}^{\mathrm{T}} F_p^{\mathrm{T}} D_p F_p \tilde{\mathbf{B}} \, dx\, dy\, dz &= \int_{\mathbb{T}_{h0}} \tilde{\mathbf{B}}^{\mathrm{T}} |J_p| F_p^{\mathrm{T}} D_p F_p \tilde{\mathbf{B}} \, d\xi\, d\eta\, d\zeta \\ &= \int_{\mathbb{T}_{h0}} \tilde{\mathbf{B}}^{\mathrm{T}} \tilde{D}_p \tilde{\mathbf{B}} \, d\xi d\eta\, d\zeta = \int_{\mathbb{T}_{h0}} \tilde{\mathbf{B}}^{\mathrm{T}} \tilde{D}_p \tilde{\mathbf{B}} \, d\sigma \\ &= \int_{\mathbb{T}_{h0}} \sum_{i,j,k,l=1}^{3} (\tilde{d}^p)_{mn} \frac{\partial \tilde{u}_i}{\partial \sigma_k} \frac{\partial \tilde{u}_j}{\partial \sigma_l} \, d\sigma, \end{aligned}$$

where $m = 3(i-1)+k; n = 3(j-1)+l,\ \sigma = (\sigma_i), i = 1,2,3$. Applying (5.47) and (5.48) we obtain

$$\begin{aligned}\int_{\mathbb{T}_{h0}} \frac{\partial \tilde{u}_i}{\partial \sigma_k} \frac{\partial \tilde{u}_j}{\partial \sigma_l} \, d\sigma &= \tilde{a}_i^{\mathrm{T}} \int_{\mathbb{T}_{h0}} \frac{\partial \tilde{G}}{\partial \sigma_k} \left(\frac{\partial \tilde{G}}{\partial \sigma_l} \right)^{\mathrm{T}} d\sigma . \tilde{a}_j \\ &= (\tilde{\Delta}_p^i)^{\mathrm{T}} S^{-1} \int_{\mathbb{T}_{h0}} \frac{\partial \tilde{G}}{\partial \sigma_k} \left(\frac{\partial \tilde{G}}{\partial \sigma_l} \right)^{\mathrm{T}} d\sigma (S^{-1})\, \tilde{\Delta}_p^j = (\tilde{\Delta}_p^i)^{\mathrm{T}} A_{kl} \tilde{\Delta}_p^j. \end{aligned}$$

Let us denote by

$$\boldsymbol{\Delta}_p = (\Delta_p^1, \Delta_p^2, \Delta_p^3)^{\mathrm{T}}, \qquad \tilde{\boldsymbol{\Delta}}_p = (\tilde{\Delta}_p^1, \tilde{\Delta}_p^2, \tilde{\Delta}_p^3)^{\mathrm{T}}, \qquad L_p = \begin{pmatrix} \Lambda_p & & \\ & \Lambda_p & \\ & & \Lambda_p \end{pmatrix},$$

where L_p is the matrix of the dimension 60×60. Relation (5.45) can be rewritten in the form

$$\tilde{\boldsymbol{\Delta}}_p = L_p \, \boldsymbol{\Delta}_p.$$

Let A^{kl} be the matrix of the dimension 60×60:

$$A^{kl} = (A^{kl}_{ij}) = \begin{pmatrix} A_{kl} & A_{kl} & A_{kl} \\ A_{kl} & A_{kl} & A_{kl} \\ A_{kl} & A_{kl} & A_{kl} \end{pmatrix}.$$

Then we obtain

$$\begin{aligned} \int_{\mathbb{T}_{h0}} \tilde{\mathbf{B}}^{\mathrm{T}} \tilde{D}_p \tilde{\mathbf{B}} \, d\sigma &= \tilde{\boldsymbol{\Delta}}_p^{\mathrm{T}} \left[\sum_{i,j,k,l=1}^{3} (\tilde{d}^p_{mn}) A^{kl}_{ij} \right] \tilde{\boldsymbol{\Delta}}_p \\ &= \boldsymbol{\Delta}_p^{\mathrm{T}} L_p^{\mathrm{T}} \left[\sum_{i,j,k,l=1}^{3} (\tilde{d}^p_{mn}) A^{kl}_{ij} \right] L_p \boldsymbol{\Delta}_p = \boldsymbol{\Delta}_p^{\mathrm{T}} K_p \boldsymbol{\Delta}_p, \end{aligned}$$

the matrix $K_p = L_p^{\mathrm{T}} [\sum_{i,j,k,l=1}^{3} (\tilde{d}^p_{mn}) A^{kl}_{ij}] L_p$ is the stiffness matrix corresponding to the element of $\mathbb{T}_{hk}$. Since $\tilde{d}^p_{mn} = \tilde{d}^p_{nm}$, $A^{kl}_{ij} = (A^{kl}_{ji})^{\mathrm{T}}$, the matrix K_p is a symmetric matrix. The six fundamental matrices A_{kl}, $k, l = 1, 2, 3$, by which the matrix K_p is generated, are computed only once and stored in the inner memory of the computer. Matrices L_p and coefficients are different for every element $\mathbb{T}_{hk}$. For computation of the integral $\int_{\mathbb{T}_{h0}} (\partial \tilde{u}_i / \partial \sigma_k)(\partial \tilde{u}_j / \partial \sigma_l) \, d\sigma$ we will use the relation

$$\int_{\mathbb{T}_{h0}} \xi^a \eta^b \zeta^c \, d\xi \, d\eta \, d\zeta = \frac{a!b!c!}{(a+b+c+3)!} \tag{5.52}$$

Construction of the Right-Hand Side Let us investigate the following:

(i) The Integral of Body Forces $\int_{\mathbb{T}_{hk}} f_i u_i \, d\mathbf{x}$ Every component of the body forces will be approximated on every tetrahedron $\mathbb{T}_{hk}$ by a linear polynomial. The component of the body forces by using the transformation (5.38) are transformed on the function

$$\tilde{f}_i(\xi, \eta, \zeta) = f_i(x(\xi, \eta, \zeta), y(\xi, \eta, \zeta), z(\xi, \eta, \zeta)), \quad i = 1, 2, 3,$$

which is also in every component of the linear polynomial. According to the theory on transformation of the integral, it holds that

$$\begin{aligned} &\int_{\mathbb{T}_{hk}} \mathbf{f}(x, y, z) \, \mathbf{u}(x, y, z) \, d\mathbf{x} \\ &= |J_p| \int_{\mathbb{T}_{h0}} \{\tilde{f}_1(\xi, \eta, \zeta) \, \tilde{u}_1(\xi, \eta, \zeta) + \tilde{f}_2(\xi, \eta, \zeta) \, \tilde{u}_2(\xi, \eta, \zeta) \\ &\quad + \tilde{f}_3(\xi, \eta, \zeta) \, \tilde{u}_3(\xi, \eta, \zeta)\} \, d\xi \, d\eta \, d\zeta. \end{aligned} \tag{5.53}$$

The function f_i will be approximated on $\mathbb{T}_{hk}$ by the linear polynomial. This polynomial is uniquely determined by the vector with components $\hat{f}_i = (f_i(P_1), f_i(P_2),$

$f_i(P_3), f_i(P_4))^{\mathrm{T}}$. After this transformation on the fundamental tetrahedron for components $\tilde{f}_i$, which are also linear polynomials, it holds that

$$\tilde{f}_i(\xi,\eta,\zeta) = \tilde{b}_i^{\mathrm{T}} H(\xi,\eta,\zeta), \qquad i = 1,2,3, \tag{5.54}$$

where $\tilde{b}_i$ are the fourth component vectors of coefficients $\tilde{b}_i = (\tilde{b}_1^i, \tilde{b}_2^i, \tilde{b}_3^i, \tilde{b}_4^i)^{\mathrm{T}}$ and $H(\xi,\eta,\zeta) = (1,\xi,\eta,\zeta)^{\mathrm{T}}$.

Let us denote by U the matrix of columns of the form

$$U = (H(R_1), H(R_2), H(R_3), H(R_4)) = \begin{pmatrix} 1 & 1 & 1 & 1 \\ 1 & 0 & 0 & 0 \\ 0 & 1 & 0 & 0 \\ 0 & 0 & 0 & 1 \end{pmatrix}.$$

Then $\hat{f}_i^{\mathrm{T}} = \tilde{b}_i^{\mathrm{T}} U$ and for the vector of the polynomial $\tilde{f}_i(\xi,\eta,\zeta)$, it holds that

$$\tilde{f}_i^{\mathrm{T}}(\xi,\eta,\zeta) = \hat{f}_i^{\mathrm{T}} U^{-1} H(\xi,\eta,\zeta).$$

Then

$$\begin{aligned} &\int_{\mathbb{T}_{hk}} \mathbf{f}(x,y,z)\,\mathbf{u}(x,y,z)\,d\mathbf{x} \\ &= |J_p|(\mathbf{\Delta}_p^i)^{\mathrm{T}} \Lambda_p^{\mathrm{T}} S^{-1} \int_{\mathbb{T}_{h0}} \tilde{G}(\xi,\eta,\zeta)\, H^{\mathrm{T}}(\xi,\eta,\zeta)\, d\xi\, d\eta\, d\zeta\, (U^{-1})^{\mathrm{T}} \hat{f}_i. \end{aligned} \tag{5.55}$$

Let us denote by $\tilde{F}_p^{(v,i)} = |J_p| S^{-1} \int_{\mathbb{T}_{h0}} \tilde{G}(\xi,\eta,\zeta) H^{\mathrm{T}}(\xi,\eta,\zeta)\, d\xi\, d\eta\, d\zeta (U^{-1})^{\mathrm{T}} \hat{f}_i$ and vector $\tilde{\mathbf{F}}_p^{(v)} = (\tilde{F}_p^{(v,1)}, \tilde{F}_p^{(v,2)}, \tilde{F}_p^{(v,3)})^{\mathrm{T}}$ of the dimension 60×1. For the mutual relation of vector $\tilde{\mathbf{F}}_p^{(v)}$ on the tetrahedron $\mathbb{T}_{hk}$ and vector $\tilde{\mathbf{F}}_p^{(v)}$ on the fundamental tetrahedron $\mathbb{T}_{h0}$, it holds that

$$\mathbf{F}_p^{(v)} = L_p^T \tilde{\mathbf{F}}_p^{(v)}, \tag{5.56}$$

where L_p is the matrix defined above. The vector $\mathbf{F}_p^{(v)}$ is the vector of the transformed body loading of the element $\mathbb{T}_{hk}$.

The integral $\int_{\mathbb{T}_{h0}} \tilde{G}(\xi,\eta,\zeta) H^{\mathrm{T}}(\xi,\eta,\zeta)\, d\xi\, d\eta\, d\zeta$ will be computed by using relation (5.52).

(ii) The Integral of Surface Forces $\int_{\Gamma_{\tau h}} P_i u_i\, ds$ Let us assume that the face $P_1P_2P_3$ of the tetrahedron $\mathbb{T}_{hk}$ is the boundary face of the region Ω_h and it belongs into the boundary Γ_τ. For simplicity we denote it by the symbol M. Let us denote the fundamental face as M_0 with vertices (1, 0, 0), (0, 1, 0), (0, 0, 0). The face M can be parametrically described by the relations

$$\begin{aligned} x &= x_1 + \overline{x}_2 w + \overline{x}_3 v, \\ y &= y_1 + \overline{y}_2 w + \overline{y}_3 v, \\ z &= z_1 + \overline{z}_2 w + \overline{z}_3 v. \end{aligned}$$

From the theory of surface integrals it holds that

$$\int_M P_i(x,y,z)\,u_i(x,y,z)\,ds = 2\text{meas}(M)\int_{M_0} \tilde{P}_i(w,v)\,\tilde{u}_i(w,v)\,dw\,dv,$$

where $\text{meas}(M) = \frac{1}{2}[(\overline{x}_2\overline{y}_3 - \overline{x}_3\overline{y}_2)^2 + (\overline{x}_2\overline{z}_3 - \overline{x}_3\overline{z}_2)^2 + (\overline{y}_2\overline{z}_3 - \overline{y}_3\overline{z}_2)^2]^{1/2}$ is the area of the triangle M and where

$$\begin{aligned}\tilde{P}_i(w,v) &= P_i(x(w,v), y(w,v), z(w,v)),\\ \tilde{u}_i(w,v) &= \tilde{u}_i(x(w,v), y(w,v), z(w,v))\end{aligned}$$

are parametrical relations of functions $u_i(x,y,z)$ and $P_i(x,y,z)$.

The functions $\tilde{u}_i(w,v)$, $i = 1,2,3$ are polynomials of the third order in coordinates w and v. Then

$$\tilde{u}_i(w,v) = \tilde{a}_i^{\mathrm{T}} g(w,v),$$

where $g(w,v) = (1, w, v, w^2, wv, v^2, w^3, w^2v, wv^2, v^3)^{\mathrm{T}}$ and $\tilde{a}_i = (\tilde{a}_1^1, \ldots, \tilde{a}_{10}^i)$ is the vector of coefficients. Let us define the vector $\tilde{\boldsymbol{\Delta}}^i$ of dimension 10×1 and matrix $\tilde{S}$ of dimension 10×10 by relations

$$\tilde{\boldsymbol{\Delta}}^i = \begin{pmatrix} \tilde{u}_i(0,0), \dfrac{\partial \tilde{u}_i(0,0)}{\partial w}, \dfrac{\partial \tilde{u}_i(0,0)}{\partial v}, \tilde{u}_i(0,1), \dfrac{\partial \tilde{u}_i(0,1)}{\partial w}, \dfrac{\partial \tilde{u}_i(0,1)}{\partial v}, \\ \tilde{u}_i(1,0), \dfrac{\partial \tilde{u}_i(1,0)}{\partial w}, \dfrac{\partial \tilde{u}_i(1,0)}{\partial v}, \tilde{u}_i\left(\dfrac{1}{3}, \dfrac{1}{3}\right) \end{pmatrix}^{\mathrm{T}}$$

$$\tilde{S} = \begin{pmatrix} g(0,0), \dfrac{\partial g(0,0)}{\partial w}, \dfrac{\partial g(0,0)}{\partial v}, g(0,1), \dfrac{\partial g(0,1)}{\partial w}, \dfrac{\partial g(0,1)}{\partial v}, \\ g(1,0), \dfrac{\partial g(1,0)}{\partial w}, \dfrac{\partial g(1,0)}{\partial v}, g\left(\dfrac{1}{3}, \dfrac{1}{3}\right) \end{pmatrix}^{\mathrm{T}}.$$

The function $\tilde{u}_i(w,v)$ after substitution from (5.48) for coefficients $\tilde{a}_i$ then has the form

$$\begin{aligned}\tilde{u}_i(w,v) &= \tilde{a}_i^{\mathrm{T}} g(w,v) = (\tilde{\boldsymbol{\Delta}}_p^i)^{\mathrm{T}}\,\tilde{S}^{-1} g(w,v)\\ &= \tilde{\boldsymbol{\Delta}}^i (\tilde{\Lambda}_p)^{\mathrm{T}}\,\tilde{S}^{-1} g(w,v),\end{aligned}$$

as $\tilde{\boldsymbol{\Delta}}_p^i = \tilde{\Lambda}_p(\Delta_p)^{\mathrm{T}}$, where Δ_p^i is the 20-component vector defined above and

$$\tilde{\Lambda}_p = \begin{pmatrix} \tilde{T}_p & \mathbf{o}_{3,4} & \mathbf{o}_{3,4} & \mathbf{o}_{3,4} & \mathbf{o}_{3,4} \\ \mathbf{o}_{3,4} & \tilde{T}_p & \mathbf{o}_{3,4} & \mathbf{o}_{3,4} & \mathbf{o}_{3,4} \\ \mathbf{o}_{3,4} & \mathbf{o}_{3,4} & \tilde{T}_p & \mathbf{o}_{3,4} & \mathbf{o}_{3,4} \\ \mathbf{o}_{1,4} & \mathbf{o}_{1,4} & \mathbf{o}_{1,4} & \mathbf{o}_{1,4} & \mathbf{c}_{1,4} \end{pmatrix}$$

$$\tilde{T}_p = \begin{pmatrix} 1 & 0 & 0 & 0 \\ 0 & \overline{x}_2 & \overline{y}_2 & \overline{z}_2 \\ 0 & \overline{x}_3 & \overline{y}_2 & \overline{z}_2 \end{pmatrix},$$

where $\mathbf{o}_{3,4}$ is the zero matrix of the dimension 3×4, $\mathbf{o}_{1,4} = (0,0,0,0)$, $\mathbf{c}_{1,4} = (0,0,0,1)$.

Let us approximate the function P_i on M by the linear polynomial. After the transformation on the fundamental face M_0 then $\tilde{P}_i$ will be also the linear polynomial of the form

$$\tilde{P}_i(w,v) = c_i^{\mathrm{T}} W(w,v),$$

where $c_i = (c_1^i, c_2^i, c_3^i)^{\mathrm{T}}$ are three component vectors of coefficients and $W(w,v) = (1,w,v)^{\mathrm{T}}$.

The linear polynomial $\tilde{P}_i(w,v)$ is uniquely determined by the vector of values in vertices of the triangle R_1, R_2, R_3:

$$\hat{P}_i^{\mathrm{T}} = (P_i(R_1), P_i(R_2), P_i(R_3)) = c_i^{\mathrm{T}} Z,$$

where

$$Z = [W(0,0), W(1,0), W(0,1)] = \begin{pmatrix} 1 & 1 & 1 \\ 0 & 1 & 0 \\ 0 & 0 & 1 \end{pmatrix}.$$

Then

$$\tilde{P}_i(w,v) = \hat{P}_i^{\mathrm{T}} Z^{-1} W(w,v),$$

and after substitution and some modifications we obtain

$$\begin{aligned} &\int_M P_i(x,y,z)\, u_i(x,y,z)\, ds \\ &= 2\text{meas}(M) \int_{M_0} \hat{P}_i^{\mathrm{T}} Z^{-1} W^{\mathrm{T}}(w,v) \mathbf{\Delta}_p^i (\tilde{\Lambda}_p)^T \tilde{S}^{-1} g(w,v)\, dw\, dv \\ &= 2\text{meas}(M) \mathbf{\Delta}_p^i (\tilde{\Lambda}_p)^{\mathrm{T}} \tilde{S}^{-1} \int_{M_0}^{-1} g(w,v) W^{\mathrm{T}}(w,v)\, dw\, dv\, (Z^{-1}) \hat{P}_i. \end{aligned} \tag{5.57}$$

Let us set

$$P^{(s,i)} = 2\text{meas}(M) \tilde{S}^{-1} \int_{M_0} g(w,v) W^{\mathrm{T}}(w,v)\, dw\, dv\, (Z^{-1})\, \hat{P}_i, \quad i = 1,2,3,$$

$$\mathbf{P}^{(s)} = (P^{(s,1)}, P^{(s,2)}, P^{(s,3)})^{\mathrm{T}},$$

$$\overline{L}_p = \begin{pmatrix} \Lambda_p & & \\ & \Lambda_p & \\ & & \Lambda_p \end{pmatrix}.$$

Then it holds that

$$\int_M P_i(x,y,z)\,u_i(x,y,z)\,ds = \mathbf{\Delta}^{\mathrm{T}}\overline{L}_p^{\mathrm{T}} P^{(s)} = \mathbf{\Delta}^{\mathrm{T}}\mathbf{F}_p^s, \tag{5.58}$$

where $\mathbf{F}_p^s$ is the vector of transformed surface loading.

The integral $\int_{M_0} g(w,v)W^{\mathrm{T}}(w,v)\,dw\,dv$ can be computed by

$$\int_{M_0} w^m v^n\,dw\,dv = \frac{m!n!}{(m+n+2)!}. \tag{5.59}$$

The vector $\mathbf{F}_p = \mathbf{F}_p^v + \mathbf{F}_p^s$ is the vector of transformed loading. Realization of nonhomogeneous boundary conditions of the Dirichlet type is simple. If a point R_j^p is situated on the boundary Γ_u, then $u_i(R_j^p) = u_{0i}(R_j^p)$. If the point R_j^p is not situated on the boundary Γ_u, then $u_i(R_j^p)$ are independent (free) parameters.

Totally for the potential energy $\mathcal{L}_p$ on the element $\mathbb{T}_{hk}$ we obtain

$$\mathcal{L}_p(\mathbf{\Delta}_p) = \tfrac{1}{2}\mathbf{\Delta}_p^{\mathrm{T}} K_p \mathbf{\Delta}_p - \mathbf{\Delta}_p^{\mathrm{T}}\mathbf{F}_p. \tag{5.60}$$

It is evident that the functional of potential energy on the whole region Ω is a sum of functionals of potentional energy of simple tetrahedrons of the used division of the region Ω, that is,

$$\mathcal{L}(\mathbf{\Delta}) = \frac{1}{2}\mathbf{\Delta}^{\mathrm{T}} K \mathbf{\Delta} - \mathbf{\Delta}^{\mathrm{T}}\mathbf{F} = \sum_{p=1}^{n} \frac{1}{2}\mathbf{\Delta}_p^{\mathrm{T}} K_p \mathbf{\Delta}_p - \sum_{p=1}^{n} \mathbf{\Delta}_p^{\mathrm{T}}\mathbf{F}_p, \tag{5.61}$$

where n is the number of all tetrahedrons of the used division. To minimize the functional $\mathcal{L}$ from (5.61) the Jacobi, Gauss–Seidel, symmetric succesive overrelaxation method (SSOR), or the conjugate gradient methods can be used.

Remark 5.2 Similar algorithms can be used in the case of contact problems in elasticity, which leads to the minimization of functional $\mathcal{L}$ on the set of admissible displacements $K = \{\mathbf{v} | \mathbf{v} \in V, v_n^k - v_n^l \leq 0 \text{ on } \Gamma_c^{kl}\}$, where V is the set of virtual displacements, Γ_c^{kl} denotes the contact boundary between two joint parts, and the condition $v_n^k - v_n^l \leq 0$ is the nonpenetration condition. These problems will be solved and discussed in detail in Chapter 6.

5.5 MATHEMATICAL MODEL OF A LOADED LONG BONE BASED ON COMPOSITE BIOMATERIALS

Generally, a bone is created from nonhomogeneous anisotropic biomaterials. A bone has the ability to be remodeled by altering its size, shape, and structure. Load on the skeleton can be accomplished by either a muscle activity or gravity. Anisotropy in a bone has some interesting consequences. In general, most of the dangerous loads in a bone are likely to be those acting along its length. This is true for bending and compression. Tension is, in some cases, applied to a bone as a whole. Another dangerous mode of loading is a torsion, which produces maximum tensile forces at roughly 45° to the long bone. The great advantage of an anisotropic composite material is that it can be also arranged such that in at least one direction the material is much stronger and stiffer than an isotropic material of the some composition could be. The structure of a bone is very complicated. To simulate the real bone structure by the other one, we will assume that the biomaterials of a bone have a periodic structure. We will assume that the bone materials are assumed to be approximated by a periodic structure of two types of isotropic biomaterials. The method of homogenization will be used (see Chapter 6).

Let the bone occupy the region $\Omega \subset \mathbb{R}^N$, $N = 2,3$, with the Lipschitz boundary $\partial\Omega$. Let the boundary $\partial\Omega = \Gamma_\tau \cup \Gamma_u$, where we denote by $\Gamma_\tau = {}^1\Gamma_\tau \cup {}^2\Gamma_\tau$ the part of the bone surface where the bone is loaded (${}^1\Gamma_\tau$) or is unloaded (${}^2\Gamma_\tau$) and by Γ_u the part of the bone boundary where the bone is fixed.

The relation between the stress and strain is defined by the generalized linear Hooke law for the periodic biomaterials as follows: Let the coefficients of elasticity $c_{ijkl}(\mathbf{x})$ be periodical functions on the interval $Y \subset \mathbb{R}^3$ (see Fig. 5.6) such that $c^\varepsilon_{ijkl}(\mathbf{x}) = c_{ijkl}(\mathbf{x}/\varepsilon)$, thus

$$\tau^\varepsilon_{ij} = \tau^\varepsilon_{ij}(\mathbf{x}) = c^\varepsilon_{ijkl}(\mathbf{x})\, e_{kl}(\mathbf{u}^\varepsilon), \tag{5.62}$$

where coefficients $c^\varepsilon_{ijkl}(\mathbf{x})$ are Y periodic in all variables (where $Y = [0, y_1] \times [0, y_2] \times [0, y_3]$, $y_i > 0$, is the unit period in the variable y) form $(3 \times 3 \times 3 \times 3)$ matrix and satisfy

$$\begin{gathered} a^\varepsilon_0 \le c^\varepsilon_{ijkl}(\mathbf{x})\xi_{ij}\xi_{kl}|\xi|^{-2} \le A^\varepsilon_0 < +\infty \quad \forall \mathbf{x} \in \Omega,\ \xi_{ij} = \xi_{ji},\ \ \xi = (\xi_{ij}) \in \mathbb{R}^9, \\ c^\varepsilon_{ijkl}(\mathbf{x}) = c^\varepsilon_{jikl}(\mathbf{x}) = c^\varepsilon_{klij}(\mathbf{x}) = c^\varepsilon_{ijlk}(\mathbf{x}), c^\varepsilon_{ijkl}(\mathbf{x}) = c_{ijkl}\left(\frac{\mathbf{x}}{\varepsilon}\right), \end{gathered} \tag{5.63}$$

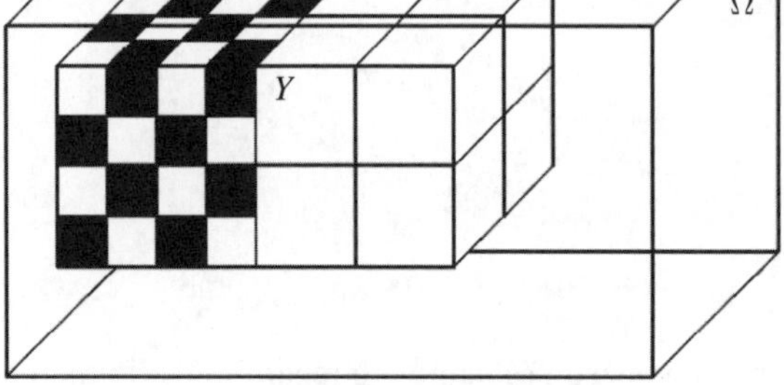

FIG. 5.6 Homogenization.

where a_0^ε, A_0^ε are constants independent of $\mathbf{x} \in \Omega$ and $a_0^\varepsilon = 2\min\{\mu^\varepsilon(\mathbf{x}), \mathbf{x} \in \Omega\}$, $A_0^\varepsilon = \max\{2\mu^\varepsilon(\mathbf{x}) + 3\lambda^\varepsilon(\mathbf{x}), \mathbf{x} \in \Omega\}$.

The stress tensor satisfies the equilibrium conditions

$$\frac{\partial \tau_{ij}^\varepsilon(\mathbf{u}(\mathbf{x}))}{\partial x_j} + f_i(\mathbf{x}) = 0 \quad \forall \mathbf{x} \in \Omega, \tag{5.64}$$

where τ_{ij}^ε is defined by (5.62) and the following boundary conditions:

$$\begin{aligned} \tau_{ij}^\varepsilon(\mathbf{u})n_j &= P_i \quad \text{on } {}^1\Gamma_\tau, \\ \tau_{ij}^\varepsilon(\mathbf{u})n_j &= 0 \quad \text{on } {}^2\Gamma_\tau, \end{aligned} \tag{5.65}$$

$$\mathbf{u}^\varepsilon(\mathbf{x}) = 0 \quad \text{on } \Gamma_u. \tag{5.66}$$

The algorithm of the used problem will be based on the homogenization method, derived in Chapter 6. We look for an asymptotic of $\mathbf{u}^\varepsilon(\mathbf{x})$ and $\tau^\varepsilon(\mathbf{x})$ in the form

$$\begin{aligned} \mathbf{u}^\varepsilon(\mathbf{x}) &= \mathbf{u}^0\left(\mathbf{x}, \frac{\mathbf{x}}{\varepsilon}\right) + \varepsilon \mathbf{u}^1\left(\mathbf{x}, \frac{\mathbf{x}}{\varepsilon}\right) + \varepsilon^2 \mathbf{u}^2\left(\mathbf{x}, \frac{\mathbf{x}}{\varepsilon}\right) + \cdots, \\ \tau^\varepsilon(\mathbf{x}) &= \tau^0\left(\mathbf{x}, \frac{\mathbf{x}}{\varepsilon}\right) + \varepsilon \tau^1\left(\mathbf{x}, \frac{\mathbf{x}}{\varepsilon}\right) + \varepsilon^2 \tau^2\left(\mathbf{x}, \frac{\mathbf{x}}{\varepsilon}\right) + \cdots, \end{aligned}$$

where $\mathbf{u}^i(\mathbf{x}, \mathbf{y})$ are functions independent of ε and periodic in the variable $\mathbf{y} = \mathbf{x}/\varepsilon$. The method used for the problem in Section 6.3 leads to the following homogenized problem:

Problem $\mathcal{P}_{\mathbf{hom}}$ Find a vector function $\mathbf{u}$ (displacement vector) satisfying

$$-\frac{\partial}{\partial x_j}\left(c_{ijkl}^0 \frac{\partial u_k^0}{\partial x_l}\right) = f_i \quad \text{in } \Omega, \tag{5.67}$$

and the following boundary conditions

$$c_{ijkl}^0 \frac{\partial u_k^0}{\partial x_l} n_j = P_i \quad \text{on } \Gamma_\tau, \tag{5.68}$$

and

$$\mathbf{u}^0 = 0 \quad \text{on } \Gamma_u, \tag{5.69}$$

representing the homogenized problem with regard to the initial one. The homogenized coefficients c_{ijkl}^0 can be found by numerical integration (see Chapter 6).

Numerically, the problem then leads to the minimization of the functional of potential energy

$$J(\mathbf{v}_h) = \frac{1}{2}\int_{\Omega} c^0_{ijkl} e_{ij}(\mathbf{v}_h) e_{kl}(\mathbf{v}_h)\, d\mathbf{x} - \int_{\Omega} f_i v_{hi}\, d\mathbf{x} - \int_{\Gamma_\tau} P_i v_{hi}\, ds$$

on the set V_h, where V_h is the finite element approximation of the space of virtual displacements $V = \{\mathbf{v} \mid \mathbf{v} \in [W^{1,2}(\Omega)]^3,\ \mathbf{v} = 0 \text{ on } \Gamma_u\}$. For more details see Nedoma (1998a) and Chapter 6.

Numerical Experiments In this section we present four model problems simulating simple biomaterials of bone tissues, based on the periodic structures of bone tissues, presented in Fig. 5.7 (a)–(d).

We will assume a piecewise isotropic biomaterial, that is, $c_{ijkl} = \lambda\delta_{ij}\delta_{kl} + \mu(\delta_{ik}\delta_{jl} + \delta_{il}\delta_{jk})$, with the following input data:

Elastic coefficients:	$\lambda_1 = 2 \times 10^{10}$ Pa,	$\mu_1 = 1 \times 10^{10}$ Pa,
	$\lambda_2 = 3 \times 10^{10}$ Pa,	$\mu_2 = 2 \times 10^{10}$ Pa,

accuracy for computation: $\varepsilon = 10^{-7}$.

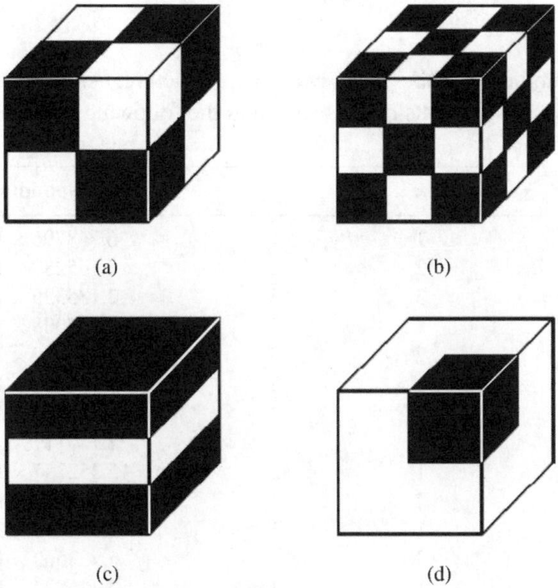

FIG. 5.7 Types of periodic structures of tissues.

Model (a) Number of cubes: 8, periodical structure. As the result we obtain anisotropic elastic coefficients c^0_{ijkl}, as shown in the following table:

i	j	k	l	Corresponding Values of c^0_{ijkl}
1	1	1	1	6.820000×10^{10}
1	1	2	2	2.100000×10^{10}
1	1	3	3	2.100000×10^{10}
1	1	1	2	0.0
1	1	2	3	0.0
1	1	1	3	0.0
2	2	2	2	6.820000×10^{10}
2	2	3	3	2.100000×10^{10}
2	2	1	2	0.0
2	2	2	3	0.0
2	2	1	3	0.0
3	3	3	3	6.820000×10^{10}
3	3	1	2	0.0
3	3	2	3	0.0
3	3	1	3	0.0
1	2	1	2	2.360000×10^{10}
1	2	2	3	0.0
1	2	1	3	0.0
2	3	2	3	2.360000×10^{10}
2	3	1	3	0.0
1	3	1	3	2.360000×10^{10}

Model (b) Number of cubes: 27, periodical structure. As the result we obtain anisotropic elastic coefficients c^0_{ijkl}, as shown in the following table:

i	j	k	l	Corresponding Values of c^0_{ijkl}
1	1	1	1	6.748796×10^{10}
1	1	2	2	2.135289×10^{10}
1	1	3	3	2.135325×10^{10}
1	1	1	2	2.947008×10^{-2}
1	1	2	3	3.027865×10^{6}
1	1	1	3	5.176286×10^{-2}
2	2	2	2	6.738705×10^{10}
2	2	3	3	2.1451179×10^{10}
2	2	1	2	-5.352257×10^{-3}
2	2	2	3	-1.507863×10^{-6}
2	2	1	3	-2.242718×10^{-2}
3	3	3	3	6.738669×10^{10}
3	3	1	2	-2.446824×10^{-2}
3	3	2	3	-1.507137×10^{6}

3	3	1	3	-3.007058×10^{-2}
1	2	1	2	2.301400×10^{10}
1	2	2	3	-1.786246×10^{-2}
1	2	1	3	1.434402×10^{6}
2	3	2	3	2.294300×10^{10}
2	3	1	3	2.154786×10^{-2}
1	3	1	3	2.301426×10^{10}

Model (c) Number of plates: 3, periodical structure. As the result we obtain anisotropic elastic coefficients c^0_{ijkl}, as shown in the following table:

i	*j*	*k*	*l*	**Corresponding Values of c^0_{ijkl}**
1	1	1	1	6.783343×10^{10}
1	1	2	2	2.063343×10^{10}
1	1	3	3	2.174778×10^{10}
1	1	1	2	0.0
1	1	2	3	$-1.180340 \times 10^{-11}$
1	1	1	3	4.809893×10^{-12}
2	2	2	2	6.783343×10^{10}
2	2	3	3	2.174780×10^{10}
2	2	1	2	0.0
2	2	2	3	$-5.545915 \times 10^{-11}$
2	2	1	3	4.809893×10^{-12}
3	3	3	3	6.667449×10^{10}
3	3	1	2	0.0
3	3	2	3	$-2.015888 \times 10^{-11}$
3	3	1	3	$-9.812182 \times 10^{-12}$
1	2	1	2	2.360000×10^{10}
1	2	2	3	0.0
1	2	1	3	$-1.455192 \times 10^{-11}$
2	3	2	3	2.115254×10^{10}
2	3	1	3	-1.509099×10^{12}
1	3	1	3	2.115254×10^{10}

Model (d) Number of cubes: 1, which is removed from the *Y* cube, periodical structure. As the result we obtain anisotropic elastic coefficients c^0_{ijkl}, as shown in the following table:

i	*j*	*k*	*l*	**Corresponding Values of c^0_{ijkl}**
1	1	1	1	7.759349×10^{10}
1	1	2	2	1.639541×10^{10}
1	1	3	3	1.639615×10^{10}
1	1	1	2	8.520586×10^{-3}

(Continued)

i	j	k	l	Corresponding Values of c^0_{ijkl}
1	1	2	3	1.019468×10^{6}
1	1	1	3	8.786604×10^{-3}
2	2	2	2	7.759573×10^{10}
2	2	3	3	1.639410×10^{10}
2	2	1	2	-7.434534×10^{-4}
2	2	2	3	-5.219826×10^{5}
2	2	1	3	4.070508×10^{-3}
3	3	3	3	7.759497×10^{10}
3	3	1	2	-7.454626×10^{-3}
3	3	2	3	-5.230976×10^{5}
3	3	1	3	-1.323510×10^{-2}
1	2	1	2	3.057976×10^{10}
1	2	2	3	5.352979×10^{-3}
1	2	1	3	-8.277078×10^{5}
2	3	2	3	3.058735×10^{10}
2	3	1	3	7.503425×10^{-3}
1	3	1	3	3.057904×10^{10}

The model with homogenized coefficients can be numerically solved by applying the finite element technique and the algorithm can be based on the polynomials of the first or third order, respectively, as in the previous section.

5.6 STOCHASTIC APPROACH

In real situations the biomaterials forming a bone have random structures, so that the mathematical description will be based on the stochastic variational principles and the stochastic finite element method. Randomness may exist in the characteristics of the structure itself (e.g., material properties of bone tissues) or in the environment to which the structure is exposed (e.g., loads, support conditions, etc.). In some structures the response is sensitive to both the material and geometric properties of the structure as well as to the applied loads. Even small uncertainties in these characteristics can affect adversely the structural performance. Since uncertainties are usually spatially distributed over the assumed structure area, thus the processes passing in it must be modeled as random fields (Soong, 1973; Kleiber and Hien, 1992). For an analysis of such problems three basic methodologies can be used to qualify structural response uncertainties, that is,

1. Taylor series expansion to formulate linear relationships among some characteristics of the random response and the random structural parameters on the basis of the perturbation approach.
2. Monte Carlo simulation in which an asset of random numbers is generated to represent the statistical uncertainties in the bone structural parameters.

Then these random parameters are substituted into the response equation to obtain a set of random numbers that reflect the uncertainty in the bone structural response and that has to be analyzed by some techniques to produce a qualification of the uncertainty.

3. The Neumann expansion technique [see Adomian (1983) and Mikhlin (1957)], which is effective when coupled with the Monte Carlo method (Yamazaki et al., 1985).

In this section the second-order version of the *i*th methodology based on the Taylor series expansion will be used because it seems both theoretically sound and computationally feasible to apply even for very large structural systems by using the techniques known from computational mechanics. We will introduce a version of the finite element method that accounts for uncertainties in both the geometry and material properties of biostructures as well as in the applied loads. This approach is known as the **stochastic finite element method (SFEM)**, and it accounts for uncertainties in the bone geometry and in material properties of the bone structure and in applied loads. In this section we present the main idea only; for more details the reader is referred to Soong (1973) and Kleiber and Hien (1992).

The idea of this approach is as follows: The analysis of stochastic partial differential equations with coefficients in the form of constant-in-time, varying-in-space random fields, say $\mathbf{b}(\mathbf{x})$, is a solution $\mathbf{u}(\mathbf{x}, t)$ of the random differential equations. The SFEM requires that the random field $\mathbf{b}(\mathbf{x})$ is spatially discretized. Then we obtain its discrete approximation represented by its nodal value vector; we denote it by $\overline{\mathbf{b}}$. The finite element discretization of displacement-type equations of motion for the linear structural dynamic problem for a loading long bone leads to a system of ordinary differential equations, which are linear in the unknown nodal displacement vector $\mathbf{X}(t)$ (a vector random process) and nonlinear in the random coefficients $\overline{\mathbf{b}}$. In certain statistical behavior the random variables $\overline{\mathbf{b}}$ are specified; then we can determine in terms of these properties certain statistical characterization of the solution $\mathbf{X}(t)$ of the dynamically loaded bone. For more details see Soong (1973) and Kleiber and Hien (1992).

In this section we limit ourselves to a statically loaded bone. Such problem leads to a static SFEM problem. The stochastic finite element methods, similar to the classical FEM, which is based on the potential energy principle, are based on the stochastic potential energy principle, which is a combination of the principle of minimum potential energy as in the deterministic case and the second-order perturbation technique.

Let us introduce some **fundamentals of stochastics**:

Let $y = f(x)$ be a real, simple-valued continuous function of a random variable x; thus y is also a random variable. The **mean value** (or statistical average or expectation) of a function f is defined as

$$E[y] = E[f(x)] = \int_{-\infty}^{+\infty} f(x)p(x)\,dx, \tag{5.70}$$

where $p(x)$ is a **probability density function** (PDF) of the random variable x. The integral of the probability density function $p(x)$, that is, $P(a) = \int_{-\infty}^{a} p(x)\,dx$, is the **cumulative distribution function** (CDF). In the special case if $f(x) = x$, then the **mean value of the random variable** x is

$$E[x] = \int_{-\infty}^{+\infty} xp(x)\,dx,$$

denoted also as x^0 or μ. For $f(x) = x^2$ is

$$E[x] = \int_{-\infty}^{+\infty} x^2 p(x)\,dx$$

the **mean-square value of the random variable** x. Its square root is called the **root-mean-square value**. The **variance** of x, denoted by Var(x), or σ_x^2, is defined as the mean-square value of x about the mean, that is,

$$\mathrm{Var}(x) = E[(x - x^0)^2] = \int_{-\infty}^{+\infty} (x - x^0)^2 p(x)\,dx = E[x^2] - (x^0)^2. \tag{5.71}$$

The positive square root of the variance

$$\sigma_x = +\sqrt{\mathrm{Var}(x)} \tag{5.72}$$

is called the **standard deviation** of x. The **coefficient of variation of a random variable** α_x is defined as

$$\alpha_x = \sqrt{\frac{\mathrm{Var}(x)}{(E[x])^2}}. \tag{5.73}$$

The mth **moment of a random variable** is defined as

$$\mu_x^m = \int_{-\infty}^{+\infty} x^m p(x)\,dx, \tag{5.74}$$

for $m = 1$ we have the mean value of the random variable x. The mth **central moment** is defined as

$$\overline{\mu}_x^m = \int_{-\infty}^{+\infty} (x - x^0)^m p(x)\,dx = E[(x - x^0)^m]. \tag{5.75}$$

Let $y=f(x_1,x_2,\ldots,x_n)$ be a real continuous function of n random variables $x_1,x_2,\ldots,x_n$, thus y is also a random function. Then the **mean value or the statistical average of the function** f is defined as

$$E[y] = E[f(x_1,x_2,\ldots,x_n)]$$
$$= \int_{-\infty}^{+\infty}\int_{-\infty}^{+\infty}\cdots\int_{-\infty}^{+\infty} f(x_1,x_2,\ldots,x_n)\,p\,(x_1,x_2,\ldots,x_n)\,dx_1\,dx_2\ldots dx_n,$$

where $p(x_1,x_2,\ldots,x_n)$ is an n-dimensional PDF of the random variables $x_1,x_2,\ldots,$ x_n. In general, the integral of the probability density function $p(x_1,x_2,\ldots,x_n)$, that is,

$$P(a_1,a_2,\ldots,a_n) = \int_{-\infty}^{a_1}\int_{-\infty}^{a_2}\cdots\int_{-\infty}^{a_n} p(x_1,x_2,\ldots,x_n)\,dx_1\,dx_2\ldots dx_n,$$

is the **multivariate (or joint) cumulative distribution function**, where $a_1,\ldots,a_n$ are constants. The joint cumulative distribution function is nondegreasing with respect to each of its arguments. Moreover,

$$p(x_1,x_2,\ldots,x_n) = \frac{\partial^n P(x_1,x_2,\ldots,x_n)}{\partial x_1\,\partial x_2\ldots\partial x_n}.$$

The **marginal probability density function** of the ith random variable is obtained from the joint probability density function $p(x_1,x_2,\ldots,x_n)$ by integrating [i.e., over $(n-1)$-fold] over the remaining variables

$$p(x_i) = \int_{-\infty}^{+\infty}\int_{-\infty}^{+\infty}\cdots\int_{-\infty}^{+\infty} p(x_1,\ldots,x_{i-1},x_i,x_{i+1},\ldots,x_n)\,dx_1\ldots dx_{i-1}\,dx_{i+1}\ldots dx_n.$$

The **cross correlation** between x_k and x_l is defined by

$$m_{kl} = E[x_k,x_l] = \int_{-\infty}^{+\infty}\int_{-\infty}^{+\infty} x_k x_l p(x_k,x_l)\,dx_k\,dx_l,$$

where [i.e., integration over $(n-2)$-fold]

$$p(x_k,x_l) = \int_{-\infty}^{+\infty}\int_{-\infty}^{+\infty}\cdots\int_{-\infty}^{+\infty} p(x_1,x_2,\ldots,x_n)\,dx_1\ldots dx_{k-1}\,dx_{k+1}\ldots dx_{l-1}\,dx_{l+1}\ldots dx_n,$$

and the **covariance** between x_k and x_l is defined by

$$\begin{aligned}
\mathrm{Cov}(x_k, x_l) &= E[(x_k - x_k^0)(x_l - x_l^0)] \\
&= \int_{-\infty}^{+\infty}\int_{-\infty}^{+\infty}\cdots\int_{-\infty}^{+\infty} (x_k - x_k^0)(x_l - x_l^0)p(x_k, x_l)\, dx_k\, dx_l \\
&= E[x_k x_l] - x_k^0 x_l^0.
\end{aligned}$$

Remark 5.3 For the dynamic problems the random processes and random fields can also be defined. A random process (random function, stochastic process are used synonymously) $x(\tau)$ is defined as an indexed family of random variables, where the index or the parameter τ belongs to some set Υ, $\tau \in \Upsilon$, and Υ is the parameter set or the index set of the process. In dynamic problems the parameter set Υ is the time interval I. In general, for an n-dimensional CDF $P_n(x_{1,}t_1; \ldots; x_n, t_n)$ then

$$p_n(x_{1,}t_1; \ldots; x_n, t_n) = \frac{\partial^n P_n(x_{1,}t_1; \ldots; x_n, t_n)}{\partial x_1 \ldots \partial x_n}.$$

The multidimensional PDF is symmetric with respect to the pairs of arguments (x_i, t_i) and (x_j, t_j). Then in a similar way we can define the mth moments of a random process $x(\tau)$ at a given t, the variance and the cross-correlation function at a given t and so forth.

Model problem Let the bone occupy the region $\Omega \subset \mathbb{R}^N$, $N = 2, 3$, with the Lipschitz boundary $\partial\Omega$. Let the boundary $\partial\Omega = \Gamma_\tau \cup \Gamma_u$, where we denote by $\Gamma_\tau = {}^1\Gamma_\tau \cup {}^2\Gamma_\tau$ the part of the bone surface where the bone is loaded (${}^1\Gamma_\tau$) or unloaded (${}^2\Gamma_\tau$) and by Γ_u the part of the bone boundary, where the bone is fixed.

Let us consider a linear structural bone system under static load

$$\frac{\partial \tau_{ij}(\mathbf{u})}{\partial x_j} + \rho f_i(\mathbf{x}) = 0 \quad \text{for all } \mathbf{x} \in \Omega, \tag{5.76}$$

where

$$\tau_{ij} = c_{ijkl} e_{kl}(\mathbf{u}),\ e_{ij} = \frac{1}{2}\left(\frac{\partial u_i}{\partial x_j} + \frac{\partial u_j}{\partial x_i}\right), \qquad i, j, k, l = 1, \ldots, N$$

(we also use the notation $u_{i,j} = \partial u_i / \partial x_j$) and the following boundary conditions:

$$\begin{aligned}
\tau_{ij}(\mathbf{u})n_j &= P_i \quad \text{on } {}^1\Gamma_\tau, \\
\tau_{ij}(\mathbf{u})n_j &= 0 \quad \text{on } {}^2\Gamma_\tau,
\end{aligned} \tag{5.77}$$

$$\mathbf{u}(\mathbf{x}) = 0 \quad \text{on } \Gamma_u. \tag{5.78}$$

Let us assume

$$\mathbf{b}(x_k) = \{b_1(x_k), b_2(x_k), \ldots, b_K(x_k)\}, \qquad k = 1, 2, 3, \tag{5.79}$$

to be a set of K random fields representing randomness of the Young's modulus, mass density of the bone biomaterials, and randomness of external load. The first two statistical moments for the random fields $b_r(x_k)$, $r = 1, 2, \ldots, K$, $k = 1, 2, 3$, are defined as

$$\begin{aligned} E[b_r] = b_r^0 &= \int_{-\infty}^{+\infty} b_r p_1(b_r)\, db_r, \\ \mathrm{Cov}(b_r, b_s) \equiv S_b^{rs} &= \int_{-\infty}^{+\infty} \int_{-\infty}^{+\infty} (b_r - b_r^0)(b_s - b_s^0) p_2(b_r, b_s)\, db_r\, db_s, \\ & \qquad r, s = 1, \ldots, K. \end{aligned} \tag{5.80}$$

Then the main idea behind the second-order perturbation to the stochastic version of the potential energy principle involves expanding all the random field variables, that is, elastic moduli $c_{ijkl}[\mathbf{b}(x_k); x_k]$, mass density $\rho[\mathbf{b}(x_k); x_k]$, body forces $\mathbf{f}[\mathbf{b}(x_k); x_k]$, surface forces $\mathbf{P}[\mathbf{b}(x_k); x_k]$, and displacement vector $\mathbf{u}[\mathbf{b}(x_k); x_k]$ about the spatial mean values of the random field variables $\mathbf{b}(x_k) = \{b_r(x_k)\}$, denoted by $\mathbf{b}^0(x_k) = \{b_r^0(x_k)\}$, in the Taylor series with a given small parameter ε and retaining terms up to the second order. Then we have (repetition of the subscript k in the arguments x_k does not imply summation)

$$\begin{aligned} c_{ijkl}[\mathbf{b}(x_k); x_k] = {} & c_{ijkl}^0[\mathbf{b}(x_k); x_k] + \varepsilon c_{ijkl}^{\prime r}[\mathbf{b}^0(x_k); x_k]\Delta b_r(x_k) \\ & + \tfrac{1}{2}\varepsilon^2 c_{ijkl}^{\prime rs}[\mathbf{b}^0(x_k); x_k]\Delta b_r(x_k)\Delta b_s(x_k), \end{aligned} \tag{5.81a}$$

$$\begin{aligned} \rho[\mathbf{b}(x_k); x_k] = {} & \rho^0[\mathbf{b}(x_k); x_k] + \varepsilon \rho^{\prime r}[\mathbf{b}^0(x_k); x_k]\Delta b_r(x_k) \\ & + \tfrac{1}{2}\varepsilon^2 \rho^{\prime rs}[\mathbf{b}^0(x_k); x_k]\Delta b_r(x_k)\Delta b_s(x_k), \end{aligned} \tag{5.81b}$$

$$\begin{aligned} f_i[\mathbf{b}(x_k); x_k] = {} & f_i^0[\mathbf{b}(x_k); x_k] + \varepsilon f_i^{\prime r}[\mathbf{b}^0(x_k); x_k]\Delta b_r(x_k) \\ & + \tfrac{1}{2}\varepsilon^2 f_i^{\prime rs}[\mathbf{b}^0(x_k); x_k]\Delta b_r(x_k)\Delta b_s(x_k), \end{aligned} \tag{5.81c}$$

$$\begin{aligned} P_i[\mathbf{b}(x_k); x_k] = {} & P_i^0[\mathbf{b}(x_k); x_k] + \varepsilon P_i^{\prime r}[\mathbf{b}^0(x_k); x_k]\Delta b_r(x_k) \\ & + \tfrac{1}{2}\varepsilon^2 P_i^{\prime rs}[\mathbf{b}^0(x_k); x_k]\Delta b_r(x_k)\Delta b_s(x_k), \end{aligned} \tag{5.81d}$$

$$\begin{aligned} u_i[\mathbf{b}(x_k); x_k] = {} & u_i^0[\mathbf{b}(x_k); x_k] + \varepsilon u_i^{\prime r}[\mathbf{b}^0(x_k); x_k]\Delta b_r(x_k) \\ & + \tfrac{1}{2}\varepsilon^2 u_i^{\prime rs}[\mathbf{b}^0(x_k); x_k]\Delta b_r(x_k)\Delta b_s(x_k), \end{aligned} \tag{5.81e}$$

where Eqs. (5.81) are symmetric with respect to r and s and where $\delta b_r(x_k) = \varepsilon \Delta b_r(x_k) = \varepsilon[b_r(x_k) - b_r^0(x_k)]$ is the first-order variation of $b_r(x_k)$ about $b_r^0(x_k)$, and

$$\delta b_r(x_k)\delta b_s(x_k) = \varepsilon^2 \Delta b_r(x_k)\Delta b_s(x_k) = \varepsilon^2[b_r(x_k) - b_r^0(x_k)][b_s(x_k) - b_s^0(x_k)]$$

denotes the second-order variation of $b_r(x_k)$ and $b_s(x_k)$ about $b_r^0(x_k)$ and $b_s^0(x_k)$, respectively, and the symbol $(\cdot)^0$ denotes the value of the function taken at b_r^0, and the symbols $(\cdot)^{\prime r}$ and $(\cdot)^{\prime rs}$ denote the first and the second (mixed) partial derivatives with respect to the random field variables $b_r(x_k)$ evaluated at their mean values, respectively. Substituting the above series (5.81) into the principle of minimum potential energy in the sense of the perturbation approach, we find

ε^0 terms:

$$\int_\Omega c_{ijkl}^0 u_{i,j}^0 \delta u_{k,l}\, d\mathbf{x} = \int_\Omega \rho^0 f_i^0 \delta u_i\, d\mathbf{x} + \int_{\Gamma_\tau} P_i^0 \delta u_i\, ds \quad \text{(one equation)}, \tag{5.82a}$$

ε^1 terms:

$$\begin{aligned}\int_\Omega c_{ijkl}^0 u_{i,j}^{\prime r} \delta u_{k,l}\, d\mathbf{x} &= \int_\Omega (\rho^{\prime r} f_i^0 + \rho^0 f_i^{\prime r}) \delta u_i\, d\mathbf{x} \\ &+ \int_{\Gamma_\tau} P_i^{\prime r} \delta u_i\, ds - \int_\Omega c_{ijkl}^0 u_{i,j}^0 \delta u_{k,l}\, d\mathbf{x} \quad (K \text{ equations}),\end{aligned} \tag{5.82b}$$

ε^2 terms:

$$\begin{aligned}\int_\Omega c_{ijkl}^0 u_{i,j}^{\prime rs} S_b^{rs} \delta u_{k,l}\, d\mathbf{x} &= \int_\Omega (\rho^{\prime r} f_i^0 + 2\rho^{\prime r} f_i^{\prime s} + \rho^{\prime rs} f_i^0) S_b^{rs} \delta u_i\, d\mathbf{x} \\ &+ \int_{\Gamma_\tau} P_i^{\prime rs} S_b^{rs} \delta u_i\, ds - \int_\Omega (2c_{ijkl}^{\prime r} u_{i,j}^{\prime s} + c_{ijkl}^{\prime rs} u_{i,j}^0) S_b^{rs} \delta u_{k,l}\, d\mathbf{x}\end{aligned} \tag{5.82c}$$

[one equation for every (rs)]

where we denoted by $f_i' = \partial f / \partial x_i$ and so forth and where $b_r^0 = E[b_r]$ and $S_b^{rs} = \int_{-\infty}^{\infty} \Delta b_r(x_k) \Delta b_s(x_k) p_K(\mathbf{b}(x_k))\, d\mathbf{b}$.

Solving (5.82) for $u_i^0(x_k)$, $u_i^{\prime r}(x_k)$, and $u_i^{\prime rs}(x_k)$ we put first $\varepsilon = 0$, which means that we solve the deterministic case and we find the zero-order displacements $u_i^0(x_k)$. The higher-order terms $u_i^{\prime r}(x_k)$ and $u_i^{\prime rs}(x_k)$ can be evaluated from (5.82b,c). Putting $\varepsilon = 1$ yields that the fluctuation of the random field variables $\overline{\mathbf{b}}(x_k)$ is small. Then introducing (5.81e) into the expression for the mean value of the random displacement field $u_i[\mathbf{b}(x_k); x_k]$, we find [for more details see Kleiber and Hien (1992)]

$$E[u_i[\mathbf{b}(x_k); x_k]] = \int_{-\infty}^{\infty} u_i[\mathbf{b}(x_k); x_k] p_K(\mathbf{b}(x_k))\, d\mathbf{b}$$

and for its second-order estimate we find

$$E[u_i[\mathbf{b}(x_k); x_k]] = u_i^0(x_k) + \tfrac{1}{2} u_i^{\prime rs}(x_k) S_b^{rs} \quad [\text{summing over } (rs)]. \tag{5.83}$$

If only the first-order accuracy of the displacement estimation is required, then (5.83) is reduced to

$$E[u_i[\mathbf{b}(x_k); x_k]] = u_i^0(x_k).$$

The first-order accurate cross covariances of $u_i[\mathbf{b}(x_k^{(1)}); x_k^{(1)}]$ and $u_j[\mathbf{b}(x_k^{(2)}); x_k^{(2)}]$ can be evaluated by substituting the second-order expansion of the random displacement field $u_i[\mathbf{b}(x_k); x_k]$ into the expression for the cross covariance, and we obtain [for the derivation see Kleiber and Hien (1992)]

$$\mathrm{Cov}[u_i[\mathbf{b}(x_k^{(1)}); x_k^{(1)}], u_j[\mathbf{b}(x_k^{(2)}); x_k^{(2)}]] = S_u^{ij}(x_k^{(1)}, x_k^{(2)}) = u_i^{\prime r}(x_k^{(1)}) u_i^{\prime s}(x_k^{(2)}) S_b^{rs}.$$

From these known values we can estimate the strain and stress probabilistic characteristics [for more details see Kleiber and Hien (1992)]:
Second-order accurate mean value:

$$\begin{aligned} E[e_{ij}[\mathbf{b}(x_k); x_k]] &= E\left[\tfrac{1}{2}(u_{i,j}[\mathbf{b}(x_k); x_k] + u_{j,i}[\mathbf{b}(x_k); x_k])\right] \\ &= \tfrac{1}{2}\left[u_{i,j}^0(x_k) + u_{j,i}^0(x_k) + \tfrac{1}{2}\left(u_{i,j}^{\prime rs}(x_k) + u_{j,i}^{\prime rs}(x_k)\right) S_b^{rs}\right] \\ &= e_{ij}^0(x_k) + \tfrac{1}{2} e_{ij}^{\prime rs}(x_k) S_b^{rs}, \end{aligned}$$

in which the following expansion has been used

$$\begin{aligned} e_{ij}[\mathbf{b}(x_k); x_k] &= e_{ij}^0[\mathbf{b}^0(x_k); x_k] + e_{ij}^{\prime r}[\mathbf{b}^0(x_k); x_k]\Delta b_r(x_k) \\ &\quad + \tfrac{1}{2} e_{ij}^{\prime rs}[\mathbf{b}^0(x_k); x_k]\Delta b_r(x_k)\Delta b_s(x_k). \end{aligned}$$

First-order accurate cross covariance:

$$\mathrm{Cov}[e_{ij}[\mathbf{b}(x_k^{(1)}); x_k^{(1)}], e_{kl}[\mathbf{b}(x_k^{(2)}); x_k^{(2)}]] = S_e^{ijkl}(x_k^{(1)}, x_k^{(2)}) = e_{ij}^{\prime r}(x_k^{(1)}) e_{kl}^{\prime s}(x_k^{(2)}) S_b^{rs}$$

and similarly for $E[\tau_{ij}[\mathbf{b}(x_k); x_k]]$ and $\mathrm{Cov}[\tau_{ij}[\mathbf{b}(x_k^{(1)}); x_k^{(1)}], \tau_{kl}[\mathbf{b}(x_k^{(2)}); x_k^{(2)}]]$, where

$$\tau_{ij} = c_{ijkl} e_{kl} = (c_{ijkl}^0 + c_{ijkl}^{\prime r}\Delta b_r + \tfrac{1}{2} c_{ijkl}^{\prime rs}\Delta b_r \Delta b_s)(e_{kl}^0 + e_{kl}^{\prime u}\Delta b_u + \tfrac{1}{2} e_{kl}^{\prime uv}\Delta b_u \Delta b_v).$$

Remark 5.4 In Kleiber and Hien (1992) the dynamic problem is also analyzed and a code for the deterministic and stochastic analyses of static and dynamic special problems are given.

In many biomechanical problems the input data (physical coefficients, right-hand sides, boundary values, friction limits, etc.) are not determined uniquely but only in some intervals with the uncertainty determined by their measurement errors. If no probability density is available such problems can be solved by the worst scenario (antioptimization) approach. Then the reliable solution of the problem represents the worst case among a set of possible solutions, where the possibility will be given by uncertain input data and the degree of badness will be measured by a certain criterion functional. This problem is discussed in Chapter 6 [see also Hlaváček and Nedoma (2004)] for the case of a contact problem, but it can be also applied for this problem of a loaded long bone.

5.7 MATHEMATICAL MODEL OF HEAT GENERATION AND HEAT PROPAGATION IN THE NEIGHBORHOOD OF THE BONE CEMENT. PROBLEMS OF BONE NECROSIS

5.7.1 Introduction

Many materials used in orthopedic practice are absorbable or nonabsorbable polymers, metals, and ceramics. The absorbable biomaterials are made from synthetic polymers that fulfill their function within the body and then are absorbed. These materials were developed and used as sutures for the fixation of bone defects or fractures. A key consideration in the design and evaluation of an absorbable polymer is its degradation mechanism. Bulk degradation and surface erosion have occurred mostly with absorbable polymers. Bulk degradation is characterized by water entering the polymer bulk where hydrolysis causes degradation; the degradation products are carboxylic acids. Surface erosion occurs in cases where the polymer is significantly hydrophobic [see Middleton and Tipton (2000) and Gopferich (1996)].

Nonabsorbable biomaterials are used for implantations of hardware such as rods, pins, screws, plates, stems of joint total replacements, or injection of cements. The bone cement is made of the nondegradable polymer, that is, polymethylmethacrylate (PMMA). The bone cement is used as a filling material for the space between the total joint replacement and the bone. Surgeons also cement internally fractures with PMMA by an in situ polymerization reaction as in the case of total joint replacements. The mechanical properties of PMMA are sufficient to stand stresses of in vivo loads and to transmit stresses from the stem of the total joint replacement into the lower part of the bone. But there are also some disadvantages, caused by difficulties in controlling the temperature rise during exothermic polymerization. Temperatures at the bone–cement interface are between 84 and 125°C and result in cell necrosis (Huiskes, 1979, 1980; Nedoma et al., 2006; Schaldach and Hochmann, 1986).

The bone cement was first applied by Haboush (in 1951) and McKee and Sir John Charnley in 1958. Charnley succeeded in anchoring femoral component in the femur with in situ autopolymerization of PMMA. His studies represent new surgical techniques in orthopedics (Charnley, 1960, 1970). The bone cement must satisfy some requirements such as biocompatibility, chemical and physical stabilities, relatively low temperature of polymerization, an easy process of preparation, and suitable mechanical properties.

The bone cement used at present is a thermoplastic material from the group of synthetic organic polymers—PMMA—and as a group they are called acrylic-based cements. These cements have been used for more than 40 years. The bone cement serves as a filling between the high-modulus metallic implant and the low-stiffness natural bone and as transfer and distribution of static and cyclic loads due to daily activities.

PMMA is a two-components thermoplastic material, one of which is in a powder state and the second one in a liquid state. The powder polymer is of granularity 20–80 μm and it is composed of about 89% PMMA, about 0.5% of an initiator such as benzol peroxide, and about 10.5% of contrastive X-ray medium [zirconium dioxine (ZrO_2) or barium sulfate ($BaSO_4$)]. The liquid part is composed of about 93% of monomer (MMA), which has a pH 6.6, about 7% of activator -*N*-dimethyl-*p*-toluidine (DMPT), and of about 0.005% of inhibitor—hydroquinone. The antibiotic additives are also applied. The polymerization reaction is in progress for a relatively small time period in situ in the layer between implant and the bone. The temperature during the polymerization is of about 84 to 125°C. The process of polymerization is of an exothermic type. It is evident that the aim of the researchers in this field of sciences is to find the bone cement with a very small temperature of polymerization. The resulting heat is transported from a greater part into the implants, a smaller part is transported by the blood into greater distances from the origin of polymerization. Results of polarization are high temperatures at the interface of cement–bone, which is characterized by the bone necrosis, and the necrosis is about 0.5 mm thick. The tissue has a property of regeneration, so that during a certain time the bone regenerates. The time of polymerization is less than about 10 min. The situation after the total joint replacement is complicated because the heat is transported into greater parts of the skeleton, for example, in the case of the hip joint into the femur and the pelvis. From the rheological point of view the bone cement is a thermoviscoelastic up to a thermoviscoplastic material. But in this section we will study only its thermal properties [for more details see Serbetci and Hasirci (2004)].

The relation between a temperature increase during the polymerization process and the bone temperature depends on many factors. If the bone cement is isolated from the neighboring medium, then the growing capacity of temperature is given by $dT = C^{-1}Q$, where C is a heat capacity ($\mathrm{J\,m^{-3}\,°C}$), and Q is a heat generation per unit volume of acrylic cement ($\mathrm{J\,m^{-3}}$). If the bone cement is in a contact with a neighboring medium (bone, implant), thermal conduction originated during and at the end of the polymerization process. On the one hand the degree of heat generation in the bone cement depends on the degree of polymerization, while on the other hand a degree of thermal conduction depends on properties of heat conduction in the bone cement, the bone, and their common interface [see Jefferis et al. (1975), Schaldach and Hohmann (1976), Swenson et al. (1976), Swanson and de Groot (1985), and D'Souza et al. (1977)].

The theory is based on the following assumptions: (i) incompressibility of biomaterials; (ii) local heat decreasing on the system surface (i.e., on the outer bone surface) is realized by conduction and convection; (iii) thermal properties of bones, implants, and bone cement are isotropic; (iv) homogeneity of bone cements; (v) thermal conductivity and heat capacity are independent on a temperature (in real situations they depend on temperature); (vi) thermal sources are functions of time and independent of temperature (in real situations they depend on temperature); (vii) interfaces between bones and bone cement are assumed to be continuous and sufficiently smooth; and (viii) local heat transport across the bone–bone cement interfaces is proportional to the difference in temperature between both materials.

In the course of polymerization the PMMA powder component is changed into a PMMA matrix. This new biomaterial has properties of a solid material. Heat originates in the bone cement during the polymerization process owing to an exoteric reaction of manometers during polymerization. The total heat Q depends on monomer quantity and on the total quantity of the originated heat Q_t of a monomer related to a mass unit, that is,

$$Q = \nu_m \rho_m Q_t \quad (\mathrm{J\,m^{-3}}), \tag{5.84}$$

where ρ_m is the density of the monomer, $\rho_m \sim 0.94 \times 10^3\,\mathrm{kg\,m^{-3}}$ [see Charnley (1970)], ν_m is a volume fraction of the actually polymerizing monomer in the cement mixture,

$$\nu_m = \frac{\rho_m}{\rho_p + 10^3 P/L} + \nu_{rm} - \nu_{am}, \tag{5.85}$$

where ρ_p is the density of the polymer, $\rho_p \sim 1.18 \times 10^3\,\mathrm{kg\,m^{-3}}$ [see Charnley (1970)], P/L is the relation polymer/monomer and $\simeq 2$ and $\nu_{rm} - \nu_{am} \simeq 0.02$; in the case if a residual monomer and additive components to the liquid component each comprise approximately 3% of the liquid component. Due to Trommsdorf (1963) $Q_t \sim 5.4 \times 10^5\,\mathrm{J\,kg^{-1}}$. In Fig. 5.8 the relation P/L and ν_m is presented. The polymerization process is described by the function of polymerization $p[t)$ $[0 \le p(t) \le 1]$, which defines a part of their monomer taking place in the polymerization process at time t. The polymerization time τ (sec) is defined as the time from the beginning of the polymerization process until 75% of the monomer has polymerized. The beginning of the process is usually somewhat retarded by the stabilizer. This interval is called retardation time t_r (sec) and for the bone cement $t_r \sim 100$–200 sec and $\tau \sim 200$–400 sec. In the dimensionless form the function of polymerization is given in Fig. 5.9 and it is of the following form:

$$p_1 = p_1\left(t - \frac{t_r}{\tau}\right), \qquad t \ge t_r,$$

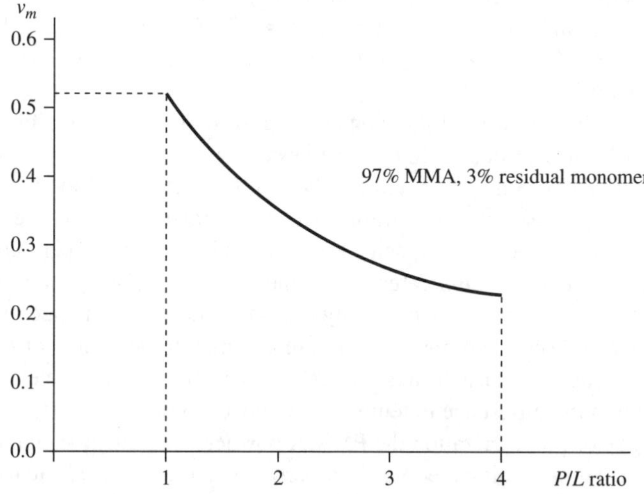

FIG. 5.8 Relation P/L and ν_m.

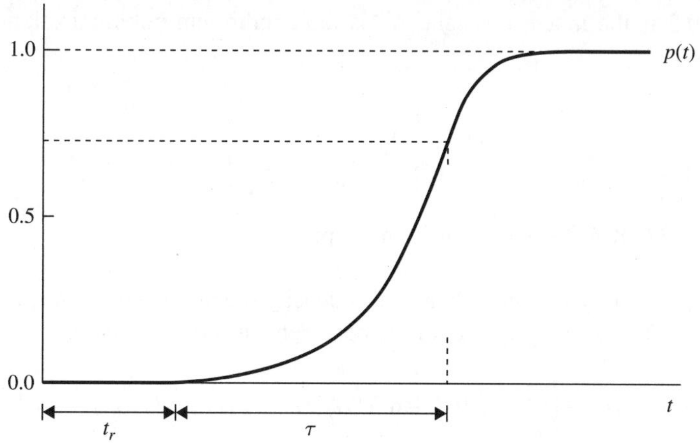

FIG. 5.9 The polymerization function $p(t)$ due to de Wijn [modified after de Wijn presented in Huiskes (1980)].

and it only depends on t_r and τ [see Huiskes (1980)]. The thermal sources $\Phi(t)$, the functions of heat generation, are proportional to the polymerization degree, that is,

$$\Phi(t) = \nu_m \rho_m Q_t \frac{dp(t)}{dt} \quad (\mathrm{J\,m^{-3}\,s}). \tag{5.86}$$

Remark 5.5 Mazzullo et al. (1991) and Stańczyk and Rietbergen (2004) define the thermal sources as

$$\Phi(\mathbf{x}, t) = \eta(\mathbf{x}, t)\frac{Q}{C(\mathbf{x})}\frac{\partial w(\mathbf{x}, t)}{\partial t},$$

where Q is the latent heat generated by cement polymerization, $\eta(\mathbf{x}) = 1$ in the bone cement and $\eta(\mathbf{x}) = 0$ otherwise, w is the polymerization fraction. Moreover,

$$\frac{\partial w(\mathbf{x}, t)}{\partial t} = a \exp\left(-\frac{E_a}{RT(\mathbf{x}, t)}\right) P(T(\mathbf{x}, t), w(\mathbf{x}, t)),$$

where $a = 2.6397 \times 10^8 (1/\mathrm{sec})$, $E_a = 62{,}866$ (J/mol), and R is the universal gas constant. The function $P(T, w)$ is defined as

$$P(T, w) = \begin{cases} \dfrac{\alpha}{w^*(T)} w^{\alpha - 1/\alpha}(w^*(T) - w)^{\alpha + 1/\alpha} & \text{if } w < w^*(T), \\ 0 & \text{if } w \geq w^*(T), \end{cases}$$

here $\alpha = 9.2$ is the model constant, w^* is the equilibrium polymerization fraction defined as

$$w^* = \begin{Bmatrix} \dfrac{T}{T_g} & \text{if } T \leq T_g, \\ 1 & \text{if } T > T_g, \end{Bmatrix}$$

where $T_g = 378$ K is the glass transition temperature.

The heat capacity $C = \rho c$, where ρ is a density (kg m^{-3}) and c is a specific heat (J kg^{-1} °C). The thermal capacity of the bone cement is defined as

$$C_c(t) = \rho_c c_c(t) \sim [1.25 \times 10^3 + 6.5T(t)]\,[\rho_{p\_}v_m(1 - p(t))\,(\rho_m - \rho_p)] \qquad (5.87)$$

where C_c is the heat capacity of the bone cement, ρ_c (kg m^{-3}) is the density of the bone cement, and c_c is its specific heat (J kg^{-1} °C). In general $\rho_c c_c$, κ_c, where κ_c is a thermal conductivity of the bone cement, depend on temperature T and on phase transitions taking place during the polymerization process. The phase transition problems are problems with free boundary between phases, and they are known as problems of the Stephan type.

If the bone cement polymerizes adiabatically (i.e., ideally isolated), temperature $T(t)$ is described by

$$C_c \frac{\partial T}{\partial t} = Q \frac{\partial p}{\partial t}, \qquad (5.88)$$

hence

$$T(t) = Q C_c^{-1} p(t) + T_{c0}, \qquad (5.89)$$

where C_c is a heat capacity, $T_{c0} \sim 25$°C. Due to Huiskes (1980) $C_c \sim 1.9 \times 10^6$ J m^{-3} °C. In Table 5.1 [after Huiskes (1980) and Nedoma et al. (2006)] the values of densities, specific heats, heat capacities, and thermal conductivities for the cancellous and compact bones, materials of implants, and the bone cement are presented. In Table 5.2 [after Huiskes (1980) and Nedoma et al. (2006)] some values for determination of thermal sources, where indices c, bk, cs denote quantities corresponding to the bone cement, compact bone and cancellous bone, T_a is a neighboring temperature, and T_{c0}, T_{i0} are initial temperatures in the bone cement and implants.

5.7.2 Mathematical Models and Their Solutions

Linear Thermal Models

A. Heat Distribution in Closed Neighborhood of the Polymerizing Bone Cement Let us assume that the acetabulum is cemented by the bone cement into the pelvis and the stem of the femoral prosthesis by the bone cement fixed into the

TABLE 5.1 Material Data of Bones and Implants

Material	Density ρ ($kg\,m^{-3}$)	Specific Heat c ($J\,kg^{-1}\,°C$)	Heat Capacity C ($J\,m^{-3}\,°C$)	Thermal Conductivity κ ($W\,m^{-1}\,K^{-1}$)
PMMA	1.19×10^3	1.60×10^3 (50°C)	1.9×10^6 (50°C)	0.17
$BaSO_4$	4.50×10^3	0.43×10^3 (50°C)	1.96×10^6 (50°C)	
ZrO_2	5.60×10^3	0.70×10^3	3.91×10^6	1.5
Al_2O_3	3.69–3.94 $\times 10^3$	0.75–0.79 $\times 10^3$		20–30
MMA	0.94×10^3	1.60×10^3 (50°C)	1.50×10^3 (50°C)	
Acryllic cement	1.10×10^3	1.70×10^3	1.8–2 $\times 10^6$	0.14–0.2
Cr–Ni steel	7.80×10^3	0.46×10^3	3.6×10^6	14
HDPE	0.96×10^3	2.22×10^3	2.06×10^6	0.29
Cancellous bone	2.1×10^3	1.15–1.73 $\times 10^3$	2.64–3.64 $\times 10^6$	0.36–0.6
Compact bone	2.1–2.9 $\times 10^3$	1.26×10^3	2.2–3.2 $\times 10^6$	0.2–0.6

TABLE 5.2 Data for Determination Thermal Sources

Parameter	Unit	Probable Value	Interval
ρ_m	$kg\,m^{-3}$	0.94×10^3	
Q_t	$J\,kg^{-1}$	5.4×10^5	
ν_m		0.35	0.27–0.42
Q	$J\,m^{-3}$	1.8×10^8	1.4–2.1 $\times 10^8$
t_r	sec	150	100–200
τ	sec	300	200–400
C_c	$J\,m^{-3}\,°C$	1.8–2 $\times 10^6$	1.7–2.1 $\times 10^6$
κ_c	$J\,m^{-1}$ sec °C	0.17	0.14–0.2
C_{bk}	$J\,m^{-3}\,°C$	2.2–3.2 $\times 10^6$	1.8–4.1 $\times 10^6$
κ_{bk}	$J\,m^{-3}\,°C$	0.3–0.5	0.2–0.6
C_{cs}	$J\,m^{-3}\,°C$	2.8–3.4 $\times 10^6$	2.6–3.6 $\times 10^6$
κ_{cs}	$J\,m^{-1}$ sec °C	0.5	0.4–0.6
β_{ci}	$J\,m^{-2}$ sec °C	1000–10,000	500– ∞
β_{cb}	$J\,m^{-2}$ sec °C	100–1000	50– ∞
α	$J\,m^{-2}$ sec °C	100–500–10,000	100– ∞
$T_\alpha = T_{b0}$	°C	32–34	30–37
T_{cb}	°C	20–25	
T_{i0}	°C	20–25	

femur, and that the total hip arthroplasty (THA) occupies a region $\Omega = \cup_{\iota=1}^{s} \Omega^\iota$ in $\mathbb{R}^N$, $N = 2, 3$, where Ω^ι denote the pelvis (Ω^1), the acetabulum (Ω^2), the femoral part of the total prosthesis (Ω^3), the bone cement in the neighborhoods of acetabulum (Ω^4), and of the femoral part of THA (Ω^5). Its boundary we denote by $\partial\Omega = \overline{\Gamma}_u \cup \overline{\Gamma}_\tau$, where $\Gamma_u = {}^1\Gamma_u \cup {}^2\Gamma_u$, $\Gamma_\tau = {}^1\Gamma_\tau \cup {}^2\Gamma_\tau$. The boundaries between subregions Ω^ι, for example, between the artificial acetabulum (Ω^k) and the joint head (Ω^l), and so forth, we denote as Γ^{kl}, $\Gamma^{kl} = \partial\Omega^k \cap \partial\Omega^l$, $k \neq l$. The time interval $I = (t_0, t_p)$, $t_p > 0$. Distribution of the temperature field $T(\mathbf{x}, t)$ satisfies the heat equation

$$C(\mathbf{x})\frac{\partial T(\mathbf{x},t)}{\partial t} - \frac{\partial}{\partial x_i}\left(\kappa_{ij}(\mathbf{x})\frac{\partial T(\mathbf{x},t)}{\partial x_j}\right) = \Phi(\mathbf{x},t),$$
$$i,j = 1,\ldots,N, N = 2,3, \quad \text{in } \Omega_t = \Omega \times I, \tag{5.90}$$

where $C(\mathbf{x})$ is a heat capacity, defined above, κ_{ij} is a thermal conductivity (for isotropic biomaterials $\kappa_{ij} = \kappa\delta_{ij}$, where δ_{ij} is the Kronecker symbol), $\Phi(\mathbf{x}, t)$ are thermal sources. The temperature on the pelvis and femur surfaces is given by the temperature of a human body T_a. Then on the boundary $\partial\Omega$ of the region Ω we prescribe the condition

$$T(\mathbf{x},t) = T_a(\mathbf{x},t) \quad \text{on } \partial\Omega \times I, \tag{5.91}$$

where T_a acquires different values on ${}^1\Gamma_u$, ${}^2\Gamma_u$, ${}^1\Gamma_\tau$, and ${}^2\Gamma_\tau$.

On the interfaces Γ^{kl} the temperature and the heat flux are continuous for all $t \in I$, that is,

$$T^k = T^l \qquad \kappa_{ij}\frac{\partial T(\mathbf{x},t)}{\partial x_j}n_i|_{(k)} = -\kappa_{ij}\frac{\partial T(\mathbf{x},t)}{\partial x_j}n_i|_{(l)}, \tag{5.92}$$

where $\mathbf{n} = (n_i)$ is the unit outward normal to the boundary $\partial\Omega^k$ or $\partial\Omega^l$, respectively. In the initial time t_0 the initial condition

$$T(\mathbf{x}, t_0) = T_0(\mathbf{x}) \tag{5.93}$$

is prescribed.

In this model the thermal sources will be evoked by the polymerization process, where $\Phi(\mathbf{x}, t)$ is given by (5.86). Our goal is to determine the temperature T in the time t, which is propagated from the place of polymerization of the bone cement. Since the assumed physical parameters are independent of the temperature, the model problem is linear. The mathematical analysis for the case, if thermal sources changed very slowly, is parallel to that of the thermal part for a quasi-static case discussed in Chapter 6, and, therefore, it is not necessary to present it here.

Next, we will investigate the case.

We say that the operator $C(\partial/\partial t) + L$ is of a parabolic type if L, defined by $LT = -(\partial/\partial x_i)[\kappa_{ij}(\mathbf{x})\partial T(\mathbf{x}, t)/\partial x_j)]$, is a second-order differential operator of an elliptic type, that is, if it satisfies

$$\sum_{i,j=1}^{N} \kappa_{ij}(\mathbf{x})\xi_i\xi_j \geq c_0|\xi^2|, \quad \forall\xi \in \mathbb{R}^N, \qquad N = 2,3, \quad \text{a.e } \mathbf{x} \in \Omega, c_0 = \text{const} > 0.$$

Let $H^1(\Omega^\iota)$, $H_0^1(\Omega^\iota)$ be the Sobolev spaces in the usual sense and let

$$V = \{z|z \in \prod_{\iota=1}^{s} H^1(\Omega^\iota),\ z = 0 \quad \text{on } \Gamma_u,\ z^k = z^l \quad \text{on } \cup\Gamma^{kl}\},$$

then $\prod_{\iota=1}^{s} H_0^1(\Omega^\iota) \subset V \subset \prod_{\iota=1}^{s} H^1(\Omega^\iota)$. Let us denote by $L^2(I; V)$ the space

$$L^2(I; V) = \left\{ z | z : I \to V | z \text{ is measurable and } \int_I \|z(t)\|_1^2 \, dt < \infty \right\},$$

where $\|z\|_1^2 = \sum_{\iota=1}^{s} \|z^\iota\|_{1,\Omega^\iota}^2$, and similarly other functional spaces for time–space functions.

To formulate the model problem in its weak (variational) formulation, we multiply (5.90) by $z \in V$, integrate over Ω, apply the Green's theorem, and the boundary conditions, and then we obtain the following problem:

Problem $(\mathcal{P})_v$ Given $\Phi(\mathbf{x}, t) \in L^2(\Omega_t)$, $\Omega_t = \Omega \times I$, $T_0(\mathbf{x}) \in L^2(\Omega)$, $\kappa_{ij} \in L^\infty(\overline{\Omega})$, $\rho, c \in \prod_{\iota=1}^{s} C(\overline{\Omega}^\iota)$. Find $T \in L^2(I; V) \cap C^0(I; L^2(\Omega))$ such that

$$\begin{aligned} \left(\rho c \frac{\partial T(t)}{\partial t}, z \right) + a(T(t), z) = (\Phi(t), z) \quad \forall z \in V, \\ T(\mathbf{x}, t_0) = T_0(\mathbf{x}), \end{aligned} \tag{5.94}$$

where $(\cdot, \cdot)$ denotes the scalar product in $L^2(\Omega)$ and $a(\cdot, \cdot)$ is the bilinear form defined by

$$\begin{aligned} a(T(t), z) &= \int_\Omega \kappa_{ij}(\mathbf{x}) \frac{\partial T(t)}{\partial x_j} \frac{\partial z}{\partial x_i} \, d\mathbf{x}, \\ \left(\rho c \frac{\partial T(t)}{\partial t}, z \right) &= \int_\Omega \rho c \frac{\partial T(t)}{\partial t} z \, d\mathbf{x}, (\Phi(t), z) = \int_\Omega \Phi(t) z \, d\mathbf{x}. \end{aligned} \tag{5.95}$$

The existence and uniqueness of the solution is given by next theorem:

Theorem 5.3 Let the bilinear form $a(\cdot, \cdot)$ be continuous on $V \times V$ and coercive, that is,

$$a(z, z) \geq c\|z\|_1^2 \quad \forall z \in V. \tag{5.96}$$

Let $\Phi(\mathbf{x}, t) \in L^2(\Omega_t)$, $T_0(\mathbf{x}) \in L^2(\Omega)$, $\kappa_{ij}(\mathbf{x}) \in L^\infty(\overline{\Omega})$, $\rho(\mathbf{x}), c(\mathbf{x}) \in C(\overline{\Omega})$. Then there exists a unique solution $T \in L^2(I; V) \cap C^0(I; L^2(\Omega))$ of (5.94). Moreover, $\partial T / \partial t \in L^2(I; V')$ and

$$\max_{t \in I} \|T(t)\|_0^2 + c \int_I \|T(t)\|_1^2 \, dt \leq \|T(t_0)\|_0^2 + c^{-1} \int_I \|\Phi(t)\|_1^2 \, dt, \tag{5.97}$$

the energy estimate, holds.

The proof is based on the Faedo–Galerkin method, and it is parallel to that of Quarteroni and Valli (1991) and Nedoma (1998a).

Semidiscrete Approximation and the FEM Method Numerically, the problem can be solved by finite difference, finite element, and finite volume methods, a wavelet method, the Faedo–Galerkin method, operator-spliting methods, decomposition methods, a spectral collocation method and their modifications, explicit, semi-implicit, and fully implicit discrete approximations, as well as the method of discretization in time, known also as the Rothe method [see Wheeler (1973), Quarteroni and Valli (1991), Ženíšek (1990), Rektorys (1983), Raviart and Thomas (1983), Křížek and Neittaanmäki (1990), Kačur (1985), Nedoma (1998a), and Zienkiewicz and Taylor (2000)].

The variational formulation of the linear thermal problem $(\mathcal{P})_v$ leads to a semidiscrete problem by approximating the space V by a finite dimensional space V_h. We will use the Faedo–Galerkin method, consisting of choosing the approximate space $V_h = V_n = \text{span}\{w_1, \ldots, w_n\}$, where $\{w_i\}$ represents a complete orthogonal basis of the space of virtual temperatures V. The space V_h can be constructed as

$$V_h = \{z_h | z_h \in X_h(\Omega), z_h = 0 \text{ on } \Gamma_u\},$$
$$X_h = \{z_h \in C^0(\overline{\Omega}); z_h|_{\mathbb{T}_{hi}} \in P_1 \ \forall \mathbb{T}_{hi} \in \mathcal{T}_h\},$$

where X_h is the space of triangular finite elements in the 2D case and of tetrahedra elements in the 3D case, and $\mathbb{T}_{hi}$ is the triangular element in the 2D case or the tetrahedra element in the 3D case of the used partition. Then problem $(\mathcal{P})_v$ in its semidiscrete version leads to the following problem:

Problem $(\mathcal{P})_{vh}$ Find $T_h(t) \in V_h$ for every $t \in I$, such that

$$\frac{d}{dt}(\rho c T_h(t), z_h) + a(T_h(t), z_h) = (\Phi_h(t), z_h) \quad \forall z_h \in V_h, t \in I,$$
$$T_h(\mathbf{x}, t_0) = T_{0h}(\mathbf{x}), \tag{5.98}$$

where $\Phi_h(t) \in L^2(\Omega_t)$, $T_{0h}(\mathbf{x}) \in V_h$ represent suitable approximations.

Next, we will assume that the domains Ω^ι are polygonal domains and their boundaries $\partial\Omega^\iota$ are Lipschitzian. Let us put

$$T_h(t) = \sum_{i=1}^{n(h)} \alpha_i(t) w_i(x), \tag{5.99}$$

where $\{w_i\}$, $i = 1, \ldots, n(h)$, is a basis of V_h, and $T_{0h} = \sum_{i=1}^{n(h)} \alpha_{0i} w_i$. Then Problem $(\mathcal{P})_{vh}$ leads to a system of ordinary differential equations

$$C\alpha'(t) + A\alpha(t) = B(t),$$
$$\alpha(t_0) = \alpha_0, \tag{5.100}$$

where $C = (C_{ij})$, $C_{ij} = (\rho c w_i, w_j)$, $A = (A_{ij})$, $A_{ij} = a(w_i, w_j)$, $B = (B_i)$, $B_i = (\Phi_h(t), w_j)$, $i, j = 1, \ldots, n(h)$. The matrices C and A are positive definite. Then there exists a unique solution $\alpha(t)$ of (5.100).

By a similar technique to that of the proof of Theorem 5.3 we can find the similar energy estimate of (5.97) with the property that T_{h0} converges to T_0 in $L^2(\Omega)$, which proves the stability of the method.

The following theorem gives a convergence of $T_h(t)$ to $T(t)$:

Theorem 5.4 Let $\{\mathcal{T}_h\}$ be a regular family of partitions. Let the bilinear form $a(\cdot,\cdot)$ be coercive in V and let $\Phi(t) \in L^2(\Omega_t)$, $T_0 \in V$, $\kappa_{ij} \in C^1(\overline{\Omega})$, $\rho, c \in C(\overline{\Omega})$. Then the solutions T and T_h of Problem $(\mathcal{P})_v$ and Problem $(\mathcal{P})$, respectively, satisfy

$$\max_{t\in I} \|T(t) - T_h(t)\|_0^2 + c\int_I \|T(t) - T_h(t)\|_1^2\, dt$$

$$\leq \|T_0 - T_{0h}\|_0^2 + c_1(c,\gamma)h^2(\|T_{0h}\|_1^2 + \|T_0\|_1^2 + \int_I \|\Phi(t)\|_0^2\, dt), \quad (5.101)$$

where c is the coerciveness constant in (5.96), γ is a constant of continuity of the bilinear form $a(\cdot,\cdot)$, and $c_1(c,\gamma) = \text{const} > 0$ is a constant independent of h. Moreover, T_{0h} can be taken such that $\|T_{0h}\|_1 \leq \|T_0\|_1$.

The proof is parallel to that of Quarteroni and Valli (1991) and Nedoma (1998a). If $T_{0h} \in V_h$ is taken in such a way that

$$\|T_0 - T_{0h}\|_0 + h\|T_0 - T_{0h}\|_1 \leq ch\|T_0\|_1,$$

then the error

$$\|T - T_h\| = O(h) \quad (5.102)$$

in the space $C^0(I;L^2(\Omega)) \cap L^2(I;H^1(\Omega))$, where [see Quarteroni and Valli (1991)]

$$\|\cdot\| = \|\cdot\|_{C^0(I;L^2(\Omega))} + \|\cdot\|_{L^2(I;H^1(\Omega))}.$$

Wheeler (1973) found the L^2–error estimate. She assumed that the bilinear form was symmetric and weakly coercive, that is, that $a(y,z) = a(z,y)$, $a(z,z) + \lambda\|z\|_0^2 \geq c\|z\|_1^2$ for all $z \in V$ (the Gårding inequality), $T_0 \in H^{k+1}\Omega$, $k \geq 1$ and the solution of Problem $(\mathcal{P})_v$ is such that $T' \in L^1(I;H^{k+1}(\Omega))$. Then, if the finite elements of V_h are defined by piecewise polynomials of degrees less than or equal to k, $\forall t \in I$ the solution T_h satisfies

$$\|T(t) - T_h(t)\|_0 \leq \|T_0 - T_{0h}\|_0 + ch^{k+1}\left(\|T_0\|_{k+1} + \int_{t_0}^t \|T'(\tau)\|_{k+1}\, d\tau\right), \quad (5.103)$$

where $c = \text{const} > 0$ independent of h [see Wheeler (1973)].

The L^∞ error estimate was studied by Schatz et al. (1980). They assumed that the bilinear form $a(\cdot,\cdot)$ was symmetric and weakly coercive and $T_0, T(t)$ sufficiently regular. Then for every $t \in I$ the error estimate

$$\|T(t) - T_h(t)\|_{L^\infty(\Omega)} \leq ch^2|\ln h|^2\left(\|T_0\|_{W^{2,\infty}(\Omega)} + \int_{t_0}^t \|T(\tau)\|_{W^{2,\infty}(\Omega)}\, d\tau\right) \quad (5.104)$$

holds.

Fully Discrete Approximation and the FEM Method The weak formulation of Problem $(\mathcal{P})_v$ needs to be discretized with respect to both the time and space variables. In the previous section we discussed the Faedo–Galerkin method, where $T_h(\mathbf{x}, t)$ was approximated by (5.99), which leads to a semidiscrete (or continuous-in-time) approximation. Next, we use the following approximation:

Let a uniform mesh for the time variable t be defined with $\Delta t = (t_p - t_0)/n$, n being integer, as the time step. Then $t_i = i\,\Delta t$, $i = 0, 1, \ldots, n$. Let us apply the Θ – scheme to Problem $(\mathcal{P})_{vh}$; then we have the following problem:

Problem $(\mathcal{P}\Theta)_{vh}$ Find $T_h^i \in V_h$ such that $T_h^0 = T_{0h}$ and

$$\begin{aligned}&\rho c\,\Delta t^{-1}(T_h^{i+1} - T_h^i, z_h) + a(\Theta T_h^{i+1} + (1-\Theta)T_h^i, z_h)\\&\quad = (\Theta\Phi_h(t_{i+1}) + (1-\Theta)\Phi_h(t_i), z_h), \quad \forall z_h \in V_h\end{aligned} \tag{5.105}$$

for every $i = 0, 1, \ldots, n-1$, and where we assume that Φ_h is everywhere defined on $\overline{I} = [t_0, t_p]$, $\Theta \in [0, 1]$.

Let us put

$$T_h^{i+1} = \sum_{j=1}^{n(h)} \alpha_j^{i+1} w_j, \tag{5.106}$$

where w_j are the basis functions of V_h, then at each time step we solve the linear system

$$(C + \Theta\,\Delta t\,A)\alpha^{i+1} = \mathbf{b}^i, \tag{5.107}$$

where $\mathbf{b}^i$ is determined from the previous steps, the matrix $C = (C_{ij})$, $C_{ij} = (\rho c w_i, w_j)$, $A = (A_{ij})$, $A_{ij} = a(w_i, w_j)$. Assuming that the form $a(\cdot,\cdot)$ is coercive, the matrix $(C + \Theta\,\Delta t\,A)$ is positive definite, so that the linear system (5.107) has a unique solution. Moreover, $(C + \Theta\,\Delta t\,A)$ is symmetric as the bilinear form $a(\cdot,\cdot)$ is symmetric, that is, $a(T, z) = a(z, T)\ \forall T, z \in V$. For $\Theta = 0$ or $\Theta = 1$ the scheme (5.105) is known as the forward Euler method or backward Euler method (or as semi-implicit method, respectively). For $\Theta = \frac{1}{2}$ is known as the Crank–Nicolson method. For $\Theta \geq \frac{1}{2}$ the scheme is A-stable[1] [see Lambert (1991, p. 244)] and $\frac{1}{2} \leq \Theta \leq 1$ is unconditionally stable with respect to the $L^2(\Omega)$ norm and convergent.

Theorem 5.5 Let us assume that the bilinear form $a(\cdot,\cdot)$ is coercive, that is, $a(z, z) \geq c_0\|z\|^2\ \forall z \in V$, $c_0 > 0$, and that the mapping $t \to \|\Phi(t)\|_0$ is bounded in

[1] A scheme is said to be absolutely stable for an assigned value $\mu\,\Delta t$ if the approximate solution y^n of $dy(t)/dt = \mu y(t)$, $\mu \in C$, generated by a discrete scheme vanishes as $n \to \infty$. The region of absolute stability is the set of all $\mu\,\Delta t$ for which the method is absolute stable. A scheme is called A-stable if its region of absolute stability includes the region $\{z \in C|\mathrm{Re}\,z < 0\}$.

$\bar{I} = [t_0, t_p]$. Then if $0 \leq \Theta < \frac{1}{2}$, assume that $\mathcal{T}_h$ creates a quasi-uniform family of triangulations[2] and the restriction on the time step

$$\Delta t(1 + ch^{-2}) \leq 2c_0(1 - 2\Theta)^{-1}c_1^{-2}, \tag{5.108}$$

the condition of stability, holds, where $c > 0$ is a constant for which $\|\nabla T_h\|_0^2 \leq ch^{-2}\|T_h\|_0^2$, c_0, c_1 are the coerciveness and continuity constants of the bilinear form $a(\cdot, \cdot)$. Then T_h^i defined by (5.105) satisfies

$$\|T_h^i\|_0 \leq c_\Theta \left(\|T_{0h}\|_0 + \max_{t \in I} \|\Phi(t)\|_0 \right), \qquad i = 0, 1, \ldots, n, \tag{5.109}$$

where $c_\Theta > 0$ is a nondecreasing function of c_0^{-1}, c_1, and $\mathbb{T}_{hi}$ and independent of i, Δt, and h.

The proof is parallel to that of Quarteroni and Valli (1991) and Nedoma (1998a).

The next theorem yields a convergence between the semidiscrete $T_h(t_i)$ and the fully discrete one T_h^i approximations for a fixed h.

Theorem 5.6 Let us assume that the bilinear form $a(\cdot, \cdot)$ is coercive and that $\partial T_h(t_0)/\partial t \in L^2(\Omega)$, $\Phi \in L^2(\Omega_t)$ with $\partial\Phi/\partial t \in L^2(\Omega_t)$. When $0 \leq \Theta < \frac{1}{2}$, assume that T_h is a quasi-uniform family of triangulation and that for the time step (5.108) holds. Then

$$\|T_h^i - T_h(t_i)\|_0 \leq c_\Theta\, \Delta t \left(\left\| \frac{\partial T_h(t_0)}{\partial t} \right\|_0^2 + \int_I \left\| \frac{\partial \Phi(\tau)}{\partial t} \right\|_0^2 d\tau \right)^{1/2}, \quad i = 0, 1, \ldots, n(h), \tag{5.110}$$

Let $\Theta = \frac{1}{2}$, $\partial^2\Phi/\partial t^2 \in L^2(\Omega_t)$, $\partial^2 T_h(t_0)/\partial t^2 \in L^2(\Omega)$, then

$$\|T_h^i - T_h(t_i)\|_0 \leq c_\Theta\, (\Delta t)^2 \left(\left\| \frac{\partial^2 T_h(t_0)}{\partial t^2} \right\|_0^2 + \int_I \left\| \frac{\partial^2 \Phi(\tau)}{\partial t^2} \right\|_0^2 d\tau \right)^{1/2}, \quad i = 0, 1, \ldots, n(h), \tag{5.111}$$

where $c_\Theta > 0$ is a nondecreasing function of c_0^{-1}, c_1, $\mathbb{T}_{hi}$, and independent of $n(h)$, Δt, and h.

The proof is similar to that of Quarteroni and Valli (1991) and Nedoma (1998a).

[2] A family of partitions $\mathcal{T}_h$, $h > 0$, is said to be quasi-uniform (strongly regular) if it is regular, and moreover, there exists a constant $c > 0$ such that $\min_{\mathbb{T}_{hi} \in \mathcal{T}_h} h_{\mathbb{T}_{hi}} \geq ch\, \forall h$.

Assuming the approximation property (Quarteroni and Valli, 1991, pp. 376–377)

$$\left\| z - \Pi^k_{1,h}(z) \right\|_0 + h \left\| z - \Pi^k_{1,h}(z) \right\|_1 \le c h^{k+1} |z|_{k+1} \quad \forall z \in H^{k+2}(\Omega),$$

where $\Pi^k_{1,h}(z)$ is the elliptic projection operator defined as

$$\Pi^k_{1,h}(z) \in V_h\colon\ a\left(\Pi^k_{1,h}(z), z_h\right) = a(z, z_h) \quad \forall z_h \in V_h, \tag{5.112}$$

then we have the following convergence result:

Theorem 5.7 Let the bilinear form $a(\cdot,\cdot)$ be symmetric and coercive. Let $T_0 \in H^{k+1}(\Omega)$ and let the solution T of Problem $(\mathcal{P})_v$ satisfy $\partial T/\partial t \in L^1(I; H^{k+1}(\Omega))$. When $0 \le \Theta < \frac{1}{2}$ let the time step Δt satisfy $\Delta t h^{-2} \le c_2/1 - 2\Theta$, where c_2 is a constant independent of h and Δt. Then T^i_h defined in Problem $(\mathcal{P}_\Theta)_{vh}$ satisfies

$$\|T^i_h - T(t_i)\|_0 \le \|T_{0h} - T_0\|_0 + c h^{k+1} \left(|T_0|_{k+1} + \int_0^{t_i} \left| \frac{\partial T(\tau)}{\partial \tau} \right|_{k+1} d\tau \right)$$
$$+ c_\Theta \,\Delta t \int_{t_0}^{t_i} \left\| \frac{\partial^2 T(\tau)}{\partial \tau^2} \right\|_0 d\tau, \qquad i = 0, 1, \ldots, n(h) \tag{5.113}$$

and for $\Theta = \frac{1}{2}$ under the additional assumption that $\partial^3 T/\partial t^3 \in L^1(I; L^2(\Omega))$ then for every $i = 0, 1, \ldots, n(h)$ we have

$$\|T^i_h - T(t_i)\|_0 \le \|T_{0h} - T_0\|_0 + c_3 h^{k+1} \left(|T_0|_{k+1} + \int_0^{t_i} \left| \frac{\partial T(\tau)}{\partial \tau} \right|_{k+1} d\tau \right)$$
$$+ \frac{(\Delta t)^2}{8} \int_{t_0}^{t_i} \left\| \frac{\partial^3 T(\tau)}{\partial \tau^3} \right\|_0 d\tau, \qquad i = 0, 1, \ldots, n(h), \tag{5.114}$$

where $c_3 > 0$ is independent of n, Δt, and h.

Remark 5.6 Other estimates can be found, for example, in Raviart and Thomas (1983) and Thomée (1984), and other numerical methods based on Galerkin and spectral FE methods can be found in Douglas and Dupont (1970), Bramble, Thomée (1974), Hackbush (1981), Dupont (1982), Baiocchi and Brezzi (1983), Baker et al. (1977), and Quarteroni and Valli (1991).

B. Heat Distribution in the Neighborhood of the Polymerizing Bone Cement with Draining It of Blood Let the region be defined as in the previous case. We will assume that on the part of the boundary ${}^1\Gamma_u$ the temperature of the human body is prescribed and on the boundary ${}^2\Gamma_u \cup {}^1\Gamma_\tau \cup {}^2\Gamma_\tau$ we will assume that temperature losses are observed. Then temperature losses originated by deliveries of heat by the blood, by the neighboring muscular and ligament, tissues, and the like. This loss of temperature, which we denote by α, depends on the surface conduction and convection coefficients on the boundary ${}^2\Gamma_u \cup {}^1\Gamma_\tau \cup {}^2\Gamma_\tau$. Experiments estimate its value as $100 < \alpha < 500\,\mathrm{J\,m^{-2}\,sec\,°C}$, but it can be greater than $500\,\mathrm{J\,m^{-2}\,sec\,°C}$ (Huiskes, 1980).

Distribution of a field of temperatures satisfies the linear heat equation

$$C(\mathbf{x},t)\frac{\partial T(\mathbf{x},t)}{\partial t} - \frac{\partial}{\partial x_i}\left(\kappa_{ij}(\mathbf{x},t)\frac{\partial T(\mathbf{x},t)}{\partial x_j}\right) = \Phi(\mathbf{x},t),$$
$$i,j = 1,\ldots,N,\ N = 2,3,\ \text{ in } \Omega_t = \Omega \times I. \tag{5.115}$$

The temperature on the boundary ${}^1\Gamma_u$ is determined as the temperature of a human body, that is,

$$T(\mathbf{x},t) = T_a(\mathbf{x},t) \quad \text{on} \quad {}^1\Gamma_u \times I. \tag{5.116}$$

On the boundaries ${}^2\Gamma_u \cup {}^1\Gamma_\tau \cup {}^2\Gamma_\tau$ different thermal losses are characterized by different changes of thermal flows, depending on an outer temperature of a human body and interface conductivity coefficients across the boundaries ${}^2\Gamma_u \cup {}^1\Gamma_\tau \cup {}^2\Gamma_\tau$, that is,

$$\kappa_{ij}(\mathbf{x},t)\frac{\partial T(\mathbf{x},t)}{\partial x_j}n_i = -\beta(\mathbf{x})\,(T(\mathbf{x},t) - T_a(\mathbf{x},t)) \equiv \varphi(\mathbf{x},t,T)$$
$$\text{on } {}^2\Gamma_u \cup {}^1\Gamma_\tau \cup {}^2\Gamma_\tau \times I, \tag{5.117}$$

where $\mathbf{n} = (n_i)$, $i = 1,\ldots,N$, $N = 2,3$, is an outward normal vector to the boundary ${}^2\Gamma_u \cup {}^1\Gamma_\tau \cup {}^2\Gamma_\tau$, β is an interface conductivity coefficient, T_a is a temperature of the neighborhood, that is, the temperature of a human body ~35–37°C.

On the inner boundaries Γ^{kl} the temperature and heat flux are continuous, that is,

$$\begin{aligned} &T^k(\mathbf{x},t) = T^l(\mathbf{x},t), \\ &\kappa_{ij}(\mathbf{x},t)\frac{\partial T(\mathbf{x},t)}{\partial x_j}n_i|_{(k)} = -\kappa_{ij}(\mathbf{x},t)\frac{\partial T(\mathbf{x},t)}{\partial x_j}n_i|_{(l)} \quad \text{on } \Gamma^{kl} \times I, \end{aligned} \tag{5.118}$$

where $\mathbf{n} = (n_i)$ is a unit outward normal to Ω^k or Ω^l, respectively.

In the initial time t_0 the initial condition

$$T(\mathbf{x},t) = T_0(\mathbf{x}) \tag{5.119}$$

is given.

Nonlinear Thermal Models

Evolution of Nonlinear Thermal Model of the Polymerizing Bone Cement with Draining by the Blood Let us assume that the part of the human skeleton with the artificial joint replacement occupies a region $\Omega = \cup_{\iota=1}^{s} \Omega^{\iota}$ in $\mathbb{R}^N$, $N = 2, 3$. About its boundary $\partial\Omega$ we assume that $\partial\Omega = \cup_{\iota=1}^{s} \partial\Omega^{\iota}$, and every $\partial\Omega^{\iota}$ is Lipschitzian and that $\partial\Omega = \overline{\Gamma}_u \cup \overline{\Gamma}_\tau \cup (\cup_{k,l} \overline{\Gamma}^{kl})$, where $\Gamma_u \cup \Gamma_\tau = {}^1\Gamma_u \cup {}^2\Gamma_u \cup {}^1\Gamma_\tau \cup {}^2\Gamma_\tau$, similar to the previous model. Let $\overline{I} = [t_0, t_p]$ be a time interval. Moreover, we will assume that the physical parameters—the thermal conductivity $\kappa = \kappa(\mathbf{x}, t, T(\mathbf{x}, t))$, heat capacity $C = C(\mathbf{x}, t, T(\mathbf{x}, t)) = \rho(\mathbf{x}, t, T(\mathbf{x}, t))\, c_e(\mathbf{x}, t, T(\mathbf{x}, t))$, depend on temperature $T = T(\mathbf{x}, t)$. Then we will solve following problem:

Problem $\mathcal{P}$ Find a temperature $T(\mathbf{x}, t)$ satisfying

$$C(\mathbf{x}, t, T)\frac{\partial T(\mathbf{x}, t)}{\partial t} = \frac{\partial}{\partial x_i}\left(\kappa(\mathbf{x}, t, T)\frac{\partial T(\mathbf{x}, t)}{\partial x_i}\right) + \Phi(\mathbf{x}, t, T),$$
$$i = 1, \ldots, N,\ N = 2, 3, \quad \text{in } \Omega_t = \Omega \times I. \tag{5.120}$$

The boundary conditions

$$T(\mathbf{x}, t) = T_a(\mathbf{x}, t) \quad \text{on} \quad {}^1\Gamma_u \times I, \tag{5.121}$$

$$\kappa(\mathbf{x}, t, T)\frac{\partial T(\mathbf{x}, t)}{\partial x_j} n_j = \varphi(\mathbf{x}, t, T) \quad \text{on} \quad {}^2\Gamma_u \cup {}^1\Gamma_\tau \cup {}^2\Gamma_\tau \times I, \tag{5.122}$$

$$\begin{aligned} &T^k(\mathbf{x}, t) = T^l(\mathbf{x}, t), \\ &\kappa(\mathbf{x}, t, T)\frac{\partial T(\mathbf{x}, t)}{\partial x_j} n_j|_{(k)} = -\kappa(\mathbf{x}, t, T)\frac{\partial T(\mathbf{x}, t)}{\partial x_j} n_j|_{(l)} \quad \text{on} \quad \Gamma^{kl} \times I \end{aligned} \tag{5.123}$$

and the initial condition

$$T(\mathbf{x}, t_0) = T_0(\mathbf{x}). \tag{5.124}$$

We will assume that $C(\mathbf{x}, t, T)$ and $\kappa(\mathbf{x}, t, T)$ are piecewise continuously differentiable functions bounded from below as well as above by positive constants, that is,

$$0 < C_1 \le C(\mathbf{x}, t, T) \le C_2,\ 0 < \kappa_1 \le \kappa(\mathbf{x}, t, T) \le \kappa_2 \quad \forall T \in (-\infty, \infty), \tag{5.125}$$

and thermal sources $\Phi(\mathbf{x}, t, T)$ satisfy

$$\begin{aligned} &|\Phi(\mathbf{x}, t, T_1) - \Phi(\mathbf{x}, t, T_2)| \le c|T_1 - T_2|, \\ &|\Phi(\mathbf{x}, t_1, T) - \Phi(\mathbf{x}, t_2, T)| \le c|t_1 - t_2|, \end{aligned} \quad \mathbf{x} \in \Omega,\ t \in I,\ T_1, T_2 \in (-\infty, \infty). \tag{5.126}$$

Moreover, we will assume that $T_0(\mathbf{x}) \in H^2(\Omega)$, $T_a(\mathbf{x}, t)$ is a continuous function on $\overline{\Omega} \times I = \overline{\Omega}_t$, $\mathbf{n}$ is outward normal to ${}^2\Gamma_u \cup {}^1\Gamma_\tau \cup {}^2\Gamma_\tau$, and $\varphi(\mathbf{x}, t, T)$ is continuous for $T \in (-\infty, \infty)$, $\mathbf{x} \in \overline{\Omega}$, $t \in I$ and

$$\varphi(\mathbf{x}, t, T_2) - \varphi(\mathbf{x}, t, T_1) \geq 0 \quad \forall T_1, T_2,\ T_1 \geq T_2. \tag{5.127}$$

We introduce the following spaces:

$$H^k(\Omega) = \{z | z \in L^2(\Omega), D^\alpha z \in L^2(\Omega),\ \forall |\alpha| \leq k\},$$

is the Sobolev space in the usual sense ,

$$W^{k,\infty}(\Omega) = \{z | z \in L^\infty(\Omega), D^\alpha z \in L^\infty(\Omega),\quad \forall |\alpha| \leq k\},$$

$$\mathbb{H}^k(\Omega) = \prod_{\iota=1}^{s} H^k(\Omega^\iota), \qquad \mathbb{W}^{k,\infty}(\Omega) = \prod_{\iota=1}^{s} \mathbb{W}^{k,\infty}(\Omega^\iota),$$

$$V = \{z | z \in \mathbb{H}^1(\Omega), z|_{{}^1\Gamma_u} = 0,\ z^k = z^l \text{ on } \cup_{k,l} \Gamma^{kl}\}.$$

Let X be a Banach space, $L^\infty(I; X)$ denote the space of functions $z(t) : t \to z(t) \in X$, $t \in I$, which are measurable and such that

$$\text{ess sup}_{t \in I} \|z(t)\|_X = \|z\|_{L^\infty(X)} < \infty.$$

We introduce the enthalpy by

$$H(T) = \int_0^T C(\tau)\, d\tau, \tag{5.128}$$

and the Kirchhoff transformation by

$$G(T) = \int_0^T \kappa(\tau)\, d\tau. \tag{5.129}$$

Let us assume $T(\mathbf{x}, t)$ to be sufficiently smooth, that is, $G(T) \in L^\infty(I; \mathbb{W}^{k,\infty}(\Omega))$, $\partial G(T)/\partial t \in L^\infty(I; \mathbb{W}^{k,\infty}(\Omega))$, $\partial^2 H(T)/\partial t^2 \in L^\infty(I; L^\infty(\Omega))$. Multiplying (5.120) by $z \in V$, applying the Green theorem and the boundary condition and (5.128) and (5.129), we obtain a weak formulation in the enthalpy formulation:

Problem $\mathcal{P}_v$ Find functions $H(\mathbf{x}, t) \in L^\infty(I; L^2(\Omega))$ and $T(\mathbf{x}, t) \in L^2(I; V)$ satisfying

$$\left(\frac{\partial H(T)}{\partial t}, z\right) + a(G(T), z) = \int_\Omega \Phi(\mathbf{x}, t, T(\mathbf{x}, t)) z\, d\mathbf{x} + \int_{{}^2\Gamma_u \cup {}^1\Gamma_\tau \cup {}^2\Gamma_\tau} \varphi(\mathbf{x}, t, T(\mathbf{x}, t)) z\, ds \quad \forall z \in V, t \in I, \tag{5.130}$$

or

$$(H'(T), z) + a(G(T), z) = (\Phi, z) + \langle \varphi, z \rangle \quad \forall z \in V, t \in I,$$

where we denoted

$$(H'(T), z) = \int_\Omega H' z \, d\mathbf{x}, a(G(T), z) = \int_\Omega \nabla G \cdot \nabla z \, d\mathbf{x},$$

$$\langle \varphi, z \rangle = \int_{{}^2\Gamma_u \cup {}^1\Gamma_\tau \cup {}^2\Gamma_\tau} \varphi z \, ds, \quad (\Phi, z) = \int_\Omega \Phi z \, d\mathbf{x}.$$

Further, we will assume that problem $(\mathcal{P})_v$ has a unique solution.

Numerical Solution Numerical solution of Problem $\mathcal{P}$ can be based on several approaches such as the predictor–corrector method, Crank–Nicolson method, the Rothe method, and the like. In the following an idea of Zlámal (1980) will be presented, as its approach gives good results for rapid variation of heat capacity and when Δt is not very small.

Let us introduce a family $\mathcal{T}_h$ of triangulations consisting of triangles $\mathbb{T}_{hi}$ in the 2D case and tetrahedra $\mathbb{T}_{hi}$ in the 3D case, with vertices lying in $\overline{\Omega}$, where we denote by $\mathbb{T}_{hi}$, a triangle in the 2D case or a tetrahedron in the 3D case, respectively, that is, an element of $\mathcal{T}_h$ for which $h_i = \text{diam}(\mathbb{T}_{hi})$, $\rho_{\mathbb{T}_{hi}} = \sup\{\text{diam}\, S\}$, S being a ball contained in $\{\mathbb{T}_{hi}\}$, $\overline{\Omega} = \cup_{\mathbb{T}_{hi} \in \mathcal{T}_h} \mathbb{T}_{hi}$, and $\partial\Omega_h = {}^1\Gamma_u \cup {}^2\Gamma_u \cup {}^1\Gamma_\tau \cup {}^2\Gamma_\tau \cup (\cup_{k,l}\Gamma^{kl})$. We assume that the family $\mathcal{T}_h$ is regular, that is: (i) there exists a constant $\beta > 0$ such that $\rho^{-1} h|_{\mathbb{T}_{hi}} \leq \beta \, \forall \mathbb{T}_{hi} \in \cup_h \mathcal{T}_h$; (ii) the quantity $h = \max_{\mathbb{T}_{hi} \in \mathcal{T}_h} h|_{\mathbb{T}_{hi}}$, $h \to 0$; and is of an acute type if (iii) all angles of the triangles and all angles made by adjacent faces and edges of the tetrahedra are not greater than $\pi/2$. To every triangulation $\mathcal{T}_h$ we introduce the finite dimensional space

$$W_h = \{z_h | z_h \in C^0(\overline{\Omega}^\iota), \ \ \iota \leq s, \ \ z_h \text{ is piecewise linear on } \mathcal{T}_h\},$$
$$V_h = \{z_h | z_h \in W_h, \ \ z_h|_{{}^1\Gamma_u} = 0, \ \ z_h^k = z_h^l \text{ on } \cup \Gamma^{kl}\}.$$

Let $\{\mathbf{x}_k\}_{k=1}^s$ be a set of all nodes of $\mathcal{T}_h$ and $\{\mathbf{x}_k\}_{k=1}^r$ the set of all nodes from $\Omega_h \cup {}^2\Gamma_u \cup {}^1\Gamma_\tau \cup {}^2\Gamma_\tau$. Let us denote by $z_k(\mathbf{x})$ the basis function associated with the node $\mathbf{x}_k$, $z_k(\mathbf{x}_i) = 0$ for $k \neq i$, $z_k(\mathbf{x}_k) = 1$. Let

$$t_i = i\Delta t, \qquad i = 1, \ldots, n, \qquad n = \Delta t^{-1}(t_p - t_0), \qquad 0 < t_1 < \cdots < t_n.$$

Let the value $T^i = T(\mathbf{x}, t_i)$ of the exact solution $T(\mathbf{x}, t)$ at $t = t_i$ be approximated by

$$\Theta^i = \sum_{j=1}^{s} T_j^i z_j(\mathbf{x}), \qquad \Theta_j^i = T_a(\mathbf{x}^j, t_i), \qquad j = r+1, \ldots, s. \tag{5.131}$$

Then for approximation of $H^i = H(T^i)$, $G^i = G(T^i)$, $\varphi^i = \varphi(\mathbf{x}, t_i, T^i)$, $\Phi^i = \Phi(\mathbf{x}, t_i, T^i)$ we take

$$\chi^i = \sum_{j=1}^{s} H(\Theta_j^i)\, z_j(\mathbf{x}), \qquad Y^i = \sum_{j=1}^{s} G(\Theta_j^i)\, z_j(\mathbf{x}),$$

$$\Psi^i = \sum_{j=1}^{s} \varphi(\mathbf{x}^j, t_i, \Theta_j^i)\, z_j(\mathbf{x}), \qquad W^i = \sum_{j=1}^{s} \Phi(\mathbf{x}^j, t_i, \Theta_j^i)\, z_j(\mathbf{x}), \tag{5.132}$$

representing interpolates of the functions H^i, G^i, φ^i, and Φ^i. For approximations of scalar products and bilinear and linear forms, we will use the following quadrature formulas:

$$\int_\Omega f(\mathbf{x})\, d\mathbf{x} = (m+1)^{-1} \operatorname{meas}(\Omega) \sum_{j=1}^{m+1} f(\mathbf{x}^j),$$

$$\int_{\partial\Omega} g(\mathbf{x})\, ds = (m+1)^{-1} \operatorname{meas}(\partial\Omega) \sum_{j=1}^{m+1} g(\mathbf{x}^j).$$

Let us put $t = t_{i+1}$, then $\partial H^{i+1}/\partial t \simeq \Delta t^{-1}(\chi^{i+1} - \chi^i)$, $G^{i+1} \simeq Y^{i+1}$, $\varphi^{i+1} \simeq \Psi^{i+1}$, $\Phi^i \simeq W^i$, $(T, z) \simeq (T, z_h)$, $a(T, z) \simeq a_h(T, z)$, $T_0 \simeq T_{0h}$. Then (5.130) yields

$$\begin{aligned} &(\chi^{i+1} - \chi^i, z)_h + \Delta t a_h(Y^{i+1}, z) = \Delta t\langle \Psi^{i+1}, z\rangle_h + \Delta t(W^i, z)_h \\ &\qquad\qquad \forall z \in V_h, i = 0, 1, \ldots, n-1, \\ &\Theta^0 = T_{0h}. \end{aligned} \tag{5.133}$$

Let us denote by $\{\mathbf{x}^i\}_{i=1}^{r'}$ nodes from Ω_h and by

$$\begin{aligned} &\mathbf{X} = \{X_j\}_{j=1}^{r}, \qquad X_j = \Theta_j^{i+1}, \\ &H(\mathbf{X}) = (H(X_1), \ldots, H(X_r))^T, \qquad G(\mathbf{X}) = (G(X_1), \ldots, G(X_r))^{\mathrm{T}}, \\ &\varphi(\mathbf{X}) = (0, \ldots, 0, \varphi(\mathbf{x}^{r'+1}, t_{i+1}, X_{r'+1}), \ldots, \varphi(\mathbf{x}^r, t_{i+1}, X_r))^{\mathrm{T}}, \\ &C = (C_{ij}) = \{(z_i, z_j)_h\}_{i,j=1}^{r}, \qquad A = (A_{ij}) = \{a_h(z_i, z_j)\}_{i,j=1}^{r}, \\ &B = (B_{ij}) = \{\langle z_i, z_j\rangle_h\}_{i,j=1}^{r}. \end{aligned}$$

The matrices C, A, B are $r \times r$ band matrices, C and A are positive definite and B is positive semidefinite. It can be shown that due to the special choice of the quadrature formulas matrices C and B are diagonal. Then (5.133) leads to the nonlinear system of equations:

$$CH(\mathbf{X}) + \Delta t\, AG(\mathbf{X}) + \Delta t\, B\varphi(\mathbf{X}) = \mathbf{f}. \tag{5.134}$$

Let us introduce new variables

$$y_j = G(X_j),\ j = 1, \ldots, r \tag{5.135}$$

and let the diagonal entries of matrices C and B be defined by

$$C_j = C_{jj} > 0, \qquad B_j = B_{jj} \geq 0, \qquad j = 1, \ldots, r.$$

Due to assumptions (5.125) the mapping in (5.135) maps $\mathbb{R}^r$ one-to-one to $\mathbb{R}^r$. Putting

$$\begin{aligned} &\sigma_j(\xi) = C_j H(\xi) + \Delta t\, B_j \varphi(x^j, t_{i+1}, \xi), \\ &\mathbf{F}(\mathbf{y}) = (F_1(y_1), \ldots, F_r(y_r))^{\mathrm{T}}, \qquad F_j(\mathbf{y}) = \sigma_j G^{-1}. \end{aligned} \tag{5.136}$$

Then from (5.134) and (5.136) we find the nonlinear system

$$\mathbf{F}(\mathbf{y}) + \Delta t\, A\mathbf{y} = \mathbf{f}. \tag{5.137}$$

According to the theory of nonlinear systems [see Ortega and Rheinboldt (1970)] it is known that (5.137) represents a necessary condition for the minimum of the functional

$$J(\mathbf{y}) = \frac{1}{2}\Delta t\, \mathbf{y}^{\mathrm{T}} A\mathbf{y} - \mathbf{f}^{\mathrm{T}}\mathbf{y} + \sum_{j=1}^{r} \int_0^{y_j} F_j(\xi)\, d\xi. \tag{5.138}$$

Theorem 5.8 The functional $J(\mathbf{y})$ is uniformly strictly convex on $\mathbb{R}^r$ and thus it has a unique global minimizer.

Proof Due to Ortega and Rheinboldt (1970, pp. 141–145) and (5.125) and (5.128) and because the matrices C and A are positive definite, B is positive semidefinite, and since due to quadrature formulas used the matrices C and B are diagonal matrices, then the G derivative of J is a uniformly monotone mapping on $\mathbb{R}^r$. Then J is convex on $\mathbb{R}^r$ if and only if $(J'(\mathbf{y}) - J'(\mathbf{x}))\,(\mathbf{y} - \mathbf{x}) \geq 0\ \forall \mathbf{x}, \mathbf{y} \in \mathbb{R}^r$. Moreover, it is uniformly convex on $\mathbb{R}^r$ as $(J'(\mathbf{y}) - J'(\mathbf{x}))\,(\mathbf{y} - \mathbf{x}) \geq 2c\|\mathbf{y} - \mathbf{x}\|^2\ \forall \mathbf{x}, \mathbf{y} \in \mathbb{R}^r$, $c > 0$. As it is uniformly convex on $\mathbb{R}^r$, then it has a unique local and global minimizer (Ortega and Rheinboldt, 1970). Hence J has a minimum at the point $\mathbf{y} = \mathbf{y}^0$, then $X_j^0 = G^{-1}(y_j^0)$ is the solution of (5.134). □

It can be found in the following estimate:

Theorem 5.9 Let us assume that the triangulations $\mathcal{T}_h$ is of an acute type. Let us assume that the solution $T(\mathbf{x}, t)$ of Problem $(\mathcal{P})$ is sufficiently smooth, that is, we assume that $G(T)$, $\partial G(T)/\partial t \in L^\infty(I; W^{2,\infty}(\Omega))$, $\partial^2 H(T)/\partial t^2 \in L^\infty(I; L^\infty(\Omega))$. Then

$$\|T^i - \Theta^i\|_{L^\infty(\Omega \cap \Omega_h)} \leq c(\vartheta(h) + \Delta t), \qquad i = 1, 2, \ldots, n, \tag{5.139}$$

where $c > 0$ does not depend on $\vartheta(h)$, Δt, and i and $\vartheta(h)$ denotes the size of the minimal angle in the trianglulation $\mathcal{T}_h$.

For the proof see Zlámal (1980) and Nedoma (1998a).

Phase Change Problems in Orthopedics Above we saw that during the polymerization process the phase changes are observed. The bone cement changes its physical properties. These problems lead to free boundary problems, the Stefan-like problems. Analyses of these problems are very difficult. Mathematically, these problems represent free boundary-type problems for parabolic equations with the initial boundary conditions, where a part of the boundary is free and where additional conditions are given at this free (unknown) boundary. A two-phase Stefan problem is such that on both sides of the free boundary there are given different parabolic equations and initial boundary conditions.

Let $\overline{I} = [t_0, t_p]$, $t_p > 0$, $\Omega = \cup_{\iota=1}^{m} \Omega^\iota \subset \mathbb{R}^N$, $N = 2, 3$. Let every domain Ω^ι be a convex polygonal in 2D and polyhedral in 3D and the bounded region occupied by an elastic or viscoelastic or viscoplastic part of the human joint system with its artificial replacement head and acetabulum fixed by the bone cement. Its boundaries are $\partial\Omega^\iota = \Gamma_u^\iota \cup \Gamma_\tau^\iota$. Let $\Omega_t = \Omega \times (t_0, t_p)$, $\Omega(t) = \Omega \times \{t\}$, $\partial\Omega_t = \partial\Omega \times (t_0, t_p)$, $\partial\Omega(t) = \partial\Omega \times \{t\}$, $t_0 < t < t_p$, their boundaries. Let us denote $T^\iota = T^\iota(\mathbf{x}, t)$ the temperature in Ω_t, let $\rho^\iota = \rho^\iota(\mathbf{x}, t, T^\iota)$ be a density in Ω^ι, $c_e^\iota = c_e^\iota(\mathbf{x}, t, T^\iota)$ be a specific heat in Ω_t, $C^\iota = \rho^\iota c_e^\iota$ a heat capacity in Ω^ι, and $\kappa^\iota = \kappa^\iota(\mathbf{x}, t, T^\iota)$ a thermal conductivity in Ω^ι.

Then we seek the distribution of the temperature $T(\mathbf{x}, t)$ in Ω_t and surfaces of phase transitions $R^{\iota,s}(t) = \{(\mathbf{x}, t) \in \Omega_t^\iota,\ \Psi^\iota(\mathbf{x}, t) = 0\}$, that is, surfaces between two phases of the bone cement during its polymerization, satisfying

$$C^\iota(\mathbf{x}, t, T^\iota)\frac{\partial T^\iota(\mathbf{x}, t)}{\partial t} = \frac{\partial}{\partial x_i}\left(\kappa^\iota(\mathbf{x}, t, T^\iota)\frac{\partial T^\iota(\mathbf{x}, t)}{\partial x_i}\right) + \Phi^\iota(\mathbf{x}, t, T^\iota),$$
$$\text{for a.e.} \quad (\mathbf{x}, t) \in \Omega_t^\iota = \Omega^\iota \times I,\ \iota = 1, \ldots, m,\ i = 1, \ldots, N,\ N = 2, 3, \qquad (5.140)$$

the boundary conditions

$$T^\iota(\mathbf{x}, t) = T_1^\iota(\mathbf{x}, t) \quad \text{for} \quad (\mathbf{x}, t) \in \Gamma_{ut}^\iota = \Gamma_u^\iota \times I, \iota = 1, \ldots, m, \qquad (5.141)$$

$$\kappa^\iota(\mathbf{x}, t, T^\iota)\frac{\partial T^\iota(\mathbf{x}, t)}{\partial x_j} n_j^\iota = \varphi^\iota(\mathbf{x}, t, T^\iota) \quad \text{for} \quad (\mathbf{x}, t) \in \Gamma_{\tau t}^\iota = \Gamma_\tau^\iota \times I, \quad \iota = 1, \ldots, m, \qquad (5.142)$$

the phase change conditions

$$T_S^{\iota,s}(\mathbf{x}, t) = T_L^{\iota,s}(\mathbf{x}, t) = T_R^{\iota,s}(\mathbf{x}, t) \quad \text{for} \quad (\mathbf{x}, t) \in R^{\iota,s}(t),$$
$$\left(\kappa^\iota(\mathbf{x}, t, T^\iota)\frac{\partial T^\iota(\mathbf{x}, t)}{\partial n}\right)_S^s - \left(\kappa^\iota(\mathbf{x}, t, T^\iota)\frac{\partial T^\iota(\mathbf{x}, t)}{\partial n}\right)_L^s \qquad (5.143)$$
$$= -\rho^{\iota,s}(\mathbf{x}, t, T^\iota)L^{\iota,s}(\mathbf{x}, t)\nu^{\iota,s} \quad \text{for} \quad (\mathbf{x}, t) \in R^{\iota,s}(t)$$

and the initial condition

$$T^\iota(\mathbf{x}, t_0) = T_0^\iota(\mathbf{x}) \quad \text{on } \Omega^\iota(t_0), \quad \iota = 1, \ldots, m, \qquad (5.144)$$

where $\rho^{\iota,s}$ is a density, $T_R^{\iota,s}$ is a temperature at the phase change boundary $R^{\iota,s}(t)$, $T_S^{\iota,s}$ and $T_L^{\iota,s}$ are temperatures in both phases of polymerized bone cement, $\nu^{\iota,s}$ is the unit normal vector to $R^{\iota,s}(t)$ pointing toward $\Omega_S^{\iota,s}$, $R^{\iota,s}(t)$ represents the hypersurface in Ω_t^ι, $\Omega_{St}^{\iota,s}$ is the domain lying in Ω_t^ι and bounded by $R^{\iota,s}(t)$ and $\partial\Omega_t^\iota$, $v_\nu^{\iota,s} = \partial R^{\iota,s}(t)/\partial t$ is the speed of $R^{\iota,s}(t)$ along $\nu^{\iota,s}$, $L^{\iota,s}$ is the latent heat of the phase change, $T_0^\iota(\mathbf{x})$ is the initial temperature, $\varphi^\iota(\mathbf{x}, t, T^\iota)$ is the heat flow, $T_1^\iota(\mathbf{x}, t)$ is the given temperature on Γ_{ut}, $\Phi^\iota(\mathbf{x}, t, T^\iota)$ are the thermal sources in $\Omega^\iota(t)$, $C^\iota(\mathbf{x}, t, T^\iota) = \rho^\iota(\mathbf{x}, t, T^\iota)c_e^\iota(\mathbf{x}, t, T^\iota)$ is a heat capacity, and $\Psi^{\iota,s}(\mathbf{x}, t)$ is a C^1 function in $\overline{\Omega} \times [t_0, t_p]$ such that

$$R^{\iota,s}(t) = \{(\mathbf{x}, t) \in \overline{\Omega}^\iota \times [t_0, t_p], \Psi^{\iota,s}(\mathbf{x}, t) = 0\}, \qquad \nabla_x \Psi^{\iota,s}(\mathbf{x}, t) \neq 0 \quad \text{on } R^{\iota,s}(t),$$

$$\Psi^{\iota,s}(\mathbf{x}, t) < 0 \quad \text{in } \Omega_S^{\iota,s} \quad \text{and} \quad \Psi^{\iota,s}(\mathbf{x}, t) > 0 \quad \text{in } \Omega_L^{\iota,s}.$$

Equations (5.143) can be rewritten in the equivalent form

$$\begin{aligned} &T_S^{\iota,s}(\mathbf{x}, t) = T_L^{\iota,s}(\mathbf{x}, t) = T_R^{\iota,s}(\mathbf{x}, t) \quad \text{for } (\mathbf{x}, t) \in R^{\iota,s}(t), \\ &(\kappa^\iota(\mathbf{x}, t, T^\iota)\nabla T^\iota \nabla \Psi^\iota)_S^s - (\kappa^\iota(\mathbf{x}, t, T^\iota)\nabla T^\iota \nabla \Psi^\iota)_L^s = \alpha^{\iota,s} \frac{\partial \Psi^{\iota,s}}{\partial t}, \\ &\qquad \text{for } (\mathbf{x}, t) \in R^{\iota,s}(t), \end{aligned}$$

where $\alpha^{\iota,s}$ are given positive constants.

The problem will be solved (i) by using assumptions (5.125) and (ii) the Kirchhoff transformation

$$y = G(\mathbf{x}, t, T) = \int_0^{T(\mathbf{x},t)} \kappa(\mathbf{x}, t, \xi)\, d\xi, \qquad (\mathbf{x}, t) \in \Omega_t.$$

The mapping $T \to y = G(T)$ is one to one. If T_R is the temperature of a phase change, then $y_R(\mathbf{x}, t) = G(\mathbf{x}, t, T_R)$. Denoting $G^{-1}(y)$ the inverse of $y = G(T)$, then $T = G^{-1}(y)$; and (iii) the enthalpy defined as

$$H(\mathbf{x}, t, T) = \int_0^T C(G^{-1}(\zeta))\, d\zeta + \begin{cases} 0 & \text{if } T < y_R(\mathbf{x}, t), \\ [0, \alpha] & \text{if } T = y_R(\mathbf{x}, t), \\ \alpha & \text{if } T > y_R(\mathbf{x}, t). \end{cases}$$

It can be shown that a weak solution of such a problem exists and it is unique. For numerical solution the semi-implicit discretization approach, which leads to a boundary value problem for nonlinear partial differential equations of an elliptic type at every time, can be used. The analysis of such problems is very complicated and is beyond the scope of this book. For more details see Glowinski (1984), Jerome (1977, 1983), Elliott (1981), Friedman (1982), Jerome and Rose (1982), White (1982a,b), Elliott and Ockendon (1982), Nedoma (1997a,b, 1998a, 2003, 2006), and Visintin (1996). The literature on these problems is very large.

5.7.3 Heat Generation

The heat in the human body is generated by muscle contractions, chemical and biological processes, as well as by mechanical work and mechanical deformations of a

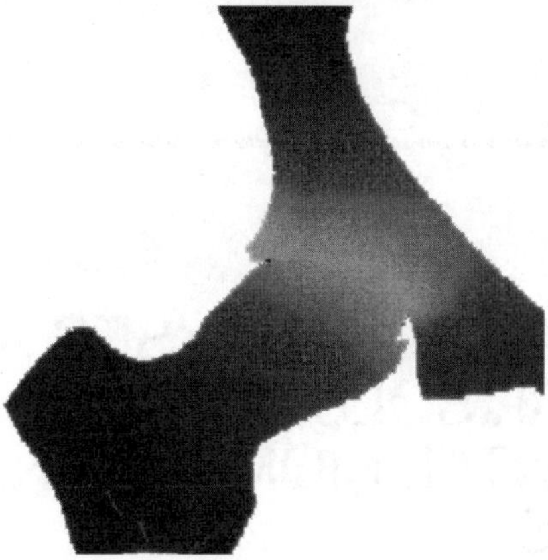

FIG. 5.10 Generation heat on the contact boundary between the head of femur and the acetabulum and in the neighboring area.

skeletal system and its parts. The heat is also generated during the exothermic polymerization reaction of the monomer in the course of surgical treatment of artificial joint replacement. A viscosity can be interpreted as an inner friction, and moreover, the corresponding term dissipates mechanical energy into the thermal energy, and therefore, into heat. On the contact surfaces of a human joint a friction exists, the coefficient of friction is small in the case of a healthy joint and is relatively high in a damaged joint, the corresponding effect contributes to the generation of heat in the human body. The coupling terms in the thermo(visco)elastic theory describe (i) the heat generated by the boundary friction and we speak about the frictional heat, (ii) the heat generated by the viscous inner friction and we speak about it as of the viscous heat. In the human body the dissipated mechanical energy cannot be lost, but it is transferred into the thermal energy and, therefore, into the heat. That is why additional terms in the coupled system of the heat equation and the equation of motion, describing the mutual exchange of energy between the thermal field in the musculoskeletal system and (visco)elastic musculoskeletal system, are introduced.

The heat in the musculoskeletal system is transferred mainly by conduction (and partly by convection). Thus the temperature distribution can be determined by solving the relevant form of the heat conduction equation. Heat propagates mainly by conduction, which is characterized by a coefficient of thermal conduction κ and by a heat flow q. The dissipative term representing the dissipation of deformation energy into heat can be taken as the thermal sources in the area of a contact zone of human joints or its artificial replacements. Fig. 5.10 [see Krejčí et al. (1997)] demonstrates the effect of heat generation on the hip joint contact surface and in the neighboring areas in the acetabulum, pelvis, and the head of the femur. The model is based on the contact problem in thermoelasticity. For numerical realization the program code based on the theory presented in Nedoma and Dvořák (1995) was used.

CHAPTER 6

MATHEMATICAL ANALYSES AND NUMERICAL SOLUTIONS OF FUNDAMENTAL BIOMECHANICAL PROBLEMS

6.1 BACKGROUND OF FUNCTIONAL ANALYSIS, FUNCTION SPACES, AND VARIATIONAL INEQUALITIES

6.1.1 Introduction to Functional Analysis

Functional analysis is one of the important mathematical disciplines, having an application in the analysis of mathematical model problems and their numerical approximations in many scientific and technological disciplines, including biomechanics, as well as in the analysis of algebraic problems. Therefore, we give a short survey of the main ideas and results of functional analysis in their generalized form. For more details see Danford and Schwartz (1958), Taylor (1967), Yosida (1974), Rudin (1991), Zeidler (1995a,b), and Atkinson and Han (2001).

Let us denote by X a set of elements. Then the set X is called a real or complex **linear or vector space** if the ordinary arithmetic rules of vector addition and scalar multiplication are valid. As an example of the linear space is the N-dimensional vector space $\mathbb{R}^N = \{u = (u_i)^{\mathrm{T}},\ i = 1, \ldots, N,\ u_i \in \mathbb{R}\}$ or the set of all real symmetric matrices of order $N \times N$, we denote it by $\mathbb{R}^{N\times N}$. A nonnegative function $\|u\| \equiv \|u\|_X$, defined on X, is called a **norm** on X if it satisfies (i) $\|u\|_X = 0$ if and only if $u = 0$, (ii) $\|\lambda u\|_X = |\lambda|\,\|u\|_X$ for all $\lambda \in \mathbb{R}$, and (iii) $\|u + v\|_X \leq \|u\|_X + \|v\|_X$, the triangle inequality. Next, we only deal with real linear spaces so that all the scalars are also real.

A vector space X with a norm $\|u\| \equiv \|u\|_X$ is called a **normed space** and the number $\|u\|_X$ is called the norm of the element $u \in X$. Let X be a normed linear space and let $u \in X$, $r > 0$. Then the set $B(u, r) = \{v \in X; \|u - v\| < r\}$ is called an open ball with its

Mathematical and Computational Methods in Biomechanics of Human Skeletal Systems: An Introduction, First Edition. Jiří Nedoma, Jiří Stehlík, Ivan Hlaváček, Josef Daněk, Taťjana Dostálová, and Petra Přečková.

center in u and radius r. A subset $M \subset X$ is called an open set in X if for every $u \in M$ there exists an $r = r(u) > 0$ such that $B(u, r) \subset M$. A subset $M \subset X$ is called a closed set in X if $X - M$ is an open set in X.

Let X be a real vector space. Then a real function defined on $X \times X$ is called an **inner or scalar product** of $u \in X$, $v \in X$, and it is denoted by $(u, v) \equiv (u, v)_X$, $u, v \in X$, if it satisfies (i) $(u + v, z)_X = (u, z)_X + (v, z)_X$, (ii) $(\lambda u, v)_X = \lambda(u, v)_X$ with $\lambda \in \mathbb{R}$, and (iii) $(u, v)_X = (v, u)_X$, $(u, u)_X > 0$ if $u \neq 0$. Equipped with such a functional, X is called an **inner product space (pre-Hilbert space)**, and the functional $\|u\|_X = (u, u)_X^{1/2}$ is a norm on X. Moreover, the inequalities $|(u, v)| \leq \|u\| \|v\|$, the Schwarz inequality, and $\|u + v\|_X \leq \|u\|_X + \|v\|_X$ hold. Then X is a **normed vector space**. We say that two vectors u and v are orthogonal if $(u, v)_X = 0$.

Let X be a normed vector space and $\{u_n\}_{n=1}^{\infty}$ a sequence in X. Then the sequence u_n converges to $u \in X$ in X, that is, $u_n \to u$ in X, if $\lim_{n\to\infty} \|u_n - u\|_X = 0$, that is, if for every $\varepsilon > 0$ there exists an $n_0 = n_0(\varepsilon) \in \mathbb{N}$ such that $\|u_n - u\|_X < \varepsilon$ holds for all $n > n_0$. We say that the sequence $\{u_n\}_{n=1}^{\infty}$ converges strongly in X or in the norm of X. If the sequence $\{u_n\}_{n=1}^{\infty}$ converges strongly in X, then it is called convergent in X and u is called its limit.

A sequence $\{u_n\}_{n=1}^{\infty}$ in X is called a **fundamental** or Cauchy sequence if $\lim_{m,n\to\infty} \|u_m - u_n\|_X = 0$, that is, if for every $\varepsilon > 0$ there exists an $n_0 = n_0(\varepsilon) \in \mathbb{N}$ such that $\|u_m - u_n\|_X < \varepsilon$ holds for all $m, n > n_0$.

Let M be a subset of a normed vector space. Then the closure of M in X, that is, $\overline{M}$, is defined as the set of all elements $u \in X$ such that there exists a sequence $\{u_n\}_{n=1}^{\infty}$ in M converging strongly in M, that is, $u_n \to u$ in X. It is evident that $M \subset \overline{M}$. A subset $M \subset X$ is said to be dense in X if $\overline{M} = X$. A subset of a normed vector space X is called a subspace of X if it is a linear set that is closed in X. A subspace M of a normed vector space X with the norm $\|\cdot\|_X$ is also a normed vector space with the norm $\|x\|_M \equiv \|x\|_X$, $x \in M$.

A set M is called **countable** if there exists a bijective [i.e., a mutually univalued (surjective and injective)] map of M onto $\mathbb{N}$, where $\mathbb{N}$ denotes the set of natural numbers $n = 1, 2, \ldots$. A normed vector space X is called **separable** if there exists a countable set $M \subset X$, which is dense in X. A subspace of a separable normed vector space X is a separable normed vector space.

A normed vector space X is said to be **complete** if every fundamental sequence $\{u_n\}_{n=1}^{\infty}$ is convergent in X, which means that it has a limit $u \in X$. A complete normed vector space is called a **Banach space**.

Let X be an inner product space with inner product (u, v) and let X be complete with respect to the norm $\|u\|_X = (u, u)^{1/2}$. Then X is the **Hilbert space**. A subspace of a Banach and/or of a Hilbert space is also a Banach or Hilbert space, respectively.

Let X be a vector space. A subset $K \subset X$ is said to be **convex** if it has the property

$$u, v \in K \Rightarrow (1 - t)u + tv \in K \quad \forall t \in [0, 1].$$

Linear Operators and Linear Functionals Let X, Y be normed linear spaces (real or complex) and let M be a set in X. Let us suppose that for every $u \in M$ there exists a uniquely determined $v \in Y$ such that $v = Au$, where A is the **operator or**

mapping function defined on M and M is the domain of operator A and is denoted by $\mathcal{D}(A)$. We also say that operator A is an operator from X into Y that maps M into Y. The set $\mathcal{R}(A)=\{w\in Y; w=Ay \text{ with } y\in M\}$ is called the range of operator A. If for all $u, v\in M$ we have $Au\neq Av$ provided $u\neq v$, then every $w\in\mathcal{R}(A)$ is assigned a uniquely determined element v such that $Av=w$. Then $v=A^{-1}w$ and A^{-1} is the inverse operator to A.

Let X, Y be normed vector spaces. An operator A from X into Y is called a linear operator if $\mathcal{D}(A)$ is a linear set and if (i) $A(u+v)=Au+Av$, and (ii) $A(\lambda u)=\lambda Au$, for all $u, v\in\mathcal{D}(A)$ and for every scalar λ.

An operator A from X into Y is said to be continuous if $v_i\to v$ in X implies $Av_i\to Av$ in Y for any sequence $\{v_i\}_{i=1}^{\infty}$ provided $v_i, v\in\mathcal{D}(A)$.

A linear operator A from X into Y is said to be bounded if $\sup\|Av\|_Y<\infty$, where the supremum is taken over all $v\in\mathcal{D}(A)$ such that $\|v\|_X\leq 1$. We see that a linear operator is bounded if and only if it is continuous.

Let A be a continuous linear operator from X into Y. The norm of the operator A is defined by $\|A\|=\sup\|Au\|_Y$, where the supremum is taken over all $v\in\mathcal{D}(A)$ such that $\|u\|_X\leq 1$ or by

$$\|A\| = \sup_{\substack{u\in\mathcal{D}(A)\\ \|u\|_X=1}} \|Au\|_Y = \sup_{\substack{u\in\mathcal{D}(A)\\ \|u\|_X\neq 0}} \frac{\|Au\|_Y}{\|u\|_X}.$$

Theorem 6.1 (Banach) Let X, Y be Banach spaces, A a linear operator from X onto Y with $\mathcal{D}(A)=X$ and $\mathcal{R}(A)=Y$. Suppose that A is continuous and that A^{-1} exists. Then the inverse operator A^{-1} is continuous.

Let A be a linear operator from a normed vector space X into a normed vector space Y with $\mathcal{D}(A)=X$. Operator A is said to be completely continuous (compact) if it maps every bounded set in X onto a relatively compact set in Y. We see that every completely continuous linear operator is continuous.

Two normed vector spaces are said to be isomorphic if there exists a continuous linear operator A such that $\mathcal{D}(A)=X$, $\mathcal{R}(A)=Y$, and A^{-1} exists and is continuous. Operator A is then called an isomorphism between X and Y or an isomorphism mapping.

Let X, Y be two normed vector spaces and let $X\subset Y$. Then the identity operator I from X into Y with $\mathcal{D}(I)=\mathcal{R}(I)=X$ is defined as the operator that maps every element $u\in X$ into itself, that is, $Iu=u$, regarded as an element of Y. Operator I is a linear operator. If it is also continuous, then it is called the **imbedding operator** or the imbedding from X into Y, and we write $X\hookrightarrow Y$. Continuity of the imbedding implies that there exists a constant $c>0$ such that $\|x\|_Y\leq c\|x\|_X$ for every $x\in X$. If the imbedding operator is completely continuous, then $X\hookrightarrow\hookrightarrow Y$.

Let X be a real normed space and Y the real line R. A linear operator from X into Y is called a real **linear functional** and it will be denoted by φ, ψ, and so forth and its value at $u\in X$ by $\varphi(u)$. The same results given above for linear continuous operators are valid also in the case of continuous linear functionals. A norm of a linear functional φ is defined by $\|\varphi\|=\sup|\varphi(u)|$, where the supremum is taken over all $u\in\mathcal{D}(\varphi)$ such that $\|u\|_X\leq 1$.

Theorem 6.2 (Hahn–Banach) Let φ be a continuous linear functional defined on a linear subset M of a normed linear space X. Then there exists a continuous linear functional ψ defined on X such that $\psi(u)=\varphi(u)$ for $u\in M$ and $\|\varphi\|=\|\psi\|$.

Let G be the set of all continuous linear functionals defined on X. Then G is a vector space as the addition of functionals and the multiplication of a functional by a scalar can be defined, that is, $(\varphi+\psi)(u)=\varphi(u)+\psi(u)$ and $(\lambda\varphi)(u)=\lambda\cdot\varphi(u)$, where φ, $\psi\in G$, and λ is a scalar. Moreover, G is a normed linear space with the norm defined by $\|\varphi\|=\sup|\varphi(u)|$, and it is called the **dual (adjoint, conjugate) space** of X and denoted by X'.

Let X be a Banach space. Then the dual space X' is a Banach space. The duality pairing between X' and X is denoted by $\langle v', v\rangle$ or $\varphi(v)$ for $v',\varphi\in X'$, and $v\in X$. The norm on X' is given by

$$\|\varphi\|_{X'} = \sup_{0\neq v\in X} \frac{|\varphi(v)|}{\|v\|_X}.$$

A normed space X is said to be **reflexive** if X may be identified with X'' by the canonical embedding χ, that is, if $\chi(X)=X''$, where X'' denotes the bidual, which is a Banach space. Each element $u\in X$ induces a linear continuous functional $\varphi_u\in X''$ by the relation $\varphi_u(v')=\langle v',u\rangle$ for any $v'\in X'$. Since the mapping $u\mapsto\varphi_u$ from X into X'' is linear and isometric, that is, $\|\varphi_u\|_{X''}=\|u\|_X$ for all $u\in X$, then the normed space X may be viewed as a linear subspace of the Banach space X'' by the embedding $u\mapsto\varphi_u=\chi(u)$. Then we can say that a normed space is reflexive if X may be identified with X'' by the canonical embedding χ, that is, if $\chi(X)=X''$. A reflexive space must be complete and thus it is a Banach space.

Let X be a normed vector space, and $\{u_i\}_{i=1}^{\infty}$ a sequence in X. We say that u_i converges weakly to $u\in X$, that is, $u_i\rightharpoonup u$ weakly, if $\lim_{i\to\infty}\varphi(u_i)=\varphi(u)$ for every $\varphi\in X'$. It is known that every weakly convergent sequence is bounded.

Theorem 6.3 If X is a reflexive Banach space, then any bounded sequence in X has a weakly convergent subsequence.

Let X be a normed space and X' be its dual. Then, we say that a sequence $\{u'_n\}\subset X'$ converges weakly star or weakly-* to $u'\in X'$ if

$$\langle u'_n, v\rangle \to \langle u', v\rangle \quad \text{as} \quad n\to\infty, \quad \text{for all } v\in X,$$

where u' is the weak-* limit of $\{u'_n\}$ and we write $u'_n \overset{*}{\rightharpoonup} u'$ as $n\to\infty$.

We say that space X is separable if there exists a set $\{v_1, v_2, \ldots,\}\subset X$ such that finite linear combinations of $\{v_i\}$ are dense in X, that is, any element of X can be approximated by a sequence of finite linear combinations of $\{v_i\}$.

Theorem 6.4 If X is a separable Banach space, then any bounded sequence in X' has a weakly star convergent subsequence.

Let X and Y be two normed spaces and $A : X \to Y$. Then we say that operator A is compact if for any bounded sequence $\{v_i\} \subset X$, the sequence $\{Av_i\} \subset Y$ has a subsequence converging in Y.

On Hilbert spaces, continuous linear functionals are limited in the forms they can take. The next theorem, the Riesz theorem, makes this more precise. This theorem identifies a Hilbert space with its dual.

Theorem 6.5 (Riesz Representation Theorem) Let X be a Hilbert space and let $\varphi \in X'$. Then there is a unique $u \in X$ for which

$$\varphi(v) = (u, v)_X \qquad \text{for all } v \in X.$$

Moreover,

$$\|\varphi\|_{X'} = \|u\|_X.$$

From this theorem it follows that a Hilbert space can be identified with its bidual, and, therefore, every Hilbert space is reflexive.

Now, we introduce the definition of **monotone operators** in inner product spaces as follows:

Definition 6.1 Let X be a space with inner product $(\cdot, \cdot)_X$ and norm $\|\cdot\|_X$. Then we say that an operator $A : X \to X$ is a monotone operator if

$$(Au - Av, u - v)_X \geq 0 \qquad \text{for all } u, v \in X.$$

The operator A is strictly monotone if

$$(Au - Av, u - v)_X > 0 \qquad \text{for all } u, v \in X, u \neq v,$$

and strongly monotone if there exists a constant $m > 0$ such that

$$(Au - Av, u - v)_X \geq m\|u - v\|_X^2 \qquad \text{for all } u, v \in X.$$

If $A : X \to X$ is a linear operator, then the monotonicity, strict monotonicity, and strong monotonicity are defined by

$$(Av, v)_X \geq 0 \qquad \text{for all } v \in X,$$

$$(Av, v)_X > 0 \qquad \text{for all } v \in X, v \neq 0,$$

$$(Av, v)_X \geq m\|v\|_X^2 \qquad \text{for all } v \in X, \text{ where } m = \text{const} > 0.$$

Definition 6.2 Let X be a reflexive Banach space with its dual X' and with the corresponding norms $\|\cdot\|_X$, $\|\cdot\|_{X'}$, respectively. The operator $A : X \to X'$ is (i) **hemicontinuous** at $u \in X$ if for any $v \in X$ the map $t \to A((1-t)u + tv)$ is continuous from $[0, 1]$ into the weak topology of X', (ii) **demicontinuous** if it is continuous from finite dimensional subspaces in X into the weak topology of X', and it is (iii) **pseudomonotone** if it is demicontinuous and if $u_n \rightharpoonup u, n \to \infty$, and $\limsup_{n\to\infty} (Au_n, u_n - u) \leq 0$ implies $(Au, u - v) \leq \liminf_{n\to\infty} (Au_n, u_n - v)$ for all $v \in X$.

Let $f : X \to \mathbb{R} \cup (\pm\infty)$. Function f is said to be proper if $f(v) > -\infty$ for all $v \in X$ and $f(u) < \infty$ for some $u \in X$ and it is convex if

$$f((1-t)u + tv) \leq (1-t)f(u) + tf(v) \tag{6.1}$$

for any $u, v \in X$, $t \in (0, 1)$ for which the right-hand side is meaningful, and strictly convex if the strict inequality holds for $u \neq v$ and $t \in (0, 1)$.

Definition 6.3 A function $f : X \to \mathbb{R} \cup (\pm\infty)$ is said to be lower semicontinuous (l.s.c.) at $u \in X$ if

$$\liminf_{n\to\infty} f(u_n) \geq f(u) \tag{6.2}$$

for any sequence $\{u_n\} \subset X$, converging to u in X, and it is weakly lower semicontinuous (weakly l.s.c) if (6.2) holds for any sequence $\{u_n\} \subset X$ converging weakly to u.

Theorem 6.6 A proper convex function $f : X \to \mathbb{R} \cup (\pm\infty)$ is lower semicontinuous if and only if it is weakly lower semicontinuous.

Theorem 6.7 Let X be a reflexive Banach space and $f : X \to \mathbb{R} \cup (\pm\infty)$ a proper convex lower semicontinuous function such that $f(u) \to +\infty$ when $\|u\|_X \to +\infty$, that is, f is coercive. Then f is bounded from below and it attains its minimum on X. The set of minimizers of f is closed and is a convex set in X; if f is a strictly convex function, then f attains its minimum on X at only one point.

Remark 6.1 The theorem is a modification of the well-known Weierstrass theorem.

Definition 6.4 Let $f : X \to \mathbb{R} \cup (\pm\infty)$ and let $u \in X$ such that $f(u) \in \mathbb{R}$. Then we say that the function f is the Gâteaux differentiable at u if there exists $u' \in X'$ such that

$$\lim_{t\to\infty} \frac{f(u+tv) - f(u)}{t} = \langle u', v \rangle \qquad \text{for all } v \in X.$$

Then the element u' is unique and is called the Gâteaux derivative of f at u, and we denote it by $f'(u)$.

In the next theorem we can show that the Gâteaux derivative characterizes the convexity of Gâteaux-differentiable functionals.

Theorem 6.8 Let X be a normed space and $K \subset X$ be a nonempty convex subset. Let $f : X \to \mathbb{R} \cup (\pm\infty)$ be Gâteaux differentiable. Then, the next statements are equivalent:

(i) f is convex,
(ii) $f(v) \geq f(u) + \langle f'(u), v-u \rangle$ for all $u, v \in X$,
(iii) $\langle f'(v) - f'(u), v-u \rangle \geq 0$ for all $u, v \in X$.

Operator A mapping a Banach space P into itself is called contractive in this space if there exists a number α, $0 < \alpha < 1$, such that for every pair of elements $u, v \in P$ we have $\|Au - Av\| \leq \alpha \|u - v\|$.

Theorem 6.9 (Banach Fixed-Point Theorem) Let A be a contractive operator in a Banach space P. Then the equation $u = Au$ has one and only one solution in the space, that is, there exists exactly one element $u \in P$ for which $u = Au$ holds. This element can be obtained as the limit of a sequence of elements $u_n \in P$, $u = \lim_{n\to\infty} u_n$, where $u_{n+1} = Au_n$, $n = 1, 2, \ldots$, while the element $u_1 \in P$ may be chosen arbitrary.

For the proof see Atkinson and Han (2001).

6.1.2 Functional Spaces and Fundamental Theorems

Let $\Omega = \cup_{\iota=1}^{s} \Omega^{\iota} \subset \mathbb{R}^N$, $N = 2, 3$, be a domain (i.e., a bounded, open connected set) in the Euclidean space $\mathbb{R}^N$, $N = 2, 3$, with a sufficiently smooth boundary $\partial\Omega = \cup_r \Gamma_r$ $(\in C^{1,1})$, $r = \tau, u, c$. Let $I = (0, t_p)$ be the time interval.

By $C(\Omega) = C^0(\Omega)$ we denote the set of all functions defined and continuous on Ω. By $C^k(\Omega)$, $k \in \mathbb{N}$, $\mathbb{N}$ is the set of all positive integers, or $C^{k,N}(\Omega) = [C^k(\Omega)]^N = C^k(\Omega, \mathbb{R}^N)$, $N = 2, 3$), we denote the set of all functions u (scalar or vector) defined on Ω, which have continuous derivatives up to the order k on Ω. The norm is defined as

$$\|u\|_{C^k(\Omega)} = \sup_{\mathbf{x}\in\Omega, |\alpha|\leq k} |D^{\alpha} u(\mathbf{x})|,$$

where $\alpha = (\alpha_1, \ldots, \alpha_N) \in \{0, 1, \ldots\}^N$ is the multi-index, $|\alpha| = \alpha_1 + \cdots + \alpha_N$ its length and

$$D^{\alpha} u(\mathbf{x}) = \frac{\partial^{|\alpha|}}{\partial x^{\alpha}} u(\mathbf{x}) = \frac{\partial^{\alpha_1 + \cdots + \alpha_N}}{\partial x_1^{\alpha_1} \ldots \partial x_N^{\alpha_N}} u(\mathbf{x}).$$

Let u be a function (scalar or vector) defined on $\Omega \subset \mathbb{R}^N$. Then the set supp $u = \overline{\{\mathbf{x} \in \Omega, u(\mathbf{x}) \neq 0\}}$, where the bar denotes the closure in the space $\mathbb{R}^N$, is called the support of the function u.

Let $\mathbb{N}_0$ be a set of all nonnegative integers and let $k \in \mathbb{N}_0 \cup \{\infty\}$. Then $C_0^k(\Omega)$ denotes the set of all functions $u \in C^k(\Omega)$, whose supports are compact subsets of Ω. By $C_0^{\infty}(\Omega)$ we denote the set of all infinitely differentiable functions with compact support.

By $C(\overline{\Omega})$ or $C^0(\overline{\Omega})$ we denote the set of all functions in $C(\Omega)$, which are continuously extendable to the boundary $\partial\Omega$ and by $C^k(\overline{\Omega})$, $k \in \mathbb{N}$, the set of all functions $u \in C^k(\Omega)$ such that $D^\alpha u \in C(\overline{\Omega})$ for all $\alpha \in \mathcal{N}_{N,k}$, where $\mathcal{N}_{N,k}$ is the set of all N-dimensional multi-indices of length k.

We denote $C^\infty(\Omega) = \bigcap_{k=0}^{\infty} C^k(\Omega)$ and $C^\infty(\overline{\Omega}) = \bigcap_{k=0}^{\infty} C^k(\overline{\Omega})$.

Let Ω be a domain in $\mathbb{R}^N$, and let $\lambda \in (0, 1]$, α be a multi-index, $|\alpha| \le k$. By $C^{k,\lambda}(\overline{\Omega})$ we denote the subset of all functions $u \in C^k(\overline{\Omega})$ such that

$$\sup_{\mathbf{x},\mathbf{y}\in\Omega,\mathbf{x}\neq\mathbf{y}} \frac{|D^\alpha u(\mathbf{x}) - D^\alpha u(\mathbf{y})|}{|\mathbf{x} - \mathbf{y}|^\lambda} < \infty \quad \text{for all } \alpha \text{ with } |\alpha| = k.$$

We say that a function u defined on Ω satisfies the **Hölder condition** with the exponent $\lambda \in (0, 1]$, if there exists a nonnegative constant $c = c(u)$ such that

$$|u(\mathbf{x}) - u(\mathbf{y})| \le c|\mathbf{x} - \mathbf{y}|^\lambda \quad \text{for all } \mathbf{x}, \mathbf{y} \in \Omega.$$

For $\lambda = 1$ we speak about the **Lipschitz condition**.

We say that the functions in $C^{0,\lambda}(\overline{\Omega})$ are Hölder continuous (or hölderian), for $\lambda = 1$ Lipschitz continuous (lipschitzian).

By $C^{0,k,\lambda}(\overline{\Omega})$ we denote the subset of all functions $u \in C^{k,\lambda}(\overline{\Omega})$ satisfying the condition: For every $\varepsilon > 0$ there exists $\delta = \delta(\varepsilon, u)$ such that for all $\mathbf{x}, \mathbf{y} \in \Omega$ with $0 < |\mathbf{x} - \mathbf{y}| < \delta$

$$\frac{|D^\alpha u(\mathbf{x}) - D^\alpha u(\mathbf{y})|}{|\mathbf{x} - \mathbf{y}|^\lambda} < \varepsilon \quad \text{for all } \alpha \text{ with } |\alpha| = k.$$

The spaces $C^k(\overline{\Omega})$ and $C^{k,\lambda}(\overline{\Omega})$ are Banach spaces.

For u_n, $u \in C^0(\overline{\Omega})$, $n = 1, 2, \ldots$, the convergence $u_n \to u$ in $C^0(\overline{\Omega})$, that is, $\lim_{n\to\infty} \|u_n - u\|_0 = 0$, is the uniform convergence of u_n to u in $\overline{\Omega}$. The convergence $u_n \to u$ in $C^k(\overline{\Omega})$, that is, $\lim_{n\to\infty} \|u_n - u\|_k = 0$, is the uniform convergence of all derivatives $D^\alpha u_n$ with $|\alpha| \le k$, to $D^\alpha u$ in $\overline{\Omega}$.

Let $k \in \mathbb{N}_0 = \{0, 1, 2, \ldots\}$, $\lambda \in (0, 1]$, and let us set

$$\|u\|_{C^k(\overline{\Omega})} = \|u\|_k = \sum_{|\alpha|\le k} \sup_{\mathbf{x}\in\Omega} |D^\alpha u(\mathbf{x})| \quad \text{for } u \in C^k(\overline{\Omega})$$

and

$$\|u\|_{C^{k,\lambda}(\overline{\Omega})} = \|u\|_{k,\lambda} = \|u\|_k + \sum_{|\alpha|\le k} \sup_{\mathbf{x},\mathbf{y}\in\Omega,\mathbf{x}\neq\mathbf{y}} \frac{|D^\alpha u(\mathbf{x}) - D^\alpha u(\mathbf{y})|}{|\mathbf{x} - \mathbf{y}|^\lambda}$$

for $u \in C^k(\overline{\Omega})$.

Then $\|u\|_k$ and $\|u\|_{k,\lambda}$ are norms in the vector spaces $C^k(\overline{\Omega})$ and $C^{k,\lambda}(\overline{\Omega})$, respectively.

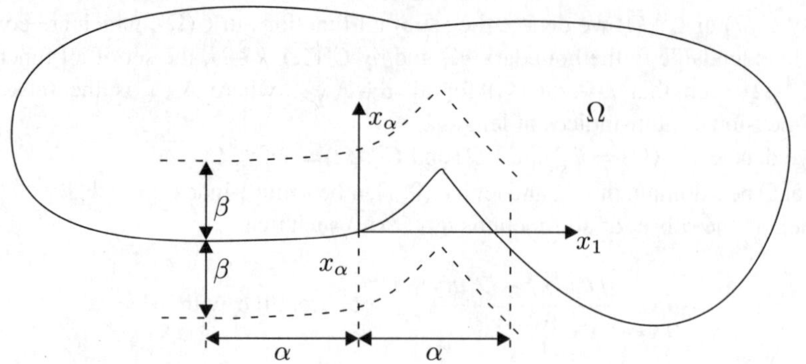

FIG. 6.1 Domain with the Lipschitz boundary.

The space $C^k(\overline{\Omega})$ is separable, while the space $C^{k,\lambda}(\overline{\Omega})$ is not separable.

Let K be a subset of $C^0(\overline{\Omega})$. Then K is said to be equicontinuous if for every $\varepsilon > 0$ there exists $\delta = \delta(\varepsilon) > 0$ such that $|u(\mathbf{x}) - u(\mathbf{y})| < \varepsilon$ holds for all $u \in K$ and all $\mathbf{x}, \mathbf{y} \in \overline{\Omega}$ for which $|\mathbf{x} - \mathbf{y}| < \delta$.

Theorem 6.10 Let $k \in \mathbb{N}_0$, $\lambda \in (0, 1]$. Then $C^{k,\lambda}(\overline{\Omega}) \hookrightarrow C^k(\overline{\Omega})$ (the imbedding) and $C^{k,\lambda}(\overline{\Omega}) \hookrightarrow\hookrightarrow C^k(\overline{\Omega})$ (the compact imbedding). (For $k = 0$ and $\lambda = 1$ it is called the **Arzelá-Ascoli theorem**.)

For classification of domains used we introduce the following definition (for illustration see Fig. 6.1).

Definition 6.5 Let $\Omega \subset \mathbb{R}^n$ be a domain with the boundary $\partial\Omega$. A boundary is said to be the Lipschitz boundary if there exist real numbers $\alpha > 0$, $\beta > 0$ such that for each $\mathbf{x} \in \partial\Omega$, the Cartesian coordinate system can be rotated and translated to $\mathbf{x}$ in such a way that it holds that: there exists a function $g(x_1, \ldots, x_{n-1})$ defined on K_{n-1}, where K_{n-1} is an $(n-1)$-dimensional open cube defined by

$$K_{n-1} = \{|x_i| < \alpha = \text{const} \quad \text{for } i = 1, \ldots, n-1\}$$

such that

$$g(x_1, \ldots, x_{n-1}) = x_n, \quad (x_1, \ldots, x_{n-1}) \in K_{n-1}$$

and satisfying the Lipschitz condition

$$|g(x_1, \ldots, x_{n-1}) - g(y_1, \ldots, y_{n-1})| < c_0 |(x_1, \ldots, x_{n-1}) - (y_1, \ldots, y_{n-1})|$$

for every $(x_1, \ldots, x_{n-1}) \in K_{n-1}$ and $(y_1, \ldots, y_{n-1}) \in K_{n-1}$. Moreover, all the points $\mathbf{x} = (x_1, \ldots, x_{n-1}, x_n)$ such that $(x_1, \ldots, x_{n-1}) \in K_{n-1}$ and $g(x_1, \ldots, x_{n-1}) < x_n < g(x_1, \ldots, x_{n-1}) + \beta$ lie inside Ω while all the points $\mathbf{x} = (x_1, \ldots, x_{n-1}, x_n)$, $(x_1, \ldots, x_{n-1}) \in K_{n-1}$, $g(x_1, \ldots, x_{n-1}, x_n) - \beta < x_n < g(x_1, \ldots, x_{n-1}, x_n)$ lie outside Ω.

Definition 6.6 We say that a real function is integrable in the Lebesgue sense if it is measurable on Ω and the Lebesgue integral $\int_\Omega f(\mathbf{x})\,d\mathbf{x}$ exists, but may be $\pm\infty$.

Theorem 6.11 (Green's Formula) Let Ω be a domain in $\mathbb{R}^N$ with the Lipschitz boundary, let u, v be two functions from $C^1(\overline{\Omega})$. Then

$$\int_\Omega \frac{\partial v(\mathbf{x})}{\partial x_i} u(\mathbf{x})\,d\mathbf{x} = -\int_\Omega v(\mathbf{x})\frac{\partial u(\mathbf{x})}{\partial x_i}d\mathbf{x} + \int_{\partial\Omega} v(\mathbf{x})u(\mathbf{x})n_i(\mathbf{x})\,ds$$

holds, where n_i is the ith component of the unit vector of the outward normal $\mathbf{n}$ to $\partial\Omega$.

As usual we define the spaces $L^p(M)$, $L^{p,N}(M) = [L^p(M)]^N$, $M = \Omega$ or Γ_r, $r = \tau, u, c$, $1 \le p < \infty$, $N = 2, 3$, as the spaces of all measurable functions such that $\|u\|_{L^p(M)} = (\int_M |u(\mathbf{x})|^p d\mathbf{x})^{1/p} = \|u\|_{0,p,M} < \infty$. By $L^\infty(M)$ we denote the set of all measurable functions u defined almost everywhere on M, such that $\|u\|_{L^\infty(M)} = \|u\|_\infty = \text{ess sup}_M |u(\mathbf{x})|$ are finite.

We denote by $W^{k,p}(M)$, $M = \Omega$ or $\Gamma_r, r = \tau, u, c$ the Sobolev space of $L^p(M)$–functions having the (fractional, if k is noninteger) derivatives (in the distributional sense) of the order at most k such that these derivatives belong to $L^p(M)$. For $p = 2$ we write $W^{k,2}(M) \equiv H^k(M)$, and we define $[H^k(M)]^N \equiv H^{k,N}(M)$, $H^{0,N}(M) = L^{2,N}(M)$; the corresponding norm is denoted by $\|\cdot\|_{k,N}$ and the scalar product and the dual pairing by $(\cdot,\cdot)_{k,M}$ and $\langle\cdot,\cdot\rangle_{k,M}$. The dual space of $H^{k,N}(M)$ we denote by $(H^{k,N}(M))' = H^{-k,N}(M)$ and the corresponding norms by $\|\cdot\|_{H^{-k,N}(M)} \equiv \|\cdot\|_{-k,N}$.

Furthermore, we introduce the Bochner spaces $L^\infty(I;X)$, $L^p(I;X)$, $W^{k,p}(I;X)$, $p \in <1, \infty)$, $\overline{I} = [0, T]$, $T < \infty$, X being Banach spaces in the usual way [see Besov et al. (1975), Adams (1975), and Adams and Fournier (2003)]. The space $L^\infty(I;X)$ is the space of all continuous and bounded functions from I into X with the sup-norm $\|v\|_{L^\infty(I;X)} = \text{ess sup}_{t\in I}\|u(t)\|_X$. The space $L^p(I;X)$ is the space of all measurable functions $v: I \to X$ such that $\int_I \|v(t)\|_X^p\,dt < \infty$, and its norm is defined by $\|v(t)\|_{L^p(I;X)} = (\int_I \|v(t)\|_X^p\,dt)^{1/p}$. The space $W^{k,p}(I;X)$, $k \in \mathbb{N}$, $1 \le p \le \infty$, is defined as $W^{k,p}(I;X) = \{v \in L^p(I;X) | \|v^{(j)}\|_{L^p(I;X)} < \infty\ \forall j \le k\}$ and its norm by $\|v\|_{W^{k,p}(I;X)} = (\int_I \sum_{0\le j\le k} \|v^{(j)}\|_{L^p(I;X)}\,dt)^{1/p}$ for $p < \infty$ and for $p = \infty$ by $\|v\|_{W^{k,\infty}(I;X)} = \max_{0\le j\le k} \text{ess sup}_{t\in I}\|v^{(j)}\|_X$. Moreover, for $p = 2$ we put $W^{k,2}(I;X) \equiv H^k(I;X)$, for $k = 0$, we have the space $L^2(I;X)$ with the norn $\|v\|_{L^2(I;X)} = (\int_I \|v(t)\|_X^2\,dt)^{1/2}$.

We also define the space $H^{1/2}(\Gamma) \equiv W^{1/2}(\Gamma)$ for $\Gamma \subset \partial\Omega$ with the norm

$$\|u\|^2_{W^{1/2,2}(\Gamma)} = \|u\|^2_{L^2(\Gamma)} + \int_\Gamma\int_\Gamma \frac{|u(\mathbf{x}) - u(\mathbf{y})|^2}{|\mathbf{x} - \mathbf{y}|^N}\,d\mathbf{x}\,d\mathbf{y}.$$

Theorem 6.12 (Trace Theorem) Let $\Omega \subset \mathbb{R}^N$, $N = 2, 3$, be a domain with the Lipschitz boundary $\partial\Omega$. If $1 \le p < N$, let

$$\frac{1}{q} = \frac{1}{p} - \frac{p-1}{(N-1)p}.$$

If $p = N$, let $q \in [1, +\infty)$ be arbitrary.

Then there exists a linear continuous mapping $\gamma: W^{1,p}(\Omega) \to L^q(\partial\Omega)$ such that if $u \in C(\overline{\Omega}) \cap W^{1,p}(\Omega)$, then $\gamma(u) = u|_{\partial\Omega}$.

If $1 < p < N$, let

$$1 \geq \frac{1}{r} > \frac{1}{p} - \frac{p-1}{(N-1)p}.$$

If $p = N$, let $r \in \{1, +\infty)$ be arbitrary.

Then the mapping $\gamma: W^{1,p}(\Omega) \to L^r(\partial\Omega)$ is compact. The function $\gamma(u)$ is called the trace of the function u.

Theorem 6.13 (Friedrich's Inequality) Let Ω be a bounded domain with the Lipschitz boundary and $\Gamma_u \subset \partial\Omega$, meas $\Gamma_u > 0$. There exists a positive constant $c \equiv c(\Omega)$, such that

$$\|u\|_{1,\Omega} \leq c \left(\int_\Omega |\nabla u|^2 \, d\mathbf{x} \right)^{1/2}$$

holds for all $u \in V = \{u \in H^1(\Omega) : \gamma(u) = 0 \text{ on } \Gamma_u\}$.

Theorem 6.14 (Sobolev Imbedding Theorem) Let $\Omega \subset \mathbb{R}^N$, $N = 2, 3$, be a bounded domain with the Lipschitz boundary.

(i) If $p \in [1, +\infty]$ and $mp > N$, then $W^{m,p}(\Omega) \hookrightarrow \hookrightarrow C(\overline{\Omega})$.

(ii) If $m > k \geq 0$, then $W^{m,p}(\Omega) \hookrightarrow \hookrightarrow W^{k,p}(\Omega)$.
If $m = 1, p = 2$ and $k = 0$, this assertion is known as the Rellich theorem.

(iii) Let $p \geq 1$, $mp < N$ and $1/q = 1/p - m/N$. Then $W^{m,p}(\Omega) \hookrightarrow L^q(\Omega)$.
If $1/r > 1/p - m/N$, then $W^{m,p}(\Omega) \hookrightarrow \hookrightarrow L^r(\Omega)$.
If $mp = N$, then $W^{m,p}(\Omega) \hookrightarrow \hookrightarrow L^q(\Omega)$ for any $q \in [1, +\infty)$.

Lemma 6.1 (Hölder Inequality) Let $f \in L^p(\Omega)$, $g \in L^{p'}(\Omega)$, $p > 1$, $1/p + 1/p' = 1$, $f \cdot g \in L^1(\Omega)$. Then

$$\left| \int_\Omega f(x)g(x) \, dx \right| \leq \int_\Omega |f(x)||g(x)| \, dx \leq \|f\|_{0,p,\Omega} \|g\|_{0,p',\Omega}.$$

Lemma 6.2 (Gronwall) (i) Continuous version: Let $f \in L^1(t_0, t_1)$ be a nonnegative function, and f and φ be continuous functions on (t_0, t_1). If φ satisfies

$$\varphi(t) \leq g(t) + \int_{t_0}^{t} f(\tau)\varphi(\tau) \, d\tau \quad \forall t \in \langle t_0, t_1 \rangle,$$

then

$$\varphi(t) \leq g(t) + \int_{t_0}^{t} f(s)\varphi(s) \exp\left(\int_s^t f(\tau) \, d\tau \right) ds \quad \forall t \in \langle t_0, t_1 \rangle.$$

Moreover, if g is nondecreasing, then

$$\varphi(t) \leq g(t)\exp\left(\int_{t_0}^{t} f(\tau)\,d\tau\right) \quad \forall t \in \langle t_0, t_1 \rangle.$$

(ii) Discrete version: Let

$$\varphi(M) \leq \psi(M) + \sum_{r=0}^{M-1} \chi(r)\varphi(r), \qquad M = 1, \ldots, m, \ \chi(r) \geq 0 \quad \forall r.$$

Then

$$\varphi(m) \leq \psi(m) + \sum_{r=0}^{m-1} \chi(r)\varphi(r) \prod_{s=r+1}^{m-1} (1 + \chi(s)).$$

For the proof see Lions and Magenes (1972) and Quarteroni and Valli (1994).

6.2 VARIATIONAL EQUATIONS AND INEQUALITIES AND THEIR NUMERICAL APPROXIMATIONS

6.2.1 Elliptic Variational Inequalities and Equations and Their Discrete Approximations

In this section we will use the following assumption:

Definition 6.7 Let $a : V \times V \to \mathbb{R}$. We say that $a(\cdot,\cdot)$ is a bilinear form on V if it is linear with respect to each argument. The bilinear form $a(\cdot,\cdot)$ is continuous or bounded if there exists a number $M > 0$ such that

$$|a(u, v)| \leq M\|u\|_V \|v\|_V \quad \forall u, v \in V, \tag{6.3}$$

and it is V–elliptic if there exists a constant $m > 0$ such that

$$a(v, v) \geq m\|v\|_V^2 \quad \forall v \in V, \tag{6.4}$$

and it is symmetric if

$$a(u, v) = a(v, u) \quad \forall u, v \in V. \tag{6.5}$$

Let us consider a real Hilbert space V with inner product $(\cdot,\cdot)$ and associated norm $\|\cdot\|$, and denote by V' the dual space of V. Let K be a closed convex nonempty subset of V. Furthermore, let $a(\cdot,\cdot) : V \times V \to \mathbb{R}$ be a bilinear continuous and V-elliptic form on $V \times V$, and $S : V \to \mathbb{R}$ a continuous linear functional and $j(\cdot) : V \to \mathbb{R} \cup \{+\infty\}$

a convex lower semicontinuous (l.s.c.) and proper functional. We say that the functional $j(\cdot): V \to \mathbb{R} \cup \{+\infty\}$ is proper if $j(\cdot) > -\infty$, $\forall v \in V$ and $j \neq +\infty$. Let there be given an operator $A: V \to V$ such that $a(u, v) = (Au, v)_V$ $\forall u, v \in V$ and the functional S such that $S(v) = (f, v)_V$ $\forall v \in V$. Next, we will distinguish two variational inequality problems of the first and second kinds.

An **elliptic variational inequality of the first kind** is the following:

Problem $\mathcal{P}_1$ Find $u \in K$ such that u is a solution of the variational inequality

$$a(u, v - u) \geq S(v - u) \quad \forall v \in K,$$

or in the operator form (6.6)

$$(Au, v - u) \geq S(v - u) \quad \forall v \in K.$$

An **elliptic variational inequality of the second kind** is the following:

Problem $\mathcal{P}_2$ Find $u \in V$ such that u is a solution of the variational inequality

$$a(u, v - u) + j(v) - j(u) \geq S(v - u) \quad \forall v \in V,$$

or in the operator form (6.7)

$$(Au, v - u) + j(v) - j(u) \geq S(v - u) \quad \forall v \in V.$$

If $K = V$ and/or $j(v) = 0$, then Problems $\mathcal{P}_1$ and $\mathcal{P}_2$ represent variational elliptic problems, that is, the classical variational equations

$$u \in V, \qquad a(u, v) = S(v) \quad \forall v \in V,$$

or in the operator form (6.8)

$$(Au, v)_V = S(v) \quad \forall v \in V.$$

From the theory of variational inequalities it is known that Problem $\mathcal{P}_1$ is a particular case of Problem $\mathcal{P}_2$ if the functional $j(\cdot)$ in Problem $(\mathcal{P}_2)_v$ it is replaced by the indicator functional I_K of K defined by $I_K(v) = 0$ for $v \in K$, and $I_k = +\infty$ for $v \neq K$. The indicator functional I_K is a convex l.s.c. and proper functional. Then Problem $\mathcal{P}_1$ is equivalent to the problem of finding $u \in V$ such that

$$a(u, v - u) + I_K(v) - I_K(u) \geq S(v - u) \quad \forall v \in V.$$

Theorem 6.15 Then Problem $\mathcal{P}_1$ has a unique solution.

For Problem $\mathcal{P}_2$ we have the following result:

Theorem 6.16 For any linear functional $S(v) = (f, v)$, $f \in V$, Problem $\mathcal{P}_2$ has a unique solution. Moreover, the solution depends Lipschitz continuously on $f \in V$.

Theorem 6.17 Let V be a Hilbert space. Let us assume that the bilinear form $a: V \times V \to \mathbb{R}$ is continuous and V–elliptic, the functional $j: V \to j(v) \geq -C_0 > -\infty$

is proper, convex, and l.s.c. on V, and the functional $S: V \to \mathbb{R}$ is a linear continuous functional. Then there exists a unique solution $u \in V$ to the elliptic variational inequality of the second kind (6.7) (Problem $\mathcal{P}_2$).

Lemma 6.3 (Lax–Milgram Lemma for Variational Equation) Let V be a Hilbert space, $a: V \times V \to \mathbb{R}$ be a bounded, V–elliptic bilinear form, and let $S: V \to \mathbb{R}$ be a linear continuous functional. Then there exists a unique solution $u \in V$ to the variational equation $a(u, v) = S(v)\ \forall v \in V$ and

$$\|u\|_V \le m^{-1} \|S\|_{V'}.$$

The proof follows from the previous theorem if $j \equiv 0$ and $K = V$.

If we assume that the bilinear form is symmetric, that is, $a(u, v) = a(v, u)\ \forall u, v \in V$, then problem (6.7) is equivalent to solving the minimum problem

$$\min_{v \in V} L(v), \qquad L(v) = \tfrac{1}{2} a(v, v) + j(v) - S(v). \tag{6.9}$$

Numerical Approximation of Elliptic Variational Inequalities and Equations Let $V_h \subset V$ be a finite element space and let $K_h \subset V_h$, $\forall h$, be a nonempty, convex, and closed set, but in general $K_h \notin K$. Then assume that for the finite element approximation the bilinear form $a(\cdot, \cdot)$ satisfies

$$\begin{aligned}
|a(u_h, v_h)| &\le M \|u_h\|_{V_h} \|v_h\|_{V_h} && \forall u_h, v_h \in V_h,\\
a_h(v_h, v_h) &\ge m \|v_h\|_V^2 && \forall v_h \in V_h,\\
a(u_h, v_h) &= a(v_h, u_h) && \forall u_h, v_h \in V_h.
\end{aligned}$$

where M and m are positive constants.

Then the finite element approximation of (6.6) and (6.7) leads to the following problem:

Problem $(\mathbf{P}_1)_h$ Find $u_h \in K_h$ such that

$$a(u_h, v_h - u_h) \ge S(v_h - u_h) \quad \forall v_h \in K_h. \tag{6.10}$$

Problem $(\mathbf{P}_2)_h$ Find $u_h \in V_h$ such that

$$a(u_h, v_h - u_h) + j(v_h) - j(u_h) \ge S(v_h - u_h) \quad \forall v_h \in V_h. \tag{6.11}$$

Theorem 6.18 Under the stated assumptions on the given data Problem $(\mathcal{P}_1)_h$ has a unique solution.

Next theorems give the weak and strong convergences and a priori estimates of u_h. The convergence results are based on the results of Falk (1974), Mosco and Strang (1974), Ciarlet (1978), Glowinski (1984), Glowinski et al. (1976, 1981),

Brezzi et al. (1977), Nedoma (1998a), Atkinson and Han (2001), and Han and Sofonea (2002).

Theorem 6.19 Let for any $v \in K$ there exist a sequence $\{v_h\}$, $h \to 0$ such that $v_h \in V_h$ and $\|v - v_h\| \to 0$. Let K, K_h be closed convex nonempty subsets of V or V_h, defined above. Let u be a solution of Problem $\mathcal{P}_1$ and u_h of Problem $(\mathcal{P}_1)_h$. Then, under the above-defined assumptions, we obtain

$$\lim_{h \to 0} \|u - u_h\| = 0. \tag{6.12}$$

Next, we introduce a generalization of the Céa inequality to the finite element approximation of elliptic variational inequalities of the first kind.

Theorem 6.20 Let u be a solution of Problem $(\mathcal{P}_1)_v$ and u_h be a solution of Problem $(\mathcal{P}_1)_h$. Then, there exists a constant $c > 0$ independent of h and u, such that

$$\|u - u_h\|_V \leq c \left\{ \inf_{v_h \in K_h} (\|u - v_h\|_V + |R(v_h - u)|^{1/2}) + \inf_{v \in K} |R(v - u_h)|^{1/2} \right\} \tag{6.13}$$

where

$$R(v) = a(u, v) - S(v)$$

is a residuum quantity.

Remark 6.2 In the case if $K_h \subset K$, the second term in (6.13) vanishes because in this case $u_h \in K$ and we have

$$\|u - u_h\|_V \leq c \inf_{v_h \in K_h} (\|u - v_h\|_V + |R(v_h - u)|^{1/2}), \tag{6.14}$$

and we speak about the internal approximation of the elliptic variational inequality of the first kind.

In the case if $K = V$, $K_h = V_h$, then (6.14) reduces to the Céa inequality

$$\|u - u_h\|_V \leq c \inf_{v_h \in V_h} \|u - v_h\|_V \tag{6.15}$$

for a finite element approximation of a variational equation.

Now we introduce an estimate of the approximation of elliptic variational inequalities of the second kind. We will assume as above that $j : V \to \mathbb{R}$ is a proper, convex, and l.s.c. functional, and, moreover, we will assume that j is proper also on V_h. Then we have the following results:

Theorem 6.21 Under the above-defined assumption, the discrete Problem $(\mathcal{P}_2)_h$ has a unique solution.

For an abstract error estimate $u - u_h$ we have the following theorem:

Theorem 6.22 Let u be a solution of Problem $\mathcal{P}_2$ and let u_h be a solution of Problem $(\mathcal{P}_2)_h$. Then, there exists a constant $c > 0$ independent of h and u such that

$$\|u - u_h\|_V \leq c \inf_{v_h \in V_h} \left\{ \|u - v_h\|_V + |R(v_h, u)|^{1/2} \right\}, \tag{6.16}$$

where

$$R(v_h, u) = a(u, v_h - u) + j(v_h) - j(u) - S(v_h - u)$$

represents a residual quantity.

For the proof see Han and Reddy (1999).

Problem $\mathcal{P}_2$ can be also solved by using the regularization method and by the method of Lagrangian multipliers, that is, methods that are very useful for our further applications onto biomechanical model problems. Many problems in the computational biomechanics are problems with friction in elasticity. Therefore, we will introduce these methods for the simplified friction problem, where in (6.7) the functional $j(v)$ will be introduced by

$$j(v) = g \int_{\partial\Omega} |v| \, ds,$$

where $g > 0$ is given.

The idea of the **regularization method** is based on an approximation of the nondifferentiable term $j(\cdot)$ by a family of differentiable ones $j_\varepsilon(\cdot)$, where $\varepsilon > 0$ is a small regularization parameter. Many possible regularization functions can be used for this method. The method will be convergent when $\varepsilon \to 0$. A general convergence result for this method can be found in Glowinski et al. (1981) or in Glowinski (1984).

Then the functional $j(v)$ will be approximated by the regularized functional $j_\varepsilon(v)$, defined as

$$j_\varepsilon(v) = g \int_{\partial\Omega} \varphi_\varepsilon(v) \, ds,$$

where the regularization function $\varphi_\varepsilon(y)$ (see Fig. 6.2) is differentiable with respect to y, and it approximates $|y|$ as $\varepsilon \to 0$ and can be introduced by

$$\varphi_\varepsilon(y) = (y^2 + \varepsilon^2)^{1/2}.$$

Then (6.7) is approximated by the regularized problem

$$u_\varepsilon \in V, \qquad a(u_\varepsilon, v - u_\varepsilon) + j_\varepsilon(v) - j_\varepsilon(u_\varepsilon) \geq S(v - u_\varepsilon) \quad \forall v \in V.$$

Since $j_\varepsilon(\cdot)$ is differentiable, then (6.7) is transformed into a nonlinear equation

$$u_\varepsilon \in V, \qquad a(u_\varepsilon, v) + \left\langle j_\varepsilon'(u_\varepsilon), v \right\rangle \geq S(v) \quad \forall v \in V.$$

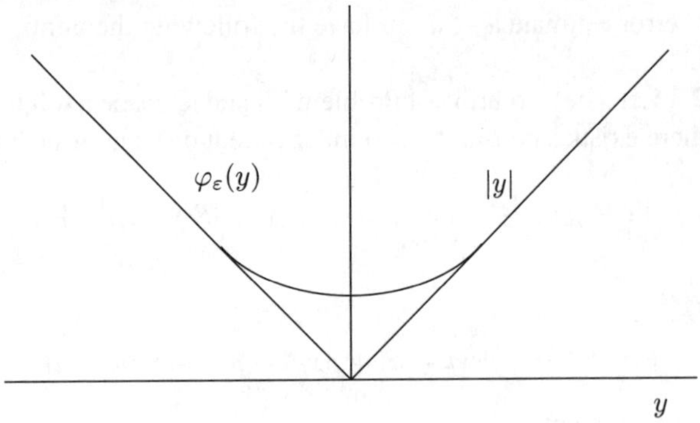

FIG. 6.2 Regularization function.

It can be shown that

$$\|u - u_\varepsilon\|_V \leq c\varepsilon^\beta, \quad \text{where } \beta > 0.$$

But the major problem with this approach is that the conditioning of a regularized problem deteriorates with $\varepsilon \to 0$, thus it is necessary to find the useful choice of the regularized parameter. If ε is too small, the numerical solution of the regularized problem cannot be computed accurately. Therefore, it is desirable to have a posteriori error estimates of the form $\|u - u_\varepsilon\|_V \leq F(u_\varepsilon)$ that can give us computable error bounds as well as a stopping criterion in our computations. On the a posteriori error analysis an adaptive algorithm can be constructed.

The **method of Lagrangian multipliers** for the simplified faction problem can be introduced as follows (Glowinski, 1984; Atkinson and Han, 2001): Let

$$\Lambda = \{\mu \in L^\infty(\partial\Omega) | \, |\mu(\mathbf{x})| \leq 1 \text{ a.e. on } \partial\Omega\}.$$

Then Problem $\mathcal{P}_2$ is equivalent to the problem of finding $u \in V$ and $\lambda \in \Lambda$ such that

$$a(u, v) + g \int_{\partial\Omega} \lambda v \, ds = S(v) \quad \text{for all } v \in V,$$

$$\lambda u = |u| \quad \text{a.e. on } \partial\Omega,$$

where λ is the so-called Lagrangian multiplier.

It can be shown that it is possible to develop an iterative solution procedure for the inequality problem (6.7). Let $\rho > 0$ be a parameter. Then we have the following algorithm:

1. Let us assume $\lambda_0 \in \Lambda$, for example, $\lambda_0 = 0$.
2. For $n = 0, 1, 2, \ldots$ we find $u_n \in V$ as the solution of the boundary value problem

$$a(u_n, v) = S(v) - g \int_{\partial\Omega} \lambda_n v \, ds \quad \text{for all } v \in V,$$

and the Lagrangian multiplier

$$\lambda_{n+1} = \mathcal{P}_\Lambda(\lambda_n + \rho g u_n),$$

where $\mathcal{P}_\Lambda$ is a projection operator to Λ defined by

$$\mathcal{P}_\Lambda(\mu) = \sup(-1, \inf(1, \mu)) \quad \text{for all } \mu \in L^\infty(\partial\Omega).$$

It can be shown that there exists $\rho_0 > 0$ such that if $\rho \in (0, \rho_0)$, then the iterative algorithm converges, that is,

$$u_n \to u \quad \text{in } V, \qquad \lambda_n \to \lambda \quad \text{in } \Lambda.$$

Remark 6.3 Both methods can be used for the numerical approximation of Problem $\mathcal{P}_2$.

Moreover, the following variational inequality problem will be also introduced.

Problem $\mathcal{P}_3$ Find $u \in K$ such that

$$a(u, v-u) + j(v) - j(u) \geq S(v-u) \qquad \forall v \in K$$

or in the operator form (6.17)

$$(Au, v-u)_V + j(v) - j(u) \geq S(v-u) \quad \forall v \in K.$$

Problem $\mathcal{P}_3$ can be solved by using the **penalty method**. Let us assume that the functional $j(v)$ has the following properties:

$$j : V \to \mathbb{R} \cup \{\pm\infty\} \text{ is convex, proper and l.s.c.,}$$
$$j(v) = 0 \iff v \in K,$$
$$j(v) \geq 0.$$

Let $\varepsilon > 0$ be given. Then we define $j_\varepsilon : V \to \mathbb{R} \cup \{\pm\infty\}$ by

$$j_\varepsilon(v) = \frac{1}{\varepsilon} j(v).$$

Then the penalized problem to Problem $\mathcal{P}_3$ is the following:

Problem $(\mathcal{P}_3)_p$ Find $u_\varepsilon \in V$ such that

$$a(u_\varepsilon, v - u_\varepsilon) + j_\varepsilon(v) - j_\varepsilon(u_\varepsilon) \geq S(v - u_\varepsilon) \quad \forall v \in V. \tag{6.18}$$

If j_ε is differentiable, then the solution of (6.18) represents a solution of a nonlinear variational equation

$$u_\varepsilon \in V, \qquad a(u_\varepsilon, v) + \langle j'_\varepsilon(u_\varepsilon), v\rangle = S(v) \quad \forall v \in V, \tag{6.19}$$

where $j'_\varepsilon(v) \in V'$ (where V' is the dual space of V) is the differential of j_ε at v and $\langle \cdot, \cdot \rangle$ is the duality pairing between V' and V.

If $a(\cdot,\cdot)$ is symmetric, then the penalized Problem $(\mathcal{P}_3)_p$ is equivalent to the minimization problem given as follows:

Problem $(\mathcal{P}_3)_{ps}$ Find $u_\varepsilon \in V$ such that

$$L_\varepsilon(u_\varepsilon) \leq L_\varepsilon(v) \quad \forall v \in V, \tag{6.20}$$

where $L_\varepsilon(v) = \frac{1}{2}a(v,v) + j_\varepsilon(v) - S(v)$.

Then, it can be proved by the following result:

Theorem 6.23 Let $V, K, a(\cdot,\cdot), S(\cdot)$ be defined as above. Then

$$\lim_{\varepsilon \to 0} \|u_\varepsilon - u\| = 0, \qquad \lim_{\varepsilon \to 0} j_\varepsilon(u_\varepsilon) = 0, \tag{6.21}$$

where u, u_ε are solutions of Problem $\mathcal{P}_3$ and Problem $(\mathcal{P}_3)_p$, respectively.

For the proof see Haslinger et al. (1996, Theorem 3.5) and also Nedoma (1998a).

Remark 6.4 We see that the penalized Problem $(\mathcal{P}_3)_p$ is transformed to the inequality problem of the second kind.

Next, we will discuss briefly the numerical solution of the above-defined inequality problem:

Problem $(\mathcal{P}_3)_h$ Find $u_h \in K_h$ such that

$$a(u_h, v_h - u_h) + j(v_h) - j(u_h) \geq S(v_h - u_h) \qquad \forall v_h \in K_h$$

or (6.22)

$$(Au_h, v_h - u_h)_V + j(v_h) - j(u_h) \geq S(v_h - u_h) \quad \forall v_h \in K_h.$$

Theorem 6.24 Problem $(\mathcal{P}_3)_h$ has a unique solution.

To estimate the error $u - u_h$, the Falk technique (Falk, 1974) can be used.

Theorem 6.25 Let u be the solution of Problem $\mathcal{P}_3$ and u_h its discrete FE approximation. Then

$$\begin{aligned}\|u - u_h\|_V &\leq c a(u - u_h, u - u_h)^{1/2} \\ &\leq c\{a(u_h - u, v_h - u) + a(u, v - u_h) + a(u, v_h - u) - S(v - u_h) \\ &\quad - S(v_h - u) + j(v_h) - j(u) + j(v) - j(u_h)\}^{1/2} \quad \forall v \in K, \quad \forall v_h \in K_h.\end{aligned} \tag{6.23}$$

Corollary 6.1 Let $K_h \subset K$. Then substituting $v = u_h$ in (6.23) we obtain

$$\|u - u_h\|_V \le c\{a(u_h - u, v_h - u) + a(u, v_h - u) - S(v_h - u) + j(v_h) - j(u)\}^{1/2}$$
$$\forall v_h \in K_h. \tag{6.24}$$

6.2.2 Time-Dependent Variational Inequalities and Their Numerical Solution

Let $\bar{I} = [0, T]$, $T > 0$, $1 < p \le \infty$. Let V be a real Hilbert space with inner product $(\cdot,\cdot)_V$ and norm $\|\cdot\|_V$, and let $a : V \times V \to \mathbb{R}$ be a bilinear form, $j : V \to \mathbb{R}$ a functional, and $f : \bar{I} \to V$. Then we will solve the following problem:

Problem $\mathcal{P}_t$ Find $u : \bar{I} \to V$ such that for a.a. $t \in I$

$$a(u(t), v - u'(t)) + j(v) - j(u'(t)) \ge (f(t), v - u'(t)) \quad \forall v \in V, \tag{6.25}$$
$$u(0) = u_0. \tag{6.26}$$

Next, the following assumptions are introduced:

Presumption Let

$$\left.\begin{array}{l}
a : V \times V \to \mathbb{R} \text{ be a symmetric bilinear form and} \\
\quad \text{(i) there exists } M > 0 \text{ such that} \\
\qquad |a(u, v)| \le M\|u\|_V \|v\|_V \quad \forall u, v \in V, \\
\quad \text{(ii) there exist } m > 0 \text{ such that} \\
\qquad a(v, v) \ge m\|v\|_V^2 \quad \forall v \in V. \\
j \text{ be a continuous seminorm on } V \text{ such that it is continuous} \\
\quad \text{with respect to the norm } \|\cdot\|_V, \\
f \in W^{1,p}(I; V), \\
u_0 \in V \quad \text{and} \quad a(u_0, v) + j(v) \ge (f(0), v)_V \quad \forall v \in V
\end{array}\right\} \tag{6.27}$$

Remark 6.5 A seminorm $|\cdot|_V$ is a nonnegative function from V to $\mathbb{R}$ satisfying the properties of the norm (ii) and (iii) of Section 6.1.1, but $|u|_V = 0$ need not imply that $u = 0$.

The following theorem gives the well-posedness of the Problem $\mathcal{P}_t$.

Theorem 6.26 Let us assume that the above-defined assumptions (6.27) hold. Then there exists a unique solution $u \in W^{1,p}(I; V)$ to Problem $\mathcal{P}_t$. Moreover, the mapping $(f, u_0) \mapsto u$ is Lipschitz continuous from $W^{1,1}(I; V) \times V$ to $L^\infty(I; V)$.

The proof is based on a time discretization (Rothe) method, compactness, and lower semicontinuity arguments. For details see Han and Sofonea (2002). It is similar to that in the hyperbolic variational inequality case, presented below.

The numerical solution can be based on the semidiscrete or fully discrete and/or Rothe methods. We limit ourselves to the the semidiscrete method only. In the case of the semidiscrete method we have two types of schemes. In the first type we replace the space V by a finite-dimensional subspace V_h. Thus we obtain a finite-dimensional system of ordinary differential equations and inequalities and we speak about the spatially semidiscrete schemes. In the second type we replace the time derivatives by finite differences. Then the problem leads to solving elliptic variational inequalities over the space V for every time level. In this case we speak about the temporally semidiscrete schemes. Analyses of these methods are close to that of the fully discrete methods. Further, Problem $\mathcal{P}_t$ will be approximated by the spatially semidiscrete method.

Let $\{V_h\}$ be a family of finite-dimensional subspaces of V such that

$$\lim_{h\to 0} \inf_{v_h \in V_h} \|v - v_h\|_V = 0 \quad \forall v \in V.$$

Then Problem $\mathcal{P}_t$ will be approximated as follows:

Problem $(\mathcal{P}_t)_h$ Find $u_h : \bar{I} \to V_h$ such that

$$u_h(0) = r_h u_0 \tag{6.28}$$

and for almost all $t \in I$,

$$a(u_h(t), v_h - u_h'(t)) + j(v_h) - j(u_h'(t)) \geq (f(t), v_h - u_h'(t)) \quad \forall v_h \in V_h, \tag{6.29}$$

where $r_h u_0 \in V_h$ is the finite element projection of u_0 on V_h such that

$$a(r_h u_0, v_h) = a(u_0, v_h) \quad \forall v_h \in V_h.$$

Note that $r_h u_0 \in V_h$ is unique because of the properties of the bilinear form $a(\cdot, \cdot)$.

Theorem 6.27 Problem $(\mathcal{P}_t)_h$ has a unique solution $u_h \in W^{1,p}(I; V_h)$.

Proof The existence of a unique solution u_h to Problem $(\mathcal{P}_t)_h$ follows from Theorem 6.26, where we replace V by V_h. For more details see Han and Sofonea (2002). □

For an estimate of the semidiscrete approximation error $u - u_h$ we have the following result:

Theorem 6.28 Let assumptions (6.27) be satisfied. Let V_h be a finite-dimensional subspace of V. Let $u \in W^{1,p}(I; V)$, $u_h \in W^{1,p}(I; V_h)$ be solutions of Problem $\mathcal{P}_t$ and Problem $(\mathcal{P}_t)_h$, respectively. Then the semidiscrete approximation error

$$\begin{aligned} \|u - u_h\|_{L^\infty(I;V)} &\leq c\|u_0 - r_h u_0\|_V \\ &+ c \inf_{v_h \in L^2(I;V_h)} \left\{ \|u' - v_h\|_{L^2(I;V)} + \|R(\cdot; v_h(\cdot), u'(\cdot))\|_{L^1(I)}^{1/2} \right\} \end{aligned} \tag{6.30}$$

holds.

For the proof see Han and Sofonea (2002).

The last theorem is a basis for a convergence analysis and optimal error estimates.

6.2.3 Hyperbolic Variational Inequalities and Equations and Their Discrete Approximations

Let $\bar{I} = [0, T]$, $T > 0$, $1 < p \leq \infty$. Let V be a real Hilbert space with inner product $(\cdot, \cdot)_V$ and norm $\|\cdot\|_V$, and let $a : V \times V \to \mathbb{R}$ be a bilinear form, $j : V \to \mathbb{R}$ a functional, and $f : \bar{I} \to V$. Then we will solve the following hyperbolic variational inequality problem:

Problem ($\mathcal{P}_{\text{hyp}}$) Find $u \in V$ such that for a.a. $t \in I$

$$(u''(t), v - u'(t)) + a(u(t), v - u'(t)) + j(v) - j(u'(t)) \geq (f(t), v - u'(t)) \quad \forall v \in V, \tag{6.31}$$

$$u(0) = u_0, \qquad u'(0) = u_d, \tag{6.32}$$

where $u_0, u_d \in V$, and let

$$\left.\begin{array}{l}
a : V \times V \to \mathbb{R} \text{ be a symmetric bilinear form and} \\
\quad \text{(i) there exist } M > 0 \text{ such that} \\
\qquad |a(u, v)| \leq M \|u\|_V \|v\|_V \quad \forall u, v \in V, \\
\quad \text{(ii) there exist } m > 0 \text{ such that} \\
\qquad a(v, v) \geq m \|v\|_V^2 \quad \forall v \in V. \\
j \text{ be a continuous seminorm on } V \text{ such that it is continuous} \\
\quad \text{with respect to the norm } \|\cdot\|_V, \\
f \in W^{1,p}(I; V).
\end{array}\right\} \tag{6.33}$$

Let the interval I be divided into n subintervals (t_i, t_{i+1}), $t_0 = 0$, $t_n = T$, $k = t_{i+1} - t_i = T/n$. Let $u_i^{(0)} \equiv u_i = u(t_i)$, $u_i^{(1)} \equiv \delta_k u_i = (u_{i+1} - u_i)/k$, $u_i^{(2)} \equiv \delta_k^2 u_i = (u_{i+1} - 2u_i + u_{i-1})/k^2$, $u_i^{(p)} \equiv \delta_k^p u_i = \delta_k(\delta_k^{p-1} u_i)$, and let us denote by $U_n^{(p)}(t) = u_i^{(p)} + 1/k(t - t_i)(u_{i+1}^{(p)} - u_i^{(p)})$ and $\overline{U}_n^{(p)}(t) = u_i^{(p)}$ for $t_i \leq t \leq t_{i+1}$, and where $u_n(t) = U_n^{(0)}(t)$ corresponds to the **Rothe method**, and $\bar{u}_n(t) = \overline{U}_n^{(0)}(t)$ is the corresponding step function [see Kačur (1985)].

To prove the existence of a regular solution of the Problem ($\mathcal{P}_{\text{hyp}}$), we will assume that there exists $z_0 \in V$ (where $\delta_k^2 u \overset{\text{def}}{\equiv} z_0$) such that

$$(z_0, v - u_d) + a(u_0, v - u_d) + j(v) - j(u_d) \geq (f(0), v - u_d) \quad \forall v \in V, \tag{6.34}$$

and the existence analysis is based on the Rothe approach.

Moreover, we introduce the operator $A_k : V \to V$ by $(A_k u, v) = 1/k^2(u, v) + a(u, v)$ $\forall u, v \in V$. If the bilinear form $a(\cdot, \cdot)$ satisfies the properties in (6.33), then the form $(A_k u, v)$ satisfies the same properties as the bilinear form $a(\cdot, \cdot)$.

For the existence proof we need the following lemmas, which are modifications of Lemmas 4.1.9 and 4.1.13 of Kačur (1985), which are valid also for the hyperbolic variational inequality in linear elasticity and which we will introduce without proofs.

Lemma 6.4 Let the interval I be divided into n subintervals (t_i, t_{i+1}). Under assumption (6.34) there exist $n_0 > 0$ and the constant c such that for $i = 1, \ldots, n-1$, and $n \geq n_0$, the following estimates

$$|\delta_k^2 u_{i+1}| \leq c, \quad \|\delta_k u_{i+1}\| \leq c, \quad \|u_{i+1}\| \leq c \tag{6.35}$$

hold.

From this lemma it follows

$$\left|\overline{U}_n^{(2)}(t)\right| \leq c, \quad \left\|\overline{U}_n^{(1)}(t)\right\| \leq c, \quad \|\overline{u}_n(t)\| \leq c,$$

and hence

$$\begin{aligned} &|U_n^{(1)}(t) - U_n^{(1)}(t')| \leq c|t-t'|, \qquad \|u_n(t) - u_n(t')\| \leq c|t-t'|, \\ &|\overline{U}_n^{(1)}(t) - U_n^{(1)}(t)| \leq \frac{c}{n}, \qquad \|\overline{u}_n(t) - u_n(t)\| \leq \frac{c}{n} \\ &\qquad\qquad \text{for all } n \geq n_0 \quad \text{and} \quad t, t' \in I. \end{aligned} \tag{6.36}$$

Lemma 6.5 There exists $u \in C(I; V)$ with $u' \in L^\infty(I; V) \cap C(I; L^2(\Omega))$ and $u'' \in L^\infty(I; L^2(\Omega))$ such that $u_n \to u$ in $C(I; V)$, $U_n^{(1)} \to u'$ in $C(I; L^2(\Omega))$. Moreover, $\overline{U}_n^{(1)}(t) \to u'$ in V for all $t \in I$ and $dU_n^{(1)}/dt \rightharpoonup u''$ in $L^2(I; L^2(\Omega))$.

As a result we have the following result:

Theorem 6.29 Let $u_0, u_1 \in V$ and let (6.33) and (6.34) hold. Then there exists a unique solution u of Problem $(\mathcal{P}_{\text{hyp}})$ such that

$$u \in C(I; V), \qquad u' \in L^\infty(I; V) \cap C(I; L^2(\Omega)), \qquad u'' \in L^2(I; \Omega).$$

Proof The proof will be based on the discretization (Rothe) method and under the assumption that the corresponding elliptic operator is pseudomonotone and coercive in $V \to V'$ for $k \leq k_0$, $k_0 > 0$. Since $a(u, v) = (A, u, v)$ $\forall u, v \in V$, t being fixed, the operator $A : V \to V'$ is a bounded pseudomonotone operator (see Definition 6.2) and is coercive, that is, $a(u, u) = (Au, u) \leq c\|u\|^2$ for all $u \in V, t \in I$.

We introduce the corresponding approximation of (6.31), where we put

$$\delta_k u_i = \frac{u_i - u_{i-1}}{k}, \qquad \delta_k^2 u_i = \frac{u_i - 2u_{i-1} + u_{i-2}}{k^2} = \frac{1}{k}(\delta_k u_i - \delta_k u_{i-1}),$$

$$f_i = \frac{1}{k}\int_{t_{i-1}}^{t_i} f(\tau)\, d\tau.$$

Then under the above-defined assumptions

$$(\delta_k^2 u_i, v - \delta_k u_i) + a(u_i, v - \delta_k u_i) + j(v) - j(\delta_k u_i) \geq (f_i, v - \delta_k u_i) \quad \forall v \in V. \quad (6.37)$$

Since $\delta_k^2 u_i = \frac{1}{k}(\delta_k u_i - \delta_k u_{i-1})$ and since

$$a(u_i, v - \delta_k u_i) = ka(\delta_k u_i, v - \delta_k u_i) + \sum_{j=0}^{i-1} ka(\delta_k u_j, v - \delta_k u_i) + a(u_0, v - \delta_k u_i),$$

then (6.37) represents an elliptic variational inequality for $\delta_k u_i$ provided $\delta_k u_j$, $j = 1, \ldots, i-1$, are known.

From (6.37) we have

$$(\overline{U}_n^{(2)}(t), v - \overline{U}_n^{(1)}(t)) + a(\overline{u}_n(t), v - \overline{U}_n^{(1)}(t)) + j(v) - j(\overline{U}_n^{(1)}(t))$$
$$\geq (\overline{f}_n(t), v - \overline{U}_n^{(1)}(t)) \quad \forall v \in V \text{ and } t \in I. \qquad (6.38)$$

Then from (6.38) and a priori estimates (6.36), we have $|j(\overline{U}_n^{(1)}(t))| \leq c$ for all $t \in I$. Then the results of the above Lemmas and the Fatou lemma yield (Kačur, 1985)

$$\int_{t_1}^{t_2} j(u'(t))\, dt \leq \liminf_{n\to\infty} \int_{t_1}^{t_2} j(\overline{U}_n^{(1)}(t))\, dt$$

since $\overline{U}_n^{(1)}(t) \rightharpoonup u'(t)$ in V for all $t \in I$ and the functional $j(\cdot)$ is weakly l.s.c on V. Integrating (6.38) over the interval (t_1, t_2) and using the limit $n \to \infty$ we obtain (6.31). The uniqueness follows from the existence and uniqueness of the solution, which completes the proof. □

Remark 6.6 Numerically Problem ($\mathcal{P}_{\text{hyp}}$) can be solved by the Rothe method with a finite element method and by the semi-implicit schemes in time and a finite element method in space.

6.2.4 Nonlinear Variational Equations and Inequalities and Their Numerical Solutions

Let the bilinear form in (6.6) or (6.7) be replaced by

$$a(u, v) = (A(u), v) \quad \text{for all } u, v \in V, \qquad (6.39)$$

where $A(u)$ is the operator $A : V \to V$ such that

(i) there exists $M > 0$ such that

$$\|A(u) - Av\| \leq M\|u - v\| \quad \forall u, v \in V;$$

(ii) there exists $m > 0$ such that

$$(A(u) - A(v), u - v) \geq m\|u - v\|^2 \quad \forall u, v \in V.$$

Then the variational inequality (6.6) with the form $a(\cdot,\cdot)$ has a unique solution and similarly for the variational inequality (6.7) provided $j: V \to \mathbb{R} \cup \{\pm\infty\}$ is a proper, convex, and lower semicontinuous functional. For the proof see Lions (1969) and Han and Sofonea (2002).

If $K = V$, then the nonlinear variational inequality with the form (6.39) leads to the variational equation and similarly for the variational inequality (6.7).

In solving nonlinear variational inequalities and variational equalities it is convenient to transform the nonlinear problem into a sequence of linear problems, which can be constructed by the method of contraction or by the secant modules method.

The method of contraction is one of simple, but very effective method for solving nonlinear inequalities and equations.

Theorem 6.30 Let the operator $\mathcal{A}: V \to V$, V being a Hilbert space, be Lipschitz continuous and suppose that

$$\|\mathcal{A}u - \mathcal{A}v\| \leq c\|u - v\|, \quad \text{where } 0 \leq c = \text{const} < 1.$$

Then there exists a unique point, the fixed point of operator $\mathcal{A}$, and this fixed point may be found by means of the iteration process $u_{n+1} = \mathcal{A}u_n$, where $u_0 \in V$ and $u_{n+1} \to u$. Moreover, the error bound

$$\|u_{n+1} - u_n\| \leq \frac{c^{n+1}}{1-c}\|u_0 - \mathcal{A}u_0\|$$

is valid for the $(n+1)$th iterate.

For the proof see Nečas and Hlaváček (1981).

The rate of convergence of this method is linear, that is,

$$\frac{\|u_{n+1} - u_n\|}{\|u_n - u_{n-1}\|} \leq q < 1.$$

Then in every step of the contraction method we solve the simpler problem of a variational inequality or variational equation, respectively, with a bilinear form $a(\cdot,\cdot)$.

The Secant-Modulus (Kachanov) Method for Variational Inequalities

Let us introduce the following assumptions:

Assumptions

(i) Let K be a nonempty closed convex set in the real Hilbert space V, let $b \in V'$, and let $u_0 \in K$ be given.

(ii) The functional $L: K \to \mathbb{R}$ is G–differentiable. There exists a functional $B: K \times V \times V \to \mathbb{R}$ such that for $u, v, w \in K$, we have the representation formula

$$\langle L'(u), v - w \rangle = B(u; u, v - w) \tag{6.40}$$

and the key inequality

$$L(v) - L(u) \leq 2^{-1}(B(u; v, v) - B(u; u, u)). \tag{6.41}$$

(iii) For each $u \in K$, the map $(v, w) \mapsto B(u; v, w)$ is bilinear, bounded, and symmetric from $V \times V$ to $\mathbb{R}$. There exist numbers $\rho > 0$ and $\delta > 0$ such that

$$|B(u; v, w)| \leq \delta \|v\| \, \|w\| \quad \text{for all } u \in K, v, w \in V$$

and

$$B(u; w - v, w - v) \geq \rho \|w - v\|^2 \quad \text{for all } u, v, w \in K.$$

(iv) The operator $L': K \to V'$ is continuous and strongly monotone, that is, there is a number $\rho_0 > 0$ such that

$$\langle L'(u) - L'(v), u - v \rangle \geq \rho_0 \|u - v\|^2 \quad \text{for all } u, v \in K.$$

Theorem 6.31 Let (i)–(iii) hold. Let $j: V \to \mathbb{R}$ be weakly lower semicontinuous

$$j(v) \geq 0 \quad \forall v \in V \quad \text{and} \quad |j(u) - j(v)| \leq C\|u - v\| \quad \forall u, v \in V,$$

Then

(a) For $k = 0, 1, \ldots$, the quadratic variational problem

$$2^{-1}B(u_k; u_{k+1}, u_{k+1}) - b(u_k + 1) \to \min \quad u_{k+1} \in K, \tag{6.42}$$

has a unique solution u_{k+1}, and

$$B(u_k; u_{k+1}, v - u_{k+1}) - j(v) - j(u_{k+1}) \geq b(v - u_{k+1}) \quad \text{for all } v \in K. \tag{6.43}$$

(b) If, in addition, (iv) holds, then the original variational problem

$$L(u) - b(u) \to \min \quad u \in K,$$

has a unique solution u. Moreover, u is the unique solution of the variational inequality

$$\langle L'(u), v - u \rangle + j(v) - j(u) \geq b(v - u) \quad \text{for all } v \in K \text{ and fixed } u \in K, \tag{6.44}$$

and the secant-modules method converges, that is,

$$u_k \to u \text{ in } V \text{ as } k \to \infty.$$

For the proof see Han et al. (1997); for $j \equiv 0$ see Zeidler (1990).

Corollary 6.2 Let (i)–(iii) hold. Suppose that the set K is compact and that the map

$$u \mapsto B(u; v, w)$$

is equicontinuous on K with respect to all $v, w \in V$.

Then the sequence (u_k) has at least one cluster point and each such cluster point u is a solution of the variational inequality (6.44) (with $j \equiv 0$).

For more details see Fučík and Kufner (1983), Nečas and Hlaváček (1983), Nečas (1983), Glowinski (1984), Kačur (1985), Feistauer and Sobotíková (1990), Ženíšek (1990), Zeidler (1990a,b), and Han et al. (1997).

6.3 BIOMECHANICAL MODELS OF HUMAN JOINTS AND THEIR TOTAL REPLACEMENTS

6.3.1 Introduction

Recent trends in total joint replacements and clinical failures have provided a strong impetus for the development of accurate stress–strain analyses in normal and prosthetic joints, which in the future will be also required in the navigated surgical techniques, where an optimal super-fast algorithm will be needed. The success of artificial replacements of human joints depend on many factors. The mechanical factor is an important one.

Mechanical loading—static and especially dynamic loadings—is presupposed to play an important role in the origin, development, and progression of joint osteoarthritis. Besides the dynamic loading (i.e., contact pressures), motion (i.e., kinematics) also contributes to an interactive effect of progression of joint osteoarthritis. The idea of a prosthesis being a device that transfers the joint loads to the bone allows one to explain the mechanical factor in terms of the load transfer mechanism. A complex relation exists between this mechanism and the magnitude and direction of the loads, the geometry of the bone/joint–prosthesis configuration, the elastic properties of the materials, and the physical connections at the material connections, where motion and loads influence wear. Similar situations are observed in the normal and artificial knee, which result in an osteolysis and in a worse-case implant failure. Therefore, knowledge of in vivo human joint motion and loading during functional activities is necessary to comprehend knowledge of human joint degeneration and restoration.

It can be shown that the static, quasi-static, and dynamic contact problems in suitable rheology, and their finite element approximations, are very useful tools for analyzing human joints and their artificial replacements. Such problems lead to solving variational inequalities, which physically describe the principle of virtual work in its inequality form.

In this section we will deal with new methodologies for simulating deformable contacts in the human joints within a mathematical model of multibody dynamics, based on the analyses of the quasi-static and dynamic contact multibody problems in (thermovisco)elastic rheology. Moreover, we will investigate the dynamic multibody contact problem with Coulomb friction in linear viscoelasticity with short and long memories and damping in N dimensions.

6.3.2 Formulation of the Model Problems

The physical settings of the investigated biomechanical problems are as follows. Let $I=(0,t_p)$ be a time interval, $t_p<+\infty$. Let $\Omega\subset\mathbb{R}^N, N=2,3$, be a region occupied by a system of bodies of arbitrary shapes Ω^ι such that $\Omega=\cup_{\iota=1}^s\Omega^\iota$. Let Ω^ι be domains with Lipschitz boundaries $\partial\Omega^\iota$, and let us assume that $\partial\Omega=\Gamma_\tau\cup\Gamma_u\cup\Gamma_0\cup\Gamma_c$, where the disjoint parts Γ_τ, Γ_u Γ_0 are open subsets. Moreover, let $\Gamma_u={}^1\Gamma_u\cup{}^2\Gamma_u$ and $\Gamma_c=\cup_{k,l}\Gamma_c^{kl}, \Gamma_c^{kl}=\partial\Omega^k\cap\partial\Omega^l, k\neq l, k,l\in\{1,\ldots,s\}$. Let $\Omega_t=I\times\Omega$ denote the time–space domain and let $\Gamma_\tau(t)=\Gamma_\tau\times I$, $\Gamma_u(t)=\Gamma_u\times I$, $\Gamma_c(t)=\Gamma_c\times I$ denote the parts of its boundary $\partial\Omega_t=\partial\Omega\times I$.

Furthermore, let $\mathbf{n}$ denote the outer normal unit vector on the boundary, $u_n=u_in_i$, $\mathbf{u}_t=\mathbf{u}-u_n\mathbf{n}$, $\tau_n=\tau_{ij}n_jn_i$, $\boldsymbol{\tau}_t=\boldsymbol{\tau}-\tau_n\mathbf{n}$ be normal and tangential components of displacement and stress vectors $\mathbf{u}=(u_i)$, $\boldsymbol{\tau}=(\tau_i)$, $\tau_i=\tau_{ij}n_j$, $i,j=1,\ldots,N$. Let $\mathbf{F}(t,\mathbf{x}),\mathbf{P}(t,\mathbf{x})$ be the body and surface forces, $\rho(\mathbf{x})$ the density, $\alpha(\mathbf{x})\geq 0$ physically means the damping, W be the thermal sources, $c_e(\mathbf{x})$ be the specific heat, $c(\mathbf{x})=\rho(\mathbf{x})c_e(\mathbf{x})$ the thermal capacity, and β_{ij} the coefficient of linear thermal expansion. The respective time derivatives are denoted by a prime. Let us denote by $\mathbf{u}'=(u_i')$ the velocity vector. Let Γ_c^{kl} the positive direction of the outer normal vector $\mathbf{n}$ be assumed with respect to Ω^s.

The stress–strain relation will be defined by the generalized Hooke or the Duhamel–Neumann law:

(a) $\tau_{ij}=\tau_{ij}(\mathbf{u})=c_{ijkl}^{(0)}(\mathbf{x})e_{kl}(\mathbf{u})$, $\quad i,j,k,l=1,\ldots,N$, in the linear elasticity,

(b) $\tau_{ij}=\tau_{ij}(\mathbf{u},T)=c_{ijkl}^{(0)}(\mathbf{x})e_{kl}(\mathbf{u})-\beta_{ij}(T-T_0)={}^e\tau_{ij}+{}^T\tau_{ij}$, $\quad i,j,k,l=1,\ldots,N$, in the linear thermoelasticity,

(c) $\tau_{ij}=\tau_{ij}(\mathbf{u},\mathbf{u}')=c_{ijkl}^{(0)}(\mathbf{x})e_{kl}(\mathbf{u})+c_{ijkl}^{(1)}(\mathbf{x})e_{kl}(\mathbf{u}')$, $\quad i,j,k,l=1,\ldots,N$, in viscoelastic rheology with short memory,

(d) $\tau_{ij}=\tau_{ij}(\mathbf{u},\mathbf{u}',T)=c_{ijkl}^{(0)}(\mathbf{x})e_{kl}(\mathbf{u})+c_{ijkl}^{(1)}(\mathbf{x})e_{kl}(\mathbf{u}')-\beta_{ij}(T-T_0)$
$={}^e\tau_{ij}+{}^v\tau_{ij}+{}^T\tau_{ij}$, $\quad i,j,k,l=1,\ldots,N$, in thermoviscoelastic rheology with short memory,

(e) $\tau_{ij}=c_{ijkl}^{(0)}e_{kl}(\mathbf{u}(t,\cdot))+\int_0^t b_{ijkl}(t-\tau)e_{kl}(\mathbf{u}(\tau,\cdot))\,d\tau-\beta_{ij}(T-T_0)$
$={}^e\tau_{ij}+{}^v\tau_{ij}+{}^T\tau_{ij}$, $\quad i,j,k,l=1,\ldots,N$, in thermoviscoelastic rheology with long memory, and where

$$e_{ij}(\mathbf{u})=\frac{1}{2}\left(\frac{\partial u_i}{\partial x_j}+\frac{\partial u_j}{\partial x_i}\right),\qquad i,j=1,\ldots,N, \tag{6.45}$$

where $c_{ijkl}^{(n)}(\mathbf{x})$, $n=0,1$, are elastic (for $n=0$) and viscous (for $n=1$) coefficients; $b_{ijkl}=b_{ijkl}(\mathbf{x},t)$ represent the material memory coefficients, depending on t and $\mathbf{x}$, bounded in $\mathbf{x}, t$ and satisfying the symmetry and regularity conditions; and $e_{ij}(\mathbf{u})$ are components of the small strain tensor, T is the temperature, and T_0 the initial temperature, and N is the space dimension. For the tensors $c_{ijkl}^{(n)}(\mathbf{x}), n=0,1$, we assume the properties of symmetry, ellipticity, and boundedness and for β_{ij} the symmetry conditions, that is, we assume that

$$c_{ijkl}^{(n)} \in L^\infty(\Omega), \qquad c_{ijkl}^{(n)} = c_{jikl}^{(n)} = c_{klij}^{(n)} = c_{ijlk}^{(n)}, \qquad n = 0, 1,\ i,j,k,l = 1,\dots,N.$$

$$c_{ijkl}^{(n)} e_{ij} e_{kl} \geq c_0^{(n)} e_{ij} e_{ij} \quad \forall e_{ij}, \qquad e_{ij} = e_{ji} \quad \text{and a.e. } \mathbf{x} \in \Omega,\ c_0^{(n)} > 0.$$

$$c_{ijkl}^{(n)} = \lambda^{(n)} \delta_{ij}\delta_{kl} + \mu^{(n)}(\delta_{ik}\delta_{jl} + \delta_{il}\delta_{jk}) \text{ for the isotropic materials,}$$

$$b_{ijkl} = b_{jikl}, \qquad b_{ijkl}, \frac{\partial b_{ijkl}}{\partial t}, \frac{\partial^2 b_{ijkl}}{\partial t^2} \in L^\infty(\Omega \times I), \qquad i,j,k,l = 1,\dots,N.$$

$$\beta_{ij} \in C^1(\overline{\Omega}), \quad \beta_{ij} = \beta_{ji}, \qquad \beta_{ij} \geq \beta_0 > 0, \qquad i,j = 1,\dots,N, \tag{6.46}$$

where $\lambda^{(n)}$, $\mu^{(n)}$ for $n=0$ are the Lamé elastic coefficients and for $n=1$ the viscous coefficients for isotropic materials, ${}^e\tau_{ij}$, ${}^v\tau_{ij}$, ${}^T\tau_{ij}$ are the elastic, viscous, and thermal stresses, and a repeated index implies the summation from 1 to N.

From the momentum conservation law and the entropy density law the equilibrium equation, the equation of motion, and the expanded equation of heat conduction

$$\frac{\partial \tau_{ij}^\iota(\mathbf{u}^\iota, \mathbf{u}^{\iota\prime}, T^\iota)}{\partial x_j} + F_i^\iota = 0, \tag{6.47a}$$

$$-\frac{\partial}{\partial x_j}\left(\kappa_{ij}^\iota \frac{\partial T^\iota}{\partial x_i}\right) = W^\iota \tag{6.47b}$$

for the static case and for the dynamic case

$$\rho^\iota \frac{\partial^2 u_i^\iota}{\partial t^2} + \alpha^\iota \frac{\partial u_i^\iota}{\partial t} = \frac{\partial \tau_{ij}^\iota(\mathbf{u}^\iota, \mathbf{u}^{\iota\prime}, T^\iota)}{\partial x_j} + F_i^\iota, \tag{6.47c}$$

$$\rho^\iota c_e^\iota \frac{\partial T^\iota}{\partial t} - \frac{\partial}{\partial x_j}\left(\kappa_{ij}^\iota \frac{\partial T^\iota}{\partial x_i}\right) = W^\iota - \rho^\iota \beta_{ij}^\iota T_0^\iota e_{ij}(\mathbf{u}^{\iota\prime}) + c_{ijkl}^{(1)\iota} e_{kl}(\mathbf{u}^{\iota\prime}) e_{ij}(\mathbf{u}^{\iota\prime}), \tag{6.47d}$$

where $i,j=1,\dots,N, \iota=1,\dots,r, (t,\mathbf{x}) \in \Omega_t^\iota = I \times \Omega^\iota, \iota=1,\dots,s,$ hold, where T_0^ι are the initial temperatures of nondeformable parts of the human skeletal system, $\kappa_{ij}^\iota(\mathbf{x})$, $\iota=1,\dots,s$, the coefficients of thermal conductivity that are symmetric and uniformly positive definite, that is,

$$\kappa_{ij}^\iota(\mathbf{x}) = \kappa_{ji}^\iota(\mathbf{x}), \qquad \kappa_{ij}^\iota(\mathbf{x})\zeta_i\zeta_j \geq \kappa_0^\iota |\zeta|^2, \qquad \kappa_0^\iota = \text{const} > 0, \qquad \zeta \in \mathbb{R}^N, \qquad N = 2,3 \tag{6.48}$$

and where the term $\rho^\iota \beta_{ij}^\iota T_0^\iota e_{ij}(\mathbf{u}^{\iota\prime})$ represents the deformation energy that changes into heat, and the term $c_{ijkl}^{(1)\iota} e_{ij}(\mathbf{u}^{\iota\prime}) e_{kl}(\mathbf{u}^{\iota\prime})$ is the additional viscous dissipation and represents one of the heat sources in the heat equation, in order to satisfy the first thermodynamic law.

We introduce the following contact problems:

A. Multibody Unilateral Contact Problems in (Thermo)elasticity, Static Case

Problem $\mathcal{P}$ Let $N=2,3,\ s\geq 2$. Find a pair of functions $(T,\mathbf{u}):\Omega\to\mathbb{R}\times\mathbb{R}^N$, $N=2,3$, satisfying (6.47ca) with (6.45b), (6.46), (6.48), and a boundary value and unilateral contact conditions with Coulombian friction of the form

$$T(\mathbf{x})=T_1(\mathbf{x}),\qquad \tau_{ij}n_j=P_i,\qquad i,j=1,\ldots,N\quad \text{on } \Gamma_\tau \tag{6.49}$$

$$T(\mathbf{x})=T_1(\mathbf{x}),\qquad \mathbf{u}=0\quad \text{on } \Gamma_u, \tag{6.50}$$

$$\begin{aligned}
&\text{(a)}\ T^k(\mathbf{x})=T^l(\mathbf{x}), && \kappa_{ij}\frac{\partial T(\mathbf{x})}{\partial x_i}n_{j|(k)}=-\kappa_{ij}\frac{\partial T(\mathbf{x})}{\partial x_i}n_{j|(l)},\\
&\text{(b)}\ u_n^k-u_n^l\leq 0, && \tau_n^k=\tau_n^l\equiv\tau_n^{kl}\leq 0,\quad (u_n^k-u_n^l)\tau_n^{kl}=0,\\
&\text{(c)}\ \boldsymbol{\tau}_t^k(\mathbf{u})=-\boldsymbol{\tau}_t^l(\mathbf{u})\equiv\boldsymbol{\tau}_t^{kl}(\mathbf{u}, && |\boldsymbol{\tau}_t^{kl}(\mathbf{u})|\leq g_c^{kl}\equiv\mathcal{F}_c^{kl}|\tau_n^{kl}(\mathbf{u})|\\
&\quad \text{if } |\boldsymbol{\tau}_t^{kl}(\mathbf{u})|<g_c^{kl}, && \text{then } \mathbf{u}_t^k(\mathbf{x})-\mathbf{u}_t^l(\mathbf{x})=0,\\
&\quad \text{if } |\boldsymbol{\tau}_t^{kl}(\mathbf{u})|=g_c^{kl}, && \text{then there exists } \Theta\geq 0 \text{ such that } \mathbf{u}_t^k(\mathbf{x})-\mathbf{u}_t^l(\mathbf{x})=-\Theta\boldsymbol{\tau}_t^{kl}(\mathbf{u}),
\end{aligned} \tag{6.51}$$

where $T(\mathbf{x})$, $T_0(\mathbf{x})$ are the temperature and the initial temperature, $\mathbf{u}(\mathbf{x})$, $\boldsymbol{\tau}(\mathbf{x})$ are the displacement and the stress vectors, τ_{ij} the stress tensor, u_n, $\mathbf{u}_t$, τ_n, $\boldsymbol{\tau}_t$ the normal and tangential components of displacement and stress vectors, $u_n^k=u_i^kn_i^k$, $u_n^l=u_i^ln_i^k$, $\mathcal{F}_c^{kl}$ the coefficient of Coulombian friction, $\mathbf{n}$ the outward normal to the boundary $\partial\Omega$, and on $\cup\Gamma_c^{kl}$ due to the equilibrium of forces $\tau_{ij}(\mathbf{u}^k)n_j^k=-\tau_{ij}(\mathbf{u}^l)n_j^l$.

Remark 6.7 If the thermal sources are equal to zero, then the thermal part of the problem can be omitted, and we have the multibody contact problem in linear elasticity only.

B. Multibody Unilateral Contact Problems in (Thermo)viscoelastic Rheology with Short Memory, Dynamic Case

Problem ($\mathcal{P}_{\text{sm}}$) Let $N=2,3,\ s\geq 2$. Find a pair of functions $(T,\mathbf{u}):\Omega\times I\to(\mathbb{R}\times\mathbb{R}^N)\times I$, $N=2,3$, satisfying (6.47c,d) with (6.45c), (6.46), (6.48), and a boundary value and unilateral contact conditions

$$\kappa_{ij}\frac{\partial T(\mathbf{x},t)}{\partial x_i}n_j=K(Y(\mathbf{x},t)-T(\mathbf{x},t)),\qquad \tau_{ij}n_j=P_i,\quad i,j=1,\ldots,N,$$
$$\text{for } (t,\mathbf{x})\in\Gamma_\tau(t)=I\times\cup_{\iota=1}^{s}(\Gamma_\tau\cap\partial\Omega^\iota), \tag{6.52}$$

$$\kappa_{ij}\frac{\partial T(\mathbf{x},t)}{\partial x_i}n_j=K(Y(\mathbf{x},t)-T(\mathbf{x},t)),\qquad u_i=u_{2i},\quad i=1,\ldots,N,$$
$$\text{for } (t,\mathbf{x})\in{}^1\Gamma_u(t)=I\times\cup_{\iota=1}^{s}({}^1\Gamma_u\cap\partial\Omega^\iota), \tag{6.53}$$

$$T(\mathbf{x},t) = T_1(\mathbf{x},t), \qquad u_i = 0,\ i = 1,\ldots,N, \text{ on } {}^2\Gamma_u(t) = I \times \cup_{\iota=1}^{s}({}^2\Gamma_u \cap \partial\Omega^\iota), \tag{6.54}$$

$$\kappa_{ij}\frac{\partial T(\mathbf{x},t)}{\partial x_i}n_j = q \left(\text{or } \kappa_{ij}\frac{\partial T(\mathbf{x},t)}{\partial x_i}n_j = K(Y(\mathbf{x},t) - T(\mathbf{x},t))\right),$$
$$u_n = 0,\ \boldsymbol{\tau}_t = 0 \text{ on } \Gamma_0(t) = I \times \cup_{\iota=1}^{s}(\Gamma_0 \cap \partial\Omega^\iota), \tag{6.55}$$

$$\left.\begin{aligned}
&\text{(a) } T^k(\mathbf{x},t) = T^l(\mathbf{x},t),\\
&\quad \kappa_{ij}\frac{\partial T(\mathbf{x},t)}{\partial x_i}n_{j|(k)} + \kappa_{ij}\frac{\partial T(\mathbf{x},t)}{\partial x_i}n_{j|(l)}\\
&\quad = \mathcal{F}_c^{kl}|\tau_n^{kl}|\,|\mathbf{u}_t'^k - \mathbf{u}_t'^l|,\\
&\text{(b) } u_n^k - u_n^l \le d^{kl},\ \tau_n^k = \tau_n^l \equiv \tau_n^{kl} \le 0,\\
&\quad (u_n^k - u_n^l - d^{kl})\tau_n^{kl} = 0,\\
&\quad \text{where } u_n^k = u_i^k n_i^k,\ u_n^l = u_i^l n_i^k\\
&\text{(c) } \mathbf{u}_t'^k - \mathbf{u}_t'^l = 0 \Longrightarrow |\boldsymbol{\tau}_t^{kl}| \le \mathcal{F}_c^{kl}|\tau_n^{kl}|,\\
&\quad \mathbf{u}_t'^k - \mathbf{u}_t'^l \ne 0 \Longrightarrow \boldsymbol{\tau}_t^{kl}\\
&\quad = -\mathcal{F}_c^{kl}(\mathbf{u}_t'^k - \mathbf{u}_t'^l)|\tau_n^{kl}|\frac{\mathbf{u}_t'^k - \mathbf{u}_t'^l}{|\mathbf{u}_t'^k - \mathbf{u}_t'^l|},
\end{aligned}\right\} (t,\mathbf{x}) \in I \times \cup_{k,l}\Gamma_c^{kl}, \tag{6.56}$$

$$T(\mathbf{x},0) = T_0(\mathbf{x}), \qquad \mathbf{u}(\mathbf{x},0) = \mathbf{u}_0(\mathbf{x}), \qquad \mathbf{u}'(\mathbf{x},0) = \mathbf{u}_1(\mathbf{x}), \qquad \mathbf{x} \in \Omega, \tag{6.57}$$

where $q = q(\mathbf{x},t)$ is the heat flux, K the heat transition coefficient, $Y(\mathbf{x},t)$ the external temperature, $T_0(\mathbf{x})$, $T(\mathbf{x},t)$ the initial and actual temperatures, and the term $\mathcal{F}_c^{kl}|\tau_n^{kl}|\ |\mathbf{u}_t'^k - \mathbf{u}_t'^l|$ represents the frictional heatings on the contact surfaces Γ_c^{kl}, $\mathcal{F}_c^{kl}$ the coefficient of friction (defined below), $\mathbf{u}(\mathbf{x},t)$, $\boldsymbol{\tau}(\mathbf{x},t)$ are the displacement and the stress vectors, τ_{ij} the stress tensor, u_n, $\mathbf{u}_t$, τ_n, $\boldsymbol{\tau}_t$ the normal and tangential components of displacement and stress vectors, $\mathbf{n}$ the outward normal to the boundary $\partial\Omega$ and where $\mathbf{u}_0$, $\mathbf{u}_1$, $\mathbf{u}_2$ are given functions, $\mathbf{u}_2$ has a time derivative $\mathbf{u}_2'$, and $\mathbf{u}_0$, $\mathbf{u}_1$ satisfy the static linear contact problem in elasticity with or without Coulombian friction and on $\cup\Gamma_c^{kl}$ due to the equilibrium of forces $\tau_{ij}(\mathbf{u}^k)n_j^k = -\tau_{ij}(\mathbf{u}^l)n_j^l$.

The coefficient of friction $\mathcal{F}_c^{kl} \equiv \mathcal{F}_c^{kl}(\mathbf{x},\mathbf{u}')$ is globally bounded, nonnegative, and it satisfies the Carathéodory property, that is, $\mathcal{F}_c^{kl}(\cdot,\mathbf{v})$ is measurable, for all $\mathbf{v} \in \mathbb{R}$, and $\mathcal{F}_c^{kl}(\mathbf{x},\cdot)$ is continuous for a.e. $\mathbf{x} \in \Gamma_c^{kl}$. Moreover, it has a compact support

$$\mathcal{SF}_c \equiv \operatorname*{supp}_x(\mathcal{F}_c) = \overline{\{(\mathbf{x},t) \in \Gamma_c(t) = \Gamma_c \times I | \exists \mathbf{u}_t' \quad \mathcal{F}_c^{kl}(\mathbf{x},\mathbf{u}_t') \ne 0\}},$$

which depends on the space variable $\mathbf{x}$, and, since we model also the difference between the coefficients of friction and of the stick as well as the slip, it depends also on the tangential displacement rate component $\mathbf{u}_t'$ (see Fig. 6.3).

Furthermore, we denote by $(y)_+ \equiv \max\{y,0\}$ the positive part of y. Let us denote by ${}^a\Gamma_c^{kl} \subset \Gamma_c^{kl}$ the actual contact set, that is, for which $u_n^k - u_n^l - d^{kl} = 0$ on ${}^a\Gamma_c^{kl}$ and

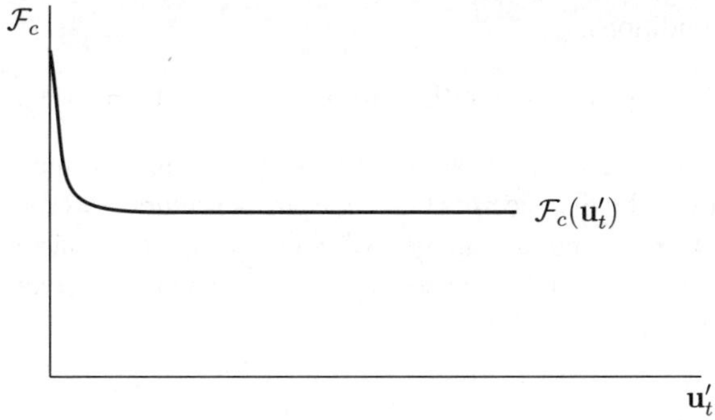

FIG. 6.3 Coefficient of Coulombian friction.

$u_n^k - u_n^l - d^{kl} < 0$ on ${}^c\Gamma_c^{kl} = \Gamma_c^{kl} \setminus {}^a\Gamma_c^{kl}$, where d^{kl} is a gap. For the continuous displacement $\mathbf{u}$ the actual contact zone ${}^a\Gamma_c^{kl}$ is well defined and a closed subset of $\cup\Gamma_c^{kl}$. Next, for simplicity, in some problems we will assume that $d^{kl} = 0$.

Remark 6.8 In case the generalized Hooke law is taken in the form (6.45a) or in the form (6.45c) and if the thermal part of the problem can be omitted, then we have the multibody contact problems in the linear elasticity or in viscoelastic rheology with short memory. Next, such a type of problem will be also investigated and discussed.

C. Multibody Bilateral Contact Problems in Thermoviscoelastic Rheology with Long Memory, Dynamic Case

Problem ($\mathcal{P}_{\mathbf{lm}}$) Find a pair of functions $(T, \mathbf{u}) : \Omega \times I \to (\mathbb{R} \times \mathbb{R}^N) \times I$, $N = 2, 3$, and a stress tensor $\tau_{ij} : \Omega \times I \to \mathbb{R}^{N \times N} \times I$ satisfying (6.47c) with (6.45e), (6.46), (6.48), and the boundary value conditions

$$T(\mathbf{x}, t) = T_1(\mathbf{x}, t) \quad (=0), \qquad \tau_{ij} n_j = P_i \quad \text{on} \quad \Gamma_\tau \times I, \tag{6.58}$$

$$\kappa_{ij} \frac{\partial T(\mathbf{x}, t)}{\partial x_i} n_j = 0, \qquad (\mathbf{u}(\mathbf{x}, t) = \mathbf{u}_2(\mathbf{x}, t) \quad (=0) \quad \text{on} \quad \Gamma_u \times I, \tag{6.59}$$

$$T^k(\mathbf{x}, t) = T^l(\mathbf{x}, t), \qquad \kappa_{ij} \frac{\partial T(\mathbf{x}, t)}{\partial x_i} n_{j|(k)} = -\kappa_{ij} \frac{\partial T(\mathbf{x}, t)}{\partial x_i} n_{j|(l)} \quad \text{on} \quad \cup_{k,l} \Gamma_c^{kl} \times I, \tag{6.60}$$

and the **bilateral** contact condition with Coulombian friction on $\Gamma_c^{kl} \times I$ of the form

$$\text{if } u_n^k - u_n^l = 0 \quad \text{and} \quad |\boldsymbol{\tau}_t^{kl}| \le \mathcal{F}_c^{kl} |\tau_n^{kl}| \equiv g_c^{kl},$$
$$\left.\begin{cases} \text{if} \quad |\boldsymbol{\tau}_t^{kl}| < g_c^{kl}, & \text{then } \mathbf{u}_t'^k - \mathbf{u}_t'^l = 0, \\ \text{if} \quad |\boldsymbol{\tau}_t^{kl}| = g_c^{kl}, & \text{then there exists } \lambda \ge 0 \\ & \text{such that } \mathbf{u}_t'^k - \mathbf{u}_t'^l = -\lambda \boldsymbol{\tau}_t^k \end{cases}\right\} \text{ on } \cup_{k,l} \Gamma_c^{kl} \times I, \tag{6.61}$$

and the initial conditions

$$T(\mathbf{x},0) = T_0(\mathbf{x}), \qquad \mathbf{u}(\mathbf{x},0) = \mathbf{u}_0(\mathbf{x}), \qquad \mathbf{u}'(\mathbf{x},\mathbf{0}) = \mathbf{u}_1(\mathbf{x}), \tag{6.62}$$

where $T_0(\mathbf{x})$, $T_1(\mathbf{x},t)$, and $T(\mathbf{x},t)$ are the initial, the given, and the actual temperatures, $\mathbf{u}(\mathbf{x},t)$, $\boldsymbol{\tau}(\mathbf{x},t)$ the displacement, and the stress vectors $\boldsymbol{\tau} = (\tau_{ij}n_j)$, τ_{ij} the stress tensor, u_n, $\mathbf{u}_t$, τ_n, $\boldsymbol{\tau}_t$ normal and tangential components of displacement and stress vectors, $\mathbf{n}$ the outward normal to the boundary $\partial\Omega$, $\mathbf{u}_1$ and $\mathbf{u}_2$ the given functions, and where $g_c^{kl}(\mathbf{x},t)$ is a slip limit.

6.4 STRESS–STRAIN ANALYSIS OF TOTAL HUMAN JOINT REPLACEMENTS IN LINEAR, NONLINEAR, ELASTICITY, AND THERMOELASTICITY: STATIC CASES, FINITE ELEMENT APPROXIMATIONS, HOMOGENIZATION AND DOMAIN DECOMPOSITION METHODS, AND ALGORITHMS

6.4.1 Formulation of the Problem

Equilibrium and Heat Equations Let $0, x_1, \ldots, x_N$ be the orthogonal Cartesian coordinate system, where N is the space dimension and let $\mathbf{x} = (x_1, \ldots, x_N)$ be a point in this Cartesian system. Let the body, being in an initial stress–strain state and created by a system of elastic anisotropic or isotropic bodies, occupy a region Ω. Let Ω be the region in $\mathbb{R}^N$, $N = 2, 3$. Moreover, we shall assume that $\Omega = \cup_{\iota=1}^{s} \Omega^\iota$, where Ω^ι are bounded domains with Lipschitz boundaries. Let the boundary $\partial\Omega$ be divided into disjoint parts $\Gamma_\tau, \Gamma_u, \Gamma_c$, and Γ_0 such that $\partial\Omega = \Gamma_\tau \cup \Gamma_u \cup \Gamma_c \cup \Gamma_0 \cup \mathcal{R}$, where the surface measure of $\mathcal{R}$ is zero and the parts Γ_τ, Γ_u, Γ_c, and Γ_0 are open sets in $\partial\Omega$. Assume that Lamé coefficients λ and μ as well as anisotropic elastic coefficients c_{ijkl} and thermal conductivity coefficients κ, κ_{ij} are bounded functions. The heat equation and the equilibrium equations for every domain Ω^ι of Ω read as follows:

$$\frac{\partial}{\partial x_i}\left(\kappa_{ij}^\iota \frac{\partial T^\iota}{\partial x_j}\right) + W^\iota = 0, \qquad \frac{\partial \tau_{ij}(\mathbf{u}^\iota)}{\partial x_j} + F_i^\iota = 0,$$
$$i,j = 1,\ldots,N, \iota = 1,\ldots,s, \text{ in } \Omega^\iota, \tag{6.63a,b}$$

where $\mathbf{F}^\iota$ are body forces and W^ι are thermal sources. A repeated index implies summation from 1 to N.

The relation between the displacement vector $\mathbf{u} = (u_i)$, $i = 1, \ldots, N$, and the small strain tensor e_{ij} is defined by

$$e_{ij} = e_{ij}(\mathbf{u}) = \frac{1}{2}\left(\frac{\partial u_i}{\partial x_j} + \frac{\partial u_j}{\partial x_i}\right), \qquad i,j = 1,\ldots,N. \tag{6.64}$$

The relation between the stress and strain tensors is defined by the generalized Hooke law (in thermoelasticity also known as the Duhamel–Neumann law):

$$\tau_{ij}^\iota = c_{ijkl}^\iota e_{kl}(\mathbf{u}^\iota) - \beta_{ij}^\iota(T^\iota - T_0^\iota), \qquad i,j,k,l = 1,\ldots,N,\ \iota = 1,\ldots,s, \tag{6.65}$$

in the anisotropic case, whereas in the isotropic case

$$c^{\iota}_{ijkl} = \mu^{\iota}(\delta_{ik}\delta_{jl} + \delta_{il}\delta_{jk}) + \lambda^{\iota}\delta_{ij}\delta_{kl}, \qquad \beta^{\iota}_{ij} = \gamma^{\iota}\delta_{ij}, \tag{6.66}$$

where λ^{ι}, μ^{ι} represent the Lamé coefficients, β^{ι}_{ij}, γ^{ι} coefficients of thermal expansion, and T^{ι}_0 is the initial temperature. The coefficients c^{ι}_{ijkl} form a matrix of the type $(N \times N \times N \times N)$ and satisfy the symmetry and Lipschitz conditions:

$$\begin{aligned} & c^{\iota}_{ijkl} = c^{\iota}_{jikl} = c^{\iota}_{klij} = c^{\iota}_{ijlk}, \\ & 0 < a^{\iota}_0 \le c^{\iota}_{ijkl}(\mathbf{x})\xi_{ij}\xi_{kl}|\xi|^{-2} \le A^{\iota}_0 < +\infty, \qquad \xi \in \mathbb{R}^{N^2}, \\ & \xi_{ij} = \xi_{ji}, \text{ for a.e. } \mathbf{x} \in \mathbf{\Omega}^{\iota}, \end{aligned} \tag{6.67}$$

where a^{ι}_0, A^{ι}_0 are constants independent of $\mathbf{x} \in \Omega^{\iota}$. For the isotropic case $a^{\iota}_0 = 2\min\{\mu^{\iota}(\mathbf{x}); \mathbf{x} \in \Omega^{\iota}\}$ and $A^{\iota}_0 = \max\{2\mu^{\iota}(\mathbf{x}) + 3\lambda^{\iota}(\mathbf{x}); \mathbf{x} \in \mathbf{\Omega}^{\iota}$. The coefficients of thermal expansion satisfy the symmetry conditions

$$\beta^{\iota}_{ij} = \beta^{\iota}_{ji}, \qquad i,j = 1,\ldots,N,\ \iota = 1,\ldots,s. \tag{6.68}$$

The thermal conductivity coefficients κ^{ι}_{ij} are symmetric and uniformly positive definite, that is,

$$\begin{aligned} & \kappa^{\iota}_{ij} = \kappa^{\iota}_{ji} \quad \text{on} \quad \Omega^{\iota}, \qquad i,j = 1,\ldots,N,\ \iota = 1,\ldots,s \\ & 0 < k^{\iota}_0 \le \kappa^{\iota}_{ij}(\mathbf{x})\zeta_i\zeta_j|\zeta|^{-2} \le k^{\iota}_1 < +\infty \text{ for a.e. } \mathbf{x} \in \Omega^{\iota},\ \zeta \in \mathbb{R}^N, \end{aligned} \tag{6.69}$$

where k^{ι}_0, k^{ι}_1 are constants independent of $\mathbf{x} \in \Omega^{\iota}$.

Boundary Conditions We shall consider the condition of heat flux and of loading on the part Γ_τ of the boundary in the form

$$\kappa_{ij}\frac{\partial T}{\partial x_i}n_j = q, \qquad \tau_{ij}n_j = P_i \quad \text{on } \Gamma_\tau. \tag{6.70}$$

Let the boundary $\Gamma_u = {}^1\Gamma_u \cup {}^2\Gamma_u$. Let us assume that a portion of the examined body is fixed at a certain boundary, which will be denoted by ${}^1\Gamma_u$. Let us denote the outward unit normal to the boundary $\partial\Omega$ by $\mathbf{n} = (n_i)$. We thus have, for example, the conditions

$$\kappa_{ij}\frac{\partial T}{\partial x_i}n_j = q, \qquad u_i = {}^1u_{0i} \quad \text{on} \quad {}^1\Gamma_u. \tag{6.71}$$

Furthermore, we can assume that the temperature and the displacement vector are given at the boundary ${}^2\Gamma_u$

$$T = T_1, \qquad u_i = {}^2u_{0i}, \qquad i = 1,\ldots,N, \quad \text{on} \quad {}^2\Gamma_u. \tag{6.72}$$

From Eqs. (6.63), (6.65), and (6.70) it follows that the effect of a change of temperature due to the deformation is equivalent to the replacement of mass forces by forces $F_i - (\partial/\partial x_j)(\beta_{ij}(T - T_0))$. Let several neighboring bodies, say Ω^k and Ω^l, be in a mutual contact. Denote the common contact boundary between both bodies Ω^k and Ω^l before deformation by Γ_c^{kl}; assume that Γ_c^{kl} has a positive $(N-1)$-dimensional measure. Further, denote by $\tau_i = \tau_{ij}n_j$ the stress vector, its normal and tangential components by $\tau_n = \tau_i n_i = \tau_{ij}n_i n_j$, $\boldsymbol{\tau}_t = \boldsymbol{\tau} - \tau_n \mathbf{n}$, and the normal and tangential components of the displacement vector by $u_n = u_i n_i$ and $\mathbf{u}_t = \mathbf{u} - u_n \mathbf{n}$. Denote by $\mathbf{u}^k$, $\mathbf{u}^l$, T^k, T^l (indices k, l correspond with the neighboring bodies in contact) the displacements and the temperatures in the neighboring bodies. All these quantities are functions of spatial coordinates. Then on the contact boundaries Γ_c^{kl} the condition of nonpenetration

$$u_n^k(\mathbf{x}) - u_n^l(\mathbf{x}) \leq 0 \quad \text{on} \quad \Gamma_c^{kl} \tag{6.73}$$

holds, where $u_n^l = u_i^l n_i^k$. For the contact forces, due to the law of action and reaction, we find

$$\tau_n^k(\mathbf{x}) = \tau_n^l(\mathbf{x}) \equiv \tau_n^{kl}(\mathbf{x}), \qquad \boldsymbol{\tau}_t^k(\mathbf{x}) = -\boldsymbol{\tau}_t^l(\mathbf{x}) \equiv \boldsymbol{\tau}_t^{kl}(\mathbf{x}). \tag{6.74}$$

Since the normal components of contact forces cannot be positive, that is, cannot be tensile forces,

$$\tau_n^k(\mathbf{x}) = \tau_n^l(\mathbf{x}) \equiv \tau_n^{kl}(\mathbf{x}) \leq 0 \quad \text{on} \quad \Gamma_c^{kl}. \tag{6.75}$$

During deformation the bodies are either in contact or they are not in contact. If they are not in contact, then $u_n^k - u_n^l < 0$, and the contact forces are equal to zero, that is, $\tau_n^k = \tau_n^l \equiv \tau_n^{kl} = 0$. If the bodies are in contact, that is, $u_n^k - u_n^l = 0$, then there may exist nonzero contact forces $\tau_n^k = \tau_n^l \equiv \tau_n^{kl} \leq 0$. These cases are included in the following condition:

$$(u_n^k(\mathbf{x}) - u_n^l(\mathbf{x}))\tau_n^{kl}(\mathbf{x}) = 0 \quad \text{on} \quad \Gamma_c^{kl}. \tag{6.76}$$

Further, if both bodies are in contact, then on the contact boundary the Coulombian type of friction acts. The frictional forces g_c^{kl} acting on the contact boundary Γ_c^{kl} are, in their absolute value, proportional to the normal stress component, where the coefficient of proportionality is the coefficient of Coulombian friction $\mathcal{F}_c^{kl}(\mathbf{x})$, that is,

$$g_c^{kl}(\mathbf{x}) = \mathcal{F}_c^{kl}(\mathbf{x})|\tau_n^{kl}(\mathbf{x})|. \tag{6.77}$$

Due to the acting and frictional forces we have the following cases:

If the absolute value of tangential forces $\boldsymbol{\tau}_t^{kl}(\mathbf{x})$ is less than the frictional forces g_c^{kl}, then the frictional forces preclude the mutual shifts of both bodies being in contact. If the tangential forces $\boldsymbol{\tau}_t^{kl}$ are equal in their absolute value to the frictional forces, then there are no forces that can preclude the mutual motion of both elastic bodies. Thus the contact points change their position in the direction opposite to that in which the tangential stress component acts. These conditions are described by the following conditions:

$$\left.\begin{array}{l}\text{(a) If } u_n^k - u_n^l = 0, \text{ then } |\boldsymbol{\tau}_t^{kl}(\mathbf{x})| \le g_c^{kl}(\mathbf{x}).\\ \text{(b) If } |\boldsymbol{\tau}_t^{kl}(\mathbf{x})| < g_c^{kl}(\mathbf{x}), \text{ then } \mathbf{u}_t^k(\mathbf{x}) - \mathbf{u}_t^l(\mathbf{x}) = 0, \text{ which means that}\\ \quad\text{the friction forces are sufficient to preclude the mutual shifting}\\ \quad\text{between the assumed bodies.}\\ \text{(c) If } |\boldsymbol{\tau}_t^{kl}(\mathbf{x})| = g_c^{kl}, \text{ then there exists a function } \vartheta \ge 0 \text{ such that}\\ \quad\mathbf{u}_t^k(\mathbf{x}) - \mathbf{u}_t^l(\mathbf{x}) = -\vartheta\boldsymbol{\tau}_t^{kl}(\mathbf{x}), \text{ which means that the friction forces}\\ \quad\text{are not sufficient to preclude the mutual shifting of both bodies.}\\ \quad\text{This shift acts in an opposite direction to the acting tangential}\\ \quad\text{forces. Then we have } \mathbf{u}_t^l = \mathbf{u}^l - u_n^l\mathbf{n}^k \text{ and } u_n^l = \mathbf{u}^l\cdot\mathbf{n}^k \equiv u_i^l n_i^k.\end{array}\right\} \quad (6.78)$$

On the contact boundary between the elastic bodies we shall assume that the temperatures and heat flow are continuous, that is,

$$T^k = T^l, \qquad \kappa_{ij}\frac{\partial T}{\partial x_j}n_{i|(k)} = -\kappa_{ij}\frac{\partial T}{\partial x_j}n_{i|(l)} \quad \text{on } \Gamma_c^{kl}. \tag{6.79}$$

If $\mathcal{F}_c^{kl} = 0$, then $g_c^{kl} = 0$ and then $\tau_t^k = \tau_t^l = 0$, and we speak about the case of contact problems without friction. In the case if $s = 1$, that is, if the second body is approximated by an absolutely rigid material and the frictional forces are equal to zero, then Eqs. (6.73), (6.75), (6.76) reduce to

$$u_n \le 0, \qquad \tau_n \le 0, \qquad u_n\tau_n = 0. \tag{6.80}$$

In some problems the bilateral conditions with or without friction can be used, for example, for the case without friction

$$u_n = 0, \qquad \tau_{tj} = 0, \qquad j = 1,\ldots,N. \tag{6.81}$$

These conditions describe the conditions on the axis Γ_0 (or plane) of symmetry.

The amplitude of the Coulombian coefficient of friction is not known, but for the existence of a solution it can be estimated [see Nečas et al. (1980), Hlaváček et al. (1988), Haslinger et al. (1996), Jarušek (1983), and Eck et al. (2005) for the elastic case and Nedoma (1987, 1998a) and Eck et al. (2005) for the thermoelastic case], for example, for the isotropic and for the anisotropic cases by

$$\|\mathcal{F}_c^{kl}\|_{L^\infty(\Gamma_c^{kl})} < (\mu/(\lambda + 2\mu))^{1/4}$$

or

$$\|\mathcal{F}_c^{kl}\|_{L^\infty(\Gamma_c^{kl})} < \left(\frac{3}{4}\frac{a_0}{A_0}\right)^{1/2} < \left(\frac{3-4\nu}{4-4\nu}\right)^{1/2} \tag{6.82}$$

for $N = 3$, where λ and μ are the Lamé coefficients, via the Poisson's ratio ν, a_0 and A_0 are constants as defined above. We see that the bound for coefficient of friction depends on the material properties only.

The problem investigated will be coercive if $\Gamma_u^\iota \neq \emptyset$ for all $\iota = 1, \ldots, s$ and semi-coercive if at least one part of $\Gamma_u = \cup_{\iota=1}^s \Gamma_u^\iota$, say Γ_u^j, is empty. The problem is coupled due to the coupling terms $(\partial/\partial x_j)(\beta_{ij}^\iota(T^\iota - T_0^\iota))$ following from Eqs. (6.63b) and (6.65) and we speak about the quasi-coupled model problem.

6.4.2 Formulation of the Contact Problem with Given Friction (the Tresca Model)

In this section we will deal with the solvability of a generalized semicoercive contact problem in linear thermoelasticity describing the global model of a human joint system, the spine and the fixation of long bones, and so forth. The problem studied is formulated as the primary variational inequality problem (Haslinger and Hlaváček, 1980; Haslinger et al., 1996; Hlaváček et al., 1988; Kikuchi and Oden, 1988; Nedoma, 1983, 1987; Panagiotopoulos, 1985; Hlaváček and Nedoma, 2002a), that is, in terms of displacements, arising from the variational formulation of the contact problem with friction in thermoelasticity. We will assume the generalized case of bodies of arbitrary shapes that are in mutual contacts. On one part of the boundary the bodies are loaded and on the second one they are fixed, and therefore, as a result, some of the bodies can shift and rotate.

Let $\Omega \subset \mathbb{R}^N$, $N = 2, 3$ be a union of domains, occupied by a body, with Lipschitz boundaries $\partial\Omega^\iota$, consisting of four parts Γ_τ, Γ_u, Γ_c, Γ_0, and $\partial\Omega = \Gamma_\tau \cup \Gamma_u \cup \Gamma_c \cup \Gamma_0 \cup \mathcal{R}$, all defined above. Let $\mathbf{x} = (x_i)$, $i = 1, \ldots, N$, be the Cartesian coordinates and let $\mathbf{n} = (n_i)$, $\mathbf{t} = (t_i)$ [for $N = 2$ $\mathbf{t} = (-n_2, n_1)$] be the outward normal and tangential vectors to $\partial\Omega$.

Let $\mathbf{F} \in [L^2(\Omega)]^N$, $\mathbf{P} \in [L^2(\Gamma_\tau)]^N$, and $\mathbf{u}_0 \in [C(\Gamma_u)]^N$. Next, we will deal with the following problem, where $g_c^{kl} \in L^2(\Gamma_c^{kl})$ are given slip limits.

Problem $\mathcal{P}$ Find a pair of functions $(T, \mathbf{u})$, a scalar function T, and a vector function $\mathbf{u}$, satisfying

$$\frac{\partial}{\partial x_i}\left(\kappa_{ij}^\iota \frac{\partial T^\iota}{\partial x_j}\right) + W^\iota = 0, \qquad \frac{\partial \tau_{ij}(\mathbf{u}^\iota)}{\partial x_j} + F_i^\iota = 0,$$
$$i,j = 1, \ldots, N,\ \iota = 1, \ldots, s \text{ in } \Omega^\iota, \tag{6.83a,b}$$

$$\tau_{ij}^\iota = c_{ijkl}^\iota e_{kl}(\mathbf{u}^\iota) - \beta_{ij}^\iota (T^\iota - T_0^\iota), \qquad i,j,k,l = 1, \ldots, N,\ \iota = 1, \ldots, s, \tag{6.84}$$

$$\kappa_{ij}\frac{\partial T}{\partial x_i} n_j = q, \qquad \tau_{ij} n_j = P_i, \qquad i,j = 1, \ldots, N \quad \text{on } \Gamma_\tau, \tag{6.85}$$

$$\kappa_{ij}\frac{\partial T}{\partial x_i} n_j = q, \qquad u_i = {}^1u_{0i}, \qquad i,j = 1, \ldots, N \quad \text{on } {}^1\Gamma_u, \tag{6.86}$$

$$T = T_1, \qquad u_i = {}^2u_{0i}, \qquad i = 1, \ldots, N \quad \text{on } {}^2\Gamma_u, \tag{6.87}$$

$$T^k = T^l, \qquad \kappa_{ij}\frac{\partial T}{\partial x_i} n_{j|(k)} = -\kappa_{ij}\frac{\partial T}{\partial x_i} n_{j|(l)}, \qquad i,j = 1, \ldots, N \quad \text{on } \Gamma_c^{kl}, \tag{6.88}$$

$$
\begin{aligned}
&u_n^k - u_n^l \le 0, \qquad \tau_n^k = \tau_n^l \equiv \tau_n^{kl} \le 0, \qquad (u_n^k - u_n^l)\tau_n^{kl} = 0,\\
&\text{if } u_n^k - u_n^l = 0, \text{ then } |\boldsymbol{\tau}_t^{kl}| \le g_c^{kl} \quad \text{on } \Gamma_c^{kl},\\
&\text{if } |\boldsymbol{\tau}_t^{kl}| < g_c^{kl}, \text{ then } \mathbf{u}_t^k - \mathbf{u}_t^l = 0,\\
&\text{if } |\boldsymbol{\tau}_t^{kl}| = g_c^{kl}, \text{ then there exists a function } \vartheta \ge 0\\
&\text{such that } \mathbf{u}_t^k - \mathbf{u}_t^l = -\vartheta \boldsymbol{\tau}_t^{kl} \quad \text{on } \Gamma_c^{kl},
\end{aligned}
\tag{6.89}
$$

where (6.74) holds for the normal and tangential components of the stress vector,

$$
u_n = 0, \qquad \tau_{tj} = 0, \qquad j = 1, \ldots, N \quad \text{on } \Gamma_0. \tag{6.90}
$$

The last condition [i.e. Eq. (6.90)] is suitable for numerical computation as it represents the bilateral condition or the condition of symmetry.

Definition 6.8 A pair of functions $(T^\iota, \mathbf{u}^\iota)$ is called a classical solution of Problem $\mathcal{P}$, if $T^\iota \in C^2(\Omega^\iota) \cap C^1(\overline{\Omega}^\iota)$, $\mathbf{u}^\iota \in [C^2(\Omega^\iota)]^N \cap [C^1(\overline{\Omega}^\iota)]^N$ and satisfy Eqs. (6.83) and (6.84) in each point of Ω, boundary conditions (6.85)–(6.87) in each point of Γ_τ or ${}^1\Gamma_u$ or ${}^2\Gamma_u$, respectively, contact conditions and conditions of the Coulombian law of friction (6.88) and (6.89) on $\cup_{k,l}\Gamma_c^{kl}$ and conditions (6.90) on Γ_0.

Variational (Weak) Solution of the Problem in Two Dimensions In the following we will consider a 2D semicoercive case with the contact conditions (6.88) and (6.89) on $\cup\Gamma_c^{kl}$, $k \ne l$ and (6.90) on Γ_0. The generalization to the 3D cases is possible and not so difficult.

Let us look for the temperature $T^\iota \in H^1(\Omega^\iota)$ and the displacement vector $\mathbf{u} = (u_i) \in \mathbf{W}(\Omega) = \prod_{\iota=1}^{s} [H^1(\Omega^\iota)]^2$, where $H^1(\Omega^\iota)$ is the Sobolev space in the usual sense. Let $e_{ij}(\mathbf{u})$, $\tau_{ij}(\mathbf{u})$ be the small strain tensor and the stress tensor, respectively, $T_0 = T_0(\mathbf{x})$ the initial temperature at which the bodies are in an initial strain and stress state, $\beta_{ij}^\iota(\mathbf{x}) \in C^1(\overline{\Omega}^\iota)$ the coefficient of thermal expansion, satisfying (6.68) and $\rho^\iota = \rho^\iota(\mathbf{x}) \in C(\overline{\Omega}^\iota)$ the density. Let $W^\iota \in L^2(\Omega^\iota)$, $F_i^\iota \in L^2(\Omega^\iota)$ be the heat sources and components of the body forces and c_{ijkl}^ι, $\kappa_{ij}^\iota \in C^1(\overline{\Omega}^\iota)$ the coefficients of elasticity and of heat conductivity, respectively, satisfying the usual symmetry conditions and the usual ellipticity conditions, that is, Eqs. (6.67) and (6.69). Let $g_c^{kl} \in L^2(\Gamma_c^{kl})$ be given slip limits.

Since Problem $\mathcal{P}$ is static, we will investigate the 2D quasi-coupled problem. We will denote $\tilde{F}_i^\iota = F_i^\iota - (\partial/\partial x_j)(\beta_{ij}^\iota(T^\iota - T_0^\iota))$, $q_0 \in L^2(\Gamma_\tau)$ is the heat flow, $\tilde{F}_i^\iota \in L^2(\Omega^\iota)$, $P_i \in L^2(\Gamma_\tau)$, $u_{0i}^\iota \in H^{1/2}(\Gamma_u^\iota)$, $T_1^\iota \in H^{1/2}({}^2\Gamma_u^\iota)$, $\iota = 1, \ldots, s$, $T_0 = \text{const}$. Let us denote by $(\cdot,\cdot)$ the scalar product in $[L^2(\Omega)]^2$, by $\langle\cdot,\cdot\rangle$ the scalar product in $[L^2(\Gamma_c)]^2$, by $\|\cdot\|_k$ the norm in $[H^k(\Omega)]^2$, k being an integer, where $H^k(\Omega)$ denotes the Sobolev space in the usual sense. Let us denote by

$$
\begin{aligned}
&{}^1V_0 = \{z | z \in {}^1W \equiv H^1(\Omega^1) \times \cdots \times H^1(\Omega^s),\ z = 0 \text{ on } \cup\, {}^2\Gamma_u, z^k = z^l \text{ on } \Gamma_c^{kl}\},\\
&{}^1V = \{z | z \in {}^1W,\ z = T_1 \text{ on } \cup\, {}^2\Gamma_u,\ z^k = z^l \text{ on } \Gamma_c^{kl}\},
\end{aligned}
$$

$$V_0 = \{\mathbf{v} | \mathbf{v} \in \mathbf{W} \equiv [H^1(\Omega^1)]^2 \times \cdots \times [H^1(\Omega^s)]^2,\ \mathbf{v} = \mathbf{0} \text{ on } {}^1\Gamma_u \cup {}^2\Gamma_u,$$
$$v_n = 0 \text{ on } \Gamma_0\},$$
$$V = \{\mathbf{v} | \mathbf{v} \in \mathbf{W}, \mathbf{v} = \mathbf{u}_0 \text{ on } {}^1\Gamma_u \cup {}^2\Gamma_u, v_n = 0 \text{ on } \Gamma_0\}$$

the spaces and sets of virtual temperatures and virtual displacements, respectively, and by

$$K = \{\mathbf{v} | \mathbf{v} \in V, v_n^k - v_n^l \leq 0 \text{ on } \cup_{k,l} \Gamma_c^{kl}\}$$

the set of all admissible displacements, which for $\mathbf{u}_0 = 0$ is a convex cone with vertex at the origin.

As our quasi-coupled problem investigated is indeed not coupled, both the problems in thermics and elasticity can be solved separately [see Nedoma (1987)] and the additional term $(\partial/\partial x_j)(\beta_{ij}^\iota(T^\iota - T_0^\iota))$ has a meaning of body forces. Since we assume that $\beta_{ij}^\iota \in C^1(\overline{\Omega}^\iota)$, $T^\iota, T_0^\iota \in H^1(\Omega^\iota)$, then $\beta_{ij}^\iota(T^\iota - T_0^\iota) \in H^1(\Omega^\iota)$, and, therefore, $(\partial/\partial x_j)(\beta_{ij}^\iota(T^\iota - T_0^\iota)) \in L^2(\Omega^\iota)$.

Definition 6.9 By a variational (weak) solution of Problem $\mathcal{P}$ we mean a pair of functions $(T, \mathbf{u})$, $T \in {}^1V$, $\mathbf{u} \in K$, such that

$$b(T, z - T) \geq s(z - T) \quad \forall z \in {}^1V, \tag{6.91}$$

$$a(\mathbf{u}, \mathbf{v} - \mathbf{u}) + \langle g_c^{kl}, |\mathbf{v}_t^k - \mathbf{v}_t^l| - |\mathbf{u}_t^k - \mathbf{u}_t^l|\rangle \geq S(T; \mathbf{v} - \mathbf{u}) \quad \forall \mathbf{v} \in K, \tag{6.92}$$

where for $T, z \in H^1(\Omega)$, $\mathbf{u}, \mathbf{v} \in \mathbf{W}$ we put

$$b(T, z) = \sum_{\iota=1}^{s} b^\iota(T^\iota, z^\iota) = \int_\Omega \kappa_{ij}(\mathbf{x}) \frac{\partial T}{\partial x_j} \frac{\partial z}{\partial x_i} d\mathbf{x},$$

$$s(z) = \sum_{\iota=1}^{s} s^\iota(z^\iota) = \int_\Omega Wz \, d\mathbf{x} + \int_{\Gamma_\tau} q_0 z \, ds,$$

$$a(\mathbf{u}, \mathbf{v}) = \sum_{\iota=1}^{s} a(\mathbf{u}, \mathbf{v}) = \int_\Omega c_{ijkl} e_{ij}(\mathbf{u}) e_{kl}(\mathbf{v}) \, d\mathbf{x},$$

$$S(T; \mathbf{v}) = \sum_{\iota=1}^{s} S^\iota(\mathbf{v}^\iota) = \int_\Omega F_i v_i \, d\mathbf{x} + \int_{\Gamma_\tau} P_i v_i \, ds + \int_\Omega \beta_{ij}(T - T_0) e_{ij}(\mathbf{v}) \, d\mathbf{x},$$

$$j_{gn}(\mathbf{v}) = \int_{\cup_{k,l}\Gamma_c^{kl}} g_c^{kl} |\mathbf{v}_t^k - \mathbf{v}_t^l| \, ds = \langle g_c^{kl}, |\mathbf{v}_t^k - \mathbf{v}_t^l|\rangle.$$

Problem $\mathcal{P}$ is equivalent to the following variational formulation:

Find a pair of functions $(T, \mathbf{u})$, $T \in {}^1V$, $\mathbf{u} \in K$, such that

$$l(T) \leq l(z) \quad \forall z \in {}^1V, \tag{6.93}$$

$$L(T; \mathbf{u}) \leq L(T; \mathbf{v}) \quad \forall \mathbf{v} \in K, \tag{6.94}$$

where $l(z)$, $L(\mathbf{v})$ are defined by

$$l(z) = \tfrac{1}{2}b(z,z) - s(z),$$
$$L(T;\mathbf{v}) = L_0(T;\mathbf{v}) + j_{gn}(\mathbf{v}), \qquad L_0(T;\mathbf{v}) = \tfrac{1}{2}a(\mathbf{v},\mathbf{v}) - S(T;\mathbf{v}).$$

It can be proved that any classical solution of Problem $\mathcal{P}$ is a weak solution and, conversely, if the weak solution is smooth enough, it represents a classical solution.

To prove the existence and uniqueness of the variational (weak) solution, we introduce the set of all rigid displacements and rotations

$$P = \cup_{\iota=1}^{s} P^{\iota}, \qquad P^{\iota} = \{\mathbf{v}^{\iota} | \mathbf{v}^{\iota} = (v_1^{\iota}, v_2^{\iota}), v_1^{\iota} = a_1^{\iota} - b^{\iota}x_2, v_2^{\iota} = a_2^{\iota} + b^{\iota}x_1\},$$

where $a_i^{\iota} \in \mathbb{R}$, $b^{\iota} \in \mathbb{R}$, $i = 1,2$, $\iota = 1,\ldots,s$, are arbitrary and the set of bilateral admissible rigid displacements

$$P_0 = \{\mathbf{v} \in K \cap P | \mathbf{v} \in P_0 \Rightarrow -\mathbf{v} \in P_0\}$$
$$= \{\mathbf{v} \in P_V \equiv P \cap V_0 | v_n^k - v_n^l = 0 \text{ on } \cup_{k,l} \Gamma_c^{kl}\}.$$

Lemma 6.6 (Korn's Inequality) Let Ω be a bounded domain with the Lipschitz boundary and $\mathbf{u} = (u_1,\ldots,u_N) \in V = \{\mathbf{u} \in H^{1,N}(\Omega) | \mathbf{u} = 0 \text{ on } \Gamma_u \subset \partial\Omega, \text{meas } \Gamma_u > 0\}$. Then there exists a positive constant $c \equiv c(\Omega)$, such that

$$|e(\mathbf{u})| \geq c\|\mathbf{u}\|_{1,N} \quad \forall \mathbf{u} \in V,$$

where

$$|e(\mathbf{u})|^2 = \int_{\Omega} e_{ij}(\mathbf{u})e_{ij}(\mathbf{u})\,d\mathbf{x}, \qquad e_{ij}(\mathbf{u}) = \frac{1}{2}\left(\frac{\partial u_i}{\partial x_j} + \frac{\partial u_j}{\partial x_i}\right),$$
$$i,j = 1,\ldots,N, \; N = 2,3.$$

For the proof see Nečas and Hlaváček (1981).

Lemma 6.7 Let $\Omega \subset \mathbb{R}^2$, $\mathbf{u}^{\iota} \in [H^1(\Omega^{\iota})]^2$. Then

$$e_{ij}(\mathbf{u}^{\iota}) = 0 \quad \forall i,j = 1,2 \iff u_1^{\iota} = a_1^{\iota} - b^{\iota}x_2, \qquad u_2^{\iota} = a_2^{\iota} + b^{\iota}x_1, \tag{6.95}$$

where a_1^{ι}, a_2^{ι}, b^{ι} are real constants.

Lemma 6.8 Let $\Gamma_u^{\iota} \neq \emptyset$, $\iota = 1,\ldots,s$. Then

$$P_V \equiv P \cap V_0 = \{0\},$$

that is, only the zero function lies in the intersection $P \cap V_0$.

Lemma 6.9 Since $e_{ij}(\mathbf{v})=0\ \forall\mathbf{v}\in P,\ \forall i,j$, then

$$a(\mathbf{u},\mathbf{v})=0 \quad \forall\mathbf{v}\in P. \tag{6.96}$$

Moreover, if $\mathbf{w}\in\mathbf{W}$, $e_{ij}(\mathbf{w})=0\ \forall i,j$, then $\mathbf{w}\in P$.

Lemma 6.10 Let there exist a weak solution of Problem $\mathcal{P}$. Then $S(T;\mathbf{w})\leq j_{gn}(\mathbf{w})$ $\forall\mathbf{w}\in K\cap P$, that is,

$$\int_\Omega F_iw_i\,d\mathbf{x}+\int_{\Gamma_\tau}P_iw_i\,ds-\int_{\cup_{k,l}\Gamma_c^{kl}}g_c^{kl}|\mathbf{w}_t^k-\mathbf{w}_t^l|\,ds\leq 0 \quad \forall\mathbf{w}\in K\cap P. \tag{6.97}$$

Proof Since the problem is quasi-coupled, it is sufficient to consider the elastic part of the problem only. Since the weak solution $\mathbf{u}$ satisfies $a(\mathbf{u},\mathbf{v}-\mathbf{u})-S(T;\mathbf{v}-\mathbf{u})+j_{gn}(\mathbf{v})-j_{gn}(\mathbf{u})\geq 0\ \forall\mathbf{v}\in K$, then putting $\mathbf{v}=\mathbf{u}+\mathbf{w}$, $\mathbf{v}-\mathbf{u}=\mathbf{w}\in K\cap P$, we find that $\mathbf{v}\in K$ and

$$a(\mathbf{u},\mathbf{u}+\mathbf{w}-\mathbf{u})-S(T;\mathbf{u}+\mathbf{w}-\mathbf{u})+j_{gn}(\mathbf{u}+\mathbf{w})-j_{gn}(\mathbf{u})\geq 0.$$

Hence

$$\begin{aligned}a(\mathbf{u},\mathbf{w})-S(T;\mathbf{w})&\geq\langle g_c^{kl},|\mathbf{u}_t^k-\mathbf{u}_t^l|-|\mathbf{v}_t^k-\mathbf{v}_t^l|\rangle\\&=\langle g_c^{kl},|\mathbf{u}_t^k-\mathbf{u}_t^l|-|(\mathbf{u}+\mathbf{w})_t^k-(\mathbf{u}+\mathbf{w})_t^l|\rangle\\&=\langle g_c^{kl},|\mathbf{u}_t^k-\mathbf{u}_t^l|-|(\mathbf{u}_t^k-\mathbf{u}_t^l)+(\mathbf{w}_t^k-\mathbf{w}_t^l)|\rangle.\end{aligned}$$

Due to Lemma 6.9 and since $|a+b|-|a|\leq|b|$, then $S(T;\mathbf{w})\leq\langle g_c^{kl},|\mathbf{w}_t^k-\mathbf{w}_t^l|\rangle$, which completes the proof. □

Remark 6.9 Equation (6.97) represents the condition of the total equilibrium, and it is the necessary condition for the existence of a solution.

Lemma 6.11 Any classical solution of Problem $\mathcal{P}$ is a weak solution. On the other hand, if the weak solution is smooth enough, it is a classical solution.

The proof is a modification of that of Hlaváček et al. (1988) and Haslinger et al. (1996).

Theorem 6.32 Assume that

$$P_0=\{0\}, \qquad P\cap K\neq\{0\} \tag{6.98}$$

and

$$S(T;\mathbf{w})<j_{gn}(\mathbf{w}) \quad \forall\mathbf{w}\in P\cap K\backslash\{0\}. \tag{6.99}$$

Then L is coercive on K and there exists a weak solution of Problem $\mathcal{P}$. If

$$S(T; \mathbf{w}) > j_{gn}(\mathbf{w}) \quad \forall \mathbf{w} \in P_V \backslash \{0\}, \tag{6.100}$$

then the solution is unique. If

$$|S(T; \mathbf{w})| \leq j_{gn}(\mathbf{w}) \quad \forall \mathbf{w} \in P_V,$$

then for any two solutions $\mathbf{u}, \mathbf{u}^*$

$$\mathbf{w} \equiv \mathbf{u}^* - \mathbf{u} \in P_V, \qquad S(T; \mathbf{w}) = j_{gn}(\mathbf{u}^*) - j_{gn}(\mathbf{u}) \tag{6.101}$$

holds.

For the proof see Hlaváček and Nedoma (2002a).

Finite Element Solution of the Problem Let every domain $\Omega^\iota \subset \mathbb{R}^N$ be approximated by a polygonal (for $N = 2$) or polyhedral (for $N = 3$) domain Ω_h^ι. Let the domain Ω_h^ι be "triangulated," that is, the domain $\overline{\Omega}_h^\iota = \Omega_h^\iota \cup \partial\Omega_h^\iota$ is divided into a system of m triangles T_{h_i} in the 2D case and into a system of m tetrahedra T_{h_i} in the 3D case, generating a triangulation $\mathcal{T}_h$ such that $\overline{\Omega}_h^\iota = \cup_{i=1}^m T_{h_i}$ and such that two neighboring triangles have only a vertex or an entire side common in the 2D case, and that two neighboring tertrahedra have only a vertex or an entire edge or an entire face common in the 3D case. Denote by $h = \max_{1 \leq i \leq m}(\text{diam } T_{h_i})$ the maximal side of the triangles T_{h_i} in the 2D case and/or the maximal edge of the tetrahedra in the 3D case in $\mathcal{T}_h$. Let ρ_{T_i} denote the radius of the maximal circle (for the 2D case) or maximal ball (for the 3D case), inscribed in the simplex T_{h_i}. A family of triangulation $\{\mathcal{T}_h\}$, $0 < h \leq h_0 < \infty$, is said to be regular if there exists a constant $\vartheta_0 > 0$ independent of h and such that $h/\rho_{T_i} \leq \vartheta_0$ for all $h \in (0, h_0)$. We will assume that the sets $\overline{\Gamma}_u \cap \overline{\Gamma}_\tau$, $\overline{\Gamma}_u \cap \Gamma_c$, $\overline{\Gamma}_u \cap \overline{\Gamma}_0$, $\Gamma_c \cap \overline{\Gamma}_\tau$, $\Gamma_c \cap \overline{\Gamma}_0$, $\overline{\Gamma}_\tau \cap \overline{\Gamma}_0$ coincide with the vertices or edges of T_{h_i}.

Let $R_i \in \Omega_h^\iota$ be an arbitrary interior vertex of the triangulation $\mathcal{T}_h$. Generally, the basis function w_h^i (where w_h^i is a scalar or vector function) is defined to be a function linear on each element $T_{h_i} \in \mathcal{T}_h$ and taking the values $w_h^i(R_j) = \delta_{ij}$ at the vertices of the triangulation, where δ_{ij} is the Kronecker symbol. The function w_h^i represents a "pyramid" of height 1 with its vertex above the point R_i and with its support (supp w_h^i) consisting of those triangles or tetrahedra that have the vertex R_i in common. The basis function has small support since diam(supp w_h^i) $\leq 2h$ and the parameter $h \to 0$. Further, for simplicity, we will discuss the 2D case only.

Let us assume that $N = 2$ and let u_{0i}^ι and T_1^ι, $\iota \leq s$, be piecewise linear continuous functions. Let ${}^1V_h$ and V_h be the spaces of linear finite elements, that is, the spaces of continuous scalar and/or vector functions in $\overline{\Omega}_h$, piecewise linear over $\mathcal{T}_h$, that is,

$${}^1V_h = \{z \in C(\overline{\Omega}_h^1) \times \cdots \times C(\overline{\Omega}_h^s) \cap {}^1V | z_{|T_{h_i}} \in P_1 \quad \forall T_{h_i} \in \mathcal{T}_h\},$$

$$V_h = \{\mathbf{v} \in [C(\overline{\Omega}_h^1)]^2 \times \cdots \times [C(\overline{\Omega}_h^s)]^2 \cap V | \mathbf{v}_{|T_{h_i}} \in [P_1]^2 \quad \forall T_{h_i} \in \mathcal{T}_h\},$$

and

$$K_h = \{\mathbf{v} \in V_h | v_n^k - v_n^l \leq 0 \text{ on } \cup \Gamma_c^{kl}\} = K \cap V_h.$$

Definition 6.10 A pair of functions $(T_h, \mathbf{u}_h)$, $T_h \in {}^1V_h$, $\mathbf{u}_h \in K_h$, is said to be a finite element approximation of Problem $\mathcal{P}$, if

$$l(T_h) \leq l(z) \quad \forall z \in {}^1V_h, \tag{6.102}$$

$$L(T_h; \mathbf{u}_h) \leq L(T_h; \mathbf{v}) \quad \forall \mathbf{v} \in K_h. \tag{6.103}$$

To find an a priori estimate for the error of the solution $(T_h - T^*, \mathbf{u}_h - \mathbf{u}^*)$, a modification of the Falk technique [see, e.g., Falk (1974) and Nedoma (1987)] will be used. Since the problem is quasi-coupled, the method will be based on the following lemma.

Lemma 6.12 Let $|\cdot|$ be the seminorm defined by

$$|\mathbf{v}|^2 = \int_{\cup_{\iota=1}^s \Omega_h^\iota} e_{ij}(\mathbf{v}) e_{ij}(\mathbf{v})\, d\mathbf{x}, \tag{6.104}$$

and $S_1(T; \mathbf{v}) = \int_\Omega \beta_{ij} T e_{ij}(\mathbf{v})\, d\mathbf{x}$. Then it holds that

$$\begin{aligned} c_0 |\mathbf{u} - \mathbf{u}_h|^2 &\leq a(\mathbf{u} - \mathbf{u}_h, \mathbf{u} - \mathbf{u}_h) \\ &\leq \{a(\mathbf{u}_h - \mathbf{u}, \mathbf{v}_h - \mathbf{u}) + a(\mathbf{u}, \mathbf{v} - \mathbf{u}_h) + a(\mathbf{u}, \mathbf{v}_h - \mathbf{u}) \\ &\quad + j_{gn}(\mathbf{v}_h) - j_{gn}(\mathbf{u}) + j_{gn}(\mathbf{v}) - j_{gn}(\mathbf{u}_h) + S(T_h; \mathbf{u} - \mathbf{v}_h) \\ &\quad + S(T_h; \mathbf{u}_h - \mathbf{v}) + S_1(T_h - T; \mathbf{v} - \mathbf{u})\} \\ &\quad \text{for any } \mathbf{v} \in K, \mathbf{v}_h \in K_h, c_0 = \text{const.} > 0. \end{aligned} \tag{6.105}$$

Proof The proof follows from the conditions

$$a(\mathbf{u}, \mathbf{v} - \mathbf{u}) + \langle g_c^{kl}, |\mathbf{v}_t^k - \mathbf{v}_t^l| - |\mathbf{u}_t^k - \mathbf{u}_t^l| \rangle - S(T; \mathbf{v} - \mathbf{u}) \geq 0 \quad \forall \mathbf{v} \in K,$$

$$a(\mathbf{u}_h, \mathbf{v}_h - \mathbf{u}_h) + \langle g_c^{kl}, |(\mathbf{v}_h^k)_t - (\mathbf{v}_h^l)_t| - |(\mathbf{u}_h^k)_t - (\mathbf{u}_h^l)_t| \rangle - S(T_h; \mathbf{v}_h - \mathbf{u}_h) \geq 0 \quad \forall \mathbf{v}_h \in K_h.$$

Adding these inequalities, adding and subtracting the term $a(\mathbf{u}, \mathbf{u}_h) - a(\mathbf{u}_h, \mathbf{u})$ to the resulting inequality, and performing some modifications, then we obtain

$$\begin{aligned} a(\mathbf{u} - \mathbf{u}_h, \mathbf{u} - \mathbf{u}_h) &\leq a(\mathbf{u}_h - \mathbf{u}, \mathbf{v}_h - \mathbf{u}) + a(\mathbf{u}, \mathbf{v} - \mathbf{u}_h) + a(\mathbf{u}, \mathbf{v}_h - \mathbf{u}) \\ &\quad - S(T_h; \mathbf{v} - \mathbf{u}_h) - S(T_h; \mathbf{v}_h - \mathbf{u}) + j_{gn}(\mathbf{v}_h) - j_{gn}(\mathbf{u}) \\ &\quad + j_{gn}(\mathbf{v}) - j_{gn}(\mathbf{u}_h) + S_1(T_h - T; \mathbf{v} - \mathbf{u}). \end{aligned}$$

Hence (6.105) follows immediately. □

Corollary 6.3 Let $K_h \subset K$. Then substituting $\mathbf{v} = \mathbf{u}_h$ in (6.105) we have

$$|\mathbf{u} - \mathbf{u}_h| \leq c\{a(\mathbf{u}_h - \mathbf{u}, \mathbf{v}_h - \mathbf{u}) + a(\mathbf{u}, \mathbf{v}_h - \mathbf{u}) + j_{gn}(\mathbf{v}_h) - j_{gn}(\mathbf{u}) \\ - S(T_h; \mathbf{v}_h - \mathbf{u}) + S_1(T_h - T; \mathbf{u}_h - \mathbf{u})\}^{1/2} \ \forall \mathbf{v}_h \in K_h, \ c = \text{const} > 0. \tag{6.106}$$

Theorem 6.33 Let $\partial\Omega$ and its parts $\Gamma_u, \Gamma_\tau, \Gamma_c, \Gamma_0$ be piecewise polygonal, let ${}^2\Gamma_u$ have a positive length. Let $T^\iota \in H^2(\Omega^\iota)$, $\mathbf{u}^\iota \in [H^2(\Omega^\iota)]^2$, $\iota = 1, \ldots, s$, be the solutions of (6.93) and (6.94) and such that the stress components $\tau_{ij}(\mathbf{u}^\iota) \in H^1(\Omega^\iota)$, $i, j = 1, 2$, $\iota = 1, \ldots, s$, $\boldsymbol{\tau} \in [L^\infty(\Gamma_c^{kl})]^2$, $u \in H^2(\Gamma_c \cap \Gamma_{ci}^{kl})$ (where Γ_{ci}^{kl} is a side of a triangle adjacent to Γ_c^{kl}) holds for any part Γ_c^{kl} of Γ_c, $g_c^{kl} \in L^\infty(\Gamma_c^{kl})$. Let $K_h \subset K$ and let the changes $u_n^k - u_n^l < 0 \rightarrow u_n^k - u_n^l = 0$ and $\mathbf{u}_t^k - \mathbf{u}_t^l = 0 \rightarrow \mathbf{u}_t^k - \mathbf{u}_t^l \neq 0$ occur at finite set of points of $\cup_{k,l}\Gamma_c^{kl}$ only. Then

$$\|T - T_h\|_1 = O(h), \qquad |\mathbf{u} - \mathbf{u}_h| = O(h). \tag{6.107}$$

Proof Since the problem is quasi-coupled, we will analyze both parts of the problem separately (Nedoma, 1987; Hlaváček and Nedoma, 2002a). For the second part of the problem the technique of the proof is based on a generalization of results of Hlaváček and Lovíšek (1980), Haslinger et al. (1996), and Nedoma (1987, 1998a). □

Thermal Part of the Problem Let $\tilde{T} \in {}^1V_h$ be a fixed element such that $\tilde{T}|_{\Omega^\iota} \in H^2(\Omega^\iota)$ $\forall \iota \leq s$. For the solution T we have $T = \tilde{T} + \zeta$, where $\zeta \in {}^1V_0 \cap \prod_{\iota=1}^s H^2(\Omega^\iota)$ by assumption. Then

$$b(\zeta, z) = s(z) - b(\tilde{T}, z) \quad \forall z \in {}^1V_0 \tag{6.108}$$

follows from definition (6.91). Note that

$$f(z) \equiv s(z) - b(\tilde{T}, z)$$

belongs to the dual space $({}^1V_0)'$ since

$$|f(z)| \leq (\|s\|_{({}^1V_0)'} + C_B\|\tilde{T}\|_1)\|z\|_1. \tag{6.109}$$

We define

$${}^1V_{0h} = \{z \in {}^1V_0 | z_{|T_{h_i}} \in P_1(T_{h_i}) \quad \forall T_{h_i} \in \mathcal{T}_h\}.$$

Then $T_h = \tilde{T} + \zeta_h$, where $\zeta_h \in {}^1V_{0h}$ and

$$b(\zeta_h, z_h) = s(z_h) - b(\tilde{T}, z_h) \quad \forall z_h \in {}^1V_{0h} \tag{6.110}$$

follows from definition (6.102).

We may use the Céa inequality (6.15)

$$\|\zeta - \zeta_h\|_1 \le C \inf \left\{ \|\zeta - z_h\|_1 ; z_h \in {}^1V_{0h} \right\}, \tag{6.111}$$

since the bilinear form $b(\cdot, \cdot)$ is continuous and ${}^1V_0$–elliptic.

The interpolation theorem yields [see Bramble and Zlámal (1970)]

$$\|z - I_h z\|_1 \le c_1 M_2 h, \quad \text{if} \quad z^\iota \in H^2(\Omega^\iota), \tag{6.112}$$

where $I_h z \in {}^1V_0$ is the Lagrange linear interpolate of $z \in {}^1V_0$ over the given triangulation $\mathcal{T}_h$.

Setting $z_h = I_h \zeta$ and using (6.111) and (6.112), we arrive at

$$\|T - T_h\|_1 = \|\zeta - \zeta_h\|_1 \le C_1 h, \tag{6.113}$$

which completes the thermal part of the proof.

Elastic Part of the Problem Using Corollary 6.3 we will estimate (6.106). This estimate can be applied, provided the solution $\mathbf{u}$ is sufficiently regular. The Green theorem implies

$$\begin{aligned} a(\mathbf{u}, \mathbf{v}_h - \mathbf{u}) - S(\boldsymbol{\tau}_h \mathbf{v}_h - \mathbf{u}) = &\int_{\cup \Gamma_c^{kl}} [\tau_n(\mathbf{u})(\mathbf{v}_h - \mathbf{u})_n + \boldsymbol{\tau}_t(\mathbf{u})(\mathbf{v}_h - \mathbf{u})_t] \, ds \\ &- \int_\Omega \beta_{ij}(T_h - T_0) e_{ij}(\mathbf{v}_h - \mathbf{u}) \, d\mathbf{x}, \end{aligned} \tag{6.114}$$

using also (6.83a,b)–(6.85). By virtue of Corollary 6.3 we have

$$\begin{aligned} |\mathbf{u} - \mathbf{u}_h| \le c_0 \{ &a(\mathbf{u}_h - \mathbf{u}, \mathbf{v}_h - \mathbf{u}) + a(\mathbf{u}, \mathbf{v}_h - \mathbf{u}) - S(T_h, \mathbf{v}_h - \mathbf{u}) + j_{gn}(\mathbf{v}_h) \\ &- j_{gn}(\mathbf{u}) + S_1(T_h - T, \mathbf{u}_h - \mathbf{u}) \}^{1/2}, \quad \forall \mathbf{v}_h \in K_h. \end{aligned}$$

The bilinear form $a(\cdot, \cdot)$ is bounded, so that

$$a(\mathbf{u}_h - \mathbf{u}, \mathbf{v}_h - \mathbf{u}) \le M |\mathbf{u}_h - \mathbf{u}| \, |\mathbf{v}_h - \mathbf{u}|.$$

Since for every $\mathbf{v} \in \prod_{\iota=1}^{s} [C(\overline{\Omega}^\iota)]^2 \cap K$ the linear interpolate over $\mathcal{T}_h$ $\mathbf{v}_{LI} \in K_h$ and since the inequality (6.106) is valid for any arbitrary $\mathbf{v}_h \in K_h$, it is valid also for $\mathbf{v}_h = \mathbf{u}_{LI}$. As $ab \le \frac{1}{2}\varepsilon a^2 + \frac{1}{2\varepsilon} b^2$, $a, b \in \mathbb{R}^1$, $\varepsilon > 0$, applying the interpolation theorem we obtain

$$\begin{aligned} a(\mathbf{u}_h - \mathbf{u}, \mathbf{v}_h - \mathbf{u}) &\le \frac{1}{2}\varepsilon M |\mathbf{u}_h - \mathbf{u}|^2 + \frac{1}{2\varepsilon} M |\mathbf{v}_h - \mathbf{u}|^2 \\ &\le \frac{1}{2}\varepsilon M |\mathbf{u}_h - \mathbf{u}^2 + c\varepsilon^{-1} h^2 \|\mathbf{u}\|_2^2. \end{aligned}$$

In a parallel way we obtain

$$|S_1(T_h - T, \mathbf{u}_h - \mathbf{u})| \leq C\|T_h - T\|_0|\mathbf{u}_h - \mathbf{u}| \leq C_1\frac{\varepsilon}{2}|\mathbf{u}_h - \mathbf{u}|^2 + C_1(2\varepsilon)^{-1}h^2.$$

To estimate the integrals

$$J_1\left(\cup_{k,l}\Gamma_c^{kl}\right) = \int_{\cup_{k,l}\Gamma_c^{kl}} \tau_n^{kl}[(v_h^k - v_h^l)_n - (u_n^k - u_n^l)]\, ds$$

and

$$J_2\left(\cup_{k,l}\Gamma_c^{kl}\right) = \int_{\cup_{k,l}\Gamma_c^{kl}} \boldsymbol{\tau}_t^{kl}[(\mathbf{v}_h^k - \mathbf{v}_h^l)_t - (\mathbf{u}_t^k - \mathbf{u}_t^l)]\, ds,$$

we assume that

$$\Gamma_c = \cup_{k,l}\Gamma_c^{kl} = \cup_{k,l}\left(\cup_i\Gamma_{ci}^{kl}\right),$$

where Γ_{ci}^{kl} is a (straight) side of a triangle T_{h_i}.

Let us put $\mathbf{v}_h = \mathbf{u}_{LI}$ in what follows.

To estimate the integral $J_1(\cup\overline{\Gamma}_c^{kl})$ we distinguish the following cases:

1. $u_n^k(\mathbf{x}) - u_n^l(\mathbf{x}) < 0, \mathbf{x} \in \overline{\Gamma}_{ci}^{kl}$: Since $(u_n^k - u_n^l)\tau_n^{kl} = 0$, then $J_1(\overline{\Gamma}_{ci}^{kl}) = 0$.
2. $u_n^k(\mathbf{x}) - u_n^l(\mathbf{x}) = 0 \ \ \forall \mathbf{x} \in \Gamma_{ci}^{kl}$. Since then also $(u_{LI}^k)_n - (u_{LI}^l)_n = 0$, we have $J_1(\overline{\Gamma}_{ci}^{kl}) = 0$.
3. Γ_{ci}^{kl} contains precisely one point of the change $u_n^k(\mathbf{x}) - u_n^l(\mathbf{x}) < 0 \rightarrow u_n^k(\mathbf{x}) - u_n^l(\mathbf{x}) = 0$. Then

$$\begin{aligned} |J_1(\overline{\Gamma}_{ci}^{kl})| \leq \|&(u_{LI}^k(\mathbf{x}))_n - (u_{LI}^l(\mathbf{x}))_n - (u_n^k(\mathbf{x}) \\ &- u_n^l(\mathbf{x}))\|_{L^\infty(\overline{\Gamma}_{ci}^{kl})} \int_{\overline{\Gamma}_{ci}^{kl}} |\tau_n^{kl}|\, ds \leq c_{si}h^2 \end{aligned}$$

holds by virtue of the assumptions and the interpolation theory. As case (3) may occur at a finite number of segments Γ_{ci}^{kl} only, we conclude that

$$|J_1(\overline{\Gamma}_{ci}^{kl})| \leq c_s h^2.$$

To estimate the integral $J_2(\cup\overline{\Gamma}_{ci}^{kl})$ we distinguish the following cases:

1. $\mathbf{u}_t^k(\mathbf{x}) - \mathbf{u}_t^l(\mathbf{x}) > 0$ for $\mathbf{x} \in \Gamma_{ci}^{kl}$: It holds that

$$\begin{aligned} J_1(\cup_{k,l}\Gamma_c^{kl}) + J_2(\cup_{k,l}\Gamma_c^{kl}) &+ \int_\Omega \beta_{ij}(T_0 - T)e_{ij}(\mathbf{v}_h - \mathbf{u})\, d\mathbf{x} \\ &+ j_{gn}(\mathbf{v}_h) - j_{gn}(\mathbf{u}) \geq 0 \end{aligned}$$

by virtue of (6.92) and (6.114) and $g_c^{kl}|\mathbf{u}_t^k - \mathbf{u}_t^l| + \boldsymbol{\tau}_t^{kl}(\mathbf{u}_t^k - \mathbf{u}_t^l) = 0$ a.e. on $\cup_{k,l}\Gamma_c^{kl}$. Hence $g_c^{kl} = -\boldsymbol{\tau}_t^{kl}$. Putting $\mathbf{v}_h = \mathbf{u}_{LI}$, then $(\mathbf{u}_{LI}^k - \mathbf{u}_{LI}^l)_t = (\mathbf{u}_t^k - \mathbf{u}_t^l)_{LI}$. Then $(\mathbf{u}_t^k - \mathbf{u}_t^l)_{LI} > 0$ on $\overline{\Gamma}_{ci}^{kl}$ and

$$\int_{\cup\Gamma_{ci}^{kl}} \{-g_c^{kl}[(\mathbf{u}_t^k - \mathbf{u}_t^l)_{LI} - (\mathbf{u}_t^k - \mathbf{u}_t^l)] + g_c^{kl}[(\mathbf{u}_t^k - \mathbf{u}_t^l)_{LI} - (\mathbf{u}_t^k - \mathbf{u}_t^l)]\}\, ds = 0.$$

2. $\mathbf{u}_t^k(\mathbf{x}) - \mathbf{u}_t^l(\mathbf{x}) = 0$ for $\mathbf{x} \in \Gamma_{ci}^{kl}$: Since $(\mathbf{u}_t^k - \mathbf{u}_t^l)_{LI} = 0$ on $\cup\overline{\Gamma}_{ci}^{kl}$, then

$$\int_{\cup\Gamma_{ci}^{kl}} \{\boldsymbol{\tau}_t^{kl}(\mathbf{u})[(\mathbf{u}_t^k - \mathbf{u}_t^l)_{LI} - (\mathbf{u}_t^k - \mathbf{u}_t^l)] + g_c^{kl}[|(\mathbf{u}_t^k - \mathbf{u}_t^l)_{LI}| - |\mathbf{u}_t^k - \mathbf{u}_t^l|]\}\, ds = 0.$$

3. $\mathbf{u}_t^k(\mathbf{x}) - \mathbf{u}_t^l(\mathbf{x}) < 0$ for $\mathbf{x} \in \Gamma_{ci}^{kl}$: As $\mathbf{u}_t^k - \mathbf{u}_t^l < \mathbf{0}$ on Γ_{ci}^{kl}, then $|\mathbf{u}_t^k - \mathbf{u}_t^l| = -(\mathbf{u}_t^k - \mathbf{u}_t^l)$ and $(\mathbf{u}_t^k - \mathbf{u}_t^l)_{LI} < 0$. Thus $g_c^{kl} = \boldsymbol{\tau}_t^{kl}$ a.e. on $\overline{\Gamma}_{ci}^{kl}$. Putting $\mathbf{v}_h = \mathbf{u}_{LI}$, then

$$\int_{\cup\overline{\Gamma}_{ci}^{kl}} \{g_c^{kl}[(\mathbf{u}_t^k - \mathbf{u}_t^l)_{LI} - (\mathbf{u}_t^k - \mathbf{u}_t^l)] + g_c^{kl}[-(\mathbf{u}_t^k - \mathbf{u}_t^l)_{LI} + (\mathbf{u}_t^k - \mathbf{u}_t^l)]\}\, ds = 0.$$

4. $\mathbf{u}_t^k(\mathbf{x}) - \mathbf{u}_t^l(\mathbf{x}) = 0$ changes to $\mathbf{u}_t^k(\mathbf{x}) - \mathbf{u}_t^l(\mathbf{x}) \neq 0$ for $\mathbf{x} \in \Gamma_{ci}^{kl}$: Let $\mathbf{v}_h = \mathbf{u}_{LI}$, under the assumptions of the theorem, then since $\mathbf{u} \in [W^{1,\infty}(\cup\overline{\Gamma}_{ci}^{kl})]^2$, $\boldsymbol{\tau} \in [L^\infty(\cup\overline{\Gamma}_{ci}^{kl})]^2$, $g_c^{kl} \in L^\infty(\cup\overline{\Gamma}_{ci}^{kl})$, $(\mathbf{u}_t^k - \mathbf{u}_t^l) \in [W^{1,\infty}(\cup\overline{\Gamma}_{ci}^{kl})]^2$, and $\boldsymbol{\tau}_t^{kl} \in [L^\infty(\cup\overline{\Gamma}_{ci}^{kl})]^2$.
Then

$$\int_{\cup\overline{\Gamma}_{ci}^{kl}} \{\boldsymbol{\tau}_t^{kl}(\mathbf{u})[(\mathbf{u}_t^k - \mathbf{u}_t^l)_{LI} - (\mathbf{u}_t^k - \mathbf{u}_t^l)] + g_c^{kl}[|(\mathbf{u}_t^k - \mathbf{u}_t^l)_{LI}| - |\mathbf{u}_t^k - \mathbf{u}_t^l|]\}\, ds$$
$$\leq \|(\mathbf{u}_t^k - \mathbf{u}_t^l)_{LI} - (\mathbf{u}_t^k - \mathbf{u}_t^l)\|_{L^\infty(\cup\overline{\Gamma}_{ci}^{kl})} \int_{\cup\overline{\Gamma}_{ci}^{kl}} (|\boldsymbol{\tau}_t^{kl}(\mathbf{u})| + g_c^{kl})\, ds \leq c_{ri}h^2.$$

Thus

$$\left| \int_{\cup\overline{\Gamma}_{ci}^{kl}} \boldsymbol{\tau}_t^{kl}(\mathbf{u})[(\mathbf{u}_t^k - \mathbf{u}_t^l)_{LI} - (\mathbf{u}_t^k - \mathbf{u}_t^l)]ds + j_{gn}(\mathbf{u}_{LI}) - j_{gn}(\mathbf{u}) \right| \leq c_r h^2.$$

Using (6.106) and the above given estimates then

$$|\mathbf{u} - \mathbf{u}_h| \leq c_0\left\{\tfrac{1}{2}\varepsilon|\mathbf{u} - \mathbf{u}_h|^2 + \varepsilon^{-1}h^2\|\mathbf{u}\|_2^2 + c_s h^2 + c_r h^2\right\}^{1/2},$$

so that (6.107) follows, choosing ε sufficiently small.

6.4.3 Algorithm

The numerical solution of problem (6.103) leads in practice to the numerical approximation of a saddle point. In fact, by a saddle point approach (i.e., a mixed formulation) one can avoid the unpleasant occurrence of the nondifferentiable term $j_{gn}(\mathbf{v})$ in the functional $L(T, \mathbf{v})$. A useful approach for a solution of the saddle point problem can be based on the Uzawa algorithm [see Céa (1971), Ekeland and Temam (1976), Nečas and Hlaváček (1981), Hlaváček et al. (1988), Nedoma (1987, 1998a), and Haslinger et al. (1996)].

The Uzawa Method Let $L(\mathbf{u})$ be a functional on a Hilbert space V, let $K \subset V$ be a closed convex set and let $\Psi(\mathbf{v}, \mu)$ be a functional defined on $V \times M$, where M is another Hilbert space. Let Λ be a closed convex cone in M with its vertex at the origin. Suppose that $\Psi(\mathbf{v}, \mu)$ is positive homogeneous in the component μ, that is, that $\Psi(\mathbf{v}, t\mu) = t\Psi(\mathbf{v}, \mu), t \geq 0$. Moreover, suppose that the set K is characterized by

$$\mathbf{v} \in K \Leftrightarrow \Psi(\mathbf{v}, \mu) \leq 0 \quad \forall \mu \in \Lambda.$$

Then $\sup_{\mu \in \Lambda} \Psi(\mathbf{v}, \mu) = \infty$ for $\mathbf{v} \notin K$ and $\sup_{\mu \in \Lambda} \Psi(\mathbf{v}, \mu) = 0$ for $\mathbf{v} \in K$. Let us introduce the Lagrangian $\mathcal{L}(\mathbf{v}, \mu) = L(\mathbf{v}) + \Psi(\mathbf{v}, \mu)$. Then

$$\inf_{\mathbf{v} \in K} L(\mathbf{v}) = \inf_{\mathbf{v} \in V} \sup_{\mu \in \Lambda} \mathcal{L}(\mathbf{v}, \mu).$$

The problem of finding $\inf_{\mathbf{v} \in V} \sup_{\mu \in \Lambda} \mathcal{L}(\mathbf{v}, \mu)$ is the primal problem while the problem of finding $\sup_{\mu \in \Lambda} \inf_{\mathbf{v} \in V} \mathcal{L}(\mathbf{v}, \mu)$ is the dual problem. The main goal is to find a pair $(\mathbf{u}, \lambda) \in V \times \Lambda$ such that

$$\mathcal{L}(\mathbf{u}, \lambda) = \inf_{\mathbf{v} \in V} \sup_{\mu \in \Lambda} \mathcal{L}(\mathbf{v}, \mu) = \sup_{\mu \in \Lambda} \inf_{\mathbf{v} \in V} \mathcal{L}(\mathbf{v}, \mu).$$

Definition 6.11 The pair $(\mathbf{u}, \lambda)$ is called a saddle point if $(\mathbf{u}, \lambda) \in V \times \Lambda$ and

$$\mathcal{L}(\mathbf{u}, \mu) \leq \mathcal{L}(\mathbf{u}, \lambda) \leq \mathcal{L}(\mathbf{v}, \lambda) \quad \text{for all } (\mathbf{v}, \mu) \in V \times \Lambda.$$

Definition 6.12 By an approximate saddle point is meant a point $(\mathbf{u}_h, \lambda_h) \in V_h \times \Lambda_h \subset V \times \Lambda$, $V_h \subset V$, $\Lambda_h \subset \Lambda$, where Λ_h is a convex cone with the vertex at the origin, if

$$\mathcal{L}(\mathbf{u}_h, \mu_h) \leq \mathcal{L}(\mathbf{u}_h, \lambda_h) \leq \mathcal{L}(\mathbf{v}_h, \lambda_h)$$

for all pairs of functions $(\mathbf{v}_h, \mu_h) \in V_h \times \Lambda_h$.

Theorem 6.34 Assume that a set $K_h \subset V_h$ is given such that $\mathbf{v}_h \in K_h \Leftrightarrow \Psi(\mathbf{v}_h, \mu_h) \leq 0$ $\forall \mu_h \in \Lambda_h$. Then the first component $\mathbf{u}_h$ of the approximate saddle point solves the problem: Find $\mathbf{u}_h \in K_h$ such that

$$\inf_{\mathbf{v}_h \in K_h} L(\mathbf{v}_h) = L(\mathbf{u}_h).$$

Proof It holds that $L(\mathbf{u}_h)+\Psi(\mathbf{u}_h, \mu_h)\leq L(\mathbf{u}_h)+\Psi(\mathbf{u}_h, \lambda_h)\leq L(\mathbf{v}_h)+\Psi(\mathbf{v}_h,\lambda_h)$. Putting $\mu_h=0$ and $\mu_h=2\lambda_h$, then $\Psi(\mathbf{u}_h,\lambda_h)=0$ and therefore $L(\mathbf{u}_h)\leq L(\mathbf{v}_h)+\Psi(\mathbf{v}_h,\lambda_h)$ for all $\mathbf{v}_h\in V_h$. If $\mathbf{v}_h\in K_h$ then $\Psi(\mathbf{v}_h,\lambda_h)\leq 0$, thus $L(\mathbf{u}_h)\leq L(\mathbf{v}_h)$ for all $\mathbf{v}_h\in K_h$ since $\Psi(\mathbf{u}_h,\mu_h)\leq 0\ \forall\mu_h\in\Lambda_h$, $\mathbf{u}_h\in K_h$. □

It can be shown that a pair of functions $(\mathbf{u},\lambda)\in V\times\Lambda$ is a saddle point if and only if

$$\max_{\mu\in\Lambda}\inf_{\mathbf{v}\in V}\mathcal{L}(\mathbf{v},\mu)=\min_{\mathbf{v}\in V}\sup_{\mu\in\Lambda}\mathcal{L}(\mathbf{v},\mu)=\mathcal{L}(\mathbf{u},\lambda) \tag{6.115}$$

and the minimum and maximum are achieved at the points $\mathbf{u}$ and λ, respectively [see Ekeland and Temam (1976) and Nečas and Hlaváček (1981)]. Similarly, the same proposition for the approximate saddle point $(\mathbf{u}_h,\lambda_h)\in V_h\times\Lambda_h$ can be proved.

To find the necessary conditions for the existence of a saddle point, the Kuhn–Tucker conditions, we will assume that the functional L and Ψ are Gâteaux differentiable with respect to both variables.

These necessary conditions are as follows:

Theorem 6.35 Let $(\mathbf{u},\lambda)\in V\times\Lambda$ be a saddle point. Let L and Ψ be differentiable, then

$$\begin{aligned} &D_vL(\mathbf{u},\mathbf{v})+D_v\Psi(\mathbf{u},\lambda,\mathbf{v})=0 && \forall\mathbf{v}\in V,\\ &D_\mu\Psi(\mathbf{u},\lambda,\mu)\leq 0 && \forall\mu\in\Lambda,\\ &D_\mu\Psi(\mathbf{u},\lambda,\lambda)=0. \end{aligned} \tag{6.116}$$

If $L(\mathbf{v})+\Psi(\mathbf{v},\mu)$ is a convex functional of a variable $\mathbf{v}$ and $\Psi(\mathbf{v},\mu)$ is a concave functional of μ, then (6.116), the **Kuhn–Tucker conditions**, are the sufficient conditions for the element $(\mathbf{u},\lambda)\in V\times\Lambda$ to be a saddle point.

For the proof see Céa (1971), Ekeland and Temam (1976) and Nečas and Hlaváček (1981).

Similarly, we can find the necessary conditions for the existence of an approximate saddle point $(\mathbf{u}_h,\lambda_h)\in V_h\times\Lambda_h$.

For finding a saddle point, the Uzawa algorithm can be used [see Ekeland and Temam (1976)].

Theorem 6.36 Let the functional $L(\mathbf{u}_h)$ possess a strictly monotone Gâteaux differential in V_h, that is,

$$DL(\mathbf{u}_h+\mathbf{h},\mathbf{h})-DL(\mathbf{u}_h,\mathbf{h})\geq m\|\mathbf{h}\|^2\quad\forall\mathbf{h}\in V_h,$$

let us suppose that $\Psi(\mathbf{v},\mu)=(\mu,\Psi(\mathbf{v}))_M$ is continuously differentiable and $D\Psi(\mathbf{u}_h,\mathbf{h})$ is bounded, that is,

$$|D\Psi(\mathbf{u}_h,\mathbf{h})|\leq c\|\mathbf{h}\|.$$

Let us assume that $\Psi(\mathbf{v}_h, \mu_h) = (\mu_h, \tilde{\Psi}(\mathbf{v}_h))$ is weakly lower semicontinuous and convex for all μ_h. Choose ρ_n in such a way that

$$2m\rho_i - c^2\rho_i^2 \geq \beta > 0.$$

Let $(\mathbf{u}_h, \lambda_h)$ be a saddle point in the sense of Definition 6.12 in its approximate version and let P be the projection operator from M into Λ ($Pw \in \Lambda$ is the element close to $w \in M$). Then the iterative Uzawa algorithm,

$$\lambda_h^{i+1} = P[\lambda_h^i + \rho_i \tilde{\Psi}(\mathbf{u}_h^i)],$$
$$DL(\mathbf{u}_h^i, \mathbf{v}) + (\lambda_h^i, D\tilde{\Psi}(\mathbf{u}_h^i, \mathbf{v}))_M = 0 \quad \text{for all } \mathbf{v} \in V_h,$$

converges, that is,

$$\mathbf{u}_h^i \to \mathbf{u}_h \quad \text{in } V, \quad \text{as } i \to \infty,$$

where $\mathbf{u}$ is the solution of the problem $\inf_{\mathbf{v}\in V} \sup_{\mu\in\Lambda} \mathcal{L}(\mathbf{v}, \mu)$. Moreover, if the saddle point is unique and $\rho_i \to \rho$, then $\lambda^i \rightharpoonup \lambda$ weakly in M.

For the proof see Nečas and Hlaváček (1981).

Problem (6.103) is equivalent to the following problem:
Find $\mathbf{u}_h \in K_h$ such that

$$L(T_h, \mathbf{u}_h) = \min_{\mathbf{v}_h\in K_h} \sup_{\mu_h\in\Lambda_h} \mathcal{L}(T_h, \mathbf{v}_h, \mu_h)$$
$$= \min_{\mathbf{v}_h\in K_h} \sup_{\mu_h\in\Lambda_h} \left\{ L_0(T_h, \mathbf{v}_h) + \int_{\cup\Gamma_c^{kl}} \mu_h^{kl} g_c^{kl} (\mathbf{v}_{ht}^k - \mathbf{v}_{ht}^l)\, ds \right\},$$

where

$$\Lambda_h = \{\mu_h | \mu_h \in L^2(\cup\Gamma_c^{kl}) | \, |\mu_h| \leq 1 \quad \text{a.e. on } \cup\Gamma_c^{kl}\}.$$

Since $\int_{\cup\Gamma_c^{kl}} g_c^{kl} |\mathbf{v}_{ht}^k - \mathbf{v}_{ht}^l|\, ds = \sup_{\mu_h^{kl}\in\Lambda_h} \int_{\cup\Gamma_c^{kl}} \mu_h^{kl} g_c^{kl} (\mathbf{v}_{ht}^k - \mathbf{v}_{ht}^l)\, ds = j_{gn}(\mathbf{v}_h)$, we define

$$\Psi(\mathbf{v}_h, \mu_h) = \int_{\cup\Gamma_c^{kl}} \mu_h^{kl} g_c^{kl} (\mathbf{v}_{ht}^k - \mathbf{v}_{ht}^l)\, ds, \quad \mathbf{v}_h \in V_h, \ \mu_h \in \Lambda_h.$$

Therefore, we can write the original (primal) problem as

$$\inf_{\mathbf{v}_h\in K_h} L(T_h, \mathbf{v}_h) = \inf_{\mathbf{v}_h\in K_h} \sup_{\mu_h\in\Lambda_h} (L_0(T_h, \mathbf{v}_h) + \Psi(\mathbf{v}_h, \mu_h))$$
$$= \inf_{\mathbf{v}_h\in K_h} \sup_{\mu_h\in\Lambda_h} \mathcal{L}(T_h, \mathbf{v}_h, \mu_h).$$

Through these means, the problem is transformed to seeking the saddle point of the Lagrangian

$$\mathcal{L}(T_h, \mathbf{v}, \mu) = L_0(T_h, \mathbf{v}) + \Psi(\mathbf{v}, \mu) \quad \text{over } K_h \times \Lambda_h.$$

Consider the inner part of the dual problem only, that is,

$$\inf_{\mathbf{v}\in K_h} \mathcal{L}(T_h, \mathbf{v}, \mu) = \inf_{\mathbf{v}\in K_h} \left\{ \frac{1}{2}\int_\Omega c_{ijkm} e_{ij}(\mathbf{v}) e_{km}(\mathbf{v})\, d\mathbf{x} - \int_\Omega F_i v_i\, d\mathbf{x} - \int_{\Gamma_\tau} P_i v_i\, ds + \int_\Omega \beta_{ij}(T_h - T_0) e_{ij}(\mathbf{v})\, d\mathbf{x} + \int_{\cup\Gamma_c^{kl}} \mu^{kl} g_c^{kl}(v_n^k - v_n^l)\, ds \right\}.$$

The Uzawa algorithm is based on the construction of two sequences $\mathbf{u}_h^i \in K_h$ and $\lambda_h^i \in \Lambda_h$, $i = 1, 2, \ldots$ as follows:

Let $\lambda_h^0 \in \Lambda_h$ be arbitrary, for example $\lambda_h^0 = 0$. Let $\lambda_h^i \in \Lambda_h$ be known, then we find $\mathbf{u}_h^i \in K_h$, solving the following problem:

$$\min_{\mathbf{v}_h \in K_h} \left\{ \mathcal{L}(T_h, \mathbf{v}) = L_0(T_h, \mathbf{v}_h) + \int_{\cup\Gamma_c^{kl}} \lambda_h^{kl} g_c^{kl}(\mathbf{v}_t^k - \mathbf{v}_t^l)\, ds \right\}$$

and λ_h^{i+1} will be found from the relation

$$\lambda_h^{i+1} = P\left[\lambda_h^i + \rho_i g_c^{kl}((\mathbf{u}_{ht}^i)^k - (\mathbf{u}_{ht}^i)^l)\right], \tag{6.117}$$

where P is a projector $P : L^2\left(\cup\Gamma_c^{kl}\right) \to \Lambda_h$ defined as follows:

$$\begin{aligned} &\mu \in L^2\left(\cup\Gamma_c^{kl}\right), \qquad y \in \cup\Gamma_c^{kl},\\ &P\mu(y) = \mu(y) \quad \text{for } |\mu(y)| \le 1,\\ &P\mu(y) = 1 \quad \text{for } \mu(y) > 1,\\ &P\mu(y) = -1 \quad \text{for } \mu(y) < -1, \end{aligned}$$

and where $0 < \overline{\rho}_1 \le \rho_i \le \overline{\rho}_2$, $\overline{\rho}_1, \overline{\rho}_2$ are sufficiently small numbers.

Then we generate a functional $f(\mathbf{x})$ by $L_0(T_h, \mathbf{v}_h)$, $\Psi(\mathbf{u}_h, \lambda_h) \approx \boldsymbol{\lambda}^{\mathrm{T}} A\mathbf{x}$. Thus, we seek the saddle point of $\mathcal{L}(T_h, \mathbf{x}, \lambda_h)$, where

$$\mathcal{L}(T_h, \mathbf{x}, \lambda_h) = f(\mathbf{x}) + \boldsymbol{\lambda}^{\mathrm{T}} A\mathbf{x} = \tfrac{1}{2}\mathbf{x}^{\mathrm{T}} C\mathbf{x} - \mathbf{x}^{\mathrm{T}}\mathbf{d} + \boldsymbol{\lambda}^{\mathrm{T}} A\mathbf{x}, \qquad \mathbf{x} \in \mathbb{R}^n,\ \boldsymbol{\lambda} \in \mathbb{R}^m, \tag{6.118}$$

where C is the stiffness matrix. The minimization of the functional in (6.118), that is,

$$f_\lambda(\mathbf{x}) = \tfrac{1}{2}\mathbf{x}^{\mathrm{T}} C\mathbf{x} - \mathbf{x}^{\mathrm{T}}(\mathbf{d} - A^{\mathrm{T}}\boldsymbol{\lambda}) \quad \text{over } K_h,$$

can be accomplished by a conjugate gradient method with constraints.

Then the problem leads to

$$f(\mathbf{x}) = \tfrac{1}{2}\mathbf{x}^{\mathrm{T}}C\mathbf{x} - \mathbf{x}^{\mathrm{T}}(\mathbf{d} - A^{\mathrm{T}}\boldsymbol{\lambda}) \to \min$$

with constraints (6.119)

$$A\mathbf{x} \le 0.$$

Remark 6.10 The global stiffness matrix C, which is generated by standard techniques, is the block diagonal of the dimension $n \times n$, where every block is a sparse, symmetric, positive semidefinite matrix and corresponds to just one body in the model. In the coercive case C is a positive definite matrix. The constraint matrix A is of the type $m \times n$, $m \ll n$; and we assume that its rows are linearly independent. The matrices C and A have a great number of zero entries. It is necessary to devote some attention to the modes of their storage in a computer memory. The stiffness matrix C can be stored in two formats — in the SKY-LINE format, where we store only the active length from each column j, and the SPARSE format, where we store only nonzero entries, which lie above the diagonal, from each column.

Remark 6.11 The usual computing criterion can be as follows: STOP if error $< \varepsilon$, where error $= \|x^{k+1} - x^k\| / \max(1.0, \|x^k\|)$, x^k is the solution in the kth iteration, $k \le$ max iter, and ε is the prescribed tolerance ($\varepsilon = 10^{-6}$). max iter is the maximum number of iterations. The overflows we define as the value greater than 10^{20}.

Conjugate Gradient (CG) Method with Constraints The problem discussed represents a problem of the minimization quadratic function with linear constraints, the problem of quadratic programming. The CG method is the gradient projection method, and it generally solves the problem

$$\begin{aligned} f(\mathbf{x}) &= \tfrac{1}{2}\mathbf{x}^{\mathrm{T}}C\mathbf{x} - \mathbf{x}^{\mathrm{T}}\mathbf{b} \to \min, \\ \mathbf{a}_i^{\mathrm{T}}\mathbf{x} - d_i &\le 0 \qquad i \in \mathcal{J}^-, \\ \mathbf{a}_i^{\mathrm{T}}\mathbf{x} - d_i &= 0 \qquad i \in \mathcal{J}^0, \end{aligned} \tag{6.120}$$

where $\mathbf{x}, \mathbf{a}_i \in \mathbb{R}^n$, $\mathbf{d} = (d_i) \in \mathbb{R}^m$, $d_i \in \mathbb{R}$, $\mathcal{J}^- \cup \mathcal{J}^0 = \{1, \ldots, m\}$, $\mathcal{J}^-$, $\mathcal{J}^0$ are finite sets of indices, and C is a symmetric, positive semidefinite matrix $n \times n$, $b \in \mathbb{R}^n$. (If Γ_0 is empty, then $\mathcal{J}^0$ is empty).

The main idea of the algorithm is the successive minimization of $f(\mathbf{x})$ on the facets created by constraints, for which the equality is satisfied. We solve the minimization problem on each of such facets by using the conjugate gradient method (CGM). Since the CGM converges after a finite number of iterate steps and since the number of facets is limited, the algorithm converges after a finite number of steps. For more details see Daniel (1971), Luenberger (1984), Pschenichny and Danilin (1978), Brouse (1988), Nedoma (1998b), Nedoma et al. (1999b) and Kestřánek (1999).

Denote by $A_{\mathcal{J}}$ the $m \times n$-dimensional matrix, where m is the number of elements of the set $\mathcal{J}$, which corresponds to the number of all nodal points on $\cup\Gamma_c^{kl}$, and n is the dimension of $\mathbf{x}$, that is, whose rows have the indices $i \in \mathcal{J} \subseteq (\mathcal{J}^- \cup \mathcal{J}^0)$.

Lemma 6.13 Let the vectors $\mathbf{a}_i$, $i \in \mathcal{J} \subseteq (\mathcal{J}^- \cup \mathcal{J}^0)$. Then the matrix $A_{\mathcal{J}} A_{\mathcal{J}}^{\mathrm{T}}$ is regular.

For the proof see Pschenichny and Danilin (1978) and Nedoma (1998a).

Let us construct the projection operator $P_{\mathcal{J}}$ as follows:

$$P_{\mathcal{J}} = A_{\mathcal{J}}^{\mathrm{T}}(A_{\mathcal{J}} A_{\mathcal{J}}^{\mathrm{T}})^{-1} A_{\mathcal{J}} \qquad \text{if } \mathcal{J} \neq \emptyset,$$
$$P_{\mathcal{J}} = 0 \qquad \text{if } \mathcal{J} = \emptyset.$$

The operator $P_{\mathcal{J}}$ is an orthogonal projection into a subspace spanned by vectors $\mathbf{a}_i$, $i \in \mathcal{J}$. Let $\mathcal{J} = \{i \in \mathcal{J}^0 \cup \mathcal{J}^-, (\mathbf{x}^0)^{\mathrm{T}}\mathbf{a}_i - b_i = 0\}$ and $\mathbf{u}^k = -(A_{\mathcal{J}} A_{\mathcal{J}}^{\mathrm{T}})^{-1} A_{\mathcal{J}} f'(\mathbf{x}^k)$, $k = 0, 1, \ldots$. Then the gradient of $f(\mathbf{x})$ is $f'(\mathbf{x}^k) = C\mathbf{x}^k - \mathbf{d}$, and $(I - P_{\mathcal{J}})f'(\mathbf{x}^k) = f'(\mathbf{x}^k) + A_{\mathcal{J}}^{\mathrm{T}}\mathbf{u}^k$.

Then the algorithm is the following:

```
Let x⁰ be given and let it=0. Then f′(x⁰) = Cx⁰ − b;
do while it<max it;
set 𝒥; and call PROJECT(𝒥,f′ (x⁰),u⁰,(I− P𝒥) f′(x⁰));
if (‖(I− P𝒥)f′(x⁰)‖ ≈ 0) then
   if (u⁰ᵢ ≥ 0 ∀i∈𝒥∩𝒥⁻) then x=x⁰ {solution}; goto 2
     else j: ={i∈𝒥∩𝒥⁻|u⁰ᵢ < 0};𝒥′=𝒥−{j} end if;
else 𝒥′=𝒥 end if;
   call CG(𝒥′,x⁰,f′(x⁰)); it=it+1;
end do { maximum number of iterations reached };
2 end;
SUBROUTINE CG(𝒥′,x,f′)
```

In the conjugate gradient method with constraints, unlike the standard CGM, we use the projection $(I - P_{\mathcal{J}'})f'(\mathbf{x}^k)$ instead of the gradient $f'(\mathbf{x}^k)$. Moreover, we check the nonactive constraints and in every iteration we correct the new step length $\alpha^{k+1} := \min(\alpha^{k+1}, \overline{\alpha}^{k+1})$, where

$$\overline{\alpha}^{k+1} = \min_{\mathcal{M}} \frac{-(\mathbf{a}_i, \mathbf{x}^k)}{(\mathbf{a}_i, \mathbf{p}^{k+1})} \quad \text{and} \quad \mathcal{M} := \{i | i \notin \mathcal{J}' \wedge (\mathbf{a}_i, \mathbf{p}^{k+1}) > 0\}.$$

We have the following algorithm:

```
Let k=0; x⁰ =x; f′(x⁰) = f′ {from the previous iteration};
do while k<max it 2;
call PROJECT(𝒥′,f′(xᵏ),u, (I− P𝒥′) f′(xᵏ));
g=−(I− P𝒥′) f′(xᵏ); rᵏ⁺¹ = ‖g‖²;
if (rᵏ⁺¹ < ε) then x=xᵏ; f′ = f′(xᵏ); return;
end if;
```

```
if k=0 then p^1=g else β^{k+1}=r^{k+1}/r^k; p^{k+1}=g+β^{k+1}p^k; end if;
α1=r^{k+1}; α2=(p^{k+1},Cp^{k+1}) {scal. product in R^N};
if (α1<min(1.0,|α2|)* max val) then α^{k+1}=α1/α2;
else α^{k+1}=max val; end if;
M:={i|i∉J' ∧ (a_i,p^{k+1})>0};
if M≠{∅} then ᾱ^{k+1}=min_M (d_i-(a_i,x^k))/(a_i,p^{k+1})   {d_i=0 in our case} (A*);
else ᾱ^{k+1}=max val end if;
if ᾱ^{k+1}<α^{k+1} then x=x^k+ᾱ^{k+1}p^{k+1}; f'=f'(x^k)+ᾱ^{k+1}Cp^{k+1};
return; else if α^{k+1}=max val then STOP;
        else x^{k+1}=x^k+α^{k+1}p^{k+1}; f'(x^{k+1})=f'(x^k)+α^{k+1}Cp^{k+1};
        end if;
dd=||x^{k+1}-x^k||/(max(1,||x^k||));
if dd<ε then x=x^{k+1}; f'=f'(x^{k+1}); return; end if;
k=k+1;
end if;
x=x^k {point obtained after max. num. of iterations};
    f'=f'(x^k);
return end;.
SUBROUTINE PROJECT(J,f'(x),u,(I-P_J)f'(x))
The calculation of u=-(A_J A_J^T)^{-1}·A_J f'(x) and (I-P_J)f'(x)=
f'(x)+A_J^T u by the CG method.
Input: J, f'(x);
Output: u, (I-P_J)f'(x);
return;
```

Remark 6.12 (i) We set $\mathbf{x}^0 = (0, \ldots, 0)$ for the initial guess. As $A_{\mathcal{J}^-}$ has a special structure, we may also choose $\mathbf{x}^0$ so that the inequalities are satisfied strictly ("inner point"). For the models, having only two bodies stuck in one point, the degree of freedom $\mathbf{x}_r$ appears at most in one constraint $\mathbf{a}_s$; we choose $\mathbf{x}_r = -\text{sign}(\mathbf{a}_{sr}) \cdot k$, $k > 0$ suitable constant not exceeding the dimension of the model. We may also choose non-constrained degrees as proportional to k. If more than two bodies stick, the restricted number of degrees of freedom may appear in more constraints. We arrive at a contradiction to the previous choice if the corresponding coefficients for $\mathbf{x}_r$ have the opposite signs. Here we choose $\mathbf{x}_r = 0$ again. (ii) Denote the value of $\|(I - P_J)f'(\mathbf{x}^0)\|$ in the nth iteration ($0 \leq n < \max it$) by pg^n. Then $pg^n \approx 0$ numerically represents the comparison $[pg^{n+1}/\max(1.0, pg^n)] < \varepsilon$. Similarly, we use the test $\mathbf{u}_i^0/\mathbf{u} > (-\varepsilon)$, where $\mathbf{u} = \max(1.0, \mathbf{u}_{\mathcal{J}}^0)$ and $\mathbf{u}_{\mathcal{J}}^0 = \max_{m \in \mathcal{J}}(0.0, \mathbf{u}_m^0)$ for the multipliers $\mathbf{u}_i^0$. It is also necessary to test the magnitudes of $\mathbf{x}^k$ and $\mathbf{p}^k$ in a semicoercive case. (iii) The value $\max it1$ depends on n. $\max it2$ is the number of iterations in CG. We would choose $n - m'$, where m' is the number of active constraints [see Pschenichny and Danilin (1978)]. However, the result will be more accurate if we choose the value slightly greater than n (e.g., $\approx 2n$). (iv) For some models, it is convenient to use the following strategy. We choose less strict tolerance for subproblems (subroutine CG – conjugate gradient algorithm) in the first several iterations within the CGC subroutine – conjugate gradient algorithm with constraints. The tolerance is set to a more strict value after a limited number of these iterations. We can get remarkable acceleration of the process. (v) If the stiff matrix C is positive definite, then

$$(f'(\mathbf{x}^k), \mathbf{p}^{k+1}) \neq 0 \quad \text{and} \quad (\mathbf{p}^{k+1}, C\mathbf{p}^{k+1}) = 0.$$

In this case $f(\mathbf{x}^k + \alpha \mathbf{p}^{k+1})$ decreases when α is increased. If $\overline{\alpha}^{k+1} = \text{max val}$, then f on K_d is not bounded from below. (vi) We may use the diagonal form of $(A_{\mathcal{J}} A_{\mathcal{J}}^{\mathrm{T}})$ in the case of "multibody contact" for the calculation of the vector $\mathbf{u}$ in the subroutine PROJECT. A more general case (when more than two bodies stick in one point or the preconditioning) can be solved as follows: $\mathbf{u}$ solves the system $(A_{\mathcal{J}} A_{\mathcal{J}}^{\mathrm{T}})\mathbf{u} = -A_{\mathcal{J}} f'(\mathbf{x})$, where $A_{\mathcal{J}} A_{\mathcal{J}}^{\mathrm{T}}$ is symmetric and positive definite. This property is due to definition and linear independence of rows $A_{\mathcal{J}}$. The minimization is carried out by the conjugate gradient method again. In this case, the dimension of the problem is far lower (contact pairs), the matrix $A_{\mathcal{J}}$ is sparse, and there are no constraints. (vii) The matrix $(A_{\mathcal{J}} A_{\mathcal{J}}^{\mathrm{T}})$ is not stored, the multiplication $\mathbf{w} = (A_{\mathcal{J}} A_{\mathcal{J}}^{\mathrm{T}})\mathbf{u}$ is gradually transformed to $\mathbf{v} = A_{\mathcal{J}}^{\mathrm{T}}\mathbf{u}$, $\mathbf{w} = A_{\mathcal{J}}\mathbf{v}$. On the basis of the fact that $\overline{\alpha}^1 > 0$ (see Subroutine CG), we can prove that the CG algorithm makes a nonzero step (i.e., does not cycle) similarly as in Pschenichny and Danilin (1978). If the implication

$$j \in J \Rightarrow (j \in J' \vee (\mathbf{a}_j, \mathbf{p}^1) \leq 0)$$

is valid, then it follows in the subroutine CG that $\overline{\alpha}^1 > 0$. Therefore, it is sufficient to focus the case $\|(I - P_J)f'(\mathbf{x}^0)\| \approx 0$ and the removed index $j \in J - J'$.

Lemma 6.14 Let $\|(I - P_{\mathcal{J}})f'(\mathbf{x}^0)\| = 0$. Let $A_{\mathcal{J}'}$ be created from $A_{\mathcal{J}}$ by removing the row with the index $j|\mathbf{u}_j^0 < 0$. Then $(\mathbf{a}_j, \mathbf{p}^1) < 0, j \in \mathcal{J} - \mathcal{J}'$.

For the proof see Pschenichny and Danilin (1978).

The condition $(\mathbf{a}_j, \mathbf{p}^1) < 0\, j \in \mathcal{J} - \mathcal{J}'$ may be satisfied even in the case where more indices $\{j|u_j^0 < 0\}$ are removed [e.g., all with $j|u_j^0 < (-\varepsilon)$]. For this see the next lemma.

Lemma 6.15 Let $\|(I - P_{\mathcal{J}})f'(\mathbf{x}^0)\| = 0$. Let $A_{\mathcal{J}'}$ be created from $A_{\mathcal{J}}$ by removing the rows with indices $j|u_j^0 < 0$. Furthermore, let the rows of $A_{\mathcal{J}}$ satisfy $(\mathbf{a}_i, \mathbf{a}_j) = 0$, $i \neq j$, $i, j \in \mathcal{J}$. Then $(\mathbf{a}_j, \mathbf{p}_1) < 0, j \in \mathcal{J} - \mathcal{J}'$.

Proof We have

$$\begin{aligned} 0 &= (I - P_J)f'(\mathbf{x}^0) = f'(\mathbf{x}^0) + A_J^{\mathrm{T}}\mathbf{u}^0 = f'(\mathbf{x}^0) + A_{J'}^{\mathrm{T}}\mathbf{u}_{\prime}^0 + A_{J-J'}^{\mathrm{T}}\mathbf{u}_{\prime\prime}^0, \\ -\mathbf{p}^1 &= (I - P_{J'})f'(\mathbf{x}^0) = f'(\mathbf{x}^0) + A_{J'}^{\mathrm{T}}\mathbf{v}_{\prime}, \end{aligned}$$

where

$$\mathbf{v}_{\prime} = -(A_{\mathcal{J}'} A_{\mathcal{J}'}^{\mathrm{T}})^{-1} A_{\mathcal{J}'} f'(\mathbf{x}^0)$$

Subtracting and multiplying by the vector $\mathbf{a}_j$, $j \in \mathcal{J} - \mathcal{J}'$, we obtain $(\mathbf{a}_j, \mathbf{p}^1) = c \cdot \mathbf{u}_{j''}^0$ where $c = (\mathbf{a}_j, \mathbf{a}_j) > 0$ and from the assumption $\mathbf{u}_{''}^0 < 0$. Thus, $(\mathbf{a}_j, \mathbf{p}^1) < 0$. □

Corollary 6.4 Let the assumptions of the previous lemma be fulfilled. Then $\overline{\alpha}^1 > 0$, and as a result the algorithm CGC does not cycle.

The condition for the rows of $A_{\mathcal{J}}$ is fulfilled in the "multibody contact." It may be slightly violated in a general case and also if the preconditioning is used. Nevertheless, for such cases we often have an acceleration as well.

Preconditioning Method Consider again problem (6.120) with $\mathbf{d} = \mathbf{0}$, that is,

$$f(\mathbf{x}) = \tfrac{1}{2}\mathbf{x}^{\mathrm{T}} C \mathbf{x} - \mathbf{x}^{\mathrm{T}} \mathbf{b} \to \min$$
$$A\mathbf{x} \leq 0.$$

Now we assume that C is positive definite. Let W be a positive definite matrix $n \times n$ in the form $W = EE^{\mathrm{T}}$. Introducing a new variable $\mathbf{y} = E^{\mathrm{T}}\mathbf{x}$ and express $(\mathcal{P}_d)$ in terms of a new variable $\mathbf{y}$, then we have

$$\overline{f}(\mathbf{y}) = \tfrac{1}{2}\mathbf{y}^{\mathrm{T}}\overline{C}\mathbf{y} - \mathbf{y}^{\mathrm{T}}\overline{\mathbf{b}} \to \min,$$
$$\overline{A\mathbf{y}} \leq 0, \tag{6.121}$$

where

$$\overline{C} = E^{-1}CE^{-\mathrm{T}}, \qquad \overline{\mathbf{b}} = E^{-1}\mathbf{b}, \qquad \overline{A} = AE^{-\mathrm{T}}.$$

As $E^{-\mathrm{T}}\overline{C}E^{\mathrm{T}} = W^{-1}C$, the matrices $\overline{C}$ and $W^{-1}C$ have the same eigenvalues. It can be shown that the convergence of CGM depends on the spectral condition number $(\lambda_{\max}/\lambda_{\min})$ of the matrix in the functional $\overline{f}(\mathbf{y})$, that is, of the matrices $C, \overline{C}$. The speed of the convergence increases when the spectral condition number decreases. The lowest spectral condition number has an identity matrix. Therefore, we try to find W, which is an easy invertible approximation of C, and for which we can show that $W^{-1}C$ has a lower spectral condition number.

Then the preconditioning CGM is as follows:

SUBROUTINE PCG$(\mathcal{J}', \mathbf{x}\{= E^{-\mathrm{T}}\ \mathbf{y}\}, E^{\mathrm{T}}, \overline{f}')$
$\mathbf{y}^0 = \mathbf{y} = E^{\mathrm{T}}\mathbf{x}$; $\overline{f}'(\mathbf{y}^0) = \overline{C}\mathbf{y}^0 - \mathbf{b}$;
for $k = 0, 1, \ldots$
$\overline{g} = (I - \overline{P}_{\mathcal{J}'})\,\overline{f}'(y^k)$; $\overline{r}^{k+1} = \|\overline{g}\|^2$;
if $k = 0$ **then** $\overline{\mathbf{p}}^1 = \overline{\mathbf{g}}$; **else** $\beta^{k+1} = \overline{r}^{k+1}/\overline{r}^k$; $\overline{\mathbf{p}}^{k+1} = \overline{\mathbf{g}} + \beta^{k+1}\overline{\mathbf{p}}^k$; **end if**;
$\alpha^{k+1} = \overline{r}^{k+1}/(\overline{\mathbf{p}}^{k+1}, \overline{C}\overline{\mathbf{p}}^{k+1})$; $\overline{\alpha}^{k+1} = \min_{\overline{\mathcal{M}}} \frac{-(\overline{\mathbf{a}}_i, \mathbf{y}^k)}{(\overline{\mathbf{a}}_i, \overline{\mathbf{p}}^{k+1})}$;
if $\overline{\alpha}^{k+1} < \alpha^{k+1}$ **then** $\mathbf{y} = \mathbf{y}^k + \overline{\alpha}^{k+1}\overline{\mathbf{p}}^{k+1}$; $\overline{f}' = \overline{f}'(\mathbf{y}^k) + \overline{\alpha}^{k+1}\overline{C}\overline{\mathbf{p}}^{k+1}$
{ and return to CGC };
else $\mathbf{y}^{k+1} = \mathbf{y}^k + \alpha^{k+1}\overline{\mathbf{p}}^{k+1}$; $\overline{f}'(\mathbf{y}^{k+1}) = \overline{f}'(\mathbf{y}^k) + \alpha^{k+1}\overline{C}\overline{\mathbf{p}}^{k+1}$; **end if**;

At the same time $\overline{P}_{J'} = \overline{A}_{J'}^{\mathrm{T}}(\overline{A}_{J'}\overline{A}_{J'}^{\mathrm{T}})^{-1}\overline{A}_{J'}$ and $\overline{\mathcal{M}}$ is connected with $\mathcal{M}$ by the transformation $\mathbf{y} = E^{\mathrm{T}}\,\mathbf{x}$.

Introducing a vector $\mathbf{v}^{k+1}$ by $\mathbf{v}^{k+1} = E^{-\mathrm{T}}\overline{\mathbf{p}}^{k+1}$ and using $\mathbf{h}^k := \overline{f}'(\mathbf{y}^k) = E^{-1}f'(\mathbf{x}^k)$, $(\overline{\mathbf{p}}^{k+1}, \overline{C}\overline{\mathbf{p}}^{k+1}) = (\mathbf{v}^{k+1}, C\,\mathbf{v}^{k+1})$ and $(\overline{\mathbf{a}}_i, \overline{\mathbf{p}}^{k+1}) = (\mathbf{a}_i,\ \mathbf{v}^{k+1})$, we can write PCG in $\mathbf{x}$ variable.

```
SUBROUTINE PCG(J', x, E^T, f')
f'(x^0) = f' {from previous iteration};
for k=0,1,...
h^k = E^-1 f'(x^k);  g¯ = -(I - P¯_J') h^k;  r¯^(k+1) = ||g¯||^2;
if k=0 then v^1 = E^-T g¯ else β^(k+1) = r¯^(k+1)/r¯^k;  v^(k+1) = E^-T g¯ + β^(k+1) v^k;
end if;
α^(k+1) = r¯^(k+1)/(v^(k+1), Cv^(k+1));  α¯^(k+1) = min_M¯ -(a_i, x^k)/(a_i, v^(k+1));
if α¯^(k+1) < α^(k+1) then x = x^k + α¯^(k+1) v^(k+1);  f' = f'(x^k) + α¯^(k+1) Cv^(k+1)
{and return to CGC };
else x^(k+1) = x^k + α^(k+1) v^(k+1);  f'(x^(k+1)) = f'(x^k) + α^(k+1) Cv^(k+1);
end if;
```

In the subroutine PROJECT, which is called in PCG (the calculation of $\overline{\mathbf{g}}$), the multiplications $A_{\mathcal{J}'}\mathbf{x}$, $A_{\mathcal{J}'}^{\mathrm{T}}\mathbf{x}$ are replaced by $\overline{A}_{\mathcal{J}'}\mathbf{y}$, $\overline{A}_{\mathcal{J}'}^{\mathrm{T}}\mathbf{y}$, that is $A_{\mathcal{J}'}E^{-\mathrm{T}}\mathbf{y}$, $E^{-1}A_{\mathcal{J}'}^{\mathrm{T}}\mathbf{y}$. As $E^{-\mathrm{T}}$ is regular, $\overline{A}_{\mathcal{J}'}$ has also linearly independent rows. The matrix $\overline{C}$ does not occur in the transformed problem.

The choice of the preconditioning matrix can be made in several ways. A simple form is represented by the diagonal of the matrix C, we denote it by D, then $W = D$, and then $E^{\mathrm{T}} = D^{1/2}$, and, therefore, it is sufficient to store it only as a vector. Another way is represented by the SSOR decomposition [see Axelsson (1994), Nedoma et al. (1999b) and Kestřánek (1999)]. Let $C = D + L + L^{\mathrm{T}}$ (L is the lower triangular matrix); then the preconditioning matrix will be of the form

$$W = \frac{1}{2-\omega}\left(\frac{1}{\omega}D + L\right)\left(\frac{1}{\omega}D\right)^{-1}\left(\frac{1}{\omega}D + L\right)^{\mathrm{T}}, \qquad 0 < \omega < 2, \tag{6.122}$$

where the factor $1/2 - \omega$ can be omitted. Thus

$$E^{\mathrm{T}} = \left(\frac{1}{\omega}D\right)^{-(1/2)}\left(\frac{1}{\omega}D + L^{\mathrm{T}}\right).$$

The spectral condition number of the matrix $\overline{C} = W^{-1}C$, which we denote as $\kappa(\overline{C})$, can be under certain assumptions smaller than the spectral condition number $\kappa(C)$ of the matrix C, that is, $\kappa(\overline{C}) < \kappa(C)$. Therefore, the rate of the convergence of the PCG method is faster than the rate of convergence of the CG method. Here $\kappa(\overline{C}) = \overline{\lambda}_n/\overline{\lambda}_1$, $\kappa(C) = \lambda_n/\lambda_1$, where $\overline{\lambda}_1$ and $\overline{\lambda}_n$ or λ_1 and λ_n are the smallest and largest eigenvalues, respectively, of $\overline{C} = W^{-1}C$ or C, respectively [see Axelsson (1994)].

Theorem 6.37 Let C be a positive definite matrix and let W be determined by (6.122). Let

$$\|D^{-(1/2)}LD^{-(1/2)}\|_\infty \leq \tfrac{1}{2}, \qquad \|D^{-(1/2)}L^{\mathrm{T}}D^{-(1/2)}\|_\infty \leq \tfrac{1}{2}.$$

Then

$$\min_{0<\omega<2} \kappa(\overline{C}) \leq \sqrt{\tfrac{1}{2}\kappa(C)} + \tfrac{1}{2}.$$

Remark 6.13 The optimal value of ω can be estimated, thus $0 < \omega < 2$.

For the preconditioning matrix the incomplete factorization and the Choleski's decomposition can be used, and we may say that they represent a more effective approach [see Axelsson (1994), Kestřánek (1999), and Nedoma et al. (1999b)]. For the incomplete factorization we consider factorization $C = LL^{\mathrm{T}}$. To analyze the preconditioning based on the incomplete factorization we introduce the following definition:

Definition 6.13 The matrix C is a $\overline{M}$ matrix, if

(i) $c_{ii} > 0$, $\quad i = 1, \ldots, n-1$,

(ii) $c_{ij} < 0$, $\quad i \neq j$,

(iii) $\max\{j \,|(i \leq j \leq n) \wedge (c_{ij} \neq 0)\} > i$, $\quad$ where $1 \leq i < n$.

Then we have the following theorem:

Theorem 6.38 The incomplete factorization is a stable process for the diagonal dominant $\overline{M}$ matrix in the following sense: the number

$$q = \frac{\max_{i,j,r} |c_{ij}^{(r)}|}{\max_{i,j} |c_{ij}|}, \qquad i,j = 1, \ldots, n, \ r = 1, \ldots, n-1,$$

is bounded from above (even $q = 1$).

The preelimination method In the previous paragraphs we have shown that in Problem $\mathcal{P}_d$ only the contact degrees of freedom, which belong to some contact pair $a_i \in \Gamma_c$, are constrained in matrix A. The number of degrees of freedom with this property is often far smaller than the total number of all degrees of freedom. By the elimination of nonconstrained degrees of freedom (Haslinger and Tvrdý, 1983; Kikuchi and Oden, 1988; Nedoma, 1993a,b, 1998a), we can reduce the number of variables in the minimized functional and therefore carry out the iterations for a smaller problem.

We will start from problem (6.121), that is, from the preconditioned problem

$$\overline{f}(\mathbf{x}) = \tfrac{1}{2}\mathbf{x}^{\mathrm{T}}\overline{C}\mathbf{x} - \mathbf{x}^{\mathrm{T}}\overline{\mathbf{b}} \rightarrow \min,$$
$$\overline{A}\mathbf{x} \leq 0.$$

In the Uzawa algorithm, if we calculate the projection operator, only variables $\mathbf{x}_1, \ldots, \mathbf{x}_m$, $m = m(h)$, corresponding to values $\mathbf{u}_{hi}$ in points of the triangulation $\mathcal{T}_h$ laying on $\cup\overline{\Gamma}_c^{kl}\backslash\overline{\Gamma}_u$, are taken into account. We will assume that unknown variables, corresponding to u_{hi} in nodes of the triangulation $\mathcal{T}_h$ laying on $\cup\overline{\Gamma}_c^{kl}\backslash\overline{\Gamma}_u$, are situated on the last $m = m(h)$ places of a vector $\mathbf{x}$. Then we may split a vector $\mathbf{x} \in \mathbb{R}^n$ in such a way that $\mathbf{x} = (\mathbf{x}_1, \mathbf{x}_2)^{\mathrm{T}} \in \mathbb{R}^{n-m} \times \mathbb{R}^m$, where $\mathbf{x}_1$ and $\mathbf{x}_2$ correspond to the free and constrained variables, respectively. Then we define the set

$$K_{h0} = \{\mathbf{x} \in \mathbb{R}^n, \mathbf{x} = (\mathbf{x}_1, \mathbf{x}_2)^{\mathrm{T}}, \mathbf{x}_1 \in \mathbb{R}^{n-m}, \mathbf{x}_2 \in \mathbb{R}^m, \overline{A}\mathbf{x}_2 \leq 0\}.$$

Due to the splitting of vector $\mathbf{x}$, we also split vector $\overline{\mathbf{d}}$, and decompose matrix $\overline{C}$ by

$$\overline{C} = \begin{pmatrix} \overline{C}_{11} & \overline{C}_{12} \\ \overline{C}_{21} & \overline{C}_{22} \end{pmatrix},$$

where $\overline{C}_{11}$ and $\overline{C}_{22}$ are $(n-m) \times (n-m)$ or $m \times m$ matrices, respectively, and $\overline{C}_{12}$ and $\overline{C}_{21}$, where $\overline{C}_{12} = \overline{C}_{21}^{\mathrm{T}}$, are rectangular $(n - m \times m)$ or $m \times (n-m)$ matrices. Then (6.121) is equivalent to the following problem: Find $\mathbf{x} = (\mathbf{x}_1, \mathbf{x}_2)^{\mathrm{T}}$, $\mathbf{x}_1 \in \mathbb{R}^{n-m}$, $\mathbf{x}_2 \in \mathbb{R}^m$ and $\overline{A}\mathbf{x}_2 \leq 0$ such that

$$(\overline{C}\mathbf{x}, \mathbf{v} - \mathbf{x}) \geq (\overline{\mathbf{b}}, v - x) \quad \forall \mathbf{v} \in K_{h0}. \tag{6.123}$$

Let $\mathbf{v} = (\mathbf{v}_1, \mathbf{v}_2)^{\mathrm{T}}$, $\overline{\mathbf{b}} = (\overline{\mathbf{b}}_1, \overline{\mathbf{b}}_2)^{\mathrm{T}}$ such that $\mathbf{v}_1, \overline{\mathbf{b}}_1 \in \mathbb{R}^{n-m}$, $\mathbf{v}_2, \overline{\mathbf{b}}_2 \in \mathbb{R}^m$, $\overline{A}\mathbf{v}_2 \leq 0$. Let us put $\mathbf{v} = (\mathbf{v}_1, \mathbf{v}_2)^{\mathrm{T}}$ in (6.123) such that $\mathbf{v}_1 = \mathbf{x}_1 \pm \mathbf{y}_1$, $\mathbf{v}_2 = \mathbf{x}_2$, where $\mathbf{y}_1 \in \mathbb{R}^{n-m}$ is arbitrary. Then from (6.123) we obtain

$$\begin{pmatrix} \mathbf{x}_1 \pm \mathbf{y}_1 - \mathbf{x}_1 \\ \mathbf{x}_2 - \mathbf{x}_2 \end{pmatrix}^{\mathrm{T}} \begin{pmatrix} \overline{C}_{11} & \overline{C}_{12} \\ \overline{C}_{21} & \overline{C}_{22} \end{pmatrix} \begin{pmatrix} \mathbf{x}_1 \\ \mathbf{x}_2 \end{pmatrix} \geq \begin{pmatrix} \mathbf{x}_1 \pm \mathbf{y}_1 - \mathbf{x}_1 \\ \mathbf{x}_2 - \mathbf{x}_2 \end{pmatrix}^{\mathrm{T}} \begin{pmatrix} \overline{\mathbf{b}}_1 \\ \overline{\mathbf{b}}_2 \end{pmatrix}.$$

Hence

$$\mathbf{y}_1^{\mathrm{T}}(\overline{C}_{11}\mathbf{x}_1 + \overline{C}_{12}\mathbf{x}_2) = \mathbf{y}_1^{\mathrm{T}}\overline{\mathbf{b}}_1, \qquad \mathbf{y}_1 \in \mathbb{R}^{n-m}$$

and thus

$$\overline{C}_{11}\mathbf{x}_1 + \overline{C}_{12}\mathbf{x}_2 = \overline{\mathbf{b}}_1. \tag{6.124}$$

Moreover, let us put in (6.123) $\mathbf{v}$ such that $\mathbf{v}_1 = \mathbf{x}_1$, $\mathbf{v}_2 = \mathbf{y}_2$, where $\mathbf{y}_2 \in \mathbb{R}^m$, $\overline{A}\mathbf{y}_2 \leq 0$. Then

$$(\mathbf{y}_2 - \mathbf{x}_2)^{\mathrm{T}}(\overline{C}_{21}\mathbf{x}_1 + \overline{C}_{22}\mathbf{x}_2) \geq (\mathbf{y}_2 - \mathbf{x}_2)^{\mathrm{T}}\,\overline{\mathbf{b}}_2.$$

Hence, using (6.124) we obtain the problem to find $\mathbf{x}_2 \in K_{hY}$ such that

$$(\mathbf{y}_2 - \mathbf{x}_2)^{\mathrm{T}} C_0 \mathbf{x}_2 \geq (\mathbf{y}_2 - \mathbf{x}_2)^{\mathrm{T}} \overline{\mathbf{b}}_{20}, \quad \forall \mathbf{y}_2 \in K_{hY}, \tag{6.125}$$

where $C_0 = \overline{C}_{22} - \overline{C}_{21}\overline{C}_{11}^{-1}\overline{C}_{12}$, $\overline{\mathbf{b}}_{20} = \overline{\mathbf{b}}_2 - \overline{C}_{21}\overline{C}_{11}^{-1}\overline{\mathbf{b}}_1$, $K_{hY} = \{\mathbf{x} | \mathbf{x} \in \mathbb{R}^m, \overline{A}\mathbf{x} \leq 0\}$.

Theorem 6.39 Let $\overline{C}$ be a symmetric, positive semidefinite matrix of the order n. Then the matrix $C_0 = \overline{C}_{22} - \overline{C}_{21}\overline{C}_{11}^{-1}\overline{C}_{12}$ is a symmetric and positive semidefinite matrix of the order m.

From (6.125) and (6.124) it follows that our original problem is equivalent to the following problem: Find $\mathbf{x} = (\mathbf{x}_1, \mathbf{x}_2)^{\mathrm{T}}$ such that

$$\tfrac{1}{2}(\mathbf{x}_2, C_0\mathbf{x}_2) - (\overline{\mathbf{b}}_{20}, \mathbf{x}_2) = \min_{\mathbf{x} \in K_{hY}} \left\{ \tfrac{1}{2}(\mathbf{x}, C_0\mathbf{x}) - (\overline{\mathbf{b}}_{20}, \mathbf{x}) \right\} \tag{6.126}$$

and $\mathbf{x}_1$ is a solution of (6.124).

It is evident that the matrix C_0 and the vector $\overline{\mathbf{b}}_{20}$ can be obtained by the Gaussian elimination from the initial matrix $\overline{C}$ and vector $\overline{\mathbf{b}}$.

From this theorem and (6.126) we see that it is sufficient to eliminate the matrix $\overline{C}$ onto C_0, then by applying the minimizing algorithm onto the form $\frac{1}{2}(\mathbf{x}, C_0\mathbf{x}) - (\overline{\mathbf{b}}_{20}, \mathbf{x})$ we find the last m unknown variables, and by back elimination we find the remaining $(n-m)$ variables.

Remark 6.14 Since the matrix C_0 is a symmetric and positive semidefinite matrix, it is evident that the LL^{T} decomposition in preconditioning can be used. Then the obtained algorithm is very fast, and, moreover, the memory requirements are lower than in the previous discussed method without a preelimination and preconditioning approach.

Remark 6.15 One can employ the second variant of the mixed variational formulation, introducing a further Lagrange multiplier for the constraint $\mathbf{u} \in K$ of nonpenetration. Such an approach is analyzed in Hlaváček (2007). See also the references therein.

6.4.4 Nonlinear Elasticity

The Model Based on anatomical and biomechanical arguments as well as CT or MRI analyses of the human joints or the fractured spine the mathematical model can be constructed.

Let $\Omega \subset R^N$, $N = 2, 3$, be a region occupied by a system of elastic bodies Ω^ι so that $\Omega = \cup_{\iota=1}^{s} \Omega^\iota$. Let Ω^ι be bounded domains with Lipschitz boundaries $\partial\Omega^\iota$. Let us assume that $\partial\Omega = \cup_{\iota=1}^{s} \partial\Omega^\iota = \Gamma_u \cup \Gamma_\tau \cup \Gamma_c \cup \mathcal{R}$, where Γ_u, Γ_τ, Γ_c are open subsets of $\partial\Omega$, $\Gamma_c \neq \emptyset$ and the surface measure of $\mathcal{R}$ is equal to zero and where Γ_u denotes the part of a boundary where the studied part of the the joint system is fixed, Γ_τ denotes the part of boundary where the studied part of the joint system is loaded (or unloaded), and $\Gamma_c = \cup_{k,l} \Gamma_c^{kl}$ represents the contact boundary between components of the joint or, for example, between different parts of the fractured vertebral body. By indices k, l we denote the neighboring components of joint or fractured parts of the vertebral body.

Let us assume that rheology of the spine can be described by a nonlinear stress–strain relation. Let the deformation energy W be the nonlinear function of a strain defined by [see Fung (1965, 1968)]

$$W = A^{\lambda}(e_{ij}) = (c_{ijkl}(\mathbf{u})e_{ij}(\mathbf{u})e_{kl}(\mathbf{u}))^{\lambda}, \qquad A(e_{ij}) = c_{ijkl}(\mathbf{u})e_{ij}(\mathbf{u})e_{kl}(\mathbf{u}), \tag{6.127}$$

where λ is a parameter. Then

$$\tau_{ij} = \frac{\partial W}{\partial e_{ij}} = 2\lambda A^{\lambda-1}(e_{ij})c_{ijkl}(\mathbf{u})e_{kl}(\mathbf{u}), \tag{6.128}$$

where $c_{ijkl}(\mathbf{u})$ are elastic coefficients satisfying the well-known symmetry and Lipschitz conditions. Hence for $0 < \lambda < 1$ the material is softened, for $\lambda = 1$ is elastic, and for $\lambda > 1$ is hardened. In the following we will assume that a repeated index implies the summation over the range $1, \ldots, N$ and

$$e_{ij}(\mathbf{u}) = \frac{1}{2}\left(\frac{\partial u_i}{\partial x_j} + \frac{\partial u_j}{\partial x_i}\right), \quad i,j = 1, \ldots, N, \tag{6.129}$$

is the small strain tensor, and $\mathbf{u} = (u_i)$ is the displacement vector. Let $u_n = u_i n_i$, and $\mathbf{u}_t = \mathbf{u} - u_n \mathbf{n}$, $\tau_n = \tau_{ij} n_j n_i$, $\boldsymbol{\tau}_t = \boldsymbol{\tau} - \tau_n \mathbf{n}$ are normal and tangential components of displacement, and stress vector $\mathbf{u} = (u_i)$, $\boldsymbol{\tau} = (\tau_i)$, $\tau_i = \tau_{ij} n_j$, $i,j = 1, \ldots, N$, $\mathbf{n} = (n_i)$ is the unit outward normal vector to $\partial\Omega$. Moreover, we shall assume the simple case without friction. Finally, let the body forces $\mathbf{f} \in [L^2(\Omega)]^N$ and surface forces $\mathbf{P} \in [L^2(\Gamma_\tau)]^N$. The model to be solved is as follows:

Problem $\mathcal{P}$ Let us seek a solution of the unilateral contact problem:

$$-\frac{\partial}{\partial x_j}(2\lambda A^{\lambda-1}(e_{ij})c_{ijkl}(\mathbf{u})e_{kl}(\mathbf{u})) = f_i, \qquad i = 1, \ldots, N \quad \text{in } \Omega, \tag{6.130}$$

$$\mathbf{u} = 0 \quad \text{on } \Gamma_u, \tag{6.131}$$

$$\tau_{ij} n_j = P_i \quad \text{on } \Gamma_\tau, \tag{6.132}$$

$$u_n^k - u_n^l \le 0, \qquad \tau_n^{kl} \equiv \tau_n^k = \tau_n^l \le 0, \qquad (u_n^k - u_n^l)\tau_n^{kl} = 0 \quad \text{on } \cup_{k,l} \Gamma_c^{kl}, \tag{6.133}$$

where $u_n^l = u_i^l n_i^k$.

Weak Solution of the Problem Let us introduce the following sets:

$$V = \left\{ \mathbf{v} | \mathbf{v} \in \prod_{\iota=1}^{s} [H^1(\Omega^\iota)]^N, \ \mathbf{v} = 0 \text{ on } \Gamma_u \text{ in the sense of traces} \right\},$$

$$K = \left\{ \mathbf{v} | \mathbf{v} \in V, \ v_n^k - v_n^l \leq 0 \text{ on } \cup_{k,l} \Gamma_c^{kl} \right\},$$

$$R = \left\{ \mathbf{v} | \mathbf{v} \in \prod_{\iota=1}^{s} [H^1(\Omega^\iota)]^N, \ e_{ij}(\mathbf{v}) = 0 \text{ a.e.} \right\}$$

$$= \left\{ \mathbf{v} | \mathbf{v} \in \prod_{\iota=1}^{s} [H^1(\Omega^\iota)]^N, \ \mathbf{v}^\iota = \mathbf{a}^\iota + \mathbf{b}^\iota \times \mathbf{x}^\iota, \text{ where } \mathbf{a}^\iota, \mathbf{b}^\iota \text{ are arbitrary real vectors for } N = 3 \text{ and } v_1^\iota = a_1^\iota - b^\iota x_2, \ v_2^\iota = a_2^\iota + b^\iota x_1 \text{ for } N = 2 \right\}.$$

It is evident that V is a Hilbert space and K a convex cone. It can be shown that K is a closed convex subset of V.

Multiplying (6.130) by $v_i - u_i$, integrating over Ω, and using boundary conditions, we obtain the following variational inequality problem: Find

$$\mathbf{u} \in K, \qquad \int_\Omega 2\lambda A^{\lambda-1}(e_{ij}) c_{ijkl}(\mathbf{u}) e_{ij}(\mathbf{u}) e_{kl}(\mathbf{v} - \mathbf{u}) \, d\mathbf{x}$$
$$- \int_{\Gamma_\tau} P_i(v_i - u_i) \, ds - \int_\Omega f_i(v_i - u_i) \, d\mathbf{x} \geq 0 \quad \forall \mathbf{v} \in K, \tag{6.134}$$

or in equivalent forms

$$\mathbf{u} \in K, \qquad a(\mathbf{u}; \mathbf{u}, \mathbf{v} - \mathbf{u}) \geq (\mathbf{F}, \mathbf{v} - \mathbf{u}) \quad \forall \mathbf{v} \in K, \tag{6.135}$$

or

$$\mathbf{u} \in K, \qquad D\mathcal{J}(\mathbf{u}, \mathbf{v} - \mathbf{u}) \geq 0 \quad \forall \mathbf{v} \in K, \tag{6.136}$$

if $a(\mathbf{u}; \mathbf{u}, \mathbf{v}) = D\mathcal{J}(\mathbf{u}, \mathbf{v}) \ \forall \mathbf{u}, \mathbf{v} \in V$, [see Nedoma and Hlaváček (2002) for a necessary condition to satisfy this assumption], where $\mathcal{J}(\mathbf{u})$ is the corresponding functional of the total potential energy defined as

$$\mathcal{J}(\mathbf{u}) = \mathcal{L}(\mathbf{u}) - \int_{\Gamma_\tau} P_i u_i \, ds - \int_\Omega f_i u_i \, d\mathbf{x},$$
$$\mathcal{L}(\mathbf{u}) = \frac{1}{2} \int_\Omega \int_0^1 a(t\mathbf{u}; t\mathbf{u}, \mathbf{u}) \, dt \, d\mathbf{x} \tag{6.137}$$

Here we denoted

$$a(\mathbf{w}; \mathbf{u}, \mathbf{v}) = \int_\Omega 2\lambda A^{\lambda-1}(e_{ij}(\mathbf{w})) c_{ijkl}(\mathbf{w}) e_{ij}(\mathbf{u}) e_{kl}(\mathbf{v}) \, d\mathbf{x}, \tag{6.138}$$

$$(\mathbf{F}, \mathbf{v}) = \int_{\Gamma_\tau} P_i v_i \, ds + \int_\Omega f_i v_i \, d\mathbf{x}. \tag{6.139}$$

The problem can be solved by the secant modules method (see Section 6.2.4). For this method in the classical form see, for example, Nečas and Hlaváček (1981), and for variational inequality problems see, for example, Zeidler (1990b), Nečas and Hlaváček (1983), and Nedoma and Hlaváček (2002). The theoretical convergence is valid for $\lambda = 1$, and the isotropic case only, for $\lambda \neq 1$, the problem represents an open mathematical problem. We see that the method of secant modules consists of solving a sequence of the following variational inequalities:

$$\mathbf{u}_{n+1} \in K, \qquad \int_{\Omega} 2\lambda A^{\lambda-1}(e_{ij}(\mathbf{u}_n))c_{ijkl}(\mathbf{u}_n)e_{ij}(\mathbf{u}_{n+1})e_{kl}(\mathbf{v}-\mathbf{u}_{n+1})\,d\mathbf{x}$$
$$-\int_{\Gamma_\tau} P_i(v_i - (\mathbf{u}_{n+1})_i)\,d\mathbf{s} - \int_{\Omega} f_i(v_i - (\mathbf{u}_{n+1})_i)\,d\mathbf{x} \geq 0, \qquad n = 1, 2, \ldots, \tag{6.140}$$

where $\mathbf{u}_n$ is the nth approximate solution of the problem investigated.

Hence the problem investigated leads to the solution of a sequence of variational inequalities of the semicoercive type with variable coefficients, which can be written as the following problem:

Find $\mathbf{u}_n \in K$, $n = 1, 2, \ldots$ be such that

$$a(\mathbf{u}_n; \mathbf{u}_{n+1}, \mathbf{v} - \mathbf{u}_{n+1}) \geq (\mathbf{F}, \mathbf{v} - \mathbf{u}_{n+1}) \quad \forall \mathbf{v} \in K. \tag{6.141}$$

Let us assume that the bilinear form $a(\mathbf{w}; \mathbf{u}, \mathbf{v})$ is symmetric in $\mathbf{u}, \mathbf{v}$ and such that

$$a(\mathbf{w}; \mathbf{u}, \mathbf{u}) \geq c_0 \|\mathbf{u}\|^2, c_0 = \text{const} > 0, \tag{6.142}$$

$$|a(\mathbf{w}; \mathbf{u}, \mathbf{v})| \leq c_1 \|\mathbf{u}\| \, \|\mathbf{v}\|, \tag{6.143}$$

$$a(\mathbf{u}; \mathbf{u}, \mathbf{v}) = D\mathcal{L}(\mathbf{u}, \mathbf{v}), \tag{6.144}$$

$$\tfrac{1}{2}a(\mathbf{u}; \mathbf{v}, \mathbf{v}) - \tfrac{1}{2}a(\mathbf{u}; \mathbf{u}, \mathbf{u}) - \mathcal{L}(\mathbf{v}) + \mathcal{L}(\mathbf{u}) \geq 0 \quad \forall \mathbf{u}, \mathbf{v} \in V, \tag{6.145}$$

and that $\mathcal{L}$ has the second Gâteaux differential $D^2\mathcal{L}(\mathbf{w}; \mathbf{u}, \mathbf{v})$; the mapping $\mathbf{w} \mapsto D^2\mathcal{L}(\mathbf{w}; \mathbf{u}, \mathbf{v})$ is continuous on every segment and that

$$D^2\mathcal{L}(\mathbf{w}; \mathbf{u}, \mathbf{u}) \geq c\|\mathbf{u}\|^2, c = \text{const} > 0 \tag{6.146}$$

holds. Let $\mathbb{R}^* = \{\mathbf{p} \in R \cap K | p_n^k - p_n^l = 0 \text{ on } \cup_{k,l}\Gamma_c^{kl}\} = \{0\}$. Moreover, assume that $(\mathbf{F}, \mathbf{p}) \neq 0 \; \forall \mathbf{p} \in R \cap V \backslash \{0\}$ and

$$(\mathbf{F}, \mathbf{p}) < 0 \quad \forall \mathbf{p} \in R \cap K \backslash \{0\} \tag{6.147}$$

holds, where $(\mathbf{F}, \mathbf{p})$ is defined by (6.139). As a result we have the following theorem, the proof of which is based on the results of Nečas and Hlaváček (1983) and Zeidler (1990b) for the abstract contact problem.

Theorem 6.40 Let $P_V = R \cap V$ and $V = P_V \oplus Q$ be the orthogonal decomposition. Let (6.142), (6.144), (6.145), and (6.146) be satisfied for $\mathbf{u}, \mathbf{v}, \mathbf{w} \in Q$, and let (6.145) hold. Then (i) the functionals

$$\mathcal{J}(\mathbf{v}) = \mathcal{L}(\mathbf{v}) - (\mathbf{F}, \mathbf{v}), \qquad \omega(\mathbf{u}) = \tfrac{1}{2}a(\mathbf{v}; \mathbf{u}, \mathbf{u}) - (\mathbf{F}, \mathbf{u}) \tag{6.148}$$

are coercive and weakly lower semicontinuous in K, (ii) the problem to find $\mathbf{u} \in K$ such that

$$D\mathcal{L}(\mathbf{u}, \mathbf{v} - \mathbf{u}) \geq (\mathbf{F}, \mathbf{v} - \mathbf{u}) \quad \forall \mathbf{v} \in K \tag{6.149}$$

has a unique solution, (iii) the problem to find $u_{n+1} \in K$ for $\mathbf{u}_n \in K$, $n = 1, 2, \ldots,$ such that

$$a(\mathbf{u}_n; \mathbf{u}_{n+1}, \mathbf{v} - \mathbf{u}_{n+1}) \geq (\mathbf{F}, \mathbf{v} - \mathbf{u}_{n+1}) \quad \forall \mathbf{v} \in K, \tag{6.150}$$

has a unique solution and

$$\lim_{n \to \infty} P_Q \mathbf{u}_n = P_Q \mathbf{u}, \tag{6.151}$$

where P_Q denotes the projection of V onto Q, and moreover, if $\mathbf{u}_{n_k} \to \mathbf{u}$ as $k \to \infty$, then $\mathbf{u}^*$ is the solution $\mathbf{u}$ of (6.149) and

$$\lim_{n \to \infty} \mathbf{u}_n = \mathbf{u}. \tag{6.152}$$

Proof Sketch of the proof. The coerciveness of $\mathcal{J}$ follows from the fact that

$$\mathcal{J}(\mathbf{v}) \geq c_0 \|\mathbf{v}\| - c_1 \quad \forall \mathbf{v} \in K, \tag{6.153}$$

where c_0, c_1 are positive constants. Hence

$$\lim_{\mathbf{v} \in K, \|\mathbf{v}\| \to \infty} \quad \inf \frac{\mathcal{J}(\mathbf{v})}{\|\mathbf{v}\|} \geq c_0 > 0.$$

Due to the definition of $\mathcal{L}(\mathbf{v})$, the functional $\mathcal{J}$ is convex and G–differentiable and, therefore, it is weakly lower semicontinuous. Hence the existence of the solution of (6.149), due to the coerciveness and weakly lower semicontinuity follows immediately. Similarly

$$\omega(\mathbf{u}) = \tfrac{1}{2} a(\mathbf{v}; \mathbf{u}, \mathbf{u}) - (\mathbf{F}, \mathbf{u}) \geq c_2 \|\mathbf{u}\|^2 - c_3 \|\mathbf{u}\| \quad \forall \mathbf{u} \in K \tag{6.154}$$

is valid uniformly with respect to $\mathbf{v}$. By a similar way we prove the existence of the solution of (6.150). We can prove the uniqueness of the solution of (6.149) by the standard technique, that is, we assume that $\mathbf{u}$ and $\mathbf{u}'$ are two solutions, then $\mathbf{u}' = \mathbf{u} + \mathbf{p}$, $\mathbf{p} \in P$, $\mathbf{u} + \mathbf{p} \in K$, and $(\mathbf{F}, \mathbf{p}) = 0$ holds so that $\mathbf{p} = 0$. By a parallel way we prove the uniqueness of the solution of (6.150). Let us denote by P_Q the projector of V onto Q. Now we shall prove the convergence of (6.151) for $n \to \infty$. According to the previous results, inserting $\mathbf{v} = 0$ in (6.150) and using (6.142) and (6.154), then there exists a sequence $\{\mathbf{u}_n\}$ such that

$$\|\mathbf{u}_n\| \leq c \quad \forall n. \tag{6.155}$$

We have

$$C\|P_Q \mathbf{u}_{n+1} - P_Q \mathbf{u}_n\|^2 \leq a(\mathbf{u}_n; \mathbf{u}_{n+1} - \mathbf{u}_n, \mathbf{u}_{n+1} - \mathbf{u}_n).$$

After some modifications and due to the convergence of $\mathcal{J}(\mathbf{u}_n)$, we find

$$\lim_{n\to\infty} \|P_Q\mathbf{u}_{n+1} - P_Q\mathbf{u}_n\| = 0. \tag{6.156}$$

Further, we have after some modifications and using (6.155) and (6.156),

$$\begin{aligned}\tfrac{1}{2}c\|P_Q\mathbf{u}_n - P_Q\mathbf{u}\|^2 &\le D\mathcal{L}(\mathbf{u}_n, \mathbf{u}_n - \mathbf{u}) - D\mathcal{L}(\mathbf{u}, \mathbf{u}_n - \mathbf{u})\\ &= a(\mathbf{u}_n; \mathbf{u}_n, \mathbf{u}_n - \mathbf{u}) - D\mathcal{L}(\mathbf{u}, \mathbf{u}_n - \mathbf{u})\\ &\le D\mathcal{L}(\mathbf{u}, \mathbf{u}_{n+1} - \mathbf{u}_n) + a(\mathbf{u}_n; \mathbf{u}_{n+1}, \mathbf{u}_n - \mathbf{u}_{n+1})\\ &+ a(\mathbf{u}_n; \mathbf{u}_n - \mathbf{u}_{n+1}, \mathbf{u}_n - \mathbf{u}) \to 0.\end{aligned}$$

Let there exist a subsequence $\mathbf{u}_{n_k} \to \mathbf{u}^*$ as $k \to \infty$. Then

$$a(\mathbf{u}_{n_k-1}; \mathbf{u}_{n_k}, \mathbf{v} - \mathbf{u}_{n_k}) \ge (\mathbf{F}, \mathbf{v} - \mathbf{u}_{n_k}) \quad \forall \mathbf{v} \in K,$$

hence

$$a(\mathbf{u}_{n_k-1}; \mathbf{u}^*, \mathbf{v} - \mathbf{u}^*) \ge (\mathbf{F}, \mathbf{v} - \mathbf{u}^*) + \varepsilon_{n_k}(\mathbf{v}), \quad \varepsilon_{n_k}(\mathbf{v}) \to 0.$$

Since

$$\begin{aligned}a(\mathbf{u}_{n_k-1}; \mathbf{u}^*, \mathbf{v} - \mathbf{u}^*) &= a(\mathbf{u}_{n_k} + \mathbf{u}_{n_k-1} - \mathbf{u}_{n_k}; \mathbf{u}^*, \mathbf{v} - \mathbf{u}^*)\\ &= a(\mathbf{u}_{n_k} + P_Q\mathbf{u}_{n_k-1} - P_Q\mathbf{u}_{n_k}; \mathbf{u}^*, \mathbf{v} - \mathbf{u}^*),\end{aligned}$$

then using (6.156) and the fact that $a(\mathbf{w}_n; \mathbf{v}, \mathbf{z}) \to a(\mathbf{w}; \mathbf{v}, \mathbf{z})\ \forall \mathbf{v}, \mathbf{z} \in V$, if $\mathbf{w}_n \to \mathbf{w}$, we find that

$$a(\mathbf{u}_{n_k-1}; \mathbf{u}^*, \mathbf{v} - \mathbf{u}^*) \to a(\mathbf{u}^*; \mathbf{u}^*, \mathbf{v} - \mathbf{u}^*) = D\mathcal{L}(\mathbf{u}^*, \mathbf{v} - \mathbf{u}^*).$$

Hence $\mathbf{u}^*$ is the solution of (6.149) and $\mathbf{u}^* = \mathbf{u}$, so that

$$\lim_{n\to\infty} \mathbf{u}_n = \mathbf{u},$$

which completes the proof. □

Numerical Solution Let the region $\Omega \subset R^N$ be triangulated. We divide $\overline{\Omega} = \Omega \cup \partial\Omega$ into the system of m triangles T_{h_i} in the 2D case and into a system of m tetrahedrons in the 3D case, generating a triangulation $\mathcal{T}_h$ such that $\overline{\Omega} = \cup_{i=1}^m T_{h_i}$ and such that two neighboring triangles have only a vertex or an entire side common in the 2D case, and that two neighboring tetrahedra have only a vertex or an entire edge or an entire face common in the 3D case. Let $h = \max_{1\le i\le m}(\operatorname{diam} T_{h_i})$ and let the family of triangulation $\{\mathcal{T}_h\}$ used be regular. We further assume that the endpoints $\overline{\Gamma}_u \cap \overline{\Gamma}_\tau$, $\overline{\Gamma}_u \cap \overline{\Gamma}_c$, $\overline{\Gamma}_\tau \cap \overline{\Gamma}_c$ coincide with vertices of T_{h_i}. Let

$$V_h = \{\mathbf{v} | \mathbf{v} \subset \prod_{\iota=1}^{s} [C(\Omega^\iota)]^N,\ \mathbf{v}|_{T_h} \in [P_1]^N,\ \forall T_h \in \mathcal{T}_h,\ \mathbf{v} = 0 \text{ on } \Gamma_u\ \forall T_h \in \mathcal{T}_h\},$$

where P_1 is the space of all linear polynomials,

$$K_h = \left\{ \mathbf{v} | \mathbf{v} \in V_h, v_n^k - v_n^l \leq 0 \text{ on } \cup_{k,l} \Gamma_c^{kl} \right\} = V_h \cap K.$$

It is evident that K_h is a convex and closed subset of V_h $\forall h$. Let $V_h = P_h \oplus Q_h$ be the orthogonal decomposition of V_h. Then we solve the following sequence of variational inequalities:

$$\mathbf{u}_h^{r+1} \in K_h, \int_\Omega 2\lambda A^{\lambda-1}(e_{ij}(\mathbf{u}_h^r))c_{ijkl}(\mathbf{u}_h^r)e_{ij}(\mathbf{u}_h^{r+1})e_{kl}(\mathbf{v} - \mathbf{u}_h^{r+1})\,d\mathbf{x}$$
$$- \int_{\Gamma_\tau} P_i(v_i - (\mathbf{u}_h^{r+1})_i)\,d\mathbf{s} - \int_\Omega f_i(v_i - (\mathbf{u}_h^{r+1})_i)\,d\mathbf{x} \geq 0$$
$$\forall \mathbf{v} \in K_h, r = 1, 2, \ldots, \tag{6.157}$$

where $\mathbf{u}_h^r$ is the rth FEM-secant modules approximate solution. The problem leads to a solution of a sequence of approximate problems of variational inequalities with variable coefficients of the semicoercive type. The analysis of such problems is parallel to that of analyses of FEM approximation of variational inequalities in linear elasticity, as the variational inequality problem (6.157) represents a linear variational inequality of the theory of elasticity, where the elastic coefficients c_{ijkl} are replaced by variable coefficients $c^*_{ijkl} = 2\lambda A^{\lambda-1}(e_{ij}(\mathbf{u}_h^r))c_{ijkl}(\mathbf{u}_h^r)$. Moreover, we can prove by the technique of the theory of contact problems that

$$\|\mathbf{u}^{r+1} - \mathbf{u}_h^{r+1}\| \to 0 \quad \text{as} \quad h \to 0_+, \tag{6.158}$$

as well as estimates based on the Falk lemma. The algorithm, which can be used for numerical realization, is parallel to that of algorithms used in the theory of contact problems in linear elasticity.

6.4.5 Homogenization Approach—Bone Tissues as Composite Materials with Periodic Structures

The structure of human bone tissue is very complicated and represents composite materials with very complicated structures. However, in the first approximation we can assume that the structure of human bone tissues can be approximated from place to place by periodic structures. Under these assumptions the homogenization method can be an important method for simulations of composite bone structures. In the biophysical sense, the homogenization approach consists of replacing the periodically heterogeneous bone tissue by an equivalent homogeneous (and anisotropic) one. In this section we will deal with homogenization of biomechanical model problems of bone tissue structures in linear elastic rheology. Mathematically, the homogenization approach means that the equation with highly oscillating periodic coefficients is approximated by a problem with constant (anisotropic) coefficients. These problems can be formulated in displacements (the primal formulation) or in stresses (the dual formulation). Above all we limit ourselves to the formulation in displacements; the formulation in stresses will be discussed only briefly.

Let us assume that the material the makes up the human skeleton is a nonhomogeneous material with elastic coefficients c_{ijkl}. For simplicity, we will assume that homogeneous parts of bone tissues, by which we simulate the bone structure, have regular periodical structures.

Let the elastic coefficients $c_{ijkl}(\mathbf{x})$ be periodic functions on the interval $Y \subset R^3$ such that $c^{\varepsilon}_{ijkl}(\mathbf{x}) = c_{ijkl}(\mathbf{x}/\varepsilon)$. The structure of such a type of biomaterial will be called Y-periodical bone material. The period Y contains groups of homogeneous subdomains Y_i, on which the coefficients c_{ijkl} are constant, while on the whole domain Y the coefficients c_{ijkl} are functions of x_i. Therefore, the Y-periodical biomaterials will be characterized by a certain anisotropy. In the real situation the bone structure of the human skeleton is much more complicated, its structure is stochastically periodical, and, moreover, in some parts of the skeleton, we denote them as Ω_i, they can be substantially different. Then the main goal will be to determine the homogenized elastic coefficients on every part of the skeleton Ω_i, that is, the elastic coefficients, which are by a certain way averaged but which on the whole domain Ω_i will be constant. These coefficients will be denoted by c^0_{ijkl}.

We introduce the Y-periodic function as follows:

Definition 6.14 Let $Y = [0, y'_1] \times [0, y'_2] \times [0, y'_3]$, $y'_i > 0$, $i = 1, 2, 3$, be the unit period. Then a function $f(x, y, z)$ is said to be Y periodic in y if

$$f(x, y_1 + k_1 y'_1, y_2 + k_2 y'_2, y_3 + k_3 y'_3, z) = f(x, y_1, y_2, y_3, z)$$

holds for all integers k_1, k_2, k_3. The integral average $M(f)$ in y will be defined by

$$M(f) = \int_Y \frac{f(x, y, z)}{\text{meas}(Y)} d\mathbf{y}.$$

If the function $f(y)$ is Y periodic in y, then $M(\partial f/\partial y_i) = 0$, $i = 1, 2, 3$, holds.

The Model Let us assume that the investigated part of the human skeleton occupies the domain $\Omega \subset R^3$ with the boundary $\partial\Omega = \Gamma_u \cup \Gamma_\tau$, where parts of the boundary Γ_u and Γ_τ are nonempty, and on Γ_u the displacements $\mathbf{u}_0$ are prescribed and on Γ_τ the surface loading $\mathbf{P}$ is prescribed. The fields of displacements and stresses in the investigated part of the human skeleton are described by the equilibrium equation. In this static case displacements, strains, and stresses are functions of coordinates only. Then

$$\frac{\partial \tau_{ij}}{\partial x_j} + f_i = 0 \quad \text{in } \Omega, \qquad i, j = 1, 2, 3, \tag{6.159}$$

where $\mathbf{f} = (f_i)$ denotes the body forces and τ_{ij} is the stress tensor, defined by the generalized Hooke law

$$\tau_{ij} = c_{ijkl}(\mathbf{x}) e_{kl}(\mathbf{u}), \qquad e_{ij}(\mathbf{u}) = \frac{1}{2}\left(\frac{\partial u_i}{\partial x_j} + \frac{\partial u_j}{\partial x_i}\right), \qquad i, j, k, l = 1, 2, 3, \tag{6.160}$$

where e_{ij} is the small strain tensor, c_{ijkl} are the elastic coefficients satisfying the conditions

$$c_{ijkl} = c_{klij} = c_{jikl} = c_{ijlk},$$
$$0 < c_0 \le c_{ijkl}\xi_{ij}\xi_{kl}|\boldsymbol{\xi}|^{-2} \le c_1 < +\infty \quad \text{for almost all } \mathbf{x} \in \Omega, \boldsymbol{\xi} \in R^9,\ \xi_{ij} = \xi_{ji}$$

where c_0, c_1 are constants. On the boundary the conditions

$$\mathbf{u} = \mathbf{u}_0 \quad \text{on } \Gamma_u, \tag{6.161}$$

and

$$\tau_{ij}n_j = P_i \quad \text{on } \Gamma_\tau, \qquad i,j = 1,2,3 \tag{6.162}$$

are prescribed.

In the dual formulation the inverse strain–stress law is defined as

$$e_{ij} = s_{ijkl}(\mathbf{x})\tau_{kl}(\mathbf{u}), \tag{6.163}$$

where coefficients s_{ijkl} are of the same type and they satisfy

$$s_{ijkl} = s_{klij} = s_{jikl} = s_{ijlk},$$
$$0 < s_0 \le s_{ijkl}\zeta_{ij}\zeta_{kl}|\zeta|^{-2} \le s_1 < +\infty \quad \text{for almost all } \mathbf{x} \in \Omega,\ \boldsymbol{\zeta} \in R^9,\ \zeta_{ij} = \zeta_{ji},$$
$$c_{ijmn}s_{mnkl} = \tfrac{1}{2}(\delta_{ik}\delta_{jl} + \delta_{il}\delta_{jk}),$$

where s_0, s_1 are constants independent of $\mathbf{x} \in \Omega$ and δ_{ij} is the Kronecker symbol. For the isotropic biomaterials with the Lamé constants λ and μ we have

$$c_{ijkl} = \lambda\delta_{ij}\delta_{kl} + \mu(\delta_{ik}\delta_{jl} + \delta_{il}\delta_{jk}),$$
$$s_{ijkl} = \frac{1}{4\mu}(\delta_{ik}\delta_{jl} + \delta_{il}\delta_{jk}) - \frac{\lambda}{2\mu(3\lambda + 2\mu)}\delta_{ij}\delta_{kl}.$$

The periodical structure of the biomaterials is expressed by the periodic coefficients c^{ε}_{ijkl} and s^{ε}_{ijkl} of the generalized Hooke law, where

$$c^{\varepsilon}_{ijkl}(\mathbf{x}) = c_{ijkl}\left(\frac{\mathbf{x}}{\varepsilon}\right), \qquad s^{\varepsilon}_{ijkl}(\mathbf{x}) = s_{ijkl}\left(\frac{\mathbf{x}}{\varepsilon}\right),$$

where $c_{ijkl}(\mathbf{y})$ and $s_{ijkl}(\mathbf{y})$ are Y-periodic functions. Further, we will consider even the problem with nonuniformly oscillating coefficients

$$c^{\varepsilon}_{ijkl}(\mathbf{x}) = c_{ijkl}\left(\frac{\mathbf{x}, \mathbf{x}}{\varepsilon}\right), \qquad s^{\varepsilon}_{ijkl}(\mathbf{x}) = s_{ijkl}\left(\frac{\mathbf{x}, \mathbf{x}}{\varepsilon}\right),$$
$$c_{ijkl}(\mathbf{x}, \mathbf{y}), \qquad s_{ijkl}(\mathbf{x}, \mathbf{y}) \text{ are } Y \text{ periodic in } \mathbf{y}.$$

Let us assume that $c_{ijkl} \in L^{\infty}(\Omega)$, $f_i \in L^2(\Omega)$, $P_i \in L^2(\Gamma_\tau)$, and $\mathbf{u}_0 \in [L^2(\Gamma_u)]^3$. Our goal will be to determine the displacement field $\mathbf{u}$ satisfying the boundary conditions (6.161) and (6.162).

Derivation of the Homogenized Problem The solution of the problem with the coefficients c^{ε}_{ijkl}, s^{ε}_{ijkl} will be denoted by $\mathbf{u}^{\varepsilon}$ and/or τ^{ε}, respectively.

Then the problem [(6.159)–(6.162)] for $\varepsilon > 0$ is as follows:

$$\frac{\partial}{\partial x_j}(c^{\varepsilon}_{ijkl}(\mathbf{x})e_{kl}(\mathbf{u}^{\varepsilon})) + f_i = 0 \quad \text{in } \Omega, \qquad i,j = 1,2,3, \tag{6.164}$$

$$\mathbf{u}^{\varepsilon} = \mathbf{u}_0 \quad \text{on } \Gamma_u, \tag{6.165}$$

$$c^{\varepsilon}_{ijkl}(\mathbf{x})e_{kl}(\mathbf{u}^{\varepsilon})n_j = P_i \quad \text{on } \Gamma_\tau, \qquad i,j = 1,2,3. \tag{6.166}$$

Let us pass to the limit $\varepsilon \to 0^+$. If coefficients $c^{\varepsilon}_{ijkl}(\mathbf{x}) = c_{ijkl}(\mathbf{x}, \mathbf{x}/\varepsilon)$ oscillate strongly on the interval Y, then the function $\mathbf{u}^{\varepsilon}(\mathbf{x})$ will change strongly its function values, but it will not be periodic. Therefore, we will introduce a new variable $\mathbf{y} = \mathbf{x}/\varepsilon$ and we define $c^{\varepsilon}_{ijkl}(\mathbf{x}) = c_{ijkl}(\mathbf{x}, \mathbf{x}/\varepsilon) = c_{ijkl}(\mathbf{x}, \mathbf{y})$. Then we find a function $\mathbf{u}^{\varepsilon}(\mathbf{x})$ in the form of asymptotic expansion of the solution in the point $\mathbf{x}/\varepsilon$, that is,

$$\mathbf{u}^{\varepsilon}(\mathbf{x}) = \mathbf{u}\left(\mathbf{x}, \frac{\mathbf{x}}{\varepsilon}\right) = \mathbf{u}^0\left(\mathbf{x}, \frac{\mathbf{x}}{\varepsilon}\right) + \varepsilon \mathbf{u}^1\left(\mathbf{x}, \frac{\mathbf{x}}{\varepsilon}\right) + \varepsilon^2 \mathbf{u}^2\left(\mathbf{x}, \frac{\mathbf{x}}{\varepsilon}\right) + \cdots, \tag{6.167}$$

where $\mathbf{u}^i(\mathbf{x}, \mathbf{x}/\varepsilon) = \mathbf{u}^i(\mathbf{x}, \mathbf{y})$, $i = 0, 1, \ldots,$ are independent of ε and Y periodic in $\mathbf{y} = \mathbf{x}/\varepsilon$. Substituting (6.167) instead of $\mathbf{u}^{\varepsilon}(\mathbf{x}) = \mathbf{u}(\mathbf{x}, \mathbf{x}/\varepsilon)$ into (6.164), we have

$$-\frac{1}{\varepsilon^2}\frac{\partial}{\partial y_j}\left(c_{ijkl}\frac{\partial u^{\varepsilon}_k}{\partial y_l}\right) - \frac{1}{\varepsilon}\left[\frac{\partial}{\partial y_j}\left(c_{ijkl}\frac{\partial u^{\varepsilon}_k}{\partial x_l}\right) + \frac{\partial}{\partial x_j}\left(c_{ijkl}\frac{\partial u^{\varepsilon}_k}{\partial y_l}\right)\right]$$
$$-\frac{\partial}{\partial x_j}\left(c_{ijkl}\frac{\partial u^{\varepsilon}_k}{\partial x_l}\right) = f_i \quad \text{in } \Omega, \qquad i = 1,2,3,$$

where

$$\frac{df(\mathbf{x}, \mathbf{x}/\varepsilon)}{dx_i} = \frac{\partial f(\mathbf{x}, \mathbf{x}/\varepsilon)}{\partial x_i} + \frac{1}{\varepsilon}\frac{\partial f(\mathbf{x}, \mathbf{x}/\varepsilon)}{\partial y_i}$$

was used and where $y_i = x_i/\varepsilon$. Let us denote by

$$(A^0\mathbf{u}^{\varepsilon})_i = -\frac{\partial}{\partial y_j}\left(c_{ijkl}\frac{\partial u^{\varepsilon}_k}{\partial y_l}\right),$$

$$(A^1\mathbf{u}^{\varepsilon})_i = -\left[\frac{\partial}{\partial y_j}\left(c_{ijkl}\frac{\partial u^{\varepsilon}_k}{\partial x_l}\right) + \frac{\partial}{\partial x_j}\left(c_{ijkl}\frac{\partial u^{\varepsilon}_k}{\partial y_l}\right)\right],$$

$$(A^2\mathbf{u}^{\varepsilon})_i = -\frac{\partial}{\partial x_j}\left(c_{ijkl}\frac{\partial u^{\varepsilon}_k}{\partial x_l}\right),$$

then

$$-\frac{\partial}{\partial x_j}\left(c^{\varepsilon}_{ijkl}(\mathbf{x})\frac{\partial u^{\varepsilon}_k}{\partial x_l}\right) = \left[\left(\frac{1}{\varepsilon^2}A^0 + \frac{1}{\varepsilon}A^1 + A^2\right)\mathbf{u}^{\varepsilon}\right]_i = f_i.$$

Inserting (6.167) for $\mathbf{u}^\varepsilon(\mathbf{x})$ we obtain

$$(\varepsilon^{-2}A^0 + \varepsilon^{-1}A^1 + A^2)_i(\mathbf{u}^0 + \varepsilon\mathbf{u}^1 + \varepsilon^2\mathbf{u}^2)_i = f_i, \tag{6.168}$$

or in vector form

$$\varepsilon^{-2}A^0\mathbf{u}^0 + \varepsilon^{-1}A^0\mathbf{u}^1 + A^0\mathbf{u}^2 + \varepsilon^{-1}A^1\mathbf{u}^0 + A^1\mathbf{u}^1 + \varepsilon A^1\mathbf{u}^2 + A^2\mathbf{u}^0 + \varepsilon A^2\mathbf{u}^1 + \varepsilon^2 A^2\mathbf{u}^2 = \mathbf{f}. \tag{6.169}$$

Comparing terms at the corresponding powers of ε^k for all integers $k \geq -2$, we obtain the following system:

$$A^0\mathbf{u}^0 = 0 \quad \text{on } \Omega \times Y, \tag{6.170}$$

$$A^0\mathbf{u}^1 + A^1\mathbf{u}^0 = 0 \quad \text{on } \Omega \times Y, \tag{6.171}$$

$$A^0\mathbf{u}^2 + A^1\mathbf{u}^1 + A^2\mathbf{u}^0 = \mathbf{f} \quad \text{on } \Omega \times Y, \tag{6.172}$$

$$A^1\mathbf{u}^2 + A^2\mathbf{u}^1 = \mathbf{0} \quad \text{on } \Omega \times Y, \tag{6.173}$$

$$A^2\mathbf{u}^2 = \mathbf{0} \quad \text{on } \Omega \times Y. \tag{6.174}$$

A similar approach leads to the following boundary conditions:

$$c_{ijkl}(\mathbf{y})\frac{\partial u_k^0}{\partial y_l}(\mathbf{x},\mathbf{y})n_j(\mathbf{x}) = 0 \quad \text{on } \Gamma_\tau \times Y, \tag{6.175}$$

$$c_{ijkl}(\mathbf{y})\left(\frac{\partial u_k^1}{\partial y_l}(\mathbf{x},\mathbf{y}) + \frac{\partial u_k^0}{\partial x_l}(\mathbf{x},\mathbf{y})\right)n_j(\mathbf{x}) = P_i \quad \text{on } \Gamma_\tau \times Y, \tag{6.176}$$

$$c_{ijkl}(\mathbf{y})\left(\frac{\partial u_k^2}{\partial y_l}(\mathbf{x},\mathbf{y}) + \frac{\partial u_k^1}{\partial x_l}(\mathbf{x},\mathbf{y})\right)n_j(\mathbf{x}) = 0 \quad \text{on } \Gamma_\tau \times Y, \tag{6.177}$$

$$c_{ijkl}(\mathbf{y})\frac{\partial u_k^2}{\partial x_l}(\mathbf{x},\mathbf{y})n_j(\mathbf{x}) = 0 \quad \text{on } \Gamma_\tau \times Y, \tag{6.178}$$

$$u_i^0 = u_{0i} \quad \text{on } \Gamma_u \times Y, \tag{6.179}$$

$$u_i^1 = 0 \quad \text{on } \Gamma_u \times Y, \tag{6.180}$$

$$u_i^2 = 0 \quad \text{on } \Gamma_u \times Y. \tag{6.181}$$

Equations (6.174), (6.178), and (6.181) define the problem with the solution $\mathbf{u}^2 = 0$, as if no outer forces act on the body and if the displacements on one part of the boundary are zero, then the resulting displacements vanish.

We introduce the following lemma:

Lemma 6.16 Let: $\mathbf{f} \in [L^2_{\text{per}}(Y)]^3$. The equation $A^0\mathbf{u} = \mathbf{f}$ has the solution $\mathbf{u} \in [W^{1,2}_{\text{per}}(Y)]^3$ if and only if $M(\mathbf{f}) = 0$. This solution is determined uniquely up to an additive constant.

Proof The proof is based on the properties of the Y-periodic function and the Lax–Milgram lemma [see Milka (1989) and Nedoma (1998a, p. 230)]. □

Equation (6.170) and the lemma yield that its solution $\mathbf{u}^0$ does not depend on the variable $\mathbf{y}$, that is, $\mathbf{u}^0 = \mathbf{u}^0(\mathbf{x})$, as from this equation

$$-\frac{\partial}{\partial y_j}\left(c_{ijkl}(\mathbf{y})\frac{\partial u_k^0}{\partial y_l}\right) = 0$$

follows. The last equality will be satisfied identically only if $\mathbf{u}^0$ will be independent of $\mathbf{y}$.

From Eq. (6.171) we obtain

$$A^0\mathbf{u}^1 = -A^1\mathbf{u}^0 = \frac{\partial}{\partial y_j}\left(c_{ijkl}(\mathbf{y})\frac{\partial u_k^0(\mathbf{x})}{\partial x_l}\right) + \frac{\partial}{\partial x_j}\left(c_{ijkl}(\mathbf{y})\frac{\partial u_k^0(\mathbf{x})}{\partial y_l}\right). \tag{6.182}$$

The last term at the right-hand side is equal to zero. Due to Lemma 6.16 the above-defined problem has a unique solution at every $\mathbf{x}$ from the space $[W^{1,2}_{\text{per}}(Y)]^3$. We will seek such a solution in the form

$$u_g^1(\mathbf{x}, \mathbf{y}) = -\chi_g^{kl}(\mathbf{y})\frac{\partial u_k^0(\mathbf{x})}{\partial x_l} + \tilde{u}_g(\mathbf{x}), \tag{6.183}$$

where $\boldsymbol{\chi}^{kl} \in [W^{1,2}_{\text{per}}(Y)]^3$.

Inserting into (6.182), we have

$$\left[A^0\left(-\chi^{kl}(\mathbf{y})\frac{\partial u_k^0(\mathbf{x})}{\partial x_l} + \tilde{\mathbf{u}}(\mathbf{x})\right)\right]_i = \frac{\partial}{\partial y_j}\left(c_{ijkl}(\mathbf{y})\frac{\partial u_k^0(\mathbf{x})}{\partial x_l}\right). \tag{6.184}$$

If the operator A^0 was expressed by using the partial derivatives, then we have

$$-\frac{\partial}{\partial y_j}\left[c_{ijgh}(\mathbf{y})\frac{\partial}{\partial y_h}\left(-\chi_g^{kl}(\mathbf{y})\frac{\partial u_k^0(\mathbf{x})}{\partial x_l} + \tilde{u}_g(\mathbf{x})\right)\right] = \frac{\partial}{\partial y_j}\left(c_{ijkl}(\mathbf{y})\frac{\partial u_k^0(\mathbf{x})}{\partial x_l}\right). \tag{6.185}$$

Applying the derivation with respect to the variable $\mathbf{y}$, then after some calculation we obtain

$$\frac{\partial}{\partial y_j}\left[c_{ijgh}(\mathbf{y})\left(\frac{\partial \chi_g^{kl}(\mathbf{y})}{\partial y_h} - \delta_{gk}\delta_{hl}\right)\right] = 0. \tag{6.186}$$

Here we used the fact that the derivative of functions, which depend on the variable **x**, are equal to zero and that $\partial u_k^0/\partial x_l \neq 0$.

Equation (6.175) is fulfilled because $\mathbf{u}^0$ is independent of a variable **y**. Due to Lemma 6.16 Eq. (6.172) has a solution in the space $[W^{1,2}_{\text{per}}(Y)]^3$ just if $M(A^1\mathbf{u}^1 + A^2\mathbf{u}^0 - \mathbf{f}) = 0$ as $\mathbf{u}^2 = \mathbf{0}$. Hence using also Eq. (6.183) we obtain

$$M\left[\frac{\partial}{\partial x_j}\left(c_{ijgh}\frac{\partial \chi_g^{kl}}{\partial y_h}\frac{\partial u_k^0}{\partial x_l}\right) - \frac{\partial}{\partial x_j}\left(c_{ijkl}\frac{\partial u_k^0}{\partial x_l}\right) - f_i\right] = 0, \tag{6.187}$$

where c_{ijgh} and χ_g^{kl} are Y-periodic functions. Denoting

$$c_{ijkl}^0 = M\left[c_{ijgh}(\mathbf{y})\left(\delta_{gk}\delta_{hl} - \frac{\partial \chi_g^{kl}(\mathbf{y})}{\partial y_h}\right)\right], \tag{6.188}$$

we find from (6.187)

$$\frac{\partial}{\partial xj}\left(c_{ijkl}^0\frac{\partial u_k^0}{\partial x_l}\right) + f_i = 0, \qquad i = 1, 2, 3. \tag{6.189}$$

Similarly, for the boundary condition on Γ_τ from (6.176) we derive

$$c_{ijkl}^0\frac{\partial u_k^0}{\partial x_l}n_j = P_i, \tag{6.190}$$

and for the boundary condition on Γ_u we obtain

$$u_i^0 = u_{0i}. \tag{6.191}$$

Equation (6.175) is satisfied as $\mathbf{u}^0$ is a function independent of the variable **y**.

As a consequence the initial problem is transformed to the homogenized problem [(6.189)–(6.191)] with homogenized elastic coefficients c_{ijkl}^o defined by (6.188) and the functions $\chi^{kl} \in [W^{1,2}_{\text{per}}(Y)]^3$ satisfy (6.186). For the functions χ^{kl} the symmetry conditions

$$\chi^{kl} = \chi^{lk} \tag{6.192}$$

hold, and for the homogenized elastic coefficients the following conditions can be obtained:

$$c_{ijkl}^0 = c_{jilk}^0 = c_{ijlk}^0 = c_{klij}^0,$$
$$0 < m_0 \leq c_{ijkl}^0\xi_{ij}\xi_{kl} \mid \boldsymbol{\xi}|^{-2} \leq m_1 < +\infty \quad \forall \mathbf{x} \in \Omega^\iota, \boldsymbol{\xi} \in R^9, \qquad \xi_{ij} = \xi_{ij}.$$

The homogenized material is a fictional material that has similar physical properties like the initial one. The next theorem (which will be presented without the proof) shows the connection between the function $\mathbf{u}^0$ solving the homogenized problem and the function $\mathbf{u}$ (resp. $\mathbf{u}^\varepsilon$), which solves the initial problem.

Theorem 6.41 The solution $\mathbf{u}^\varepsilon$ of problem (6.164)–(6.166) converges weakly to the solution $\mathbf{u}^0$ of the homogenized problem (6.189)–(6.191) in the space $[W^{1,2}(\Omega)]^3$ for $\varepsilon \to 0_+$.

For the proof see, for example, Birolli (1977), Bensoussan et al. (1978), and Franců (1982).

Let us recall that the homogenized coefficients can be computed from (6.188), that is, from the relation

$$c^0_{ijkl} = M\left[c_{ijgh}(\mathbf{y})\left(\delta_{gk}\delta_{hl} - \frac{\partial \chi_g^{kl}(\mathbf{y})}{\partial y_h}\right)\right].$$

To compute the derivative of $\chi_g^{kl}(\mathbf{y})$ we start from the following information: We have [see (6.186)]

$$\frac{\partial}{\partial y_j}\left[c_{ijgh}(\mathbf{y})\left(\frac{\partial \chi_g^{kl}(\mathbf{y})}{\partial y_h} - \delta_{gk}\delta_{hl}\right)\right] = 0, \qquad i = 1,2,3.$$

We know that the functions $\boldsymbol{\chi}^{kl}(\mathbf{y})$ are in the space $[W^{1,2}_{\text{per}}(Y)]^3$ uniquely defined up to an additive constant.

Let us multiply these equations by an arbitrary function $\boldsymbol{\varphi} \in [W^{1,2}_{\text{per}}(Y)]^3$ and integrate over the domain Y. Then we have

$$\int_Y \frac{\partial}{\partial y_j}\left[c_{ijgh}(\mathbf{y})\left(\frac{\partial \chi_g^{kl}(\mathbf{y})}{\partial y_h} - \delta_{gk}\delta_{hl}\right)\right]\varphi_i\, d\mathbf{y} = 0 \quad \forall \boldsymbol{\varphi} \in [W^{1,2}_{\text{per}}(Y)]^3.$$

Let Y_i, $i = 1,\ldots,n$, be subdomains of Y, where the elastic coefficients are given. The analysis of this problem shows that the resulting equation after a discretization leads to a matrix that need not be a square matrix, and, moreover, we do not know whether the obtained matrix is regular. Therefore, we will solve another problem.

Instead of the problem in Eq. (6.186) we will solve the following problem: Find $\overline{\chi}^{kl} \in V_a$ such that

$$\int_Y \frac{\partial}{\partial y_j}\left[c_{ijgh}(\mathbf{y})\left(\frac{\partial \overline{\chi}_g^{kl}(\mathbf{y})}{\partial y_h} - \delta_{gk}\delta_{hl}\right)\right]\varphi_i\, d\mathbf{y} = 0 \quad \forall \boldsymbol{\varphi} \in V_a = [W_0^{1,2}(Y)]^3, \quad (6.193)$$

where $\overline{\boldsymbol{\chi}}^{kl}$ approximates the function $\boldsymbol{\chi}^{kl}$. Applying the Green theorem we obtain

$$\int_Y \left[c_{ijgh}(\mathbf{y})\left(\frac{\partial \overline{\chi}_g^{kl}(\mathbf{y})}{\partial y_h} - \delta_{gk}\delta_{hl}\right)\right]\frac{\partial \varphi_i}{\partial y_j} d\mathbf{y} = 0 \quad \forall \boldsymbol{\varphi} \in V_a. \quad (6.194)$$

Let $\mathbf{u}, \mathbf{v} \in V_a$ and let us define

$$a(\mathbf{u}, \mathbf{v}) = \int_Y c_{ijgh}(\mathbf{y}) \frac{\partial u_i}{\partial y_j} \frac{\partial v_g}{\partial y_h} d\mathbf{y}, \qquad f^{kl}(\mathbf{v}) = \int_Y c_{ijkl}(\mathbf{y}) \frac{\partial v_i}{\partial y_j} d\mathbf{y}, \quad \text{for fixed } k, l. \tag{6.195}$$

Then problem (6.193) can be rewritten in the form

$$\overline{\chi}^{kl} \in V_a, \qquad a(\overline{\chi}^{kl}, \varphi) = f^{kl}(\varphi) \quad \forall \varphi \in V_a. \tag{6.196}$$

The proof of the existence and uniqueness is similar to that of the previous lemma.

Lemma 6.17 Problem (6.196) has just one solution $\overline{\chi}^{kl}$ in the space $V_a = [W_0^{1,2}(Y)]^3$.

The proof is based on the Lax–Milgram theorem.

Let us investigate the relation between χ^{kl} and $\overline{\chi}^{kl}$. Since every function from the space $[W_{\text{per}}^{1,2}(Y)]^3$ can be written as the sum of the function from the space $[W_0^{1,2}(Y)]^3$ and the function from the space $[W_{\text{per}}^{1,2}(Y)]^3$, which is nonzero in a small zone near the boundary of the interval Y only, the function χ^{kl} can be replaced by the function $\overline{\chi}^{kl}$ almost on the whole domain Y. Near the boundary of the domain Y the difference between both functions will be substantial.

In what follows we will show an algorithm for the computation of the function $\overline{\chi}^{kl}$ [for more details see Nedoma (1998a)].

We will start from problem (6.196). The problem can be solved by the finite element method.

Let $\{\varphi_i\}_{i=1}^{I}$ be a basis of the finite-dimensional subspace $V_{ah} \subset V_a$. Let $\overline{\chi}_h^{kl}$ be approximated by using the basis functions φ_i^h in the form $\overline{\chi}_h^{kl} = \sum_{i=1}^{n} \alpha_i^{kl} \varphi_i^h$, $h = 1, 2, 3$. We obtain a system of linear equations of the form $A\boldsymbol{\alpha}^{kl} = \mathbf{b}^{kl}$ for unknown α_i^{kl}, with the matrix $A = (A_{mn})$, and with the right-hand-side vector $\mathbf{b}^{kl} = (b_n^{kl})$, where

$$A_{mn} = \int_Y c_{ijgh}(\mathbf{y}) \frac{\partial \varphi_m^i}{\partial y_j} \frac{\partial \varphi_n^g}{\partial y_h} d\mathbf{y}, \qquad b_n^{kl} = \int_Y c_{ijkl}(\mathbf{y}) \frac{\partial \varphi_n^i}{\partial y_j} d\mathbf{y}.$$

The matrix A is regular, symmetric, and positive definite. The construction of the matrix $A = (A_{mn})$ and the vector of the right-hand-side $\mathbf{b}^{kl} = (b_n^{kl})$ are similar as in the classical finite element technique. The system of linear equations $A\boldsymbol{\alpha}^{kl} = \mathbf{b}^{kl}$ for unknown coefficients α_i^{kl}, can be solved, for example, by the Choleski method or by the conjugate gradient method. The homogenized coefficients are then computed by using relation (6.188). The problem then leads to the minimization of the functional of the potential energy

$$J(\mathbf{v}) = \frac{1}{2} \int_\Omega c_{ijkl}^0 e_{ij}(\mathbf{v}) e_{kl}(\mathbf{v})\, d\mathbf{x} - \int_\Omega f_i v_i \, d\mathbf{x} - \int_{\Gamma_\tau} P_i v_i \, ds$$

on the set $V = \{\mathbf{v} | \mathbf{v} \in [W^{1,2}(\Omega)]^3, \mathbf{v} = \mathbf{u}_0 \text{ na } \Gamma_u\}$.

6.4.6 Domain Decomposition

Introduction In biomechanics, there are problems whose investigations lead to solving model numerical problems based on variational formulations. Such problems are frequently formulated by variational inequalities that physically describe the principle of virtual work in its inequality form. In this section we will deal with the numerical solution of a biomechanical problem based on the generalized semicoercive contact problem with the given friction in linear elasticity for the case that s bodies of arbitrary shapes are in mutual contacts and they are loaded by external forces. The problem will be formulated as the primary variational inequality problem, that is, in displacements.

Modern computers are constructed as parallel computers with a greater number of processors. Therefore, the general trend for accurately solving larger and more complicated problems brings more attention to the numerical procedures for solving such biomechanical problems. The purpose of this section is mainly to describe some of the directions of the recent progress in the numerical methods and algorithms for the solution of unilateral problems in linear elasticity, that is, problems of biomechanical analyses of human joints and their artificial replacements. From the above discussion these algorithms will be based on domain decomposition approaches and finite element techniques. The parallel algorithm, employing properties of modern parallel computers with a greater number of processors, is based on the nonoverlapping domain decomposition method and the finite element method proposed by Le Tallec (1994) for linear problems. This approach was generalized in Daněk et al. (2005a) and Hlaváček et al. (2006) for a unilateral contact problem with the given friction and the finite element method.

Formulation of the Model Let the investigated part of the human skeleton occupy a union Ω of s-bounded domains Ω^ι, $\iota = 1, \ldots, s$ in $\mathbb{R}^N$, $N = 2, 3$, denoting separate components of human joints, with Lipschitz boundaries $\partial\Omega^\iota$. Let the boundary $\partial\Omega = \cup_{\iota=1}^{s} \partial\Omega^\iota$ consist of four disjoint parts such that $\partial\Omega = \Gamma_\tau \cup \Gamma_u \cup \Gamma_c \cup \Gamma_0$. Based on knowledge of the physiological distribution of insertions in a bone tissue and skeletal sites through which the loading forces are transmitted due to the weight of the human body and due to acting muscular forces, we eliminate this portion of the skeleton boundary as $\Gamma_\tau = {}^1\Gamma_\tau \cup {}^2\Gamma_\tau$, where on ${}^1\Gamma_\tau$ the part of the human joint system is loaded by muscular forces or by a weight of the body (or of its part) and is unloaded on ${}^2\Gamma_\tau$. By Γ_u we denote the part of the skeletal boundary, where we simulate its fixation or the surgical osteotomy technique, respectively. The common contact boundary between both joint components Ω^k and Ω^l before deformation we denote by $\Gamma_c^{kl} = \partial\Omega^k \cap \partial\Omega^l$, $k, l = 1, \ldots, s$, $k \neq l$, and by $\Gamma_c = \cup_{k,l} \Gamma_c^{kl}$ the whole contact boundary. Moreover, the boundary Γ_0 simulates, for example, the symmetry of the pelvis.

Let body forces $\mathbf{F}$, surface tractions $\mathbf{P}$, and slip limits g^{kl} be given. We have the following problem:

Problem $\mathcal{P}$ Find the displacements $\mathbf{u}^\iota$ in all Ω^ι such that

$$\frac{\partial}{\partial x_j}\tau_{ij}(\mathbf{u}^\iota) + F_i^\iota = 0 \quad \text{in } \Omega^\iota, \qquad 1 \le \iota \le s, i = 1, \ldots, N, \tag{6.197}$$

where the stress tensor τ_{ij} is defined by

$$\tau_{ij}(\mathbf{u}^\iota) = c^\iota_{ijkl} e_{kl}(\mathbf{u}^\iota) \quad \text{in } \Omega^\iota, \qquad 1 \le \iota \le s, i = 1, \ldots, N, \tag{6.198}$$

with boundary conditions

$$\tau_{ij}(\mathbf{u})n_j = P_i \quad \text{on } {}^1\Gamma_\tau, \qquad i = 1, \ldots, N, \tag{6.199}$$

$$\tau_{ij}(\mathbf{u})n_j = 0 \quad \text{on } {}^2\Gamma_\tau, \qquad i = 1, \ldots, N, \tag{6.200}$$

$$\mathbf{u} = \mathbf{u}_0 \quad \text{on } \Gamma_u, \tag{6.201}$$

$$u^k_n - u^l_n \le 0, \qquad \tau^{kl}_n \le 0, \qquad (u^k_n - u^l_n)\tau^{kl}_n = 0 \quad \text{on } \cup_{k,l} \Gamma^{kl}, \qquad 1 \le k, l \le s, \tag{6.202}$$

$$\begin{aligned} &|\boldsymbol{\tau}^{kl}_t| \le g^{kl} \quad \text{on } \cup_{k,l} \Gamma^{kl}, \qquad 1 \le k, l \le s, \\ &|\boldsymbol{\tau}^{kl}_t| < g^{kl} \Longrightarrow \mathbf{u}^k_t - \mathbf{u}^l_t = 0, \\ &|\boldsymbol{\tau}^{kl}_t| = g^{kl} \Longrightarrow \text{there exists } \vartheta \ge 0 \text{ such that } \mathbf{u}^k_t - \mathbf{u}^l_t = -\vartheta \boldsymbol{\tau}^{kl}_t, \end{aligned} \tag{6.203}$$

$$u_n = 0, \qquad \tau_{ti} = 0, \quad \text{on } \Gamma_0. \tag{6.204}$$

Here $e_{ij}(\mathbf{u}) = \frac{1}{2}(\partial u_i/\partial x_j + \partial u_j/\partial x_i)$ is the small strain tensor, normal, and tangential components of the displacement vector $\mathbf{u}$ $[\mathbf{u} = (u_i), i = 1, 2]$ and the stress vector $\boldsymbol{\tau}$ $[\boldsymbol{\tau} = (\tau_i)]$, $u^k_n = u^k_i n^k_i$, $u^l_n = u^l_i n^k_i$ (no sum over k or l), $\mathbf{u}^k_t = (u^k_{ti})$, $u^k_{ti} = u^k_i - u^k_n n^k_i$, $\mathbf{u}^l_t = (u^l_{ti})$, $u^l_{ti} = u^l_i - u^l_n n^l_i$, $i = 1, \ldots, N$, $\tau^k_n = \tau^k_{ij} n^k_i n^k_j$, $\boldsymbol{\tau}^k_t = (\tau^k_{ti})$, $\tau^k_{ti} = \tau^k_{ij} n^k_j - \tau^k_n n^k_i$, $\tau^l_n = \tau^l_{ij} n^l_i n^l_j$, $\tau^{kl}_n = \tau^k_n$, $\boldsymbol{\tau}^l_t = (\tau^l_{ti})$, $\tau^l_{ti} = \tau^l_{ij} n^l_j - \tau^l_n n^l_i$, $\boldsymbol{\tau}^{kl}_t \equiv \boldsymbol{\tau}^k_t$.

Assume that c^ι_{ijkl} are positive definite symmetric matrices such that

$$0 < c^\iota_0 \le c^\iota_{ijkl} \xi_{ij} \xi_{kl} |\boldsymbol{\xi}|^{-2} \le c^\iota_1 < +\infty \quad \text{for a.a. } \mathbf{x} \in \Omega^\iota, \boldsymbol{\xi} \in R^{N^2}, \xi_{ij} = \xi_{ij},$$

where c^ι_0, c^ι_1 are constants independent of $\mathbf{x} \in \boldsymbol{\Omega}^\iota$.

Let us introduce $W = \prod_{\iota=1}^{s} [H^1(\Omega^\iota)]^2$, $\|l\mathbf{v}\|_W = (\sum_{\iota \le s} \sum_{i \le N} \|v^\iota_i\|^2_{1,\Omega^\iota})^{1/2}$ and the sets of virtual and admissible displacements $V_0 = \{\mathbf{v} \in W | \mathbf{v} = 0 \text{ on } \Gamma_u \text{ and } v_n = 0 \text{ on } \Gamma_0\}$, $V = \mathbf{u}_0 + V_0$, $K = \{\mathbf{v} \in V \mid v^k_n - v^l_n \le 0 \text{ on } \cup_{k,l} \Gamma^{kl}_c\}$. Assume that $u^k_{0n} - u^l_{0n} = 0$ on $\cup_{k,l} \Gamma^{kl}_c$. Let $c^\iota_{ijkl} \in L^\infty(\Omega^\iota)$, $F^\iota_i \in L^2(\Omega^\iota)$, $P_i \in L^2({}^1\Gamma_\tau)$, and $\mathbf{u}^\iota_0 \in [H^1(\Omega^\iota)]^2$.

Multiplying Eq. (6.197) by a test function, integrating per partes over the domain Ω^ι, and using the boundary conditions, we obtain the following variational problem $(\mathcal{P})_v$:

Definition 6.15 We say that the function $\mathbf{u}$ is a weak solution of Problem $\mathcal{P}$, if $\mathbf{u} \in K$ and

$$a(\mathbf{u}, \mathbf{v} - \mathbf{u}) + j_{gn}(\mathbf{v}) - j_{gn}(\mathbf{u}) \ge L(\mathbf{v} - \mathbf{u}) \quad \forall \mathbf{v} \in K, \tag{6.205}$$

where

$$a(\mathbf{u},\mathbf{v}) = \sum_{\iota=1}^{s} \int_{\Omega^\iota} c^\iota_{ijkl} e_{ij}(\mathbf{u}^\iota) e_{kl}(\mathbf{v}^\iota)\, d\mathbf{x},$$

$$j_{gn}(\mathbf{v}) = \sum_{k,l} \int_{\Gamma^{kl}} g^{kl} |\mathbf{v}^k_t - \mathbf{v}^l_t|\, ds,$$

$$L(\mathbf{v}) = \sum_{\iota=1}^{s} \int_{\Omega^\iota} F^\iota_i v^\iota_i \, d\mathbf{x} - \sum_{\iota \le s} \int_{\Gamma^\iota_\tau} P^\iota_i v^\iota_i \, ds. \tag{6.206}$$

Let us define the sets of rigid displacements and rotations

$$P = \prod_{\iota=1}^{s} P^\iota, \qquad P^\iota = \{\mathbf{v}^\iota = (v^\iota_1, v^\iota_2) | v^\iota_1 = a^\iota_1 - b^\iota x_2, v^\iota_2 = a^\iota_2 + b^\iota x_1\}$$

where a^ι_i, $i = 1, 2$ and b^ι are arbitrary real constants.

It can be shown that problem (6.205) has a unique solution. The condition $S(\mathbf{v}) \le j_g(\mathbf{v})\ \forall \mathbf{v} \in P \cap K$ is necessary for the existence of a weak solution of Problem $\mathcal{P}$ [see Hlaváček and Nedoma (2002a)].

Finite Element Approximation For simplicity, we restrict ourselves to the 2D case only. Let the region $\Omega = \cup_{\iota=1}^{s} \Omega^\iota$ be approximated by $\Omega_h = \cup_{\iota=1}^{s} \Omega^\iota_h$ with the polygonal boundary $\partial\Omega_h = \overline{\Gamma}_{uh} \cup \overline{\Gamma}_{\tau h} \cup \overline{\Gamma}_{ch} \cup \overline{\Gamma}_{0h}$, where $\overline{\Gamma}_{uh}, \overline{\Gamma}_{\tau h}, \overline{\Gamma}_{ch}, \overline{\Gamma}_{0h}$ are piecewise linear. Let $\Omega_h = \cup_{\iota=1}^{s} \Omega^\iota_h$ be triangulated, let q_i be nodes of used triangulation. Let $\mathcal{T}^\iota_h$, $\iota = 1, \ldots, s$, denote triangulations of polygonal domains Ω^ι_h, $\iota = 1, \ldots, s$, and $\mathcal{T}_h = \{\mathcal{T}^\iota_h, \iota = 1, \ldots, s\}$. We assume that $\mathcal{T}^\iota_h$, $\iota = 1, \ldots, s$, are consistent with the respective decompositions of the boundaries $\partial\Omega^\iota_h$, $\iota = 1, \ldots, s$, and let the nodes lie on Γ^{kl}_c belonging to the triangulations corresponding to the neighboring subdomains Ω^k and Ω^l being in a mutual contact. The triangulation $\mathcal{T}_h$ is said to be regular; and if all $\mathcal{T}^\iota_h$, $\iota = 1, \ldots, s$, are regular, h is the maximal side of the triangulation. For every node q_i of the triangulation $\mathcal{T}_h$ on Γ^{kl}_c and Γ_0 we define the set of indices $\mathcal{N}^{kl}_i = \{j \in \{1, \ldots, r\} | q_i \in \Gamma^{kl}_{cj}\}$ and $\mathcal{N}_i = \{j \in \{1, \ldots, r'\} | q_i \in \Gamma_{0j}\}$, where $\Gamma^{kl}_c = \cup_{j=1}^{r} \Gamma^{kl}_{cj}$, $\Gamma_0 = \cup_{j=1}^{r'} \Gamma_{0j}$, Γ^{kl}_{cj}, Γ_{0j} denote segments on Γ^{kl}_c, Γ_0, and r, r' the number of segments on Γ^{kl}_c and Γ_0, respectively.

Let us define a finite dimensional space V_h by

$$V_h = \{\mathbf{v}_h | \mathbf{v}_h \in [C(\Omega^1)]^2 \times \cdots \times [C(\Omega^s)]^2, \mathbf{v}_{h|T_{hi}} \in [P_1(T_{hi})]^2, \quad \forall T_{hi} \in \mathcal{T}_h;$$
$$v_{hn}(q_i) = 0, q_i \in \Gamma_0; \mathbf{v}_h(q_i) = \mathbf{u}_0(q_i), q_i \in \Gamma_u\},$$

and a finite-dimensional set of admissible displacements

$$K_h = \{\mathbf{v}_h | \mathbf{v}_h \in V_h, (v^k_{hn} - v^l_{hn})(q_i) \le 0, \quad q_i \in \Gamma^{kl}_c, 1 \le k, l \le s\}.$$

Definition 6.16 We say that the function $\mathbf{u}_h \in K_h$ is a solution of Problem $(\mathcal{P})_h$ if

$$a(\mathbf{u}_h, \mathbf{v} - \mathbf{u}_h) + j_{gn}(\mathbf{v}_h) - j_{gn}(\mathbf{u}_h) \geq L(\mathbf{v}_h - \mathbf{u}_h) \quad \forall \mathbf{v}_h \in K_h. \tag{6.207}$$

The next theorem gives the connection between Problem $\mathcal{P}_{sg}$ and Problem $(\mathcal{P}_{sg})_h$ if $h \to 0_+$ under the assumption that the solution of the problem is sufficiently smooth.

Theorem 6.42 Let $\partial\Omega$ and its parts Γ_u, Γ_τ, Γ_c, Γ_0 be piecewise polygonal, and let Γ_u have the positive length. Let $\mathbf{u}^\iota \in H^{2,2}(\Omega^\iota)$, $\iota = 1, \ldots, s$, be the solution of Problem $\mathcal{P}$ such that the stress components $\tau_{ij}(\mathbf{u}^\iota) \in H^1(\Omega^\iota)$, $i,j = 1,2$, $\iota = 1, \ldots, s$, $\boldsymbol{\tau} \in [L^\infty(\Gamma_c^{kl})]^2$, $\mathbf{u} \in H^{2,2}(\Gamma_c \cap \Gamma_{ci}^{kl})$ (where Γ_{ci}^{kl} side of triangle adjacent to Γ_c^{kl}) holds for any part Γ_c^{kl} of Γ_c, $g_c^{kl} \in L^\infty(\Gamma_c^{kl})$. Let $K_h \subset K$ and let the changes $u_n^k - u_n^l < 0 \to u_n^k - u_n^l = 0$ and $\mathbf{u}_t^k - \mathbf{u}_t^l = 0 \to \mathbf{u}_t^k - \mathbf{u}_t^l \neq 0$ occur at finite set of points of $\cup_{k,l}\Gamma_c^{kl}$ only. Then for the semicoercive case

$$|\mathbf{u} - \mathbf{u}_h| = O(h), \qquad \text{where} \quad |\mathbf{w}| = \left(\sum_{\iota=1}^{s} \int_{\Omega_h^\iota} e_{ij}(\mathbf{w}) e_{ij}(\mathbf{w})\, d\mathbf{x} \right)^{1/2}, \tag{6.208}$$

and for the coercive case

$$\|\mathbf{u} - \mathbf{u}_h\| = O(h). \tag{6.209}$$

For the proof see Nedoma (1987, 1998a), Haslinger et al. (1996), and Hlaváček and Nedoma (2002a).

Domain Decomposition Algorithm Let every region $\overline{\Omega}^\iota = \cup_{i=1}^{J(\iota)} \overline{\Omega}_i^\iota$ in $\mathbb{R}^2$, where $J(\iota)$ is a number of subdomains of Ω^ι. Let $\Gamma_i^\iota = \partial\Omega_i^\iota \backslash \partial\Omega^\iota$, $\iota \in \{1, \ldots, s\}$, $i \in \{1, \ldots, J(\iota)\}$, be a part of the dividing line (boundary line) and let $\Gamma = \cup_{\iota=1}^{s} \cup_{i=1}^{J(\iota)} \Gamma_i^\iota$ be the whole interface boundary. Let

$$T^\iota = \{j \in \{1, \ldots, J(\iota)\} : \overline{\Gamma}_c \cap \partial\overline{\Omega}_j^\iota = 0\}, \qquad \iota = 1, \ldots, s, \tag{6.210}$$

be the set of all indices of subdomains of the domain Ω^ι that are not adjacent to a contact, and let

$$\Omega^{*j} = \cup_{[i,\iota]\in\vartheta} \Omega_i^\iota, \tag{6.211}$$

where $\vartheta = \{[i, \iota] : \partial\Omega_i^\iota \cap \Gamma_c \neq \emptyset\}$ represent subdomains in unilateral contact. Suppose that $\Gamma \cap \Gamma_c = \emptyset$. Then for the trace operator $\gamma : [H^1(\Omega_i^\iota)]^2 \to [L^2(\partial\Omega_i^\iota)]^2$ we have

$$V_\Gamma = \gamma K_{|\Gamma} = \gamma V_{|\Gamma}. \tag{6.212}$$

Let $\gamma^{-1} : V_\Gamma \to V$ be an arbitrary linear inverse mapping satisfying

$$\gamma^{-1}\overline{\mathbf{v}} = 0 \quad \text{on } \cup_{k,l} \Gamma_c^{kl} \quad \forall \overline{\mathbf{v}} \in V_\Gamma. \tag{6.213}$$

Let us introduce restrictions $\overline{R}_i^{\iota} : V_\Gamma \to \Gamma_i^{\iota}$; $L_i^{\iota} : L^{\iota} \to \Omega_i^{\iota}$; $j_i^{\iota} : j_{gn}^{\iota} \to \Gamma_c^{kl}$; $a_i^{\iota}(.,.)$: $a_i^{\iota}(.,.) \to \Omega_i^{\iota}$; $V(\Omega_i^{\iota}) \to \Omega_i^{\iota}$, and let $V^0(\Omega_i^{\iota}) = \{\mathbf{v} \in V_0 | \mathbf{v} = 0 \text{ on } \overline{(\cup_{\iota=1}^{s} \Omega^{\iota}) \backslash \Omega_i^{\iota}}\}$ be the space of functions with zero traces on Γ_i^{ι}.

Theorem 6.43 A function $\mathbf{u}$ is a solution of a global problem $(\mathcal{P})$, if and only if its trace $\overline{\mathbf{u}} = \gamma \mathbf{u}_{|\Gamma}$ on the interface Γ satisfies the condition

$$\sum_{\iota=1}^{s} \sum_{i=1}^{J(\iota)} [a_i^{\iota}(\mathbf{u}_i^{\iota}(\overline{\mathbf{u}}), \gamma^{-1}\overline{\mathbf{w}}) - L_i^{\iota}(\gamma^{-1}\overline{\mathbf{w}})] = 0 \quad \forall \overline{\mathbf{w}} \in V_\Gamma, \overline{\mathbf{u}} \in V_\Gamma \tag{6.214}$$

and its restrictions $\mathbf{u}_i^{\iota}(\overline{\mathbf{u}}) \equiv \mathbf{u}_{|\Omega_i^{\iota}}$ satisfy:

(i) the condition

$$a_i^{\iota}(\mathbf{u}_i^{\iota}(\overline{\mathbf{u}}), \varphi_i^{\iota}) = L_i^{\iota}(\varphi_i^{\iota}) \quad \forall \varphi_i^{\iota} \in V^0(\Omega_i^{\iota}),\ \mathbf{u}_i^{\iota}(\overline{\mathbf{u}}) \in V(\Omega_i^{\iota}),$$
$$\gamma \mathbf{u}_i^{\iota}(\overline{\mathbf{u}})_{|\Gamma_i^{\iota}} = \overline{R}_i^{\iota}\overline{\mathbf{u}}, i \in T^{\iota},\ \iota = 1, \ldots, s, \tag{6.215}$$

(ii) the condition

$$\sum_{[i,\iota] \in \vartheta} a_i^{\iota}(\mathbf{u}_i^{\iota}(\overline{\mathbf{u}}), \varphi_i^{\iota}) + j_{gn}^{\iota}(\mathbf{u}_i^{\iota}(\overline{\mathbf{u}}) + \varphi_i^{\iota}) - j_{gn}^{\iota}(\mathbf{u}_i^{\iota}(\overline{\mathbf{u}}))$$
$$\geq \sum_{[i,\iota] \in \vartheta} L_i^{\iota}(\varphi_i^{\iota}) \quad \forall \varphi \in (\varphi_i^{\iota}, [i, \iota] \in \vartheta, \varphi_i^{\iota} \in V^0(\Omega_i^{\iota}), \tag{6.216}$$

and such that

$$\mathbf{u} + \varphi \in K, \quad \gamma \mathbf{u}_i^{\iota}(\overline{\mathbf{u}})_{|\Gamma_i^{\iota}} = \overline{R}_i^{\iota}\overline{\mathbf{u}} \quad \text{for } [i, \iota] \in \vartheta. \tag{6.217}$$

For the proof see Daněk et al. (2005a).

Local and Global Operators of the Schur Complement To analyze condition (6.214) we will introduce the concept of the **local and global Schur complements**. Let us denote

$$V_i^{\iota} = \{\gamma \mathbf{v}_{|\Gamma_i^{\iota}} | \mathbf{v} \in K\} = \{\gamma \mathbf{v}_{|\Gamma_i^{\iota}} | \mathbf{v} \in V\}$$

and define a particular case of the restriction of the inverse mapping $\gamma^{-1}(\cdot)_{|\Omega_i^{\iota}}$ by

$$\mathrm{Tr}_{i\iota}^{-1} : V_i^{\iota} \to V(\Omega_i^{\iota}), \qquad \gamma(\mathrm{Tr}_{i\iota}^{-1}\overline{\mathbf{u}})_{|\Gamma_i^{\iota}} = \overline{\mathbf{u}}_i^{\iota}, \qquad i = 1, \quad J(\iota), \quad \iota = 1, \ldots, s,$$
$$a_i^{\iota}(\mathrm{Tr}_{i\iota}^{-1}\overline{\mathbf{u}}_i^{\iota}, \mathbf{v}_i^{\iota}) = 0 \quad \forall \mathbf{v}_i^{\iota} \in V_0(\Omega_i^{\iota}),$$
$$\mathrm{Tr}_{i\iota}^{-1}\overline{\mathbf{u}}_i^{\iota} \in V(\Omega_i^{\iota}), \quad \text{for } i \in T^{\iota}, \iota = 1, \ldots, s. \tag{6.218}$$

For $[i, \iota] \in \vartheta$ we complete the definition by the boundary condition (6.213), that is,

$$\mathrm{Tr}_{i\iota}^{-1}\overline{\mathbf{u}}_i^{\iota} = 0 \quad \text{on } \cup_{k,l} \Gamma^{kl}. \tag{6.219}$$

Definition 6.17 By the local Schur complement for $i \in T^{\iota}$ it is meant the operator $\mathcal{S}_i^{\iota} : V_i^{\iota} \to (V_i^{\iota})^*$ defined by

$$\langle \mathcal{S}_i^{\iota}\overline{\mathbf{u}}_i^{\iota}, \overline{\mathbf{v}}_i^{\iota}\rangle = a_i^{\iota}(\mathrm{Tr}_{i\iota}^{-1}\overline{\mathbf{u}}_i^{\iota}, \mathrm{Tr}_{i\iota}^{-1}\overline{\mathbf{v}}_i^{\iota}) \quad \forall \overline{\mathbf{u}}_i^{\iota}, \overline{\mathbf{v}}_i^{\iota} \in V_i^{\iota}, \tag{6.220}$$

and in the matrix form by

$$\mathcal{S}_i^{\iota}\overline{\mathbf{U}}_i^{\iota} = (\overline{A}_{i\iota} - \overline{B}_{i\iota}^{T}\,{}^{\circ}A_{i\iota}^{-1}B_{i\iota})\overline{\mathbf{U}}_i^{\iota}, \tag{6.221}$$

where

$$A_{i\iota} \equiv a_i^{\iota}(.,.) = \begin{pmatrix} {}^{\circ}A_{i\iota} & B_{i\iota} \\ B_{i\iota}^{T} & \overline{A}_{i\iota} \end{pmatrix} \quad \mathbf{U}_i^{\iota} = \begin{pmatrix} {}^{\circ}\mathbf{U}_i^{\iota} \\ \overline{\mathbf{U}}_i^{\iota} \end{pmatrix},$$

$$\mathrm{Tr}_{i\iota}^{-1} = \begin{pmatrix} -{}^{\circ}A_{i\iota} & B_{i\iota} \\ 0 & I \end{pmatrix}, \qquad i \in T^{\iota},\ \iota = 1, \ldots, s, \tag{6.222}$$

where the nodes of $\overline{\mathbf{U}}_i^{\iota}$ belong to Γ_i^{ι} and the internal degrees of freedom are ${}^{\circ}\mathbf{U}_i^{\iota}$.

For subdomains that are in contact we will define a **common local Schur complement** as follows:

Definition 6.18 The common local Schur complement for the union $\Omega_i^k \cup \Omega_j^l$ (where $\Gamma_c^{kl} \subset \Gamma_c$ and $[i,k] \in \vartheta$, $[j,l] \in \vartheta$) is the operator

$$\mathcal{S}^{kl} : (V_i^k \times V_j^l) \to (V_i^k \times V_j^l)' = (V_i^k)' \times (V_j^l)'$$

defined by the relation

$$\langle \mathcal{S}^{kl}(\overline{\mathbf{y}}_i^k, \overline{\mathbf{y}}_j^l), \overline{\mathbf{v}}_i^k, \overline{\mathbf{v}}_j^l)\rangle = a_i^k(\mathbf{u}_i^k(\overline{\mathbf{y}}_i^k), \mathrm{Tr}_{ik}^{-1}\overline{\mathbf{v}}_i^k) + a_j^l(\mathbf{u}_j^l(\overline{\mathbf{y}}_j^l), \mathrm{Tr}_{jl}^{-1}\overline{\mathbf{v}}_j^l)$$
$$\forall(\overline{\mathbf{v}}_i^k, \overline{\mathbf{v}}_j^l) \in V_i^k \times V_j^l, \tag{6.223}$$

where Tr_{ik}^{-1} and Tr_{jl}^{-1} are defined by means of (6.218) and (6.219).

Condition (6.214) can be expressed by means of local Schur complements. Then we have the following lemma:

Lemma 6.18 The trace $\overline{\mathbf{u}} = \gamma\mathbf{u}_{|\Gamma}$ of the weak solution satisfies the following condition:

$$\sum_{\iota=1}^{s}\sum_{i\in T^{\iota}}\langle \mathcal{S}_i^{\iota}\overline{\mathbf{u}}_i^{\iota}, \overline{\mathbf{v}}_j^{\iota}\rangle + \sum_{k,l}\langle \mathcal{S}^{kl}(\overline{\mathbf{u}}_i^k, \overline{\mathbf{u}}_j^l), (\overline{\mathbf{v}}_i^k, \overline{\mathbf{v}}_j^l)\rangle = \sum_{\iota=1}^{s}\sum_{i=1}^{J(\iota)} L_i^{\iota}(\mathrm{Tr}_{i\iota}^{-1}\overline{\mathbf{v}}_i^{\iota})$$
$$\forall \overline{\mathbf{v}} \in V_{\Gamma}, [i,k] \in \vartheta, [j,l] \in \vartheta, \Gamma_c^{kl} \subset \Gamma_c, \tag{6.224}$$

where $\overline{\mathbf{v}}_i^{\iota} = \overline{R}_i^{\iota}\overline{\mathbf{v}}, \overline{\mathbf{u}}_i^{\iota} = \overline{R}_i^{\iota}\overline{\mathbf{u}}$.

Then we will solve Eq. (6.224) on the interface Γ in the dual space $(V_\Gamma)'$. We rewrite (6.224) into the following form:

$$\mathcal{S}_0\overline{\mathbf{U}} + \mathcal{S}_{\text{KON}}\overline{\mathbf{U}} = \mathbf{F}, \tag{6.225}$$

where

$$\mathcal{S}_0 = \sum_{\iota=1}^{s}\sum_{i\in T^\iota}(\overline{R}_i^\iota)^{\text{T}}\mathcal{S}_i^\iota\overline{R}_i^\iota, \mathcal{S}_{\text{KON}} = \sum_{k,l}\overline{R}_{kl}^{\text{T}}\mathcal{S}^{kl}\overline{R}_{kl},$$
$$\mathbf{F} = \sum_{\iota=1}^{s}\sum_{i=1}^{J(\iota)}(\overline{R}_i^\iota)^{\text{T}}(Tr_{i\iota}^{-1})^{\text{T}}\mathcal{S}_i^\iota, \tag{6.226}$$

and $\overline{R}_{kl}(\overline{\mathbf{u}}) = (\overline{R}_i^k(\overline{\mathbf{u}}), \overline{R}_j^l(\overline{\mathbf{u}}))^{\text{T}}, \overline{\mathbf{u}} \in V_\Gamma, [i,k] \in \vartheta, [j,l] \in \vartheta, \Gamma_c^{kl} \subset \Gamma_c$.

Equation (6.225) will be solved by **successive approximations** because the operators $\mathcal{S}^{kl}$ and therefore $\mathcal{S}_{\text{KON}}$ are nonlinear. We choose a suitable initial approximation $\overline{\mathbf{U}}^0$, for instance, the solution of the global primal problem, where the boundary conditions on Γ_c are replaced by the linear "classical" bilateral conditions (which correspond with $g^{kl} \equiv 0$ and $j(\mathbf{u}) \equiv 0$)

$$u_n^k - u_n^l = 0, \qquad \boldsymbol{\tau}_t^{kl} = 0 \quad \text{on } \Gamma_{c0} \equiv \cup_{k,l}\Gamma_{c0}^{kl}, \tag{6.227}$$

where Γ_{c0}^{kl} are parts of Γ_c^{kl}, meas $\Gamma_{c0}^{kl} > 0$, chosen a priori (e.g., $\Gamma_{c0}^{kl} = \Gamma_c^{kl}$). On $\Gamma_c^{kl}\setminus\Gamma_{c0}^{kl}$ we consider homogeneous conditions of zero surface load $P_j^k = P_j^l = 0, j = 1, 2$.

Then we replace the set K by $K^0 = \{\mathbf{v} \in V_0 | v_n^k - v_n^l = 0 \text{ on } \cup_{k,l}\Gamma_{c0}^{kl}\}$ and, therefore, we solve the following problem:

$$\mathbf{u}^0 = \arg\min_{\mathbf{v}\in K^0}\left\{\tfrac{1}{2}a(\mathbf{v},\mathbf{v}) - L(\mathbf{v})\right\} \tag{6.228}$$

and we set $\overline{\mathbf{U}}^0 = \gamma\mathbf{u}^0|_\Gamma$. The auxiliary problem (6.228) represents a linear elliptic boundary value problem of the system of s elastic bodies with bilateral contact, and it can be solved by the domain decomposition method again.

We will assume that the approximation $\overline{\mathbf{U}}^{k-1}$ is known and the next approximation $\overline{\mathbf{U}}^k$ we find as the solution of the following linear problem:

$$\mathcal{S}_0\overline{\mathbf{U}}^k = \mathbf{F} - \mathcal{S}_{\text{KON}}\overline{\mathbf{U}}^{k-1}, \qquad k = 1, 2, \ldots, \tag{6.229}$$

for which a suitable preconditioning of the Neumann–Neumann type can be applied (Daněk et al., 2005a).

Solution of the Auxiliary Problem Instead of (6.207) we will solve the variational equation for $\mathbf{u}^0 \in K^0$:

$$a(\mathbf{u}^0, \mathbf{v}) = L(\mathbf{v}) \quad \forall\mathbf{v} \in K^0. \tag{6.230}$$

Thus an analog of Theorem 6.43 can be derived, where condition (6.216) is replaced by the corresponding variational equality and where a mapping $\gamma_0^{-1} : V_\Gamma \to V$ satisfies conditions $(\gamma_0^{-1}\overline{\mathbf{v}})_n^k - (\gamma_0^{-1}\overline{\mathbf{v}})_n^l = 0$ on $\cup_{k,l}\Gamma_{c0}^{kl}$.

We introduce operators of the Schur complements. For $i \in T^\iota$, $\iota = 1, \dots, s$, we define the mappings $\mathrm{Tr}_{i\iota}^{-1}$ according to (6.218) and the local Schur complements $\mathcal{S}_i^{0\iota}$ by (6.220).

Definition 6.19 The common local Schur complement for the union $\Omega_i^k \cup \Omega_j^l$, where $\Gamma_{c0}^{kl} \subset \Gamma_c$ and $[i,k] \in \vartheta$, $[j,l] \in \vartheta$,

$$\mathcal{S}^{0kl} : (V_i^k \times V_j^l) \to (V_i^k)' \times (V_j^l)'$$

is defined by the following relation

$$\langle \mathcal{S}^{0kl}(\overline{\mathbf{u}}_i^{0k}, \overline{\mathbf{u}}_j^{0l}), (\overline{\mathbf{v}}_i^k, \overline{\mathbf{v}}_j^l) \rangle = a_i^k(\mathbf{u}_i^k(\overline{\mathbf{u}}_i^k), \mathrm{Tr}_{ik}^{-1}\overline{\mathbf{v}}_i^k) + a_j^l(\mathbf{u}_j^l(\overline{\mathbf{u}}_j^l), \mathrm{Tr}_{jl}^{-1}\overline{\mathbf{v}}_j^l)$$

$$\forall (\overline{\mathbf{v}}_i^k, \overline{\mathbf{v}}_j^l) \in V_i^k \times V_j^l, \tag{6.231}$$

where Tr_{ik}^{-1} and Tr_{jl}^{-1} are defined by means of $(\mathrm{Tr}_{ik}^{-1}\overline{\mathbf{v}}_i^k)_n - (\mathrm{Tr}_{jl}^{-1}\overline{\mathbf{v}}_j^k)_n = 0$ on Γ_{c0}^{kl} and

$$a_i^k(\mathrm{Tr}_{ik}^{-1}\overline{\mathbf{v}}_i^k, \mathbf{w}_i^k) + a_j^l(\mathrm{Tr}_{jl}^{-1}\overline{\mathbf{v}}_j^l, \mathbf{w}_j^l) = 0 \quad \forall \mathbf{w}_i^k \in V^0(\Omega_i^k), \mathbf{w}_j^l \in V^0(\Omega_j^l)$$

such that

$$(\mathbf{w}_i^k)_n - (\mathbf{w}_j^l)_n = 0 \quad \text{on } \Gamma_{c0}^{kl}.$$

A global Schur complement $\mathcal{S}$ is defined by

$$\mathcal{S} = \mathcal{S}_0 + \sum_{k,l} (\overline{R}_{kl})^{\mathrm{T}} \mathcal{S}^{0kl} \overline{R}_{kl}, \tag{6.232}$$

where $\mathcal{S}_0$ is defined in (6.226).

Then the condition corresponding to (6.224) of the auxiliary problem on the interface implies the equation

$$\mathcal{S}\overline{\mathbf{U}} = \mathbf{F} \quad \text{in the dual space } (V_\Gamma)'. \tag{6.233}$$

Let $\mathcal{M}$ be a suitable matrix of preconditioning.

The PCG1 Algorithm Choose $\overline{\mathbf{U}}^0$, $\mathbf{H}^0$ and let $\mathbf{P}^0 = 0$:

(i) Compute the preconditioned direction of descent

$$\mathbf{G}^m = \mathcal{M}^{-1}\mathbf{H}^m, \qquad m = 0, 1, \dots .$$

(ii) Compute

$$\mathbf{P}^m = \mathbf{G}^m + \frac{\langle \mathbf{H}^m, \mathbf{G}^m \rangle}{\langle \mathbf{H}^{m-1}, \mathbf{G}^{m-1} \rangle} \mathbf{P}^{m-1}, \qquad m = 1, 2, \ldots .$$

(iii) On each subdomain Ω_i^ι, $i \in T^\iota$, $\iota = 1, \ldots, s$, solve the Dirichlet problem

$$\mathcal{A}_{i\iota}\, {}^{\circ}\mathbf{U}_i^\iota = B_{i\iota} \overline{R}_i^\iota \mathbf{P}^m .$$

(iv) On every subdomain $\Omega_i^k \cup \Omega_j^l$, (where $\Gamma_{c0}^{kl} \subset \Gamma_c$) solve the following problem:

$$({}^{\circ}\mathbf{U}_i^k, {}^{\circ}\mathbf{U}_j^l) = \arg \min_{E^k\, {}^{\circ}\mathbf{V}^k + E^l\, {}^{\circ}\mathbf{V}^l = 0} \boldsymbol{\Psi}({}^{\circ}\mathbf{V}^k, {}^{\circ}\mathbf{V}^l), \tag{6.234}$$

where

$$\begin{aligned} \boldsymbol{\Psi}({}^{\circ}\mathbf{V}^k, {}^{\circ}\mathbf{V}^l) &= ({}^{\circ}\mathbf{V}^k)^T \mathcal{A}_{ik}\, {}^{\circ}\mathbf{V}^k + 2(\overline{R}_i^k \mathbf{P}^m)^T B_{ik}^T\, {}^{\circ}\mathbf{V}^k \\ &\quad + ({}^{\circ}\mathbf{V}^l)^T \mathcal{A}_{jl}\, {}^{\circ}\mathbf{V}^l + 2(\overline{R}_j^l \mathbf{P}^m)^T B_{jl}^T\, {}^{\circ}\mathbf{V}^k \end{aligned}$$

and the condition $E^k\, {}^{\circ}\mathbf{V}^k + E^l\, {}^{\circ}\mathbf{V}^l = 0$ is equivalent with $(\mathbf{v}^k)_n - (\mathbf{v}^l)_n = 0$ on Γ_{c0}^{kl}. Then using the decompositions $\mathbf{U}^\iota = ({}^{\circ}\mathbf{U}^\iota, \overline{\mathbf{U}}^\iota)^{\mathrm{T}}$, $\iota = k, l$, and inserting $\overline{\mathbf{U}}^k = \overline{R}_i^k \mathbf{P}^m$, $\overline{\mathbf{U}}^l = \overline{R}_j^l \mathbf{P}^m$, we have

$$\begin{aligned} (\mathcal{S}^{0kl} \overline{\mathbf{U}})_i^k &= B_{ik}^{\mathrm{T}}\, {}^{\circ}\mathbf{U}^k + A_{ik} \overline{R}_i^k \mathbf{P}^m , \\ (\mathcal{S}^{0kl} \overline{\mathbf{U}})_j^l &= B_{jl}^{\mathrm{T}}\, {}^{\circ}\mathbf{U}^l + A_{jl} \overline{R}_j^l \mathbf{P}^m . \end{aligned}$$

(v) Compute

$$\begin{aligned} \mathbf{Z}^m &= \mathcal{S}\mathbf{P}^m = \sum_{k,l} (\overline{R}_{kl})^{\mathrm{T}} \mathcal{S}^{0kl} \overline{R}_{kl} \mathbf{P}^m + \sum_{\iota=1}^{s} \sum_{i \in T^\iota} (\overline{R}_i^\iota)^{\mathrm{T}} (\overline{A}_{i\iota} \overline{R}_i^\iota \mathbf{P}^m - B_{i\iota}^{\mathrm{T}}\, {}^{\circ}\mathbf{U}_i^\iota), \\ \alpha_m &= \frac{\langle \mathbf{H}^m, \mathbf{G}^m \rangle}{\langle \mathbf{Z}^m, \mathbf{P}^m \rangle}, \\ \mathbf{H}^{m+1} &= \mathbf{H}^m - \alpha_m \mathbf{Z}^m , \\ \overline{\mathbf{U}}^{m+1} &= \overline{\mathbf{U}}^m + \alpha_m \mathbf{P}^m , \end{aligned}$$

goto (i).

In Daněk et al. (2005a) the Neumann–Neumann preconditioner is derived.

Solution of the Original Problem on the Interface Recall that we have to solve problem (6.225) by successive approximations, that is, by means of a sequence of problems (6.229), which can be written in the form

$$\mathcal{S}_0 \overline{\mathbf{U}}^k = \mathbf{b}^k, \qquad k = 1, 2, \ldots , \tag{6.235}$$

where $\mathbf{b}^k = \mathbf{F} - \mathcal{S}_{\mathrm{KON}}\overline{\mathbf{U}}^{k-1}$. Now $\overline{\mathbf{U}}^0$ is the solution of auxiliary problem, that is, $\overline{\mathbf{U}}^0 = \gamma \mathbf{u}^0|_\Gamma$, where $\mathbf{u}^0$ is a solution of problem (6.228).

To solve problem (6.235) the method of preconditioned conjugate gradients can be used. Let a suitable matrix of preconditioning $\mathcal{M}_0$ be given. Then we have the following algorithm:

The PCG2 Algorithm Choose $\boldsymbol{\omega}_0, \boldsymbol{\rho}_0 = \mathbf{b}^k - \mathcal{S}_0\boldsymbol{\omega}_0, \boldsymbol{\pi}_0 = 0$.

(i) Compute the preconditioned direction of descent

$$\mathbf{g}_m = \mathcal{M}_0^{-1}\boldsymbol{\rho}_m,$$

where $\mathcal{M}_0^{-1}$ is a “reduced” preconditioner.

(ii) Compute

$$\boldsymbol{\pi}_m = \mathbf{g}^m + \frac{\langle \boldsymbol{\rho}_m, \mathbf{g}_m \rangle}{\langle \boldsymbol{\rho}_{m-1}, \mathbf{g}_{m-1} \rangle}\boldsymbol{\pi}_{m-1}.$$

(iii) On every subdomain Ω_i^ι, $\iota = 1, \ldots, s$ and $i \in T^\iota$ solve (parallel) the system (Dirichlet problem)

$$\mathring{A}_{i\iota}\,{}^\circ\boldsymbol{\omega}_i^\iota = B_{i\iota}\overline{R}_i^\iota\boldsymbol{\pi}_m.$$

(iv) Compute

$$\boldsymbol{\xi}_m = \mathcal{S}_0\boldsymbol{\pi}^m = \sum_{\iota=1}^{s}\sum_{i\in T^\iota}(\overline{R}_i^\iota)^T(\overline{A}_{i\iota}\overline{R}_i^\iota\boldsymbol{\pi}_m - B_{i\iota}^T\,{}^\circ\boldsymbol{\omega}_i^\iota).$$

(v) Compute

$$\alpha_m = \frac{\langle \boldsymbol{\rho}_m, \mathbf{g}_m \rangle}{\langle \boldsymbol{\xi}_m, \boldsymbol{\pi}_m \rangle},$$

$$\boldsymbol{\rho}_{m+1} = \boldsymbol{\rho}_m - \alpha_m\boldsymbol{\xi}_m,$$

$$\omega_{m+1} = \boldsymbol{\omega}_m + \alpha_i\boldsymbol{\pi}_m,$$

goto (i).

A preconditioner of the Neumann–Neumann type $\mathcal{M}_0^{-1}$ was derived in Daněk et al. (2005a).

In Daněk, et al. (2005a) the convergence of the method of successive approximation (6.235) to the solution of the original problem (6.225) is proved for $g^{kl} \equiv 0$ on $\cup_{k,l}\Gamma_c^{kl}$.

6.5 STRESS–STRAIN ANALYSES OF HUMAN JOINTS AND THEIR REPLACEMENTS BASED ON QUASI-STATIC AND DYNAMIC MULTIBODY CONTACT PROBLEMS IN VISCOELASTIC RHEOLOGIES

6.5.1 Introduction

Recent trends in total joint replacement and clinical failures have provided a strong impetus for the development of accurate stress analyses in normal and prosthetic joints. The success of artificial replacements of human joints depends on many factors. The mechanical factor is an important one. Moreover, the bone tissue has properties of viscoelastic materials. Single parts of the human joint system as well as its replacements represent a system, where its parts are in mutual contact. Similarly, the loosened total joint replacements represent also the system, where its parts, including loosened parts of the total joint replacement, are in mutual contact.

Mechanical loading, static and especially dynamic loadings, is presupposed to play an important role in the origin, development, and progression of joint osteoarthritis. Besides the loading (i.e., the contact pressures) also motion (i.e., kinematics) contributes to an interactive effect of progression of joint osteoarthritis. The idea of a prosthesis being a device that transfers the joint loads to the bone allows one to explain the mechanical factor in terms of the load transfer mechanism. A complex relation exists between this mechanism and the magnitude and direction of the loads, the geometry of the bone/joint–prosthesis configuration, the elastic properties of the materials, and the physical connections at the material connections, where motion and loads influence wear. Therefore, knowledge of in vivo joint motion and loading during functional activities is necessary to comprehend knowledge of human joint degeneration and restoration.

Optimal division of loads acting in the joint components after implantation of total joint replacement is of primary importance in orthopedic surgery. For example, the pressure ratios in the knee joint after total knee replacement (TKR), the soft tissue tension (i.e., capsules of joints, ligaments, muscular insertions) in the vicinity of the replacement and the resulting axial position of the whole limb are determined for optimal distribution of stresses in the replacement knee joint. These factors are influenced especially by the technique of implantation, and, in a decisive way, it determines the survival time of the implant. Nonobservance of the balance of both compartments (medial and lateral) or possible overloading of the posterior part of the tibia plate leads to wear of the polyethylene inlay. Asymmetrical overloading leads to the premature abrasion of a plastic chirulen insert with production of a great amount of polyethylene elements, which instigate a complicated inflammatory reaction leading to loosening of metallic components of the total replacement from the bone. Therefore, the optimal navigated surgery technique must be based not only on the kinematic factors of the joint but on the optimal distribution of the stresses inside the whole joint system, on the contact stresses on the contact surfaces, and mainly on the optimal distribution of the stresses inside the joint system during static and dynamic loading. These problems will be discussed in this and the next chapter of the book. From the orthopedic point of view, total joint replacements are studied by Périé and Hobartho (1998), Hungerford (1995), and Sparmann et al. (2003).

It is evident that complex relations exist between this mechanism, the magnitude and direction of the loads, the geometry of the bone/joint–prosthesis configuration, the elastic properties of the materials, and the physical connections at the material connections. The contact problems in suitable rheology and their finite element approximations are very useful tools for analyzing these relations for joints and their artificial replacements (Nedoma, 1993b; Nedoma et al., 1999a,b, 2000, 2003a,b, 2006; Rojek et al., 2001; Daněk et al., 2004). Therefore, the aim of the book is to analyze these problems and to give the optimal methods and algorithms not only for the analyses of human joints and their replacements but also to give optimal methods and algorithms that can be applied for their use in the navigated orthopedic surgery. The last goal is very difficult because of the very quick convergence of derived methods and algorithms.

Therefore, this study proposes a computationally efficient methodology for combining multibody quasi-static and dynamic simulation methods with a deformable contact in human joint models.

Formulation of the Problem to Be Investigated Here we will consider a dynamic multibody contact problem with Coulomb friction in linear viscoelasticity with a short memory and damping in N dimensions formulated in velocities.

Let $I=(0,t_p)$ be the time interval. Let $\Omega \subset \mathbb{R}^N$, $N=2,3$, be the region occupied by a system of bodies of arbitrary shapes Ω^ι such that $\Omega = \cup_{\iota=1}^r \Omega^\iota$. Let Ω^ι be bounded domains with Lipschitz boundaries $\partial\Omega^\iota$ and let us assume that $\partial\Omega = \Gamma_\tau \cup \Gamma_u \cup \Gamma_c$, where the disjoint parts Γ_τ, Γ_u, Γ_c are open subsets. Moreover, let $\Gamma_u = {}^1\Gamma_u \cup {}^2\Gamma_u$ and $\Gamma_c = \cup_{k,l}\Gamma_c^{kl}$, $\Gamma_c^{kl} = \partial\Omega^k \cap \partial\Omega^l$, $k \neq l$, $k,l \in \{1,\ldots,r\}$. Let $Q = I \times \Omega$ denote the time–space region and let $\Gamma_\tau(t)=\Gamma_\tau \times I$, $\Gamma_u(t)=\Gamma_u \times I$, $\Gamma_c(t)=\Gamma_c \times I$ denote the parts of its boundary $\partial Q = \partial\Omega \times I$.

Furthermore, let $\mathbf{n}$ denote the outer normal vector of the boundary, $u_n = u_i n_i$, $\mathbf{u}_t = \mathbf{u} - u_n\mathbf{n}$, $\tau_n = \tau_{ij}n_j n_i$, $\boldsymbol{\tau}_t = \boldsymbol{\tau} - \tau_n\mathbf{n}$ be normal and tangential components of displacement and stress vectors $\mathbf{u}=(u_i)$, $\boldsymbol{\tau}=(\tau_i)$, $\tau_i = \tau_{ij}n_j$, $i,j=1,\ldots,N$. Let $\mathbf{F},\mathbf{P}$ be the body and surface forces, ρ the density, $\alpha(\mathbf{x}) \geq 0$ physically have a meaning of damping. Let us denote by $\mathbf{u}'=(u_k')$ the velocity vector. Let on Γ_c^{kl} the positive direction of the outer normal vector $\mathbf{n}$ be assumed with respect to Ω^k. Let us denote by $[u_n']^{kl} = u_n'^k - u_n'^l$ the jump of the normal velocity across the contact zone between neighboring bodies Ω^k and Ω^l. Similarly, we denote $[\mathbf{u}_t']^{kl} \equiv \mathbf{u}_t'^k - \mathbf{u}_t'^l$. If one of the colliding bodies say Ω^l is absolutely rigid, then $[u_n']^{kl} \equiv u_n'^k$ and $[\mathbf{u}_t']^{kl} \equiv \mathbf{u}_t'^k$.

The stress–strain relation will be defined by Hooke's law

$$\begin{aligned} \tau_{ij} &= \tau_{ij}(\mathbf{u},\mathbf{u}') = c_{ijkl}^{(0)}(\mathbf{x})e_{kl}(\mathbf{u}) + c_{ijkl}^{(1)}(\mathbf{x})e_{kl}(\mathbf{u}'), \\ e_{ij}(\mathbf{u}) &= \frac{1}{2}\left(\frac{\partial u_i}{\partial x_j} + \frac{\partial u_j}{\partial x_i}\right), \qquad i,j,k,l = 1,\ldots,N, \end{aligned} \tag{6.236}$$

where $c_{ijkl}^{(n)}(\mathbf{x})$, $n=0,1$, are elastic and viscous coefficients, respectively, and $e_{ij}(\mathbf{u})$ are components of the small strain tensor. For the tensors $c_{ijkl}^{(n)}(\mathbf{x})$, $n=0,1$, we assume

$$\begin{aligned} &c_{ijkl}^{(n)} \in L^{\infty}(\Omega), \qquad n=0,1, \qquad c_{ijkl}^{(n)} = c_{jikl}^{(n)} = c_{klij}^{(n)} = c_{ijlk}^{(n)}, \\ &c_{ijkl}^{(n)} e_{ij} e_{kl} \geq c_0^{(n)} e_{ij} e_{ij} \quad \forall e_{ij},\ e_{ij}=e_{ji} \quad \text{and} \quad \text{a.e.}\ \mathbf{x} \in \Omega,\ c_0^{(n)} > 0. \end{aligned} \tag{6.237}$$

A repeated index implies the summation from 1 to N.

The dynamic contact problems with Coulombian friction, where the nonpenetration conditions are formulated in displacements, are up-to-date unsolved in general. Therefore, for these mathematical analyses conditions will be formulated in velocities, that is, we will assume the following contact conditions:

$$[u_n']^{kl} \leq 0, \qquad \tau_n^k = \tau_n^l \equiv \tau_n^{kl} \leq 0, \qquad [u_n']^{kl} \tau_n^{kl} = 0 \quad \text{on } \cup \Gamma_c^{kl}. \tag{6.238}$$

If one of the colliding bodies is absolutely rigid, say Ω^l, then $[u_n']^{kl} \equiv u_n'^k \equiv u_n'$ and the Signorini conditions are as follows:

$$u_n' \leq 0, \qquad \tau_n \leq 0, \qquad u_n' \tau_n = 0 \quad \text{on } \Gamma_c^k.$$

Remark 6.16 Conditions (6.238) can be interpreted as first-order approximations (with respect to the time variable) of the original contact conditions. They are physically realistic for a short time interval only.

The problem to be solved has the following classical formulation:

Problem $\mathcal{P}$ Let $N=2,3$, $s \geq 2$. Find a vector function $\mathbf{u} : \overline{\Omega} \times I \to \mathbb{R}$, satisfying

$$\begin{aligned} &\rho^{\iota} \frac{\partial^2 u_i^{\iota}}{\partial t^2} + \alpha^{\iota} \frac{\partial u_i^{\iota}}{\partial t} = \frac{\partial \tau_{ij}^{\iota}(\mathbf{u}^{\iota}, \mathbf{u}^{\iota\prime})}{\partial x_j} + F_i^{\iota}, \\ &\qquad i,j=1,\ldots,N,\ \iota=1,\ldots,r, (t,\mathbf{x}) \in Q^{\iota} = I \times \Omega^{\iota}, \end{aligned} \tag{6.239}$$

$$\begin{aligned} &\tau_{ij}^{\iota} = \tau_{ij}^{\iota}(\mathbf{u}^{\iota}, \mathbf{u}'^{\iota}) = c_{ijkl}^{(0)\iota}(\mathbf{x}) e_{kl}(\mathbf{u}^{\iota}) + c_{ijkl}^{(1)\iota}(\mathbf{x}) e_{kl}(\mathbf{u}'^{\iota}), \\ &\qquad i,j,k,l=1,\ldots,N,\ \iota=1,\ldots,r, \end{aligned} \tag{6.240}$$

$$\begin{aligned} &\tau_{ij} n_j = P_i, \qquad i,j=1,\ldots,N, \\ &\qquad (t,\mathbf{x}) \in \Gamma_{\tau}(t) = I \times \cup_{\iota=1}^{r} (\Gamma_{\tau} \cap \partial\Omega^{\iota}), \end{aligned} \tag{6.241}$$

$$\begin{aligned} &u_i = u_{2i}, \qquad i=1,\ldots,N, \\ &\qquad (t,\mathbf{x}) \in {}^1\Gamma_u(t) = I \times \cup_{\iota=1}^{r} ({}^1\Gamma_u \cap \partial\Omega^{\iota}), \end{aligned} \tag{6.242}$$

$$u_i = 0, \qquad i=1,\ldots,N, \quad \text{on } {}^2\Gamma_u(t) = I \times \cup_{\iota=1}^{r} ({}^2\Gamma_u \cap \partial\Omega^{\iota}), \tag{6.243}$$

and one of the following conditions:

(a) The nonpenetration contact conditions in velocities:

$$[u_n']^{kl} \le 0, \qquad \tau_n^k = \tau_n^l \equiv \tau_n^{kl} \le 0, \qquad [u_n']^{kl}\tau_n^{kl} = 0; \tag{6.244a}$$

(b) the Signorini conditions for the case with a rigid body Ω^l:

$$u_n'^k \le 0, \qquad \tau_n^k \le 0, \qquad u_n'^k \tau_n^k = 0; \tag{6.244b}$$

and

(c) the Coulombian law of friction,

$$\begin{aligned} [\mathbf{u}_t']^{kl} &= 0 \Longrightarrow |\boldsymbol{\tau}_t^{kl}| \le \mathcal{F}_c^{kl}(0)|\tau_n^{kl}|, \\ [\mathbf{u}_t']^{kl} &\ne 0 \Longrightarrow \boldsymbol{\tau}_t^{kl} = -\mathcal{F}_c^{kl}([\mathbf{u}_t']^{kl})|\tau_n^{kl}|\frac{[\mathbf{u}_t']^{kl}}{|[\mathbf{u}_t']^{kl}|}, \end{aligned} \tag{6.244c}$$

and the initial conditions

$$\mathbf{u}(\mathbf{x},0) = \mathbf{u}_0(\mathbf{x}), \qquad \mathbf{u}'(\mathbf{x},0) = \mathbf{u}_1(\mathbf{x}), \qquad \mathbf{x} \in \Omega, \tag{6.245}$$

where $\mathbf{u}_0$, $\mathbf{u}_1$, $\mathbf{u}_2$ are the given functions, we assume that $\mathbf{u}_2$ has a time derivative $\mathbf{u}_2'$, and $\mathbf{u}_0$, $\mathbf{u}_1$ satisfy the static linear contact problems in elasticity with or without Coulombian friction $\mathcal{F}_c^{kl}$ [see Nedoma (1983, 1987, 1998a)] or can be equal to zero.

The coefficient of friction $\mathcal{F}_c^{kl} \equiv \mathcal{F}_c^{kl}(\mathbf{x}, \mathbf{u}')$ is globally bounded, nonnegative, and satisfies the Carathéodory property, that is, $\mathcal{F}_c^{kl}(\cdot, \mathbf{v})$ is measurable for any $\mathbf{v} \in \mathbb{R}$, and $\mathcal{F}_c^{kl}(\mathbf{x}, \cdot)$ is continuous for a.e. $\mathbf{x} \in \Gamma_c^{kl}$. Moreover, it has a compact support $\mathcal{SF}_c$ defined in Section 6.3, and it depends also on the tangential displacement rate component $\mathbf{u}_t'$ (see Fig. 6.3).

6.5.2 Weak Solutions of Problems in Viscoelastic Rheology with Short Memory

Since the dynamic contact problems with Coulombian friction formulated in displacements are up to now unsolved in general, we will formulate the nonpenetration conditions in velocities.

Variational Formulation of the Problem and the Penalty Approximation

The existence of the solution will be proved under the following assumptions:

Presumption 6.1 The given input data satisfy $\rho \in C(\Omega)$, $\mathbf{F}$, $\mathbf{F}' \in L^2(I; L^{2,N}(\Omega))$ $\mathbf{P}$, $\mathbf{P}' \in L^2(I; L^{2,N}(\Gamma_\tau))$, $\mathbf{u}_1 \in \prod_{\iota=1}^r H^{1,N}(\Omega^\iota)$, $u_{1n}^s - u_{1n}^m = 0$ on $\cup_{s,m}\Gamma_c^{sm}$, $\mathbf{u}_0 \in \prod_{\iota=1}^r H^{1,N}(\Omega^\iota)$, and $\mathbf{u}_2' \in L^2\left(I; \prod_{\iota=1}^r H^{1,N}(\Omega^\iota)\right)$ such that trace $\mathbf{u}_2'|_{^1\Gamma_u} = \mathbf{u}_2'$ on $I \times {}^1\Gamma_u$ and $\mathbf{u}_2'|_{^2\Gamma_u} = 0$ on $I \times {}^2\Gamma_u$, the coefficient of friction $\mathcal{F}_c^{sm} \equiv \mathcal{F}_c^{sm}(\mathbf{x}, \mathbf{u}_t')$ is assumed to be bounded, nonnegative, and satisfies the Carathéodory property. Let n be an integer; then $k = t_p/n$ is a time subinterval.

Contact Problem $\mathcal{P}$ has a weak formulation in terms of a variational inequality. Let us introduce the set of virtual displacements and the set of admissible displacements by

$$V_0 = \left\{ \mathbf{v} | \mathbf{v} \in \prod_{\iota=1}^{r} H^{1,N}(\Omega^\iota),\ \mathbf{v} = 0 \quad \text{on } \Gamma_u = {}^1\Gamma_u \cup {}^2\Gamma_u \right\},$$

$$\mathcal{V}_0 = \left\{ \mathbf{v} | \mathbf{v} \in L^2\left(I; \prod_{\iota=1}^{r} H^{1,N}(\Omega^\iota) \right),\ \mathbf{v} = 0 \quad \text{on } \Gamma_u(t), \quad \text{a.e. on } I \times \cup_{s,m} \Gamma_c^{sm} \right\}$$
$$= L^2(I; V_0),$$
$$V = \mathbf{u}_2 + V_0, \mathcal{V} = \mathbf{u}_2' + \mathcal{V}_0 = L^2(I; V),$$
$$\mathcal{K} = \left\{ \mathbf{v} | \mathbf{v} \in L^2\left(I; \prod_{\iota=1}^{s} H^{1,N}(\Omega^\iota) \right),\ \mathbf{v} = \mathbf{u}_2' \text{ on } \Gamma_u(t), v_n^s - v_n^m \le 0 \right.$$
$$\left. \text{a.e. on } I \times \cup_{s,m} \Gamma_c^{sm} \right\}.$$

Multiplying (6.239) by $\mathbf{v} - \mathbf{u}'$, where $\mathbf{v}$ are suitable test functions, integrating over $\Omega \times I$, using Green's theorems and boundary and contact conditions, we obtain the following variational problem:

Problem $(\mathcal{P})_v$ Find a vector function $\mathbf{u} \in \mathcal{K}$ with $\mathbf{u}' \in \mathcal{K} \cap B(I; L^{2,N}(\Omega))$ and $\mathbf{u}(0, \cdot) = \mathbf{u}_0$, $\mathbf{u}'(0, \cdot) = \mathbf{u}_1$, such that for all $t \in I$

$$\int_I \{(\mathbf{u}''(t), \mathbf{v} - \mathbf{u}'(t)) + (\alpha \mathbf{u}'(t), \mathbf{v} - \mathbf{u}'(t)) + a^{(0)}(\mathbf{u}(t), \mathbf{v} - \mathbf{u}'(t))$$
$$+ a^{(1)}(\mathbf{u}'(t), \mathbf{v} - \mathbf{u}'(t)) + j(\mathbf{v}) - j(\mathbf{u}'(t))\}\, dt$$
$$\ge \int_I (\mathbf{f}(t), \mathbf{v} - \mathbf{u}'(t))\, dt \quad \forall \mathbf{v} \in \mathcal{K}. \tag{6.246}$$

Here we assume that the initial data $\mathbf{u}_0$, $\mathbf{u}_1$ satisfy the static contact multibody linear elastic problem, and where

$$(\mathbf{u}'', \mathbf{v}) = \sum_{\iota=1}^{r} (\mathbf{u}''^\iota, \mathbf{v}^\iota) = \int_\Omega \rho u_i'' v_i \, d\mathbf{x},$$

$$(\alpha \mathbf{u}', \mathbf{v}) = \sum_{\iota=1}^{r} (\alpha^\iota \mathbf{u}'^\iota, \mathbf{v}^\iota) = \int_\Omega \alpha u_i' v_i \, d\mathbf{x},$$

$$a^{(n)}(\mathbf{u}, \mathbf{v}) = \sum_{\iota=1}^{r} a^{(n)\iota}(\mathbf{u}^\iota, \mathbf{v}^\iota) = \int_\Omega c_{ijkl}^{(n)} e_{kl}(\mathbf{u}) e_{ij}(\mathbf{v}) \, d\mathbf{x}, \qquad n = 0, 1,$$

$$(\mathbf{f}, \mathbf{v}) = \sum_{\iota=1}^{r} (\mathbf{f}^\iota, \mathbf{v}^\iota) = \int_\Omega F_i v_i \, d\mathbf{x} + \int_{\Gamma_\tau} P_i v_i \, ds,$$

$$j(\mathbf{v}) = \sum_{\iota=1}^{r} j^\iota(\mathbf{v}^\iota) = \int_{\cup_{s,m} \Gamma_c^{sm}} \mathcal{F}_c^{sm}(\mathbf{u}_t'^s - \mathbf{u}_t'^m) |\tau_n^{sm}(\mathbf{u}, \mathbf{u}')| |\mathbf{v}_t^s - \mathbf{v}_t^m| \, ds$$
$$\equiv \langle \mathcal{F}_c^{sm}(\mathbf{u}_t'^s - \mathbf{u}_t'^m) |\tau_n^{sm}(\mathbf{u}, \mathbf{u}')|, |\mathbf{v}_t^s - \mathbf{v}_t^m| \rangle_{\Gamma_c},$$

where the bilinear forms $a^{(n)\iota}(\mathbf{u}^\iota, \mathbf{v}^\iota)$, $n=0,1$, are symmetric in $\mathbf{u}^\iota$, $\mathbf{v}^\iota$ and satisfy $a^{(n)\iota}(\mathbf{u}^\iota, \mathbf{u}^\iota) \geq c_0^{(n)} \|\mathbf{u}^\iota\|_{1,N}^2$, $c_0^{(n)} = \text{const} > 0$, $|a^{(n)\iota}(\mathbf{u}^\iota, \mathbf{v}^\iota)| \leq c_1^{(n)} \|\mathbf{u}^\iota\|_{1,N} \|\mathbf{v}^\iota\|_{1,N}$, $c_1^{(n)} = \text{const} > 0$, $\mathbf{u}, \mathbf{v} \in V_0$ and where $B(I; L^{2,N}(\Omega))$ is the space from I into $L^{2,N}(\Omega)$ with the norm $\|\mathbf{v}\| = \|\mathbf{v}\|_{L^\infty(I;L^{2,N}(\Omega))} = \sup_{t\in I} \|\mathbf{v}(t)\|_0$ [see Adams (1975)].

It can be shown that every solution $\mathbf{u} \in C^{2,N}(I \times \overline{\Omega})$ of (6.239)–(6.244a,c) and (6.245) is a solution of (6.246) and that every sufficiently smooth solution $\mathbf{u} \in C^{2,N}(I \times \overline{\Omega})$ of (6.246) is also a solution of (6.239)–(6.244a,c) and (6.245).

As a result we have the following theorem:

Theorem 6.44 Let the assumptions concerning Ω, Γ_τ, Γ_u, Γ_c and physical data given in Presumption 6.1 be satisfied. Then there exists at least one weak solution of the dynamic contact problem with friction and damping in viscoelasticity with a short memory.

Proof The proof of the existence of the solution is based on the penalization and regularization techniques. In the first step, the penalty method will be used. This leads to a contact problem of the normal compliance type. The contact condition (6.244a) will be replaced by the nonlinear boundary condition putting $\tau_n(\mathbf{u}, \mathbf{u}') = -(1/\delta)[u_n'^s - u_n'^m]_+$ with $[\cdot]_+ := \max\{\cdot, 0\}$, $\delta > 0$. The similar problem was analyzed in Nedoma (2009). □

Remark 6.17 It can be shown similarly as in Duvaut and Lions (1976) that the contact problem in the classical linear elasticity is the limit case of the problem in the linear viscoelasticity with short memory, where $\mathbf{u}_\lambda(t)$ and $\mathbf{u}(t)$ are the solutions of the investigated Problem $(\mathcal{P})_v$ with

$$\tau_{ij}(t) = c_{ijkl}^{(0)} e_{kl}(\mathbf{u}_\lambda(t)) + \lambda c_{ijkl}^{(1)} e_{kl}(\mathbf{u}_\lambda(t)), \qquad \lambda \to 0, \tag{6.247}$$

and/or with

$$\tau_{ij}(t) = c_{ijkl}^{(0)} e_{kl}(\mathbf{u}(t)), \tag{6.248}$$

respectively, that is, $\mathbf{u}_\lambda \to \mathbf{u}$ in $\mathcal{K}$, if $\lambda \to 0$.

6.5.3 Quasi-Static Contact Model Problem in Biomechanics of Human Joints

In the case when the skeletal part of the human body is assumed to deform very slowly, then the inertia forces in the initial dynamic formulation can be neglected. Moreover, if we assume that the damping is also very small, then the damping term in dynamic formulation can be also neglected. Moreover, we will formulate the problem in displacements because it approximates the reality better. Then we solve the following problem:

Problem $\mathcal{P}$ Let $N = 2, 3$. Find a vector function $\mathbf{u} : \overline{\Omega} \times I \to \mathbb{R}^N$ satisfying

$$\frac{\partial \tau_{ij}^{\iota}}{\partial x_j} + F_i^{\iota} = 0, \qquad i,j = 1,\ldots,N, \qquad \iota = 1,\ldots,r, \qquad (t,\mathbf{x}) \in I \times \Omega^{\iota} = \Omega_t^{\iota}, \tag{6.249}$$

$$\tau_{ij}^{\iota} = \tau_{ij}^{\iota}(\mathbf{u}^{\iota}, \mathbf{u}'^{\iota}) = c_{ijkl}^{(0)\iota}(\mathbf{x}) e_{kl}(\mathbf{u}^{\iota}) + c_{ijkl}^{(1)\iota}(\mathbf{x}) e_{kl}(\mathbf{u}'^{\iota}), \tag{6.250}$$

or

$$\tau_{ij}^{\iota} = \tau_{ij}^{\iota}(\mathbf{u}^{\iota}) = c_{ijkl}^{(0)\iota}(\mathbf{x}) e_{kl}(\mathbf{u}^{\iota}), \qquad (t,\mathbf{x}) \in I \times \Omega^{\iota} = \Omega_t^{\iota},$$

$$\tau_{ij} n_j = P_i, \qquad i,j = 1,\ldots,N, \qquad (t,\mathbf{x}) \in \Gamma_\tau(t) = I \times \cup_{\iota=1}^{r} (\Gamma_\tau \cap \partial\Omega^{\iota}), \tag{6.251}$$

$$u_i = u_{2i}, \qquad i = 1,\ldots,N, \qquad (t,\mathbf{x}) \in {}^{1}\Gamma_u(t) = I \times \cup_{\iota=1}^{r} ({}^{1}\Gamma_u \cap \partial\Omega^{\iota}), \tag{6.252}$$

$$u_i = 0, \qquad i = 1,\ldots,N, \qquad (t,\mathbf{x}) \in {}^{2}\Gamma_u(t) = I \times \cup_{\iota=1}^{r} ({}^{2}\Gamma_u \cap \partial\Omega^{\iota}), \tag{6.253}$$

$$\begin{aligned}
& u_n^s - u_n^m \le d^{sm}, \qquad \tau_n^s = \tau_n^m \equiv \tau_n^{sm}, \qquad (u_n^s - u_n^m - d^{sm})\tau_n^{sm} = 0, \\
& \mathbf{u}_t'^s - \mathbf{u}_t'^m = 0 \Rightarrow |\boldsymbol{\tau}_t^{sm}| \le \mathcal{F}_c^{sm}(0)|\tau_n^{sm}|, \qquad (t,\mathbf{x}) \in \Gamma_c(t) = I \times \cup_{s,m} \Gamma_c^{sm}, \\
& \mathbf{u}_t'^s - \mathbf{u}_t'^m \ne 0 \Rightarrow \boldsymbol{\tau}_t^{sm} = -\mathcal{F}_c^{sm}(\mathbf{u}_t'^s - \mathbf{u}_t'^m)|\tau_n^{sm}| \frac{\mathbf{u}_t'^s - \mathbf{u}_t'^m}{|\mathbf{u}_t'^s - \mathbf{u}_t'^m|},
\end{aligned} \tag{6.254}$$

$$\mathbf{u}(0,\mathbf{x}) = \mathbf{u}_0 \quad \text{for } \mathbf{x} \in \Omega. \tag{6.255}$$

In this formulation the nonpenetration conditions are formulated in displacements, while the Coulombian law is formulated in velocities, that is, the time derivative $\mathbf{u}'$ occurs in the friction law only. Moreover, the initial condition is formulated in displacements only. Let us introduce

$$V_0 = V_0(\Omega) = \left\{ \mathbf{v} \in \prod_{\iota=1}^{r} H^{1,N}(\Omega^{\iota}),\ \mathbf{v} = 0 \text{ on } \Gamma_u \right\} \quad \text{and} \quad \mathcal{V} = \mathbf{u}_2' + V_0,$$

$$\mathcal{K}(t) = \left\{ \mathbf{v} \in \prod_{\iota=1}^{r} H^{1,N}(\Omega^{\iota}), \mathbf{v} = \mathbf{u}_2(t,\cdot) \quad \text{on } \Gamma_u \quad \text{and} \quad v_n^s - v_n^m \le d^{sm} \quad \text{on} \cup \Gamma_c^{sm} \right\}.$$

Let us multiply (6.249) by $\mathbf{w} - \mathbf{u}$, and using the Green theorem and boundary conditions, we arrive at the following variational problem:

Problem $\mathcal{P}$ Find a function $\mathbf{u}(t,\cdot) \in \mathcal{K}(t)$ for a.e. $t \in I$ such that for a.e. $t \in I$ and every $\mathbf{w} \in \mathcal{K}(t)$

$$A(\mathbf{u}; \mathbf{u}', \mathbf{v} - \mathbf{u}') + \int_{\cup\Gamma_c^{sm}} \mathcal{F}_c^{sm} |\tau_n^{sm}(\mathbf{u}, \mathbf{u}')| (|[\mathbf{v}_t]^{sm}| - |[\mathbf{u}_t']^{sm}|)\, ds \ge (\mathbf{f}, \mathbf{v} - \mathbf{u}') \tag{6.256}$$

holds, where

$$A(\mathbf{u};\mathbf{u}',\mathbf{w}) = \int_\Omega \tau_{ij}(\mathbf{u},\mathbf{u}')e_{ij}(\mathbf{w})\,d\mathbf{x},$$

$$(\mathbf{f},\mathbf{w}) = \int_\Omega \mathbf{F}\cdot\mathbf{w}\,d\mathbf{x} + \int_{\Gamma_\tau} \mathbf{P}\cdot\mathbf{w}\,ds \quad \text{and} \quad [\mathbf{v}_t]^{sm} \equiv \mathbf{v}_t^s - \mathbf{v}_t^m.$$

We used the decomposition

$$\mathbf{w} - \mathbf{u} = (\mathbf{w} - \mathbf{u} + \mathbf{u}') - \mathbf{u}' = \mathbf{v} - \mathbf{u}'.$$

The existence of proof for such a problem was given by Andersson (2000) and Eck et al. (2005). The proof is based on the Rothe method. The frictionless case in viscoelasticity formulated in displacements is analyzed in Han and Sofonea (2002). The analysis of solvability is based on the theory of monotone operators. Numerical approximations can be based on the semidiscrete and the fully discrete techniques as well as the Rothe method.

6.5.4 Dynamic Contact Problem Formulated in Displacements and Its Numerical Solution

In this section we will deal with a new methodology for simulating deformable contact in the human joints and their total replacements within multibody dynamics, based on the analyses of the quasi-static and dynamic contact multibody problems in (visco)elastic rheology.

Next, we will solve the problem (6.239)–(6.243), (6.244c), and (6.245), and with

$$[u_n]^{sm} \leq d^{sm}, \qquad \tau_n^{sm} \leq 0, \qquad ([u_n]^{sm} - d^{sm})\tau_n^{sm} = 0. \tag{6.257}$$

Variational (Weak) Solution of the Model Problem The dynamic problem with damping and Coulombian friction (Problem $\mathcal{P}$), where the nonpenetration conditions are formulated in velocities is analyzed in Nedoma (2005, 2010). The dynamic contact problem with Coulombian friction, where the nonpenetration conditions are formulated in displacements is up to now unsolved in general. In what follows, we will solve another problem in which at every time level we will solve the dynamic contact problem with the Tresca model of friction, that is, we will assume that the Coulombian law of friction depends at every time level on its value g_c^{sm} from the previous time level(s). For simplicity, we can write $g_c^{sm} \equiv (\mathcal{F}_c^{sm}(\mathbf{u}_t'^s - \mathbf{u}_t'^m)|\,\tau_n^{sm}(\mathbf{u},\mathbf{u}')|)(t - \delta t)$. As a consequence g_c^{sm} is a nonnegative function, and it has a meaning of a given friction limit (or a given friction bound, i.e., the magnitude of the limiting friction traction at which the slip originates) and δt is a time step. Then such biomechanical models are called Tresca-type models. In this section the problem will be solved by using the semi-implicit scheme in time and the FEM method in space. The other possibilities are represented by the explicit and fully implicit schemes in time and the FEM method in space or by the Rothe method.

Multiplying (6.239) by $\mathbf{v}-\mathbf{u}$, integrating the result over Ω, using the Green theorem, the boundary conditions, and contact conditions in displacements, we obtain:

Problem $(\mathcal{P})_v$ Find a displacement field $\mathbf{u}:\bar{I}\to V$ such that $\mathbf{u}(t)\in K$ for a.e. $t\in I$, and

$$(\mathbf{u}''(t),\mathbf{v}-\mathbf{u}(t))+(\alpha\mathbf{u}'(t),\mathbf{v}-\mathbf{u}(t))+a^{(0)}(\mathbf{u}(t),\mathbf{v}-\mathbf{u}(t))$$
$$+a^{(1)}(\mathbf{u}'(t),\mathbf{v}-\mathbf{u}(t))+j(\mathbf{v})-j(\mathbf{u}(t))\geq(\mathbf{f}(t),\mathbf{v}-\mathbf{u}(t))$$
$$\forall\mathbf{v}\in K,\ \text{a.e. } t\in I, \tag{6.258}$$

$$\mathbf{u}(\mathbf{x},0)=\mathbf{u}_0(\mathbf{x}),\mathbf{u}'(\mathbf{x},0)=\mathbf{u}_1(\mathbf{x}), \tag{6.259}$$

where

$$V_0=\{\mathbf{v}|\mathbf{v}\in\prod_{\iota=1}^{r}H^{1,N}(\Omega^\iota),\mathbf{v}=0 \text{ on } \Gamma_u={}^1\Gamma_u\cup{}^2\Gamma_u\},\qquad V=\mathbf{u}_2+V_0,$$
$$K=\{\mathbf{v}|\mathbf{v}\in\prod_{\iota=1}^{r}H^{1,N}(\Omega^\iota),\mathbf{v}=\mathbf{u}_2 \text{ on } \Gamma_u, v_n^s-v_n^m\leq 0 \text{ a.e. on } I\times\cup_{s,m}\Gamma_c^{sm}\},$$
$$\mathcal{K}=L^2(I;K),$$

and where we assume that the initial data $\mathbf{u}_0$, $\mathbf{u}_1$ satisfy the static contact multibody linear elastic problem [see, e.g., Nedoma (1983, 1987) and Hlaváček and Nedoma (2002a)], where

$$(\mathbf{u}'',\mathbf{v})=\sum_{\iota=1}^{r}(\mathbf{u}''^\iota,\mathbf{v}^\iota)=\int_\Omega\rho u_i''v_i\,d\mathbf{x},$$

$$(\alpha\mathbf{u}',\mathbf{v})=\sum_{\iota=1}^{r}(\alpha^\iota\mathbf{u}'^\iota,\mathbf{v}^\iota)=\int_\Omega\alpha u_i'v_i\,d\mathbf{x},$$

$$a^{(n)}(\mathbf{u},\mathbf{v})=\sum_{\iota=1}^{r}a^{(n)\iota}(\mathbf{u}^\iota,\mathbf{v}^\iota)=\int_\Omega c_{ijkl}^{(n)}e_{kl}(\mathbf{u})e_{ij}(\mathbf{v})\,d\mathbf{x},\qquad n=0,1,$$

$$(\mathbf{f},\mathbf{v})=\sum_{\iota=1}^{r}(\mathbf{f}^\iota,\mathbf{v}^\iota)=\int_\Omega F_iv_i\,d\mathbf{x}+\int_{\Gamma_\tau}P_iv_i\,ds,$$

$$j(\mathbf{v})=\int_{\cup_{k,l}\Gamma_c^{kl}}g_c^{kl}|\mathbf{v}_t^k-\mathbf{v}_t^l|\,ds\equiv\langle g_c^{kl}|\mathbf{v}_t^k-\mathbf{v}_t^l|\rangle_{\cup\Gamma_c^{kl}},$$

where the bilinear forms $a^{(n)\iota}(\mathbf{u}^\iota,\mathbf{v}^\iota)$, $n=0,1$, are symmetric in $\mathbf{u}^\iota,\mathbf{v}^\iota$ and satisfy $a^{(n)\iota}(\mathbf{u}^\iota,\mathbf{u}^\iota)\geq c_0^{(n)}\|\mathbf{u}^\iota\|_{1,N}^2$, $c_0^{(n)}=\text{const}>0$, $a^{(n)\iota}(\mathbf{u}^\iota,\mathbf{v}^\iota)\leq c_1^{(n)}\|\mathbf{u}^\iota\|_{1,N}\|\mathbf{v}^\iota\|_{1,N}$, $c_1^{(n)}=\text{const}>0$, $\mathbf{u},\mathbf{v}\in V_0$.

Problem $(\mathcal{P})_v$ is equivalent to the following weak formulation:

Problem $(\mathcal{P})_W$ A weak solution of (6.239)–(6.245) with (6.257) instead of (6.244a) is a function $\mathbf{u}\in B(I;\prod_{\iota=1}^{r}H^{1,N}(\Omega^\iota))$ with $\mathbf{u}(t,\cdot)\in\mathcal{K}$, for a.e. $t\in I$,

$\mathbf{u}' \in L^2(I; \prod_{\iota=1}^r H^{1,N}(\Omega^\iota)) \cap L^\infty(I; \prod_{\iota=1}^r L^{2,N}(\Omega^\iota))$, $\mathbf{u}'(t_p, \cdot) \in L^{2,N}(\Omega)$ such that for all $\mathbf{v} \in \prod_{\iota=1}^r H^{1,N}(\Omega^\iota(t))$ with $\mathbf{v}(t, \cdot) \in \mathcal{K}$ a.e. in I the following inequality holds

$$\int_I \{(\mathbf{u}''(t), \mathbf{v} - \mathbf{u}(t)) + (\alpha\mathbf{u}'(t), \mathbf{v} - \mathbf{u}(t)) + a^{(0)}(\mathbf{u}(t), \mathbf{v} - \mathbf{u}(t))$$

$$+ a^{(1)}(\mathbf{u}'(t), \mathbf{v} - \mathbf{u}(t)) + j(\mathbf{v}) - j(\mathbf{u}(t))\}\, dt \geq \int_I (\mathbf{f}(t), \mathbf{v} - \mathbf{u}(t))\, dt,$$

$$\mathbf{u}(\mathbf{x}, 0) = \mathbf{u}_0(\mathbf{x}), \qquad \mathbf{u}'(\mathbf{x}, 0) = \mathbf{u}_1(\mathbf{x}), \tag{6.260}$$

where $B(I; X)$, X being a Banach space, denotes the space of all continuous and bounded functions from I into X with subnorm $\sup_{t \in I} \|\mathbf{u}(t)\|_X$, $X = \prod_{\iota=1}^r H^{1,N}(\Omega^\iota)$ [see, e.g., Adams (1975)].

To prove the existence of the solution the following decomposition

$$\mathbf{v} - \mathbf{u} = \mathbf{v} - \mathbf{u} + \mathbf{u}' - \mathbf{u}' = \mathbf{w} - \mathbf{u}'$$

or, if the physical units are considered, the decomposition

$$\mathbf{v} - \mathbf{u} = \mathbf{v} - \mathbf{u} + c\mathbf{u}' - c\mathbf{u}' = c\mathbf{w} - c\mathbf{u}',$$

are used, where $c > 0$ is an arbitrary parameter that may be used for transformation of the physical units. Then the proof is similar to that of Eck et al. (2005, Chapter 4). For the proof the techniques of penalization and regularization were used.

Numerical Solution Let Ω be approximated by Ω_h, being a polygon for the 2D case and a polyhedron for the 3D case, with the boundary $\partial\Omega_h = \Gamma_{\tau h} \cup \Gamma_{uh} \cup \Gamma_{ch}$. Let $I = (0, t_p)$, $t_p > 0$, let $m > 0$ be an integer, then $\Delta t = t_p/m$, $t_i = i\Delta t$, $i = 0, \ldots, m$. Let $\{\mathcal{T}_h\}$ be a regular family of finite element partitions $\mathcal{T}_h$ of $\overline{\Omega}_h$ compatible with the boundary subsets $\overline{\Gamma}_{\tau h}$, $\overline{\Gamma}_{uh}$, and $\overline{\Gamma}_{ch}$. Let $V_h \subset V$ be the finite element space of linear elements corresponding to the partition $\mathcal{T}_h$. Then $K_h = V_h \cap K$ is the set of continuous piecewise linear functions. It is evident that K_h is a nonempty, closed, convex subset of $V_h \subset V$. Let $\mathbf{u}_{0h} \in K_h$, $\mathbf{u}_{1k} \in K_h$ be approximations of $\mathbf{u}_0$ or $\mathbf{u}_1$. Let us assume that $\mathbf{u}_2 = 0$. Further, we assume that the end points $\overline{\Gamma}_{\tau h} \cup \overline{\Gamma}_{uh}$, $\overline{\Gamma}_{uh} \cup \overline{\Gamma}_{ch}$, and $\overline{\Gamma}_{\tau h} \cup \overline{\Gamma}_{ch}$ coincide with the vertices of T_{hi}. Then we have the following semidiscrete problem:

Problem $(\mathcal{P})_h$ Find a displacement field $\mathbf{u}_h : \overline{I} \to V_h$ with

$$\mathbf{u}_h(0) = \mathbf{u}_{0h}, \qquad \mathbf{u}_h'(0) = \mathbf{u}_{1h}, \tag{6.261}$$

such that for a.e. $t \in I$, $\mathbf{u}_h(t) \in K_h$ and

$$(\mathbf{u}_h''(t), \mathbf{v}_h - \mathbf{u}_h(t)) + (\alpha\mathbf{u}_h'(t), \mathbf{v}_h - \mathbf{u}_h(t)) + a^{(0)}(\mathbf{u}_h(t), \mathbf{v}_h - \mathbf{u}_h(t))$$

$$+ a^{(1)}(\mathbf{u}_h'(t), \mathbf{v}_h - \mathbf{u}_h(t)) + j(\mathbf{v}_h) - j(\mathbf{u}_h(t)) \geq (\mathbf{f}(t), \mathbf{v}_h - \mathbf{u}_h(t))$$

$$\forall \mathbf{v}_h \in K_h, \tag{6.262}$$

holds.

To prove the existence of the discrete (FEM) solution $\mathbf{u}_h$ of Problem $(\mathcal{P})_h$ the decomposition

$$\mathbf{v}_h - \mathbf{u}_h = \mathbf{v}_h - \mathbf{u}_h + \mathbf{u}'_h - \mathbf{u}'_h = \mathbf{w}_h - \mathbf{u}'_h$$

and the penalty and regularization techniques are used. Then the proof will be similar to that in the continuous case [see Eck et al. (2005)]. We remark that the test function $\mathbf{v}$ will correspond with the test function $\mathbf{v}+\mathbf{u}'-\mathbf{u}$ in (6.258) and $\mathbf{v}_h$ with the test function $\mathbf{v}_h+\mathbf{u}'_h-\mathbf{u}_h$ in (6.262) after the used decompositions.

6.6 ALGORITHMS

6.6.1 Consistency, Stability, and the Lax Condition

The requirement of the algorithms discussed in this section is that they converge and they are stable. We will assume that problems studied in the book are well-posed. The problems are well-posed if for small changes in the data they will make only small changes in the solution. This is essential in engineering problems where the data can frequently be measured only very roughly. Therefore, we will analyze these type of problems in the section concerning the worst scenario method in biomechanics of human joint replacements.

We say that an **algorithm is convergent** if for t_i fixed and $\Delta t = t_p/m$, $t_i = i\,\Delta t$, $i=0,\ldots,m$, $\mathbf{u}_h^i \to \mathbf{u}(t_i)$ as $\Delta t \to 0$. An algorithm convergence is connected with consistency, completeness, and stability (Wood, 1990; Bathe, 1982, 1996; Zienkiewicz and Taylor, 2000; Belytschko et al., 2000).

Consistency is defined in the context of the finite approximation methods. A discrete approximation $\mathcal{L}_h(\mathbf{u})$ of the continuous problem $\mathcal{L}(\mathbf{u})$ is consistent if the error is of the order of the mesh size, that is, if

$$\mathcal{L}(\mathbf{u}) - \mathcal{L}_h(\mathbf{u}) = O(h^p), \quad \text{with } p \geq 1.$$

The above states that the truncation error of a consistent discrete approximation must tend to zero as the step of the mesh size tends to zero. For the time-dependent problems, the discretization error will be a function of the time step Δt and the mesh size h, and the truncation error of the time and spatial discretization must tend to zero [see, e.g., Belytschko et al. (2000)]. The condition of consistency is a test for continuity (Oden and Reddy, 1976).

Stability is defined as follows: We say that a discrete procedure is stable if small perturbations in initial data result in small changes in the discrete solution, that is,

$$\|\mathbf{u}_{ha}^i - \mathbf{u}_{hb}^i\| \leq c\varepsilon \quad \forall i > 0 \quad \text{for all } \mathbf{u}_{ha}^0 \text{ such that } \|\mathbf{u}_{ha}^0 - \mathbf{u}_{hb}^0\| \leq c, \qquad c > 0, \tag{6.263}$$

where $\mathbf{u}_{ha}^i$ corresponds to initial data and $\mathbf{u}_{hb}^i$ corresponds to perturbed data. An algorithm that yields a stable discrete solution is said to be stable. It states that for a well-posed problems, a discretization that is stable and consistent is convergent (the **Lax equivalence condition**).

Considering the stability of integration schemes of two procedures, which are unconditionally stable and conditionally stable, can be used. An integration scheme is **unconditionally stable** if the solution for any initial conditions does not grow without bound for any time step Δt, in particular when $\Delta t/T$ is large, $T = 2\pi/\omega$, where ω is a frequency, T is the natural period. The schemes that require the use of a time step Δt smaller than a critical time step Δt_{crit} (such as the central difference scheme) are said to be **conditionally stable**. For these schemes $\Delta t/T$ is smaller than or equal to a certain value, the stability limit.

6.6.2 Approximation of Biomechanic Models Based on the Central Difference Method, an Explicit Scheme

In the present dynamic problems of biomechanics and orthopedics, the explicit scheme (often in a penalized approach) is frequently used because of its relatively simple algorithm. In what follows, we shortly present derivation of an algorithm for the dynamic problem with friction in elasticity.

The algorithm will be based on the explicit scheme and on the assumption that the slip limit functional is regularized and approximated by its value at the previous time level(s). Such a problem leads to the case with regularized slip limit.

Let $m > 0$ be an integer, then $\Delta t = t_p/m$, $t_i = i\,\Delta t$, $i = 0, 1, \ldots, m$. Then Problem $(\mathcal{P})_h$ (6.261) and (6.262) can be discretized by the following scheme:

$$
\begin{aligned}
&(\Delta t^{-2}(\mathbf{u}_h^{i+1} - 2\mathbf{u}_h^i + \mathbf{u}_h^{i-1}), \mathbf{v}_h - \mathbf{u}_h^{i+1}) + ((2\Delta t)^{-1}\alpha(\mathbf{u}_h^{i+1} - \mathbf{u}_h^{i-1}), \mathbf{v}_h - \mathbf{u}_h^{i+1}) \\
&\quad + a^{(0)}(\mathbf{u}_h^i, \mathbf{v}_h - \mathbf{u}_h^{i+1}) + \int_{\bigcup_{s,m}\Gamma_c^{sm}} g_{ch}^{sm} |\mathbf{v}_{th}^s - \mathbf{v}_{th}^m|\, ds \\
&\quad - \int_{\bigcup \Gamma_c^{sm}} g_{ch}^{sm} |\mathbf{u}_{th}^s(t_{i+1}) - \mathbf{u}_{th}^m(t_{i+1})|\, ds \geq (\mathbf{f}(t_{i+1}), \mathbf{v}_h - \mathbf{u}_h^{i+1}) \quad \forall \mathbf{v}_h \in K_h,
\end{aligned}
$$

where the velocity vector $\mathbf{v} = \mathbf{u}'$ and the acceleration vector $\mathbf{a} = \mathbf{v}' = \mathbf{u}''$ were approximated by

$$
\mathbf{v}_h^i = \mathbf{u}_h^{\prime i} = \frac{\mathbf{u}_h^{i+1} - \mathbf{u}_h^{i-1}}{2\Delta t}, \qquad \mathbf{a}_h^i = \mathbf{u}_h^{\prime\prime i} = \frac{\mathbf{u}_h^{i+1} - 2\mathbf{u}_h^i + \mathbf{u}_h^{i-1}}{\Delta t^2},
$$

and where we denoted by

$$
\begin{aligned}
&\mathbf{u}_h^i = \mathbf{u}_h(t_i), \qquad \Delta\mathbf{u}_h^i = \mathbf{u}_h(t_i) - \mathbf{u}_h(t_{i-1}), \qquad \mathbf{u}_h^0 = \mathbf{u}_{0h}, \\
&\mathbf{u}_h^{-1} = \mathbf{u}_{0h} - \Delta t \cdot \mathbf{u}_{1h} + \frac{\Delta t^2}{2}\mathbf{a}_h^0, \\
&g_{ch}^{sm} = g_{ch}^{sm}(t_i) = \mathcal{F}_c^{sm}(\Delta t^{-1}(\Delta\mathbf{u}_{th}^s(t_i) - \Delta\mathbf{u}_{th}^m(t_i))) \left| \tau_n^{sm}\left(\mathbf{u}_h(t_i), \frac{\Delta\mathbf{u}_h(t_i)}{\Delta t}\right)\right|.
\end{aligned}
$$

Under the above-defined assumption we obtain the following scheme: Find $\mathbf{u}_h^{i+1} \in K_h$ such that

$$(\mathbf{u}_h^{i+1}, \mathbf{v}_h - \mathbf{u}_h^{i+1}) + \frac{\Delta t}{2}(\alpha \mathbf{u}_h^{i+1}, \mathbf{v}_h - \mathbf{u}_h^{i+1}) + \Delta t^2 a^{(0)}(\mathbf{u}_h^i, \mathbf{v}_h - \mathbf{u}_h^{i+1})$$
$$\geq (\mathbf{F}(t_{i+1}), \mathbf{v}_h - \mathbf{u}_h^{i+1}), \qquad \forall \mathbf{v}_h \in K_h,\ t = t_{i+1} \in I, \tag{6.264}$$

where

$$(\mathbf{F}(t_{i+1}), \mathbf{v}_h) = \Delta t^2(\mathbf{f}(t_{i+1}), \mathbf{v}_h) + (2\mathbf{u}_h^i - \mathbf{u}_h^{i-1}, \mathbf{v}_h) + \frac{\Delta t}{2}(\alpha \mathbf{u}_h^{i-1}, \mathbf{v}_h)$$
$$- \Delta t^2 \int_{\bigcup_{s,m} \Gamma_c^{sm}} g_{ch}^{sm} \varphi_\delta(\mathbf{v}_{th}^s - \mathbf{v}_{th}^m)\, ds,$$

where for a numerical solution we set, for example, $\varphi_\delta(\lambda) = (1/1+\delta)|\lambda|^{1+\varepsilon}$, $\delta > 0$, that is, $j(\mathbf{v}_h) \sim j_\delta(\mathbf{v}_h)$, where $j_\delta(\mathbf{v}_h) = \int_{\bigcup_{s,m}\Gamma_c^{sm}} g_{ch}^{sm} \varphi_\delta(\mathbf{v}_{th}^s - \mathbf{v}_{th}^m)\, ds$ is a convex regularization of $j(\mathbf{v}_h)$.

To prove the convergence of the algorithm (6.264), we need to prove the stability and the consistency of the scheme. Most of the theory of stability of the numerical method is concerned with linear and linearized problems. If a numerical algorithm is unstable for a linear problem, it will, of course, be unstable for the nonlinear problem because linear problems are a subset of nonlinear problems. Therefore, the stability of numerical procedures for linear problems provides a useful approach in both linear and nonlinear problems. The set K_h is a subset of V_h. Therefore, it is sufficient to investigate the following problem:

Find a displacement field $\mathbf{u}_h : \bar{I} \to V_h$ with

$$\mathbf{u}_h(0) = \mathbf{u}_{0h},\ \mathbf{u}_h'(0) = \mathbf{u}_{1h}$$

and such that for a.e. $t \in I$, $\mathbf{u}_h \in V_h$

$$(\mathbf{u}_h''(t), \mathbf{v}_h) + (\alpha \mathbf{u}_h'(t), \mathbf{v}_h) + a^{(0)}(\mathbf{u}_h(t), \mathbf{v}_h) = (\mathbf{f}(t), \mathbf{v}_h) \quad \forall \mathbf{v}_h \in V_h \tag{6.265}$$

holds.

Let us derive finite element (FE) approximate equations. By the usual way the displacements measured in a local coordinate system x, y, z within each element T_{hi} are assumed to be a function of the displacements at the n finite element nodal points. Since in general the displacement vector includes the displacements and rotations at the supports of the element assemblage, we need to impose the known values of the displacement vector prior to solving for the unknown nodal point displacement, thus in our use $\mathbf{u}_h^{(k)}$ will represent the virtual displacements.

By the usual way (see the Appendix) for the kth element we have

$$\mathbf{u}_h^{(k)}(x, y, z) = H^{(k)}(x, y, z)\mathbf{U}_h,$$

where $H^{(k)}$ is a displacement interpolation matrix corresponding to the kth element and $\mathbf{U}_h$ is a vector of the three global displacement components U_j, V_j, W_j at all nodal points, that is,

$$\mathbf{U}_h = (U_1, V_1, W_1, \dots, U_n, V_n, W_n),$$

that is, $\mathbf{U}_h$ is a $3n$-dimensional vector.

The corresponding element strains are evaluated by

$$\mathbf{e}^{(k)}(x, y, z) = B^{(k)}(x, y, z)\mathbf{U}_h,$$

where $B^{(k)}$ is a strain–displacement matrix, where the rows of $B^{(k)}$ are obtained by differentiating and combining rows of the matrix $H^{(k)}$. The mass matrix M is defined as follows:

$$M = \sum_{k=1}^{n} M^{(k)} = \sum_{k=1}^{n} \int_{T_{h_k}} \rho^{(k)} (H^{(k)})^{\mathrm{T}} H^{(k)} d\mathbf{x},$$

where $\rho^{(k)}$ is a mass density property parameter of the kth element. The damping matrix C is defined by

$$C = \sum_{k=1}^{n} C^{(k)} = \sum_{k=1}^{n} \int_{T_{h_k}} \alpha^{(k)} (H^{(k)})^{\mathrm{T}} H^{(k)}\, d\mathbf{x},$$

where $\alpha^{(k)}$ is a damping property parameter of the kth element. The stiffness matrix K is defined by

$$K = \sum_{k=1}^{n} K^{(k)} = \sum_{k=1}^{n} \int_{T_{h_k}} (B^{(k)})^{\mathrm{T}} \overline{C}^{(k)} B^{(k)}\, d\mathbf{x},$$

where $\overline{C}^{(k)}$ is the elasticity matrix of the kth element. The load vector $\mathbf{F}$ includes the effect of the element body and surface forces as

$$\mathbf{F} = \sum_{k=1}^{n} \mathbf{F}^{(k)} = \sum_{k=1}^{n} \int_{T_{h_k}} (H^{(k)})^{\mathrm{T}} \mathbf{f}^{(k)}\, d\mathbf{x} + \sum_{k=1}^{n_s} \int_{S^{(k)}} (H^{S^{(k)}})^{\mathrm{T}} \mathbf{P}^{(k)}\, ds,$$

where $S^{(k)}$ denotes a part of the element boundary, the surface force $\mathbf{P}^{(k)}$ is prescribed, n_s is a number of corresponding nodal points, and $\mathbf{f}^{(k)}$ are body forces acting on the kth element. Since the total internal virtual work is equal to the total external virtual work over all finite elements, then under the assumption that the element acceleration and element velocities are approximated in the same way as the element displacements, we obtain

$$M\mathbf{U}_h'' + C\mathbf{U}_h' + K\mathbf{U}_h = \mathbf{F}_h. \tag{6.266}$$

To apply the definition of stability and putting $\mathbf{d}_h = \mathbf{U}_{ha} - \mathbf{U}_{hb}$, where $\mathbf{U}_{ha}$ corresponds to the solution of the initial problem and $\mathbf{U}_{hb}$ corresponds to the solution of the problem with perturbed data, we obtain

$$M\mathbf{d}_h'' + C\mathbf{d}_h' + K\mathbf{d}_h = \mathbf{0}. \tag{6.267}$$

Stability according to (6.263) then requires that $\mathbf{d}_h(t)$ does not grow.

We require that using the central difference method a time step Δt will be smaller than the critical time step Δt_{crit}. For the analysis of a direct integration method it is necessary to find the relation between mode superposition and direct integration. In the mode superposition analysis a change of the basis form the finite element nodal displacements to the basis of eigenvalues of the generalized eigenproblem

$$K\boldsymbol{\varphi} = \omega^2 M\boldsymbol{\varphi}$$

is performed to the time integration. Putting

$$\mathbf{d}(t) = \boldsymbol{\varphi}\mathbf{X}(t),$$

where the columns in $\boldsymbol{\varphi}$ are the M-orthonormalized eigenvectors $\boldsymbol{\varphi}_1, \ldots, \boldsymbol{\varphi}_n$, into (6.267), then we obtain

$$\mathbf{X}''(t) + \overline{P}\mathbf{X}'(t) + \varpi^2\mathbf{X}(t) = \mathbf{0}, \tag{6.268}$$

where $\overline{P}$ is a diagonal matrix, $\overline{P} = \operatorname{diag}(2\omega_j\xi_j)$, where $\xi_j = (a_1/2\omega_j) + (a_2\omega_j/2)$ is the damping ratio in the jth mode, and ϖ^2 is a diagonal matrix listing the eigenvalues $\omega_1^2, \ldots, \omega_n^2$. If all n equations in (6.267) are integrated using the same time step Δt, then the mode superposition analysis is completely equivalent to a direct integration analysis in which the same integration scheme and the same time step Δt are used. Thus, the variables to be considered in the stability and accuracy analyses are Δt, ω_j, ξ_j, $j = 1, \ldots, n$, and not all elements of the mass, damping, and stiffness matrices. Because all n equations in (6.268) are similar, we will analyze only one typical row of (6.268), that is, the equation

$$x'' + 2\xi\omega x' + \omega^2 x = 0. \tag{6.269}$$

Using the central difference integration scheme, where

$$x'' = \frac{1}{\Delta t^2}(x^{i+1} - 2x^i - x^{i-1}), \qquad x' = \frac{1}{2\Delta t}(x^{i+1} - x^{i-1}),$$

then after substituting into (6.269) we obtain

$$x^{i+1} = \frac{2 - \omega^2\,\Delta t^2}{1 + \xi\omega\,\Delta t}x^i - \frac{1 - \xi\omega\,\Delta t}{1 + \xi\omega\,\Delta t}x^{i-1}$$

or in the matrix form

$$\begin{pmatrix} x^{i+1} \\ x^i \end{pmatrix} = \begin{pmatrix} \dfrac{2-\omega^2\,\Delta t^2}{1+\xi\omega\,\Delta t} & -\dfrac{1-\xi\omega\,\Delta t}{1+\xi\omega\,\Delta t} \\ 1 & 0 \end{pmatrix} \begin{pmatrix} x^i \\ x^{i-1} \end{pmatrix}.$$

Let $\xi = 0$. Then we will solve the eigenvalue problem

$$\mathbf{Ax} \equiv \begin{pmatrix} 2-\omega^2\Delta t^2 & -1 \\ 1 & 0 \end{pmatrix} \mathbf{x} = \lambda \mathbf{x}.$$

The eigenvalues are the roots of the characteristic polynomial

$$\begin{aligned} p(\lambda) &= \det(\mathbf{A} - \lambda\mathbf{I}) \equiv (2-\omega^2\,\Delta t^2 - \lambda)(-\lambda) + 1 \\ &= 1 - (2-\omega^2\,\Delta t^2)\lambda + \lambda^2 = 0, \end{aligned}$$

thus

$$\lambda_{1,2} = \frac{2-\omega^2\,\Delta t^2}{2} \pm \sqrt{\frac{(2-\omega^2\,\Delta t^2)^2}{4} - 1}.$$

For stability we need that the spectral radius $\rho(\mathbf{A}) = \max_j |\lambda_j| \le 1$, that is, $|\lambda_1|$ and $|\lambda_2|$ are smaller than or equal to 1. Since $\omega = 2\pi/T$, we find $\Delta t \le T/\pi$. Thus, the central difference method is stable provided that $\Delta t \le \Delta t_{\mathrm{crit}} = T_n/\pi = 2/\omega_n$.

Let ω_n be the largest frequency of the used finite element mesh and $\omega_n^{(k)}$ be the largest frequency of the kth element. Then it can be shown that $\omega_n \le \max_{(k)} \omega_n^{(k)}$, where $\max_{(k)} \omega_n^{(k)}$ is the largest element frequency of all elements in the used mesh. Then

$$\Delta t = \frac{2}{\max_{(k)} \omega_n^{(k)}} \le \Delta t_{\mathrm{crit}} = \frac{T_n}{\pi},$$

where $\max_{(k)} \omega_n^{(k)} \ge \left(\sum_k \mathcal{K}^{(k)} / \sum_k \mathcal{M}^{(k)}\right)$ and where $\mathcal{K}^{(k)} = \boldsymbol{\varphi}_n^{\mathrm{T}} K^{(k)} \boldsymbol{\varphi}_n$, $\mathcal{M}^{(k)} = \boldsymbol{\varphi}_n^{\mathrm{T}} M^{(k)} \boldsymbol{\varphi}_n$. The number $(\Delta t/\Delta t_{\mathrm{crit}})$ is the Courant number, which can be used to indicate the size of the time step for a dynamic finite element solution.

Remark 6.18 Using the Rayleigh quotient $\rho(\boldsymbol{\varphi}) = (\boldsymbol{\varphi}^{\mathrm{T}} K \boldsymbol{\varphi} / \boldsymbol{\varphi}^{\mathrm{T}} M \boldsymbol{\varphi})$, where $K\boldsymbol{\varphi} = \lambda M \boldsymbol{\varphi}$, K and M are positive definite matrix, then $\rho(\boldsymbol{\varphi})$ has finite values for all $\boldsymbol{\varphi}$, and therefore, we obtain its bounds, that is, $0 < \lambda_1 \le \rho(\boldsymbol{\varphi}) \le \lambda_n < \infty$. For a single kth element

$$\rho^{(k)} = \frac{\boldsymbol{\varphi}^{\mathrm{T}} K^{(k)} \boldsymbol{\varphi}}{\boldsymbol{\varphi}^{\mathrm{T}} M^{(k)} \boldsymbol{\varphi}} = \frac{\mathcal{K}^{(k)}}{\mathcal{M}^{(k)}}, \qquad \text{hence} \quad \mathcal{K}^{(k)} \le (\omega_n^{(k)})^2 \mathcal{M}^{(k)}$$

as the size of K, $\mathcal{K}^{(k)}$, and $\mathcal{M}^{(k)}$ are the same, and therefore some of $\mathcal{K}^{(k)}$, $\mathcal{M}^{(k)}$ can be zero. Since

$$\omega_n = \left[\frac{\boldsymbol{\varphi}^{\mathrm{T}} \sum_k K^{(k)} \boldsymbol{\varphi}}{\boldsymbol{\varphi}^{\mathrm{T}} \sum_k M^{(k)} \boldsymbol{\varphi}}\right]^{1/2} = \left[\frac{\sum_k \mathcal{K}^{(k)}}{\sum_k \mathcal{M}^{(k)}}\right]^{1/2} \le \left[\frac{\sum_k \left(\omega_n^{(k)}\right)^2 \mathcal{M}^{(k)}}{\sum_k \mathcal{M}^{(k)}}\right]^{1/2}$$

$$\le \left[\max_k \left(\omega_n^{(k)}\right)^2 \frac{\sum_k \mathcal{M}^{(k)}}{\sum_k \mathcal{M}^{(k)}}\right]^{1/2} = \max_k \left(\omega_n^{(k)}\right).$$

The same results are valid for $\xi > 0$.

From the theory it follows that for linear problems in the use of the central difference scheme the time step Δt is smaller than the critical value Δt_{crit}, which can be calculated from the mass and stiffness properties of the complete element construction. The stability analysis shows that the following estimates are valid for the critical time step size:

$$\Delta t \le \Delta t_{\text{crit}} = \frac{T_n}{\pi} \quad \text{(the Courant criterion)},$$

where T_n is the smallest period of the finite element discretization with n degrees of freedom, where T_n can be calculated by using the solution method for eigenproblems or a lower bound on T_n may be evaluated using norms and based on the fact that $\|K\| \; \|\boldsymbol{\varphi}\| \ge |\lambda| \; \|M\| \; \|\boldsymbol{\varphi}\|$ (Bathe, 1982, 1996). For nonlinear problems Belytschko et al. (2000) find the Courant criterion in the form

$$\Delta t \le \Delta t_{\text{crit}} = \delta \frac{h}{c_L},$$

where h/c_L is the time for the pressure wave to cross the element, $c_L = 3k(1-\nu)/\rho(1-\nu)$, k is the modulus of compression, ν is the Poisson ration, ρ the density, h the mesh step, and $\delta \in (0.2, 0.9)$ is a reduction factor determined experimentally, which is necessary because of the destabilizing effect of round-off and the possibility of rapidly varying material properties.

To prove the consistency of the scheme, the approach introduced by Richtmayer and Morton (1967) will be used. They introduced the local truncation error for the pth step algorithms, which for the equation

$$M\mathbf{x}'' + C\mathbf{x}' + K\mathbf{x} = \mathbf{f} \tag{6.270}$$

is the following:

$$\sum_{j=0}^{p} (\alpha_j \, \Delta t^{-2} M + \gamma_j \, \Delta t^{-1} C + \beta_j K)\mathbf{x}_{i+j} - \beta_j \mathbf{f}_{i+j} = 0. \tag{6.271}$$

Then the aim is to measure how closely the classical solution of (6.270) satisfies the difference scheme (6.271). The local truncation error is as follows:

$$\begin{aligned}\epsilon_{\Delta t}(\mathbf{x}(i\,\Delta t)) &= \sum_{j=0}^{p}\{(\alpha_j\,\Delta t^{-2}M + \gamma_j\,\Delta t^{-1}C + \beta_j K)\mathbf{x}((i+j)\,\Delta t) - \beta_j \mathbf{f}_{i+j}\}\\ &= O(\Delta t^q), \qquad q \geq 1,\end{aligned}$$

as $\Delta t \to 0$, for all sufficiently smooth classical solutions of the differential equation (6.270). If the local truncation error tends to zero as $\Delta t \to 0$, the algorithm is consistent.

For the central difference algorithm we have

$$\frac{1}{\Delta t^2}M(\mathbf{U}_h^{i+1} - 2\mathbf{U}_h^i + \mathbf{U}_h^{i-1}) + \frac{1}{2\Delta t}C(\mathbf{U}_h^{i+1} - \mathbf{U}_h^{i-1}) + K\mathbf{U}_h^i = \mathbf{F}_h^i. \tag{6.272}$$

To find the local truncation error expanding by the Taylor series about $i\,\Delta t$ will be applied. Thus we obtain

$$\begin{aligned}\epsilon_{\Delta t}(\mathbf{U}_h(i\,\Delta t)) = M&\left\{\mathbf{U}_h''(i\,\Delta t) + \frac{\Delta t^2}{12}\mathbf{U}_h''(i\,\Delta t) + \ldots\right\}\\ &+ C\left\{\mathbf{U}_h'(i\,\Delta t) + \frac{\Delta t^2}{3}\mathbf{U}_h'(i\,\Delta t) + \ldots\right\}\\ &+ K\mathbf{U}_h(i\,\Delta t) - \mathbf{F}_h(i\,\Delta t).\end{aligned}$$

Hence

$$\epsilon_{\Delta t}(\mathbf{U}_h(i\,\Delta t)) = O(\Delta t^2), \tag{6.273}$$

that is, the central difference algorithm is consistent of order 2.

Remark 6.19 Some relatively simple **explicit algorithm**, which can be also used for contact problems with friction, is based on the penalty and regularization approaches. The penalized and regularized terms are referred to $t = t_i$, so that we have to solve a system of equation for the unknown displacements $\mathbf{u}_h^{i+1}$ at the time t_{i+1}:

$$\left(\frac{1}{\Delta t^2}M + \frac{1}{2\Delta t}C\right)\mathbf{u}_{\varepsilon h}^{i+1} = \tilde{\mathbf{F}}_\varepsilon(t_i), \tag{6.274}$$

where

$$\begin{aligned}\tilde{\mathbf{F}}_\varepsilon(t_i) = \mathbf{F}(t_i) &- \int_{\bigcup_{s,m}\Gamma_c^{sm}} g_{ch}^{sm}(t_i)\varphi_\delta(\mathbf{u}_{ht}^s(t_i) - \mathbf{u}_{ht}^m(t_i))\,ds\\ &- \frac{1}{2\varepsilon}\int_{\bigcup\Gamma_c^{sm}}([u_n^s(t_i) - u_n^m(t_i)]_+)^2\,ds = \mathbf{F}(t_i) - j_\delta(\mathbf{u}_{\delta h}^i) - \frac{1}{2\varepsilon}\bar{j}_\varepsilon(\mathbf{u}_{\varepsilon h}^i).\end{aligned}$$

In this system the matrices M and C as well as $[(1/\Delta t^2)M + 1/(2\Delta t)C]$ do not change, and hence they can be computed once. All nonlinearities only enter on the right-hand side. Since matrices M and C are diagonal, the factorization of $[(1/\Delta t^2)M + 1/(2\Delta t)C]$ is trivial, so that only vector operations are needed for calculation. The right-hand side of (6.274) depends, besides on the loading term $\mathbf{f}(t_{i+1})$, upon vectors that are computed at the time t_i. Thus the initial conditions can be used directly when the algorithm starts. The integrals in $j_{gn\varepsilon}$ and j_ε are computed in the usual way.

For a solution a small time step size must be applied. To analyze the central difference method from its integration point of view, we need to calculate the spectral radius of the approximation operator. From the theory it follows that for linear problems in the use of the central difference scheme, the time step Δt is smaller than the critical value Δt_{crit}, which can be calculated from the mass and stiffness properties of the complete element construction.

The algorithm as follows:

Initial Calculation

1. Form the mass matrix M, the damping matrix C, and the stiffness matrix K.
2. Initialize $\mathbf{U}_h^0$, $(\mathbf{U}_h')^0$, $(\mathbf{U}_h'')^0$.
3. Select the time step Δt such that $\Delta t \le \Delta t_{\text{crit}}$.
4. Calculate integration constants

$$a_0 = \frac{1}{\Delta t^2}, \qquad a_1 = \frac{1}{2\Delta t}, \qquad a_2 = 2a_0, \qquad a_3 = \frac{1}{a_2}.$$

5. Calculate $\mathbf{U}_h^{-1} = \mathbf{U}_h^0 - \Delta t(\mathbf{U}_h')^0 + a_3(\mathbf{U}_h'')^0$.
6. Form the effective mass matrix $\overline{M} = a_0 M + a_1 C$.
7. Triangularize the effective mass matrix: $\overline{M} : \overline{M} = LDL^{\mathrm{T}}$.

At each time step:

1. Calculate effective loads at the time $t = t_i$:

$$\mathbf{F}_\varepsilon(t_i) = \mathbf{f}(t_i) - (K - a_2 M)\mathbf{U}_h^i - (a_0 M - a_1 C)\mathbf{U}_h^{i-1}.$$

2. Calculate the effective penalty and regularized terms at the time $t = t_i$:

$$\bar{j}_\varepsilon(\mathbf{U}_h) = \frac{1}{2\varepsilon}\int_{\bigcup_{s,m}\Gamma_c^{sm}} ([\mathbf{U}_{hn}^i]_+^{sm})^2\, ds,$$

$$j_\delta(\mathbf{U}_h) = \int_{\bigcup_{s,m}\Gamma_c^{sm}} g_{ch}^{sm}\varphi_\delta([\mathbf{U}_{ht}^i]^{sm})\, ds.$$

3. Calculate effective loads together with penalized and regularized terms:

$$\overline{\mathbf{F}}(t_i) = \mathbf{F}_\varepsilon(t_i) - j_\delta(\mathbf{U}_h) - \bar{j}_\varepsilon(\mathbf{U}_h).$$

4. Solve displacements $\mathbf{U}_h$ at the time $t = t_{i+1} = t_i + \Delta t$:

$$LDL^{\mathrm{T}}\mathbf{U}_h^{i+1} = \overline{\mathbf{F}}(t_i).$$

5. If required, we can compute accelerations and velocities at the time $t = t_i$:

$$\mathbf{U}_h'' = a_0(\mathbf{U}_h^{i+1} - 2\mathbf{U}_h + \mathbf{U}_h^{i-1}), \qquad \mathbf{U}_h' = a_1(\mathbf{U}_h^{i+1} - \mathbf{U}_h^{i-1}).$$

6.6.3 Biomechanical Model of Human Joints Based on the Semi-implicit Scheme

Next, we briefly present the derivation of the algorithm [see also Nedoma (2009)].

The algorithm is based on the semi-implicit scheme and on the assumption that the frictional term is approximated by its value in the previous time level (s). Such a problem leads to the case with given friction.

Let $m > 0$ be an integer. Then $\Delta t = t_p/m$, $t_i = i\,\Delta t$, $i = 0, 1, \ldots, m$. Then

$$\begin{aligned}
&(\Delta t^{-2}(\mathbf{u}_h^{i+1} - 2\mathbf{u}_h^i + \mathbf{u}_h^{i-1}), \mathbf{v}_h - \mathbf{u}_h^{i+1}) + (\Delta t^{-1}\alpha(\mathbf{u}_h^{i+1} - \mathbf{u}_h^i), \mathbf{v}_h - \mathbf{u}_h^{i+1}) \\
&\quad + a^{(0)}(\mathbf{u}_h^{i+1}, \mathbf{v}_h - \mathbf{u}_h^{i+1}) + a^{(1)}(\Delta t^{-1}(\mathbf{u}_h^{i+1} - \mathbf{u}_h^i), \mathbf{v}_h - \mathbf{u}_h^{i+1}) \\
&\quad + \int_{\bigcup_{s,m}\Gamma_c^{sm}} g_{ch}^{sm}|\mathbf{v}_{th}^s - \mathbf{v}_{th}^m|\,ds - \int_{\bigcup_{s,m}\Gamma_c^{sm}} g_{ch}^{sm}|\mathbf{u}_{th}^s(t_{i+1}) - \mathbf{u}_{th}^m(t_{i+1})|\,ds \\
&\quad \geq (\mathbf{f}(t_{i+1}), \mathbf{v}_h - \mathbf{u}_h^{i+1}) \quad \forall \mathbf{v}_h \in K_h,
\end{aligned}$$

where

$$\begin{aligned}
&\mathbf{u}_h^i = \mathbf{u}_h(t_i), \qquad \mathbf{u}_h^0 = \mathbf{u}_{0h}, \qquad \mathbf{u}_h^{-1} = \mathbf{u}_{0h} - \Delta t \cdot \mathbf{u}_{1h}, \\
&g_{ch}^{sm} = g_{ch}^{sm}(t_i) = \mathcal{F}_c^{sm}(\Delta t^{-1}(\Delta\mathbf{u}_{th}^s(t_i) - \Delta\mathbf{u}_{th}^m(t_i)))\left|\tau_n^{sm}\left(\mathbf{u}_h(t_i), \frac{\Delta\mathbf{u}_h(t_i)}{\Delta t}\right)\right|, \\
&\Delta\mathbf{u}_h^i = \Delta\mathbf{u}_h(t_i) = \mathbf{u}_h(t_i) - \mathbf{u}_h(t_{i-1}),
\end{aligned}$$

Then

$$\begin{aligned}
&(\mathbf{u}_h^{i+1}, \mathbf{v}_h - \mathbf{u}_h^{i+1}) + \Delta t(\alpha\mathbf{u}_h^{i+1}, \mathbf{v}_h - \mathbf{u}_h^{i+1}) + \Delta t^2 a^{(0)}(\mathbf{u}_h^{i+1}, \mathbf{v}_h - \mathbf{u}_h^{i+1}) \\
&\quad + \Delta t a^{(1)}(\mathbf{u}_h^{i+1}, \mathbf{v}_h - \mathbf{u}_h^{i+1}) \\
&\quad + \Delta t^2 \int_{\bigcup_{s,m}\Gamma_c^{sm}} g_{ch}^{sm}(t_i)(|\mathbf{v}_{th}^s - \mathbf{v}_{th}^m| - |\mathbf{u}_{th}^s(t_{i+1}) - \mathbf{u}_{th}^m(t_{i+1})|)\,ds \\
&\quad \geq (\mathbf{F}(t_{i+1}), \mathbf{v}_h - \mathbf{u}_h^{i+1}),
\end{aligned}$$

where

$$\begin{aligned}
(\mathbf{F}(t_{i+1}), \mathbf{v}_h) &= \Delta t^2(\mathbf{f}(t_{i+1}), \mathbf{v}_h) + (2\mathbf{u}_h^i - \mathbf{u}_h^{i-1}, \mathbf{v}_h) \\
&\quad + \Delta t(\alpha\mathbf{u}_h^{i-1}, \mathbf{v}_h) + \Delta t a_h^{(1)}(\mathbf{u}_h^{i-1}, \mathbf{v}_h). \qquad (6.275)
\end{aligned}$$

For the convergence of the finite element discretization, it is necessary and sufficient that it will be consistent and stable. To prove the stability we are concerned with the linearized problem based on the penalty and regularity approaches such as in the previous explicit case. The model problem in elasticity is a special case of the model problem formulated in viscoelasticity. Thus, it is sufficient to analyze problem formulated in elasticity, and, moreover, from the same reasons as in the previous explicit case, it is sufficient to analyze problem (6.265). Corresponding to the fact that when the scheme corresponding to (6.265) will be stable and consistent, the scheme corresponding to the problem formulated in viscoelasticity will be also stable and consistent, and therefore it will be convergent.

As the variables, which will be considered in the stability analysis of the integration method, are $\Delta t, \omega_j, \xi_j, j = 1, \ldots, n$, and not all elements of the mass, damping, and stiffness matrices, where Δt is a time step, ω_j are the eigenvalues of $K\mathbf{y}_k = \omega^2 M\mathbf{y}_k$, $\mathbf{y}_k$ are the M-orthonormalized eigenvectors (free-vibration modes), and ξ_j are the damping ratio in the jth mode. Then the critical time step is given in terms of the maximal eigenvalue (i.e., of the spectral radius) of the system $K\mathbf{y} = \omega^2 M\mathbf{y}$.

Applying the definition of stability and putting $\mathbf{d}_h = \mathbf{U}_{ha} - \mathbf{U}_{hb}$, where $\mathbf{U}_{ha}$ and $\mathbf{U}_{hb}$ are solutions corresponding to the initial and perturbed problems, respectively, then the semi-implicit scheme leads to the linear discrete equation of motion for a damped system of the form

$$M\mathbf{d}_h'' + C\mathbf{d}_h' + K\mathbf{d}_h = \mathbf{0}, \tag{6.276}$$

where $C = a_1 M + a_2 K$, a_1, a_2 are arbitrary parameters. Let us put $\mathbf{d}_h = \sum_j \alpha_j(t)\mathbf{y}_j$, where $\mathbf{y}_j$ are eigenvectors of the associated eigenproblem

$$K\mathbf{y}_k = \lambda_k M\mathbf{y}_k,$$

which are orthogonal with respect to M and K and the eigenvalues λ_k are real. The orthogonality conditions can be written as

$$\mathbf{y}_j M\mathbf{y}_k = \delta_{jk}, \qquad \mathbf{y}_j K\mathbf{y}_k = \lambda_k \delta_{jk} \quad \text{(no sum on } k\text{)}.$$

Then (6.276) gives

$$\alpha'' + (a_1 + a_2\omega_j^2)\alpha' + \omega_j^2\alpha = 0, \quad \text{where } \omega_j^2 = \lambda_j, \qquad j = 1, \ldots, n.$$

The damping is usually given as a fraction of critical damping ξ, so that (the index j can be omitted)

$$\alpha'' + 2\xi\omega\alpha' + \omega^2\alpha = 0,$$

where

$$\xi = \frac{a_1}{2\omega} + \frac{a_2\omega}{2}.$$

Applying the difference approximations we have

$$\frac{\alpha_{n+1} - 2\alpha_n + \alpha_{n-1}}{\Delta t^2} + 2\xi\omega\frac{\alpha_{n+1} - \alpha_n}{\Delta t} + \omega^2\alpha_{n+1} = 0. \tag{6.277}$$

Since this difference equation is linear, the solution is of the form $\alpha_{n+1} = \lambda\alpha_n = \lambda^2\alpha_{n-1}$. Substituting this solution into (6.277) then we obtain

$$(1 + 2\xi\omega\,\Delta t + \omega^2\,\Delta t^2)\lambda^2 + (-2\xi\omega\,\Delta t - 2)\lambda + 1 = 0.$$

Setting $\mu = 1 + z/1 - z$, the z-transform [see Belytscko et al. (2000)], we obtain a quadratic equation in z of the form

$$(4 + 4\xi\omega\,\Delta t + \omega^2\,\Delta t^2)z^2 + 2(2\xi\omega\,\Delta t + \omega^2\,\Delta t^2)z + \omega^2\,\Delta t^2 = 0,$$

that is, $c_0z^2 + c_1z + c_2 = 0$. Comparing the coefficients at powers of z with the elements of the Hurwicz matrix of the polynomial equation of order p

$$\sum_{i=0}^{p} c_i z^{p-i} = 0 \quad \text{with} \quad c_0 > 0,$$

where the Hurwitz matrix is defined by

$$H_{ij} = \begin{Bmatrix} c_{2j-i} & \text{if } 0 \le 2j - i \le p \\ 0 & \text{otherwise,} \end{Bmatrix},$$

we find the stability conditions:

$$\begin{aligned} &4 - 4\xi\omega\,\Delta t - \omega^2\,\Delta t^2 \ge 0, \\ &2\xi\omega\,\Delta \ge 0, \\ &\omega^2\,\Delta t^2 \ge 0. \end{aligned} \tag{6.278}$$

The third condition is directly satisfied, the second one is satisfied for $\xi \ge 0$, and the first one gives

$$\omega\,\Delta t = -2\xi \pm 2(\xi^2 - 1)^{1/2}.$$

For a positive root and between points, where the first inequality of (6.278) vanishes, the critical time step is as follows:

$$\Delta t_{\text{crit}} = \max_k \frac{2}{\omega_k}[(\xi_k^2 - 1)^{1/2} - \xi_k] \equiv \max_k \frac{2}{\lambda_k^{1/2}}[(\xi_k^2 - 1)^{1/2} - \xi_k],$$

as $\omega_k^2 = \lambda_k$. The factor in the bracket is equal to 1 for $\xi_k = 0$ and less than 1 for $\xi_k > 0$. Hence, the delay in the velocity decreases the stable time step when the damping is assumed. The critical time step is also called the stable time step. Since the time step corresponds to the stability limit T_n/π, then similarly as in the previous case

$$\Delta t \le \Delta t_{\text{crit}} = \frac{T_n}{\pi}[(\xi_k^2 - 1)^{1/2} - \xi_k] = \frac{h^{(n)}}{\pi}\left(\frac{\rho^{(n)}}{E^{(n)}}\right)[(\xi_k^2 - 1)^{1/2} - \xi_k].$$

It can be shown, similarly as in the previous explicit case, that the algorithm is also consistent of order 2 because the central difference method is of the second order in time, that is, the truncation error is of order Δt^2 in the displacements. We see that the used algorithm is convergent.

The scheme can be simplified. Let us set $\mathbf{u}_h^{i+1} \equiv \mathbf{u}_h$, $\tilde{g}_{ch}^{sm} = g_{ch}^{sm}(t_i)$, $\mathbf{F}(t_{i+1}) \equiv \mathbf{f}_h$, then (6.275) can be rewritten as

$$\begin{aligned}
&(\mathbf{u}_h, \mathbf{v}_h - \mathbf{u}_h) + \Delta t(\alpha \mathbf{u}_h, \mathbf{v}_h - \mathbf{u}_h) + \Delta t^2 a^{(0)}(\mathbf{u}_h, \mathbf{v}_h - \mathbf{u}_h) \\
&\quad + \Delta t a^{(1)}(\mathbf{u}_h, \mathbf{v}_h - \mathbf{u}_h) + \Delta t^2 \int_{\bigcup_{s,m}\Gamma_c^{sm}} \tilde{g}_{ch}^{sm}(|\mathbf{v}_{th}^s - \mathbf{v}_{th}^m| - |\mathbf{u}_{th}^s - \mathbf{u}_{th}^m|)\, ds \\
&\ge (\mathbf{f}_h, \mathbf{v}_h - \mathbf{u}_h), \qquad t = t_{i+1} \in I.
\end{aligned} \tag{6.279}$$

Let us define the bilinear form $A(\mathbf{u}_h, \mathbf{v}_h)$ and the functional $j(\mathbf{v}_h)$ by

$$\begin{aligned}
A(\mathbf{u}_h, \mathbf{v}_h) &= (\mathbf{u}_h, \mathbf{v}_h) + \Delta t(\alpha \mathbf{u}_h, \mathbf{v}_h) + \Delta t^2 a^{(0)}(\mathbf{u}_h, \mathbf{v}_h) + \Delta t a^{(1)}(\mathbf{u}_h, \mathbf{v}_h), \\
j(\mathbf{v}_h) &= \Delta t^2 \int_{\bigcup_{s,m}\Gamma_c^{sm}} \tilde{g}_{ch}^{\text{sm}} |\mathbf{v}_{th}^s - \mathbf{v}_{th}^m|\, ds,
\end{aligned} \tag{6.280}$$

where $\tilde{g}_{ch}^{\text{sm}}$ represents the approximate given frictional limit.

Since we assumed that $\rho \ge \rho_0 > 0, \alpha \ge \alpha_0 > 0$ and since bilinear forms $a_h^{(n)}(\mathbf{u}_h, \mathbf{v}_h)$, $n = 0, 1$, are symmetric in $\mathbf{u}_h$ and $\mathbf{v}_h$ and satisfy $a^{(n)}(\mathbf{u}_h, \mathbf{u}_h) \ge c_0^{(n)} \|\mathbf{u}_h\|_{1,N}^2$, $c_0^{(n)} = \text{const} > 0$, $|a^{(n)}(\mathbf{u}_h, \mathbf{v}_h)| \le c_0^{(n)} \|\mathbf{u}_h\|_{1,N} \|\mathbf{v}_h\|_{1,N}$, $c_1^{(n)} = \text{const} > 0$, $n = 0, 1$, $\mathbf{u}_h, \mathbf{v}_h \in V_h$, then the bilinear form $A(\mathbf{u}_h, \mathbf{v}_h)$ is also symmetric in $\mathbf{u}_h$ and $\mathbf{v}_h$ and

$$\begin{aligned}
&A(\mathbf{u}_h, \mathbf{v}_h) \ge a_0 \|\mathbf{u}_h\|_{1,N}^2, \qquad c_0^{(n)} = \text{const} > 0, \\
&|A(\mathbf{u}_h, \mathbf{v}_h)| \le a_1 \|\mathbf{u}_h\|_{1,N} \|\mathbf{v}_h\|_{1,N}, \qquad a_1 = \text{const} > 0, \qquad \mathbf{u}_h, \mathbf{v}_h \in V_{0h},
\end{aligned} \tag{6.281}$$

hold.

Then we have to solve in every time level the equivalent problem:

Problem $(\mathcal{P}_A)_h$ Find $\mathbf{u}_h \in K_h$, such that

$$A(\mathbf{u}_h, \mathbf{v}_h - \mathbf{u}_h) + j(\mathbf{v}_h) - j(\mathbf{u}_h) \ge (\mathbf{f}_h, \mathbf{v}_h - \mathbf{u}_h), \quad \forall \mathbf{v}_h \in K_h, \qquad t = t_{i+1} \in I. \tag{6.282}$$

Let us introduce the functional $\mathcal{L}(\mathbf{v}_h)$ by

$$\mathcal{L}(\mathbf{v}_h) = \mathcal{L}_0(\mathbf{v}_h) + j(\mathbf{v}_h),$$

where

$$\mathcal{L}_0(\mathbf{v}_h) = \tfrac{1}{2}A(\mathbf{v}_h, \mathbf{v}_h) - (\mathbf{f}_h, \mathbf{v}_h).$$

Then Problem $(\mathcal{P}_A)_h$ is equivalent to the following problem: Find $\mathbf{u}_h \in K_h$ such that

$$\mathcal{L}(\mathbf{u}_h) \le \mathcal{L}(\mathbf{v}_h), \quad \forall \mathbf{v}_h \in K_h, \quad t = t_{i+1} \in I. \tag{6.283}$$

We arrived at a variational inequality of the second kind; see Section 6.2.

6.6.4 Biomechanical Model of Human Joints Based on the Approximate Mixed Variational Formulation of the Frictional Tresca Model

In what follows the model of human joints and their replacements will be based on a simpler friction model introduced by Tresca. The main feature of the Tresca friction law is the assumption that on the contact surfaces between both joint components the friction bound g_c^{sm}, that is, the magnitude of the limiting friction traction at which slip begins, is given. Next, the discussed semi-implicit scheme will be used.

Saddle Point–Uzawa/CGM Approach for the Human Joint Model in Elastic Rheology—The Matching Case Since the term $j(\mathbf{v}_h)$ is not smooth, we introduce an approximate mixed variational formulation of the problem discussed, which employs Lagrange multipliers. Lagrange multipliers will be identified with normal and tangential components of the contact stress vector. The problem (6.283) can be reformulated [see Haslinger et al. (1996), Kestřánek (1999), Hlaváček (2007), Nedoma (2010), and Nedoma et al. (1999b, 2006)] in the form of an approximate saddle point problem. Piecewise linear finite elements will be used for the approximation of displacements, and the Lagrange multipliers will be approximated by piecewise linear functions.

We introduce the standard spaces of linear elements on the regular partition $\mathcal{T}_h = \cup_{\iota=1}^{r} \mathcal{T}_h^{\iota}$ of polygonal (for $N=2$) or polyhedral (for $N=3$) domains $\overline{\Omega}^{\iota}$, $\iota = 1, \dots, r$. Let

$$V_h = V_h^1 \times \cdots \times V_h^r = \prod_{\iota=1}^{r} V_h^{\iota},$$

where

$$V_h^{\iota} = \{\mathbf{v}_h \in [C(\overline{\Omega}^{\iota})]^N : \mathbf{v}_h|_{T_{hi}} \in [P_1(T_{hi})]^N\ \forall T_{hi} \in \mathcal{T}_h^{\iota},\ \mathbf{v}_h = 0 \text{ on } \Gamma_u \cap \partial\Omega^{\iota}\},$$
$$K_h^{\iota} = \{\mathbf{v}_h \in V_h^{\iota} : [\mathbf{v}_h \cdot \mathbf{n}]^{sm}(a_i) \le 0 \text{ on } \Gamma_c^{sm},\ \forall i\},$$

where by T_{hi} we denote any triangle (for $N=2$) or any tetrahedron (for $N=3$) of the partition $\mathcal{T}_h^{\iota}$, $\iota = 1, \dots, r$, and a_i, $i = 1, \dots, q$, denote the nodes of $\mathcal{T}_h^{\iota}$ laying on Γ_c^{sm}. When Γ_c^{sm} is curved the normal $\mathbf{n}^s$ at a point a_i is approximated as an arithmetic mean of values of both normals to the neighboring segments at the point a_i.

Let us introduce the approximate Lagrange multiplier spaces W_h and M_h as follows:

$$W_h = W_h(\cup\Gamma_c^{sm}) = \{(\boldsymbol{\varphi}_t) | \exists \mathbf{v}_h \in V_h \text{ such that } [\mathbf{v}_{ht}] = \boldsymbol{\varphi}_t\},$$
$$M_h = \{\boldsymbol{\mu}_h \in W_h \mid (\boldsymbol{\mu}_{ht}, \boldsymbol{\varphi}_{ht})_{0,\cup\Gamma_c^{sm}} \le (1, |\boldsymbol{\varphi}_{ht}|)_{0,\cup\Gamma_c^{sm}} \ \forall \boldsymbol{\varphi}_{ht} \in W_h\}.$$

We will assume that any partition $\mathcal{T}_h$ is compatible with the endpoints of ${}^1\Gamma_u$, ${}^2\Gamma_u$, Γ_τ, Γ_c and that the nodes of $\mathcal{T}_h^s$ and $\mathcal{T}_h^m$ coincide on Γ_c^{sm}. We observe that,

$$j(\mathbf{v}_h) = \sup_{M_h} (\tilde{g}_c^{sm} \boldsymbol{\mu}_{ht}, [\mathbf{v}_{ht}]^{sm})_{0,\cup\Gamma_c^{sm}}.$$

Then problem (6.282) leads to the following saddle point problem: Find a pair $(\mathbf{u}_h, \boldsymbol{\lambda}_h) \in K_h \times M_h$ such that

$$\mathcal{H}_h(\mathbf{u}_h, \boldsymbol{\mu}_h) \le \mathcal{H}_h(\mathbf{u}_h, \boldsymbol{\lambda}_h) \le \mathcal{H}_h(\mathbf{v}_h, \boldsymbol{\lambda}_h) \quad \forall (\mathbf{v}_h, \boldsymbol{\mu}_h) \in K_h \times M_h,$$

where

$$\mathcal{H}_h(\mathbf{v}_h, \boldsymbol{\mu}_h) = \tfrac{1}{2} A(\mathbf{v}_h, \mathbf{v}_h) - (\mathbf{f}_h, \mathbf{v}_h) + b_h(\boldsymbol{\mu}_h, \mathbf{v}_h),$$
$$b_h(\boldsymbol{\mu}_h, \mathbf{v}_h) = (\tilde{g}_c^{sm} \boldsymbol{\mu}_{ht}, [\mathbf{v}_{ht}]^{sm})_{0,\cup\Gamma_c^{sm}}, \tag{6.284}$$

or in an equivalent form: Find $(\mathbf{u}_h, \boldsymbol{\lambda}_h) \in K_h \times M_h$ such that

$$A(\mathbf{u}_h, \mathbf{v}_h - \mathbf{u}_h) + \left(\tilde{g}_c^{sm} \boldsymbol{\lambda}_h, [\mathbf{v}_{ht}]^{sm} - [\mathbf{u}_{ht}]^{sm}\right)_{0,\cup\Gamma_c^{kl}}$$
$$\ge (\mathbf{f}_h, \mathbf{v}_h - \mathbf{u}_h), \quad \forall \mathbf{v}_h \in K_h, \qquad t = t_{i+1} \in I. \tag{6.285}$$
$$\left(\tilde{g}_c^{sm} (\boldsymbol{\mu}_h - \boldsymbol{\lambda}_h), [\mathbf{u}_{ht}]^{sm}\right)_{0,\cup\Gamma_c^{sm}} \le 0 \quad \forall \boldsymbol{\mu}_h \in M_h.$$

For the local matrix representation then for every element T_{hi} (with appropriate boundary segments—sides for $N=2$ and faces for $N=3$), we find the local functional of the form

$$f_n(\mathbf{y}_n, \boldsymbol{\mu}_n) = \tfrac{1}{2} \mathbf{y}_n^{\mathrm{T}} C_n \mathbf{y}_n - \mathbf{y}_n^{\mathrm{T}} \mathbf{d}_n + \mathbf{y}_n^{\mathrm{T}} G_n^{\mathrm{T}} \boldsymbol{\mu}_n,$$

where in the 3D case C_n is the local stiffness matrix of dimension 12×12, $\mathbf{y}_n$ is the vector (a representation of $\mathbf{v}_h$ on T_{hi}) of the dimensions 12×1, and $\mathbf{d}_n$ is the local load vector of the dimensions 12×1. Dimensions of G_n and $\boldsymbol{\mu}_n$ depend on the number of faces, which we denote as p_n. Then $\boldsymbol{\mu}_n$ is the vector of the dimension $3p_n \times 1$ and G_n is the constraint matrix of the dimension $3p_n \times 12$. The same holds for $N=2$.

The global matrix representation of the problem (6.284) for every $t = t_{i+1} \in I$ leads to the following problem: Find $(\mathbf{x}, \boldsymbol{\lambda}) \in K_d \times M_d$ such that

$$\begin{aligned}
&\mathcal{H}(\mathbf{x}, \boldsymbol{\mu}) \le \mathcal{H}(\mathbf{x}, \boldsymbol{\lambda}) \le \mathcal{H}(\mathbf{y}, \boldsymbol{\lambda}) \quad \forall (\mathbf{y}, \boldsymbol{\mu}) \in K_d \times M_d, \\
&\mathcal{H}(\mathbf{y}, \boldsymbol{\mu}) = \tfrac{1}{2}\mathbf{y}^{\mathrm{T}} C \mathbf{y} - \mathbf{y}^{\mathrm{T}}\mathbf{d} + \mathbf{y}^{\mathrm{T}} G^{\mathrm{T}} \boldsymbol{\mu}, \\
&\quad K_d = \{\mathbf{y} \in \mathbb{R}^N \mid A\mathbf{y} \le 0\} \\
&\quad M_d = \left\{\boldsymbol{\mu} \in \mathbb{R}^{3\overline{P}} \mid \mu_{3i-2}^2 + \mu_{3i-1}^2 + \mu_{3i}^2 \le 1,\ i = 1, \ldots, \overline{P}\right\} \quad \text{for } N = 3,
\end{aligned} \tag{6.286}$$

where C is the global stiffness matrix ($\dim n \times n$), block diagonal, every block of which is a sparse, symmetric positive (semi)definite matrix, A is the constraint matrix representing the nonpenetration condition ($\dim m \times n$), G is the matrix representing the friction forces ($\dim 3\overline{P} \times n$), and $\mathbf{d}$ is the global load vector ($\dim n \times 1$); moreover, it holds $m \ll n$, $3\overline{P} \ll n$.

Problem (6.285) for every $t = t_{i+1} \in I$ will be solved by the modified Uzawa algorithm with the preconditioning conjugate gradient method (CGM) with constraints [see Kestřánek and Nedoma (1996, 1998), Nedoma (1998b), Nedoma et al. (1999b), Kestřánek (1999), and Nedoma (2010)]. The algorithm for every $t = t_{i+1} \in I$ will be the following:

Let $\boldsymbol{\lambda}_H^0$ be an initial approximation; then, knowing $\boldsymbol{\lambda}_H^k \in M_d$, we calculate $\mathbf{x}^k$ as a solution of the minimization problem

$$\begin{aligned}
&\mathcal{H}(\mathbf{x}, \boldsymbol{\lambda}_H^k) \to \min, \\
&\mathbf{x} \in \mathbf{K}_d.
\end{aligned} \tag{6.287}$$

Then we replace $\boldsymbol{\lambda}_H^k$ by $\boldsymbol{\lambda}_H^{k+1}$ by using the process

$$\boldsymbol{\lambda}_H^{k+1} = \Pi_{M_d}(\boldsymbol{\lambda}_H^k + \rho(\mathbf{x}^k)^{\mathrm{T}} G^{\mathrm{T}}), \quad \rho > 0 \tag{6.288}$$

and return to (6.287). The value of $\boldsymbol{\lambda}_H^0 \in M_d$ is chosen arbitrarily. The symbol Π_{M_d} is the projection of $\mathbb{R}^{3\overline{P}}$ onto the set M_d, that is, $\Pi_{M_d}(\mathbf{y}) = [(\Pi_{M_d})_1^{\mathrm{T}}, \ldots, (\Pi_{M_d})_{\overline{P}}^{\mathrm{T}}]^{\mathrm{T}}$, where for $i = 1, \ldots, \overline{P}$

$$(\Pi_{M_d}(\mathbf{y}))_i = \left\{\begin{array}{ll} \mathbf{y}_i = (\mathbf{y}_{3i-2}^2, \mathbf{y}_{3i-1}^2, \mathbf{y}_{3i}^2)^{\mathrm{T}} & \text{for } |\mathbf{y}_i| \le 1, \\ \mathbf{y}_i / |\mathbf{y}_i| & \text{for } |\mathbf{y}_i| > 1, \end{array}\right\} \quad \text{for } N = 3,$$

where $|\mathbf{y}_i| = (\mathbf{y}_{3i-2}^2 + \mathbf{y}_{3i-1}^2 + \mathbf{y}_{3i}^2)^{1/2}$,

$$(\Pi_{M_d}(\mathbf{y}))_i = \left\{\begin{array}{ll} \mathbf{y}_i & \text{for } |\mathbf{y}|_i \le 1, \\ \operatorname{sgn} \mathbf{y}_i & \text{for } |\mathbf{y}|_i > 1, \end{array}\right\} \quad \text{for } N = 2,$$

where $\mathbf{y}_i = (\mathbf{y}_{2i-1}, \mathbf{y}_{2i})^{\mathrm{T}}$, $\quad |\mathbf{y}_i| = (\mathbf{y}_{2i-1}^2 + \mathbf{y}_{2i}^2)^{1/2}$.

From the proof of the convergence of the Uzawa algorithm we know that there exists a suitable parameter ρ such that $\rho \in (\rho_1, \rho_2) \sim (10^3, 10^4)$, for instance.

The minimization problem (6.287) leads to

$$\begin{aligned}
&f(\mathbf{x}) = \tfrac{1}{2}\mathbf{x}^{\mathrm{T}} C \mathbf{x} - \mathbf{x}^{\mathrm{T}}\mathbf{d}_0 \to \min \\
&\mathbf{x}^{\mathrm{T}}\mathbf{a}_i \le \mathbf{0}, \qquad 1 \le i \le m,
\end{aligned}$$

where $\mathbf{x}, \mathbf{a}_i, \mathbf{d}_0 \in \mathbb{R}^N$. The idea of the algorithm is based on the sequential minimization of $f(\mathbf{x})$ on sides, determined by such rows $\mathbf{a}_i$ in which the equality is set. On every side we solve the minimization process by the conjugate gradient method with constraints [for details see Nedoma et al. (1999a), Nedoma (1998a), and Kestřánek (1999)]. Since the number of steps in the conjugate gradient method is finite, and, moreover, since the number of sides is also finite, the algorithm converges after a finite step of Uzawa's iterations. For a better convergence the preconditioned Uzawa/CGM techniques can be used [see, e.g., Nedoma (1998a) and Nedoma et al. (1999b)], where it is shown that the preconditioning incomplete and complete Choleski method are the best methods and that they converge quickly.

Remark 6.20 Since the stiffness matrix remains the same during the whole iteration process, the number of the approximated displacement vector $\mathbf{u}_h$ subject to the condition on $\cup\Gamma_c^{sm}$ is small when compared to the total number of components of $\mathbf{u}_h$, we can use the effective algorithm of the problem discussed. The linear term of the Lagrangian $\mathcal{H}$ changes during the iteration process only those components that correspond to the components of $\mathbf{u}_h$ on $\cup\Gamma_c^{sm}$. Therefore, it is possible to eliminate the free components of $\mathbf{u}_h$ corresponding to the nodes not belonging to $\cup\Gamma_c^{sm}$, and to carry out the iteration process with only the other components of $\mathbf{u}_h$ [for details see Haslinger and Tvrdý (1983), Hlaváček et al. (1988), and Nedoma and Stehlík (1995)].

The Mortar Approach—The Nonmatching Case We formulate the problem as a saddle point (mixed) problem. To define a saddle point formulation, we introduce a Lagrange multiplier space M. Let us assume that Ω^ι, $\iota = 1, \ldots, r$, are domains with sufficiently smooth boundaries $\partial\Omega^\iota$, then we can introduce the trace space $W = \prod_{s,m} [H^{1/2}(\Gamma_c^{sm})]^N = \prod_{s,m} H^{1/2,N}(\Gamma_c^{sm})$, being the trace space of V_0 restricted to $\cup_{s,m}\Gamma_c^{sm}$, and its dual W'.

Remark 6.21 In the general case, if $\overline{\Gamma}_c^s = \partial\Omega^s \cap (\Gamma_u \cap \partial\Omega^s)$, we will use $H_{00}^{1/2,N}(\Gamma_c^{sm}) \equiv [H_{00}^{1/2}(\Gamma_c^{sm})]^N$ instead of $H^{1/2,N}(\Gamma_c^{sm})$, where $H_{00}^{1/2,N}(\Gamma_c^{sm})$ is the Hilbertian interpolation space between $L^{2,N}(\Gamma_c^{sm})$ and $H_0^{1,N}(\Gamma_c^{sm})$. Its dual is $H_{00}^{-1/2,N}(\Gamma_c^{sm}) = (H_{00}^{1/2,N}(\Gamma_c^{sm}))'$ and we note that $H_{00}^{1/2,N}(\Gamma_c^{sm})$ is a proper and continuously embedded subspace of $H^{1/2,N}(\Gamma_c^{sm})$ [see Lions and Magenes (1972)], and it is assigned with the norm $\|\mathbf{z}\|_{H_{00}^{1/2,N}(\Gamma_c^{sm})}$.

Elastic Case, the Second Variant of the Mixed Formulation Let $\mathbf{u}, \mathbf{v} \in V_0 = \{\mathbf{v} \in \prod_{\iota=1}^r H^{1,N}(\Omega^\iota) | \mathbf{v} = 0 \text{ on } \Gamma_u,\ v_n = 0 \text{ on } \Gamma_0\}$, $\boldsymbol{\mu} = (\mu_n, \boldsymbol{\mu}_t) \in W' \times L^{2,N-1}(\Gamma_c^s)$. We define

$$A(\mathbf{u}, \mathbf{v}) = a^{(0)}(\mathbf{u}, \mathbf{v}) = \int_\Omega c_{ijkl}(\mathbf{x}) e_{kl}(\mathbf{x}) e_{ij}(\mathbf{x})\, d\mathbf{x},$$

$$(\mathbf{f}, \mathbf{v}) = \int_\Omega \mathbf{F} \cdot \mathbf{v}\, d\mathbf{x} + \int_{\Gamma_\tau} \mathbf{P} \cdot \mathbf{v}\, ds,$$

$$j(\mathbf{v}) - \int_{\cup_s \Gamma_c^s} g_c^{sm} |\mathbf{v}_t^s - \mathbf{v}_t^m|\, ds \equiv \int_{\cup_s \Gamma_c^s} g_c^{sm} |[\mathbf{v}_t]^{sm}|\, ds,$$

and

$$b(\boldsymbol{\mu}, \mathbf{v}) = \langle \mu_n, \ [\mathbf{v} \cdot \mathbf{n}]^s - d^{sm} \rangle_{\cup_s \Gamma_c^s} + \int_{\cup_s \Gamma_c^s} g_c^{sm} \boldsymbol{\mu}_t \cdot [\mathbf{v}_t]^s \, ds,$$

where

$$[\mathbf{v} \cdot \mathbf{n}]^s(\mathbf{x}) = v_n^s(\mathbf{x}) - v_n^m(\mathcal{R}^{sm}(\mathbf{x})), \qquad [\mathbf{v}_t]^s(\mathbf{x}) = \mathbf{v}_t^s(\mathbf{x}) - \mathbf{v}_t^m(\mathcal{R}^{sm}(\mathbf{x})), \quad (6.289)$$

where $\mathcal{R}^{sm} : \Gamma_c^s(t) \mapsto \Gamma_c^m(t)$, at $t \in I$, is a bijective map satisfying $\Gamma_c^m(t) \subset \mathcal{R}^{sm}(\Gamma_c^s(t))$, $t \in I$, that is, $\mathcal{R}^{sm}$ at $t \in I$ may be defined by mapping any $\mathbf{x}(t) \in \Gamma_c^s(t)$ to the intersection of the normal to $\Gamma_c^s(t)$ at $\mathbf{x}(t)$ with $\Gamma_c^m(t)$, and therefore, $\Gamma_c^m(t)$ is defined as the image of $\mathcal{R}^{sm}(\mathbf{x}(t))$. Here $\langle \cdot, \rangle_{\Gamma_c^s}$ denotes the duality pairing between W and W'. Moreover, we introduce

$$M_n = \{\mu_n \in W'; \mu_n \geq 0\},$$
$$M_t = \{\boldsymbol{\mu}_t \in L^{2,N-1}(\cup_{sm} \Gamma_c^{sm}) | \|\boldsymbol{\mu}_t\| \leq 1 \text{ a.e.}, \boldsymbol{\mu}_t = 0 \text{ on } \cup_{sm} \Gamma_c^{sm} \setminus \operatorname{supp} g_c^{sm}\}.$$

Here $\mu_n \geq 0$ means that $\langle \mu_n, \mathbf{w}^s \cdot \mathbf{n}^s \rangle \geq 0$ for all $\mathbf{w} \in W$ such that $\mathbf{w}^s \cdot \mathbf{n}^s \geq 0$.

We will assume that $\mathbf{F} \in L^2(I; L^{2,N}(\Omega))$, $\mathbf{P} \in L^2(I; L^{2,N}(\Gamma_\tau))$, $c_{ijkl}(\mathbf{x}) \in L^\infty(\Omega)$, and that $\Gamma_u, \Gamma_\tau, \Gamma_0, \Gamma_c$ have positive surface measures.

Then at every time level the mixed formulation of the contact problem with given friction consists of finding $\mathbf{u} \in V = \mathbf{u}_2 + V_0$ and $\boldsymbol{\lambda} = (\lambda_n, \boldsymbol{\lambda}_t) \in M = M_n \times M_t$ such that

$$A(\mathbf{u}, \mathbf{v}) + b(\boldsymbol{\lambda}, \mathbf{v}) = (\mathbf{f}(t), \mathbf{v}) \quad \forall \mathbf{v} \in V_0, \ t \in I,$$
$$b(\boldsymbol{\mu} - \boldsymbol{\lambda}, \mathbf{u}) \leq 0 \quad \forall \boldsymbol{\mu} \in M, \ t \in I. \qquad (6.290)$$

The existence and uniqueness of $(\boldsymbol{\lambda}, \mathbf{u})$ of this saddle point problem have been analyzed, for example, in Haslinger et al. (1996), Haslinger and Hlaváček (1982), and Hlaváček (2007) even in a semicoercive case for $N = 2$. Thus, we have:

Proposition 6.1 Let $-\tau_n(\mathbf{u}) \in M_n$. Then problem (6.290) has a unique solution $(\boldsymbol{\lambda}, \mathbf{u}) \in V \times M$, a.e. $t \in I$. Moreover, we have

$$\lambda_n^s = -\tau_n^s(\mathbf{u}) \quad \text{and} \quad g_c^s \boldsymbol{\lambda}_t^s = -\boldsymbol{\tau}_t^s(\mathbf{u}),$$

where $\mathbf{u} = \mathbf{u}(t)$ is the solution of the primal problem (6.292), where $g_c^s = \mathcal{F}_c^s(\mathbf{u}^s)|\tau_n(\mathbf{u}^s)|$.

The problem is equivalent to that of the Lagrangian formulation: Find a pair $(\mathbf{u}, \boldsymbol{\lambda}) \in V_0 \times (M_n \times M_t)$, a.e. $t \in I$, such that

$$\mathcal{H}(\mathbf{u}, \boldsymbol{\mu}) \leq \mathcal{H}(\mathbf{u}, \boldsymbol{\lambda}) \leq \mathcal{H}(\mathbf{v}, \boldsymbol{\lambda}) \quad \forall (\mathbf{v}, \boldsymbol{\mu}) \in V_0 \times (M_n \times M_t), \ t \in I, \qquad (6.291)$$

where

$$\mathcal{H}(\mathbf{u}, \boldsymbol{\mu}) = \tfrac{1}{2} A(\mathbf{u}, \mathbf{v}) + b(\boldsymbol{\mu}, \mathbf{v}) - (\mathbf{f}(t), \mathbf{v}).$$

Remark 6.22 It can be verified that $g_c^s \boldsymbol{\lambda}_t^s = -\boldsymbol{\tau}_t^s(\mathbf{u}^s)$ on Γ_c^s. If more regularity for the solution $\mathbf{u}$ is assumed, for example, $\mathbf{u}^s \in H^{q,N}(\Omega^s)$, $q > \frac{3}{2}$, then $\tau_n^s(\mathbf{u}^s)$ can be expressed pointwise and $\lambda_n^s = -\tau_n^s(\mathbf{u}^s)|_{\Gamma_c^s}$.

Remark 6.23 Problem (6.290) or (6.291) is equivalent to the following primal problem:

$$\mathbf{u} \in K, a(\mathbf{u}, \mathbf{v} - \mathbf{u}) + j(\mathbf{v}) - j(\mathbf{u}) \geq (\mathbf{f}(t), \mathbf{v} - \mathbf{u}) \quad \forall \mathbf{v} \in K,\ t \in I, \tag{6.292}$$

where $K = \{\mathbf{v} \in V, [\mathbf{v} \cdot \mathbf{n}]^{sm} \leq d^{sm} \text{ on } \cup_{sm} \Gamma_c^{sm}\}$.

Remark 6.24 If $\gamma_a^{sm} \subset \Gamma_c^{sm}$ is the actual set and if the displacement $\mathbf{u}$ is a continuous function, then the actual contact zone γ_a^{sm} is a well-defined and closed subset of Γ_c^{sm}.

Viscoelastic Case, the Second Variant of the Mixed Formulation We formulate the problem via its mixed (i.e., a saddle point) formulation. Therefore, we introduce a Lagrange multiplier space M. We assume that Ω^ι, $\iota = 1, \ldots, r$, are domains with sufficiently smooth boundaries $\partial\Omega^\iota$; then we introduce the trace space $W = \prod_{s,m} H^{1/2,N}(\Gamma_c^{sm})$, being the normal trace space of $V_0 = \prod_{\iota=1}^r V_0^\iota$ restricted to $\cup_{s,m}\Gamma_c^{sm}$ and its dual W'. Then we introduce the sets M_n, M_t, the mapping $\mathcal{R}^{sm}$, and the form $b(\boldsymbol{\mu}, \mathbf{v})$ as in the previous elastic case.

Let $M_{hH} = M_{hn} \times M_{Ht}$ be a discrete approximation of the Lagrange multiplier space $M = M_n \times M_t$. We then introduce the spaces describing the degree of the polynomial approximations as follows:

$$\begin{aligned} W_{hH}^1(\cup_s\Gamma_c^s) &= W_{hn}^1(\cup_s\Gamma_c^s) \times W_{Ht}^1(\cup_s\Gamma_c^s) \\ &= \{\mathbf{v}_h^s \cdot \mathbf{n}^s|_{\cup\Gamma_c^{sm}}, \mathbf{v}_h \in V_h\} \times \{\mathbf{v}_h^s \cdot \mathbf{t}^s|_{\cup\Gamma_c^{sm}}, \mathbf{v}_h \in V_h\}, \end{aligned}$$

$$M_{hn}^1 = \left\{\mu_{hn} \in W_{hn}^1(\cup_s\Gamma_c^s), \int_{\Gamma_c^s} \mu_{hn}\psi_h\, ds \geq 0, \quad \forall \psi_h \in W_{hn}^1, \psi_h \geq 0 \right.$$
$$\left. \text{a.e. on every } \Gamma_c^s \right\},$$

$$M_{Ht}^1 = \left\{\boldsymbol{\mu}_{Ht} \in W_{Ht}^1(\cup_s\Gamma_c^s), \int_{\Gamma_c^s} \boldsymbol{\mu}_{Ht}\boldsymbol{\psi}_H\, ds - \int_{\Gamma_c^s} g_c^{sm}|\boldsymbol{\psi}_H|\, ds \leq 0, \right.$$
$$\left. \forall \boldsymbol{\psi}_H \in W_{Ht}^1(\cup_s\Gamma_c^s) \right\},$$

or

$$\begin{aligned} W_{hH}^0(\cup_s\Gamma_c^s) &= W_{hn}^0(\cup_s\Gamma_c^s) \times W_{Ht}^0(\cup_s\Gamma_c^s) \\ &= \{\boldsymbol{\mu}_{hH}|_{\Delta_r} \in [P_0(\Delta_r)]^N,\ 0 \leq r \leq n(h^s)\}, \end{aligned}$$

$$M_{hn}^0 = \{\mu_{hn} \in W_{hn}^0(\cup_s\Gamma_c^s), \mu_{hn} \geq 0, \text{ a.e. on } \Gamma_c^s\},$$

$$M_{Ht}^0 = \left\{\boldsymbol{\mu}_{hH} \in W_{hH}^0(\cup_s\Gamma_c^s), \int_{\Gamma_c^s} \boldsymbol{\mu}_{Ht}\boldsymbol{\psi}_H\, ds - \int_{\Gamma_c^s} g_c^{sm}|\boldsymbol{\psi}_H|\, ds \leq 0, \right.$$
$$\left. \forall \boldsymbol{\psi}_H \in W_{Ht}^0(\cup_s\Gamma_c^s) \right\}$$

and where Γ_c^s is created by a finite number of parts Γ_c^{sp}, $p=1,\ldots,\bar{p}$, $\Gamma_c^{sp}=\cup_r \Delta_r$, where $\Delta_r \equiv q_r=(s_r, s_r+h_p)$, $r=0,1,\ldots,m-1$, $m=m(h_p)$, for $N=2$, and $\Gamma_c^{sp}=\cup_r \Delta_r$, where Δ_r are faces of polyhedra of the given partition, $r=0,1,\ldots,m-1$, $m=m(h_p)$, for $N=3$.

For a suitable partition Γ_c^{sp} in the 2D case, see for example, Baillet and Sassi (2002) or Hlaváček (2007).

In the following discussion we will consider the problem without friction. Then we have the following problem:

$$A(\mathbf{u}_h,\mathbf{v}_h)+b(\boldsymbol{\lambda}_{hH},\mathbf{v}_h)=(\mathbf{f},\mathbf{v}_h) \quad \forall \mathbf{v}_h \in V_h \equiv \prod_{\iota=1}^{r} V_h^{\iota},$$

$$b(\boldsymbol{\mu}_{hH}-\boldsymbol{\lambda}_{hH},\mathbf{v}_h) \leq \langle d^{sm},(\mu_{hn}-\lambda_{hn})\rangle_{\cup_{s,m}\Gamma_c^{sm}} \quad \forall \boldsymbol{\mu}_{hH} \in M_{hH}, \tag{6.293}$$

where $M_{hH}=M_{hn}^1$ or $M_{hH}=M_{hn}^0$ and where

$$A(\mathbf{u}_h,\mathbf{v}_h)=(\rho\mathbf{u}_h,\mathbf{v}_h)+\Delta t(\alpha\mathbf{u}_h,\mathbf{v}_h)+\Delta t^2 a^{(0)}(\mathbf{u}_h,\mathbf{v}_h)+\Delta t a^{(1)}(\mathbf{u}_h,\mathbf{v}_h),$$

for the dynamic viscoelastic case and

$$A(\mathbf{u}_h,\mathbf{v}_h)=\Delta t a^{(0)}(\mathbf{u}_h,\mathbf{v}_h)+a^{(1)}(\mathbf{u}_h,\mathbf{v}_h)$$

for the quasi-static viscoelastic case, and, moreover,

$$A(\mathbf{u}_h,\mathbf{v}_h)=(\rho\mathbf{u}_h,\mathbf{v}_h)+\Delta t(\alpha\mathbf{u}_h,\mathbf{v}_h)+\Delta t^2 a^{(0)}(\mathbf{u}_h,\mathbf{v}_h),$$

$$A(\mathbf{u}_h,\mathbf{v}_h)=\Delta t a^{(0)}(\mathbf{u}_h,\mathbf{v}_h)$$

for the elastic dynamic and quasi-static cases.

To ensure the existence and uniqueness of a solution to (6.293), it is necessary to verify that

$$\{\boldsymbol{\mu}_{hH} \in M_{hH}, b(\boldsymbol{\mu}_{hH},\mathbf{v}_h)=0,\ \forall \mathbf{v}_h \in V_h\}=\{0\},$$

which is obvious if $M_{hH}=M_{hH}^1$. For $M_{hH}=M_{hH}^0$, see Haslinger et al. (1996), where the discrete Babuška–Brezzi "inf-sup" condition

$$\inf_{\boldsymbol{\mu}_{hH}\in M_{hH}(\cup_{s,m}\Gamma_c^{sm})} \sup_{\mathbf{v}_h \in V_h} \frac{b(\boldsymbol{\mu}_{hH},\mathbf{v}_h)}{\|\boldsymbol{\mu}_{hH}\|_{-1/2,\cup_{s,m}\Gamma_c^{sm}}\|\mathbf{v}_h\|_1} \geq \beta,$$

where

$$\|\mathbf{v}_h\|_1=\left(\sum_{\iota=1}^{r}\|\mathbf{v}_h^{\iota}\|_{1,\Omega^{\iota}}^2\right)^{1/2} \quad \text{and} \quad \beta>0,$$

must be satisfied.

Proposition 6.2 Let $-\tau_n(\mathbf{u}_h) \in M_{hn}$. Then problem (6.293) has a unique solution $(\boldsymbol{\lambda}_{hH},\mathbf{u}_h) \in V_h \times M_{hH}$, a.e. $t \in I$. Moreover, we have

$$\lambda_{hn}^s=-\tau_n^s(\mathbf{u}_h),$$

where $\mathbf{u}_h$ is the solution of the discrete primal problem.

We define the Lagrangian

$$\mathcal{H}(\boldsymbol{\mu}, \mathbf{v}) = \tfrac{1}{2}A(\mathbf{v}_h, \mathbf{v}_h) + b(\boldsymbol{\mu}_{hH}, \mathbf{v}_h) - (\mathbf{f}, \mathbf{v}_h).$$

Problem (6.293) is equivalent to that of the Lagrangian formulation: Find for every $t \in I$ a pair $(\boldsymbol{\lambda}_{hH}, \mathbf{v}_h) \in M_{hn} \times V_h$, such that

$$\mathcal{H}(\boldsymbol{\mu}_{hH}, \mathbf{u}_h) \leq \mathcal{H}(\boldsymbol{\lambda}_{hH}, \mathbf{u}_h) \leq \mathcal{H}(\boldsymbol{\lambda}_{hH}, \mathbf{v}_h) \quad \forall(\boldsymbol{\mu}_{hH}, \mathbf{v}_h) \in M_{hn} \times V_h, t \in I. \tag{6.294}$$

Matrix Formulation and the Primal–Dual Active Set (PDAS) Strategy Method for the Model Problem in Viscoelastic Rheology In this section, we will derive the matrix formulation of (6.292) for the case without friction. Then we will apply the primal-dual active set strategy to the mixed formulation of the multibody contact problem.

The interface Γ_c^{sm} has two "sides"; the side from Ω^s will be called the "slave" side and the side from Ω^m will be denoted as the "master" side.

In Wolhmuth and Krause (2003) it was shown that the standard basis of the space V_h is not a good choice. The acceptable idea is to modify the nodal basis functions on the mortar side in such a way that they will be biorthogonal with respect to the standard piecewise linear basis on the slave side.

The N = 2 Case A suitable "dual" basis for $N = 2$ can be defined by discontinuous piecewise linear functions as follows:

$$\begin{aligned} &\psi_i(s_i) = 2 \text{ (at the node } s_i \in \Gamma_c^{sm}), \\ &\psi_i(s_{i-1}) = \psi_i(s_{i+1}) = -1, \qquad \psi_i(s_i) = 0 \quad \text{for } s \in \Gamma_c^{sm}\backslash[s_{i-1}, s_i], \\ &\qquad 2 \leq i \leq m-1. \end{aligned} \tag{6.295}$$

Let s_0, s_m be the endpoints. Then functions ψ_1 and ψ_m are modified in the intervals $[s_0, s_1]$ and $[s_{m-1}, s_m]$, respectively, where they are equal to the constant 1.

Denoting $\{\varphi_j\}, j = 1, \ldots, m$, the standard piecewise linear basis on the slave side, that is, the basis of a component of $W_{hH}^1(\Gamma_c^{sm})$, we have

$$\int_{\Gamma_c^{sm}} \varphi_j \psi_i \, ds = \delta_{ij} \int_{\Gamma_c^{sm}} \varphi_j \, ds, \qquad 1 \leq i, j \leq m, \tag{6.296}$$

where δ_{ij} is the Kronecker delta.

Let us define

$$M_h^s = \{\psi_i e_k, \ i = 1, \ldots, m; \ k = 1, 2\}, \quad s \in \{1, \ldots, r\},$$

where e_k are components of the unit Cartesian basis, and

$$M_h^{s+} = \{\mu_h \in M_h^s | \langle \mu_h, \mathbf{v}_h \rangle_{\cup \Gamma_c^{sm}} \geq 0 \quad \forall v_h \in W_h^{s+}\}, \tag{6.297}$$

where

$$W_h^{s+} = \{\mathbf{v}_h \in W_{hH}^1(\Gamma_c^s) | \mathbf{v}_h \cdot \mathbf{n}^s \geq 0\}. \tag{6.298}$$

It is proved in Hüeber and Wolhmuth (2005a) that

$$M_h^{s+} = \left\{ \mu_h = \sum_{i=1}^{m} \alpha_i \psi_i | \alpha_i = \alpha_i^n n^s, \alpha_i^n \in \mathbb{R}, \alpha_i^n \geq 0,\ i \leq m \right\} \tag{6.299}$$

provided Γ_c^{sm} is a straight segment.

Finally, we define

$$M_h^+ = \prod_s M_h^{s+}, \tag{6.300}$$

$$b(\mu_h, \mathbf{v}_h) = \int_{\cup\Gamma^{sm}} \mu_h \cdot \mathbf{n}^s [\mathbf{v}_h \cdot \mathbf{n}]^s \, ds. \tag{6.301}$$

The discrete mortar formulation of the saddle point problem (6.294) for every time level is defined as follows:

Problem $(\mathcal{P}_{sp})_{dm}$ In every time level find $\mathbf{u}_h \in V_h$, $\lambda_h \in M_h^+$, satisfying

$$\begin{aligned} &A(\mathbf{u}_h, \mathbf{v}_h) + b(\lambda_h, \mathbf{v}_h) = (\mathbf{f}, \mathbf{v}_h), && \mathbf{v}_h \in V_h,\ t \in I \\ &b(\mu_h - \lambda_h, \mathbf{v}_h) \leq \langle d^{sm}, (\mu_h - \lambda_h) \rangle_{\cup\Gamma_c^{sm}} && \forall \mu_h \in M_h^+,\ t \in I. \end{aligned} \tag{6.302a,b}$$

It is readily seen that the inequality (6.302) implies the two conditions:

$$(\lambda_h, \mathbf{u}_h) = \langle d^{sm}, \lambda_h \rangle_{\cup\Gamma_c^{sm}}, \tag{6.303}$$

$$b(\mu_h, \mathbf{u}_h) \leq \langle d^{sm}, \mu_h \rangle_{\cup\Gamma_c^{sm}} \quad \forall \mu_h \in M_h^+. \tag{6.304}$$

If Γ_c^{sm} are straight segments, we can employ (6.299) to infer that condition (6.304) implies a weak nonpenetration condition (6.305). It means that the strong form of the nonpenetration condition $[\mathbf{u} \cdot \mathbf{n}]^s \leq d^{sm}$ will be replaced by its weak discrete form

$$\int_{\cup\Gamma_c^s} [\mathbf{u}_h \cdot \mathbf{n}]^s \psi_p \, ds \leq \int_{\cup\Gamma_c^s} d_p^s \psi_p \, ds, \quad p \in S, \tag{6.305}$$

where S is the set of all vertices in the potential contact part on the slave side. The constraints (6.305) give a coupling between the vertices on the slave side and the master side. We introduce now a transformation of the basis of V_h in such a way that the weak nonpenetration condition (6.295) in the new basis deals only with the vertices on the slave side.

The Lagrange multipliers λ_h are approximations of the contact stress vector $-\boldsymbol{\tau}(\mathbf{u}_h^s)$. For example, in Fig. 6.4 the 2D linear trial functions are shown.

Let us denote the nodal parameters $\mathbf{u}_h$ by $\mathbf{U}$, of λ_h by $\boldsymbol{\Lambda}_h$, and since for the frictionless case the tangential stress component on the contact boundary $\cup\Gamma_c^{sm}$ is

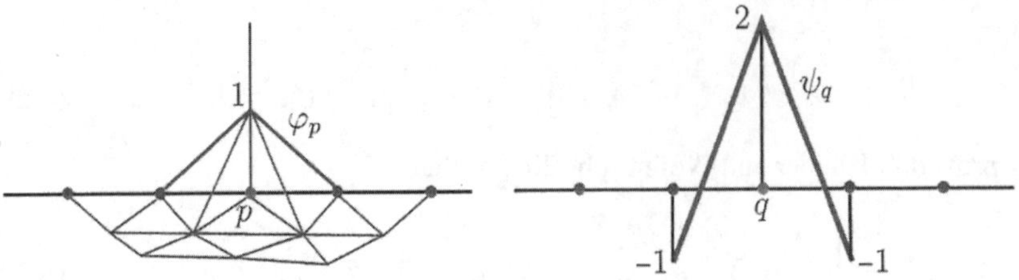

FIG. 6.4 Linear trial functions of the dual basis.

zero, then $\boldsymbol{\lambda}_{Ht} = 0$, and thus $\boldsymbol{\Lambda}_{Ht} = 0$. The dual Lagrange multipliers will be used [see Wolhmuth (2000), Hüeber and Wolhmuth (2005a,b), and Brunsson et al. (2007)].

For every $t \in I$ in the matrix form Eq. (6.302a) is as follows:

$$\mathbb{A}_h \mathbf{U} + \mathbb{B}_h \boldsymbol{\Lambda}_h = \mathbb{F}_h. \tag{6.306}$$

To examine the structure of $\mathbb{B}_h$, we introduce three sets of indices $\mathcal{N}$, $\mathcal{M}$, and $\mathcal{S}$. We decompose the set of all vertices in every time $t \in I$ into three disjoint parts $\mathcal{N}$, $\mathcal{M}$, and $\mathcal{S}$, where by $\mathcal{S}(\dim n_{cn})$ we denote all vertices on the possible contact parts on the slave side, by $\mathcal{M}(\dim n_{cn})$ all vertices of the possible contact part on the master side, and by $\mathcal{N}$ all the other ones. Then the strong formulation of the nonpenetration condition (i.e., $[\mathbf{u}_h \cdot \mathbf{n}]^{sm} \leq d^{sm}$ on $\cup\Gamma_c^{sm}$) will be replaced by its weak discrete form (6.305).

This condition connects points of sets $\mathcal{S}$ and $\mathcal{M}$. We introduce a transformation of the basis of the space V_h in such a way that the weak nonpenetration condition (6.305) in the new basis deals only with the vertices on the slave side [see Hüeber and Wolhmuth (2005a) and Wolhmuth and Krause (2003)].

Let us introduce the transformation of the basis $\varphi = (\varphi_{\mathcal{N}}, \varphi_{\mathcal{M}}, \varphi_{\mathcal{S}})^{\mathrm{T}}$. The matrices and vectors in (6.306) can be decomposed in the sense of decomposition of the set of all vertices into three disjoint parts $\mathcal{N}$, $\mathcal{M}$, and $\mathcal{S}$, that is,

$$\begin{bmatrix} \mathbb{A}_{\mathcal{N}\mathcal{N}} & \mathbb{A}_{\mathcal{N}\mathcal{M}} & \mathbb{A}_{\mathcal{N}\mathcal{S}} & \mathbb{O} \\ \mathbb{A}_{\mathcal{M}\mathcal{N}} & \mathbb{A}_{\mathcal{M}\mathcal{M}} & \mathbb{A}_{\mathcal{M}\mathcal{S}} & -\mathbb{M}^{\mathrm{T}} \\ \mathbb{A}_{\mathcal{S}\mathcal{N}} & \mathbb{A}_{\mathcal{S}\mathcal{M}} & \mathbb{A}_{\mathcal{S}\mathcal{S}} & \mathbb{D} \end{bmatrix} \begin{bmatrix} \mathbb{U}_{\mathcal{N}} \\ \mathbb{U}_{\mathcal{M}} \\ \mathbb{U}_{\mathcal{S}} \\ \boldsymbol{\Lambda}_{\mathcal{S}} \end{bmatrix} = \begin{bmatrix} \mathbb{F}_{\mathcal{N}} \\ \mathbb{F}_{\mathcal{M}} \\ \mathbb{F}_{\mathcal{S}} \end{bmatrix}. \tag{6.307}$$

Matrix $\mathbb{M}$ represents the coupling between the traces of the finite element shape functions on the master side m and the shape functions for the Lagrange multiplier on the slave side s, defined by

$$\mathbb{M}[p,q] = \int_{\cup\Gamma_c^{sm}} \varphi_p \psi_q \, ds \mathbb{I}_2, \qquad p \in \mathcal{S}, q \in \mathcal{M}, \tag{6.308}$$

where $\mathbb{I}_2$ denotes the identity matrix in $\mathbb{R}^{2\times 2}$. The matrix $\mathbb{M}$ is the block matrix and $\mathbb{D}$ is the block diagonal matrix. Due to (6.296), the entries of $\mathbb{D}$ are given by

$$\mathbb{D}[p,q] = \delta_{pq} \mathbb{I}_2 \cdot \int_{\cup\Gamma_c^{sm}} \varphi_p \psi_p \, ds, \qquad p \in \mathcal{S} \equiv \{1, \ldots, n_{cn}\}. \tag{6.309}$$

The structure of matrix $\mathbb{B}_h$ in every time level $t \in I$ is thus of the form $\mathbb{B}_h = (\mathbb{O}, -\mathbb{M}^{\mathrm{T}}, \mathbb{D})^{\mathrm{T}}$. The block matrices associated with the basis functions of the free structure nodes (i.e., $\mathcal{N}$), the potential contact nodes of the master side (i.e., $\mathcal{M}$), and the potential contact nodes on the slave side (i.e., $\mathcal{S}$) are denoted by $\mathbb{A}_{k,l}$, $k, l \in \{\mathcal{N}, \mathcal{M}, \mathcal{S}\}$. The entries of vectors $\mathbb{U}$ and $\mathbb{F}$ for $k \in \{\mathcal{N}, \mathcal{M}, \mathcal{S}\}$ are denoted by $\mathbb{U}_k$ and $\mathbb{F}_k$, respectively.

Now, our goal is to introduce a suitable new modified basis $\Phi = (\Phi_{\mathcal{N}}, \Phi_{\mathcal{M}}, \Phi_{\mathcal{S}})^{\mathrm{T}}$ instead of the basis $\varphi = (\varphi_{\mathcal{N}}, \varphi_{\mathcal{M}}, \varphi_{\mathcal{S}})^{\mathrm{T}}$. If we introduce the matrix $\hat{\mathbb{M}} = \mathbb{D}^{-1}\mathbb{M}$, then the matrix $\hat{\mathbb{M}}$ is sparse and Φ can be defined by

$$\Phi = (\Phi_{\mathcal{N}}, \Phi_{\mathcal{M}}, \Phi_{\mathcal{S}})^{\mathrm{T}} = \begin{bmatrix} \mathbb{I}_2 & \mathbb{O} & \mathbb{O} \\ \mathbb{O} & \mathbb{I}_2 & \hat{\mathbb{M}}^{\mathrm{T}} \\ \mathbb{O} & \mathbb{O} & \mathbb{I}_2 \end{bmatrix} \begin{bmatrix} \varphi_{\mathcal{N}} \\ \varphi_{\mathcal{M}} \\ \varphi_{\mathcal{S}} \end{bmatrix} = Q\varphi. \tag{6.310}$$

Then

$$\mathbb{U} = Q^{\mathrm{T}}\hat{\mathbb{U}},$$

where $\hat{\mathbb{U}}$ is the vector of coefficients with respect to the transformed basis Φ.

Let us give an algebraic representation of (6.310). The modified stiffness matrix $\hat{\mathbb{A}}_h$ associated with the transformed basis Φ is as follows:

$$\begin{aligned} \hat{\mathbb{A}}_h &= \begin{bmatrix} \mathbb{I}_2 & \mathbb{O} & \mathbb{O} \\ \mathbb{O} & \mathbb{I}_2 & \hat{\mathbb{M}}^{\mathrm{T}} \\ \mathbb{O} & \mathbb{O} & \mathbb{I}_2 \end{bmatrix} \begin{bmatrix} \mathbb{A}_{\mathcal{NN}} & \mathbb{A}_{\mathcal{NM}} & \mathbb{A}_{\mathcal{NS}} \\ \mathbb{A}_{\mathcal{MN}} & \mathbb{A}_{\mathcal{MM}} & \mathbb{A}_{\mathcal{MS}} \\ \mathbb{A}_{\mathcal{SN}} & \mathbb{A}_{\mathcal{SM}} & \mathbb{A}_{\mathcal{SS}} \end{bmatrix} \begin{bmatrix} \mathbb{I}_2 & \mathbb{O} & \mathbb{O} \\ \mathbb{O} & \mathbb{I}_2 & \mathbb{O} \\ \mathbb{O} & \hat{\mathbb{M}} & \mathbb{I}_2 \end{bmatrix} \\ &= Q\mathbb{A}_h Q^{\mathrm{T}} \\ &= \begin{bmatrix} \mathbb{A}_{\mathcal{NN}} & \mathbb{A}_{\mathcal{NM}} + \mathbb{A}_{\mathcal{NS}}\hat{\mathbb{M}} & \mathbb{A}_{\mathcal{NS}} \\ \mathbb{A}_{\mathcal{MN}} + \hat{\mathbb{M}}^{\mathrm{T}}\mathbb{A}_{\mathcal{SN}} & \mathbb{A}_{\mathcal{MM}} + \mathbb{A}_{\mathcal{MS}}\hat{\mathbb{M}} + & \mathbb{A}_{\mathcal{MS}} + \hat{\mathbb{M}}^{\mathrm{T}}\mathbb{A}_{\mathcal{SS}} \\ & +\hat{\mathbb{M}}^{\mathrm{T}}\mathbb{A}_{\mathcal{SM}} + \hat{\mathbb{M}}^{\mathrm{T}}\mathbb{A}_{\mathcal{SS}}\hat{\mathbb{M}} & \\ \mathbb{A}_{\mathcal{SN}} & \mathbb{A}_{\mathcal{SM}} + \mathbb{A}_{\mathcal{SS}}\hat{\mathbb{M}} & \mathbb{A}_{\mathcal{SS}} \end{bmatrix}, \end{aligned} \tag{6.311}$$

and the modified vector of the right-hand side $\hat{\mathbb{F}}_h$ is as follows:

$$\hat{\mathbb{F}}_h = Q\mathbb{F}_h = \begin{bmatrix} \mathbb{F}_{\mathcal{N}} \\ \mathbb{F}_{\mathcal{M}} + \hat{\mathbb{M}}^{\mathrm{T}}\mathbb{F}_{\mathcal{S}} \\ \mathbb{F}_{\mathcal{S}} \end{bmatrix}. \tag{6.312}$$

For the algebraic representation of the weak nonpenetration condition we shall need the following formula:

$$\begin{aligned} [\mathbf{u}_h \cdot \mathbf{n}]^s &= \sum_{p \in \mathcal{S}} \hat{\mathbb{U}}_p [\Phi_{pn}]^s + \sum_{q \in \mathcal{M}} \hat{\mathbb{U}}_q [\Phi_{qn}]^s \\ &= \left(\sum_{p \in \mathcal{S}} \hat{\mathbb{U}}_p \varphi_p \right) \cdot \mathbf{n}_p^s + \sum_{q \in \mathcal{M}} \hat{\mathbb{U}}_q \left(-\varphi_q + \sum_{p' \in \mathcal{S}} \hat{\mathbb{M}}[p', q] \varphi_{p'} \right) \cdot \mathbf{n}_p^s. \end{aligned}$$

By multiplying this equation with ψ_p, $p \in \mathcal{S}$ and integrating the resulting equation over $\cup_s \Gamma_c^s$, due to the biorthogonality condition (6.296) the definition of matrix $\hat{\mathbb{M}}$ and (6.310), the weak nonpenetration condition (6.305) will imply that

$$\hat{\mathbb{U}}_{n,p} \equiv (\mathbf{n}_p^s)^{\mathrm{T}} \mathbb{D}[p,p] \hat{\mathbb{U}}_p \leq d_p^{sm} \quad \forall p \in \mathcal{S}, \tag{6.313}$$

where $d_p^{sm} = \int_{\cup_s \Gamma_c^s} d_h^{sm} \psi_p \, ds$, $p \in \mathcal{S}$, as the coefficients at $\hat{\mathbb{U}}_q$, $q \in \mathcal{M}$, are nullified. This basis transformation glues the vertices on the slave side and on the master side together. The displacements of the glued vertices are given by $\hat{\mathbb{U}}_{\mathcal{M}}$ and the relative displacements between the vertices on the slave side and on the master side are given by $\hat{\mathbb{U}}_{\mathcal{S}}$.

The modified matrix $\hat{\mathbb{B}}_h$ associated with the transformed basis Φ has the form

$$\hat{\mathbb{B}}_h = Q\mathbb{B}_h = \begin{bmatrix} \mathbb{I}_2 & \mathbb{O} & \mathbb{O} \\ \mathbb{O} & \mathbb{I}_2 & \mathbb{M}^{\mathrm{T}} \\ \mathbb{O} & \mathbb{O} & \mathbb{I}_2 \end{bmatrix} \begin{bmatrix} \mathbb{O} \\ -\mathbb{M}^{\mathrm{T}} \\ \mathbb{D} \end{bmatrix} = (\mathbb{O}, \mathbb{O}, \mathbb{D})^{\mathrm{T}}.$$

Then we will solve the following problem in every time level:

$$\begin{aligned} &\hat{\mathbb{A}}_h \hat{\mathbb{U}} + \hat{\mathbb{B}}_h \mathbf{\Lambda}_{hH} = \hat{\mathbb{F}}_h, \\ &\hat{\mathbb{U}}_{n,p} \leq d_p^{sm}, \mathbf{\Lambda}_{hn,p} \geq 0, (\hat{\mathbb{U}}_{n,p} - d_p^{sm})\mathbf{\Lambda}_{hn,p} = 0, \quad \forall p \in \mathcal{S},\ t \in I, \\ &\mathbf{\Lambda}_{Ht,p} = 0, \quad \forall p \in \mathcal{S},\ t \in I, \end{aligned} \tag{6.314a–e}$$

where in (6.314) the second line represents the Karush–Kuhn–Tucker conditions of a constrained optimization problem for inequality constraints,

$$\begin{aligned} &\mathbf{\Lambda}_{hn,p} = \mathbf{n}_p^{s\mathrm{T}} \hat{\mathbb{D}}[p,p] \mathbf{\Lambda}_{hH}(p), \qquad \mathbf{\Lambda}_{hH}(p) \in \mathbb{R}^2, \\ &\mathbf{\Lambda}_{Ht,p} = \mathbf{\Lambda}_{hH}(p) - (\mathbf{\Lambda}_{hH}(p) \cdot \mathbf{n}_p^s)\mathbf{n}_p^s = (\mathbf{\Lambda}_{hH}(p) \cdot \mathbf{t}_p^s)\mathbf{t}_p^s. \end{aligned}$$

The Primal–Dual Active Set Strategy In Hintermüller et al. (2003, 2004, 2005) a method on how to find the correct "active" subset $\mathcal{A}$ of vertices from $\mathcal{S}$, where the bodies Ω^s, Ω^m are in mutual contact, is presented. To this aim, we decompose the set $\mathcal{S}$ as $\mathcal{S} = \mathcal{A} \cup \mathcal{I}$, where $\mathcal{A}$ is the active set and $\mathcal{I}$ is the inactive set. Let us define the function

$$C(\hat{\mathbb{U}}_{n,p}, \mathbf{\Lambda}_{hn,p}) = \mathbf{\Lambda}_{hn,p} - \max\left\{0, \mathbf{\Lambda}_{hn,p} + c_1(\hat{\mathbb{U}}_{n,p} - d_p^{sm})\right\}, \ c_1 = \text{const} > 0.$$

Then (6.314b–d) can be expressed as

$$C(\hat{\mathbb{U}}_{n,p}, \mathbf{\Lambda}_{hn,p}) = 0, \qquad p \in \mathcal{S},$$

and, therefore, at every time level (6.314) can be rewritten as

$$\begin{aligned} \hat{\mathbb{A}}_h \hat{\mathbb{U}} + \hat{\mathbb{B}}_h \mathbf{\Lambda}_{hH} &= \hat{\mathbb{F}}_h, \\ C(\hat{\mathbb{U}}_{n,p}, \mathbf{\Lambda}_{hn,p}) &= 0, \\ \mathbf{\Lambda}_{Ht,p} &= \mathbf{0} \end{aligned} \tag{6.315}$$

for all vertices $p \in \mathcal{S}$ and $t \in I$.

This leads to **the primal–dual active set algorithm** as follows:

The PDAS Algorithm

Step 1 Initiate the sets $\mathcal{A}_1$ (active set) and $\mathcal{I}_1$ (inactive set), such that $\mathcal{S} = \mathcal{A}_1 \cup \mathcal{I}_1$ and $\mathcal{A}_1 \cap \mathcal{I}_1 = \emptyset$, put the initial value $(\hat{\mathbb{U}}^0, \mathbf{\Lambda}_{hH}^0)$, $c_1 \in (10^3, 10^4)$ and set $k = 1$.

Step 2 If $(\hat{\mathbb{U}}^{k-1}, \mathbf{\Lambda}_{hH}^{k-1})$ and $\mathcal{A}_k, \mathcal{I}_k$ are known, find the primal-dual pair $(\hat{\mathbb{U}}^k, \mathbf{\Lambda}_{hH}^k)$ such that

$$
\begin{aligned}
\hat{\mathbb{A}}_h \hat{\mathbb{U}}^k + \hat{\mathbb{B}}_h \mathbf{\Lambda}_{hH}^k &= \hat{\mathbb{F}}_h, \\
\hat{\mathbb{U}}_{n,p}^k &= d_p^{sm} \quad \text{for all } p \in \mathcal{A}_k, \\
\mathbf{\Lambda}_{hn,p}^k &= \mathbf{0} \quad \text{for all } p \in \mathcal{I}_k, \\
\mathbf{\Lambda}_{Ht,p}^k &= \mathbf{0} \quad \text{for all } p \in \mathcal{S}.1
\end{aligned}
\tag{6.316}
$$

Step 3 Set $\mathcal{A}_{k+1}$ and $\mathcal{I}_{k+1}$ to

$$
\begin{aligned}
\mathcal{A}_{k+1} &= \left\{ p \in \mathcal{S} : \mathbf{\Lambda}_{hn,p}^k + c_1(\hat{\mathbb{U}}_{n,p}^k - d_p^{sm}) > 0 \right\}, \\
\mathcal{I}_{k+1} :&= \left\{ p \in \mathcal{S} : \mathbf{\Lambda}_{hn,p}^k + c_1(\hat{\mathbb{U}}_{n,p}^k - d_p^{sm}) \le 0 \right\}.
\end{aligned}
$$

Step 4 **If** $\mathcal{A}_{k+1} = \mathcal{A}_k$ **then** STOP **else** $k = k + 1$; **goto** Step 2.

System (6.316) can be rewritten if we decompose the set of vertices $\mathcal{S}$ on the slave side in each step k of the PDAS algorithm into the disjoint active and inactive sets $\mathcal{S} = \mathcal{A}_k \cup \mathcal{I}_k$. Since $\hat{\mathbb{B}}_h = (\mathbb{O}, \mathbb{O}, \mathbb{D})^{\mathrm{T}}$, we decompose the diagonal matrix $\mathbb{D}$ into

$$
\mathbb{D} = \begin{bmatrix} \mathbb{D}_{\mathcal{I}_k} & \mathbb{O} \\ \mathbb{O} & \mathbb{D}_{\mathcal{A}_k} \end{bmatrix}.
$$

According to the definition of $\hat{\mathbb{U}}_{n,p}$ in (6.313), we introduce the matrix $\mathbb{N}_{\mathcal{A}_k} \in \mathbb{R}^{|\mathcal{A}_k| \times 2|\mathcal{A}_k|}$, where $|\mathcal{A}_k|$ denotes the number of vertices in $\mathcal{A}_k$, by

$$
\mathbb{N}_{\mathcal{A}_k} = \begin{bmatrix} \ddots & & & \\ & w_{pp}\mathbf{n}_{p,1}^s & w_{pp}\mathbf{n}_{p,2}^s & \\ & & & \ddots \end{bmatrix}, \qquad p \in \mathcal{A}_k,
$$

where $N = 2|\mathcal{A}_k|$, w_{pp} denotes an abbreviation for $\mathbb{D}[p,p]_{1,1} = \cdots = \mathbb{D}[p,p]_{2,2} = \int_{\Gamma_c^s} \varphi_p \, ds$, having the meaning of a weighting factor. The associated tangential vectors are then given by $\mathbf{t}_p^s \perp \mathbf{n}_p^s$, that is, $\mathbf{t}_p^s = (-n_{p2}^s, n_{p1}^s)^{\mathrm{T}}$. We define the matrix $\mathbb{T}_{\mathcal{A}_k} \in \mathbb{R}^{|\mathcal{A}_k|\times 2|\mathcal{A}_k|}$ by

$$\mathbb{T}_{\mathcal{A}_k} = \begin{bmatrix} \ddots & & & \\ & & \mathbb{O}_{1\times 2} & \\ & \mathbb{O}_{1\times 2} & \mathbf{t}_p^{(s)} & \\ & & & \ddots \end{bmatrix} = \begin{bmatrix} \ddots & \ddots & & \\ & -n_{p,2}^s & n_{p,1}^s & \\ & & \ddots & \ddots \end{bmatrix}, \qquad p \in \mathcal{A}_k.$$

Then the algebraic representation of (6.316) is given by

$$\begin{bmatrix} \hat{\mathbb{A}}_{\mathcal{N}\mathcal{N}} & \hat{\mathbb{A}}_{\mathcal{N}\mathcal{M}} & \hat{\mathbb{A}}_{\mathcal{N}\mathcal{I}_k} & \hat{\mathbb{A}}_{\mathcal{N}\mathcal{A}_k} & 0 & 0 \\ \hat{\mathbb{A}}_{\mathcal{M}\mathcal{N}} & \hat{\mathbb{A}}_{\mathcal{M}\mathcal{M}} & \hat{\mathbb{A}}_{\mathcal{M}\mathcal{I}_k} & \hat{\mathbb{A}}_{\mathcal{M}\mathcal{A}_k} & 0 & 0 \\ \hat{\mathbb{A}}_{\mathcal{I}_k\mathcal{N}} & \hat{\mathbb{A}}_{\mathcal{I}_k\mathcal{M}} & \hat{\mathbb{A}}_{\mathcal{I}_k\mathcal{I}_k} & \hat{\mathbb{A}}_{\mathcal{I}_k\mathcal{A}_k} & \mathbb{D}_{\mathcal{I}_k} & 0 \\ \hat{\mathbb{A}}_{\mathcal{A}_k\mathcal{N}} & \hat{\mathbb{A}}_{\mathcal{A}_k\mathcal{M}} & \hat{\mathbb{A}}_{\mathcal{A}_k\mathcal{I}_k} & \hat{\mathbb{A}}_{\mathcal{A}_k\mathcal{A}_k} & 0 & \mathbb{D}_{\mathcal{A}_k} \\ 0 & 0 & 0 & 0 & \mathbb{I}_{\mathcal{I}_k} & 0 \\ 0 & 0 & 0 & \mathbb{N}_{\mathcal{A}_k} & 0 & 0 \\ 0 & 0 & 0 & 0 & 0 & \mathbb{T}_{\mathcal{A}_k} \end{bmatrix} \cdot \begin{bmatrix} \hat{\mathbb{U}}_{\mathcal{N}}^k \\ \hat{\mathbb{U}}_{\mathcal{M}}^k \\ \hat{\mathbb{U}}_{\mathcal{I}_k}^k \\ \hat{\mathbb{U}}_{\mathcal{A}_k}^k \\ \hat{\mathbf{\Lambda}}_{\mathcal{I}_k}^k \\ \hat{\mathbf{\Lambda}}_{\mathcal{A}_k}^k \end{bmatrix} = \begin{bmatrix} \hat{\mathbb{F}}_{\mathcal{N}} \\ \hat{\mathbb{F}}_{\mathcal{M}} \\ \hat{\mathbb{F}}_{\mathcal{I}_k} \\ \hat{\mathbb{F}}_{\mathcal{A}_k} \\ 0 \\ d_{\mathcal{A}_k}^{sm} \\ 0 \end{bmatrix}, \tag{6.317}$$

where k denotes the index of the active set step, and $d_{\mathcal{A}_k}^{sm}$ denotes the vector containing the entries d_p^{sm} associated with the active vertex $p \in \mathcal{A}_k$. System (6.317) cannot be solved directly. Due to the dual biorthogonal Lagrange multiplier space, $\hat{\mathbf{\Lambda}}_{hH}$ can be locally eliminated. Since $\hat{\mathbb{B}}_h = Q\mathbb{B}_h = (0, 0, \mathbb{D})^{\mathrm{T}}$,

$$\hat{\mathbb{B}}_h \mathbf{\Lambda}_{hH}^k = (0, 0, \mathbb{D}\mathbf{\Lambda}_{hH\mathcal{S}}^k)^{\mathrm{T}},$$

then $\hat{\mathbf{\Lambda}}_{hH\mathcal{S}}^k$ can be eliminated locally, that is,

$$\mathbf{\Lambda}_{hH\mathcal{S}}^k = \mathbb{D}^{-1}\left[\hat{\mathbb{F}}_{\mathcal{S}} - \hat{\mathbb{A}}_{\mathcal{S}\mathcal{N}}\hat{\mathbb{U}}_{\mathcal{N}}^k - \hat{\mathbb{A}}_{\mathcal{S}\mathcal{S}}\hat{\mathbb{U}}_{\mathcal{S}}^k - (\hat{\mathbb{A}}_{\mathcal{S}\mathcal{M}} + \hat{\mathbb{A}}_{\mathcal{S}\mathcal{S}}\hat{\mathbb{M}})\hat{\mathbb{U}}_{\mathcal{M}}^k\right]. \tag{6.318}$$

System (6.317) can be reduced. We can eliminate the fifth row and column of (6.317) since $\mathbf{\Lambda}_{\mathcal{I}_k}^k = 0$. Applying the matrix $\mathbb{T}_{\mathcal{A}_k}$ to the fourth row, then $\mathbb{T}_{\mathcal{A}_k}\mathbb{D}_{\mathcal{A}_k}\mathbf{\Lambda}_{\mathcal{A}_k} = 0$ since $\mathbb{T}_{\mathcal{A}_k}\mathbf{\Lambda}_{\mathcal{A}_k} = 0$ and $\mathbb{D}_{\mathcal{A}_k}$ is a diagonal matrix for all $p \in \mathcal{A}_k$. Hence $\mathbf{\Lambda}_{\mathcal{A}_k} = 0$. Now we can rewrite the sixth row and then we obtain

$$\begin{bmatrix} \hat{\mathbb{A}}_{\mathcal{N}\mathcal{N}} & \hat{\mathbb{A}}_{\mathcal{N}\mathcal{M}} & \hat{\mathbb{A}}_{\mathcal{N}\mathcal{I}_k} & \hat{\mathbb{A}}_{\mathcal{N}\mathcal{A}_k} \\ \hat{\mathbb{A}}_{\mathcal{M}\mathcal{N}} & \hat{\mathbb{A}}_{\mathcal{M}\mathcal{M}} & \hat{\mathbb{A}}_{\mathcal{M}\mathcal{I}_k} & \hat{\mathbb{A}}_{\mathcal{M}\mathcal{A}_k} \\ \hat{\mathbb{A}}_{\mathcal{I}_k\mathcal{N}} & \hat{\mathbb{A}}_{\mathcal{I}_k\mathcal{M}} & \hat{\mathbb{A}}_{\mathcal{I}_k\mathcal{I}_k} & \hat{\mathbb{A}}_{\mathcal{I}_k\mathcal{A}_k} \\ 0 & 0 & 0 & \mathbb{N}_{\mathcal{A}_k} \\ \mathbb{T}_{\mathcal{A}_k}\hat{\mathbb{A}}_{\mathcal{A}_k\mathcal{N}} & \mathbb{T}_{\mathcal{A}_k}\hat{\mathbb{A}}_{\mathcal{A}_k\mathcal{M}} & \mathbb{T}_{\mathcal{A}_k}\hat{\mathbb{A}}_{\mathcal{A}_k\mathcal{I}_k} & \mathbb{T}_{\mathcal{A}_k}\hat{\mathbb{A}}_{\mathcal{A}_k\mathcal{A}_k} \end{bmatrix} \cdot \begin{bmatrix} \hat{\mathbb{U}}_{\mathcal{N}}^k \\ \hat{\mathbb{U}}_{\mathcal{M}}^k \\ \hat{\mathbb{U}}_{\mathcal{I}_k}^k \\ \hat{\mathbb{U}}_{\mathcal{A}_k}^k \end{bmatrix} = \begin{bmatrix} \hat{\mathbb{F}}_{\mathcal{N}} \\ \hat{\mathbb{F}}_{\mathcal{M}} \\ \hat{\mathbb{F}}_{\mathcal{I}_k} \\ d_{\mathcal{A}_k}^{sm} \\ \mathbb{T}_{\mathcal{A}_k}\hat{\mathbb{F}}_{\mathcal{A}_k} \end{bmatrix}. \tag{6.319}$$

Hence, we determine $(\hat{\mathbb{U}}_{\mathcal{N}}^k, \hat{\mathbb{U}}_{\mathcal{M}}^k, \hat{\mathbb{U}}_{\mathcal{I}_k}^k, \hat{\mathbb{U}}_{\mathcal{A}_k}^k)^{\mathrm{T}}$ and by (6.318) we determine $\hat{\mathbf{\Lambda}}_{hH}^k$, where $\hat{\mathbb{A}}_h$ and $\hat{\mathbb{F}}_h$ were defined above.

The N = 3 Case Derivation of the 3D problem is parallel to that of the 2D case. Therefore, we will restrict ourselves to different relations only.

The discrete mortar formulation of problem (6.294) for every time level leads to Problem $(\mathcal{P}_{sp})_{dm}$ of (6.302), where

$$M_h^+ = \prod_s M_h^{s+} = \prod_s \left\{\mu_h \in M_h^s | \langle \mu_h, \mathbf{v}_h \rangle_{\cup \Gamma_c^{sm}} \geq 0 \ \forall v_h \in W_h^{s+}\right\},$$

where W_h^{s+} is defined by (6.298) and

$$M_h^s = \{\psi_i \mathbf{e}_k,\ i = 1, \ldots, m;\ k = 1, 2, 3\}, \qquad s \in \{1, \ldots, r\},$$

and $\mathbf{e}_k$ are components of the unit Cartesian basis and $b(\mu_h, \mathbf{v}_h)$ is defined by (6.301). In what follows, the strong form of the nonpenetration condition $[\mathbf{u} \cdot \mathbf{n}]^s \leq d^{sm}$ is replaced by its weak discrete form (6.305).

By a similar way as in the 2D case (6.302a) for every $t \in I$ leads to (6.306) with the similar structure. Applying the transformation of the basis $\varphi = (\varphi_{\mathcal{N}}, \varphi_{\mathcal{M}}, \varphi_{\mathcal{S}})^{\mathrm{T}}$, then the matrices and vectors in (6.306) are decomposed in the decomposition of the set of all vertices in the form (6.307), where the matrix $\mathbb{M}$, representing the coupling between the traces of the FE shape functions on the master side m and the shape functions for the Lagrange multiplier on the slave side s is of the form

$$\mathbb{M} = (\mathbb{M}[p, q]), \qquad \mathbb{M}[p, q] = \mathbb{I}_3 \cdot \int_{\cup \Gamma_c^{sm}} \varphi_p \varphi_q \, ds, \qquad p \in \mathcal{S}, \qquad q \in \mathcal{M},$$

where $\mathbb{I}_3$ is the identity matrix in $\mathbb{R}^{3\times 3}$. The diagonal matrix

$$\mathbb{D} = (\mathbb{D}[p, q]) = \delta_{pq} \mathbb{I}_3 \int_{\cup \Gamma_c^{sm}} \varphi_p \psi_p \, ds, \qquad p \in \mathcal{S} \equiv \{1, \ldots, n_c\}.$$

The matrix $\mathbb{B}_h = (\mathbb{O}, -\mathbb{M}, \mathbb{D})^{\mathrm{T}}$ has the similar structure as in the 2D case.

We introduce a suitable new modified basis $\Phi = (\Phi_{\mathcal{N}}, \Phi_{\mathcal{M}}, \Phi_{\mathcal{S}})^{\mathrm{T}}$ instead of the basis $\varphi = (\varphi_{\mathcal{N}}, \varphi_{\mathcal{M}}, \varphi_{\mathcal{S}})^{\mathrm{T}}$. Then

$$\Phi = (\Phi_{\mathcal{N}}, \Phi_{\mathcal{M}}, \Phi_{\mathcal{S}})^{\mathrm{T}} = \begin{bmatrix} \mathbb{I}_3 & \mathbb{O} & \mathbb{O} \\ \mathbb{O} & \mathbb{I}_3 & \hat{\mathbb{M}}^T \\ \mathbb{O} & \mathbb{O} & \mathbb{I}_3 \end{bmatrix} \begin{bmatrix} \varphi_{\mathcal{N}} \\ \varphi_{\mathcal{M}} \\ \varphi_{\mathcal{S}} \end{bmatrix} = Q\varphi,$$

where $\hat{\mathbb{M}} = \mathbb{D}^{-1}\mathbb{M}$, then $\mathbb{U} = \mathbb{Q}^{\mathrm{T}}\hat{\mathbb{U}}$, where $\hat{\mathbb{U}}$ is the vector of coefficients with respect to the transformed basis Φ. The modified stiffness matrix associated with the transformed basis Φ is the following:

$$\begin{aligned} \hat{\mathbb{A}}_h &= \begin{bmatrix} \mathbb{I}_3 & \mathbb{O} & \mathbb{O} \\ \mathbb{O} & \mathbb{I}_3 & \hat{\mathbb{M}}^{\mathrm{T}} \\ \mathbb{O} & \mathbb{O} & \mathbb{I}_3 \end{bmatrix} \begin{bmatrix} \mathbb{A}_{\mathcal{NN}} & \mathbb{A}_{\mathcal{NM}} & \mathbb{A}_{\mathcal{NS}} \\ \mathbb{A}_{\mathcal{MN}} & \mathbb{A}_{\mathcal{MM}} & \mathbb{A}_{\mathcal{MS}} \\ \mathbb{A}_{\mathcal{SN}} & \mathbb{A}_{\mathcal{SM}} & \mathbb{A}_{\mathcal{SS}} \end{bmatrix} \begin{bmatrix} \mathbb{I}_3 & \mathbb{O} & \mathbb{O} \\ \mathbb{O} & \mathbb{I}_3 & \mathbb{O} \\ \mathbb{O} & \hat{\mathbb{M}} & \mathbb{I}_3 \end{bmatrix} \\ &= Q\mathbb{A}_h Q^{\mathrm{T}} \\ &= \begin{bmatrix} \mathbb{A}_{\mathcal{NN}} & \mathbb{A}_{\mathcal{NM}} + \mathbb{A}_{\mathcal{NS}}\hat{\mathbb{M}} & \mathbb{A}_{\mathcal{NS}} \\ \mathbb{A}_{\mathcal{MN}} + \hat{\mathbb{M}}^{\mathrm{T}}\mathbb{A}_{\mathcal{SN}} & \mathbb{A}_{\mathcal{MM}} + \mathbb{A}_{\mathcal{MS}}\hat{\mathbb{M}} + \hat{\mathbb{M}}^{\mathrm{T}}\mathbb{A}_{\mathcal{SM}} + \hat{\mathbb{M}}^{\mathrm{T}}\mathbb{A}_{\mathcal{SS}}\hat{\mathbb{M}} & \mathbb{A}_{\mathcal{MS}} + \hat{\mathbb{M}}^{\mathrm{T}}\mathbb{A}_{\mathcal{SS}} \\ \mathbb{A}_{\mathcal{SN}} & \mathbb{A}_{\mathcal{SM}} + \mathbb{A}_{\mathcal{SS}}\hat{\mathbb{M}} & \mathbb{A}_{\mathcal{SS}} \end{bmatrix}, \end{aligned}$$

the right-hand side $\hat{\mathbb{F}}$ is given by (6.312).

The weak nonpenetration condition (6.305) under this basis transformation gives a similar expression as (6.313), which glues the vertices on the slave side and on the master side together. The modified matrix $\hat{\mathbb{B}} = (\mathbb{O}, \mathbb{O}, \mathbb{D})^{\mathrm{T}}$. Then we will solve the corresponding problem (6.314) in every time level, where

$$\mathbf{\Lambda}_{hn,p} = \mathbf{n}_p^{s\mathrm{T}} \hat{\mathbb{D}}[p,p] \mathbf{\Lambda}_{hH}(p), \qquad \mathbf{\Lambda}_{hH}(p) \in \mathbb{R}^3,$$

$$\mathbf{\Lambda}_{Ht,p} = \mathbf{\Lambda}_{hH}(p) - (\mathbf{\Lambda}_{hH}(p) \cdot \mathbf{n}_p^s)\mathbf{n}_p^s = (\mathbf{\Lambda}_{hH}(p) \cdot \mathbf{t}_p^s)\mathbf{t}_p^s.$$

The primal-dual active set algorithm is similar to that of the 2D case. The diagonal matrix is decomposed similarly as in the 2D case into

$$\mathbb{D} = \begin{bmatrix} \mathbb{D}_{\mathcal{I}_k} & \mathbb{O} \\ \mathbb{O} & \mathbb{D}_{\mathcal{A}_k} \end{bmatrix}, \quad \text{since } \mathcal{S} = \mathcal{A}_k \cup \mathcal{I}_k.$$

The normal and tangential vectors of all active vertices are merged in the matrices $\mathbb{N}_{\mathcal{A}_k} \in \mathbb{R}^{|\mathcal{A}_k| \times 3|\mathcal{A}_k|}$, where $|\mathcal{A}_k|$ denotes the number of vertices in $\mathcal{A}_k$, and $\mathbb{T}_{\mathcal{A}_k} \in \mathbb{R}^{|\mathcal{A}_k| \times 3|\mathcal{A}_k|}$, which are the following:

$$\mathbb{N}_{\mathcal{A}_k} = \begin{bmatrix} \ddots & & \mathbb{O} & & \\ & w_{pp}\mathbf{n}_{p,1}^s & w_{pp}\mathbf{n}_{p,2}^s & w_{pp}\mathbf{n}_{p,3}^s & \\ & & \mathbb{O} & & \ddots \end{bmatrix}, \qquad p \in \mathcal{A}_k,$$

and

$$\mathbb{T}_{\mathcal{A}_k} = \begin{bmatrix} \ddots & & \mathbb{O} & & \\ & t_{p,1}^{(m)} & t_{p,2}^{(m)} & t_{p,3}^{(m)} & \\ & & \mathbb{O} & & \ddots \end{bmatrix} \in \mathbb{R}^{|\mathcal{A}_k| \times 3|\mathcal{A}_k|}, \qquad m = \xi, \eta, \ p \in \mathcal{A}_k,$$

is the matrix containing the entries associated with the corresponding tangential vectors. Herein, $\mathbf{t}_p^{(m)}$ are associated, normalized tangential vectors given by $\mathbf{t}_p^{\xi} \perp \mathbf{n}_p^s$ and $\mathbf{t}_p^{\eta} = \mathbf{t}_p^{\xi} \times \mathbf{n}_p^s$ with $\|\mathbf{n}_p^s\| = \|\mathbf{t}_p^{\xi}\| = \|\mathbf{t}_p^{\eta}\| = 1$.

Then the algebraic representation (6.316) is given by (6.317) with $\mathbb{T}_{\mathcal{A}_k} = (\mathbb{T}_{\mathcal{A}_k}^{\xi}, \mathbb{T}_{\mathcal{A}_k}^{\eta})^{\mathrm{T}}$ and where k denotes the index of the active set step, and $d_{\mathcal{A}_k}^{sm}$ denotes the vector containing the entries d_p^{sm} associated with the active vertex $p \in \mathcal{A}_k$. System (6.317) cannot be also solved directly. Due to the dual Lagrange multiplier space, $\hat{\mathbf{\Lambda}}_{hH}$ can be eliminated locally. Since $\hat{\mathbb{B}} = (0, 0, \mathbb{D})^{\mathrm{T}}$,

$$\hat{\mathbb{B}}_h \mathbf{\Lambda}_{hH}^k = (0, 0, \mathbb{D}\mathbf{\Lambda}_{hH\mathcal{S}}^k)^{\mathrm{T}},$$

and $\hat{\mathbf{\Lambda}}_{hH\mathcal{S}}^k$ can be eliminated locally, that is, we can use (6.318); then system (6.317) can be reduced similarly as in the previous 2D case.

6.6.5 Newmark Method

Since the semidiscrete approximation of the dynamic problem leads to a system of ordinary differential equations, the problem can be solved numerically by using some of methods solving systems of ordinary equations. Moreover, the primal–dual active set algorithm can be extended to these cases with zero friction. In practice the Newmark method is frequently used.

The discrete (spatial) approximation of the mixed frictionless Problem $(\mathcal{P})_h$ [i.e., problem (6.261) and (6.262)] for $t \in I$ leads to the system

$$\overline{\mathbb{M}}_h \mathbf{u}_h''(t) + \mathbb{C}_h \mathbf{u}_h'(t) + \mathbb{A}_h \mathbf{u}_h(t) + \mathbb{B}_h(t)\mathbf{\Lambda}_h(t) = \mathbb{F}_h(t), \tag{6.320}$$

where $\overline{\mathbb{M}}_h$ is the mass matrix having a connection with the density ρ of the material, $\mathbb{C}_h$ is a damping matrix, $\mathbb{A}_h$ is a stiffness matrix, $\mathbb{B}_h$, $\mathbf{\Lambda}_h$, and $\mathbb{F}_h$ are similar as in (6.306).

Let the time interval $\overline{I} = [0, t_p]$ be divided into m subintervals (t_{n-1}, t_n), $n = 1, \ldots, N_0$, $t_n = n\Delta t$, $\Delta t = t_p/N_0$. Let us denote by $\mathbf{u}^n = \mathbf{u}_h(t_n)$, $(\mathbf{u}')^n = \mathbf{u}_h'(t_n)$, $(\mathbf{u}'')^n = \mathbf{u}_h''(t_n)$. Further, we use the Newmark scheme [see Bathe (1982, 1996) and Wood (1990)] defined by

$$\begin{aligned}
\mathbf{u}^n &= \mathbf{u}^{n-1} + \Delta t(\mathbf{u}')^{n-1} + \tfrac{1}{2}(\Delta t)^2 \left((1-2\beta)(\mathbf{u}'')^{n-1} + 2\beta(\mathbf{u}'')^n\right), \\
(\mathbf{u}')^n &= (\mathbf{u}')^{n-1} + \Delta t \left((1-\gamma)(\mathbf{u}'')^{n-1} + \gamma(\mathbf{u}'')^n\right), \qquad n = 1, \ldots, N_0,
\end{aligned} \tag{6.321}$$

where β, γ are free parameters, $0 \le \beta \le \frac{1}{2}$, $0 \le \gamma \le 1$. Let $\mathbf{u}_h(0) \in V_h$, $\mathbf{u}_h'(0) \in V_h$ be a suitable approximations of initial data and let

$$\mathbf{u}_h^0 = \mathbf{u}_h(0) = \mathbf{u}_0, (\mathbf{u}_h')^0 = \mathbf{u}_h'(0) = \mathbf{u}_1. \tag{6.322}$$

Assuming that $\mathbf{\Lambda}_h(0) = \mathbf{0}$, it follows from (6.320) that

$$(\mathbf{u}_h'')^0 = (\overline{\mathbb{M}}_h)^{-1}(\mathbb{F}_h(0) - \mathbb{C}_h \mathbf{u}_h'(0) - \mathbb{A}_h \mathbf{u}_h(0)). \tag{6.323}$$

The stability condition of the Newmark scheme yields that $\gamma \ge \frac{1}{2}$, $\beta \ge \frac{1}{4}(\gamma + \frac{1}{2})^2$, restricting the time step Δt (Hüeber et al., 2005). The acceptable values are $\gamma = 1$, $\beta \ge \frac{9}{16}$.

Denoting $\mathbf{\Lambda}_h^n = \mathbf{\Lambda}_h(t_n)$, from (6.321) and (6.322), we find

$$\begin{aligned}
&(a_1\overline{\mathbb{M}}_h + a_4\mathbb{C}_h + \mathbb{A}_h)\mathbf{u}_h^n + \mathbb{B}_h(t_n)\mathbf{\Lambda}_h^n = \overline{\mathbb{F}}_h(t_n), \\
&(\mathbf{u}_h'')^n = a_1(\mathbf{u}_h^n - \mathbf{u}_h^{n-1}) - a_2(\mathbf{u}_h')^{n-1} - a_3(\mathbf{u}_h'')^{n-1}, \\
&(\mathbf{u}_h')^n = a_4(\mathbf{u}_h^n - \mathbf{u}_h^{n-1}) + a_5(\mathbf{u}_h')^{n-1} + a_6(\mathbf{u}_h'')^{n-1}, \qquad n = 1, \ldots, N_0,
\end{aligned} \tag{6.324a–c}$$

where

$$\overline{\mathbb{F}}_h(t_n) = (a_1\overline{\mathbb{M}}_h + a_4\mathbb{C}_h)\mathbf{u}_h^{n-1} + (a_2\overline{\mathbb{M}}_h - a_5\mathbb{C}_h)(\mathbf{u}'_h)^{n-1}$$
$$+ (a_3\overline{\mathbb{M}}_h - a_6\mathbb{C}_h)(\mathbf{u}''_h)^{n-1} + \mathbb{F}_h(t_n),$$

$$a_1 = \frac{1}{\beta(\Delta t)^2}, \quad a_2 = \frac{1}{\beta\Delta t}, \quad a_3 = \frac{1-2\beta}{2\beta}, \quad a_4 = \frac{\gamma}{\beta\Delta t},$$

$$a_5 = 1 - \frac{\gamma}{\beta}, \quad a_6 = \left(1 - \frac{\gamma}{2\beta}\right)\Delta t. \tag{6.325}$$

Since in every time step $t=t_n$, due to deformation of the body, the geometry of the contact boundary changed from time to time, we reassemble $\mathbb{B}_h(t_n)$ with respect to the changes of geometry and the weak distance $d_h^{sm}(t_n)$ in every $t=t_n$ as the distance between the colliding bodies may change due to the change of the normal direction and displacements in tangential direction. For typical values $\beta=0.25$, $\gamma=0.5$ the Newmark algorithm is of order 2, for $\beta \geq 0.25$, $\gamma \geq 0.5$ the Newmark algorithm produces a "numerical damping" (Bathe, 1996; Wood, 1990).

Then the algorithm using the PDAS method will be the following:

The D-PDAS Algorithm

(i) Initiate $\mathbf{u}_h^0$ and $(\mathbf{u}'_h)^0$, $(\mathbf{u}''_h)^0$ defined by (6.322) and (6.323) and set $n=0$.

(ii) Set $n=n+1$, determine $\overline{\mathbb{F}}_h(t_n)$ from (6.325), determine the matrix $\mathbb{B}_h(t_n)$ and the weak distance $d_h^{sm}(t_n)$.

(iii) Compute $\mathbf{u}_h^n$ from (6.324a) by using the PDAS algorithm, where by $\mathcal{A}^n$ we denote the correct active set of $\mathbf{u}_h^n$, and for the initialization of $\mathcal{A}_1^n$ in the PDAS algorithm we use the active set of the previous time step $t=t_{n-1}$, that is, $\mathcal{A}_1^n = \mathcal{A}^{n-1}$.

(iv) Compute $(\mathbf{u}'_h)^n$, $(\mathbf{u}''_h)^n$ by using (6.324b,c).

(v) **If** $n=N_0$, **then** STOP **else** go to (ii).

Remark 6.25 For numerical solution of (6.320) several others methods, such as the central difference method, the Houbolt method, or the Wilson Θ method, can be used.

6.6.6 Biomechanic Models in Orthopedy Based on Thermoviscoelastic Rheology with Short Memory

Formulation of the Problem The model of human joints and their replacements being in mutual contact will be investigated. Let the set $\Omega \subset \mathbb{R}^N$, $N=2,3$, $\Omega = \cup_{\iota=1}^r \Omega^\iota$, where Ω^ι are domains with the Lipschitz boundary $\partial\Omega^\iota$ and let

$\partial\Omega = \Gamma_u \cup \Gamma_\tau \cup \Gamma_c$ be defined similarly as in previous cases. Let $t \in I = (0, t_p)$, $t_p > 0$. Then we will solve the following problem:

Problem ($\mathcal{P}_{tsm}$) Find a pair of functions $(T, \mathbf{u}) : \Omega \times I \to (\mathbb{R} \times \mathbb{R}^N) \times I$, $N = 2, 3$, and a stress tensor $\tau_{ij} : \Omega \times I \to \mathbb{R}^{N\times N} \times I$ satisfying

$$\rho^\iota \frac{\partial^2 u_i^\iota}{\partial t^2} + \alpha^\iota \frac{\partial u^\iota}{\partial t} = \frac{\partial}{\partial x_j} \tau_{ij}(\mathbf{u}^\iota, \mathbf{u}'^\iota) + f_i^\iota \quad \text{in } \Omega^\iota \times I; \tag{6.326}$$

$$\rho^\iota c_e^\iota \frac{\partial T^\iota}{\partial t} - \rho\beta_{ij} T_0 e_{ij}(\mathbf{u}'^\iota) = \frac{\partial}{\partial x_j}\left(\kappa_{ij}^\iota \frac{\partial T^\iota}{\partial x_i}\right) + W^\iota \quad \text{in } \Omega^\iota \times I; \tag{6.327}$$

$$\begin{aligned} \tau_{ij}^\iota &= \tau_{ij}^\iota(\mathbf{u}^\iota, \mathbf{u}'^\iota, T_\iota) = c_{ijkl}^{(0)\iota}(\mathbf{x}) e_{kl}(\mathbf{u}^\iota) + c_{ijkl}^{(1)\iota}(\mathbf{x}) e_{kl}(\mathbf{u}'^\iota) - \beta_{ij}^\iota (T^\iota - T_0) \\ &= {}^e\tau_{ij}^\iota + {}^v\tau_{ij}^\iota + {}^T\tau_{ij}^\iota; \qquad i, j, k, l = 1, \ldots, N, \iota = 1, \ldots, r, \end{aligned} \tag{6.328}$$

$$\left.\begin{aligned} &\tau_{ij} n_j = P_i, \qquad i, j = 1, \ldots, N \\ &\kappa_{ij} \frac{\partial T}{\partial x_i} n_i = q \end{aligned}\right\} (t, \mathbf{x}) \in \Gamma_\tau(t) = I \times \cup_{\iota=1}^r (\Gamma_\tau \cap \partial\Omega^\iota) \tag{6.329}$$

$$\left.\begin{aligned} &u_i = u_{2i}, \ i = 1, \ldots, N \\ &\kappa_{ij} \frac{\partial T}{\partial x_i} n_j = K(Y(\mathbf{x}, t) - T(\mathbf{x}, t)), \ i, j = 1, \ldots, N \end{aligned}\right\} \begin{aligned} &(t, \mathbf{x}) \in {}^1\Gamma_u(t) \\ &= I \times \cup_{\iota=1}^r ({}^1\Gamma_u \cap \partial\Omega^\iota) \end{aligned}, \tag{6.330}$$

$$\left.\begin{aligned} &u_i = 0, \ i = 1, \ldots, N \\ &T(\mathbf{x}, t) = T_1(\mathbf{x}, t) \end{aligned}\right\} \text{on } {}^2\Gamma_u(t) = I \times \cup_{\iota=1}^r ({}^2\Gamma_u \cap \partial\Omega^\iota), \tag{6.331}$$

the nonpenetration contact conditions in displacements

$$[u_n]^{sm} \le d^{sm}, \qquad \tau_n^s \equiv \tau_n^{sm} \le 0, \qquad ([u_n]^{sm} - d^{sm})\tau_n^{sm} = 0,$$

and the Coulomb law of friction

$$\left.\begin{aligned} &[\mathbf{u}_t']^{sm} = 0 \Longrightarrow |\boldsymbol{\tau}_t^{sm}| \le \mathcal{F}_c^{sm} |\tau_n^{sm}| \\ &[\mathbf{u}_t']^{sm} \ne 0 \Longrightarrow \boldsymbol{\tau}_t^{sm} = -\mathcal{F}_c^{sm} |\tau_n^{sm}| \frac{[\mathbf{u}_t']^{sm}}{|[\mathbf{u}_t']^s|} \end{aligned}\right. \qquad (t, \mathbf{x}) \in \Gamma_c(t) = I \times \cup_{s,m} \Gamma_c^{sm},$$

and (6.332a–g)

$$T^s = T^m, \qquad \left[\kappa_{ij} \frac{\partial T}{\partial x_j} n_i\right]^{sm} = \mathcal{F}_c^{sm} |\tau_n^{sm}| |[\mathbf{u}_t']^{sm}|$$

and the initial conditions

$$T^\iota(\mathbf{x}, 0) = T_0^\iota(\mathbf{x}), \qquad \mathbf{u}^\iota(\mathbf{x}, 0) = \mathbf{u}_0^\iota(\mathbf{x}), \qquad \mathbf{u}'^\iota(\mathbf{x}, 0) = \mathbf{u}_1^\iota(\mathbf{x}) \quad \text{in } \Omega^\iota \tag{6.333}$$

where $\rho^\iota(\mathbf{x})$ is the density, $c_e^\iota(\mathbf{x})$ the specific heat, $c^\iota(\mathbf{x}) = \rho^\iota(\mathbf{x})c_e^\iota(\mathbf{x})$ represents the thermal capacity, $q = q(\mathbf{x}, t)$ is the heat flux, K the heat transition coefficient, $Y(\mathbf{x}, t)$ the external temperature, $\kappa_{ij}^\iota(\mathbf{x})$ the coefficients of thermal conductivity that are symmetric and uniformly positive definite, that is,

$$\kappa_{ij}^\iota(\mathbf{x}) = \kappa_{ji}^\iota(\mathbf{x}), \quad \kappa_{ij}^\iota(\mathbf{x})\zeta_i\zeta_j \geq \kappa_0^\iota|\zeta|^2, \quad \kappa_0^\iota = \text{const} > 0, \quad \zeta \in \mathbb{R}^N, \iota = 1, \ldots, r, \tag{6.334}$$

where $\beta_{ij}^\iota(\mathbf{x})$ is the coefficient of linear thermal expansion, W^ι the thermal sources, and $\mathbf{f}^\iota$, $\mathbf{P}^\iota$ the body and surface forces. The elastic coefficients $c_{ijkl}^\iota(\mathbf{x})$ may possibly depend on $\mathbf{x}$, but they are bounded in $\mathbf{x}$ and satisfy

$$c_{ijkl}^\iota = c_{jikl}^\iota = c_{ijlk}^\iota = c_{klij}^\iota, \quad c_{ijkl}^\iota e_{kl} e_{ij} \geq c_0^\iota e_{ij} e_{ij}, \quad c_0^\iota > 0,$$
$$\text{for all symmetric matrices } \{e_{ij}\}. \tag{6.335}$$

$T_0^\iota(\mathbf{x})$, $T_1^\iota(\mathbf{x})$, $T^\iota(\mathbf{x}, t)$ are the initial, the given, and the actual temperatures, $\mathbf{u}^\iota(\mathbf{x}, t)$, $\boldsymbol{\tau}^\iota(\mathbf{x}, t)$ are the displacement and the stress vectors, τ_{ij}^ι the stress tensor, u_n, $\mathbf{u}_t$, τ_n, $\boldsymbol{\tau}_t$ the normal and tangential components of displacement and stress vectors, $\mathbf{n}^\iota$ the outward normal to the boundary $\partial\Omega^\iota$, $\mathbf{u}_2^\iota$ the given function, $\mathbf{u}_2^\iota$ has a time derivative, $\mathbf{u}_2^{\prime\iota}$, $\mathbf{u}_0^\iota$, $\mathbf{u}_1^\iota$ are the given functions, and where $\mathcal{F}_c^{sm}$ is a coefficient of friction. The coefficient of friction $\mathcal{F}_c^{sm} \equiv \mathcal{F}_c^{sm}(\mathbf{x}, \mathbf{u}')$ is assumed to be globally bounded, nonnegative, and satisfies the Carathéodory property, that is, $\mathcal{F}_c^{sm}(\cdot, \mathbf{v})$ is measurable, $\mathbf{v} \in \mathbb{R}$, and $\mathcal{F}_c^{sm}(\mathbf{x}, \cdot)$ is continuous for a.e. $\mathbf{x} \in \Gamma_c^{sm}$.

To prove the existence theorem of this problem, the nonpenetration contact conditions will be reformulated in velocities, that is,

$$[u_n']^{sm} \leq 0, \quad \tau_n^s \equiv \tau_n^{sm} \leq 0, \quad [u_n']^{sm}\tau_n^{sm} = 0,$$

and the Coulomb law of friction

$$[\mathbf{u}_t']^{sm} = 0 \Longrightarrow |\boldsymbol{\tau}_t^{sm}| \leq \mathcal{F}_c^{sm}|\tau_n^{sm}|, \tag{6.336}$$

$$[\mathbf{u}_t']^{sm} \neq 0 \Longrightarrow \boldsymbol{\tau}_t^{sm} = -\mathcal{F}_c^{sm}|\tau_n^{sm}|\frac{[\mathbf{u}_t']^{sm}}{|[\mathbf{u}_t']^s|}, \quad (t, \mathbf{x}) \in \Gamma_c(t) = I \times \cup_{s,m}\Gamma_c^{sm}.$$

and

$$T^s = T^m, \quad \left[\kappa_{ij}\frac{\partial T}{\partial x_j}n_i\right]^{sm} = \mathcal{F}_c^{sm}|\tau_n^{sm}|\,|[\mathbf{u}_t']^{sm}|.$$

Variational Formulation of the Problem and the Penalty Approximation
The existence of the solution will be proved under the following assumptions:

Presumption 6.2 Let $N \geq 2$ and let $\Omega = \cup_{\iota=1}^r \Omega^\iota$, where Ω^ι are domains bounded with the Lipschitz boundary $\partial\Omega = \Gamma_\tau \cup \Gamma_u \cup \Gamma_c$, where the disjoint parts Γ_τ, Γ_u, Γ_c are open subsets. Let $\Gamma_u = {}^1\Gamma_u \cup {}^2\Gamma_u$ and $\Gamma_c = \cup_{s,m}\Gamma_c^{sm}$, $\Gamma_c^{sm} = \partial\Omega^s \cap \partial\Omega^m$, $s \neq m$, $s, m \in \{1, \ldots, r\}$. Let $\text{meas}({}^2\Gamma_u \cap \partial\Omega^\iota) > 0$ hold for all $\iota \leq r$. Let $I = (0, t_p)$, t_p being small.

Let $\Omega_t = \Omega \times I$ denote the time–space domain and let $\Gamma_\tau(t) = \Gamma_\tau \times I$, $\Gamma_u(t) = \Gamma_u \times I$, $\Gamma_c(t) = \Gamma_c \times I$ denote the parts of its boundary $\partial\Omega_t = \partial\Omega \times I$. The coefficients $c_{ijkl}^{(n)}(\mathbf{x})$, $n = 0, 1$, satisfy (6.237). The given input data satisfy $\rho \in C(\Omega)$, $\mathbf{F}, \mathbf{F}' \in L^2(I; L^{2,N}(\Omega))$ $\mathbf{P}, \mathbf{P}' \in L^2(I; L^{2,N}(\Gamma_\tau))$, $\mathbf{u}_1^\iota \in H^{1,N}(\Omega^\iota)$, $u_{1n}^s - u_{1n}^m = 0$ on $\cup_{s,m}\Gamma_c^{sm}$, $\mathbf{u}_0^\iota \in H^{1,N}(\Omega^\iota)$, the coefficient of friction $\mathcal{F}_c^{sm} \equiv \mathcal{F}_c^{sm}(\mathbf{x}, \mathbf{u}_t')$ is assumed to be bounded, nonnegative, and satisfies the Carathéodory property.

The contact problem $(\mathcal{P}_{tsm})$ has a weak formulation in terms of a variational inequality. Let us introduce the set of virtual velocities and the set of admissible velocities by

$$V_0 = \{\mathbf{v} \mid \mathbf{v} \in \prod_{\iota=1}^{r} H^{1,N}(\Omega^\iota), \mathbf{v} = 0 \text{ on } \Gamma_u = {}^1\Gamma_u \cup {}^2\Gamma_u \text{ and } v_n^s - v_n^m = 0$$
$$\text{a.e. on } \Gamma_c\},$$

$$K = \{\mathbf{v} \mid \mathbf{v} \in \prod_{\iota=1}^{r} H^{1,N}(\Omega^\iota), \mathbf{v} = \mathbf{u}_2 \text{ on } \Gamma_u,\ v_n^s - v_n^m \leq 0 \text{ a.e. on } \cup_{s,m} \Gamma_c^{sm}\},$$

$$\mathcal{K} = \{\mathbf{v} | \mathbf{v} \in L^2(I; \prod_{\iota=1}^{r} H^{1,N}(\Omega^\iota)), \mathbf{v} = \mathbf{u}_2' \text{ on } \Gamma_u(t),\ v_n^s - v_n^m \leq 0$$
$$\text{a.e. on } I \times \cup_{s,m}\Gamma_c^{sm}\},$$

$${}^1V_0 = \{z | z \in \prod_{\iota=1}^{r} H^1(\Omega^\iota), z = 0 \text{ on } {}^2\Gamma_u\},$$

$${}^1\mathcal{V}_0 = L^2(I; \prod_{\iota=1}^{r} H^1(\Omega^\iota)), \qquad {}^1\mathcal{V} = T_1 +^1 \mathcal{V}_0.$$

Multiplying (6.326) by $\mathbf{v} - \mathbf{u}'$, where $\mathbf{v} \in K$ are suitable test functions, and (6.327) by $z - T$, where $z \in {}^1V_0$ are suitable test functions, integrating over $\Omega \times I$, using the Green theorems and boundary and contact conditions, then we obtain the following variational problem:

Problem $(\mathcal{P}_{tsm})_v$ Find a pair of functions $(\mathbf{u}, T)$, where a vector function $\mathbf{u}$ with $\mathbf{u}' \in \mathcal{K} \cap B(I; L^{2,N}(\Omega))$ and $\mathbf{u}(0, \cdot) = \mathbf{u}_0$, $\mathbf{u}'(0, \cdot) = \mathbf{u}_1$, and $T \in {}^1\mathcal{V}$ such that

$$\int_I \{(\mathbf{u}''(t), \mathbf{v} - \mathbf{u}'(t)) + (\alpha\mathbf{u}'(t), \mathbf{v} - \mathbf{u}'(t)) + a^{(0)}(\mathbf{u}(t), \mathbf{v} - \mathbf{u}'(t))$$
$$+ a^{(1)}(\mathbf{u}'(t), \mathbf{v} - \mathbf{u}'(t)) + j(\mathbf{v}) - j(\mathbf{u}'(t))\}\, dt$$
$$\geq \int_I (\mathbf{f}(t), \mathbf{v} - \mathbf{u}'(t))\, dt \quad \forall \mathbf{v} \in \mathcal{K},$$

$$\int_I \{(T'(t), z - T(t)) + b(T'(t), z - T(t)) \geq (\mathbb{W}(t), z - T(t)), \ \forall z \in {}^1\mathcal{V}, \tag{6.337}$$

where we assume that the initial data $\mathbf{u}_0$, $\mathbf{u}_1$ satisfy the static contact multibody linear elastic problem, and where

$$(\mathbf{u}'', \mathbf{v}) = \sum_{\iota=1}^{r} (\mathbf{u}''^{\iota}, \mathbf{v}^{\iota}) = \int_{\Omega} \rho u_i'' v_i \, d\mathbf{x},$$

$$(\alpha \mathbf{u}', \mathbf{v}) = \sum_{\iota=1}^{r} (\alpha^{\iota} \mathbf{u}'^{\iota}, \mathbf{v}^{\iota}) = \int_{\Omega} \alpha u_i' v_i \, d\mathbf{x},$$

$$a^{(n)}(\mathbf{u}, \mathbf{v}) = \sum_{\iota=1}^{r} a^{(n)\iota}(\mathbf{u}^{\iota}, \mathbf{v}^{\iota}) = \int_{\Omega} c_{ijkl}^{(n)} e_{kl}(\mathbf{u}) e_{ij}(\mathbf{v}) \, d\mathbf{x}, \qquad n = 0, 1,$$

$$(\mathbf{f}, \mathbf{v}) = \sum_{\iota=1}^{r} (\mathbf{f}^{\iota}, \mathbf{v}^{\iota}) = \int_{\Omega} F_i v_i \, d\mathbf{x} + \int_{\Gamma_\tau} P_i v_i \, ds,$$

$$j(\mathbf{v}) = \sum_{\iota=1}^{r} j^{\iota}(\mathbf{v}^{\iota}) = \int_{\bigcup_{s,m} \Gamma_c^{sm}} \mathcal{F}_c^{sm}(\mathbf{u}_t'^s - \mathbf{u}_t'^m) |\tau_n^{sm}(\mathbf{u}, \mathbf{u}')| |\mathbf{v}_t^s - \mathbf{v}_t^m| \, ds$$

$$\equiv \langle \mathcal{F}_c^{sm}(\mathbf{u}_t'^s - \mathbf{u}_t'^m) |\tau_n^{sm}(\mathbf{u}, \mathbf{u}')|, |\mathbf{v}_t^s - \mathbf{v}_t^m| \rangle_{\Gamma_c},$$

$$(T', z) = \sum_{\iota=1}^{r} (T'^{\iota}, z^{\iota}) = \int_{\Omega} \rho c_e T' z \, d\mathbf{x},$$

$$b(T, z) = \sum_{\iota=1}^{r} b^{\iota}(T^{\iota}, z^{\iota}) = \int_{\Omega} \kappa_{ij} \frac{\partial T}{\partial x_i} \frac{\partial z}{\partial x_j} \, d\mathbf{x},$$

$$(\mathbb{W}, z) = \sum_{\iota=1}^{r} (\mathbb{W}^{\iota}, z^{\iota}) = \int_{\Omega} (Wz + \rho \beta_{ij} T_0 e_{ij}(\mathbf{u}') z) \, d\mathbf{x}$$

$$+ \int_{{}^1\Gamma_u} (K(Y(\mathbf{x}, t) - T(\mathbf{x}, t)), z) \, ds$$

$$+ \int_{\bigcup_{sm} \Gamma_c^{sm}} \mathcal{F}_c^{sm} |\tau_n^{sm}| \, |\mathbf{u}_t'^s - \mathbf{u}_t'^m| z \, ds$$

and where the bilinear forms $a^{(n)}(\mathbf{u}, \mathbf{v})$, $n = 0, 1$, are symmetric in $\mathbf{u}$, $\mathbf{v}$ and satisfy $a^{(n)\iota}(\mathbf{u}, \mathbf{u}) \geq c_0^{(n)\iota} \|\mathbf{u}^{\iota}\|_{1,N}^2$, $c_0^{(n)\iota} = \text{const} > 0$, provided meas $({}^2\Gamma_u \cap \partial\Omega^{\iota}) > 0$, $|a^{(n)\iota}(\mathbf{u}^{\iota}, \mathbf{v}^{\iota})| \leq c_1^{(n)\iota} \|\mathbf{u}^{\iota}\|_{1,N} \|\mathbf{v}^{\iota}\|_{1,N}$, $c_1^{(n)\iota} = \text{const} > 0$, $\mathbf{u}, \mathbf{v} \in V_0$, and where the space $B(I; L^{2,N}(\Omega))$ was defined in Section 6.5.2.

It can be shown that every solution $\mathbf{u} \in \prod_{\iota=1}^{r} C^{2,N}(I \times \overline{\Omega}^{\iota})$ of (6.326)–(6.333) is a solution of (6.337) and that every sufficiently smooth solution $\mathbf{u} \in \prod_{\iota=1}^{r} C^{2,N}(I \times \overline{\Omega}^{\iota})$ of (6.337) is also a solution of (6.326)–(6.332).

As a result we have the following theorem:

Theorem 6.45 Let the assumptions concerning Ω, Γ_r, $r = \tau$, u, c and the physical data given in Presumption 6.2 be satisfied. Then there exists at least one weak solution of the dynamic contact problem with friction and damping in thermoviscoelasticity with short memory.

The proof of the existence of the solution is based on the penalization and regularization techniques. In the first step, the penalty method will be used. This leads to a contact problem of a normal compliance type. The contact condition (6.332) will be replaced by the nonlinear boundary condition putting $\tau_n(\mathbf{u}, \mathbf{u}') = -\frac{1}{\delta}[u_n'^s - u_n'^m]_+$ with $[\cdot]_+ := \max\{\cdot, 0\}$, $\delta > 0$. The problem is parallel to that of Eck et al. (2005) and Nedoma (2009).

6.6.7 Biomechanic Models in Orthopedy Based on Viscoelastic Rheology with Long Memory

Viscoelastic biomaterials can be also assumed as materials endowed with a memory in the sense that the state of stress at the instant time t depends on all the deformations undergone by the biomaterial in previous times. Then we speak about the viscoelastic biomaterials with a long memory. The corresponding constituent law

$$\tau_{ij}(t) = c_{ijkl}^{(0)} e_{kl}(\mathbf{u}(t)) + \int_0^t b_{ijkl}(t-\tau) e_{kl}(\mathbf{u}(\tau))\, d\tau$$

represents the stress–strain relation for materials for which the state of stresses at the time t depends on the deformation at the instant t and on deformations at the times preceding t, the history of deformation, where $c_{ijkl}^{(0)}$ are elastic coefficients of instantaneous elasticity, which may depend possibly on $\mathbf{x}$, but they are bounded in $\mathbf{x}$ and satisfy the conditions of symmetry and ellipticity defined above and repeated below. The coefficients $b_{ijkl} = b_{ijkl}(\mathbf{x}, t)$, representing the memory of the biomaterials, which depend on t and possibly on $\mathbf{x}$, are bounded in $\mathbf{x}$, t, and they satisfy the symmetry as well as the regularity conditions introduced above and repeated below.

Formulation of a Model Problem in the Viscoelastic Rheology with Long Memory and Its Weak Solution

Formulation of the Problem The model of a system of biological bodies being in mutual bilateral contact with given friction will be investigated. Let the set $\Omega \subset \mathbb{R}^N$, $N = 2, 3$, $\Omega = \cup_{\iota=1}^{r} \Omega^\iota$, where Ω^ι are domains with the Lipschitz boundary $\partial\Omega^\iota$ and let $\partial\Omega = \Gamma_u \cup \Gamma_\tau \cup \Gamma_c$ be defined similarly as in previous cases. Let $t \in I = (0, t_p)$, $t_p > 0$. Then we will solve the following problem:

Problem ($\mathcal{P}_{\mathbf{lm}}$) Find a function $\mathbf{u} : \Omega \times I \to \mathbb{R}^N \times I$, $N = 2, 3$, and a stress tensor $\tau_{ij} : \Omega \times I \to \mathbb{R}^{N\times N} \times I$ satisfying

$$\rho \frac{\partial^2 u_i}{\partial t^2} = \frac{\partial}{\partial x_j} \tau_{ij} + f_i \quad \text{in } \Omega \times I; \tag{6.338}$$

$$\tau_{ij}(t) = c_{ijkl} e_{kl}(\mathbf{u}(t)) + \int_0^t b_{ijkl}(t-\tau) e_{kl}(\mathbf{u}(\tau))\, d\tau = {}^e\tau_{ij} + {}^v\tau_{ij}; \tag{6.339}$$

with the boundary value and contact conditions

$$\tau_{ij} n_j = P_i \quad \text{on } ({}^1\Gamma_\tau \cup {}^2\Gamma_\tau) \times I, \qquad \text{meas}\, {}^i\Gamma_\tau > 0, \qquad i = 1, 2, \tag{6.340}$$

$$\mathbf{u}(x, t) = \mathbf{u}_2(\mathbf{x}, t) \quad \text{on } ({}^1\Gamma_u \cup {}^2\Gamma_u) \times I, \qquad \text{meas}\, {}^i\Gamma_u \geq 0, \qquad i = 1, 2, \tag{6.341}$$

where $P_i = 0$ on ${}^2\Gamma_\tau \times I$ and $\mathbf{u}_2 = 0$ on ${}^2\Gamma_u \times I$ and the **bilateral contact condition** with the Tresca model of friction on $\Gamma_c^{kl} \times I$, where $\Gamma_c^{kl} = \partial\Omega^k \cap \partial\Omega^l$, $k \neq l$, $k, l \in \{1, \ldots, r\}$, of the form

$$\left.\begin{aligned} &[u_n]^{kl} \equiv u_n^k - u_n^l = 0 \quad \text{and} \quad |\boldsymbol{\tau}_t^k| \le g_c^{kl}(\mathbf{x}, t), \\ &\begin{cases} \text{if } |\boldsymbol{\tau}_t^k| < g_c^{kl}(\mathbf{x}, t), & \text{then } [\mathbf{u}_t']^{kl} \equiv \mathbf{u}_t'^k - \mathbf{u}_t'^l = 0, \\ \text{if } |\boldsymbol{\tau}_t^k| = g_c^{kl}(\mathbf{x}, t), & \text{then there exists } \lambda \ge 0 \\ & \text{such that } [\mathbf{u}_t']^{kl} \equiv \mathbf{u}_t'^k - \mathbf{u}_t'^l = -\lambda \boldsymbol{\tau}_t^k, \end{cases} \end{aligned}\right\} \text{ on } \Gamma_c^{kl} \times I, \tag{6.342}$$

and the initial conditions

$$\mathbf{u}(\mathbf{x}, 0) = \mathbf{u}_0(\mathbf{x}), \qquad \mathbf{u}'(\mathbf{x}, 0) = \mathbf{u}_1(\mathbf{x}), \tag{6.343}$$

where $\rho(\mathbf{x})$ is the density, $\mathbf{f}, \mathbf{P}$ the body and surface forces, the elastic coefficients $c_{ijkl}(\mathbf{x})$ may depend possibly on $\mathbf{x}$, but they are bounded in $\mathbf{x}$ and satisfy

$$c_{ijkl} = c_{jikl} = c_{ijlk} = c_{klij}, \qquad c_{ijkl}e_{kl}e_{ij} \ge c_0 e_{ij}e_{ij}, \qquad c_0 > 0,$$
$$\text{for all symmetric matrices } \{e_{ij}\}, \tag{6.344}$$

the coefficients $b_{ijkl} = b_{ijkl}(\mathbf{x}, t)$, representing the material memory, depend on $\mathbf{x}$ and t, are bounded in $\mathbf{x}, t$ and they satisfy the symmetry conditions and the following regularity assumption, that is,

$$b_{ijkl} = b_{jikl}, \qquad b_{ijkl}, \frac{\partial b_{ijkl}}{\partial t}, \frac{\partial^2 b_{ijkl}}{\partial t^2} \in L^\infty(\Omega \times I), \tag{6.345}$$

$\mathbf{u}(\mathbf{x}, t)$, $\boldsymbol{\tau}(\mathbf{x}, t)$ the displacement and the stress vectors, u_n, $\mathbf{u}_t$, τ_n, $\boldsymbol{\tau}_t$ the normal and tangential components of displacement and stress vectors, $\mathbf{n}$ the outward normal to the boundary $\partial\Omega$, $\mathbf{u}_2$ the given function, and where $g_c^{kl}(\mathbf{x}, t)$ are slip limits.

Variational (Weak) Solution of the Problem Let us denote by $(\mathbf{v}, \mathbf{w})_0$ the inner product in $[L^2(\Omega)]^N$, by $\|\cdot\|_k$, $k \in \mathbb{R}^1$, the norm in $\mathbb{H}^{k,N}(\Omega) = \Pi_{\iota=1}^s [H^k(\Omega^\iota)]^N$, where $H^k(\Omega^\iota)$ denotes the standard Sobolev space and by $|\cdot|_0$ the norm corresponding to the inner product $(\mathbf{v}, \mathbf{w})_0$, $\mathbf{v}, \mathbf{w} \in H^{1,N}(\Omega)$, $N = 2, 3$.

Let us assume that $\rho^\iota \in C(\overline{\Omega}^\iota)$, $\rho^\iota \ge \rho_0^\iota > 0$, $c_{ijkl}^\iota \in L^\infty(\Omega^\iota)$, $\mathbf{f}^\iota \in L^2(I; [L^2(\Omega^\iota)]^N)$, $\mathbf{P} \in L^2(I; [L^2(\Gamma_\tau)]^N)$, $b_{ijkl}^\iota \in L^\infty(\Omega^\iota \times I)$, $g_c^{kl} \in L^\infty(\Gamma_c^{kl})$, $g_c^{kl} \ge 0$ on $\cup_{k,l}\Gamma_c^{kl}$, and $\mathbf{u}_0^\iota$, $\mathbf{u}_1^\iota \in H^{1,N}(\Omega^\iota)$.

For $\mathbf{u}, \mathbf{v} \in \mathbb{H}^{k,N}(\Omega)$ let us introduce

$$a(\mathbf{u}, \mathbf{v}) = \sum_{\iota=1}^r a^\iota(\mathbf{u}^\iota, \mathbf{v}^\iota) = \int_\Omega c_{ijkl}e_{kl}(\mathbf{u})e_{ij}(\mathbf{v})\,d\mathbf{x},$$

$$a_1(t; \mathbf{u}, \mathbf{v}) = \sum_{\iota=1}^r a_1^\iota(t; \mathbf{u}^\iota, \mathbf{v}^\iota) = \int_\Omega b_{ijkl}(t)e_{kl}(\mathbf{u})e_{ij}(\mathbf{v})\,d\mathbf{x},$$

$$(\mathbf{u}'', \mathbf{v}) = \sum_{\iota=1}^{r} (\mathbf{u}^{\iota\prime\prime}, \mathbf{v}^{\iota}) = \int_{\Omega} \rho \mathbf{u}'' \mathbf{v}\, d\mathbf{x},$$

$$(\mathbf{F}, \mathbf{v}) = \sum_{\iota=1}^{r} (\mathbf{F}^{\iota}, \mathbf{v}^{\iota}) = \int_{\Omega} f_i v_i\, d\mathbf{x} + \int_{\Gamma_\tau} P_i v_i\, ds,$$

$$j_g(\mathbf{v}) = \sum_{\iota=1}^{r} j_g^{\iota}(\mathbf{v}^{\iota}) = \int_{\cup_{kl} \Gamma_c^{kl}} g_c^{kl} |\mathbf{v}_t^k - \mathbf{v}_t^l|\, ds. \tag{6.346}$$

Let

$$\mathcal{U}_{ad}(t) = \{\mathbf{v} | \mathbf{v} \in \mathbb{H}^{1,N}(\Omega), \qquad \mathbf{v} = \mathbf{u}_2'(\mathbf{x}, t) \text{ on } {}^1\Gamma_u \quad \text{and } v_n^k - v_n^l = 0; \text{ on } \cup \Gamma_c^{kl}\},$$

Multiplying (6.338) by $\mathbf{v} - \mathbf{u}'(t)$, integrating over Ω, using the boundary value conditions and the contact conditions (6.340)–(6.342) and the Green theorem, then we have to solve the following problem:

Problem $(\mathcal{P}_{lm})_v$ Find a function $\mathbf{u}(t) \in V$, $\mathbf{u}(t) = \mathbf{u}_2$ on ${}^1\Gamma_u$, and $\mathbf{u}'(t) \in \mathcal{U}_{ad}(t)$, such that for a.a. $t \in I$,

$$(\mathbf{u}''(t), \mathbf{v} - \mathbf{u}'(t)) + a(\mathbf{u}, \mathbf{v} - \mathbf{u}'(t)) + \int_0^t a_1(t - \tau; \mathbf{u}(\tau), \mathbf{v} - \mathbf{u}'(t))\, d\tau$$
$$+ j_g(\mathbf{v}) - j_g(\mathbf{u}'(t)) \geq (\mathbf{F}(t), \mathbf{v} - \mathbf{u}'(t)) \quad \forall \mathbf{v} \in \mathcal{U}_{ad}(t), \tag{6.347}$$

and the initial conditions (6.343).

Now let us introduce $\boldsymbol{\varphi}(t)$ satisfying $\boldsymbol{\varphi}(t) \in \mathbb{H}^{1,N}(\Omega)$, $\boldsymbol{\varphi}(t) = \mathbf{u}_2(t)$ on ${}^1\Gamma_u$, $\varphi_n^k - \varphi_n^l = 0$ on $\cup \Gamma_c^{kl}$, $\boldsymbol{\varphi}'(t) \in \mathbb{H}^{1,N}(\Omega)$ ($\boldsymbol{\varphi} = 0$, if $\Gamma_u = \emptyset$), substitute $\mathbf{v} - \boldsymbol{\varphi}'$ for $\mathbf{v}$, keeping the same notation $\mathbf{u}$ and introduce the spaces

$$V = \{\mathbf{v} | \mathbf{v} \in \mathbb{H}^{1,N}(\Omega), v_n^k - v_n^l = 0 \quad \text{on } \cup \Gamma_c^{kl}, \mathbf{v} = 0 \text{ on } \Gamma_u\},$$

then we have to solve the following problem:

Problem $(\mathcal{P}_{lm})_0$ Find a function $\mathbf{u}$ such that $\mathbf{u}(t) \in V$, for a.a. $t \in I$,

$$(\mathbf{u}''(t), \mathbf{v} - \mathbf{u}'(t)) + a(\mathbf{u}(t), \mathbf{v} - \mathbf{u}'(t)) + \int_0^t a_1(t - \tau; \mathbf{u}(\tau), \mathbf{v} - \mathbf{u}'(t))\, d\tau$$
$$+ j_g(\mathbf{v} + \boldsymbol{\varphi}'(t)) - j_g(\mathbf{u}'(t) + \boldsymbol{\varphi}'(t)) \geq (\mathbf{F}_0(t), \mathbf{v} - \mathbf{u}'(t)) \quad \forall \mathbf{v} \in V, \tag{6.348}$$

where

$$(\mathbf{F}_0(t), \mathbf{v}) = (\mathbf{F}(t), \mathbf{v}) - (\boldsymbol{\varphi}''(t), \mathbf{v}) - a(\boldsymbol{\varphi}(t), \mathbf{v}) - \int_0^t a_1(t - \tau; \boldsymbol{\varphi}(\tau), \mathbf{v})\, d\tau, \tag{6.349}$$

with the initial conditions

$$\mathbf{u}(\mathbf{x}, 0) = \mathbf{u}_0(\mathbf{x}) - \boldsymbol{\varphi}(\mathbf{x}, 0),$$
$$\mathbf{u}'(\mathbf{x}, 0) = \mathbf{u}_1(\mathbf{x}) - \boldsymbol{\varphi}'(\mathbf{x}, 0). \tag{6.350}$$

Remark 6.26 It is evident that from $\mathbf{u}'(t) \in \mathcal{U}_{ad}(t)$ and from the initial conditions it follows that $\mathbf{u}(\mathbf{x}, t) = \mathbf{u}_2(\mathbf{x}, t) + (\mathbf{u}_0(\mathbf{x}) - \mathbf{u}_2(\mathbf{x}, 0))$ on ${}^1\Gamma_u$, from which the boundary condition $\mathbf{u}(\mathbf{x}, t) = \mathbf{u}_2(\mathbf{x}, t)$ follows if $\mathbf{u}_0(\mathbf{x}) = \mathbf{u}_2(\mathbf{x}, 0)$ on Γ_u and similarly for the boundary condition on Γ_τ.

Then we have the following main result:

Theorem 6.46 Let us assume that

$$\mathbf{f}, \mathbf{f}', \mathbf{f}'' \in L^2(I; [L^2(\Omega)]^N), \qquad \mathbf{P}, \mathbf{P}', \mathbf{P}'' \in L^2(I; [L^2(L_\tau)]^N),$$
$$\rho \in C(\overline{\Omega}), \qquad \rho \geq \rho_0 > 0,$$

$c_{ijkl} \in L^\infty(\Omega)$ satisfy the usual conditions of symmetry and ellipticity (6.344), b_{ijkl} satisfy the symmetry conditions and the regularity assumption (6.345), g_c^{kl} does not depend on t, $g_c^{kl} \in L^\infty(\Gamma_c^{kl})$, $g_c^{kl} \geq 0$ on $\cup_{k,l}\Gamma_c^{kl}$

$$\varphi, \varphi', \varphi'' \in L^2(I; \mathbb{H}^{1,N}(\Omega)), \varphi''', \varphi^{(iv)} \in L^2(I; [L^2(\Omega)]^N), \varphi(0) \in [H^2(\Omega)]^N,$$
$$\mathbf{u}_2(\mathbf{x}, t) \in L^2(I; \mathbb{H}^{1,N}(\Omega)),$$
$$\mathbf{u}_0 \in V, \qquad (\mathbf{F}_0(0), \mathbf{v}) - a(\mathbf{u}_0, \mathbf{v}) = (\mathbf{u}_p, \mathbf{v}), \qquad \mathbf{u}_p \in [L^2(\Omega)]^N,$$
$$\mathbf{u}_1 \in V, \qquad \mathbf{u}_{1t} + \varphi_t'(0) = 0 \quad \text{on } \cup_{k,l}\Gamma_c^{kl}.$$

Then there exists one and only one function $\mathbf{u}$ such that

$$\mathbf{u}, \mathbf{u}' \in L^\infty(I; V), \qquad \mathbf{u}'' \in L^2(I; V') \tag{6.351}$$

and satisfying (6.348)–(6.350).

Proof The method of the proof is the following:

(i) The uniqueness of the solution of (6.348)–(6.350) will be proved.
(ii) The problem in eqs. (6.348)–(6.350) will be regularized.
(iii) The existence of the regularized problem based on the Galerkin approximation will be proved.
(iv) A priori estimates I and II independent of ε will be derived.
(v) Limitation processes over m (Galerkin) and ε (regularization) will be derived.

The proof is parallel to that of Duvaut and Lions (1976).

Uniqueness Let $\mathbf{u}^1(t)$, $\mathbf{u}^2(t)$ be two solutions of the problem discussed. Let us denote $\mathbf{u}^*(t)) = \mathbf{u}^1(t) - \mathbf{u}^2(t)$. Let us put $\mathbf{v}(t) = \mathbf{u}^2(t)$ into Eq. (6.348) for $\mathbf{u}^1(t)$ and $\mathbf{v}(t) = \mathbf{u}^1(t)$ for $\mathbf{u}^2(t)$ and add. Then after some modifications, we obtain

$$\frac{1}{2}\frac{d}{dt}\{|\mathbf{u}^{*\prime}(t)|_0^2 + a(\mathbf{u}^*(t), \mathbf{u}^*(t))\} \leq -\int_0^t a_1(t-\tau; \mathbf{u}^*(\tau), \mathbf{u}^{*\prime}(t))\, d\tau,$$

and then

$$|\mathbf{u}^{*\prime}(t)|_0^2 + a(\mathbf{u}^*(t), \mathbf{u}^*(t)) \leq -2 \int_0^t \int_0^\tau a_1(\tau - \tau_1; \mathbf{u}^*(\tau_1), \mathbf{u}^{*\prime}(\tau))\, d\tau_1\, d\tau.$$

Since the form $\mathbf{v} \to a_1(t; \mathbf{u}, \mathbf{v})$ is continuous on V, then

$$a_1(t; \mathbf{u}, \mathbf{v}) = (B(t)\mathbf{u}, \mathbf{v}), B(t)\mathbf{u} \in V', B(t) \in \mathcal{L}(V; V') \tag{6.352}$$

and

$$\begin{aligned}
&\left| \int_0^t \int_0^\tau a_1(\tau - \tau_1; \boldsymbol{\varphi}(\tau_1), \boldsymbol{\varphi}'(\tau))\, d\tau_1\, d\tau \right| \\
&= \left| \int_0^t \left(\int_0^\tau B(\tau - \tau_1)\boldsymbol{\varphi}(\tau_1)\, d\tau_1, \boldsymbol{\varphi}'(\tau) \right) d\tau \right| \\
&= \left| \int_0^t \int_\tau^t (B(\tau - \tau_1)\boldsymbol{\varphi}(\tau_1), \boldsymbol{\varphi}'(\tau))\, d\tau\, d\tau_1 \right| \\
&= \left| \int_0^t (B(t - \tau_1)\boldsymbol{\varphi}(\tau_1), \boldsymbol{\varphi}(t))\, d\tau_1 - \int_0^t (B(0)\boldsymbol{\varphi}(\tau_1), \boldsymbol{\varphi}(\tau_1))\, d\tau_1 \right. \\
&\quad \left. - \int_0^t \int_\tau^t (B'(\tau - \tau_1)\boldsymbol{\varphi}(\tau_1), \boldsymbol{\varphi}(\tau))\, d\tau\, d\tau_1 \right| \\
&\leq c \left[\int_0^t \|\boldsymbol{\varphi}(\tau)\|_1^2 d\tau + \|\boldsymbol{\varphi}(t)\|_1 \int_0^t \|\boldsymbol{\varphi}(\tau)\|_1\, d\tau \right] \\
&\leq C_1 \int_0^t \|\boldsymbol{\varphi}(\tau)\|_1^2\, d\tau + C_2 \left(\frac{\varepsilon}{2} \|\boldsymbol{\varphi}(t)\|_1^2 + \frac{1}{2\varepsilon} \int_o^t \|\boldsymbol{\varphi}(\tau)\|_1^2\, d\tau \right), \qquad \varepsilon > 0.
\end{aligned} \tag{6.353}$$

Since the bilinear form $a(\mathbf{v}, \mathbf{v})$ is continuous, bounded, and coercive, that is, there exist constants $c > 0$ such that $|a(\mathbf{u}, \mathbf{v})| \leq c\|\mathbf{u}\|_1 \|\mathbf{v}\|_1$ $\forall \mathbf{u}, \mathbf{v} \in \mathbb{H}^{1,N}(\Omega)$, $a(\mathbf{v}, \mathbf{v}) \geq c\|\mathbf{v}\|_1^2$ $\forall \mathbf{v} \in \mathbb{H}^{1,N}(\Omega)$ hold, since $\mathbf{u}^*(t) = \int_0^t \mathbf{u}^{*\prime}(\tau)\, d\tau$ we obtain

$$|\mathbf{u}^{*\prime}(t)|_0^2 + \|\mathbf{u}^*(t)\|_1^2 \leq c \int_0^t [|\mathbf{u}^{*\prime}(\tau)|_0^2 + \|\mathbf{u}^*(\tau)\|_1^2 + \|\mathbf{u}^{*\prime}(\tau)\|_1^2]\, d\tau.$$

By a suitable choice of a small parameter ε we arrive at the inequality

$$|\mathbf{u}^{*\prime}(t)|_0^2 + \|\mathbf{u}^*(t)\|_1^2 \leq C \int_0^t [|\mathbf{u}^{*\prime}(\tau)|_0^2 + \|\mathbf{u}^*(\tau)\|_1^2]\, d\tau.$$

so that the Gronwall lemma (see the Appendix) yields $\mathbf{u}^*(t) = 0$.

Existence To prove the existence of the solution, the regularization technique will be used. For this we introduce the convex function $\psi_\varepsilon : \mathbb{R} \to R$ defined by

$$\psi_\varepsilon(x) = \frac{|x|^{1+\varepsilon}}{1+\varepsilon}, \tag{6.354}$$

which regularizes the function $x \to |x|$ and which is differentiable and satisfies the following inequality

$$|\,|x| - \psi_\varepsilon(|x|)| < \varepsilon, \quad \forall x \in \mathbb{R},\ \varepsilon \geq 0. \tag{6.355}$$

Define a family of regularized friction functionals $j_{g\varepsilon} : V \to \mathbb{R}$, depending on the parameter $\varepsilon > 0$

$$j_{g\varepsilon}(\mathbf{v}) = \int_{\cup_{k,l}\Gamma_c^{kl}} g_c^{kl}(\mathbf{x},t)\psi_\varepsilon(|\mathbf{v}_t^k - \mathbf{v}_t^l|)\,ds \equiv \langle g_c^{kl}, \psi_\varepsilon(|\mathbf{v}_t^k - \mathbf{v}_t^l|)\rangle_{\cup_{k,l}\Gamma_c^{kl}}, \tag{6.356}$$

which are well-defined, convex, and Gâteaux-differentiable with respect to the argument $\mathbf{v}$. The partial Gâteaux-derivative of $j_{g\varepsilon}$ with respect to the argument $\mathbf{v}$ in the direction $\mathbf{w} \in V$ is given by

$$(j'_{g\varepsilon}(\mathbf{v}), \mathbf{w}) = \langle g_c^{kl}\chi_\varepsilon(\mathbf{v}_t^k - \mathbf{v}_t^l), \mathbf{w}_t^k - \mathbf{w}_t^l\rangle_{\cup_{k,l}\Gamma_c^{kl}}, \qquad \mathbf{v}, \mathbf{w} \in V, \tag{6.357}$$

where we set $\chi_\varepsilon(\mathbf{v}_t^k - \mathbf{v}_t^l) = \psi'_\varepsilon(|\mathbf{v}_t^k - \mathbf{v}_t^l|)$.

Then the regularized problem is the following: Find $\mathbf{u}_\varepsilon(t) \in V$ such that for all $t \in I$ $\mathbf{u}'_\varepsilon(t) \in V$, $\mathbf{u}''_\varepsilon(t) \in V$

$$\begin{aligned}(\mathbf{u}''_\varepsilon(t), \mathbf{v}) + a(\mathbf{u}_\varepsilon(t), \mathbf{v}) + \int_0^t a_1(t-\tau; \mathbf{u}_\varepsilon(\tau), \mathbf{v})\,d\tau \\ + (j'_{g\varepsilon}(\mathbf{u}'_\varepsilon(t) + \boldsymbol{\varphi}'(t), \mathbf{v}) = (\mathbf{F}_0(t), \mathbf{v}), \qquad \mathbf{v} \in V,\end{aligned} \tag{6.358}$$

with the initial conditions

$$\mathbf{u}_\varepsilon(\mathbf{x}, 0) = \mathbf{u}_0(\mathbf{x}), \mathbf{u}'_\varepsilon(\mathbf{x}, 0) = \mathbf{u}_1(\mathbf{x}). \tag{6.359}$$

The existence of $\mathbf{u}_\varepsilon(t)$ will be proved by means of the finite-dimensional Galerkin approximation. Let $\{\mathbf{w}_j\}$ be a countable basis of the space V, that is, each finite subset of $\{\mathbf{w}_j\}$ is linearly independent and span$\{\mathbf{w}_j | j = 1, 2, \cdots\}$ is dense in V, as V is a separable space. Let V_m be spanned by $\{\mathbf{w}_j | 1 \leq j \leq m\}$. Then the approximate solution $\mathbf{u}_m(t)$ of the order m satisfies the system:

$$\begin{aligned}(\mathbf{u}''_m(t), \mathbf{w}_j) + a(\mathbf{u}_m(t), \mathbf{w}_j) + \int_0^t a_1(t-\tau; \mathbf{u}_m(\tau), \mathbf{w}_j)\,d\tau \\ + (j'_{g\varepsilon}(\mathbf{u}'_m(t) + \boldsymbol{\varphi}'(t), \mathbf{w}_j) = (\mathbf{F}_0(t), \mathbf{w}_j), \qquad \mathbf{w}_j \in V_m, 1 \leq j \leq m,\end{aligned} \tag{6.360}$$

$$\mathbf{u}_m(\mathbf{x}, 0) = \mathbf{u}_0(\mathbf{x}), \mathbf{u}'_m(\mathbf{x}, 0) = \mathbf{u}_{1m}(\mathbf{x}),$$
$$\mathbf{u}_{1m} \in V_m, \mathbf{u}_{1m} \to \mathbf{u}_1 \text{ in } [L^2(\Omega)]^N, \text{ as } m \to \infty. \tag{6.361}$$

Since $\{\mathbf{w}_j\}_{j=1}^m$ are linearly independent, the system (6.360) and (6.361) is a regular system of ordinary differential equations of the second order, and therefore (6.360) and (6.361) define $\mathbf{u}_m(t)$ uniquely on an interval $I_m = [0, t_m]$. By passing to the limit with $m \to \infty$, we prove the existence of solution $\mathbf{u}_\varepsilon$ of the regularized problem (6.358) and (6.359).

A Priori Estimates I To obtain the first estimates, we put $\mathbf{v} = \mathbf{u}'_\varepsilon(t) + \varphi'(t)$ in (6.358). Then

$$\begin{aligned}&(\mathbf{u}''_\varepsilon(t), \mathbf{u}'_\varepsilon(t) + \varphi'(t)) + a(\mathbf{u}_\varepsilon(t), \mathbf{u}'_\varepsilon(t) + \varphi'(t))\\ &\quad + \int_0^t a_1(t-\tau; \mathbf{u}_\varepsilon(\tau), \mathbf{u}'_\varepsilon(t) + \varphi'(t))\, d\tau\\ &\quad + (j'_{g\varepsilon}(\mathbf{u}'_\varepsilon(t) + \varphi'(t)), \mathbf{u}'_\varepsilon(t) + \varphi'(t)) = (\mathbf{F}_0(t), \mathbf{u}'_\varepsilon(t) + \varphi'(t)).\end{aligned} \tag{6.362}$$

Since the form $a_1(t; \mathbf{u}, \mathbf{v})$ is continuous on V and since $(j'_{g\varepsilon}(\mathbf{w}), \mathbf{w}) \geq 0$, $\forall \mathbf{w} \in V$,

$$\begin{aligned}&\frac{1}{2}\frac{d}{dt}[|\mathbf{u}'_\varepsilon(t)|_0^2 + a(\mathbf{u}_\varepsilon(t), \mathbf{u}_\varepsilon(t))] = (\mathbf{F}_0(t), \mathbf{u}'_\varepsilon(t) + \varphi'(t))\\ &\quad - \int_0^t a_1(t-\tau; \mathbf{u}_\varepsilon(\tau), \mathbf{u}'_\varepsilon(t) + \varphi'(t))\, d\tau - (\mathbf{u}''_\varepsilon(t), \varphi'(t)) - a(\mathbf{u}_\varepsilon(t), \varphi'(t)).\end{aligned} \tag{6.363}$$

Since $|a_1(t-\tau; \mathbf{u}, \mathbf{v})| \leq C(t, \tau)\|\mathbf{u}\|_1 \|\mathbf{v}\|_1$ $\forall \mathbf{u}, \mathbf{v} \in V$, with $C(t, \tau) \in L^\infty(I \times I)$, and according to (6.353)

$$\begin{aligned}&\left|\int_0^t \int_0^\tau a_1(\tau - \tau_1; \mathbf{u}_\varepsilon(\tau_1), \mathbf{u}'_\varepsilon(\tau))\, d\tau_1\, d\tau\right|\\ &\quad \leq c\left[\int_0^t \|\mathbf{u}_\varepsilon(\tau)\|_1^2\, d\tau + \|\mathbf{u}_\varepsilon(t)\|_1 \int_0^t \|\mathbf{u}_\varepsilon(\tau)\|_1 d\tau\right],\end{aligned} \tag{6.364}$$

then from (6.363), after integration over t and by using the coerciveness of the bilinear form $a(\mathbf{v}, \mathbf{v})$, we obtain

$$\begin{aligned}&|\mathbf{u}'_\varepsilon(t)|_0^2 + c\|\mathbf{u}_\varepsilon(t)\|_1^2 \leq |\mathbf{u}_1|_0^2 + c\|\mathbf{u}_0\|_1^2 + 2\int_0^t (\mathbf{F}_0(\tau), \mathbf{u}'_\varepsilon(\tau) + \varphi'(\tau))\, d\tau\\ &\quad - 2\int_0^t (\mathbf{u}''_\varepsilon(\tau), \varphi'(\tau))\, d\tau - 2\int_0^t a(\mathbf{u}_\varepsilon(\tau), \varphi'(\tau))\, d\tau\\ &\quad - 2\int_0^t \int_0^\tau a_1(\tau - \tau_1; \mathbf{u}_\varepsilon(\tau_1), \mathbf{u}'_\varepsilon(\tau) + \varphi'(\tau))\, d\tau_1\, d\tau\end{aligned}$$

$$\leq |\mathbf{u}_1|_0^2 + c\|\mathbf{u}_0\|_1^2 + 2\int_0^t (\mathbf{F}_0(\tau), \mathbf{u}_\varepsilon'(\tau) + \boldsymbol{\varphi}'(\tau))\, d\tau$$
$$+ 2c\left[\int_0^t \|\mathbf{u}_\varepsilon(\tau)\|_1^2\, d\tau + \|\mathbf{u}_\varepsilon(t)\|_1 \int_0^t \|\mathbf{u}_\varepsilon(\tau)\|_1\, d\tau\right]$$
$$+ 2c\left[\int_0^t \|\mathbf{u}_\varepsilon(\tau)\|_1^2\, d\tau + \|\mathbf{u}_\varepsilon(t)\|_1 \int_0^t \|\boldsymbol{\varphi}_\varepsilon(\tau)\|_1\, d\tau\right]. \qquad (6.365)$$

According to the assumptions made on $\mathbf{f}, \mathbf{P}, \boldsymbol{\varphi}$,

$$\mathbf{F}_0, \mathbf{F}_0' \in L^2(I; V').$$

Since $\mathbf{u}_\varepsilon(t) = \mathbf{u}_0 + \int_0^t \mathbf{u}_\varepsilon'(\tau) d\tau$, we have

$$|\mathbf{u}_\varepsilon(t)|_0^2 \leq c\int_0^t |\mathbf{u}_\varepsilon'(\tau)|_0^2 d\tau + 2|\mathbf{u}_0|_0^2,$$
$$2\int_0^t (\mathbf{F}_0(\tau), \mathbf{u}_\varepsilon'(\tau) + \boldsymbol{\varphi}'(\tau))\, d\tau = 2(\mathbf{F}_0(t), \mathbf{u}_\varepsilon(t)) - 2(\mathbf{F}_0(0), \mathbf{u}_\varepsilon(0))$$
$$- 2\int_0^t (\mathbf{F}_0'(\tau), \mathbf{u}_\varepsilon(\tau))\, d\tau + 2(\mathbf{F}_0(t), \boldsymbol{\varphi}(t))$$
$$- 2(\mathbf{F}_0(0), \boldsymbol{\varphi}(0)) - 2\int_0^t (\mathbf{F}_0'(\tau), \boldsymbol{\varphi}(\tau))\, d\tau,$$
$$2|(\mathbf{F}_0(t), \mathbf{u}_\varepsilon(t))| \leq c|\mathbf{u}_\varepsilon(t)|_0^2 + c\|\mathbf{F}_0(t)\|_*^2.$$

Then (6.365) yields

$$|\mathbf{u}_\varepsilon'(t)|_0^2 + \|\mathbf{u}_\varepsilon(t)\|_1^2 \leq c + c\int_0^t |\mathbf{u}_\varepsilon'(\tau)|_0^2\, d\tau + c\int_0^t \|\mathbf{F}_0(\tau)\|_*^2\, d\tau$$
$$+ 2c\left(\int_0^t \|\mathbf{u}_\varepsilon(\tau)\|_1^2\, d\tau + \|\mathbf{u}_\varepsilon(t)\|_1 \int_0^t \|\mathbf{u}_\varepsilon(\tau)\|_1\, d\tau\right).$$

Hence

$$|\mathbf{u}_\varepsilon'(t)|_0^2 + \|\mathbf{u}_\varepsilon(t)\|_1^2 \leq c\int_0^t (1 + |\mathbf{u}_\varepsilon'(\tau)|_0^2 + \|\mathbf{u}_\varepsilon(\tau)\|_1^2 + \|\mathbf{u}_\varepsilon'(\tau)\|_1^2)\, d\tau.$$

Using the similar approach as in the uniqueness case, we obtain

$$|\mathbf{u}_\varepsilon'(t)|_0^2 + \|\mathbf{u}_\varepsilon(t)\|_1^2 \leq c\int_0^t (|\mathbf{u}_\varepsilon'(\tau)|_0^2 + \|\mathbf{u}_\varepsilon(\tau)\|_1^2)\, d\tau.$$

so that the Gronwall lemma yields

$$|\mathbf{u}_\varepsilon'(t)|_0^2 + \|\mathbf{u}_\varepsilon(t)\|_1^2 \leq c,$$

that is,

$$|\mathbf{u}_\varepsilon'(t)|_0 \leq c, \qquad \|\mathbf{u}_\varepsilon(t)\|_1 \leq c, \qquad t \in I, \tag{6.366}$$

where $c = \text{const} > 0$ denotes various constants independent of ε.

Let us set in (6.358) $t = 0$. Then

$$(\mathbf{u}_\varepsilon''(0), \mathbf{v}) + a(\mathbf{u}_0, \mathbf{v}) + (j_{g\varepsilon}'(\mathbf{u}_1 + \boldsymbol{\varphi}'(0), \mathbf{v}) = (\mathbf{F}_0(0), \mathbf{v}), \qquad \mathbf{v} \in V.$$

According to the assumption of Theorem 6.46, that is, $\mathbf{u}_{1t} + \boldsymbol{\varphi}_t'(0) = 0$ on $\cup_{k,l}\Gamma_c^{kl}$, we have $(j_{g\varepsilon}'(\mathbf{u}_1 + \boldsymbol{\varphi}'(0)), \mathbf{v}) = 0$ and since

$$(\mathbf{F}_0(0), \mathbf{v}) - a(\mathbf{u}_0, \mathbf{v}) = (\mathbf{u}_p, \mathbf{v}), \qquad \mathbf{u}_p \in [L^2(\Omega)]^N,$$

then $\mathbf{u}_\varepsilon''(0) = \mathbf{u}_p$, $\mathbf{u}_\varepsilon''(0) \in [L^2(\Omega)]^N$.

A Priori Estimates II Differentiating (6.358) with respect to t we have

$$\begin{aligned}(\mathbf{u}_\varepsilon'''(t), \mathbf{v}) &+ a(\mathbf{u}_\varepsilon'(t), \mathbf{v}) + (B(0)\mathbf{u}_\varepsilon(t), \mathbf{v}) + \left(\int_0^t B'(t-\tau)\mathbf{u}_\varepsilon(\tau)\,d\tau, \mathbf{v}\right) \\ &+ \left(\frac{d}{dt}j_{g\varepsilon}'(\mathbf{u}_\varepsilon'(t) + \boldsymbol{\varphi}'(t)), \mathbf{v}\right) = (\mathbf{F}_0'(t), \mathbf{v}), \qquad \mathbf{v} \in V.\end{aligned} \tag{6.367}$$

Hence, putting $\mathbf{v} = \mathbf{u}_\varepsilon''(t) + \boldsymbol{\varphi}''(t)$, $\mathbf{u}_\varepsilon''(t) \in V$, we obtain

$$\begin{aligned}&(\mathbf{u}_\varepsilon'''(t), \mathbf{u}_\varepsilon''(t) + \boldsymbol{\varphi}''(t)) + a(\mathbf{u}_\varepsilon'(t), \mathbf{u}_\varepsilon''(t) + \boldsymbol{\varphi}''(t)) \\ &\quad + (B(0)\mathbf{u}_\varepsilon(t), \mathbf{u}_\varepsilon''(t) + \boldsymbol{\varphi}''(t)) + \left(\int_0^t B'(t-\tau)\mathbf{u}_\varepsilon(\tau)\,d\tau, \mathbf{u}_\varepsilon''(t) + \varphi''(t)\right) \\ &\quad + \left(\frac{d}{dt}j_{g\varepsilon}'(\mathbf{u}_\varepsilon'(t) + \boldsymbol{\varphi}'(t)), \mathbf{u}_\varepsilon''(t) + \boldsymbol{\varphi}''(t)\right) = (\mathbf{F}_0'(t), \mathbf{u}_\varepsilon''(t) + \boldsymbol{\varphi}''(t)).\end{aligned} \tag{6.368}$$

Hence

$$\begin{aligned}&\frac{1}{2}\frac{d}{dt}[|\mathbf{u}_\varepsilon''(t)|_0^2 + a(\mathbf{u}_\varepsilon'(t), \mathbf{u}_\varepsilon'(t))] + (B(0)\mathbf{u}_\varepsilon(t), \mathbf{u}_\varepsilon''(t) + \boldsymbol{\varphi}''(t)) \\ &\quad + \left(\int_0^t B'(t-\tau)\mathbf{u}_\varepsilon(\tau)d\tau, \mathbf{u}_\varepsilon''(t) + \boldsymbol{\varphi}''(t)\right) \\ &\quad + \left(\frac{d}{dt}j_{g\varepsilon}'(\mathbf{u}_\varepsilon'(t) + \boldsymbol{\varphi}'(t)), \mathbf{u}_\varepsilon''(t) + \boldsymbol{\varphi}''(t)\right) = (\mathbf{F}_0'(t), \mathbf{u}_\varepsilon''(t) + \boldsymbol{\varphi}''(t)) \\ &\quad - (\mathbf{u}_\varepsilon'''(t), \boldsymbol{\varphi}''(t)) - a(\mathbf{u}_\varepsilon'(t), \boldsymbol{\varphi}''(t)),\end{aligned} \tag{6.369}$$

where

$$\left(\frac{d}{dt}j'_{g\varepsilon}(\mathbf{u}'_\varepsilon(t)+\boldsymbol{\varphi}'(t)), \mathbf{u}''_\varepsilon(t)+\boldsymbol{\varphi}''(t)\right)$$
$$=\left\langle\frac{d}{dt}g_c^{kl}(\mathbf{x},t)\psi'_\varepsilon(|[\mathbf{u}'_{\varepsilon t}(t)]^{kl}+[\boldsymbol{\varphi}'_t(t)]^{kl}|), [\mathbf{u}''_\varepsilon(t)]^{kl}+[\boldsymbol{\varphi}''(t)]^{kl}\right\rangle_{\cup_{k,l}\Gamma_c^{kl}},$$

where $[\mathbf{w}_t]^{kl}=\mathbf{w}_t^k-\mathbf{w}_t^l$. Since [see (6.357)]

$$(j'_{g\varepsilon}(\mathbf{w}),\mathbf{v})=\int_{\cup_{k,l}\Gamma_c^{kl}} g_c^{kl}(\mathbf{x},t)\psi'_\varepsilon(|[\mathbf{w}_t]^{kl}|)[\mathbf{v}_t]^{kl}ds$$
$$=\int_{\cup_{k,l}\Gamma_c^{kl}} g_c^{kl}(\mathbf{x},t)\chi_\varepsilon(\mathbf{w}_t)\mathbf{v}_t\,ds,$$

then

$$\left(\frac{d}{dt}j'_{g\varepsilon}(\mathbf{w}(t),\mathbf{v}\right)=\int_{\cup_{k,l}\Gamma_c^{kl}} g_c^{kl}(\mathbf{x},t)\lim_{k\to 0}\frac{\chi_\varepsilon(\mathbf{w}_t(t+k))-\chi_\varepsilon(\mathbf{w}_t(t))}{k}\mathbf{v}_t\,ds$$

since $\partial g_c^{kl}/\partial t=0$ by the assumption. Thus

$$\left(\frac{d}{dt}j'_{g\varepsilon}(\mathbf{w}(t)),\mathbf{w}'(t)\right)$$
$$=\int_{\cup_{k,l}\Gamma_c^{kl}} g_c^{kl}(\mathbf{x},t)\lim_{k\to 0}\frac{\chi_\varepsilon(\mathbf{w}_t(t+k))-\chi_\varepsilon(\mathbf{w}_t(t))}{k}\frac{\mathbf{w}_t(t+k)-\mathbf{w}_t(t)}{k}\,ds\geq 0$$

due to the property of monotony. Hence and (6.369)

$$\frac{1}{2}\frac{d}{dt}[|\mathbf{u}''_\varepsilon(t)|_0^2+a(\mathbf{u}'_\varepsilon(t),\mathbf{u}'_\varepsilon(t))]+(B(0)\mathbf{u}_\varepsilon(t),\mathbf{u}''_\varepsilon(t)+\boldsymbol{\varphi}''(t))$$
$$+\left(\int_0^t B'(t-\tau)\mathbf{u}_\varepsilon(\tau)d\tau,\mathbf{u}''_\varepsilon(t)+\boldsymbol{\varphi}''(t)\right)$$
$$\leq(\mathbf{F}'_0(t),\mathbf{u}''_\varepsilon(t)+\boldsymbol{\varphi}''(t))-(\mathbf{u}'''_\varepsilon(t),\boldsymbol{\varphi}''(t))-a(\mathbf{u}'_\varepsilon(t),\boldsymbol{\varphi}''(t)). \tag{6.370}$$

Then we integrate (6.370) from 0 to t. Since

$$\int_0^t(\mathbf{F}'_0(\tau),\mathbf{u}''_\varepsilon(\tau))\,d\tau=(\mathbf{F}'_0(t),\mathbf{u}'_\varepsilon(t))-(\mathbf{F}'_0(0),\mathbf{u}_1)-\int_0^t(\mathbf{F}''_0(\tau),\mathbf{u}'_\varepsilon(\tau))\,d\tau$$

and according to the assumptions on $\mathbf{f},\mathbf{P},\boldsymbol{\varphi}$ it follows that

$$\mathbf{F}''_0\in L^2(I;V'),$$

and, therefore, since $y(t) = \int_0^t y'(\tau)\,d\tau + y_0$, $|y|_0^2 \le c\|y\|_1^2$,

$$\left|\int_0^t (\mathbf{F}_0'(\tau), \mathbf{u}_\varepsilon''(\tau))\,d\tau\right| \le c\|\mathbf{F}_0'(t)\|_*\|\mathbf{u}_\varepsilon'(t)\|_1 + c + \int_0^t \|\mathbf{F}_0''(\tau)\|_*\|\mathbf{u}_\varepsilon'(\tau)\|_1\,d\tau, \tag{6.371}$$

$$\int_0^t (\mathbf{u}_\varepsilon'''(\tau), \boldsymbol{\varphi}''(\tau))\,d\tau = (\mathbf{u}_\varepsilon''(t), \boldsymbol{\varphi}''(t)) - (\mathbf{u}_\varepsilon''(0), \boldsymbol{\varphi}''(0)) - \int_0^t (\mathbf{u}_\varepsilon''(\tau), \boldsymbol{\varphi}'''(\tau))\,d\tau, \tag{6.372}$$

$$\begin{aligned}
&\int_0^t \left(\int_0^\tau B'(\tau-\tau_1)\mathbf{u}_\varepsilon(\tau_1)\,d\tau_1, \mathbf{u}_\varepsilon''(\tau)\right)d\tau \\
&= \int_0^t \int_{\tau_1}^t (B'(\tau-\tau_1)\mathbf{u}_\varepsilon(\tau_1), \mathbf{u}_\varepsilon''(\tau))\,d\tau\,d\tau_1 \\
&= \int_0^t [(B'(\tau-\tau_1)\mathbf{u}_\varepsilon(\tau_1), \mathbf{u}_\varepsilon'(t)) - (B'(0)\mathbf{u}_\varepsilon(\tau_1), \mathbf{u}_\varepsilon'(\tau_1))]\,d\tau_1 \\
&- \int_0^t \int_{\tau_1}^t (B''(\tau-\tau_1)\mathbf{u}_\varepsilon(\tau_1), \mathbf{u}_\varepsilon'(\tau))\,d\tau\,d\tau_1,
\end{aligned}$$

from which, using (6.345) and (6.366)

$$\begin{aligned}
&\left|\int_0^t \left(\int_0^\tau B'(\tau-\tau_1)\mathbf{u}_\varepsilon(\tau_1)d\tau_1, \mathbf{u}_\varepsilon''(\tau)\right)d\tau\right| \\
&\le c\|\mathbf{u}_\varepsilon'(t)\|_1 + c + c\int_0^t \|\mathbf{u}_\varepsilon'(\tau)\|_1 d\tau.
\end{aligned} \tag{6.373}$$

Further, we have

$$|I| = \left|\int_0^t \int_{\cup_{k,l}\Gamma_c^{kl}} \frac{\partial g_c^{kl}(\mathbf{x},\tau)}{\partial \tau} \frac{\partial \chi_\varepsilon(\mathbf{u}_{\varepsilon t}''^k(\tau) - \mathbf{u}_{\varepsilon t}''^l(\tau) + \varphi_t''^k(t) - \varphi_t''^l(\tau))}{\partial \tau}\,ds\,d\tau\right|$$

and assuming that $\partial g_c^{kl}/\partial t = 0$, that is, that g_c^{kl} does not depend on t, we have $|I| = 0$.
To estimate $|\mathbf{u}_\varepsilon''(0)|$ from (6.358), we put $t = 0$, and we obtain

$$(\mathbf{u}_\varepsilon''(0), \mathbf{v}) + a(\mathbf{u}_\varepsilon(0), \mathbf{v}) + (j_{g\varepsilon}'(\mathbf{u}_\varepsilon'(0) + \boldsymbol{\varphi}_\varepsilon'(0)), \mathbf{v}) = (\mathbf{F}_0(0), \mathbf{v}) \quad \forall \mathbf{v} \in V.$$

Hence and since $\mathbf{u}_{1t} + \boldsymbol{\varphi}_t'(0) = 0$ due to the assumption of Theorem 6.46, we infer that

$$\begin{aligned}
(\mathbf{u}_\varepsilon''(0), \mathbf{v}) &= (\mathbf{F}_0(0), \mathbf{v}) - a(\mathbf{u}_0, \mathbf{v}) \\
&= (\mathbf{F}(0), \mathbf{v}) - (\boldsymbol{\varphi}''(0), \mathbf{v}) - a(\boldsymbol{\varphi}(0), \mathbf{v}) - a(\mathbf{u}_0, \mathbf{v}).
\end{aligned} \tag{6.374}$$

Since

$$a(\mathbf{u}_0, \mathbf{v}) = (A\mathbf{u}_0, \mathbf{v}) \quad \forall \mathbf{v} \in V,$$

$$a(\boldsymbol{\varphi}(0), \mathbf{v}) = (A\boldsymbol{\varphi}(0), \mathbf{v}) + \int_{\partial\Omega} \tau_{ij}(\boldsymbol{\varphi}(0)) n_j v_i \, ds, \quad \forall \mathbf{v} \in V,$$

then we obtain from (6.374) after some modification

$$\mathbf{u}_\varepsilon''(0) = \mathbf{F}(0) - \boldsymbol{\varphi}''(0) - A(\mathbf{u}_0 + \boldsymbol{\varphi}(0)) - c\|\boldsymbol{\varphi}(0)\|_1 \in V',$$

from which, using the assumptions of the theorem and since $|\mathbf{v}|_0^2 \leq c\|\mathbf{v}\|_1^2$,

$$\|\mathbf{u}_\varepsilon''(0)\|_1 \leq c.$$

Integrating (6.370) from 0 to t, using all estimates and the above-obtained results, and estimating some terms like in the uniqueness proof, we obtain

$$\begin{aligned} |\mathbf{u}_\varepsilon''(t)|_0^2 + \|\mathbf{u}_\varepsilon'(t)\|_1^2 &\leq c\Bigg[|\mathbf{u}_\varepsilon''(0)|_0^2 + \|\mathbf{u}_1\|_1^2 + \int_0^t |\mathbf{u}_\varepsilon'(\tau)|_0^2 \, d\tau \\ &\quad + \int_0^t \|\mathbf{F}_0'(\tau)\|_*^2 \, d\tau + \|\mathbf{u}_\varepsilon'(t)\|_1 + c + \int_0^t \|\mathbf{u}_\varepsilon'(\tau)\|_1^2 \, d\tau \Bigg] \\ &\leq c\left(1 + \int_0^t \left[|\mathbf{u}_\varepsilon''(\tau)|_0^2 + \|\mathbf{u}_\varepsilon'(\tau)\|_1^2\right] d\tau\right), \end{aligned}$$

and applying the Gronwall lemma, we find that

$$|\mathbf{u}_\varepsilon''(t)|_0^2 + \|\mathbf{u}_\varepsilon'(t)\|_1^2 \leq c,$$

from which

$$|\mathbf{u}_\varepsilon''(t)|_0^2 \leq c \qquad \|\mathbf{u}_\varepsilon'(t)\|_1^2 \leq c. \tag{6.375}$$

To prove the limit process with ε, we select, according to (6.366) and (6.375), from the sequences $\{\mathbf{u}_\varepsilon\}$ the subsequences, we denote them again by $\{\mathbf{u}_\varepsilon\}$, such that

$$\begin{aligned} &\mathbf{u}_\varepsilon \to \mathbf{u} \text{ weakly-star in } L^\infty(I; V), \\ &\mathbf{u}_\varepsilon' \to \mathbf{u}' \text{ weakly-star in } L^\infty(I; V), \\ &\mathbf{u}_\varepsilon'' \to \mathbf{u}'' \text{ weakly-star in } L^\infty(I; L^{2,N}(\Omega)). \end{aligned} \tag{6.376}$$

From (6.358) we obtain

$$\begin{aligned} &(\mathbf{u}_\varepsilon''(t), \mathbf{v} - \mathbf{u}_\varepsilon'(t)) + a(\mathbf{u}_\varepsilon(t), \mathbf{v} - \mathbf{u}_\varepsilon'(t)) + \int_0^t a_1(t - \tau; \mathbf{u}_\varepsilon(\tau), \mathbf{v} - \mathbf{u}_\varepsilon'(t)) \, d\tau \\ &\quad + j_{g\varepsilon}(\mathbf{v} + \varphi'(t)) - j_{g\varepsilon}(\mathbf{u}_\varepsilon'(t) + \boldsymbol{\varphi}'(t)) - (\mathbf{F}_0(t), \mathbf{v} - \mathbf{u}_\varepsilon'(t)) \\ &\quad = j_{g\varepsilon}(\mathbf{v} + \boldsymbol{\varphi}'(t)) - j_{g\varepsilon}(\mathbf{u}_\varepsilon'(t) + \boldsymbol{\varphi}'(t)) - (j_{g\varepsilon}'(\mathbf{u}_\varepsilon'(t) + \boldsymbol{\varphi}'(t)), \mathbf{v} - \mathbf{u}_\varepsilon'(t)) \geq 0. \end{aligned}$$

Hence, putting $\mathbf{v} = \mathbf{v}(t)$, $\mathbf{v} \in L^2(I; V)$, and integrating over $t \in \overline{I}$, we have

$$\int_0^{t_p} \Big[(\mathbf{u}_\varepsilon''(t), \mathbf{v}(t)) + a(\mathbf{u}_\varepsilon(t), \mathbf{v}(t)) + \int_0^t a_1(t-\tau; \mathbf{u}_\varepsilon(\tau), \mathbf{v}(t))\, d\tau$$

$$+ j_{g\varepsilon}(\mathbf{v}(t) + \boldsymbol{\varphi}'(t)) - (\mathbf{F}_0(t), \mathbf{v}(t) - \mathbf{u}_\varepsilon'(t)) \Big]\, dt$$

$$\geq (\mathbf{u}_\varepsilon''(t), \mathbf{u}_\varepsilon'(t)) + a(\mathbf{u}_\varepsilon(t), \mathbf{u}_\varepsilon'(t))$$

$$+ \int_0^t a_1(t-\tau; \mathbf{u}_\varepsilon(\tau), \mathbf{u}_\varepsilon'(t))\, d\tau + j_{g\varepsilon}(\mathbf{u}_\varepsilon'(t) + \boldsymbol{\varphi}'(t))\, dt$$

$$= \frac{1}{2}[|\mathbf{u}_\varepsilon'(t_p)|_0^2 + a(\mathbf{u}_\varepsilon(t_p), \mathbf{u}_\varepsilon(t_p))] + \int_0^{t_p}\int_0^t a_1(t-\tau; \mathbf{u}_\varepsilon(\tau), \mathbf{u}_\varepsilon'(t))\, d\tau\, dt$$

$$+ \int_0^{t_p} j_{g\varepsilon}(\mathbf{u}_\varepsilon'(t) + \boldsymbol{\varphi}'(t))\, dt - \frac{1}{2}a(\mathbf{u}_0, \mathbf{u}_0) - \frac{1}{2}|\mathbf{u}_1|_0^2.$$

Since

$$\liminf_{\varepsilon\to 0} \Big\{ \frac{1}{2}[|\mathbf{u}_\varepsilon'(t_p)|_0^2 + a(\mathbf{u}_\varepsilon(t_p), \mathbf{u}_\varepsilon(t_p))]$$

$$+ \int_0^{t_p}\int_0^t a_1(t-\tau; \mathbf{u}_\varepsilon(\tau), \mathbf{u}_\varepsilon'(t))\, d\tau\, dt$$

$$+ \int_0^{t_p} j_{g\varepsilon}(\mathbf{u}_\varepsilon'(t) + +\varphi'(t))\, dt \Big\}$$

$$\geq \frac{1}{2}[|\mathbf{u}'(t_p)|_0^2 + a(\mathbf{u}(t_p), \mathbf{u}(t_p))]$$

$$+ \int_0^{t_p}\int_0^t a_1(t-\tau; \mathbf{u}(\tau), \mathbf{u}'(t))\, d\tau\, dt + \int_0^{t_p} j_g(\mathbf{u}'(t) + \boldsymbol{\varphi}'(t))\, dt$$

$$\geq \int_0^{t_p} [(\mathbf{u}''(t), \mathbf{u}'(t)) + a(\mathbf{u}(t), \mathbf{u}'(t))]\, dt$$

$$+ \int_0^{t_p}\int_0^t a_1(t-\tau; \mathbf{u}(\tau), \mathbf{u}'(t))\, d\tau\, dt + \int_0^{t_p} j_g(\mathbf{u}'(t) + \boldsymbol{\varphi}'(t))\, dt.$$

Therefore,

$$\int_0^{t_p} \{(\mathbf{u}''(t), \mathbf{v} - \mathbf{u}'(t)) + a(\mathbf{u}'(t), \mathbf{v} - \mathbf{u}'(t))$$

$$+ \int_0^t a_1(t-\tau; \mathbf{u}(\tau), \mathbf{v} - \mathbf{u}'(t))\, d\tau + j_g(\mathbf{v} + \boldsymbol{\varphi}'(t))$$

$$- j_g(\mathbf{u}'(t) + \boldsymbol{\varphi}'(t)) - (\mathbf{F}_0(t), \mathbf{v} - \mathbf{u}'(t))\}\, dt \geq 0, \qquad \mathbf{v} \in L^2(I; V),$$

and then (6.348) follows, which completes the proof. □

Remark 6.27 It can be shown that the elastic case in the classical linear elasticity is the special case of the problem in the linear viscoelasticity with a long memory [see, e.g., Duvaut and Lions (1976)].

If the biomaterials are loaded very slowly, then the investigated parts of human bodies are deformed also very slowly; and/or if the investigated parts of human bodies move by an uniform velocity, then the inertia forces in both cases can be neglected.

Therefore, we will consider the following variational problem: Find $\mathbf{u} \in L^\infty(I; V)$ such that

$$a(\mathbf{u}(t), \mathbf{v}) + \int_0^t a_1(t-\tau; \mathbf{u}(\tau), \mathbf{v})\, d\tau = (\mathbf{F}(t), \mathbf{v}) \quad \forall \mathbf{v} \in V, \qquad \text{a.e. } t \in I, \quad (6.377)$$

where the bilinear forms $a(\cdot,\cdot)$ and $a_1(t;\cdot,\cdot)$ are bilinear forms, mapping $V \times V$ into $\mathbb{R}$, defined above.

It can be shown that $a(\cdot,\cdot) \equiv \bar{a}(t,t;\cdot,\cdot)$ is continuous and V-coercive, and $a_1(t-\tau;\cdot,\cdot) \equiv \bar{a}_1(t,\tau;\cdot,\cdot)$ is continuous and satisfies

$$|a_1(t-\tau; \mathbf{u}, \mathbf{v})| \le C(t,\tau)\|\mathbf{u}\|_1\|\mathbf{v}\|_1 \quad \forall \mathbf{u}, \mathbf{v} \in V, \quad \text{with } C(t,\tau) \in L^\infty(I \times I),$$

and due to the assumed regularity of **f** and **P**, we have

$$|(\mathbf{F}(t), \mathbf{v})| \le c(t)\|\mathbf{v}\|_1 \quad \forall \mathbf{v} \in V, \quad \text{with } c(t) = \|\mathbf{f}\|_0 + \|\mathbf{P}\|_0. \qquad (6.378)$$

Theorem 6.47 Under the above-defined assumptions, the solution $\mathbf{u} \in L^\infty(I; V)$ exists, and it is unique and depends continuously on **f** and **P**.

The proof is parallel to that of Linz (1985) and it is based on the Picard method.

6.6.8 Numerical Solution of the Problem

In addition to an instantaneous elastic response, biomaterials under load can display significant creep during sustained loading. The creep is a viscous effect. Therefore, the viscous biomaterials with a long time memory can be used for analyses of human joints. Analytical solutions of such a type of problems are rare, and then we must discretize the continuous problem. According to the properties of the viscoelastic biomaterials, the relaxation functions are introduced. Thus, concerning the nature of the viscoelastic functions, we have the following assumptions:

Assumptions (About Relaxation Functions) Let $I = [0, t_p]$, $t_p > 0$. We will assume that the time-dependent part of the generic stress relaxation function $\varphi(t,\tau)$ defined for $t, \tau \in I$ satisfies the following:

(i) regularity: $\varphi(t,\tau) \in C^\infty(I \times I)$,
(ii) positivity: $\varphi(t,\tau) > 0 \ \forall \tau < t$, such that $\tau, t \in I$,

(iii) causality: $\varphi(t,\tau) \equiv 0$ $\forall \tau > t$, such that $\tau, t \in I$,

(iv) fading memory: for fixed $t \in I$ $\varphi_{,\tau}(t,\tau) \leq 0$ when $\tau \in [0, t_p]$, $\varphi_{,\tau} = \partial\varphi/\partial\tau$.

Remark 6.28 These assumptions hold for the time-dependent parts of the specific relaxation functions like $\lambda(t,\tau)$, $\mu(t,\tau)$, $E(t,\tau)$, and so forth.

First, we will discuss the viscoelastic part of the quasi-static problem only. The method of discretization can be based on the semidiscrete and fully discrete approximation approaches.

Discrete Approximation of the Quasi-Static Problem in Viscoelasticity with a Long Memory The method of discretization proceeds in two steps. First, we utilize the Galerkin finite element method for discretization in the space domain. One of the procedures is a semidiscrete approximation of the problem, producing a system of Volterra second-kind equations. The second procedure is a fully discrete approximation based on replacing the Volterra operator with a discrete representation as a quadrature rule such as the trapezoidal or the Simpson rules. The numerical scheme based on the trapezoidal rule is numerically stable [see Linz (1985, Chapter 7)] and does not require starting values as the numerical scheme based on the Simpson rule. Below we construct the semi- and fully discrete schemes and briefly discuss the possibility conditions of finding semiexact solutions by treating the semidiscrete solution as an exact elastic solution.

First, let us consider a classical displacement problem for a single body occupying the domain Ω, loaded on the part Γ_τ and fixed on part $\Gamma_u = \partial\Omega \backslash \Gamma_\tau$. Let us assume that the domain Ω has a polygonal or polyhedral boundary. Let $\{\mathcal{T}_h\}$ be a regular family of finite element partitions $\mathcal{T}_h$ of $\overline{\Omega}_h$ compatible to the boundary subsets $\overline{\Gamma}_{\tau h}$ and $\overline{\Gamma}_{uh}$. We introduce the standard spaces of linear elements on the regular division $\mathcal{T}_h$ of the polygonal (for $N=2$) or polyhedral (for $N=3$) domain. We will also set $h := \max_\iota (\text{diam}\, T_{h_i})$. Let

$$V_h = \{\mathbf{v}_h \in [C(\overline{\Omega})]^N : \mathbf{v}_h|_{T_{hi}} \in [P_1(T_{hi})]^N \quad \forall T_{hi} \in \mathcal{T}_h,\ \mathbf{v}_h = 0 \text{ on } \Gamma_u \cap \partial\Omega\},$$

where we denote by T_{hi} any triangle (for $N=2$) and any tetrahedron (for $N=3$) of the division $\mathcal{T}_h$.

In the finite element theory the well-known approximation property of the space V_h is the following: Let $m \geq 0$ be a positive integer and let $\mathbf{u} \in V \cap H^{2,N}(\Omega)$, then we have

$$\inf_{\mathbf{v} \in V_h} \|\mathbf{u} - \mathbf{v}\|_{m,N,\Omega} \leq ch^{2-m} |\mathbf{u}|_{2,N,\Omega} \quad \text{for } 2 > m, \tag{6.379}$$

where c is some positive constant depending upon Ω but not upon $\mathbf{u}$ [see Ciarlet (1978)].

The semidiscrete finite element approximation to the quasi-static displacement problem is the following: Find $\mathbf{u}_h \in L(I; V_h)$ such that

$$a(\mathbf{u}_h(t), \mathbf{v}_h) + \int_0^t a_1(t-\tau; \mathbf{u}_h(\tau), \mathbf{v}_h)\, d\tau = (\mathbf{F}(t), \mathbf{v}_h) \quad \forall \mathbf{v}_h \in V_h, \text{ a.e. } t \in I. \tag{6.380}$$

Now we construct the finite-dimensional subspaces $V_h^\iota := \text{span}\{\varphi_j(\mathbf{x})\}_{j=1}^n \subset V^\iota$, $\iota = 1, \ldots, r$, where n is the number of nodes of the used partition, $\varphi_j \in P_1$, where P_1 is a finite-dimensional vector space of linear functions and the set $\{\varphi_j(\mathbf{x})\}$ is linearly independent. Then from the φ_j we may construct a basis for the space V_h, so for arbitrary $\mathbf{u}_h \in V_h$ we select the degrees of freedom for the finite element space to be the values of $\mathbf{u}_h$ at the nodes. Specifying the design of φ_j, then $\mathbf{u}_h$ can be uniquely represented by

$$\mathbf{u}_h(\mathbf{x}, t) = \sum_{i=1}^{n} \varphi_i(\mathbf{x})(\mathbf{u}_{1h}(N^i, t), \ldots, \mathbf{u}_{mh}(N^i, t))^{\mathrm{T}},$$

where N^i are the coordinates of the node labeled by the index i.

Then (6.380) leads to the Volterra system

$$A(t,t)\mathbf{U}(t) + \int_o^t B(t,\tau)\mathbf{U}(\tau)\,d\tau = F(t), \tag{6.381}$$

where $A(t,t)$ represents the stiffness matrix, $B(t,\tau)$ the history matrix of the materials, which can be derived from the bilinear forms $a(\cdot,\cdot)$ and $a_1(\cdot,\cdot)$, the time-dependent $F(t)$ contains the body forces and surface tractions, and $\mathbf{U}(t)$ is a time-dependent array of coordinates in the space V_h.

Due to the Korn inequality and the above given assumptions, $A(t,t)$ is positive definite and therefore by premultiplying (6.381) through $(A(t,t))^{-1}$ we obtain a standard form of the Volterra equation of the second order, which admits a unique solution [see, e.g., Linz (1985)].

The fully discrete finite element approximation to the quasi-static displacement problem is the following: The fully discrete problem follows from (6.381) by replacing the hereditary with a suitable numerical quadrature rule.

Let n be a positive integer and let the interval $I = [0, t_p]$ be divided into subintervals by the set $I_k = \{0 = t_0, \ldots, t_n = t_p \in \mathbb{R} : t_{i-1} < t_i,\ i = 1, \ldots, n\} \subset I$ and use the trapezoidal rule. Let us set

$$\mathbf{u}_i := m\,\mathbf{u}(t_i) \quad \text{for } t_i \in I_k,$$
$$k_i := t_i - t_{i-1} \quad \text{for } 1 \le i \le n,\ k := \max_{1\le i\le n} k_i,$$

then

$$\delta_i(\mathbf{u}) := \int_0^{t_i} \mathbf{u}(\tau)\,d\tau - \sum_{j=1}^{i} \frac{k_j}{2}(\mathbf{u}_{j-1} + \mathbf{u}_j), \qquad i > 0,$$

and then the trapezoidal rule approximates the integral and for the error we have

$$|\delta_i(\mathbf{u})| \le ck^2 \int_0^{t_i} |\mathbf{u}''(\tau)|\,d\tau, \qquad c = \text{const} > 0.$$

Then the fully discrete approximation leads to the following problem: For each $t_i \in I_k$ find a function $\mathbf{u}_{hi}^k \cong \mathbf{u}_h^k(t_i) \in V_h$ such that

$$a(t_i, t_i; \mathbf{u}_{hi}^k, \mathbf{v}_h) + \sum_{j=0}^{i} w_{ij} a_1(t_i, t_j; \mathbf{u}_{hj}^k, \mathbf{v}_h) = (\mathbf{F}(t_i), \mathbf{v}_h) \quad \forall \mathbf{v}_h \in V_h, \tag{6.382}$$

where w_{ij} denote the weights implied by the approximation of the integral by the trapezoidal rule and $w_{00} = 0$, and where $\mathbf{u}_{hi}^k \cong \mathbf{u}_h(t_i)$.

Then this problem leads to the following problem: For each $t_i \in I_k$ find the vector $\mathbf{U}_i$ of coordinates of V_h

$$A(t_i,t_i)\mathbf{U}_i + \sum_{j=0}^{i} w_{ij} B(t_i,t_j)\mathbf{U}_j = \mathbf{F}_i, \tag{6.383}$$

where $\mathbf{U}_i = \mathbf{U}(t_i)$, $\mathbf{F}_i = \mathbf{F}(t_i)$. Hence, for finding $\{\mathbf{U}_i\}_{i=1}^n$, it follows that

$$\begin{aligned}\left(A(t_i,t_i) + \frac{k_i}{2}B(t_i,t_i)\right)\mathbf{U}_i &= \mathbf{F}_i - \frac{k_i}{2}B(t_i,t_{i-1})\mathbf{U}_{i-1}\\ &\quad - \sum_{j=1}^{i-1}\frac{k_i}{2}(B(t_i,t_j)\mathbf{U}_j + B(t_i,t_{j-1})\mathbf{U}_{j-1}),\end{aligned} \tag{6.384}$$

where the starting value $\mathbf{U}_0$ can be found as the solution of the static linear elastic problem

$$A(t_0,t_0)\mathbf{U}_0 = \mathbf{F}_0. \tag{6.385}$$

Remark 6.29 If we use the higher order Simpson quadrature rule, we need as additional starting values $\mathbf{U}_0$ and $\mathbf{U}_1$.

Next, we will present the semidiscrete error estimate and the fully discrete error estimate. The estimates are based on the following abstract theorems (Linz, 1985; Shaw et al., 1994a,b, 1997), mentioned below.

Theorem 6.48 Let $\mathcal{H}$ be a Hilbert space with the norm $\|\cdot\|$ and $\mathcal{H}_h \subset \mathcal{H}$ be an appropriate finite element space. For almost every $t \in I = [0, t_p]$, $t_p > 0$, and with respect to this norm let $a(t,t;\cdot,\cdot) : \mathcal{H} \times \mathcal{H} \mapsto \mathbb{R}$ be a continuous and coercive bilinear form with constants c_0 and c_1, respectively. Additionally, for almost every $t - \tau \in I$ let $a_1(t,\tau;\cdot,\cdot) : \mathcal{H} \times \mathcal{H} \mapsto \mathbb{R}$ and $L(t;\cdot) : \mathcal{H} \mapsto \mathbb{R}$ be, respectively, a continuous bilinear form, with the constant c_2 and a continuous linear form with the constant c_3. Then, for the approximation of the abstract Volterra problem: Find $\mathbf{u} \in L^\infty(I;\mathcal{H})$ such that

$$a(t,t;\mathbf{u}(t),\mathbf{v}) + \int_0^t a_1(t-\tau;(\mathbf{u}),\mathbf{v})\,d\tau = L(t;\mathbf{v}) \quad \forall \mathbf{v} \in \mathcal{H},\ \text{a.e. } t \in I, \tag{6.386}$$

and by the semidiscrete Galerkin scheme: Find $\mathbf{u}_h \in L^\infty(I;\mathcal{H}_h)$ such that

$$a(t,t;\mathbf{u}_h(t),\mathbf{v}_h) + \int_0^t a_1(t-\tau;\mathbf{u}_h(\tau),\mathbf{v}_h)\,d\tau = L(t;\mathbf{v}_h) \quad \forall \mathbf{v}_h \in \mathcal{H}_h,\ \text{a.e. } t \in I, \tag{6.387}$$

we have,

$$\|\mathbf{u}(t) - \mathbf{u}_h(t)\| \le c\|\mathbf{u}(t) - \mathbf{v}_h\|_{L^\infty(I;\mathcal{H})} \quad \forall \mathbf{v}_h \in \mathcal{H}_h, \text{ a.e. } t \in I, \tag{6.388}$$

where c is a positive constant independent of $\mathbf{u}$.

For the proof see Shaw et al. (1994a,b, 1997).

Putting $\mathcal{H} = V$, $\mathcal{H}_h = V_h$, $L(t; \mathbf{v}) = (\mathbf{F}(t), \mathbf{v})$, $L(t; \mathbf{v}_h) = (\mathbf{F}(t), \mathbf{v}_h)$, we find the following error estimate:

Theorem 6.49 Let the conditions of the previous abstract theorem hold, let the space V_h satisfy the approximation property (6.378). Then for the semidiscrete finite element approximation of (6.377) by (6.380) under the above conditions, the following estimate

$$\|\mathbf{u}(t) - \mathbf{u}_h(t)\|_{1,N,\Omega_h} \le ch|\mathbf{u}|_{L^\infty(I;\mathbb{H}^{2,N}(\Omega_h))}, \quad \forall t \in I, \tag{6.389}$$

holds, where c is a positive constant independent of $\mathbf{u}$ and h, whenever $\mathbf{u}(t) \in V \cap \mathbb{H}^{2,N}(\Omega_h)$ for all $t \in I$.

For the fully discrete estimate we can use the following abstract theorem:

Theorem 6.50 Let the notation of the previous abstract theorem hold and let $\mathbf{u}'' \in L^1(I, \cdot)$ and $a_1(t, \tau; \cdot, \cdot)$ be twice differentiable with respect to τ. Then, the fully discrete trapezoidal approximation to the semidiscrete problem (6.387) is the following: Find $\mathbf{u}_{hi}^k \in \mathcal{H}_h$ for each $t_i \in I_k$ such that

$$a(t_i, t_i; \mathbf{u}_{hi}^k, \mathbf{v}_h) + \sum_{j=0}^{i} w_{ij} a_1(t_i, t_j; \mathbf{u}_{hj}^k, \mathbf{v}_h) = L(t_i, \mathbf{v}_h) \quad \forall \mathbf{v}_h \in \mathcal{H}_h, \tag{6.390}$$

t holds, where $I_k = \{0 = t_0, \ldots, t_n = t_p \in \mathbb{R}; t_{i-1} < t_i, i = 1, \ldots, n\} \subset I$, n integer constant, w_{ij} are the weights associated with the trapezoidal rule with $w_{00} = 0$. Then there exists for $k = \max_i (t_i - t_{i-1})$ small enough a positive constant $c = c(\mathbf{u})$ independent of k such that

$$\max_{t_i \in I_k} \|\mathbf{u}(t_i) - \mathbf{u}_{hi}^k\| \le c(\|\mathbf{u}(t) - \mathbf{v}_h\|_{L^\infty(I;\mathcal{H})} + k^2) \quad \forall \mathbf{v}_h \in \mathcal{H}_h. \tag{6.391}$$

For the proof see Linz (1985) and Shaw et al. (1994a,b, 1997).

Using this theorem onto our investigated problem and taking $\mathcal{H} = V$, $\mathcal{H}_h = V_h$, we have the following result:

Theorem 6.51 Let the assumptions about the data be as above. Then there exists the constant $c = c(\mathbf{u})$, independent of h and k such that

$$\max_{t_i \in I_k} \|\mathbf{u}(t_i) - \mathbf{u}_{hi}^k\|_{1,N,\Omega_h} \le c(h + k^2) \tag{6.392}$$

provided $\mathbf{u}'' \in L^1(I, \cdot)$.

Discrete Approximation of the Dynamic Bilateral Contact Problem in Viscoelasticity with Long Memory Let $\Omega \subset R^N$, $N = 2, 3$ be approximated by its polyhedral approximation and let $\partial\Omega_h = \Gamma_{uh} \cup \Gamma_{\tau h} \cup (\cup_{k,l}\Gamma^{kl}_{ch})$. Let $\mathcal{T}_h$ be a partition of $\overline{\Omega}_h$ by tetrahedra (for $N = 3$) and triangles (for $N = 2$) T_{hi} and let $h = h(\mathcal{T}_h)$ be the maximum diameter of the tetrahedra (triangle) element T_{hi}. We assume that the used family of finite element partitions $\{\mathcal{T}_h\}, h \to 0$, is regular and, moreover, that it also induces a regular family of finite partitions on $\cup_{k,l}\Gamma^{kl}_{ch}$.

Let the space V be approximated by the finite element space V_h, generated by linear polynomial functions, $V_h \subset V$, where

$$V = \{\mathbf{v} \in \mathbb{H}^{1,N}(\Omega) | \mathbf{v} = \mathbf{0} \text{ on } \Gamma_u, v^k_n - v^l_n = 0 \text{ on } \cup_{k,l}\Gamma^{kl}_c\}.$$

Then the discrete problem is as follows [cf. (6.348)]:

Problem $(\mathcal{P})_d$ Find $\mathbf{u}_h \in L^2(I; V_h)$ such that

$$\begin{aligned}
&(\mathbf{u}''_h(t), \mathbf{v}_h - \mathbf{u}'_h(t)) + a(\mathbf{u}_h(t), \mathbf{v}_h - \mathbf{u}'_h(t)) \\
&\quad + \int_0^t a_1(t-\tau; \mathbf{u}_h(\tau), \mathbf{v}_h - \mathbf{u}'_h(t))\, d\tau + j_g(\mathbf{v}_h + \boldsymbol{\varphi}'_h(t)) - j_g(\mathbf{u}'_h(t) + \boldsymbol{\varphi}'_h(t)) \\
&\quad \geq (\mathbf{F}_0(t), \mathbf{v}_h - \mathbf{u}'_h(t)) \quad \forall \mathbf{v}_h \in V_h,
\end{aligned} \tag{6.393}$$

with initial conditions

$$\mathbf{u}_h(0) = \mathbf{u}_{0h}, \mathbf{u}'_h(0) = \mathbf{u}_{1h}, \tag{6.394}$$

where $\mathbf{u}_{0h} \in V_h$, $\mathbf{u}_{1h} \in V_h$ are discrete approximates of $\mathbf{u}_0 - \boldsymbol{\varphi}(0)$, $\mathbf{u}_1 - \boldsymbol{\varphi}'(0)$.

It can be proved that the discrete problem $(\mathcal{P})_d$ has a unique solution $\mathbf{u}_h \in L^2(I; V_h)$.

Since the functional $j_g(\cdot)$ is not differentiable, for a numerical solution we will introduce the regularization (6.354), that is, $\psi_\varepsilon = |z|^{1+\varepsilon}/(1+\varepsilon)$, and the regularized functional $j_{g\varepsilon}$ by (6.356), which will be well-defined, convex and Gâxteaux-differentiable with respect to the argument $\mathbf{v}$, and then it can be approximated, for example, by the trapezoidal rule. The functional $j_{g\varepsilon}$ has a meaning of the frictional forces. Then the problem leads to the following [cf. (6.358) and (6.359)]:

$$\begin{aligned}
&(\mathbf{u}''_h(t), \mathbf{v}_h) + a(\mathbf{u}_h(t), \mathbf{v}_h) + \int_0^t a_1(t, \tau; \mathbf{u}_h(\tau), \mathbf{v}_h)\, d\tau \\
&= (\mathbf{F}_1(t), \mathbf{v}_h) \quad \forall \mathbf{v}_h \in V_h,\ t \in I, \\
&\mathbf{u}_h(\mathbf{x}, 0) = \mathbf{u}_{0h}(\mathbf{x}), \qquad \mathbf{u}'_h(\mathbf{x}, 0) = \mathbf{u}_{1h}(\mathbf{x}),
\end{aligned} \tag{6.395}$$

where we put $\mathbf{u}_h(t) \equiv \mathbf{u}_{h\varepsilon}(t)$ and

$$(\mathbf{F}_1(t), \mathbf{v}_h) = (\mathbf{F}_0(t), \mathbf{v}_h) - (j'_{g\varepsilon}(\mathbf{v}_h + \boldsymbol{\varphi}'_h(t)), \mathbf{v}_h).$$

Putting

$$\mathbf{u}''_h = k^{-2}(\mathbf{u}^{i+1}_h - 2\mathbf{u}^i_h + \mathbf{u}^{i-1}_h),$$

and using the notation $\mathbf{u}_h^{i+1} \equiv \mathbf{u}_h$, then $\mathbf{u}_h$ have the unique representations

$$\mathbf{u}_h(\mathbf{x}) = \sum_{i=1}^{n} \varphi_i(\mathbf{x})(\mathbf{u}_{1h}(N^i), \ldots, \mathbf{u}_{mh}(N^i))^{\mathrm{T}},$$

where $N^i \in \mathbb{R}^N$ are the coordinates of the node labeled i.

Then the semi-implicit scheme in the time and the Galerkin finite element approximation in the space variable will be used. The problem for every $t = t^{i+1} \in I$ then leads to solving the Volterra system:

$$\mathbb{A}(t,t)\mathbf{U}(t) + \int_o^t \mathbb{B}(t,\tau)\mathbf{U}(\tau)\,d\tau = \mathbb{F}(t), \qquad t = t^{i+1} \in I, \tag{6.396}$$

where $\mathbb{A}(t,t)$ is the stiffness matrix, which is positive definite, $\mathbb{B}(t,\tau)$ is the history matrix of the viscoelastic materials, which can be derived from the bilinear forms $a(t,t;\cdot,\cdot)$, $a_1(t,\tau;\cdot,\cdot)$ and the inner products $(\cdot,\cdot)$, $\mathbb{F}(t)$ contains the body and friction forces, the surface tractions, as well as the inner product $k^{-2}(-2\mathbf{u}_h^i + \mathbf{u}_h^{i-1}, \mathbf{v}_h)$. Moreover, $\mathbf{U}(t)$ is the time-dependent array of coordinates in the space V_h. The Eq. (6.396) is similar to (6.381) so that the derivation of the error estimate is also similar as above.

6.7 VISCOPLASTIC MODEL OF TOTAL HUMAN JOINT REPLACEMENTS

6.7.1 Introduction

In this section we will investigate a simple model of the human joint, say the knee joint, which is loaded by a vertical loading $\mathbf{P}(t)$, and, moreover, we will assume that the contact is frictionless because of the lubricated properties of the synovial liquid. The simulation of the contact conditions on the contact between the femoral and tibial parts of the knee joint are modeled by the classical Signorini conditions without friction.

6.7.2 Formulation of the Problem

Let us consider the total knee replacement as five elasticviscoplastic bodies in mutual contact, whose separate particles occupy open, bounded, connected domains $\Omega^\iota \subset \mathbb{R}^N$, $\iota = 1, \ldots, r$, $r = 5$, $N = 2, 3$. We will assume that every domain Ω^ι has the boundary $\partial\Omega^\iota$, which is Lipschitz continuous, and it is partitioned into three parts $\overline{\Gamma}_u, \overline{\Gamma}_\tau = {}^1\overline{\Gamma}_\tau \cup {}^2\overline{\Gamma}_\tau$, and $\overline{\Gamma}_c = \overline{\Gamma_c^{23}} \cup \overline{\Gamma_c^{34}} \equiv \cup_{k,l} \overline{\Gamma}_c^{kl}$, $k \neq l$, $k, l = \{2, 3, 4\}$, with mutually disjoint open sets Γ_u, Γ_τ, and Γ_c such that meas $\Gamma_u > 0$. Let the unit outward normal $\mathbf{n}$ to $\partial\Omega^\iota$ be defined almost everywhere, and, moreover, let its positive direction to the contact boundary Γ_c^{kl} be taken to the domain Ω^k. Let $\overline{I} = [0, T]$, $T > 0$, be the time interval, let $\Omega(t) = \Omega \times I$, $\partial\Omega(t) = \partial\Omega \times I$, $\Gamma_u(t) = \Gamma_u \times I$, $\Gamma_\tau(t) = \Gamma_\tau \times I$, and $\Gamma_c(t) = \Gamma_c \times I$. Let us assume that body forces $\mathbf{F} = \mathbf{F}(t)$ act in $\Omega = \cup_\iota \Omega^\iota$ and that the contacts are frictionless and simulated by the so-called nonpenetration conditions with zero gap functions.

We will solve the following problem:

Problem ($\mathcal{P}_{vp}$) Find a displacement field $\mathbf{u} = (\mathbf{u}^\iota)$, $\iota = 1, \ldots, 5$, $\mathbf{u}^\iota : \Omega^\iota \times \bar{I} \to \mathbb{R}^N$ and stress fields $\boldsymbol{\tau} = (\tau_{ij})$, $i, j = 1, \ldots, N$, $\boldsymbol{\tau}^\iota = (\tau_{ij}^\iota) : \Omega^\iota \times \bar{I} \to \mathbb{S}^N$ such that

$$\frac{\partial \tau_{ij}^\iota}{\partial x_j} + F_i^\iota = 0 \quad \text{or} \quad \text{Div}\, \boldsymbol{\tau}^\iota + \mathbf{F}^\iota = 0 \quad \text{in } \Omega^\iota \times I,\ \iota = 1, \ldots, r, \tag{6.397}$$

$$\boldsymbol{\tau}^{\iota\prime} = (\tau^{\iota\prime}) = C^\iota \mathbf{e}(\mathbf{u}^{\iota\prime}) + G^\iota(\tau^\iota, \mathbf{e}(\mathbf{u}^\iota)) \quad \text{in } \Omega^\iota \times I,\ \iota = 1, \ldots, r, \tag{6.398}$$

$$\mathbf{u}^\iota = \mathbf{0} \quad \text{on } \Gamma_u \times I, \tag{6.399}$$

$$\tau_{ij} n_j = P_i \quad \text{on } {}^1\Gamma_\tau \times I, \tag{6.400}$$

$$\tau_{ij} n_j = 0 \quad \text{on } {}^2\Gamma_\tau \times I, \tag{6.401}$$

$$\begin{aligned} &u_n^k - u_n^l \le 0, \qquad \tau_n^k = \tau_n^l \equiv \tau_n^{kl} \le 0, \\ &(u_n^k - u_n^l)\tau_n^{kl} = 0, \qquad \boldsymbol{\tau}_t^{kl} = 0 \quad \text{on } \Gamma_c^{kl} \times I, \end{aligned} \tag{6.402}$$

$$\mathbf{u}^\iota(0) = \mathbf{u}_0, \qquad \boldsymbol{\tau}^\iota(0) = \boldsymbol{\tau}_0^\iota \quad \text{in } \Omega^\iota, \qquad \iota = 1, \ldots, r, \tag{6.403}$$

where $u_n^\iota = u_i^\iota n_i^\iota$, $\tau_n^\iota = \tau_{ij}^\iota n_i^\iota n_j^\iota$, $\boldsymbol{\tau}_t^\iota = (\tau_{ti}^\iota)$, and $\tau_{ti}^\iota = \tau_{ij}^\iota n_j^\iota - \tau_n^\iota n_i^\iota$ are the normal displacement, normal stress, and tangential stress, and $\mathbf{u}_0^\iota$ and $\boldsymbol{\tau}_0^\iota$ are initial displacement and initial stress, respectively.

Let us denote by $\mathbb{S}^N$ the space of second-order symmetric tensors on $\mathbb{R}^N$ and let us introduce the following Hilbert space:

$$V = \prod_{\iota=1}^{r} V^\iota = \{\mathbf{v} = (v_i) | \mathbf{v} \in \mathbb{H}^{1,N}(\Omega), \mathbf{v} = 0 \text{ on } \Gamma_u,\ i = 1, \ldots, N\},$$

$$Q = \prod_{\iota=1}^{r} Q^\iota = \{\boldsymbol{\tau} = (\tau_{ij}) | \tau_{ij} = \tau_{ji} \in L^2(\Omega),\ 1 \le i, j \le N\},$$

$${}^1Q = \prod_{\iota=1}^{r} {}^1Q^\iota = \{\boldsymbol{\tau} \in \mathbf{Q} |\ \text{Div}\, \boldsymbol{\tau} \in \mathbb{L}^{2,N}(\Omega)\},$$

and the set of admissible displacement and stress fields

$$K = \{\mathbf{v} = (\mathbf{v}_i^\iota) \in V, v_n^k - v_n^l \le 0 \text{ on } \cup_{k,l} \Gamma_c^{kl}\},$$

and, for all $t \in \bar{I}$,

$$\Sigma(t) = \{\boldsymbol{\sigma} = (\boldsymbol{\sigma}^1, \ldots, \boldsymbol{\sigma}^r) \in Q : (\boldsymbol{\sigma}, \mathbf{e}(\mathbf{v}))_Q \ge (\mathbf{f}(t), \mathbf{v})_V \quad \forall \mathbf{v} \in K\}.$$

where $(\mathbf{f}(t), \mathbf{v})_V = (\mathbf{F}(t), \mathbf{v}) + (\mathbf{P}(t), \mathbf{v})$.

The inner product in the space V is defined as follows:

$$(\mathbf{u}, \mathbf{v})_V = (\mathbf{e}(\mathbf{u}), \mathbf{e}(\mathbf{v}))_Q = \sum_{\iota=1}^{r} \int_{\Omega^\iota} \mathbf{e}(\mathbf{u})\mathbf{e}(\mathbf{v})\, d\mathbf{x} \quad \forall \mathbf{u}, \mathbf{v} \in V.$$

Moreover, since $\text{meas}(\Gamma_u \cap \partial\Omega^\iota) > 0$ for all $\iota \leq r$, the Korn inequality

$$\|\mathbf{e}(\mathbf{v})\|_Q \geq c\|\mathbf{v}\|_{H^{1,N}(\Omega)} \quad \forall \mathbf{v} \in V$$

holds, and, moreover, the norm is defined as $\|\mathbf{v}\|_V = \|\mathbf{e}(\mathbf{v})\|_Q$, so that it is equivalent to the norm $\|\mathbf{v}\|_{H^{1,N}(\Omega)}$.

Presumption 6.3 Let the elastic coefficients $C^\iota = (c^\iota_{ijkl})$ and the operator G^ι, $\iota = 1, \ldots, r$, satisfy

$$\begin{aligned}
&\text{(i)} \quad C^\iota = (c^\iota_{ijkl}) : \Omega^\iota \times \mathbb{S}^N \to \mathbb{S}^N,\\
&\text{(ii)} \quad c^\iota_{ijkl} \in L^\infty(\Omega^\iota), \qquad 1 \leq i,j,k,l \leq N,\\
&\text{(iii)} \quad C^\iota \boldsymbol{\tau} \cdot \boldsymbol{\sigma} = \boldsymbol{\sigma} \cdot C^\iota \boldsymbol{\tau} \quad \forall \boldsymbol{\tau}, \boldsymbol{\sigma} \in \mathbb{S}^N, \quad \text{a.e. in } \Omega^\iota,\\
&\text{(iv)} \quad \text{there exists } c^\iota_0 > 0 \text{ such that } C^\iota \boldsymbol{\sigma} \cdot \boldsymbol{\sigma} \geq c^\iota_0 \|\boldsymbol{\sigma}\|^2 \ \forall \boldsymbol{\sigma} \in \mathbb{S}^N, \quad \text{a.e. in } \Omega^\iota.
\end{aligned} \tag{6.404}$$

(a) $G^\iota : \Omega^\iota \times \mathbb{S}^N \times \mathbb{S}^N \to \mathbb{S}^N$,
(b) there exist a constants $G^\iota_0 > 0$ such that

$$\|G^\iota(\mathbf{x}, \boldsymbol{\tau}_1, \mathbf{e}_1) - G^\iota(\mathbf{x}, \boldsymbol{\tau}_2, \mathbf{e}_2)\| \leq G^\iota_0(\|\boldsymbol{\tau}_1 - \boldsymbol{\tau}_2\| + \|\mathbf{e}_1 - \mathbf{e}_2\|)$$
$$\forall \boldsymbol{\tau}_1, \boldsymbol{\tau}_2, \mathbf{e}_1, \mathbf{e}_2 \in \mathbb{S}^N, \quad \text{a.e. in } \Omega^\iota, \tag{6.405}$$

(c) for any $\boldsymbol{\tau}, \mathbf{e} \in \mathbb{S}^N$, $\mathbf{x} \mapsto G^\iota(\mathbf{x}, \boldsymbol{\tau}, \mathbf{e})$ is measurable,
(d) the mapping $\mathbf{x} \mapsto G^\iota(\mathbf{x}, \mathbf{0}, \mathbf{0}) \in Q^\iota$.

Let the body and surface forces satisfy

$$\mathbf{F}^\iota \in W^{1,\infty}(I; L^{2,N}(\Omega^\iota)), \qquad \mathbf{P}^\iota \in W^{1,\infty}(I; L^{2,N}({}^1\Gamma_\tau)), \tag{6.406}$$

and the initial data satisfy

$$\mathbf{u}_0 \in K, \qquad \boldsymbol{\tau}_0 \in \Sigma(0), \qquad (\boldsymbol{\tau}_0, \mathbf{e}(\mathbf{u}_0))_Q = (\mathbf{f}(0), \mathbf{u}_0)_V. \tag{6.407}$$

Then by the usual technique, we derive the weak formulation as follows:

Problem $(\mathcal{P}_{vp})_v$ Find a displacement and stress fields $\mathbf{u} : \bar{I} \to V$, $\boldsymbol{\tau} : \bar{I} \to {}^1Q$ such that

$$\mathbf{u}(t) \in K, \qquad (\boldsymbol{\tau}(t), \mathbf{e}(\mathbf{v}) - \mathbf{e}(\mathbf{u}(t)))_Q \geq (\mathbf{f}(t), \mathbf{v} - \mathbf{u}(t))_V \quad \forall \mathbf{v} \in K,\ t \in \bar{I}, \tag{6.408}$$

$$\mathbf{u}(0) = \mathbf{u}_0, \boldsymbol{\tau}(0) = \boldsymbol{\tau}_0, \tag{6.409}$$

where $\boldsymbol{\tau}'(t) = C\mathbf{e}(\mathbf{u}'(t)) + G(\boldsymbol{\tau}(t), \mathbf{e}(\mathbf{u}(t)))$, for a.e. $t \in I$ and where $\mathbf{f}(t) \in W^{1,\infty}(I, V)$.

By a standard technique it can be shown that, if $(\mathbf{u}, \boldsymbol{\tau})$ is a regular solution of Problem $(\mathcal{P}_{vp})_v$, then $(\mathbf{u}, \boldsymbol{\tau})$ satisfies (6.397)–(6.403).

We have the following existence and uniqueness results:

Theorem 6.52 Let us assume that assumptions (6.404)–(6.407) are satisfied. Then there exists a unique solution $(\mathbf{u}, \boldsymbol{\tau})$ of Problem $(\mathcal{P}_{vp})_v$. Moreover, the solution satisfies

$$\mathbf{u} \in W^{1,\infty}(I; K), \qquad \boldsymbol{\tau} \in W^{1,\infty}(I; {}^1Q). \tag{6.410}$$

Furthermore, let us assume that (6.404) and (6.405) hold and let $(\mathbf{u}_k, \boldsymbol{\tau}_k)$, $k = 1, 2$, be the solution of Problem $(\mathcal{P}_{vp})_v$ with data $\mathbf{F}_k$, $\mathbf{P}_k$, $\mathbf{u}_{0k}$, $\boldsymbol{\tau}_{0k}$, $k = 1, 2$, satisfying (6.406) and (6.407). Then there exists constant $c > 0$, depending on Ω, Γ_u, C, G, and T only, such that

$$\begin{aligned} &\|\mathbf{u}_1 - \mathbf{u}_2\|_{L^\infty(I;V)} + \|\boldsymbol{\tau}_1 - \boldsymbol{\tau}_2\|_{L^\infty(I;\, {}^1Q)} \\ &\quad \le c(\|\mathbf{u}_{01} - \mathbf{u}_{02}\|_V + \|\boldsymbol{\tau}_{01} - \boldsymbol{\tau}_{02}\|_{{}^1Q} + \|\mathbf{f}_1 - \mathbf{f}_2\|_{L^\infty(I;V)}). \end{aligned} \tag{6.411}$$

The proof is parallel to that of Theorems 16.3 and 16.7 of Han and Sofonea (2002).

6.7.3 Numerical Solution—The Semidiscrete Approximation

We will assume that every domain Ω^ι is a polyhedron in the 3D case. Let $\{\mathcal{T}_h\}$ be a regular family of finite element partition of $\overline{\Omega}$ compatible to the boundary subsets $\overline{\Gamma}_u$, $\overline{\Gamma}_\tau$, and $\overline{\Gamma}_c$. Let $V_h \subset V$ be the finite element space of piecewise linear elements over the partition $\mathcal{T}_h$, and let Q_h be the space of piecewise constant functions approximating the stress variable. Then $K_h = K \cap V_h$ consists of continuous piecewise linear functions that vanish at the nodes on $\overline{\Gamma}_u$ and where normal components satisfy the nonpenetration at the nodes on $\overline{\Gamma}_c$, that is,

$$K_h = \{\mathbf{v}_h | \mathbf{v}_h \in V_h, v_{hn}^k - v_{hn}^l \le 0 \text{ on } \cup_{k,l} \Gamma_c^{kl}\}, \qquad K_h \subset K.$$

We assume that $\mathbf{u}_{0h} \in K_h$ is an approximate of $\mathbf{u}_0$. Now we introduce the orthogonal projection $P_{Q_h} : Q \to Q_h$ by

$$(P_{Q_h}\mathbf{w}, \mathbf{w}_h)_Q = (\mathbf{w}, \mathbf{w}_h)_Q \quad \forall \mathbf{w} \in Q, \mathbf{w}_h \in Q_h.$$

Applying $\mathbf{w}_h = P_{Q_h}\mathbf{w}$ we infer that

$$\|P_{Q_h}\mathbf{w}\|_Q \le \|\mathbf{w}\|_Q \quad \forall \mathbf{w} \in Q. \tag{6.412}$$

Next, the semidiscrete or the fully discrete approximation can be used. We restrict ourselves to the semidiscrete approximation only. Then, we have the following discrete problem:

Problem $(\mathcal{P}_{vp})_h$ Find a couple of fields $(\mathbf{u}_h, \boldsymbol{\tau}_h)$, $\mathbf{u}_h : \overline{I} \to K_h$, $\boldsymbol{\tau}_h : \overline{I} \to Q_h$ such that

$$(\boldsymbol{\tau}_h(t), \mathbf{e}(\mathbf{v}_h - \mathbf{u}_h(t)))_Q \ge (\mathbf{f}(t), \mathbf{v}_h - \mathbf{u}_h(t))_V \quad \forall \mathbf{v}_h \in K_h, \qquad t \in I, \tag{6.413}$$

$$\mathbf{u}_h(0) = \mathbf{u}_{0h}, \qquad \boldsymbol{\tau}_h(0) = \boldsymbol{\tau}_{0h}, \tag{6.414}$$

$$\boldsymbol{\tau}_h'(t) = P_{Q_h} C\mathbf{e}(\mathbf{u}_h'(t)) + P_{Q_h} G(\boldsymbol{\tau}_h(t), \mathbf{e}(\mathbf{u}_h(t))) \quad \text{a.e. } t \in I, \tag{6.415}$$

where $\mathbf{u}_{0h} \in K_h$, $\boldsymbol{\tau}_{0h} \in Q_h$ are appropriate approximations of $\mathbf{u}_0$ and $\boldsymbol{\tau}_0$.

Similarly, as in the continuous case, it can be shown that Problem $(\mathcal{P}_{vp})_h$ has a unique solution $(\mathbf{u}_h, \boldsymbol{\tau}_h)$, $\mathbf{u}_h \in W^{1,\infty}(I; K_h)$, $\boldsymbol{\tau}_h \in W^{1,\infty}(I; Q_h)$. To estimate the errors $\mathbf{u} - \mathbf{u}_h$ and $\boldsymbol{\tau} - \boldsymbol{\tau}_h$, we have the following result:

Theorem 6.53 Let $\mathbf{u} \in W^{1,\infty}(I; K)$, $\boldsymbol{\tau} \in W^{1,\infty}(I;^1 Q)$ be the solution of Problem $(\mathcal{P}_{vp})_v$ and $\mathbf{u}_h \in W^{1,\infty}(I; K_h)$, $\boldsymbol{\tau}_h \in W^{1,\infty}(I; Q_h)$ the solution of Problem $(\mathcal{P}_{vp})_h$. Then we have the error estimate:

$$\begin{aligned} \|\boldsymbol{\tau} - \boldsymbol{\tau}_h\|_{L^\infty(I;Q)} + \|\mathbf{u} - \mathbf{u}_h\|_{L^\infty(I;V)} &\le c\|\boldsymbol{\tau}_0 - \boldsymbol{\tau}_{0h}\|_Q + \|\mathbf{u}_0 - \mathbf{u}_{0h}\|_V \\ &+ c \inf_{\mathbf{v}_h \in L^\infty(I;K_h)} \left\{ \|\mathbf{u} - \mathbf{v}_h\|_{L^\infty(I;V)} + \sup_{t\in I} |R(t; \mathbf{u} - \mathbf{v}_h)|^{1/2} \right\} \\ &+ \|(I_Q - P_{Q_h})(\boldsymbol{\tau} - \boldsymbol{\tau}_0)\|_{L^\infty(I;Q)}, \end{aligned} \tag{6.416}$$

where $R(t; \mathbf{v})$ is a residual quantity,

$$R(t; \mathbf{v}) = (\mathbf{f}(t), \mathbf{v}) - (\boldsymbol{\tau}(t), \mathbf{e}(\mathbf{v}))_Q. \tag{6.417}$$

Proof The proof is similar to that of Theorem 16.9 of Han and Sofonea (2002), therefore, the proof will be sketched in several steps only, that is,

(i) We integrate (6.408) and (6.413) from 0 to t and use the initial conditions (6.409) and (6.414), and then compute $\boldsymbol{\tau}(t) - \boldsymbol{\tau}_h(t)$.

(ii) In (6.408) we put $\mathbf{v} = \mathbf{u}_h(t)$ and add to (6.413), then after some modification we obtain

$$\begin{aligned} (\boldsymbol{\tau}(t) - \boldsymbol{\tau}_h(t), \mathbf{e}(\mathbf{u}(t) - \mathbf{u}_h(t)))_Q &\le (\boldsymbol{\tau}(t) - \boldsymbol{\tau}_h(t), \mathbf{e}(\mathbf{u}(t) - \mathbf{v}_h))_Q \\ &+ R(t; \mathbf{u}(t) - \mathbf{v}_h). \end{aligned} \tag{6.418}$$

(iii) We estimate $\|\boldsymbol{\tau}(t) - \boldsymbol{\tau}_h(t)\|_Q$ using the result of point (i) and (6.404) and (6.405).

(iv) Using $\boldsymbol{\tau}(t) - \boldsymbol{\tau}_h(t)$ from point (i), (6.418), (6.404), and (6.405), we estimate $\|\mathbf{u}(t) - \mathbf{u}_h(t)\|_V$.

(v) Combining results of points (iii) and (iv), putting $\mathbf{v}_h = \mathbf{v}_h(t) \in K_h$, using the Gronwall lemma, estimate (6.416) follows. □

Estimate (6.416) is the basis for convergence analysis. Similarly, as in Han and Sofonea (2002) we can derive the following convergence results:

Lemma 6.19 Let us assume that $\overline{\Gamma}_u \cap \overline{\Gamma}_c^{kl} = \emptyset$, $k, l \in \{2, 3, 4\}$ and there is only a finite number of boundary points of $\overline{\Gamma}_\tau \cap (\cup_{k,l} \overline{\Gamma}_c^{kl})$, $\overline{\Gamma}_u \cap \overline{\Gamma}_\tau$. Moreover, let the initial values $\mathbf{u}_{0h} \in K_h$, $\boldsymbol{\tau}_{0h} \in Q_h$ satisfy

$$\|\mathbf{u}_0 - \mathbf{u}_{oh}\|_V \to 0, \qquad \|\boldsymbol{\tau}_0 - \boldsymbol{\tau}_{0h}\|_Q \to 0 \quad \text{as } h \to 0. \tag{6.419}$$

Then the semidiscrete method converges, that is,

$$\|\mathbf{u} - \mathbf{u}_h\|_{L^\infty(I;V)} + \|\boldsymbol{\tau} - \boldsymbol{\tau}_h\|_{L^\infty(I;Q)} \to 0 \quad \text{as } h \to 0. \tag{6.420}$$

To derive an order error estimate we introduce the additional solution regularity assumption

$$\boldsymbol{\tau}^\iota \mathbf{n}^\iota \in L^\infty(I; L^2(\Gamma_\tau)), \qquad \iota = 1, \dots, r. \tag{6.421}$$

Then, we can estimate the residuum $R(t; \mathbf{u}(t) - \mathbf{v}_h(t))$ defined in (6.417). A direct estimate yields

$$|R(t; \mathbf{u}(t) - \mathbf{v}_h(t))| \leq c(\|\mathbf{f}(t)\|_V + \|\tau(t)\|_Q)\|\mathbf{u}(t) - \mathbf{v}_h(t)\|_V, \tag{6.422}$$

but under the regularity assumption (6.421) we have

$$\begin{aligned} |R(t; \mathbf{u}(t) - \mathbf{v}_h(t))| &= \left| \int_{\cup \Gamma_c^{kl}} \tau_n^{kl}(t)([v_n]^{kl}(t) - [u_n]^{kl}(t)) \right| ds \\ &\leq \|\tau_n^{kl}(t)\|_{L^2(\Gamma_c)} \|[v_n]^{kl}(t) - [u_n]^{kl}(t)\|_{L^2(\Gamma_c)}, \qquad \Gamma_c = \cup \Gamma_c^{kl}. \end{aligned} \tag{6.423}$$

Then

$$\begin{aligned} \|\mathbf{u} - \mathbf{u}_h\|_{L^\infty(I;V)} + \|\boldsymbol{\tau} - \boldsymbol{\tau}_h\|_{L^\infty(I;Q)} &\leq c \left(\|\mathbf{u}_0 - \mathbf{u}_{0h}\|_V + \|\boldsymbol{\tau}_0 - \boldsymbol{\tau}_{0h}\|_Q \right) \\ &+ c \inf_{\mathbf{v}_h \in L^\infty(I;K_h)} \left\{ \|\mathbf{u} - \mathbf{v}_h\|_{L^\infty(I;V)} + \|u_n - v_n\|^{\frac{1}{2}}_{L^\infty(I;L^2(\Gamma_c))} \right\} \\ &+ c\|(I_Q - P_{Q_h})(\boldsymbol{\tau}(t) - \boldsymbol{\tau}_0)\|_Q, \end{aligned}$$

where $v_n - u_n = [v_n]^{kl} - [u_n]^{kl}$, $[v_n]^{kl} = v_n^k - v_n^l$.

Let us assume that $\mathbf{u} \in L^\infty(I; \mathbb{H}^{2,N}(\Omega))$, $u_{n|\Gamma_{c,i}} \in L^\infty(I; \mathbb{H}^2(\Gamma_{c,i}))$, $1 \leq i \leq i_0$, where i_0 represents the number of segments in the case $N = 2$ or polygonals in the case $N = 3$, $\boldsymbol{\tau} \in L^\infty(I; H^{1,N\times N}(\Omega))$ and let the initial values $\mathbf{u}_{0h} \in K_h$, $\boldsymbol{\tau}_{0h} \in Q_h$ be chosen such that

$$\|\mathbf{u}_0 - \mathbf{u}_{0h}\|_V + \|\boldsymbol{\tau}_0 - \boldsymbol{\tau}_{0h}\|_Q \leq ch, \qquad c = \text{const} > 0.$$

Let us introduce the piecewise linear interpolant $r_h\mathbf{u}(t) \in V_h$ of $\mathbf{u}(t)$. Then the interpolation error estimates yield

$$\begin{aligned} \|\mathbf{u} - r_h\mathbf{u}\|_{L^\infty(I;V)} &\leq ch|\mathbf{u}|_{L^\infty(I;H^{2,N}(\Omega))}, \\ \|u_n - r_h u_n\|^{1/2}_{L^\infty(I;L^2(\Gamma_c))} &\leq ch \sum_{i=1}^{i_0} |u_n|^{1/2}_{L^\infty(I;H^2(\Gamma_{c,i}))}. \end{aligned} \tag{6.424}$$

Hence

$$\|\mathbf{u} - \mathbf{u}_h\|_{L^\infty(I;V)} + \|\boldsymbol{\tau} - \boldsymbol{\tau}_h\|_{L^\infty(I;Q)} = O(h). \tag{6.425}$$

6.8 OPTIMAL SHAPE DESIGN IN BIOMECHANICS OF HUMAN JOINT REPLACEMENTS

6.8.1 Introduction

At present, new types of total joint replacements are searched for a better application in orthopedic surgery and namely for their better function and extension of their length of life. Moreover, present efforts are directed to the construction of compatible systems that make possible to solve a greater number of alternative solutions in a consequence on the character of joint deformations as well as on the possibility of the revised operations of joint replacements and on the greater power of resistance to infections. The main goal of the construction of total joint replacements is to regenerate functions of the joint. In the case of the knee joint the main goal of the construction of the knee joint replacements (TKAs) is to extend the length of their life, reduce the risk of their luxation, and to extend the range of their functions. All these facts depend on the optimal shape of total joint replacement and on the optimal distribution of stresses on the contact boundary between the femoral and tibial components of the TKA as well as on the optimal distribution of stresses inside the whole system of the total joint replacement and the femoral and tibial bones. Therefore, in the next section we will discuss a method for such analyses.

6.8.2 Main Idea of the Optimal Shape Design Problem

The main idea of the investigated problem is the following: Let O be a family of admissible regions $\Omega \subset \mathbb{R}^N$, $N = 2, 3$, that is, $O = \{\Omega(\alpha), \alpha \in U_{\text{ad}}\}$, where U_{ad} is a set of admissible controls and α is a parameter, a function describing the shape of the contact part $\Gamma_c = \cup_{k,l} \Gamma_c^{kl}$ of the boundary $\partial\Omega$. Let us suppose that it can be a part of a larger family of admissible regions $\tilde{O}$, that is, $O \subseteq \tilde{O}$. Then the main aim is to find an optimal pair, the optimal region Ω_{opt} and the corresponding solution $\mathbf{u}(\Omega_{\text{opt}})$ obtained by a minimization process of an objective functional, the so-called cost functional.

We will briefly describe the derivation of the method. We associate with any $\Omega = \cup_{\iota=1}^{r} \Omega^{\iota}$ the Hilbert space $V(\Omega)$ of functions, defined in Ω. Let $\{\Omega_n\}$ be a sequence of domains Ω_n, where $\Omega_n, \Omega \in \tilde{O}, n \in \mathbb{N}$ and let us denote by $\Omega_n \overset{O}{\rightarrow} \Omega$, for $n \to \infty$, the convergence of $\{\Omega_n\}$ to Ω, and by $\mathbf{u}_n \to \mathbf{u}$, as $n \to \infty$ the convergence of the sequence $\{\mathbf{u}_n\}$, $\mathbf{u}_n \in V(\Omega_n)$ to $\mathbf{u} \in V(\Omega)$.

Let us denote by

$$\mathbf{u} : \Omega \in O \to \mathbf{u}(\Omega) \in V(\Omega)$$

a mapping that associates with any domain $\Omega \in O$ the solution $\mathbf{u}$ of the investigated problem and let

$$G = \{(\Omega, \mathbf{u}(\Omega)) : \Omega \in O\} \quad \text{and} \quad D = \{(\Omega, \mathbf{w}) : \Omega \in O, \mathbf{w} \in V(\Omega)\}.$$

Then we introduce an objective functional, the so-called cost functional $I : D \to \mathbb{R}$, the values of which over G will be denoted by $J(\Omega)$, that is, $J(\Omega) \equiv I(\Omega, \mathbf{u}(\Omega))$. Then the optimal shape design problem is the following:

$$J(\Omega_{\text{opt}}) = \min_{\alpha \in U_{\text{ad}}} J(\Omega(\alpha)),$$

which is equivalent to the following problem Find $\Omega_{\text{opt}} \in O$ such that

$$J(\Omega_{\text{opt}}) \leq J(\Omega) \quad \text{for all } \Omega \in O. \tag{6.426}$$

We then say that $(\Omega_{\text{opt}}, \mathbf{u}(\Omega_{\text{opt}}))$ represent an optimal pair of (6.426).

Under some assumptions we can prove that there exists at least one solution $\Omega_{\text{opt}} \in O$ of (6.426). For more details see Nečas and Hlaváček (1982), Hlaváček (1986, 1989, 1994), Haslinger (1991), and Haslinger and Neittaanmäki (1996).

Numerical Solution Let $h \to 0$ be a finite element mesh step. With $h > 0$ we associate $O_h \subset \tilde{O}$. Let $\Omega_h \subset O_h$ be a discrete approximation of the region Ω and let $\partial\Omega_h$ be its boundary. With $\Omega_h \in O_h$ we associate a finite-dimensional space $V_h(\Omega_h) \subset V(\Omega_h)$ of functions defined on Ω_h. We define the convergence $\Omega_h \overset{\tilde{O}}{\to} \Omega$, $h \to 0_+$, where $\Omega_h \in O_h$, $\Omega \in \tilde{O}$, and $\mathbf{u}_{nh} \to \mathbf{u}$, $h \to 0_+$, $\mathbf{u}_h \in V_h(\Omega_h)$, $\mathbf{u} \in V(\Omega)$ similarly as in the continuous case.

Let us denote by

$$\mathbf{u}_h : \Omega_h \to \mathbf{u}_h(\Omega_h) \in V_h(\Omega_h)$$

the solution of an appropriate problem and let

$$G_h = \{(\Omega_h, \mathbf{u}_h(\Omega_h)) | \Omega_h \in O_h\}$$

and

$$D_h = \{(\Omega_h, \mathbf{y}_h) | \Omega_h \in \Omega_h, \mathbf{y}_h \in V_h(\Omega_h)\}.$$

Let $I_h : D_h \to \mathbb{R}$ be a discretization of the cost functional I and let us put

$$J_h(\Omega_h) \equiv I_h(\Omega_h, \mathbf{u}_h(\Omega_h)) \quad \text{for } \Omega_h \in O_h.$$

Then the discrete optimal shape design problem is as follows:

Problem $(\mathcal{P})_h$ Find $\Omega_{h\,\text{opt}} \in O_h$ such that

$$J_h(\Omega_{h\,\text{opt}}) \leq J_h(\Omega_h) \quad \text{for all } \Omega_h \in O_h.$$

Let us assume that:

(i)$_h$ For any $\Omega \in O$ there exists a sequence $\{\Omega_h\}$, $\Omega_h \in O_h$ such that

$$\Omega_h \overset{O}{\to} \Omega \quad \text{as } h \to 0_+;$$

$(\text{ii})_h$ for every sequence $\{(\Omega_h, \mathbf{u}_h(\Omega_h))\}$, where $\Omega_h, \mathbf{u}_h(\Omega_h) \in G_h$, there exists a subsequence $\{(\Omega_{h_j}, \mathbf{u}_h(\Omega_{h_j}))\} \subset \{\Omega_h, \mathbf{u}_h(\Omega_h)\}$ and an element $(\Omega, \mathbf{u}(\Omega)) \in G$ such that

$$\Omega_{h_j} \overset{O}{\to} \Omega \quad \text{as } j \to \infty,$$
$$\mathbf{u}_{h_j}(\Omega_{h_j}) \to \mathbf{u}(\Omega) \quad \text{as } j \to \infty;$$

$(\text{iii})_h$ if $\{\Omega_h\}$, $\Omega_h \in O_h$, and $\Omega \in O$ are such that

$$\Omega_h \overset{O}{\to} \Omega \quad \text{as } h \to 0_+;$$

and if $\mathbf{u}_h(\Omega_h) \in V_h(\Omega_h)$, $\mathbf{u}(\Omega) \in V(\Omega)$ are solutions of Problem $(\mathcal{P})_h$ and Problem $\mathcal{P}$, respectively, such that

$$\mathbf{u}_h(\Omega_h) \to \mathbf{u}(\Omega) \quad \text{as } h \to 0_+,$$

then

$$\lim_{h\to 0_+} J_h(\Omega_h) = J(\Omega).$$

We obtain the following result, which will be formulated in the next theorem.

Theorem 6.54 Let assumptions $(\text{i})_h$–$(\text{iii})_h$ be satisfied. Let for any $h > 0$ there exists a solution $(\Omega_{h\,\text{opt}}, \mathbf{u}_h(\Omega_{h\,\text{opt}}))$ of Problem $(\mathcal{P})_h$. Then there exists a subsequence $\{(\Omega_{h_j\,\text{opt}}, \mathbf{u}_h(\Omega_{h_j\,\text{opt}}))\} \subset \{(\Omega_{h\,\text{opt}}, \mathbf{u}_h(\Omega_{h\,\text{opt}}))\}$ and an element $(\Omega_{\text{opt}}, \mathbf{u}(\Omega_{\text{opt}})) \in G$ such that

$$\Omega_{h_j\,\text{opt}} \overset{O}{\to} \Omega_{\text{opt}} \quad \text{as } j \to \infty,$$
$$\mathbf{u}_{h_j}(\Omega_{h_j\,\text{opt}}) \to \mathbf{u}(\Omega_{\text{opt}}) \quad \text{as } j \to \infty,$$

where $(\Omega_{\text{opt}}, \mathbf{u}(\Omega_{\text{opt}}))$ is an optimal pair of Problem $\mathcal{P}$.

6.8.3 Optimal Shape Joint Design, Description of the Method

Next, we will apply the above presented idea to total joint replacements, which will be presented on the case of total knee joint replacement in 3D.

Let us assume that the knee joint arthroplasty (TKA) is modeled by the elastic bodies Ω^ι, $\iota = 1, \ldots, r$, being in mutual contact and that the TKA occupies the body $\Omega = \cup_{\iota=1}^r \Omega^\iota$. Let us assume that the boundary of Ω, denoted by $\partial\Omega$, be decomposed into three parts Γ_u, $\Gamma_\tau = {}^1\Gamma_\tau \cup {}^2\Gamma_\tau$, and $\Gamma_c = \cup_{k,l} \Gamma_c^{kl}$, $\Gamma_c^{kl} = \partial\Omega^k \cap \partial\Omega^l$, $k \neq l$, $k, l \in \{1, \ldots, r\}$, that is, $\partial\Omega = \overline{\Gamma}_u \cup \overline{\Gamma}_\tau \cup \overline{\Gamma}_c$. Furthermore, we will assume that on the part of the boundary $\partial\Omega$, we denote it as Γ_u, the TKA is fixed, that is, $\mathbf{u} = 0$, on the part ${}^1\Gamma_\tau$ the TKA is loaded by a load $\mathbf{P}$, that is, $\tau_i \equiv \tau_{ij} n_j = P_i$, while on the part ${}^2\Gamma_\tau$ is unloaded, that is, $\tau_{ij} n_j = 0$, and the rest of the boundary Γ_c represents the contact

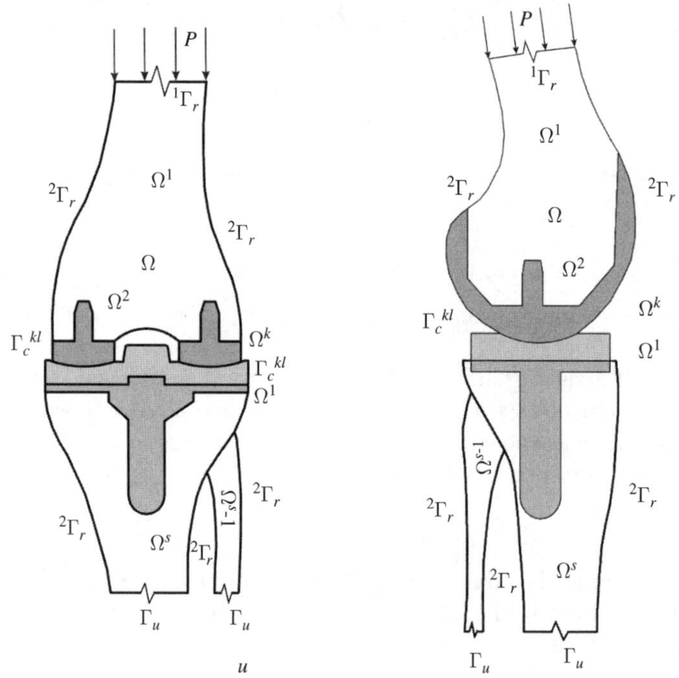

FIG. 6.5 Model of knee joint.

boundary between components of the TKA, which are in mutual contact. We assume that Γ_u, Γ_τ, and Γ_c are nonempty and open in $\partial\Omega$. Let $\text{meas}(\Gamma_u \cap \partial\Omega^\iota) > 0$ for all $\iota \leq r$.

Now, we will formulate the optimal shape design problem of the total knee joint replacements. For simplicity the model is presented in Fig. 6.5 only as the 2D case. Such a choice of the cost functional leads to a differentiable (i.e., smooth) problem. The classical formulation of the N-dim model problem corresponding to Fig. 6.5 is the following:

Problem $\mathcal{P}$ Find a displacement field $\mathbf{u} = (\mathbf{u}^\iota)$, $\iota = 1, \ldots, r$, $\mathbf{u}^\iota : \Omega^\iota \to \mathbb{R}^N$, $N = 3$, and stress fields $\boldsymbol{\tau} = (\tau_{ij})$, $i, j = 1, \ldots, N$, $\boldsymbol{\tau}^\iota = (\tau_{ij}^\iota) : \Omega^\iota \to \mathbb{S}^N$ such that

$$\frac{\partial \tau_{ij}^\iota}{\partial x_j} + F_i^\iota = 0 \quad \text{or} \quad \text{Div}\, \tau^\iota + \mathbf{F}^\iota = 0 \quad \text{in } \Omega^\iota,\ \iota = 1, \ldots, r, \tag{6.427}$$

$$\tau_{ij}^\iota = c_{ijkl} e_{kl}(\mathbf{u}^\iota) \quad \text{in } \Omega^\iota,\ \iota = 1, \ldots, r,\ i, j, k, l = 1, \ldots, N, \tag{6.428}$$

$$\mathbf{u}^\iota = \mathbf{0} \quad \text{on } \Gamma_u, \tag{6.429}$$

$$\tau_{ij} n = P_i \quad \text{on } {}^1\Gamma_\tau,\ i, j = 1, \ldots, N, \tag{6.430}$$

$$\tau_{ij} n = 0 \quad \text{on } {}^2\Gamma_\tau,\ i, j = 1, \ldots, N, \tag{6.431}$$

$$u_n^k - u_n^l \le 0, \qquad \tau_n^k = \tau_n^l \equiv \tau_n^{kl} \le 0, \qquad (u_n^k - u_n^l)\tau_n^{kl} = 0$$
$$\text{if } u_n^k - u_n^l = 0, \quad \text{then} \quad |\boldsymbol{\tau}_t^{kl}(\mathbf{x})| \le g_c^{kl}(\mathbf{x}),$$
$$\text{if } |\boldsymbol{\tau}_t^{kl}(\mathbf{x})| < g_c^{kl}(\mathbf{x}), \quad \text{then} \quad \mathbf{u}_t^k(\mathbf{x}) - \mathbf{u}_t^l(\mathbf{x}) = 0,$$
$$\text{if } |\boldsymbol{\tau}_t^{kl}(\mathbf{x})| = g_c^{kl}(\mathbf{x}), \quad \text{then there exists a function } \theta \ge 0$$
$$\text{such that } \mathbf{u}_t^k(\mathbf{x}) - \mathbf{u}_t^l(\mathbf{x}) = -\theta\boldsymbol{\tau}_t^{kl}(\mathbf{x}), \quad \text{on } \cup \Gamma_c^{kl}, \tag{6.432}$$

where $\mathbf{F}^\iota \in L^{2,N}(\Omega^\iota)$, $\mathbf{P} \in L^{2,N}(\Gamma_\tau)$, $g_c^{kl} \in L^2(\Gamma_c^{kl})$, $g_c^{kl} > 0$ and where c_{ijkl} are elastic coefficients satisfying the symmetry and the algebraic elasticity conditions as in the previous cases.

Now, we will assume a family of contact problems, where as a control parameter we will consider a function α from the set U_{ad} of an admissible control, which describes the shape of the contact boundary $\Gamma_c^{\kappa\lambda} \subset \cup_{k,l} \Gamma_c^{kl}$. We will assume that $\Gamma_c^{\kappa\lambda} \subset \cup_{k,l} \Gamma_c^{kl}$ is described by a function $\alpha \ge 0$, so that

$$\Gamma_c^{\kappa\lambda} \equiv \Gamma_c^{\kappa\lambda}(\alpha) = \{\mathbf{x} = (x_1, x_2, x_3) | x_3 = \alpha(x_1, x_2),\ x_1 \in (a, b), x_2 \in (c, d)\}.$$

The displacement field $\mathbf{u}$ corresponding to the equilibrium state minimizes the functional of the total potential energy $J(\mathbf{v})$ over the set K of kinematically admissible displacements, where

$$V = \{\mathbf{v} | \mathbf{v} \in \mathbb{H}^{1,N}(\Omega), \mathbf{v} = 0 \text{ on } \Gamma_u\}, \qquad \mathbb{H}^{1,N}(\Omega) = \prod_\iota [H^1(\Omega^\iota)]^N,$$
$$K = \{\mathbf{v} \in V | [v_n]^{kl} \le 0 \text{ on } \cup \Gamma_c^{kl}\}, \tag{6.433}$$
$$J(\mathbf{v}) = \tfrac{1}{2} a(\mathbf{v}, \mathbf{v}) + j(\mathbf{v}) - (\mathbf{f}, \mathbf{v}),$$

where we denote $[v_n]^{kl} \equiv v_n^k - v_n^l$ and $j(\mathbf{v}) = \int_{\cup \Gamma_c^{kl}} g_c^{kl} |\mathbf{v}_t^k - \mathbf{v}_t^l|\, ds$ as in the previous sections.

Then the variational solution leads to the variational inequality problem: Find

$$\mathbf{u} \in K, \qquad J(\mathbf{u}) \le J(\mathbf{v}) \quad \forall \mathbf{v} \in K, \tag{6.434}$$

which is equivalent to finding

$$\mathbf{u} \in K, \qquad a(\mathbf{u}, \mathbf{v} - \mathbf{u}) + j(\mathbf{v}) - j(\mathbf{u}) \ge (\mathbf{f}, \mathbf{v} - \mathbf{u}) \quad \forall \mathbf{v} \in K, \tag{6.435}$$

where $(\mathbf{f}, \mathbf{v}) = (\mathbf{F}, \mathbf{v})_{0,\Omega} + (\mathbf{P}, \mathbf{v})_{0,\Gamma_\tau}$.

Since $a(\mathbf{v}, \mathbf{v}) \ge c\|\mathbf{v}\|_{1,\Omega}^2 \quad \forall \mathbf{v} \in V$, (6.434) has a unique solution $\mathbf{u}$.

We assume that $\alpha \in U_{\text{ad}}$, where

$$U_{\text{ad}} = \{\alpha \in C^{0,1}([a, b] \times [c, d]) | -\gamma_2 < c_0 \le \alpha \le c_1 < \gamma_1,$$
$$|\alpha(x_1, x_2) - \alpha(\bar{x}_1, \bar{x}_2)| \le c_2 |x_1 - \bar{x}_1|\, |x_2 - \bar{x}_2|$$
$$\forall (x_1, x_2),\ (\bar{x}_1, \bar{x}_2) \in [a, b][c, d]\}, \tag{6.436}$$

where constants c_k, $k = 0, 1, 2, 3$, are chosen in such a way that $U_{\text{ad}} \ne \emptyset$.

Now, we will define for any $\alpha \in U_{\text{ad}}$ the set $\Omega(\alpha)$ as

$$\begin{aligned}
\Omega^\kappa &\equiv \Omega^\kappa(\alpha) = \{\mathbf{x} \in \mathbb{R}^3,\ (x_1,x_2) \in (a,b)\times(c,d), -\gamma_2 < x_3 < \alpha(x_1,x_2)\},\\
\Omega^\lambda &\equiv \Omega^\lambda(\alpha) = \{\mathbf{x} \in \mathbb{R}^3,\ (x_1,x_2) \in (a,b)\times(c,d),\ \alpha(x_1,x_2) < x_3 < \gamma_1\},\\
\Omega(\alpha) &= \cup_{\iota=1}^{r}\Omega^\iota, \quad \text{where } \Omega^\iota = \Omega^\iota(\alpha) \text{ for } \iota = \kappa,\lambda,\\
\Gamma_c^{kl}(\alpha) &= \Gamma_c^{kl} \text{ for } (k,l) \neq (\kappa,\lambda) \quad \text{and} \quad \Gamma_c^{\kappa\lambda}(\alpha) = \partial\Omega^\kappa(\alpha)\cap\partial\Omega^\lambda(\alpha). \qquad (6.437)
\end{aligned}$$

Thus $\hat{\Omega} \supset \Omega(\alpha)\ \forall\alpha \in U_{\text{ad}}$. For the existence and the uniqueness of the solution of the optimal shape design problem, we will assume that there exists $\delta \geq 0$, such that meas $\Gamma_u \cap \partial\Omega^\iota \geq \delta$ for any $\alpha \in U_{\text{ad}}$ and $\iota = \kappa, \lambda$, that is, $\Gamma_u \cap \partial\Omega^\iota$ cannot shrink.

Then we have the following problem:

Problem $\mathcal{P}_\alpha$ Find $\mathbf{u}(\alpha) \in K(\alpha)$ such that

$$J(\alpha,\mathbf{u}(\alpha)) \leq J(\alpha,\mathbf{v}) \quad \forall\mathbf{v} \in K(\alpha), \qquad (6.438)$$

where

$$\begin{aligned}
J(\alpha,\mathbf{v}) &= \tfrac{1}{2}a_{\Omega(a)}(\mathbf{v},\mathbf{v}) + j_\alpha(\mathbf{v}) - \langle\mathbf{f},\mathbf{v}\rangle_\alpha,\\
V(\alpha) &= \{\mathbf{v} \in \mathbb{H}^{1,N}(\Omega(\alpha)); \mathbf{v} = 0 \text{ on } \Gamma_u\ \forall\alpha\},\\
K(\alpha) &= \{\mathbf{v} \in V(\alpha) | [v_n]^{kl} \leq 0 \text{ on } \Gamma_c^{kl}\ \forall\alpha\}.
\end{aligned}$$

Next, we formulate the problem using the general cost functional $I : D \to \mathbb{R}$, where $D = \{(\alpha,\mathbf{y}) | \alpha \in U_{ad}, \mathbf{y} \in V(\alpha)\}$. Then we have the following problem:

Problem $\mathcal{P}_\mathbf{I}$ Find $\hat{\alpha} \in U_{\text{ad}}$ such that

$$I(\hat{\alpha},\mathbf{u}(\hat{\alpha})) \leq I(\alpha,\mathbf{u}(\alpha)) \quad \forall\alpha \in U_{\text{ad}}$$

with $\mathbf{u}(\alpha) \in K(\alpha)$ being the solution of Problem $\mathcal{P}(\alpha)$.

Under the assumption of the lower semicontinuity of I, that is,

$$\left\{\begin{array}{l}
\alpha_n \underset{n\to\infty}{\overset{\text{uniformly}}{\rightarrow}} \alpha \text{ in } [a,b]\times[c,d],\ \alpha_n,\alpha \in U_{\text{ad}}\\
\mathbf{y}_n \underset{n\to\infty}{\rightharpoonup} \mathbf{y} \text{ (weakly) in } \mathbb{H}^{1,N}(\hat{\Omega}),\ \mathbf{y}_n,\mathbf{y} \in \mathbb{H}^{1,N}(\hat{\Omega})
\end{array}\right\}$$

$$\Rightarrow \liminf_{n\to\infty} I(\alpha_n, \mathbf{y}_n|_{\Omega_n}) \geq I(\alpha, \mathbf{y}|_{\Omega(\alpha)}), \qquad (6.439)$$

where $\Omega_n = \Omega(\alpha_n)$, $\hat{\Omega} \supset \Omega(\alpha)$, $\forall\alpha \in \cup_{\text{ad}}$, we can prove the following result:

Theorem 6.55 Under the lower semicontinuity (6.439) there exists at least one solution of Problem $\mathcal{P}_I$.

The proof is parallel to that of Haslinger and Neittaanmäki (1996).

Numerical Solution Numerically, the optimal shape design problems of the total joint replacements will be solved by the finite element technique.

Let the rectangle $\Omega^0 \equiv [a, b] \times [c, d]$ be partitioned by a triangulation $\mathcal{T}_h^0(\Omega^0)$. We define

$$U_{\text{ad}}^h = \{\alpha_h \in C^{0,1}(\Omega^0) |\ \alpha_{|T_{hi}} \in P_1(T_{hi}) \quad \forall T_{hi} \in \mathcal{T}_h^0(\Omega^0)\},$$

that is, we approximate α by a piecewise linear function. For any $\alpha_h \in U_{\text{ad}}^h$ we define the approximation of $\Omega^\iota(\alpha)$, $\iota = \kappa, \lambda$, as

$$\Omega_h^\iota(\alpha_h) = \begin{cases} \mathbf{x} \in \mathbb{R}^3 |\ (x_1, x_2) \in (a, b) \times (c, d), & \\ \qquad -\gamma_2 < x_3 < \alpha_h(x_1, x_2) & \text{for } \Omega_h^\kappa(\alpha_h), \\ \qquad \alpha_h(x_1, x_2) < x_3 < \gamma_1 & \text{for } \Omega_h^\lambda(\alpha_h), \end{cases}$$

where $\Omega_h^\iota(\alpha_h)$ is a polyhedral domain and the contact boundary $\Gamma_c^{\kappa\lambda}(\alpha)$ is replaced by the boundary $\Gamma_c^{\kappa\lambda}(\alpha_h)$.

We introduce a family of partitions of $\overline{\Omega^\iota(\alpha_h)}$, $\iota = \kappa, \lambda$, we denote it by $\{\mathcal{T}^\iota(h, \alpha_h)\}$, $h \to 0_+$, satisfying the following:

(i) For any $h > 0$ fixed, $\mathcal{T}^\iota(h, \alpha_h)$ depend continuously on $\alpha_h \in U_{\text{ad}}^h$.

(ii) For any $h > 0$ fixed, $\mathcal{T}^\iota(h, \alpha_h)$ are topologically equivalent for $\alpha_h \in U_{\text{ad}}^h$, that is, the number of the nodes from $\mathcal{T}^\iota(h, \alpha_h)$ is the same for all $\alpha_h \in U_{\text{ad}}^h$ and the nodes still have the same neighbors.

(iii) The family $\mathcal{T}^\iota(h, \alpha_h), h \to 0_+$, is regular with respect to $h > 0$, there exists a constant $\nu > 0$ such that $h_T / \rho_T \leq \nu\ \forall T_{hi} \in \cup_{h>0} \mathcal{T}(h, \alpha_h)$ where $h_T = \text{diam}(T_{hi})$ and ρ_T is the radius of the maximal sphere inscribed in the tetrahedron T_{hi}.

For $\iota \neq \kappa, \lambda$ let the domains Ω^ι be polyhedral and $\mathcal{T}^\iota(h, \alpha_h)$ coincide with the standard regular partitions $\mathcal{T}_h^\iota(\Omega^\iota)$.

Let us introduce

$$\mathcal{T}(h, \alpha_h) = \prod_{\iota=1}^{r} \mathcal{T}^\iota(h, \alpha_h)$$

and the set

$$\Omega_h(\alpha_h) = \cup_{\iota=1}^{r} \Omega_h^\iota(\alpha_h), \tag{6.440}$$

where

$$\Omega_h^\iota(\alpha_h) = \begin{cases} \Omega^\iota & \text{for } \iota \neq \kappa, \lambda, \\ \Omega_h^\iota(\alpha_h) & \text{for } \iota = \kappa, \lambda. \end{cases}$$

Let us introduce with any $\mathcal{T}(h, \alpha_h)$ the set of virtual displacements and the set of admissible displacements:

$$V_h(\alpha_h) = \{\mathbf{v}_h \in \prod_{\iota=1}^{r} [C(\overline{\Omega_h^\iota})]^3 | \mathbf{v}_{h|T_h} \in [P_1(T_h)]^3$$

$$\forall T_h \in \mathcal{T}(h, \alpha_h), \mathbf{v}_h = \mathbf{0} \text{ on } \Gamma_u\},$$

$$K_h(\alpha_h) = V_h(\alpha_h) \cap K(\alpha_h).$$

Then we define the following approximate problem:

Problem $(\mathcal{P})_h$ Find $\alpha_{h\,\text{opt}} \in U_{\text{ad}}^h$ such that

$$I_h(\alpha_{h\,\text{opt}}, \mathbf{u}_h(\alpha_{h\,\text{opt}})) \leq I_h(\alpha_h, \mathbf{u}_h(\alpha_h)) \quad \forall \alpha_h \in U_{\text{ad}}^h, \tag{6.441}$$

where $I_h : D_h \to \mathbb{R}$ is the approximation of I; $D_h = \{(\alpha_h, \mathbf{y}_h) | \alpha_h \in U_{\text{ad}}^h, \mathbf{y}_h \in V_h(\alpha_h)\}$ and $\mathbf{u}_h \equiv \mathbf{u}_h(\alpha_h) \in K_h(\alpha_h)$ is the solution of the inequality:

$$a_{\alpha h}(\mathbf{u}_h, \mathbf{v}_h - \mathbf{u}_h) + j(\mathbf{v}_h) - j(\mathbf{u}_h) \geq (\mathbf{f}, \mathbf{v}_h - \mathbf{u}_h)_{\alpha_h} \quad \forall \mathbf{v}_h \in K_h(\alpha_h), \tag{6.442}$$

where $a_{\alpha h}(\cdot,\cdot) \equiv a_{\Omega_h(\alpha_h)}(\cdot,\cdot)$.

As the result we obtain the following theorems:

Theorem 6.56 For any $h > 0$ there exists a solution $\alpha_{h\,\text{opt}}$ of Problem $(\mathcal{P})_h$.

The proof follows from the application of the classical compactness arguments and the assumptions concerning $\mathcal{T}(h, \alpha_h)$, $h \to 0_+$.

Theorem 6.57 Let $\alpha_{h\,\text{opt}} \in U_{\text{ad}}^h$ be a solution of Problem $(\mathcal{P})_h$ and $\mathbf{u}_h(\alpha_{h\,\text{opt}})$ be the corresponding solution of (6.442). Then there exist a subsequence $\{\alpha_{h_j\,\text{opt}}, \mathbf{u}_{h_j}(\alpha_{h_j\,\text{opt}})\} \subset \{(\alpha_{h\,\text{opt}}, \mathbf{u}_h(\alpha_{h\,\text{opt}}))\}$, $h \to 0_+$ and elements $\alpha_{\text{opt}} \in U_{\text{ad}}$, $\mathbf{u}_{\text{opt}} \in K(\alpha_{\text{opt}})$ such that

$$\begin{aligned} &\alpha_{h_j\,\text{opt}} \overset{\text{uniformly}}{\rightrightarrows} \alpha_{\text{opt}} \quad \text{as } j \to \infty, \\ &\mathbf{u}_{h_j}(\alpha_{h_j\,\text{opt}}) \rightharpoonup \mathbf{u}_{\text{opt}} \quad \text{(weakly) in } \mathbb{H}^{1,3}(\hat{\Omega}) \quad \text{as } j \to \infty, \end{aligned}$$

provided the approximate cost functional I_h, $h \to 0_+$, satisfy

$$\left\{ \begin{array}{l} \alpha_h \overset{\text{uniformly}}{\rightarrow} \alpha \text{ in } [a,b] \times [c,d], \ \alpha_h \in U_{\text{ad}}^h, \qquad \alpha \in U_{\text{ad}}, \\ \mathbf{u}_h \rightharpoonup \mathbf{u} \text{ (weakly)}, \quad \text{where } \mathbf{u}_h, \mathbf{u} \text{ solve Problem } \mathcal{P}(\alpha_h) \\ \quad \text{or Problem } \mathcal{P}(\alpha), \text{ resp.} \end{array} \right\}$$

$$\Rightarrow \lim_{h \to 0_+} I_h(\alpha_h, \mathbf{u}_h(\alpha_h)) = I(\alpha, \mathbf{u}(\alpha)).$$

Moreover, α_{opt} is a solution of Problem $\mathcal{P}_I$ and $\mathbf{u}_{\text{opt}}$ of (6.438) on $\Omega(\alpha_{\text{opt}})$.

6.9 WORST-SCENARIO METHOD IN BIOMECHANICS OF HUMAN JOINT REPLACEMENTS

6.9.1 Introduction

Many mathematical problems in biomechanics of human joint replacements (Nedoma, 1987, 1993; Nedoma et al., 2003a,b, 2006; Eck et al., 1995; Penrose et al., 2002) lead to solving unilateral contact problems with friction in elasticity, thermoelasticity, thermoviscoelasticity, plasticity, and viscoplasticity. Such problems are frequently formulated by variational inequalities representing the principle of virtual displacements in its inequality form. Since the input data (physical coefficients, right-hand sides, boundary values, friction limits, etc.) cannot be determined uniquely, but only in some intervals with the uncertainty determined by their measurement errors, the problem can be solved by the worst scenario (antioptimization) approach. Then the reliable solution of the problem represents the worst case among a set of possible solutions, where possibility will be given by uncertain input data and the degree of badness will be measured by a certain criterion functional. We will formulate and analyze the corresponding worst-scenario problem and discuss its solvability.

6.9.2 Formulation of the Thermoelastic Contact Problem

Let $\Omega = \cup_{\iota=1}^{s} \Omega^{\iota} \subset R^N$, $N = \{2, 3\}$ be a bounded region with the boundary $\partial\Omega = \cup_{\iota=1}^{s} \partial\Omega^{\iota}$. Let the boundary $\partial\Omega$ consist of three disjoint parts Γ_τ, Γ_u, and Γ_c such that $\partial\Omega = \overline{\Gamma_\tau} \cup \overline{\Gamma_u} \cup \overline{\Gamma_c}$. Let $(N-1)$-dimensional measures of Γ_τ, Γ_u, and Γ_c^{kl} be positive, where $\Gamma_c = \cup_{k,l} \Gamma^{kl}$, $\Gamma^{kl} = \partial\Omega^k \cap \partial\Omega^l$, $1 \le k, l \le s$, $k \ne l$, and $\overline{\Gamma}_\tau$, $\overline{\Gamma}_u$, $\overline{\Gamma}_c$ denote the closures in $\partial\Omega$.

Since we asume that the problem is quasi-coupled, we can solve the thermal and elastic parts of the problem separately.

Stationary Heat Conduction Problem Let W^ι and T_1 be given functions, then we will investigate the following problem:

Problem $\mathcal{P}_1$ Find the function of the temperature $T = \prod_{\iota=1}^{s} T^\iota$ such that

$$\frac{\partial}{\partial x_i}\left(\kappa_{ij}^{\iota}\frac{\partial T^{\iota}}{\partial x_j}\right) + W^{\iota} = 0 \quad \text{in } \Omega^{\iota}, \qquad 1 \le \iota \le s,\ i,j = 1, \ldots, N \tag{6.443}$$

$$\kappa_{ij}\frac{\partial T}{\partial x_j} n_i = 0 \quad \text{on } \Gamma_u, \tag{6.444}$$

$$T = T_1 \quad \text{on } \Gamma_\tau, \tag{6.445}$$

$$T^k = T^l, \qquad \left(\kappa_{ij}\frac{\partial T}{\partial x_j} n_i\right)^k + \left(\kappa_{ij}\frac{\partial T}{\partial x_j} n_i\right)^l = 0 \quad \text{on } \cup_{k,l}\Gamma^{kl},\ 1 \le k,\ l \le s \tag{6.446}$$

where $\mathbf{n}^k = (n_i^k)$, $i = 1, \ldots, N$, $1 \le k \le s$, denotes the unit outward normal with respect to $\partial\Omega^k$, $\mathbf{n}^k = -s\mathbf{n}^l$ on Γ^{kl}; (κ_{ij}^{ι}) is the matrix of thermal conductivities. Assume that κ^ι are positive definite symmetric matrices,

$$0 < \kappa_0^{\iota} \le \kappa_{ij}^{\iota}\zeta_i\zeta_j|\zeta|^{-2} \le \kappa_1^{\iota} < +\infty \quad \text{for a.a. } \mathbf{x} \in \Omega^{\iota}, \zeta \in R^N,$$

where κ_0^ι, κ_1^ι are constants independent of $\mathbf{x} \in \Omega^\iota$. Let $\kappa_{ij}^\iota \in L^\infty(\Omega^\iota)$, $W^\iota \in L^2(\Omega^\iota)$, $T_1 \in \prod_{\iota=1}^{s} H^1(\Omega^\iota)$, $T_1^k = T_1^l$ on $\cup_{k,l}\Gamma^{kl}$.

Throughout the text we use the summation convention, that is, a repeated index implies summation from 1 to N.

Definition 6.20 We say that the function T is a weak solution of Problem $\mathcal{P}_1$, if $T - T_1 \in V_1$ and

$$b(T, z) = s(z) \quad \forall z \in V_1, \tag{6.447}$$

where

$$b(T, z) = \sum_{\iota=1}^{s} \int_{\Omega^\iota} \kappa_{ij}^\iota \frac{\partial T^\iota}{\partial x_i} \frac{\partial z^\iota}{\partial x_j} \, d\mathbf{x}, \qquad s(z) = \sum_{\iota=1}^{s} \int_{\Omega^\iota} W^\iota z^\iota \, d\mathbf{x},$$

$$V_1 = \left\{ z \in W_1 = \prod_{\iota=1}^{s} H^1(\Omega^\iota) | z = 0 \text{ on } \Gamma_\tau, z^k = z^l \text{ on } \cup_{k,l} \Gamma_c^{kl} \right\}.$$

Remark 6.30 Multiplying Eq. (6.443) by a test function, integrating per parts over the domain Ω^ι, and using the boundary conditions, we obtain (6.447).

Unilateral Contact Problem with Friction We will deal with the following problem:

Problem $\mathcal{P}_2$ Find a displacement vector $\mathbf{u} = (u_i)$, $i = 1, \ldots, N$, in Ω satisfying

$$\frac{\partial}{\partial x_j} \tau_{ij}(\mathbf{u}^\iota, T^\iota) + F_i^\iota = 0 \quad \text{in } \Omega^\iota, \ 1 \le \iota \le s, \ i = 1, \ldots, N, \tag{6.448}$$

$$\tau_{ij}(\mathbf{u}^\iota, T^\iota) = c_{ijkl}^\iota e_{kl}(\mathbf{u}^\iota) - \beta_{ij}^\iota (T^\iota - T_0^\iota) \quad \text{in } \Omega^\iota, \ 1 \le \iota \le s, \ i = 1, \ldots, N, \tag{6.449}$$

$$\mathbf{u} = \mathbf{u}_0 \quad \text{on } \Gamma_u, \tag{6.450}$$

$$\tau_{ij}(\mathbf{u}, T) n_j = P_i \quad \text{on } \Gamma_\tau, \ i = 1, \ldots, N, \tag{6.451}$$

$$u_n^k - u_n^l \le 0, \qquad \tau_n^k \le 0, \qquad (u_n^k - u_n^l)\tau_n^k = 0 \quad \text{on } \cup_{k,l} \Gamma^{kl}, \qquad 1 \le k, \ l \le s, \tag{6.452}$$

$$|\boldsymbol{\tau}_t^{kl}| \le g^{kl} \quad \text{on } \cup_{k,l} \Gamma^{kl}, \qquad 1 \le k, \ l \le s, \tag{6.453}$$

$$|\boldsymbol{\tau}_t^{kl}| < g^{kl} \Longrightarrow \mathbf{u}_t^k - \mathbf{u}_t^l = 0, \tag{6.454}$$

$$|\boldsymbol{\tau}_t^{kl}| = g^{kl} \Longrightarrow \text{there exists } \vartheta \ge 0 \text{ such that } \mathbf{u}_t^k - \mathbf{u}_t^l = -\vartheta \boldsymbol{\tau}_t^{kl}, \tag{6.455}$$

where $\mathbf{F} = (\mathbf{F}_i)$ are body forces, $\mathbf{P}$ the surface tractions, $\mathbf{u}_0$ boundary displacement, c_{ijkl}^ι elastic coefficients, β_{ij}^ι coefficients of thermal expansion, g_c^{kl} the slip limits, and T^ι, T_0^ι are the temperature and the reference temperature, respectively. Furthermore, we put $e_{ij}(\mathbf{u}) = \frac{1}{2}(\partial u_i/\partial x_j + \partial u_j/\partial x_i)$, $u_n^k = u_i^k n_i^k$, $u_n^l = u_i^l n_i^k = -u_i^l n_i^l$ (no sum over k or l), $u_t^k = (u_{ti}^k)$, $u_{ti}^k = u_i^k - u_n^k n_i^k$, $\mathbf{u}_t^l = (u_{ti}^l)$, $u_{ti}^l = u_i^l - u_n^l n_i^k$, $i = 1, \ldots, N$, $\tau_n^k = \tau_{ij}^k n_i^k n_j^k$, $\boldsymbol{\tau}_t^k = (\tau_{ti}^k)$, $\tau_{ti}^k = \tau_{ij}^k n_j^k - \tau_n^k n_i^k$, $\boldsymbol{\tau}_t^{kl} \equiv \boldsymbol{\tau}_t^k$.

Let $c^{\iota}_{ijkl} \in L^{\infty}(\Omega^{\iota})$, $F^{\iota}_i \in L^2(\Omega^{\iota})$, $P_i \in L^2(\Gamma_{\tau})$, $\beta^{\iota}_{ij} \in L^{\infty}(\Omega^{\iota})$, and $\mathbf{u}^{\iota}_0 \in [H^1(\Omega^{\iota})]^N$.

We will introduce the following notation: For the entries of any symmetric $(N \times N)$ matrix $\{\tau_{ij}\}$ we use the *vector notation* $\{\tau_j\}, j = 1, \ldots, j_N$, where $j_N = N(N+1)/2$ and

$$\tau_i = \tau_{ii} \quad \text{for } 1 \leq i \leq N,$$

$$\tau_3 = \tau_{12} \quad \text{for } N = 2,$$

$$\tau_4 = \tau_{23}, \qquad \tau_5 = \tau_{31}, \qquad \tau_6 = \tau_{12}, \text{ for } N = 3.$$

In a similar way we replace the symmetric matrices $\{e_{ij}(\mathbf{u})\}$, $\{\beta_{ij}\}$ by vectors $\{e_j(\mathbf{u})\}$, $\{\beta_j\}$. The stress–strain relation (6.449) will be of the following form:

$$\tau_i(\mathbf{u}^{\iota}, T) = \sum_{j=1}^{j_N} \mathcal{A}^{\iota}_{ij} e_j(\mathbf{u}^{\iota}) - \beta^{\iota}_i (T^{\iota} - T^{\iota}_0), \;\; 1 \leq i, j \leq j_N, \;\; 1 \leq \iota \leq s,$$

where $\mathcal{A}^{\iota}$ is a symmetric $(j_N \times j_N)$ matrix, $\mathcal{A}^{\iota}_{ik} \in L^{\infty}(\Omega^{\iota})$, $\iota = 1, \ldots, s$; assume that $\mathcal{A}^{\iota}_{ik} = 0$ for $i \leq N$ and $k > N$.

Since $\tau : e \equiv \tau_{ij} e_{ij} = \sum_{i=1}^{N} \tau_i e_i + 2 \sum_{i=N+1}^{j_N} \tau_i e_i$,

$$c^{\iota}_{ijkl} e_{ij} e_{kl} = \sum_{i,j=1}^{j_N} B^{\iota}_{ij} e_i e_j, \tag{6.456}$$

where B^{ι} is a symmetric $(j_N \times j_N)$ matrix such that $B^{\iota}_{ij} = \mathcal{A}^{\iota}_{ij}$ for $1 \leq i, j \leq N$, $B^{\iota}_{ij} = 0$ for $1 \leq i \leq N$, $N+1 \leq j \leq j_N$, and $B^{\iota}_{ij} = 2\mathcal{A}^{\iota}_{ij}$ for $N+1 \leq i, j \leq j_N$.

Let us assume that $\text{meas}_{N-1}(\Gamma_u \cap \partial\Omega^{\iota}) > 0$ and $\text{meas}_{N-1}(\Gamma_{\tau} \cap \partial\Omega^{\iota}) > 0$ for all $\iota = 1, \ldots, s$. Let

$$W_1 = \prod_{\iota=1}^{s} H^1(\Omega^{\iota}), \; \|w\|_{W_1} = \left(\sum_{\iota \leq s} \|w^{\iota}\|^2_{1,\Omega^{\iota}} \right)^{1/2},$$

$$W = \prod_{\iota=1}^{s} [H^1(\Omega^{\iota})]^N, \; \|\mathbf{v}\|_W = \left(\sum_{\iota \leq s} \sum_{i \leq N} \|v^{\iota}_i\|^2_{1,\Omega^{\iota}} \right)^{1/2}.$$

Let us introduce the space of virtual displacements

$$V = \{\mathbf{v} \in W : \mathbf{v} = 0 \quad \text{on } \Gamma_u\}$$

and the set of admissible displacements

$$K = \{\mathbf{v} \in V : v^k_n - v^l_n \leq 0 \quad \text{on } \cup_{k,l} \Gamma^{kl}_c\}.$$

Assume that the matrices B^{ι} are positive definite, so that

$$0 < a^{\iota}_0 \leq \sum_{i,j=1}^{j_N} B^{\iota}_{ij} \xi_i \xi_j |\xi|^{-2} \leq a^{\iota}_1 < +\infty \quad \text{for a.a. } \mathbf{x} \in \Omega^{\iota}, \xi \in R^{j_N},$$

where the constants a^{ι}_0, a^{ι}_1 are independent of $\mathbf{x} \in \Omega^{\iota}$.

Moreover, let $\mathbf{u}_0 \in W$, $T_0 \in W_1$, $g_c^{kl} \in L^\infty(\Gamma_c^{kl})$, and $\beta_j^\iota \in L^\infty(\Omega^\iota)$.

Multiplying (6.448) by a test function, integrating per parts over the domain Ω, applying the boundary conditions, assuming that $\mathbf{u}_0$ satisfies conditions $u_{0n}^k - u_{0n}^l = 0$ on $\cup_{k,l}\Gamma^{kl}$, we obtain the following result:

Definition 6.21 We say that the function $\mathbf{u}$ is a weak solution of Problem $\mathcal{P}_2$, if $\mathbf{u} - \mathbf{u}_0 \in K$ and

$$a(\mathbf{u}, \mathbf{v} - \mathbf{u}) + j_g(\mathbf{v}) - j_g(\mathbf{u}) \geq S(\mathbf{v} - \mathbf{u}, T) \quad \forall \mathbf{v} \in \mathbf{u}_0 + K, \tag{6.457}$$

where

$$a(\mathbf{u}, \mathbf{v}) = \sum_{\iota=1}^{s} \int_{\Omega^\iota} \sum_{i,j=1}^{j_N} B_{ij}^\iota e_i(\mathbf{u}^\iota) e_j(\mathbf{v}^\iota)\, d\mathbf{x},$$

$$j_g(\mathbf{v}) = \sum_{k,l} \int_{\Gamma^{kl}} g^{kl} |\mathbf{v}_t^k - \mathbf{v}_t^l|\, ds,$$

$$S(\mathbf{v}, T) = \sum_{\iota=1}^{s} \int_{\Omega^\iota} (F_i^\iota v_i^\iota + (T^\iota - T_0^\iota)\beta^\iota : e(\mathbf{v}^\iota))\, d\mathbf{x} + \int_{\Gamma_\tau} P_i v_i\, ds,$$

and where we insert the weak solution T of Problem $\mathcal{P}0_1$ in $S(\mathbf{v}, T)$.

6.9.3 Uncertain Input Data and the Worst Scenario Method

By uncertain data we mean data that cannot be determined uniquely and they are determined in some intervals given by the measurements.

Let us assume that the input data

$$A = \{B^\iota, \kappa^\iota, W^\iota, T_1, \mathbf{F}^\iota, \beta^\iota, \mathbf{P}, \mathbf{u}_0, g^{kl}, \ \iota = 1, \ldots, s, \ \forall k, l\}$$

are uncertain and belong to some sets of admissible data, that is,

$$A \in U_{\text{ad}} \Leftrightarrow B^\iota \in U_{\text{ad}}^{B^\iota}, \kappa^\iota \in U_{\text{ad}}^{\kappa^\iota}, \ W^\iota \in U_{\text{ad}}^{W^\iota}, \ T_1 \in U_{\text{ad}}^{T_1},$$
$$\mathbf{F}^\iota \in U_{\text{ad}}^{F^\iota}, \ \beta^\iota \in U_{\text{ad}}^{\beta^\iota}, \ \mathbf{P} \in U_{\text{ad}}^{P}, \mathbf{u}_0 \in U_{\text{ad}}^{u_0}, \ g^{kl} \in U_{\text{ad}}^{g^{kl}}.$$

We will assume that all the bodies Ω^ι are piecewise homogeneous, so that partitions of $\overline{\Omega}^\iota$ exist such that $\overline{\Omega}^\iota = \cup_{j=1}^{J^\iota} \overline{\Omega}_j^\iota$, $\Omega_j^\iota \cap \Omega_k^\iota = \emptyset$ for $j \neq k$, $1 \leq \iota \leq s$, $\Gamma^{kl} = \cup_{q=1}^{Q_{kl}} \overline{\Gamma}_q^{kl}$, $\Gamma_q^{kl} \cap \Gamma_p^{kl} = \emptyset$ for $q \neq p$, $\forall k, l$ and let us denote $\Gamma_u \cap \partial\Omega^\iota = \Gamma_u^\iota$, $\iota = 1, \ldots, s$, $\Gamma_\tau \cap \partial\Omega^\iota = \Gamma_\tau^\iota$, $\iota \leq s$ and let the data B^ι, κ^ι, $\mathbf{F}^\iota$, W^ι, β^ι be the *piecewise constant* with respect to the above-defined partitioning.

Let us denote by $C^{(0),1}$ the space of Lipschitz-continuous functions. We define the sets of admissible matrices by

$$U_{\text{ad}}^{B^\iota} = \{(j_N \times j_N) \text{ symmetric matrices } B^\iota : \underline{B}_{ik}^\iota(j) \leq B_{ik}^\iota{}_{|\Omega_j^\iota} = \text{const} \leq \overline{B}_{ik}^\iota(j),\ j \leq J^\iota,\ i,k = 1,\ldots,j_N\}, \tag{6.458}$$

where $\underline{B}^\iota(j)$ and $\overline{B}^\iota(j)$ are given $(j_N \times j_N)$ symmetric matrices, $\iota = 1,\ldots,s$ and let positive constants $c_B^\iota(j)$ exist such that

$$\lambda_{\min}\left(\tfrac{1}{2}(\underline{B}^\iota(j) + \underline{B}^\iota(j))\right) - \rho\left(\tfrac{1}{2}(\overline{B}^\iota(j) - \overline{B}^\iota(j))\right) \equiv c_B^\iota(j) \quad \text{for } j = 1,\ldots,J^\iota,\ \iota = 1,\ldots,s, \tag{6.459}$$

where $\lambda_{\min}$ and ρ denotes the minimal eigenvalue and the spectral radius, respectively.

Let us introduce

$$U_{\text{ad}}^{\kappa^\iota} = \{(N \times N) \text{ symmetric matrices } \kappa^\iota : \underline{\kappa}_{ik}^\iota(j) \leq \kappa_{ik|\Omega_j^\iota}^\iota = \text{const} \leq \overline{\kappa}_{ik}^\iota(j),\ j \leq J^\iota,\ i,k \leq N\} \tag{6.460}$$

where $\underline{\kappa}^\iota(j)$ and $\overline{\kappa}^\iota(j)$ are given $(N \times N)$ symmetric matrices, $j = 1,\ldots,J^\iota, \iota = 1,\ldots,s$ and let positive constants $c_\kappa^\iota(j)$ exist such that

$$\lambda_{\min}\left(\tfrac{1}{2}\left(\underline{\kappa}^\iota(j) + \overline{\kappa}^\iota(j)\right)\right) - \rho\left(\tfrac{1}{2}\left(\overline{\kappa}^\iota(j) - \underline{\kappa}^\iota(j)\right)\right) \equiv c_\kappa^\iota(j) \quad \text{for } j \leq J^\iota,\ \iota \leq s, \tag{6.461}$$

where $\lambda_{\min}$ and ρ denotes the minimal eigenvalue and the spectral radius, respectively.

Remark 6.31 The matrices $B^\iota(j) \equiv B^\iota|_{\Omega_j^\iota}$ are positive definite for any $B^\iota \in U_{\text{ad}}^{B^\iota}$, $\iota = 1,2,\ldots,s$ and any $j \leq J^\iota$ and, furthermore, the matrices $\kappa^\iota(j) = \kappa_{|\Omega_j^\iota}^\iota$ are positive definite for any $\kappa^\iota \in U_{\text{ad}}^{\kappa^\iota}$, $\iota \leq s, j \leq J^\iota$.

Next, let us introduce

$$U_{\text{ad}}^{F_i^\iota} = \{f \in L^\infty(\Omega) : \overline{F}_i^\iota(j) \leq f_{|\Omega_j^\iota} = \text{const} \leq \underline{F}_i^\iota(j),\ j \leq J^\iota\}, \quad \text{for } i \leq N,\ \iota \leq s, \tag{6.462}$$

where $\underline{F}_i^\iota(j)$ and $\overline{F}_i^\iota(j)$ are given constants;

$$U_{\text{ad}}^{W^\iota} = \{w \in L^\infty(\Omega) : \underline{W}^\iota(j) \leq w_{|\Omega_j^\iota} = \text{const} \leq \overline{W}^\iota(j),\ j \leq J^\iota\}, \quad \text{for } \iota \leq s, \tag{6.463}$$

where $\underline{W}^\iota(j)$ and $\overline{W}^\iota(j)$ are given constants;

$$U_{\text{ad}}^{T_1} = \{\mathcal{T} \in L^\infty(\Gamma_\tau) : \underline{T}_1(\iota) \leq \mathcal{T}_{|\Gamma_\tau^\iota} = \text{const} \leq \overline{T}_1(\iota),\ \iota \leq s\}, \tag{6.464}$$

where $\underline{T}_1(\iota)$ and $\overline{T}_1(\iota)$ are given constants;

$$U_{\text{ad}}^{u_{0i}} = \{u \in L^\infty(\Gamma_u) : \underline{u}_{0i}(\iota) \leq u_{|\Gamma_u^\iota} = \text{const} \leq \overline{u}_{0i}(\iota),\ \iota \leq s\}, \tag{6.465}$$

where $\underline{u}_{0i}(\iota)$ and $\overline{u}_{0i}(\iota)$, $i = 1, \ldots, N$, are given constants;

$$U_{\text{ad}}^{P_i} = \{p \in L^\infty(\Gamma_\tau) : \underline{P}_i(\iota) \leq p_{|\Gamma_\tau^\iota} = \text{const} \leq \overline{P}_i(\iota),\ \iota \leq s\}, \tag{6.466}$$

where $\underline{P}_i(\iota)$ and $\overline{P}_i(\iota)$, $i = 1, \ldots, N$, are given constants;

$$U_{\text{ad}}^{\beta_i^\iota} = \{b \in L^\infty(\Omega) : \underline{\beta}_i^\iota(j) \leq b_{|\Omega_j^\iota} = \text{const} \leq \overline{\beta}_i^\iota(j),\ j \leq J^\iota\} \quad \text{for } i \leq j_N,\ \iota \leq s, \tag{6.467}$$

where $\underline{\beta}_i^\iota(j)$ and $\overline{\beta}_i^\iota(j)$ are given constants;

$$U_{\text{ad}}^{g^{kl}} = \{g \in L^\infty(\Gamma^{kl}) \ :\ g_{|\overline{\Gamma}_q^{kl}} \in C^{(0),1}(\overline{\Gamma}_q^{kl});\ 0 \leq g(s) \leq \overline{g}_q^{kl},\ \left|\frac{dg}{ds}\right| \leq C_g^{kl} \quad \text{a.e. in } \Gamma_q^{kl},\ q \leq Q_{kl}\}, \tag{6.468}$$

for all pairs k, l under consideration, where $\overline{g}_q^{kl}$ and C_g^{kl} are given positive constants.

Finally, we introduce the set of admissible data, as follows:

$$U_{\text{ad}} = \prod_{\iota \leq s} U_{\text{ad}}^{B^\iota} \times \prod_{\iota \leq s} U_{\text{ad}}^{K^\iota} \times \prod_{\iota \leq s, i \leq N} U_{\text{ad}}^{F_i^\iota} \times \prod_{\iota \leq s} U_{\text{ad}}^{W^\iota} \times U_{\text{ad}}^{T_1} \times \prod_{\iota \leq s, i \leq j_N} U_{\text{ad}}^{\beta_i^\iota} \times \prod_{i \leq N} U_{\text{ad}}^{P_i} \times \prod_{i \leq N} U_{\text{ad}}^{u_{0i}} \times \prod_{k,l} U_{\text{ad}}^{g^{kl}}. \tag{6.469}$$

Instead of the bilinear forms and functionals $b(T, z)$, $a(\mathbf{u}, \mathbf{v})$, $j_g(\mathbf{v})$, $s(z)$, and $S(\mathbf{v}, T)$ introduced above, we will write $b(A; T, z)$, $a(A; \mathbf{u}, \mathbf{v})$, $j_g(A; \mathbf{v})$, $s(A; z)$, and $S(A; \mathbf{v}, T)$ for any $A \in U_{\text{ad}}$.

Lemma 6.20 There exist positive constants C_i, $i = 0, 1, \ldots, 6$ independent of $A \in U_{\text{ad}}$, such that

$$b(A; z, z) \geq C_0 \|z\|_{W_1}^2 \quad \forall z \in V_1, \tag{6.470}$$

$$|b(A; z, y)| \leq C_1 \|z\|_{W_1} \|y\|_{W_1} \quad \forall z, y \in W_1, \tag{6.471}$$

$$a(A; \mathbf{v}, \mathbf{v}) \geq C_2 \|\mathbf{v}\|_W^2 \quad \forall \mathbf{v} \in V, \tag{6.472}$$

$$|a(A; \mathbf{v}, \mathbf{w})| \leq C_3 \|\mathbf{v}\|_W \|\mathbf{w}\|_W \quad \forall \mathbf{v}, \mathbf{w} \in V, \tag{6.473}$$

$$|s(A; z)| \leq C_4 \|z\|_{0,\Omega} \quad \forall z \in V_1, \tag{6.474}$$

$$|S(A; \mathbf{v}, T)| \leq C_5 \big(\|\mathbf{v}\|_{0,\Omega} + \|\mathbf{v}\|_{0,\Gamma_\tau} + \|T - T_0\|_{0,\Omega} \|\mathbf{v}\|_W\big) \quad \forall \mathbf{v} \in W, \tag{6.475}$$

$$|j_g(A; \mathbf{u}) - j_g(A; \mathbf{v})| \leq C_6 \sum_{\iota \leq s} \|\mathbf{u}^\iota - \mathbf{v}^\iota\|_{0,\partial\Omega^\iota}, \quad \forall \mathbf{u}, \mathbf{v} \in W. \tag{6.476}$$

For the proof see Hlaváček and Nedoma (2004).

As the result we have the following theorem:

Theorem 6.58 There exists a unique weak solution $T(A)$ and $\mathbf{u}(A)$, respectively, of Problems $\mathcal{P}_1$ and $\mathcal{P}_2$ for any $A \in U_{\text{ad}}$.

For the proof see Hlaváček and Nedoma (2004).

To find the worst input data A in the set U_{ad}, we will introduce a criterion, a certain functional, that depends on the solutions $(T(A), \mathbf{u}(A))$ of Problems $\mathcal{P}_1$ and $\mathcal{P}_2$, respectively. Such criteria can be the following:

Let $G_r \subset \cup_{\iota \le s} \Omega^\iota$, $r = 1, \dots, \bar{r}$, be subdomains, adjacent to the boundaries $\partial\Omega^\iota$, for example.

Then we define

$$\Phi_1(T) = \max_{r \le \bar{r}} \varphi_r(T) \tag{6.477}$$

where $\varphi_r(T) = (\text{meas}_N\ G_r)^{-1} \int_{G_r} T\, d\mathbf{x}$; let $G'_r \subset \Gamma_u$, $r \le \bar{r}$ and

$$\Phi_2(T) = \max_{r \le \bar{r}} \psi_r(T); \tag{6.478}$$

where $\psi_r(T) = (\text{meas}_{N-1}\ G'_r)^{-1} \int_{G'_r} T\, ds$;

$$\Phi_3(\mathbf{u}) = \max_{r \le \bar{r}} \chi_r(\mathbf{u}), \tag{6.479}$$

where $\chi_r(\mathbf{u}) = (\text{meas}_N\ G_r)^{-1} \int_{G_r} u_i n_i(X_r)\, d\mathbf{x}$; and where $\mathbf{n}(X_r)$ is the unit outward normal at a fixed point $X_r \in \partial\Omega^\iota \cap \partial G_r$ (if $G_r \subset \Omega^\iota$) to the boundary $\partial\Omega^\iota$;

$$\Phi_4(\mathbf{u}) = \max_{r \le \bar{r}} \chi'_r(\mathbf{u}); \tag{6.480}$$

where $\chi'_r(\mathbf{u}) = (\text{meas}_{N-1}\ G'_r)^{-1} \int_{G'_r} u_i n_i(X_r)\, ds$; $G'_r \subset \cup_{\iota \le s} \partial\Omega^\iota \setminus \Gamma_u$.

Since the weak solution $\mathbf{u}(A)$ of the discussed problem $(\mathcal{P}_2)$ depends on $T(A)$, then $\mathbf{u}(A) = \mathbf{u}(A; T(A))$. Hence instead of $\Phi_i(\mathbf{u})$ we will write $\Phi_i(A; \mathbf{u}, T)$.

Another choice is

$$\Phi_5(A; \mathbf{u}, T) = \max_{r \le \bar{r}} \omega_r(A; \mathbf{u}, T)\,; \tag{6.481}$$

where $\omega_r(A; \mathbf{u}, T) = (\text{meas}_N\ G_r)^{-1} \int_{G_r} I_2^2(\tau(A; \mathbf{u}, T))\, d\mathbf{x}$, and where $I_2(\tau)$ is the intensity of shear stress defined as the second fundamental invariant of the stress tensor deviator τ_{ij}^D, that is,

$$I_2^2(\tau) = \sum_{i,j=1}^{3} \tau_{ij}^D \tau_{ij}^D, \qquad \tau_{ij}^D = \tau_{ij} - \frac{1}{3}\tau_{kk}\delta_{ij}\,;$$

that is,

$$I_2^2 = \tfrac{2}{3}[\tau_{11}^2 + \tau_{22}^2 + \tau_{33}^2 - (\tau_{11}\tau_{22} + \tau_{11}\tau_{33} + \tau_{22}\tau_{33}) + 3(\tau_{12}^2 + \tau_{13}^2 + \tau_{23}^2)]$$

and $\tau(A; \mathbf{u}, T)$ is defined by the formula (6.449) if $N = 3$.

In the case if the friction can be neglected, we set $g_c^{kl} \equiv, 0$. Then

$$\Phi_6(A; \mathbf{u}, T) = \max_{r \le \bar{r}} \mu_r(A; \mathbf{u}, T)\,; \tag{6.482}$$

where $\mu_r(A; \mathbf{u}, T) = (\text{meas}_N\ G_r)^{-1} \int_{G_r} (-\tau_n(A; \mathbf{u}, T))\, d\mathbf{x}$, $\tau_n = \tau_{ij} n_i(X_r) n_j(X_r)$, $X_r \in \Gamma_c \cap \partial G_r$, and G_r is a small subdomain adjacent to Γ_c.

Then we can formulate the **worst-scenario problems** as follows: Find

$$A^{0i} = \arg \max_{A \in U_{\text{ad}}} \Phi_i(T(A)), \qquad i = 1, 2, \tag{6.483}$$

$$A^{0i} = \arg \max_{A \in U_{\text{ad}}} \Phi_i(A, \mathbf{u}(A), T(A)), \qquad i = 3, 4, 5, 6, \tag{6.484}$$

where $T(A)$ and $\mathbf{u}(A)$ are weak solutions of Problems $\mathcal{P}_1$ and $\mathcal{P}_2$, respectively.

To analyze the solvability of the worst-scenario problems (6.483) and (6.484), we have to study the mapping $A \mapsto T(A)$ and $A \mapsto \mathbf{u}(A, T(A))$. We introduce the decomposition of $A \in U_{\text{ad}}$ as $A = \{A', A''\}$, where

$$A' = \left\{ \prod_{\iota \le s} \prod_{j \le J^\iota} \kappa^\iota(j), \prod_{\iota \le s} \prod_{j \le J^\iota} W^\iota(j), \prod_{\iota \le s} T_1^\iota \right\},$$

$$A' \in R^{p_1}, \qquad p_1 = (j_N + 1) \sum_{\iota \le s} J^\iota + s,$$

$$A'' = \left\{ \prod_{\iota \le s} \prod_{j \le J^\iota} B^\iota(j), \prod_{\iota \le s} \prod_{j \le J^\iota} \mathbf{F}^\iota(j), \prod_{\iota \le s} \mathbf{P}^\iota, \prod_{\iota \le s} \mathbf{u}_0^\iota, \prod_{\iota \le s} \prod_{j \le J^\iota} \beta^\iota(j), \prod_{k,l} \prod_{q \le Q_{kl}} g^{kl}(q) \right\},$$

$$A'' \in R^{p_2} \times \prod_{k,l} \prod_{q \le Q_{kl}} C(\overline{\Gamma}_q^{kl}),$$

$$p_2 = \left(\sum_{\iota \le s} J^\iota \right) \left[\frac{(3 + j_N) j_N}{2 + N(1 + 2s)} \right]. \tag{6.485}$$

We will show the continuity of the mappings $A' \mapsto T(A')$ for $A' \in U'_{\text{ad}} = \prod_{\iota \le s} U_{\text{ad}}^{\kappa^\iota} \times \prod_{\iota \le s} U_{\text{ad}}^{W^\iota} \times U_{\text{ad}}^{T_1^\iota}$ and $A \mapsto \mathbf{u}(A, T(A'))$ for $A \in U_{\text{ad}}$, respectively. Since the problem is quasi-coupled we can prove the following theorems and lemma [see Hlaváček and Nedoma (2004)].

Lemma 6.21 If $A_n \in U_{\text{ad}}, A_n \to A$ in U, where $U = R^{p_1 + p_2} \times \prod_{k,l} \prod_{q \le Q_{kl}} C(\overline{\Gamma}_q^{kl})$, and $\mathbf{u}_n \rightharpoonup \mathbf{u}$ weakly in W, then

$$a(A_n; \mathbf{u}_n, \mathbf{v}) \to a(A; \mathbf{u}, \mathbf{v}) \quad \forall \mathbf{v} \in W, \tag{6.486}$$

$$S(A_n; \mathbf{u}_n, T) \to S(A; \mathbf{u}, T) \quad \forall T \in W_1, \tag{6.487}$$

$$j_g(A_n; \mathbf{u}_n) \to j_g(A; \mathbf{u}). \tag{6.488}$$

For the proof see Hlaváček and Nedoma (2004).

Theorem 6.59 Let $A' \in U'_{\text{ad}}$, $A'_n \to A'$ in R^{p_1} as $n \to \infty$. Then $T(A'_n) \to T(A')$ in W_1. Let $A_n \in U_{\text{ad}}$, $A_n \to A$ in U. Then $\mathbf{u}(A_n) \to \mathbf{u}(A)$ in W.

For the proof see Hlaváček and Nedoma (2004).

To prove the existence of a solution of the worst-scenario problem, the following lemma will be used.

Lemma 6.22 Let $\Phi_i(T)$, $i = 1, 2$, be defined by (6.477) and (6.478). Let $\Phi_i(\mathbf{u})$, $i = 3, 4$, be defined by (6.479) and (6.480). Let $\Phi_i(A; \mathbf{u}, \mathbf{T})$, $i = 5, 6$, be defined by (6.481) and (6.482). Let $A_n \to A$ in U, $A_n \in U_{\text{ad}}$, $\mathbf{u}_n \to \mathbf{u}$ in W and $T_n \to T$ in W_1. Then

$$\begin{aligned} \lim_{n\to\infty} \Phi_i(T_n) &= \Phi_i(T), \qquad i = 1, 2, \\ \lim_{n\to\infty} \Phi_i(\mathbf{u}_n) &= \Phi_i(\mathbf{u}), \qquad i = 3, 4, \\ \lim_{n\to\infty} \Phi_i(A_n, \mathbf{u}_n, T_n) &= \Phi_i(A, \mathbf{u}, T), \qquad i = 5, 6. \end{aligned} \tag{6.489}$$

For the proof see Hlaváček and Nedoma (2004).

The main result of this section is represented by the next theorem:

Theorem 6.60 There exists at least one solution of the worst-scenario problems (6.483) and (6.484), $i = 1, \ldots, 6$.

For the proof see Hlaváček and Nedoma (2004, 2002b) for the case without any friction.

Remark 6.32 For a related analysis we refer to Hlaváček (2003), where the worst-scenario method has been applied to the unilateral contact with the Coulomb friction model and uncertain input data.

6.10 BIOMECHANICAL MODELS OF HUMAN JOINT REPLACEMENTS COUPLING BI- AND UNILATERAL CONTACTS, FRICTION, ADHESION, AND WEAR

Simulation of a function of human joint replacements and wear of their parts leads to solving complicated contact problems. More realistic modeling of the interface of bone–implants requires taking into account the influence of wear debris on the adhesion. **Wear** is a primary factor that limits the life span of total joint replacements in humans. Wear particles (e.g., ceramics particles) have been reported to be in a nanometric size range, but larger particles up to 7 μm have been also identified in ex vivo specimens (Hatton et al., 2000; Yoon et al., 1998). The mean (±95% confidence limits) sizes of the debris were 300 ± 200 nm for the UHMWPE particles, 30 ± 2.25 nm for the metal particles, and 9 ± 0.5 nm for the alumina-ceramic wear particles generated under the standard simulation condition. The UHMWPE particles are larger than the metal and ceramics particles, and the metal particles are larger than the ceramics particles (Tipper et al., 2003). Goldsmith et al. (2000) show that while wear rates in the range from 30 $\text{mm}^3/10^6$ cycles to 100 $\text{mm}^3/10^6$ cycles are associated with the more conventional metal-on-polymer points of diameters 22–28 mm, the mean wear rates for the 28 and 36 mm diameter metal-on-metal joints were only 0.45 $\text{mm}^3/10^6$ cycles and 0.36 $\text{mm}^3/10^6$ cycles, respectively. Therefore, computational wear prediction represents an attractive concept for evaluating new total joint replacement designs before their physical testing and then their implementation. Released polyethylene particles build a wear debris that can initiate a bone destruction, the so-called osteolysis, resulting in implant loosening and great pain for patients. To study the effects of wear one can analyze (i) implants retrieved post mortem, (ii) implants retrieved after a failure, and (iii) implant wear tests. Physical wear testing is a fundamental present approach because recent simulator designs are more successful at reproducing the wear patterns observed in retrievals (Walker et al., 1997; Giddings et al., 2001). Modern dynamic FE methods are very attractive approaches for wear-and-tear predictions, but they take a great deal CPU time. While revision and post-mortem retrievals are valuable for studying insert damage modes, which can be obtained in several years, numerical results can be obtained during several hours or days (Bartel et al., 2006). Modern methods based on bilateral and unilateral contact problems with friction, adhesion, and wear are very effective methods for such studies. The first results in this field can be found in the works of Onsager (1931), Archard (1953), Ziegler (1958, 1963), Burwell (1958), Moreau (1974), Frémond (1987, 1988a,b), Rabinowicz (1995), Strömberg et al. (1996), Raous et al. (1999), Cocu and Rocca (2000), and Rojek and Telega (2001). The study by Strömberg et al. (1996) is mainly extended for the study of fretting, that is, for a wear phenomenon arising when contacting surfaces undergo oscillatory displacements with small amplitudes.

6.10.1 Biomechanical Model Based on Quasi-Static and Dynamic (Visco)elastic Contacts with Adhesion and Friction

Adhesion between two neighboring bodies being in contact may be caused by metallic, covalent, ionic, hydrogen, or van der Waals forces [see Maugis (1990), Possart (1998), and Breme et al. (1998)]. The model will be based on the results of Raous et al. (1999). To separate these colliding bodies, one needs energies γ_k and γ_l to create the unit areas (k) and (l), whereas the interfacial energy γ_{kl} is recovered. Then the thermodynamic work of adhesion (Dupré energy of adhesion) is given by

$$w = \gamma_k + \gamma_l - \gamma_{kl}.$$

The work of adhesion is a useful quantity because it distinguishes the two states, contact and separation. Since two neighboring bodies are in contact, they can be either bounded, that is, adhesion being present, or they can be without any bounding, that is, no adhesion being present. In both cases the displacements are the same but mechanical situations are different. Therefore, a new variable to describe the state of contact will be introduced, the adhesion intensity β representing active bounds. The **intensity of adhesion** β is defined as follows:

1. If $\beta = 1$, all the bonds are active.
2. If $\beta = 0$, all the bonds are broken or no adhesion exists.
3. If $0 < \beta < 1$, one part of the bonds remains active, the remaining are broken, and the adhesion is partial.

The intensity of adhesion $\beta = \beta(\mathbf{x}, t)$ is a function of $\mathbf{x} \in \Gamma_c^{kl}$, where Γ_c^{kl}—the contact boundary, will be defined below, and of time t. The velocity β' describes the manner in which the bonds evolve. This velocity is a macroscopic representation of the velocities of the microscopic displacements that form and destroy the bonds. If the adhesion is described with a damage parameter D, we have $\beta = 1 - D$. The theory is constructed using the virtual power principle.

Let there be a joint structure, occupying at time t_0, that is, in their undeformed state, a closed region $\Omega = \cup_{\iota=1}^{s} \Omega^\iota$, $\Omega^\iota \in \mathbb{R}^N$, $N = 2, 3$, fixed along a part of the boundary $\partial\Omega$, we denote it as Γ_u, (meas $\partial\Omega^\iota \cap \Gamma_u > 0$ for all $\iota \le s$) and loaded by a surface traction $\mathbf{P}$ along a part Γ_τ and $\Gamma_c^{kl} = \Gamma_c^k \cap \Gamma_c^l$ is the common contact part of the boundary $\partial\Omega^k$ and $\partial\Omega^l$. We assume that meas $\Gamma_c^{kl} \neq 0$. The boundaries of the deformed bodies $\partial\Omega^\iota$, $\iota = 1, \ldots, s$, possess a unique outward normal $\mathbf{n}^\iota$ at each of their points. At the common contact boundary Γ_c^{kl}

$$\mathbf{n}^{kl} \equiv \mathbf{n}^k = -\mathbf{n}^l$$

holds. Let $\bar{I} = [t_0, t_p]$ be a time interval.

The motion of the joint system is described by the following equations:

$$\frac{\partial \tau_{ij}^\iota}{\partial x_j} + f_i^\iota = \rho^\iota \frac{\partial^2 u_i^\iota}{\partial t^2} \quad \text{in } \Omega^\iota \times I, \qquad \iota = 1, \ldots, s, \tag{6.490}$$

$$u_i^\iota = u_2^\iota \quad \text{on } \Gamma_u^\iota \times I, \tag{6.491}$$

$$\tau_{ij}^\iota n_j = P_i^\iota \quad \text{on } \Gamma_\tau^\iota \times I, \tag{6.492}$$

with contact conditions, which will be introduced later, and with initial conditions

$$\begin{aligned} &\mathbf{u}^\iota(t_0, \mathbf{x}) = \mathbf{u}_0^\iota(\mathbf{x}), \\ &\mathbf{u}'^\iota(t_0, \mathbf{x}) = \mathbf{u}_1(\mathbf{x}), \qquad \mathbf{x} \in \Omega^\iota,\ \iota = 1, \ldots, s, \end{aligned} \tag{6.493}$$

and with Hooke's law

$$\tau_{ij}^\iota = c_{ijkl}^\iota(\mathbf{x}) e_{kl}(\mathbf{u}^\iota), \qquad \mathbf{x} \in \Omega^\iota. \tag{6.494}$$

All quantities (static and kinematic) are functions of spatial variables and the time $t \in I$. The elastic coefficients c_{ijkl}^ι are functions from $L^\infty(\Omega^\iota)$, and they satisfy the conditions

$$\begin{aligned} &c_{ijkl}^\iota = c_{jikl}^\iota = c_{klij}^\iota = c_{ijlk}^\iota, \\ &c_0^\iota \xi_{ij}\xi_{ij} \le c_{ijkl}^\iota(\mathbf{x})\xi_{ij}\xi_{kl} \le c_1^\iota \xi_{ij}\xi_{ij}, \\ &\xi \in \mathbb{R}^{N^2}, \qquad \xi_{ij} = \xi_{ji}, \qquad c_0^\iota = \text{const} > 0, \qquad c_1^\iota = \text{const} > 0 \end{aligned} \tag{6.495}$$

for almost every $\mathbf{x} \in \Omega^\iota$, $\iota = 1, \ldots, s$, where

$$e_{ij} = \frac{1}{2}\left(\frac{\partial u_i}{\partial x_j} + \frac{\partial u_j}{\partial x_i}\right), \qquad i,j = 1,\ldots,N,$$

and where $\mathbf{u}_0$, $\mathbf{u}_1$, $\mathbf{u}_2$, $\mathbf{f}$ and $\mathbf{P}$ are prescribed functions.

To define the contact conditions, we introduce the relative displacement $[\mathbf{u}] \equiv [\mathbf{u}]^{kl} = \mathbf{u}^l - \mathbf{u}^k$ on the contact boundary Γ_c^{kl}, which can be decomposed into its normal and tangential components, u_n and $\mathbf{u}_t$, respectively, defined by

$$[u_n] \equiv [u_n]^{kl} = [\mathbf{u}]^{kl} \cdot \mathbf{n}^{kl}, \qquad [\mathbf{u}_t] \equiv [\mathbf{u}_t]^{kl} = [\mathbf{u}]^{kl} - [u_n]^{kl} \cdot \mathbf{n}^{kl},$$

where $\mathbf{n}^{kl} \equiv \mathbf{n}^k = -\mathbf{n}^l$.

Let contact forces on Γ_c^{kl} be denoted by $R_i^k = \tau_{ij}^k n_j^k$, $\mathbf{R}^k = \mathbf{R}^l$, $R_i^l = \tau_{ij}^l n_j^{kl}$ and let us decompose it into the normal and tangential components, respectively, $R_n^{kl} \equiv R_i^k n_i^k$, $\mathbf{R}_t^k = \mathbf{R}^k - \mathbf{R}_n^{kl} \cdot \mathbf{n}^k$. If no tensile normal forces on Γ_c^{kl} are allowed, then

$$R_n^{kl} \leq 0. \qquad (6.496)$$

But in this formulation contact forces can be either compressive or due to the adhesion they can be also tensile. The adhesion can generate reactions in normal as well as tangential directions. Initial adhesive bonds are given by

$$\beta(0, \mathbf{x}) = \beta_0(\mathbf{x}), \qquad \mathbf{x} \in \cup\Gamma_c^{kl}. \qquad (6.497)$$

Considering the contact area $\cup\Gamma_c^{kl}$ as a material boundary (Strömberg et al., 1996; Frémond, 1987), we introduce a surface density of internal energy E and a density of entropy S associated to the pseudodomain $\cup\Gamma_c^{kl}$. Then, the Helmholtz free energies can be written as $\Psi = E - ST$ on $\cup\Gamma_c^{kl}$ and $\psi = e - sT$ in $\cup(\Omega^k \cup \Omega^l)$, where e is the specific internal energy, s the specific entropy, and T is the absolute temperature. Further, we introduce the intensity of adhesion defined above.

According to Raous et al. (1999) and Rojek and Telega (2001), we define the generalized potential $\varphi(\beta, u_n, \mathbf{u}_t)$ by

$$\varphi(\beta, u_n, \mathbf{u}_t) = \tfrac{1}{2}\left(c_n[u_n]^2\beta^2 + c_t\left\|[\mathbf{u}_t]\right\|^2\beta^2\right) - w\beta + I_K(u_n) + I_P(\beta), \qquad (6.498)$$

where c_n and c_t are nonnegative constants characterizing the interface stiffness, w is the thermodynamic work of adhesion (the Dupré's energy), I_K and I_P are the indicatrix of the closed convex sets $K = \{\mathbf{v}|\mathbf{v} \geq 0\} \equiv \mathbb{R}^+$ and $P = \{\gamma|0 \leq \gamma \leq 1\}$, which are defined as $I_K(\mathbf{v}) = 0$ for $\mathbf{v} \in K$ and $I_K(\mathbf{u}) = \infty$ for which $\mathbf{v} \notin K$, $I_P(\gamma) = 0$ for $\gamma \in P$ and $I_P(\gamma) = \infty$ for $\gamma \notin P$ and where $\mathbb{R}^+$ denotes the set of nonnegative reals.

Let us consider a viscous contact with the adhesion and friction. Let $\mathbf{R}$ be the contact reaction. Let the contact traction be divided on Γ_c^{kl} into reversible parts R_n^r and $\mathbf{R}_t^r$ and irreversible parts R_n^i and $\mathbf{R}_t^i$. The state variables on the contact boundary are u_n, $\mathbf{u}_t$, and β. The function φ is obviously nondifferentiable. Due to Panagiotopoulos (1993) and Raous et al. (1999) the state laws are written in the form of subdifferentials:

$$\begin{aligned} R_n^r &\in \partial_{u_n}\varphi(\beta, u_n, \mathbf{u}_t)\,, \\ \mathbf{R}_t^r &\in \partial_{\mathbf{u}_t}\varphi(\beta, u_n, \mathbf{u}_t), \\ -G_\beta &\in \partial_\beta\varphi(\beta, u_n, \mathbf{u}_t), \end{aligned} \tag{6.499}$$

where $\partial_{u_n}\varphi, \partial_{\mathbf{u}_t}\varphi, \partial_\beta\varphi$ are the subdifferentials of the potential function φ with respect to the variable $u_n, \mathbf{u}_t, \beta$, respectively and G_β is the thermodynamic force. Hence, we obtain

$$\mathbf{R}_t^r = c_t[\mathbf{u}_t]\beta^2, \tag{6.500}$$

$$[u_n] \geq 0, \qquad -R_n^r + c_n[u_n]\beta^2 \geq 0, \qquad (-R_n^r + c_n[u_n]\beta^2)[u_n] = 0, \tag{6.501}$$

$$\begin{aligned} &G_\beta \geq w && \text{if } \beta = 0, \\ &G_\beta = w - (c_n[u_n]^2 + c_t\|[\mathbf{u}_t]\|^2)\beta && \text{if } 0 < \beta < 1, \\ &G_\beta \leq w - (c_n[u_n]^2 + c_t\|[\mathbf{u}_t]\|^2) && \text{if } \beta = 1, \end{aligned} \tag{6.502}$$

where (6.501) represents the generalized nonpenetration conditions for the contact with the adhesion.

The irreversible part of the contact forces $R_n^i = R_n - R_n^r$ and $\mathbf{R}_t^i = \mathbf{R}_t - \mathbf{R}_t^r$ will be derived from the pseudopotential of dissipation because the only dissipative processes under consideration are the friction and adhesion. Due to Raous et al. (1999)

$$\Phi(R_n, u_n, \beta; \mathbf{u}_t', \beta') = \mathcal{F}_c^{kl}|R_n - c_n[u_n]\beta^2|\|[\mathbf{u}_t']\| + \frac{b}{p+1}|\beta'|^{p+1} + I_{\mathbb{R}^-}(\beta'), \tag{6.503}$$

where $\mathcal{F}_c^{kl}$ is the coefficient of Coulomb friction, $\mathbb{R}^- = \{\gamma \in P | \gamma \leq 0\}$, and parameters b and p characterize viscous properties of adhesive bonds, $-1 < p \leq 1$, b characterizes a time-dependent evolution of the adhesion, and where the indicatrix $I_{\mathbb{R}^-}(\beta')$ imposes that $\beta' \leq 0$, which means that adhesive bounds can only be weaken and cannot be restituted.

The complementary laws are then written as

$$\begin{aligned} R_n^i &= 0, \\ \mathbf{R}_t^i &\in \partial_{\mathbf{u}_t'}\Phi(R_n, u_n, \beta; \mathbf{u}_t', \beta')\,, \\ G_\beta &\in \partial_{\beta'}\Phi(R_n, u_n, \beta; \mathbf{u}_t', \beta'). \end{aligned} \tag{6.504a–c}$$

The normal behavior has been supposed to be totally elastic. Making explicit the subdifferentials in (6.504),we obtain on $\cup\Gamma_c^{kl}$

$$\|\mathbf{R}_t^i - c_t[\mathbf{u}_t]\beta^2\| \leq \mathcal{F}_c^{kl}|R_n - c_n[u_n]\beta^2| \tag{6.505}$$

with two situations

$$\begin{aligned}
&\|\mathbf{R}_t^i - c_t[\mathbf{u}_t]\beta^2\| < \mathcal{F}_c^{kl}|R_n - c_n[u_n]\beta^2| \Rightarrow [\mathbf{u}_t'] = 0, \\
&\|\mathbf{R}_t^i - c_t[\mathbf{u}_t]\beta^2\| = \mathcal{F}_c^{kl}|R_n - c_n[u_n]\beta^2| \Rightarrow \text{there exists } \lambda \geq 0 \\
&\qquad \text{such that } [\mathbf{u}_t'] = \lambda(\mathbf{R}_t^i - c_t[\mathbf{u}_t]\beta^2),
\end{aligned} \tag{6.506}$$

and with regard to (6.504c) the partial subdifferential $\partial_{\beta'}\Phi$ can be written as the sum of two subdifferentials

$$\partial_{\beta'}\Phi(R_n, u_n, \beta; \mathbf{u}_t', \beta') = b|\beta'|^p\partial|\beta'| + \partial I_{\mathbb{R}^-}(\beta'),$$

where

$$\partial I_{\mathbb{R}^-}(\beta') = \begin{cases} 0 & \text{if } \beta' < 0, \\ \mathbb{R}^- & \text{if } \beta' = 0, \\ \emptyset & \text{otherwise } (\emptyset \text{ denotes the empty set}). \end{cases}$$

Then

$$\begin{aligned}
&G_\beta = -b|\beta' t|^p && \text{if } \beta' < 0, \\
&G_\beta \in \mathbb{R}^- && \text{if } \beta' = 0, \\
&\text{or} \\
&\beta' = -\left(\frac{G_\beta^-}{b}\right),
\end{aligned} \tag{6.507}$$

where G_β^- denotes the negative part of G_β because

$$\partial|\beta'| = \begin{matrix} -1 & \text{if } \beta' < 0, \\ [-1, 1] & \text{if } \beta' = 0, \\ +1 & \text{if } \beta' > 0. \end{matrix}$$

For the local model coupling adhesion, friction, and unilateral contact the contact conditions are written as follows:

1. Unilateral conditions with adhesion

$$[u_n] \geq 0, \qquad -R_n^r + c_n[u_n]\beta^2 \geq 0, \qquad (-R_n^r + c_n[u_n]\beta^2)[u_n] = 0, \tag{6.508}$$

2. Coulombian law of friction with adhesion (because $\mathbf{R}_t = \mathbf{R}_t^r + \mathbf{R}_t^i$)

$$\mathbf{R}_t^r = c_t[\mathbf{u}_t]\beta^2, \qquad \|\mathbf{R}_t^i\| \leq \mathcal{F}_c^{kl}|R_n - c_n[u_n]\beta^2|$$

a. Stick

$$\|\mathbf{R}_t^i\| < \mathcal{F}_c^{kl}|R_n - c_n[u_n]\beta^2| \Rightarrow [\mathbf{u}_t'] = 0, \tag{6.509}$$

b. Slip

$$\|\mathbf{R}_t^i\| = \mathcal{F}_c^{kl}|R_n - c_n[u_n]\beta^2| \Rightarrow \text{there exists } \lambda \geq 0$$
$$\text{such that } [\mathbf{u}_t'] = -\lambda \mathbf{R}_t^i.$$

These conditions can be written in the equivalent form

$$\Phi \leq 0, \qquad \lambda \geq 0, \qquad \Phi\lambda = 0,$$
$$[\mathbf{u}_t'] + \lambda \mathbf{R}_t^i = 0, \tag{6.510a–d}$$

where

$$\Phi = \|\mathbf{R}_t^i\| - \mathcal{F}_c^{kl}|R_n - c_n[u_n]\beta^2|.$$

From (6.507) using (6.502) we eliminate the thermodynamic force G_β, so that we obtain

3. Conditions for evolution of adhesion intensity

$$\beta' = -\left\{-\frac{1}{b}[w - (c_n[u_n]^2 + c_t\|[\mathbf{u}_t]\|^2)\beta]^-\right\}^{1/p} \qquad \text{if } \beta \in [0, 1),$$
$$\beta' \leq -\left\{-\frac{1}{b}[w - (c_n[u_n]^2 + c_t\|[\mathbf{u}_t]\|^2)]^-\right\}^{1/p} \qquad \text{if } \beta = 1. \tag{6.511}$$

The contact variables are u_n, $\mathbf{u}_t$, β, R_n, and $\mathbf{R}_t$. The model is characterized by six parameters: parameters c_n, c_t, that is, the initial normal and tangential stiffness of the interface if the adhesion is complete; the coefficient of Coulombian friction $\mathcal{F}_c$; the viscosity of the adhesion evolution b; the limit of the decohesion energy w, and the power coefficient p. If $[u_n] \geq 0$, that is, under traction, then $R_n = c_n[u_n]\beta^2$, that is, an adhesive resistance is active (elasticity with damage). The intensity of adhesion starts to decrease if the displacement is sufficiently large such that the elastic energy becomes larger than the limit of the adhesion energy w. The evolution of the adhesion is then given by (6.511). If the adhesion is totally broken, the classical Signorini conditions are obtained. On the contact boundary the friction acts if a normal compression is applied. If a normal traction is applied, that is, if $[u_n] > 0$, then the sliding limit $\mathcal{F}_c^{kl}|R_n - c_n[u_n]\beta^2|$ is zero according to conditions (6.508), and the tangential behavior is elastic with the damage $\mathbf{R}_t = c_t[\mathbf{u}_t]\beta^2$. Under compression, because $[u_n] = 0$, the sliding limit is $\mathcal{F}_c^{kl}|R_n|$. If the norm of the tangential force $\|\mathbf{R}_t\|$ is smaller than the sliding limit, then $[\mathbf{u}_t] = 0$ with respect to the assumed initial condition and $[\mathbf{u}_t'] = 0$ with respect to (6.509), and therefore, sliding does not start. If the sliding limit is reached, then according to (6.509) an elastic tangential displacement occurs. The adhesion starts to decrease if the adhesive limit is reached, and evolution of the intensity of adhesion β is then given by (6.511). If the intensity of adhesion β tends to zero, then the Coulomb friction conditions are obtained. If the adhesion limit is over, that is, if the loading is constant, the adhesion starts to decrease by relaxation. If the tangential loading is now backward, then an opposite tangential displacement occurs only if the other side of the Coulomb cone is reached [see Raous et al. (1999)].

A. Modeling Quasi-Static Unilateral Contact with Adhesion, Friction, and Wear In case that the loading slowly depends on the time, the inertial term can be omitted, and then we have the quasi-static model problem. We have the following problem:

Problem $\mathcal{P}_{qs}$ Find the displacements $\mathbf{u}^\iota$, the stresses $\boldsymbol{\tau}^\iota$, $\iota = 1, \ldots, s$, the strains $\mathbf{e} = (e_{ij})$, the contact forces $\mathbf{R}$ such that

$$\frac{\partial \tau_{ij}^\iota}{\partial x_j} + f_i^\iota = 0 \quad \text{in } \Omega^\iota \times I,\ \iota = 1, \ldots, s, \tag{6.512}$$

$$\tau_{ij}^\iota = c_{ijkl}^\iota(\mathbf{x}) e_{kl}(\mathbf{u}^\iota), \qquad e_{ij}(\mathbf{u}^\iota) = \frac{1}{2}\left(\frac{\partial u_i^\iota}{\partial x_j} + \frac{\partial u_j^\iota}{\partial x_i}\right) \quad \text{in } \Omega^\iota \times I,\ \iota = 1, \ldots, s, \tag{6.513}$$

$$u_i^\iota = u_2^\iota \quad \text{on } \Gamma_u^\iota \times I, \tag{6.514}$$

$$\tau_{ij}^\iota n_j = P_i^\iota \quad \text{on } \Gamma_\tau^\iota \times I, \tag{6.515}$$

$$[u_n] \geq 0, \qquad -R_n^r + c_n[u_n]\beta^2 \geq 0, \qquad (-R_n^r + c_n[u_n]\beta^2)[u_n] = 0$$
$$\text{on } \cup \Gamma_c^{kl} \times I,$$

$$\mathbf{R}_t^r = c_t[\mathbf{u}_t]\beta^2, \qquad \|\mathbf{R}_t^i\| \leq \mathcal{F}_c^{kl}|R_n - c_n[u_n]\beta^2|.$$

(i) Stick

$$\|\mathbf{R}_t^i\| < \mathcal{F}_c^{kl}|R_n - c_n[u_n]\beta^2| \Rightarrow [\mathbf{u}_t'] = 0,$$

(ii) Slip

$$\|\mathbf{R}_t^i\| = \mathcal{F}_c^{kl}|R_n - c_n[u_n]\beta^2| \Rightarrow \text{there exists } \lambda \geq 0$$

$$\text{such that } [\mathbf{u}_t'] = -\lambda \mathbf{R}_t^i, \tag{6.516}$$

where $[\mathbf{u}] \equiv [\mathbf{u}]^{kl} = \mathbf{u}^l - \mathbf{u}^k = [u_n \cdot \mathbf{n}]^{kl} + [\mathbf{u}_t]^{kl}$, $\boldsymbol{\tau}^k \mathbf{n}^k = -\boldsymbol{\tau}^l \cdot \mathbf{n}^l = R_n^k \cdot \mathbf{n}^k + \mathbf{R}_t$ on $\cup \Gamma_c^{kl} \times I$,

$$\beta' = -\left\{-\frac{1}{b}\left[w - \left(c_n[u_n]^2 + c_t\|[\mathbf{u}_t]\|^2\right)\beta\right]^-\right\}^{1/p} \quad \text{if } \beta \in [0, 1),$$

$$\beta' \leq -\left\{-\frac{1}{b}\left[w - \left(c_n[u_n]^2 + c_t\|[\mathbf{u}_t]\|^2\right)\beta\right]^-\right\}^{1/p} \quad \text{if } \beta = 1. \tag{6.517}$$

$$\mathbf{u}^\iota(t_0, \mathbf{x}) = \mathbf{u}_0^\iota(\mathbf{x}), \qquad \beta^\iota(t_0, \mathbf{x}) = \beta_0^\iota(\mathbf{x}), \qquad \mathbf{x} \in \Omega^\iota, \quad \iota = 1, \ldots, s. \qquad (6.518)$$

Let

$$H = L^\infty(\Gamma_c),$$

$$V^\iota = \{\mathbf{v}^\iota \in H^{1,N}(\Omega^\iota); \mathbf{v}^\iota = 0 \text{ a.e. on } \Gamma_u^\iota\}, \qquad \iota = 1, \ldots, s, \ V = \prod_{\iota=1}^{s} V^\iota,$$

$$K = \{\mathbf{v} \in V; [v_n] \geq 0 \text{ a.e. on } \cup \Gamma_c^{kl}\}$$

with the norm $\|\mathbf{v}\|_V = \sum_\iota \|\mathbf{v}^\iota\|_{V^\iota}$ and the duality pairing $\langle \cdot, \cdot \rangle_{kl}$ on $H^{1/2}(\Gamma_c^{kl}) \times H^{-1/2}(\Gamma_c^{kl})$. Let us suppose that

$$\mathbf{f} \in W^{1,2}\left(I; \prod_{\iota=1}^{s} L^{2,N}(\Omega^\iota)\right), \quad \mathbf{P} \in W^{1,2}\left(I; \prod_{\iota=1}^{s} L^{2,N}(\Gamma_\tau^\iota)\right),$$

then

$$(\mathbf{F}, \mathbf{v}) = \sum_{\iota=1}^{s} \left[\int_{\Omega^\iota} \mathbf{f}^\iota \cdot \mathbf{v}^\iota \, d\mathbf{x} + \int_{\Gamma_\tau^\iota} \mathbf{P}^\iota \cdot \mathbf{v}^\iota \, ds\right] \quad \forall \mathbf{v} = (\mathbf{v}^\iota)_{\iota=1}^s \in V,$$

thus $\mathbf{F} \in W^{1,2}(I; V)$. Let us introduce the bilinear form

$$a(\mathbf{u}, \mathbf{v}) = \sum_{\iota=1}^{s} a^\iota(\mathbf{u}^\iota, \mathbf{v}^\iota) = \sum_{\iota=1}^{s} \int_{\Omega^\iota} c_{ijkl}^\iota e_{ij}(\mathbf{u}^\iota) e_{kl}(\mathbf{v}^\iota) \, d\mathbf{x}, \qquad \mathbf{v}^\iota = (v_i^\iota)_{\iota=1}^s \in V^\iota,$$

the functional $j : H \times V \times V \to \mathbb{R}$

$$j(\beta, \mathbf{u}, \mathbf{v}) = \int_{\cup\Gamma_c^{kl}} \mathcal{F}_c^{kl} |\tau_n^*(P\mathbf{u}^k) + c_n \beta^2 [u_n]| \|[\mathbf{v}_t]\| \, ds,$$

where $\mathcal{F}_c^{kl} \in L^\infty(\Gamma_c^{kl})$, $\mathcal{F}_c^{kl} \geq 0$, $(\,\cdot\,)^* : H^{-1/2}(\Gamma_c^{kl}) \to L^2(\Gamma_c^{kl})$ is a linear and compact mapping, P is the projection of $W^{1,2}(I; V^k)$ onto V_0^k, where

$$V_0^\iota = \left\{\mathbf{w}^\iota \in W^{1,2}(I; V^\iota); \int_{t_0}^{t_p} a^\iota(\mathbf{w}^\iota, \boldsymbol{\psi}^\iota) \, dt = \int_{t_0}^{t_p} (\mathbf{f}^\iota, \boldsymbol{\psi}^\iota)_0 \, dt \right.$$

$$\left. + \int_{t_0}^{t_p} (\mathbf{P}^\iota, \boldsymbol{\psi}^\iota)_0 \, dt \ \forall \boldsymbol{\psi}^\iota \in L^2(I; V^\iota), \ \boldsymbol{\psi}^\iota = 0 \text{ a.e. on } \cup \Gamma_c^{kl} \times I \right\},$$

and $C_n, C_t : H \times V \times V \to \mathbb{R}$, where

$$C_n(\beta, \mathbf{u}, \mathbf{v}) = \int_{\cup\Gamma_c^{kl}} c_n \beta^2 [u_n][v_n] \, ds,$$

$$C_t(\beta, \mathbf{u}, \mathbf{v}) = \int_{\cup\Gamma_c^{kl}} c_t \beta^2 [\mathbf{u}_t] \cdot [\mathbf{v}_t] \, ds,$$

and

$$y(\beta, \mathbf{u}) = -\frac{1}{b}[w - \left(c_n[u_n]^2 + c_t \|[\mathbf{u}_t]\|^2\right)\beta]^-.$$

The variational formulation of Problem $\mathcal{P}_{qs}$ is the following:

Problem $(\mathcal{P}_{qs})_v$ Find $(\mathbf{u}, \beta) \in W^{1,2}(I; V) \times W^{1,2}(I; H)$ such that $\mathbf{u}(t_0) = \mathbf{u}_0$, $\beta(t_0) = \beta_0$ and for almost all $t \in \overline{I}$, $\mathbf{u}(t) \in K$ and

$$\begin{aligned}
&a(\mathbf{u}, \mathbf{v} - \mathbf{u}') + j(\beta, \mathbf{u}, \mathbf{v}) - j(\beta, \mathbf{u}, \mathbf{u}') + C_t(\beta, \mathbf{u}, \mathbf{v} - \mathbf{u}') \\
&\quad \geq (\mathbf{F}, \mathbf{v} - \mathbf{u}') + \left\langle \tau_n(\mathbf{u}^k), [v_n] - [u_n'] \right\rangle_{\cup\Gamma_c^{kl}} \quad \forall \mathbf{v} \in V, \\
&\langle \tau_n(\mathbf{u}^k), [z_n] - [u_n] \rangle_{\cup\Gamma_c^{kl}} + C_n(\beta, \mathbf{u}, \mathbf{z} - \mathbf{u}) \geq 0 \quad \forall \mathbf{z} \in K, \\
&\beta' = y(\beta, \mathbf{u}) \quad \text{a.e. on} \cup \Gamma_c^{kl},
\end{aligned} \tag{6.519}$$

where the initial conditions $\mathbf{u}_0 \in K$, $\beta_0 \in H$, $\beta_0 \in [0, 1)$ on $\cup\Gamma_c^{kl}$, satisfy the following compatibility condition:

$$\begin{aligned}
&a(\mathbf{u}_0, \mathbf{w} - \mathbf{u}_0) + j(\beta_0, \mathbf{u}_0, \mathbf{w} - \mathbf{u}_0) + C_t(\beta_0, \mathbf{u}_0, \mathbf{w} - \mathbf{u}_0 t) \\
&\quad \geq (\mathbf{F}(t_0), \mathbf{w} - \mathbf{u}_0) \quad \forall \mathbf{w} \in K.
\end{aligned} \tag{6.520}$$

The bilinear form $a(\cdot, \cdot)$ is continuous on $V \times V$ and coercive, that is,

$$\begin{aligned}
&|a(\mathbf{u}, \mathbf{v})| \leq c_0 \|\mathbf{u}\|_V \|\mathbf{v}\|_V, \qquad c_0 = \text{const} > 0, \quad \forall \mathbf{u}, \mathbf{v} \in V, \\
&a(\mathbf{u}, \mathbf{u}) \geq c_1 \|\mathbf{u}\|_V^2, \qquad c_1 = \text{const} > 0, \quad \forall \mathbf{u} \in V
\end{aligned} \tag{6.521}$$

and the mapping $j(\cdot, \cdot, \cdot)$ satisfies

$$\begin{aligned}
&|j(\beta, \mathbf{u}, \mathbf{v}) - j(\beta, \mathbf{u}, \overline{\mathbf{v}}) - j(\beta, \overline{\mathbf{u}}, \mathbf{v}) + j(\beta, \overline{\mathbf{u}}, \overline{\mathbf{v}})| \\
&\leq \overline{\mathcal{F}}_c c_2 \|\mathbf{u} - \overline{\mathbf{u}}\|_V \|\mathbf{v} - \overline{\mathbf{v}}\|_V \quad \forall \mathbf{u}, \overline{\mathbf{u}}, \mathbf{v}, \overline{\mathbf{v}} \in V,
\end{aligned} \tag{6.522}$$

where $c_2 = \text{const} > 0$, $\overline{\mathcal{F}}_c = \mathcal{F}_c^{kl}|_{L^\infty(\cup\Gamma_c^{kl})}$.

To prove the existence of the solution of Problem $(\mathcal{P}_{qs})_v$ a time discretization and a fixed point method will be used. Let us set $\Delta t = t_p/n$, $t_i = i\Delta t$, $i = 1, \ldots, n$. Using an implicit scheme, we obtain the sequence of the following problems, that is, Problem $(\mathcal{P}_{qs})_v^i$, $i = 0, \ldots, n-1$:

Problem $(\mathcal{P}_{qs})_v^i$ Let $(\mathbf{u}^0, \beta^0) \in K \times H$. Find $(\mathbf{u}^{i+1}, \beta^{i+1}) \in K \times H$ such that

$$\begin{aligned}
&a(\mathbf{u}^{i+1}, \mathbf{v} - \mathbf{u}^{i+1}) + j(\beta^{i+1}, \mathbf{u}^{i+1}, \mathbf{v} - \mathbf{u}^i) - j(\beta^{i+1}, \mathbf{u}^{i+1}, \mathbf{u}^{i+1} - \mathbf{u}^i) \\
&\quad + C(\beta^{i+1}, \mathbf{u}^{i+1}, \mathbf{v} - \mathbf{u}^{i+1}) \geq (\mathbf{F}^{i+1}, \mathbf{v} - \mathbf{u}^{i+1}) \quad \forall \mathbf{v} \in K, \\
&\beta^{i+1} = \beta^i + \Delta t y(\beta^{i+1}, \mathbf{u}^{i+1}) \quad \text{a.e. on} \cup \Gamma_c^{kl},
\end{aligned} \tag{6.523}$$

where $C(\cdot, \cdot, \cdot) = C_n(\cdot, \cdot, \cdot) + C_t(\cdot, \cdot, \cdot)$.

The problem $(\mathcal{P}_{qs})_v^i$ will be solved by using a fixed point method as follows: For every $\overline{\mathbf{u}} \in K$ we denote $s(\overline{\mathbf{u}}) = \beta$ the solution of

$$\beta = \beta^i + \Delta t \cdot y(\beta, \overline{\mathbf{u}}). \tag{6.524}$$

For every $\beta \in H$ let $\mathbf{u}(\beta)$ be the solution of

$$\begin{aligned} \mathbf{u} \in K,\ a(\mathbf{u}, \mathbf{v} - \mathbf{u}) + j(\beta, \mathbf{u}, \mathbf{v} - \mathbf{u}^i) - j(\beta, \mathbf{u}, \mathbf{u} - \mathbf{u}^i) \\ + C(\beta, \mathbf{u}, \mathbf{v} - \mathbf{u}) \geq (\mathbf{F}^{i+1}, \mathbf{v} - \mathbf{u}) \quad \forall \mathbf{v} \in K. \end{aligned} \tag{6.525}$$

The solutions of both (6.524) and (6.525) exist (Cocu, 1984; Cocu et al., 1995), and the existence and uniqueness of (6.525) exist if the friction coefficient $\overline{\mathcal{F}}_c$ is sufficiently small.

Let $T : K \to K$ be the mapping defined by

$$\forall \overline{\mathbf{u}} \in K \quad T(\overline{\mathbf{u}}) = \mathbf{u}(s(\overline{\mathbf{u}})).$$

Then, there exists $k_1 > 0$ such that for all $\overline{\mathbf{u}}_1, \overline{\mathbf{u}}_2 \in K$

$$|s(\overline{\mathbf{u}}_1) - s(\overline{\mathbf{u}}_2)| \leq k_1 \Delta t(\|\overline{\mathbf{u}}_1\| + \|\overline{\mathbf{u}}_2\|)|[\overline{\mathbf{u}}_1 - \overline{\mathbf{u}}_2]| \quad \text{a.e. on } \cup \Gamma_c^{kl}. \tag{6.526}$$

Setting $\mathbf{u}_1 = \mathbf{u}(s(\overline{\mathbf{u}}_1))$, $\mathbf{u}_2 = \mathbf{u}(s(\overline{\mathbf{u}}_2))$, and putting $\mathbf{u} = \mathbf{u}_1$, $\mathbf{v} = \mathbf{u}_2$ and $\mathbf{u} = \mathbf{u}_1$, $\mathbf{v} = \mathbf{u}_1$ into (6.525) and adding, we obtain

$$\|\mathbf{u}_1 - \mathbf{u}_2\|^2 \leq k_0 \int_{\cup \Gamma_c^{kl}} |\mathbf{u}_2|\, |\mathbf{u}_1 - \mathbf{u}_2|\, |s(\overline{\mathbf{u}}_1) - s(\overline{\mathbf{u}}_2)|\, ds \quad \forall \overline{\mathbf{u}}_1, \overline{\mathbf{u}}_2 \in K, \tag{6.527}$$

where $k_0 > 0$, and where the properties of $a(\cdot, \cdot)$, $j(\cdot, \cdot, \cdot)$ were used. Then (6.526) and (6.527) and $\|\mathbf{u}\| \leq k_2$, where k_2 is independent of $\overline{\mathbf{u}}$, admit that T is a contraction mapping for sufficiently small Δt and that T has a unique fixed point $\mathbf{u}$. Then $(\mathbf{u}, s(\mathbf{u}))$ is the solution of Problem $(\mathcal{P}_{qs})_v^i$ [see Cocu et al. 1995, 1999) and Raous et al. (1999)]. Similar problems are also discussed in Sofonea et al. (2006), Han and Sofonea (2002), and Shillor et al. (2004).

Numerical Solution To solve Problem $\mathcal{P}_{qs}$ numerically the incremental formulation and the finite element approximation will be used. Let V_h be a finite element approximation of the space V and $K_h = \{\mathbf{v}_h \in V_h | [v_{hn}] \geq 0 \text{ on } \cup \Gamma_c^{kl}\}$ be a set of discrete admissible displacements, Π_h denotes the projection in the FE discretization.

Then we have the following problem:

Problem $(\mathcal{P}_{qs})_h^i$ For each time step t^{i+1} find $(\overline{\beta}_h, \mathbf{u}_h(\beta_h))$, where $\overline{\beta}_h$ is a fixed point of the application $s(\cdot)$

$$\beta_h \to s(\mathbf{u}_h(\beta_h)) = \beta_h^i + \Delta t y(\beta_h, \mathbf{u}_h(\beta_h)), \tag{6.528}$$

and where $\mathbf{u}_h(\beta_h) \in K_h$ is the solution of

$$a(\mathbf{u}_h, \mathbf{v}_h - \mathbf{u}_h) + \Pi_h j_h(\beta_h, \mathbf{u}_h, \mathbf{v}_h - \mathbf{u}_h^i) - \Pi_h j_h(\beta_h, \mathbf{u}_h, \mathbf{u}_h - \mathbf{u}_h^i)$$
$$+ C(\beta_h, \mathbf{u}_h, \mathbf{v}_h - \mathbf{u}_h) \geq (\mathbf{F}_h^{i+1}, \mathbf{v}_h - \mathbf{u}_h) \quad \forall \mathbf{v}_h \in K_h, \tag{6.529}$$

where β_h^i, $\mathbf{u}_h^i$ are the discrete solutions in $t = t_i$ and

$$j_h(\beta_h, \mathbf{u}_h, \mathbf{v}_h - \mathbf{u}_h^i) = \int_{\cup \Gamma_c^{kl}} \mathcal{F}_c^{kl} |R_h(\mathbf{u}_h) - c_n \beta_h^2 [u_{hn}]| \cdot \|[\mathbf{v}_{ht}] - [\mathbf{u}_{ht}^i]\| \, ds,$$

$$C(\beta_h, \mathbf{u}_h, \mathbf{v}_h - \mathbf{u}_h) = \int_{\cup \Gamma_c^{kl}} c_n \beta_h^2 [u_{hn}]([v_{hn}] - [u_{hn}]) \, ds$$
$$+ \int_{\cup \Gamma_c^{kl}} c_t \beta_h^2 [\mathbf{u}_{ht}]([\mathbf{v}_{ht}] - [\mathbf{u}_{ht}]) \, ds,$$

$$y(\beta_h, \mathbf{u}_h(\beta_h)) = -\frac{1}{b}\left[w - (c_n [u_{hn}]^2 + c_t \|[\mathbf{u}_{ht}]\|^2)\beta_h\right]^- . \tag{6.530}$$

To solve (6.529) we introduce a fixed point method on the sliding limit g_c^{kl}. Then Problem $(\mathcal{P}_{qs})_h^i$ will be equivalent to the following problem:

Problem $(\mathcal{P}_{qs})_{gh}^i$ For each time step t^{i+1}, find $(\overline{g}_c, \mathbf{u}_h(g_c))$, where $\overline{g}_c$ is a fixed point of the application $S_h^i(\cdot)$

$$g_c \to S_h^i(\mathbf{u}_h(g_c)) = \mathcal{F}_c^{kl} |R_n(\mathbf{u}_h(g_c)) - c_n \beta_h^2 [u_{hn}](g_c)|, \tag{6.531}$$

where $\mathbf{u}_h(g_c) \in K_h$ is the solution of

$$a(\mathbf{u}_h(g_c), \mathbf{v}_h - \mathbf{u}_h) + \Pi_h j_g^*(\mathbf{v}_h - \mathbf{u}_h^i) - \Pi_h j_g^*(\mathbf{u}_h(g_c) - \mathbf{u}_h^i)$$
$$+ C(\beta_h, \mathbf{u}_h(g_c), \mathbf{v}_h - \mathbf{u}_h(g_c)) \geq (\mathbf{F}_h^{i+1}, \mathbf{v}_h - \mathbf{u}_h(g_c)) \quad \forall \mathbf{v}_h \in K_h, \tag{6.532}$$

where $j_g^*(\mathbf{v}) = \int_{\cup \Gamma_c^{kl}} g_c^{kl} \|[\mathbf{v}_t]\| \, ds$, $\mathbf{u}_h^i$ is the solution at $t = t_i$.

We see that (6.532) represents a classical Tresca model with a given sliding limit g_c^{kl}. Therefore, it is evident that for the solution all methods known from the theory of contact problems of the Tresca type (Hlaváček et al., 1988; Haslinger et al., 1996; Nedoma and Stehlík, 1995; Nedoma et al., 1999b) can be used. Algorithms used for the Tresca contact problem are robust, and the convergence needs a reasonable number of iterations. The convergence of the fixed point of the sliding limit g_c^{kl} is very fast and it needs only a small number of iterations.

B. Modeling Dynamic Unilateral Contact with Adhesion, Friction, and Wear

When the loading depends on time, the inertial term cannot be omitted. Then we have the following problem:

Problem $(\mathcal{P}_d)$ Find the displacement $\mathbf{u}^\iota$, the stresses $\boldsymbol{\tau}^\iota$, $\iota = 1, \ldots, s$, the strain $\mathbf{e}$, the contact force $\mathbf{R}$ satisfying (6.490) and (6.513) and (6.518).

Let

$$H = L^{\infty}(\Gamma_{\varepsilon}),$$

$$V^{\iota} = \{\mathbf{v}^{\iota} \in H^{1,N}(\Omega^{\iota}); \mathbf{v}^{\iota} = 0 \text{ a.e. on } \Gamma_u\}, \qquad \iota = 1, \ldots, s; \qquad V = \prod_{\iota=1}^{s} V^{\iota},$$

$$K = \{\mathbf{v}^{\iota} \in V; [v_n] \geq 0 \text{ a.e. on } \cup \Gamma_c^{kl}\}; \qquad \mathcal{V} = L^2(I; V); \qquad \mathcal{K} = L^2(I; K).$$

Let the bilinear form $a(\cdot, \cdot)$, the functional $(\mathbf{F}, \mathbf{v})$, the functionals $j(\cdot, \cdot, \cdot)$, $C_n(\cdot, \cdot, \cdot)$, $C_t(\cdot, \cdot, \cdot)$, and $y(\cdot, \cdot)$ be defined similarly as in the quasi-static case.

Then the variational formulation of the Problem $\mathcal{P}_d$ is the following:

Problem $(\mathcal{P}_d)_v$ Find $(\mathbf{u}, \beta) \in W^{1,2}(I; V) \times W^{1,2}(I; H)$ such that $\mathbf{u}(t_0) = \mathbf{u}_0$, $\mathbf{u}'(t_0) = \mathbf{u}_1$, $\beta(t_0) = \beta_0$ and for almost all $t \in \bar{I}$, $\mathbf{u}(t) \in K$:

$$\begin{aligned} &(\mathbf{u}''(t), \mathbf{v} - \mathbf{u}'(t)) + a(\mathbf{u}(t), \mathbf{v} - \mathbf{u}'(t)) + j(\beta, \mathbf{u}, \mathbf{v}) - j(\beta, \mathbf{u}(t), \mathbf{u}'(t)) \\ &\quad + C_t(\beta, \mathbf{u}(t), \mathbf{v} - \mathbf{u}'(t)) \geq (\mathbf{F}(t), \mathbf{v} - \mathbf{u}'(t)) \\ &\quad + \langle \tau_n(\mathbf{u}^k(t)), [v_n] - [u_n'(t)] \rangle_{\cup \Gamma_c^{kl}} \quad \forall \mathbf{v} \in V, \\ &\langle \tau_n(\mathbf{u}^k), [z_n] - [u_n] \rangle_{\cup \Gamma_c^{kl}} + C_n(\beta, \mathbf{u}, \mathbf{z} - \mathbf{u}) \geq 0 \quad \forall \mathbf{z} \in K, \\ &\beta' = y(\beta, \mathbf{u}) \quad \text{a.e. on } \cup \Gamma_c^{kl}, \end{aligned} \tag{6.533a,b,c}$$

where $\mathbf{u}_0 \in K$, $\mathbf{u}_1 \in K$, $\beta \in H = L^{\infty}(\cup \Gamma_c^{kl})$, $\beta_0 \in [0, 1]$ on $\cup \Gamma_c^{kl}$ and satisfies (6.520).

Numerically, the problem can be solved by the explicit, semi-implicit, or Newark schemes in time and the standard finite element method based on other on the Galerkin method in space [see Zienkiewicz and Taylor (1989)]. Then the algorithms are parallel to that of Section 6.6. A simple method is based on the explicit scheme in time, the finite element method in space, and the penalty approach to enforce the constraints for normal and tangential constraints, which simplifies the variational formulation of the problem. Thus we will change the conditions for the normal contact with adhesion (6.508), that is, the conditions

$$[u_n] \geq 0, \qquad -R_n^r + c_n[u_n]\beta^2 \geq 0, \qquad [u_n](-R_n^r + c_n[u_n]\beta^2) = 0, \tag{6.534a,b,c}$$

or in the equivalent form

$$\begin{aligned} &\mathbf{R}_t^r = c_t[\mathbf{u}_t]\beta^2, \\ &\Phi \leq 0, \qquad \lambda \geq 0, \qquad \Phi\lambda = 0, \\ &[\mathbf{u}_t'] + \lambda \mathbf{R}_t^i = 0, \end{aligned} \tag{6.535(a–e))}$$

where

$$\Phi = \|\mathbf{R}_t^i\| - \mathcal{F}_c^{kl} |R_n - c_n[u_n]\beta^2|, \tag{6.536}$$

where the evolution law for the adhesion intensity β is given by (6.511), by the penalized condition [see Rojek et al. (2001)].

Introducing the penalty coefficient c_n^-, then the nonpenetration condition is enforced by

$$R_n^- = c_n^- [u_n]^-, \tag{6.537}$$

where $(y)^-$ denotes the negative part of y (zero normal traction is taken into account here). Since $R_n = R_n^- + R_n^+$, then $R_n^+ = c_n^+ [u_n]^+ \beta^2$, where $(y)^+$ denotes the nonnegative parts of y. The relation (6.537) applies to compression only, the relations for adhesive traction $R_n = c_n [u_n] \beta^2$ if $[u_n] > 0$ remains unchanged.

Then (6.536) can be rewritten as

$$\Phi = \|\mathbf{R}_t^i\| - \mathcal{F}_c^{kl} |R_n^-|. \tag{6.538}$$

Thus (6.535c) will be of the following form

$$[\mathbf{u}_t'] + \lambda \mathbf{R}_t^i = \frac{1}{\varepsilon_t} (\mathbf{R}_t^i)', \varepsilon_t \to 0, \tag{6.539}$$

where $(\mathbf{R}_t^i)'$ is the time derivative of $\mathbf{R}_t^i$. Then the variational inequality (6.533a) changes to the variational equation.

The algorithm will be based on the standard Galerkin approximation. Let us introduce the space discretization of displacements by

$$\mathbf{u}(\mathbf{x}, t) = \mathbf{H}(\mathbf{x})\mathbf{U}(t), \tag{6.540}$$

where $\mathbf{H}$ is the matrix of interpolation functions. By the standard way we obtain the system of semidiscrete equations of motion

$$M\mathbf{U}''(t) = \mathbf{F}_{\text{ext}}(t) + \mathbf{F}_c(t) - \mathbf{F}_{\text{int}}(t), \tag{6.541}$$

where M is the mass matrix, $\mathbf{F}_{\text{ext}}$, $\mathbf{F}_c$, and $\mathbf{F}_{\text{int}}$ are the vectors of external, contact and internal forces, respectively, and $\mathbf{U}$, $\mathbf{U}''$ are the vectors of nodal displacements and accelerations, respectively, where

$$\begin{aligned} M &= \int_\Omega \rho \mathbf{H}^T \mathbf{H}\, d\mathbf{x}, \\ \mathbf{F}_{\text{ext}} &= \int_\Omega \rho \mathbf{H}^T \mathbf{f}\, d\mathbf{x} + \int_{\Gamma_\tau} \mathbf{H}^T \cdot \mathbf{P}\, d\mathbf{x}, \\ \mathbf{F}_{\text{int}} &= \int_\Omega \mathbf{B}^T \boldsymbol{\tau}\, d\mathbf{x} = \int_\Omega \mathbf{B}^T C \mathbf{B}\, d\mathbf{x}, \end{aligned} \tag{6.542a,b,c}$$

where $\boldsymbol{\tau}$ is the Cauchy stress tensor, C is the elasticity matrix, $\mathbf{B}$ is the strain–displacement operator matrix, $\mathbf{P}$ is the given traction vector, $\mathbf{f}$ is the vector of given body forces. The term for the nodal contact force $\mathbf{F}_c$ will be discussed later in connection with evaluation of contact forces. The regularized contact conditions constitute the basis of the central difference algorithm for the calculation of contact interaction forces.

The equation of motion (6.541) will be solved by using the explicit finite element method and the PDAS algorithm, in which the displacements $\mathbf{U}^{i+1}$ at the time $t = t_{i+1}$ are obtained from the known values at the previous time $t = t_i$ and $t = t_{i-1}$.

Let $m > 0$ be an integer and $t_0 = 0$, then $\Delta t = t_p/m$, $t_i = i\,\Delta t$, $i = 0, 1, \ldots, m$. Then

$$M\frac{\mathbf{U}^{i+1} - 2\mathbf{U}^{i} + \mathbf{U}^{i-1}}{\Delta t^2} = \mathbf{F}_{\mathrm{ext}}(t_i) + \mathbf{F}_c(t_i) - \mathbf{F}_{\mathrm{int}}(t_i)\,.$$

Hence

$$\mathbf{U}^{i+1} = \Delta t^2 M^{-1}(\mathbf{F}_{\mathrm{ext}}(t_i) + \mathbf{F}_c(t_i) - \mathbf{F}_{\mathrm{int}}(t_i)) + 2\mathbf{U}^{i} - \mathbf{U}^{i-1}. \tag{6.543}$$

Such an algorithm, similarly as in Section 6.6.2, is stable and consistent and therefore convergent. The effectiveness of the dynamic formulation is based on the use of a diagonal mass matrix $M_{\mathrm{diag}} = \mathrm{diag}\, M$, because then we do not need to solve a system of equations. The main problem of the algorithm, similarly as in the previously discussed algorithms, consists of two basic tasks: (i) determination of the contact area and (ii) evaluation of contact forces. Similarly as in Section 6.6.4 the PDAS algorithm and the mortar finite elements can be used with an advantage. In the PDAS algorithm two sides for the contact boundary Γ_c^{kl} are assumed; one is the slave side (i.e., from Ω^k) and the second one is the master side (i.e., from Ω^l). The algorithm is similar to that of Section 6.6.2 and of Rojek et al. (2001). In the contact problem without any adhesion, the contact is established for the slave nodes penetrating through the master surface, that is, $[u_n] < 0$. In the contact algorithm with the adhesion the contact search has been modified to include the possibility of an adhesive contact, that is, $[u_n] > 0$.

At the time t_i we establish the contact between a certain slave node S and a master segment, the relative displacements $[u_n(t_i)]$, $[\mathbf{u}_t(t_i)]$ and the tangential relative velocity $[\mathbf{u}_t'(t_i)]$ are then obtained. We calculate the contact interactions $R_n^-(t_i)$ and $\mathbf{R}_t^-(t_i)$ and the intensity of adhesion $\beta^{(i)}$ by using $[u_n(t_i)]$, $[\mathbf{u}_t(t_i)]$, $[\mathbf{u}_t'(t_i)]$, interface constitutive parameters c_n^-, c_n^+, c_t, w, $\mathcal{F}_c^{kl}$, b and p, where b and p characterize viscous properties of adhesive bonds [see in (6.503)]. The values of $\mathbf{R}_t^i(t_i)$, the irreversible tangential force, and the intensity of adhesion $\beta(t_i)$ are calculated from their previous values at the time t_{i-1}.

The contact force in the node S is calculated by integrating contact traction $R_n(t_i)$ and $\mathbf{R}_t(t_i)$ over the area (segment) Γ_S (see Fig. 6.6) in the neighborhood of node S, that is,

$$(\mathbf{F}_c(t_i)) = (R_n(t_i)\mathbf{n}^{kl} + \mathbf{R}_t(t_i))|\Gamma_S|, \tag{6.544}$$

where $|\Gamma_S| =$ meas Γ_S, that is, the area of the master segment.

The corresponding value on the master side is

$$(\mathbf{F}_c(t_i))_l = -N_l^{kl}(\mathbf{F}_c(t_i))_S, \qquad l = 1, \ldots, 4, \tag{6.545}$$

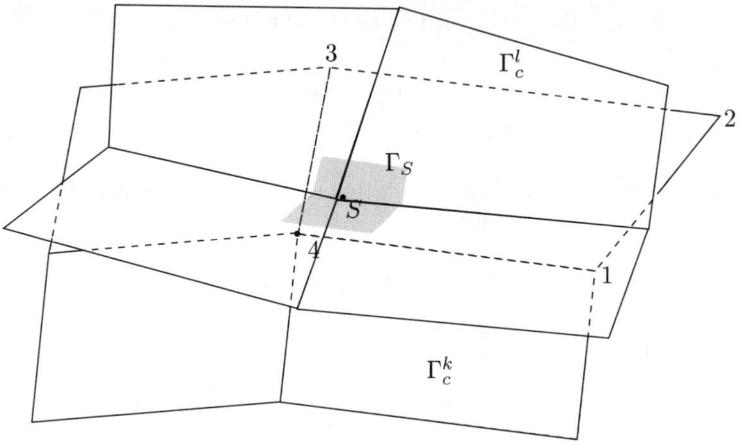

FIG. 6.6 Discretized contact surfaces Γ_c^k and Γ_c^l.

where N_l^{kl} represents the shape function corresponding to the mth node at the projection of the slave nodes S on the master segment, characterized by l nodes (in Fig. 6.6 $l=1,\ldots,4$).

The algorithm is as follows:

1. We modify the algorithm of Section 6.6.2 and execute initial calculations.
2. At each time step we calculate the effective loads at $t=t_i$:

   a. We determine the evolution of the intensity of adhesion: if adhesion bonds exist, that is, if $\beta(t_{i-1})>0$, we calculate a trial value of $\beta'(t_i)$ from (6.511a), that is,

$$\beta'_{tr}(t_i)=-\left\{-\frac{1}{b}\left[w-\left(c_n^+([u_n(t_i)]^+)^2+c_t\|[\mathbf{u}_t(t_i)]\|^2\right)\beta(t_{i-1})\right]^-\right\}^{1/p}; \tag{6.546}$$

   if $\beta(t_{i-1})<0$, that is, if a decohesion process exists, then

$$\beta(t_i)=\beta(t_{i-1})+\beta'_{tr}(t_i)\Delta t,$$

otherwise

$$\beta(t_i)=\beta(t_{i-1}). \tag{6.547}$$

   b. We determine a normal contact force: If $[u_n(t_i)]\le 0$ the compressive traction is according to (6.537):

$$R_n(t_i)=c_n^-[u_n(t_i)],$$

otherwise the tensile traction due to adhesion

$$R_n(t_i) = c_n^+[u_n(t_i)]\beta^2(t_i) \quad \text{if } \beta(t_i) > 0,$$
$$R_n(t_i) = 0 \quad \text{if } \beta(t_i) = 0. \tag{6.548}$$

c. We determine an adhesive tangential contact force: An adhesive tangential contact traction is due to (6.535):

$$\mathbf{R}_t^r(t_i) = c_t[\mathbf{u}_t(t_i)]\beta^2(t_i) \quad \text{if } \beta(t_i) > 0$$

or if adhesion is absent

$$\mathbf{R}_t^r(t_i) = \mathbf{0} \quad \text{if } \beta(t_i) = 0. \tag{6.549}$$

d. We determine a tangential contact force: If $R_n > 0$, that is, for tensile normal contact traction, the friction force is set to zero, that is,

$$\mathbf{R}_t^i(t_i) = \mathbf{0}. \tag{6.550}$$

If the normal contact traction is compressive, the friction contact traction is calculated as follows:

i. We calculate a trial state due to (6.539), with $\lambda = 0$, that is,

$$(\mathbf{R}_t^i(t_i))_{\text{tr}} = \mathbf{R}_t^i(t_{i-1}) + \varepsilon_t[\mathbf{u}_t'(t_i)]\ \Delta t. \tag{6.551}$$

ii. We check the slip condition (6.538), that is,

$$\Phi_{\text{tr}} = \|(\mathbf{R}_t^i(t_i))_{\text{tr}}\| - \mathcal{F}_c^{kl}|R_n(t_i)|. \tag{6.552}$$

iii. If $\Phi_{\text{tr}} \leq 0$, that is, the case of the stick contact, the frictional traction is

$$\mathbf{R}_t^i(t_i) = (\mathbf{R}_t^i(t_i))_{\text{tr}},$$

otherwise, that is, for the slip contact,

$$\mathbf{R}_t^i(t_i) = \mathcal{F}_c^{kl}|R_n(t_i)|\frac{(\mathbf{R}_t^i(t_i))_{\text{tr}}}{\|(\mathbf{R}_t^i(t_i))_{\text{tr}}\|}. \tag{6.553}$$

3. Determine displacements $\mathbf{U}_h$ at the time $t = t_{i+1}$.

This algorithm takes into account all possible loading, unloading, and reloading conditions. The intensity of adhesion β in the present formulation tends to zero, but it never will be zero (because it is asymptotic). The initial value of β depends on the previous solicitations applied to the interface. Usually, the total adhesion is considered (i.e., $\beta = 1$) as an initial value.

The parameters c_n, c_t, and w are evoked by the stiffness of the materials (the Young modul E) and the maximum deformation before collapse to identify the contact stiffness c_n, c_t. The thermodynamic work of adhesion w (or the adhesive energy or Dupré energy) is a given constitutive parameter of the model, but its value is difficult to choose, and we can only find an order of magnitude of its value. We usually adjust this value during the identification of the various parameters on a preliminary experiment. The set of parameters is then used on other experiments in order to validate the model (Raous and Monerie, 2002; Raous, 2009).

Under traction ($[u_n(t_i)] \geq 0$), an adhesive resistance [$R_n(t_i) = c_n^+[u_n(t_i)]\beta^2$] is active (elasticity with damage). The intensity of adhesion starts to decrease when the displacement is sufficiently large such that the elastic energy becomes larger than the limit of the adhesion energy w. Evolution of w is given by (6.517). If the adhesion is totally broken, we obtain the classical unilateral contact problem. The friction increases if the adhesion decreases. Instead of $\mathcal{F}_c^{kl}$ we put $(1-\beta)\mathcal{F}_c^{kl}$. If the adhesion is total, that is, $\beta = 1$, then friction is not active. If the adhesion decreases, the friction starts to increase, and, if $\beta \to 0$ (i.e., without adhesion), then friction is total and the classical Coulomb friction behavior is obtained (the limit is asymptotic) with the coefficient $\mathcal{F}_c^{kl}$. The interface energy is dissipated by viscosity and friction or adhesion (Raous et al., 1999; Raous and Monerie, 2002; Raous, 2009). For more details see Raous et al. (1999), Raous and Monerie (2002), and Maugis (1990).

6.10.2 Wear Models

To analyze wear mechanisms of artificial human joints, the Archard adhesive wear model, which predicts the wear rate to be proportional to the frictional dissipative rate, will be useful for our consideration. Let us assume that wear particles have similar shapes of the diameter r (representing single real contact areas) and that the volume of the wear particle is approximately r^3. Moreover, let the volumetric wear rate V_{wear} defined as the volume worn per unit sliding length be proportional to the real contact area A. The real contact area is proportional to the ratio of the normal load F_n and the hardness p of the softer body surface, which will be worn away. Then the volume of material lost by the wear process, the wear rate per unit sliding length, is given by the Holm–Archard law in the form (see Chapter 4)

$$V_{\text{wear}} = K_{\text{ad}} \frac{F_n g_t}{p}, \tag{6.554}$$

where K_{ad} is the wear coefficient depending upon the materials in contact, and, therefore, it is in relation to the friction coefficients; F_n is the normal loading, g_t is the relative sliding distance between materials, and p is the hardness of the surface, which will be worn away.

In continuum mechanics it is more convenient to use rates with respect to time rather than to the sliding length and to use local variables rather than the global variables. Therefore, we will use the contact pressure rather than the normal force F_n. Thus the Holm–Archard wear law (6.554) can be rewritten in a local form as

$$w' = k_{\text{ad}} \frac{f_n v}{p}, \tag{6.555}$$

where w' represents the depth of wear per unit time, and v is the relative sliding velocity. Since the range of pressures for which the real area of contact is proportional to the normal pressure p_n, then similar proportionality exists between the friction stress p_t and the normal pressure p_n. Hence the Holm–Archard law can be rewritten as

$$w' = \frac{\overline{k}_{\text{ad}}}{p} d', \tag{6.556}$$

where $d' = p_t v$ is the friction dissipation rate and $\overline{k}_{\text{ad}} = k_{\text{ad}} \mathcal{F}_c^{kl}$ is the modified wear coefficient, and $\mathcal{F}_c^{kl}$ is the coefficient of friction.

Thus following Strömberg et al. (1996), we can enhance conditions (6.509) and (6.510c) setting

$$\Phi = \|\mathbb{R}_t^i\| - \mathcal{F}_c^{kl}|R_n - c_n[u_n]\beta^2| + k_{\text{ad}}|R_n - \frac{c_n[u_n]\beta^2|^2}{3p}.$$

Obviously, conditions (6.516) have to be enhanced then, as well.

Remark 6.33 The wear coefficient K_{ad} can be predicted, based on the delamination theory of wear; see Suh (1973) and Lim and Ashby (1987).

Remark 6.34 In a more precise analysis in orthopedic biomechanics of artificial human joints the models with friction, adhesion, and wear must be taken into account, and because wear debris is produced between different components of artificial prostheses and the bone–implant interface, the transport of wear debris in the joint replacements must be also taken into account. Such a problem is described by the transport equation

$$\frac{\partial c(\mathbf{x}, t)}{\partial t} = \nabla \cdot [B(\mathbf{x})\nabla c(\mathbf{x}, t) - c(\mathbf{x}, t)\mathbf{v}(\mathbf{x}, t)] + R(c(\mathbf{x}, t)) = 0, \tag{6.557}$$

together with the boundary and initial conditions, where $c(\mathbf{x}, t)$ represents the concentration of wear debris on the interface bone–implant, B is the matrix of diffusion coefficients, $\mathbf{v}$ is the small velocity of wear debris, R is the reaction term, and t is the time.

PART III

BIOMECHANICAL ANALYSES OF PARTICULAR PARTS OF THE HUMAN SKELETON, JOINTS, AND THEIR REPLACEMENTS

CHAPTER 7

BIOMECHANICAL MODELS BASED ON CONTACT PROBLEMS AND BIOMECHANICAL ANALYSES OF SOME HUMAN JOINTS, THEIR TOTAL REPLACEMENTS, AND SOME OTHER PARTS OF THE HUMAN SKELETON

7.1 INTRODUCTION TO THE BIOMECHANICS OF STATICALLY LOADED AND OF MOVING LOADED HUMAN BODY

Kinematics of joints of the lower limbs are studied by gait analysis because gait has a great influence on the loading of human joints or of their artificial replacements. Gait is a determining factor affecting the character of joint loading of the lower limbs. Gait is a functional task requiring complex interaction and coordination among most of the major points of the body. Gait is a forward motion in the direction perpendicular to the frontal plane. Therefore, the acting forces in the plane in the direction of motion—the sagittal plane—are analyzed. The motion of the body when walking is assumed to be uniform, which is evoked and retained by force impulses of elastic elements—the muscles. During walking, motion about the hip joint acts in the sagittal plane (flexion–extension), in the frontal plane (adduction–abduction), and, moreover, also rotations, which is triaxial: flexion–extension occurs about a mediolateral axis; adduction–abduction occurs about an anteroposterior axis; and internal–external rotation occurs about a longitudinal axis. Flexion–extension movements are of the highest amplitude. Motions in the other two planes are substantial and consistent both within and between individuals. During walking three degrees of freedom of rotation are observed in the knee joint. The primary motion is knee flexion–extension about a mediolateral axis. Knee internal–external rotation and adduction–abduction (varus–valgus)

Mathematical and Computational Methods in Biomechanics of Human Skeletal Systems: An Introduction, First Edition. Jiří Nedoma, Jiří Stehlík, Ivan Hlaváček, Josef Daněk, Taťjana Dostálová, and Petra Přečková.

may also occur but with less consistency and amplitude among healthy individuals owing to soft tissue and bone constraints with these motions. Ankle motion is restricted by the morphological constraints of the talocrural joint, which admits only extension and flection. The ankle is a complex of three bones—the tibia, the fibula, and the talus of the foot. The ankle is responsible for load bearing and the kinematic function. The ankle joint consists of the tibiotalar, fibulotalar, and distal tibiofibular articulations, and it is a hinge type of a joint. The ankle admits only flexion–extension movement of the foot in the sagittal plane. Other movements of the foot are inversion and eversion, inward and outward rotation, and pronation and supination. These movements occur about the foot joints such as the subtalar and transverse tarsal joints between the tales and calcaneus. The foot is required to act as both a semirigid structure and a rigid structure that permits adequate stability to support body weight (Nordin and Frankel, 2001).

Human bipedal locomotion is a cyclic activity consisting of walking and running. In walking the gait is more or less symmetrical with regard to angular motions of the major joints, muscle activation patterns, and the load on the lower extremities. As a result the gait is efficient in translating the center of mass of the human body in the direction of locomotion. Usually gait is defined as the period from a heel contact of one foot to the next heel contact of the same foot. This cycle consists of two parts, a stance phase and a swing phase. The stance phase is divided into heel strike, foot flat, heel rise, push-off, and toe-off. The swing phase is divided into acceleration, toe clearance, and deceleration phases. The gait cycle is, on the average, about 1 sec in duration, the stance phase occupies 60% of the stride, and the swing phase occupies 40%. The stance phase consists of two periods of double limb support (initial and terminal), when the contralateral foot is in contact with the ground, and an intermediate period of a single limb support, when the contralateral limb is engaged in the swing phase. The stance phase is further divided into an initial double stance, followed by a period of a single stance, and then a final period of a double stance. During the double stance, both feet are in contact with the ground. During the single stance, only one foot is in contact with the ground.

The swing phase occupies 40% of the gait cycle, and it consists of three periods—an initial swing lasts approximately 60–73% of the stride (i.e., approximately one-third of the swing phase); a midswing ends when the tibia of the swinging limb is vertically oriented, and it lasts from 73 to 87% of the stride; and a terminal swing lasts from 87 to 100% of the stride, and it ends at the moment of initial contact with the ground.

During the gait cycle, the different muscles involved absorb or deliver mechanical power, which is defined as the product of the muscle force and the contraction velocity.

The motion of the human body during walking is assumed to be uniform. From the experimental measurements it follows that the reactive force is greater than the force evoked by the weight of the human body. The force operating on the head of the hip joint in the gait cycle corresponds to the course of the force registered on the tread desk, but it is higher in order, which is a result of acting muscle groups functioning during "heel-strike" and "push-off." For mathematical modeling and simulations of these motions, it is necessary to determine magnitudes of these acting forces. Thus we can deduce the following:

1. The reaction forces at the moment of heel-strike and push-off are always higher than the forces evoked by the weight of the human body. Its growth is proportional to an acceleration at thread-down or push-off.
2. Reaction forces at push-off are higher than reaction forces at thread-down.
3. A force, by which the returned lower limb acts in the hip joint, must be for the continuation of forward movement higher than a reaction force in the hip joint at heel-strike.
4. The direction of these reaction forces is close to the situation of the axis of the femur at push-off and heel-strike.
5. At the moment, if the center of mass of the body goes above on the supporting limb, the force acting on all joints of the lower limb is the smallest because muscles of the supporting limb are not in action and the swing acts as a balancing element.

We think that in contradiction to the static position on one leg, where the force loading of the head of an endoprosthesis is increased by actions of muscles retained the equilibrium, the situation during walking is reverse. The muscles increasing the force of the load on joints of the lower limb act during walking at the moment of push-off and heel-strike, and in the phase of full support the swing leg operates as a balanced element. Vertical force resulting an affect of the human body weight on the support decreases owing to a horizontal component of force of the moving human body.

From here it follows that the direction that acts on all joints of the lower limb during walking without any support will go through the sagittal plane. The highest reaction force will act on the joints at the moment of heel-strike and push-off. The direction of this force due to the sagittal plane is, in the case of the hip joint, determined by the gait length and in the case of the knee joint by the position of the femur axis.

After the implantation of an artificial hip replacement the biomechanical relation in the joint sector is changed, namely in the dependence on the patient activity. From the moving activity follows not only the value of the resulting force acting on the head of the hip joint but also the extent of wobble of the pelvis in the frontal plane. With greater patient activity the direction of the resulting force, in relation to the vertical axis of the pelvis, declines laterally. With less patient activity wear is also less and a reaction force acting on the acetabulum is closer to the ideal vertical axis of the pelvis.

Total hip joint replacement is the most extreme loading joint in the human body from the point of view of a bending movement. The big problem is the manner of fixation of the stem. Here, the problem is a fixation of the stem of the joint replacement in the marrow channel of the femur. The stem has a wedge-shaped form. This form is selected because in the course of the application the stem of the femur, which is filled with the bone cement PMMA (cemented THA), penetrated through the bone cement and at the same time was pressed down into the spongious bone on the inner channel surface. Given that the PMMA filled in all the unevenness in the channel of the upper part of the femur, it creates the ideal bed for the joint replacement stem.

Another reason for its form is its relatively simple loosening at reoperation. From a stability point of view, the wedge-shaped stem has a critical cross section. In case the stem is strongly fixed and by its collar it leans on the plane of the resected femur neck, then the total hip arthroplasty (THA) behaves like a rigid two-arm lever, and a strain at the bending moment is observed on the THA neck only. On the stem of the THA the bending moment is manifested as a resulting compression force (pressure), which acts at the end of stem, by a pressure on the inner surface of the bone channel. In case the stem is partially loosened, then the THA is manifested as a fixed slender prismatic rod. This problem will be modeled and discussed in Section 7.4.

After implantation the bending moment of an extremely loaded artificial hip joint replacement sometimes produces disorders in fixation of the PMMA. It is closely connected to the different elasticity of the femur, the stem, and the PMMA. Therefore, new solutions in the construction of the stem and its fixation are expected to be found. The goals are not only for new solutions that may render the stem fixation without PMMA and with very long functional reliability, but also provide a reachability of such an elasticity of a hip replacement stem to be close to the elasticity of the femur bone.

Mathematical simulation of the function of an arbitrary human joint based on the mathematical theory of evolution of unilateral contact problems of the Signorini type must study stress–strain as well as force situations on the contact boundaries, the magnitude of deformation on contact boundaries, transmission of loading forces, which are transmitted from the acetabulum onto the head of the femur during the time-dependent loading. Comparison of results of the mathematical simulation of the healthy and artificial joint then make it possible to estimate if the function of an artificial joint replacement will fully satisfy the requirements of its function and make a decision about the technical design of the THA and to estimate if such a clearance is optimal in providing a fully functional artificial joint replacement.

The analysis of gait and of human data is a complex and time-consuming task. Chan (2001a,b) shows that the data are multivariable, time dependent, with a highly intertrial variability and nonlinear relationship between variables. To reduce variability Winter and Yack (in 1987) used average strides. As the stride durations are not equal, the time axis is usually normalized to a stride percentage. Hausdorff et al. (1997) and Dingwell and Cusumano (2000) showed that the time variability between consecutive strides was recognized as an indicator of several pathologies. Recent studies suggest that the time fluctuation in the strides is not noise (Hausdorff et al., 1995; Dingwell and Cusumano, 2000). Traditionally, gait analysis has been based on normalizing the stride time to a percentage and then averaging several strides measured under the same conditions. The stride time average to a percentage assumes a uniform time scaling of strides with different durations. Former-Cordero et al. (2006) describe gait as a sequence of states. They analyze the knee joint angle (Figs. 1 and 2 of their study) and hip and knee joint angles in the sagittal plane of the reference averaged cycle, computed with the sequence of the states method and the normalization to stride percentage. Anderson and Pandy (2001) and Fernandez and Pandy (2006) published ground reaction forces and the muscle activity for the gait cycle and joint angles predicted by the walking model compared with experimental data for the hip, knee, and ankle joints, which are useful for mathematical modeling of the contact finite element modeling of moving human joints of the lower limbs.

The early failures of human joint replacements were due to the lack of suitable preclinical tests to adequately assess likely performance. Among the factors that affect this performance, the kinematics of a replaced joint are important as the implant components need to allow a functional range of motion, while at the same time providing the necessary level of a constraint to achieve good stability. The contact pressure acting on the articulating surfaces is also of interest, particularly as the expected relatively high loads and with the small size of the head of joints, which may be the causes of high wear rates. Practice shows that extensive contact pressures lead to a catastrophic failure.

In comparison with a total hip replacement, and to a lesser extent with a total knee replacement, only a little knowledge of the biomechanical analysis of total ankle replacements during the gait is available.

Dynamic optimizations, which are based on the forward dynamic methods in mechanics (see dynamic methods and algorithms in Chapter 6), are more powerful approaches for estimating muscle forces during movements. These approaches solve the corresponding optimization problems over the entire time intervals. The serious limitation of dynamic optimizations are these high computational costs. The methods are computationally expensive because the governing equations of motion must be integrated many times for each iterate step of the algorithm used (Anderson et al., 1995). In Section 7.5 we will analyze the time-dependent loading after the application of artificial knee joint replacement under several axial angles, the axial angle changes of the knee joint, which has a great influence on the useful knee joint replacement applications in orthopedic practice. For more details see, for example, Nordin and Frankel (2001).

7.2 BONE REMODELING AND THE CORRESPONDING MATHEMATICAL MODEL

Bone is a nonhomogeneous biomaterial that consists of various cells and organic and inorganic substances with different material properties. Moreover, a bone is characterized by anisotropic properties, that is, its mechanical properties are different in different directions. The mechanical response of a bone is dependent upon the magnitude and direction of the acting loads. Inside of anisotropic properties a bone possesses viscoelastic material properties. The mechanical response of a bone is dependent on the rate at which the loads are operated. From the macroscopic point of view all bones consist of (i) cortical (compact) bone tissue, which is a dense biomaterial forming the outer layers of bones; and (ii) spongy (cancellous or trabecular) bone tissue, which consists of thin plates, the trabeculae, in a loose mesh structure that is covered by the compact bone. Thus, bone is a complex of two-phase composite biomaterials, one of which is composed of inorganic mineral salts, the second one is composed of the organic matrices of collagen and ground substances. A similar problem was discussed in Chapters 5 and 6 where it was based on the homogenization method.

Bone fractures can be produced by a single load that exceeds the ultimate strength of the bone or by repeated loading of a lower magnitude, which causes fatigue fractures, which can be evoked by few repetitions of a high load on a great number of normal loads. The bone has the ability to be remodeled. During the remodeling process, its size, shape, and structure are changed. Load on the skeleton can be accomplished by the influence of gravity and of muscle activities. Forces acting on the contact surfaces between both parts of a fractured bone and microscopic movements on the contact surfaces are fundamental because the healing of a fracture is strongly dependent on such activity. If overloaded, long bones can be fractured. Fractures can be simple or they can be communicative fractures. In 1984 the Association Orthopaedic AO-ASIF in Bern, Switzerland, prepared the AO classification of fractures. In Fig. 7.1 the AO

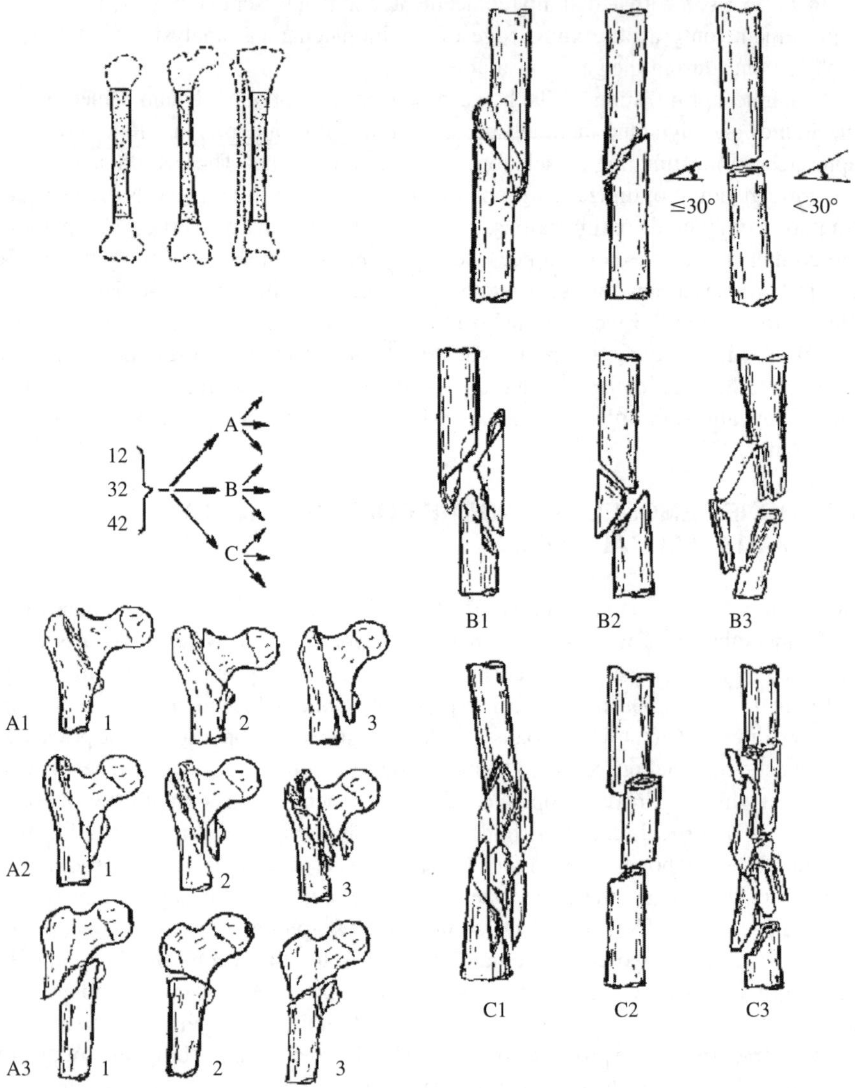

FIG. 7.1 AO classification of the femur. [After Nedoma et al. (2006).]

classification of the femur, based on the monography by Müller et al. (1990) or its computer version Classification of Fractures, is presented.

In orthopedics there are two types of fracture therapies, the internal fixation and the external fixation. Figure 7.2 shows some internal and Fig. 7.3 shows some external types.

The fractured bone [see Fig. 7.3(c) and (d)] can be simulated by the model based on contact model problems with friction, discussed in Chapter 6, because the fractured long bone can be considered two loaded bodies in contact, where the contact boundary

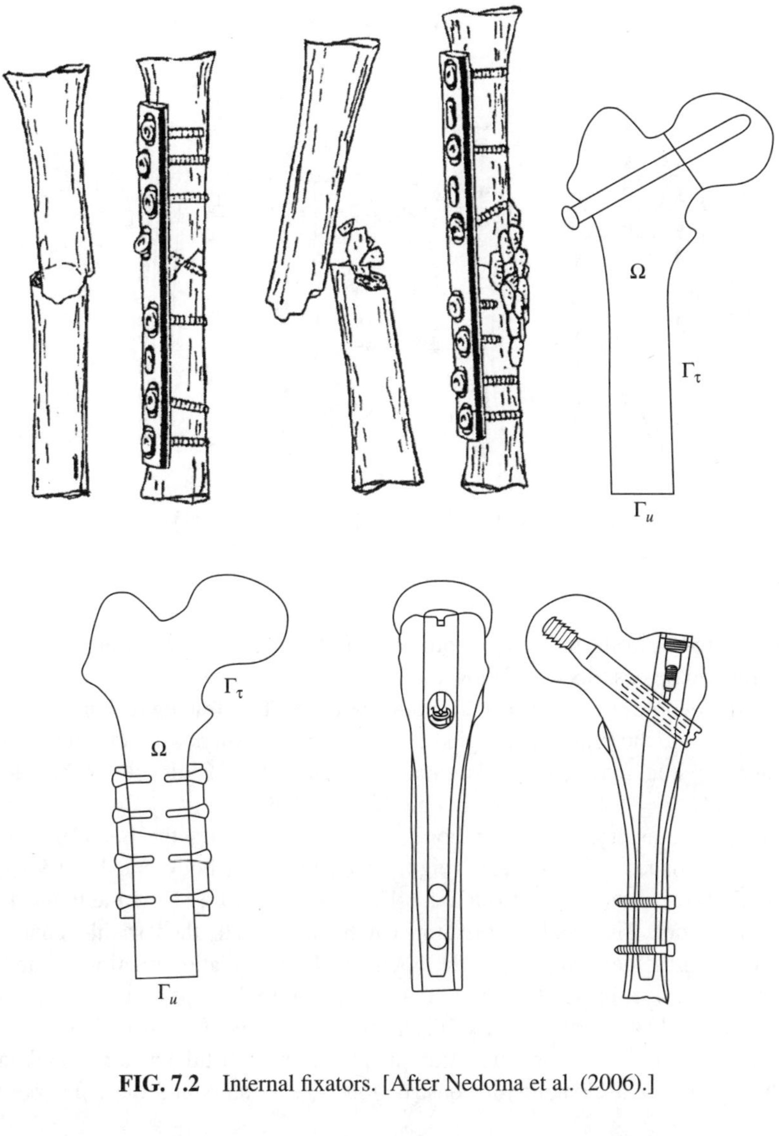

FIG. 7.2 Internal fixators. [After Nedoma et al. (2006).]

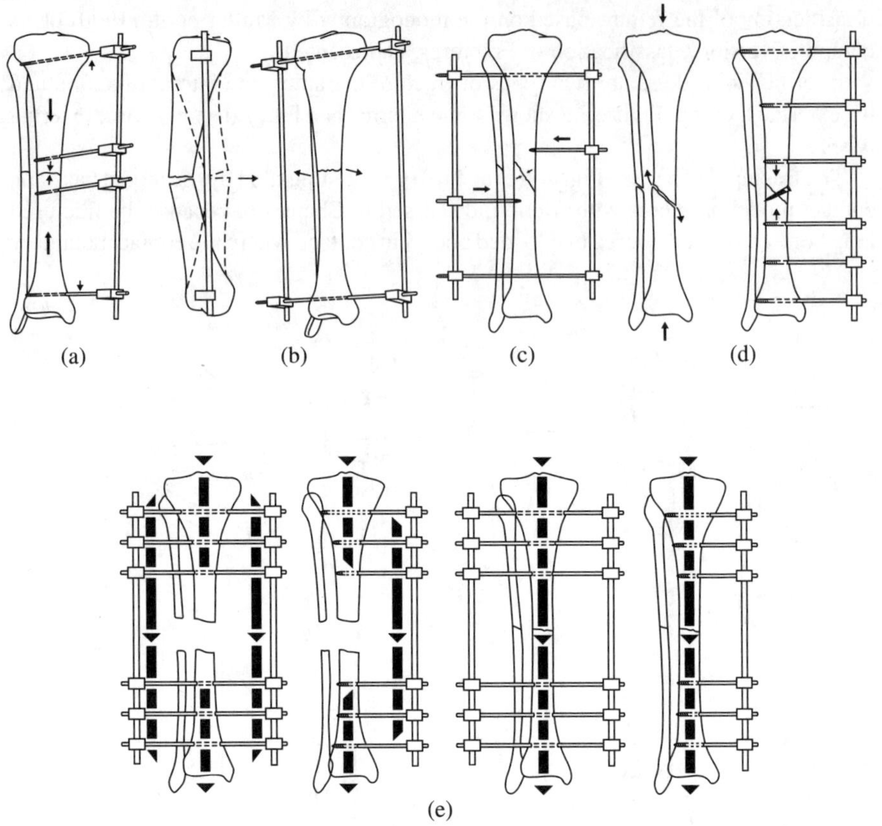

FIG. 7.3 External fixators. [After Stehlík (1995).]

is represented by both surfaces on both parts of the fractured long bone. Such model problems can be described as follows:

Let us denote by $\Omega^\iota, \iota = 1, 2$, both parts of the fractured bone, and by $\Omega^\iota, \iota = 3, \ldots, s$, the regions occupied by the fixators, where s is an integer and it is greater or equal to the maximal number of fixators. Let $\partial\Omega^\iota$ denote the boundaries of regions $\Omega^\iota, \iota = 1, 2, \ldots, s$. The boundary $\partial\Omega^1 \cup \partial\Omega^2 = \bar{\Gamma}_u \cup \bar{\Gamma}_\tau \cup \bar{\Gamma}_c$, where Γ_u denotes the boundary, where the bone is fixed, Γ_τ denotes the boundary, where the bone is loaded (${}^1\Gamma_\tau$) by the weight of the patient's body and by the weight of muscle forces, and it is unloaded on ${}^2\Gamma_\tau$, and $\Gamma_c = {}^1\Gamma_c \cap {}^2\Gamma_c$ is the boundary of the fracture zone. The number of fixators can be modeled by the Dirichlet conditions prescribed on the defined parts of the boundary Γ_u (simulating fixation of the bone in a certain number of points or areas corresponding to the type of a fixator), or they can be modeled as a certain number of different elements of construction—the fixators, then the problem will be more complicated as we must define a relatively great number of different elements of construction—the fixators and the corresponding boundary conditions. We restrict ourselves to the simpler model problem, which is as follows:

Problem $\mathcal{P}_1$ Find a displacement vector $\mathbf{u}$ satisfying the equilibrium equation

$$-\frac{\partial \tau_{ij}}{\partial x_j} = \rho f_i, \qquad i = 1, 2, \quad \text{in } \Omega, \tag{7.1}$$

the boundary conditions

$$\begin{aligned} \tau_{ij} n_j &= P_i \quad \text{on } {}^1\Gamma_\tau, \\ \tau_{ij} n_j &= 0 \quad \text{on } {}^2\Gamma_\tau, \\ \mathbf{u} &= \mathbf{u}_0 \quad \text{on } \Gamma_u, \end{aligned} \tag{7.2}$$

and the unilateral contact conditions (of the Signorini type) with the given friction:

$$u_n^1 - u_n^2 \leq 0, \qquad \tau_n^{12} \leq 0, \qquad (u_n^1 - u_n^2)\tau_n^{12} = 0,$$

and

$$\begin{aligned} &\text{if } u_n^1 - u_n^2 = 0 \quad \text{then} \quad |\tau_t^{12}| \leq g_c, \qquad \text{on } \Gamma_c, \\ &\text{if } |\tau_t^{12}| < g_c \quad \text{then} \quad \mathbf{u}_t^1 - \mathbf{u}_t^2 = 0, \\ &\text{if } |\tau_t^{12}| = g_c \quad \text{then} \quad \mathbf{u}_t^1 - \mathbf{u}_t^2 = -\vartheta \tau_t^{12}, \ \vartheta \geq 0, \end{aligned} \tag{7.3}$$

where g_c is the given nonnegative function and where the stress tensor is given by Hooke's law:

$$\tau_{ij} = c_{ijkl}(\mathbf{x}) e_{kl}(\mathbf{u}), \tag{7.4}$$

where the coefficients c_{ijkl} satisfy

$$\begin{aligned} &c_{ijkl} = c_{jikl} = c_{ijlk} = c_{klij}, \\ &c_{ijkl} e_{ij}(\mathbf{u}) e_{kl}(\mathbf{u}) \geq c_0 e_{ij}(\mathbf{u}) e_{ij}(\mathbf{u}), \qquad c_0 = \text{const} > 0, \end{aligned} \tag{7.5}$$

and where u_n, $\mathbf{u}_t$, τ_n, and τ_t are normal and tangential components of the displacement vector $\mathbf{u} = (u_i)$ and the stress vector $\tau = (\tau_i)$ defined by the relations $u_n = u_i n_i$, $\mathbf{u}_t = \mathbf{u} - u_n \mathbf{n}$, $\tau_i = \tau_{ij} n_j$, $\tau_n = \tau_i n_i = \tau_{ij} n_j n_i$, $\tau_t = \tau - \tau_n.\mathbf{n}$, that is, $\tau_{ti} = \tau_i t_i = \tau_{ij} n_j t_i$ and where $\mathbf{n} = (n_i)$ and $\mathbf{t} = (t_i)$ are normal and tangential unit vectors to the contact boundary Γ_c and it holds $u_n^2 = u_i^2 n_i^1$ and ρ is density.

Numerical Results The treatment of the fractured long bone, in our study the fractured tibia, modeled by 2D models, will be presented. Two types of relatively simple treatment techniques are given in Figs. 7.4 and 7.5. The treatment techniques based on the possibility of lateral compressions for a reposition and stability of the fractured

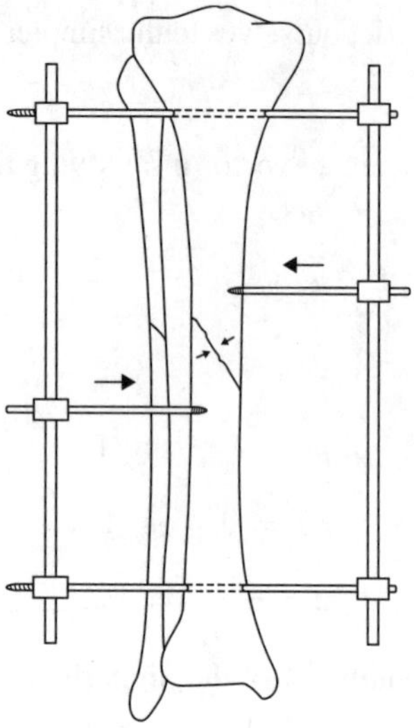

FIG. 7.4 Treatment technique based on the application of lateral compressions for the reposition and stability of the fractured tibia. [After Stehlík (1995).]

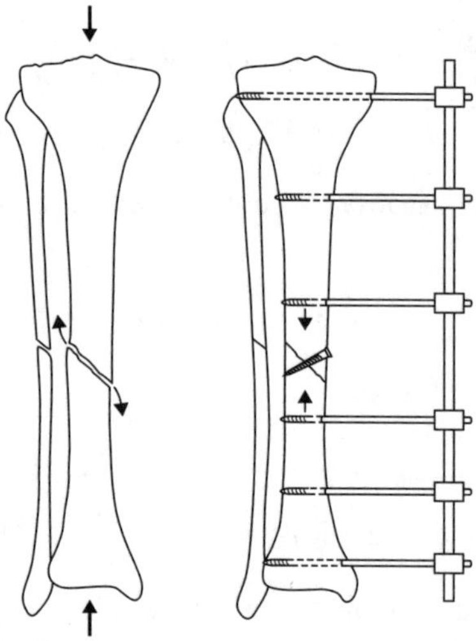

FIG. 7.5 Treatment technique based on the internal miniosteosynthesis combined with the external fixation. [After Stehlík (1995).]

tibia (see Fig. 7.4) and the internal miniosteosynthesis combined with the external fixation (see Fig. 7.5) are useful surgical techniques for treatments of the fractured tibia.

These relatively simple models of surgical treatments of long bones were used as special specimens for mathematical models of external fixation techniques in orthopedic practice. The models of the tibia were 2D models, fixed in the lower

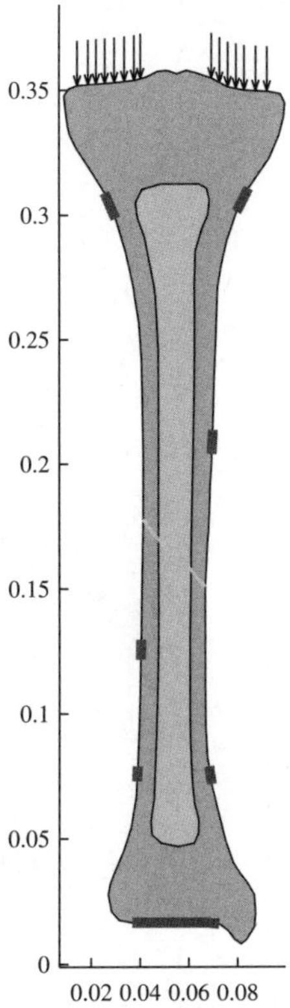

FIG. 7.6 Model of a surgical technique based on the application of lateral compressions in order to reposition and stabilize the fractured tibia.

part of the tibia, that is, in the area of the internal ankle and loaded on the upper part of the tibia, that is, in the internal and external condyles of the tibia. The long bone has a load of 1.75×10^6 Pa. The bone tissue is approximated by the cortical bone tissue with a Young's modulus of $E = 1.71 \times 10^{10}$ Pa and Poisson's ratio of 0.25 and by the marrow tissue with the Young modulus $E = 2 \times 10^6$ Pa and Poisson ratio of 0.45. The fixators are simulated by Dirichlet conditions on small areas described by a yellow color and similarly the prestressed fixators are simulated by acting surface forces, corresponding to forces the evoked prestressing, in places modeled by small areas as in the previous case. The fracture is described by an oblique line.

First, the model of a surgical technique based on the possibility of the application of lateral compressions for a reposition and stability of the fractured tibia [see Fig. 7.6(a)] will be discussed. Numerical results presented in Figs. 7.6(b)–(g) (see the accompanying DVD disk) demonstrate the distributions of deformation of the loaded fractured tibia and of displacement and stress components in the corresponding 2D model, describing the useful possibility of lateral compressions for its successful

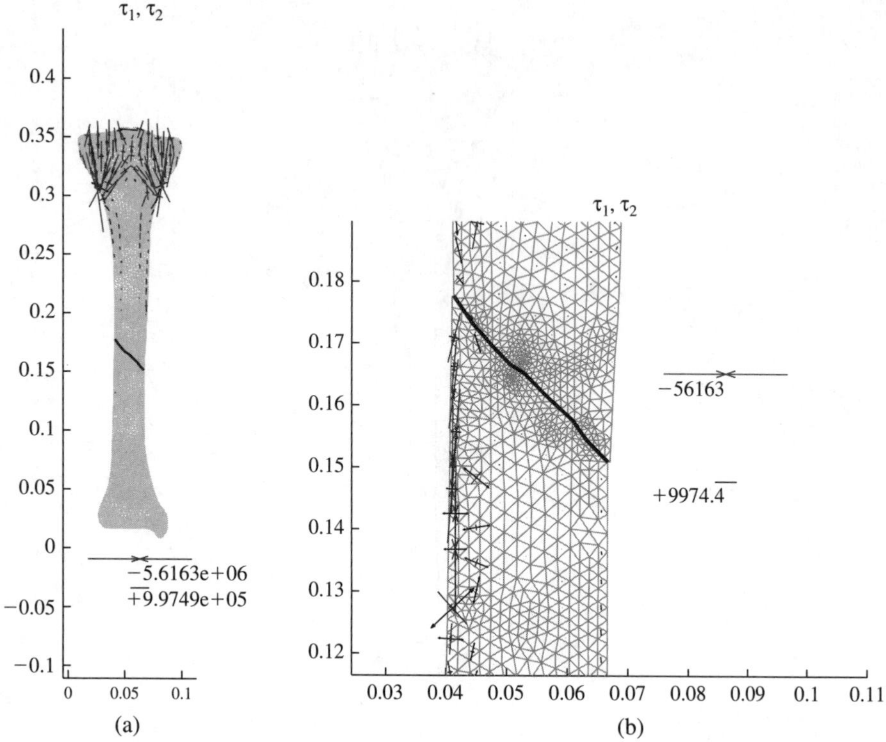

FIG. 7.7 Model of a surgical technique based on the application of lateral compressions in order to reposition and stabilize the fractured tibia—the principal stresses in the tibia and in the detailed area near the fracture.

reposition and stability. In Figs. 7.7(a) and (b) the principal stresses in the whole tibia and in the details are presented, where the symbols ←—→ and —→←— denote the compressive and tensile stresses. Numerical results show that the load is predominantly observed in the upper part of the tibia; nevertheless it shows also a small shift with small separations on the contact boundary of both parts of the fractured tibia [see Fig. 7.7(b)]. It indicates that the lateral compression must be more intensive. Numerical results also indicate the small separations in the area of the marrow tissue and at the end of the fracture.

The internal miniosteosynthesis combined with the external fixation (Fig. 7.5) is presented in Figs. 7.8 and 7.9(a) and (b). As in the previous case numerical results indicate similar distributions of displacement and stress fields, but nevertheless the values of compressive and tensile stresses are rather higher than in the previous case. In the fracture zone small separations are also observed. In both cases these effects are small, so they can have a positive influence on the treatments of the fractures.

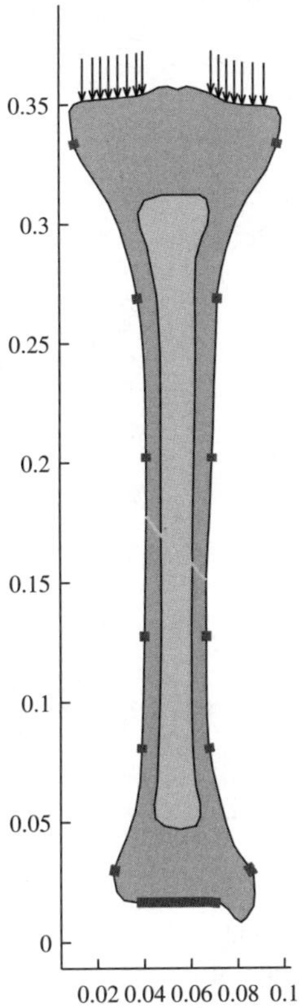

FIG. 7.8 Model of a surgical technique based on the internal miniosteosynthesis combined with the external fixation.

7.3 BIOMECHANICAL STUDIES OF CYSTS, OSTEOPHYTES, AND OF INTER- AND SUBTROCHANTERIC OSTEOTOMY OF THE FEMUR AND THE KNEE JOINT

7.3.1 Biomechanics of Cysts and Osteophytes

The occurrence of cysts is observed in all large human joints, such as the hip or the shoulder joints. The cysts occur either single or in pairs. Their shapes are of spherical, pearical, or irregular forms. The size of the cysts depends on the magnitude of operated forces as well as on the size of the joint. The tissues of the cysts are of a different character, that is, of a liquid, the gelatinous ligament character, frequently with hyaline joint cartilage and bone beam bits. The origin of the cysts is discovered by several types of hypotheses; see, for example, Collins (1949), Landells (1957), Guest and Bernes (1959), Bombelli (1983), Petrtýl et al. (1985), Nedoma (1991, 1993b), and Stehlík and Nedoma (1989).

The hip joint, in addition to the knee joint, is one of the most loaded joints. With respect to the erectile stance of the human body, the shape and extent of movements,

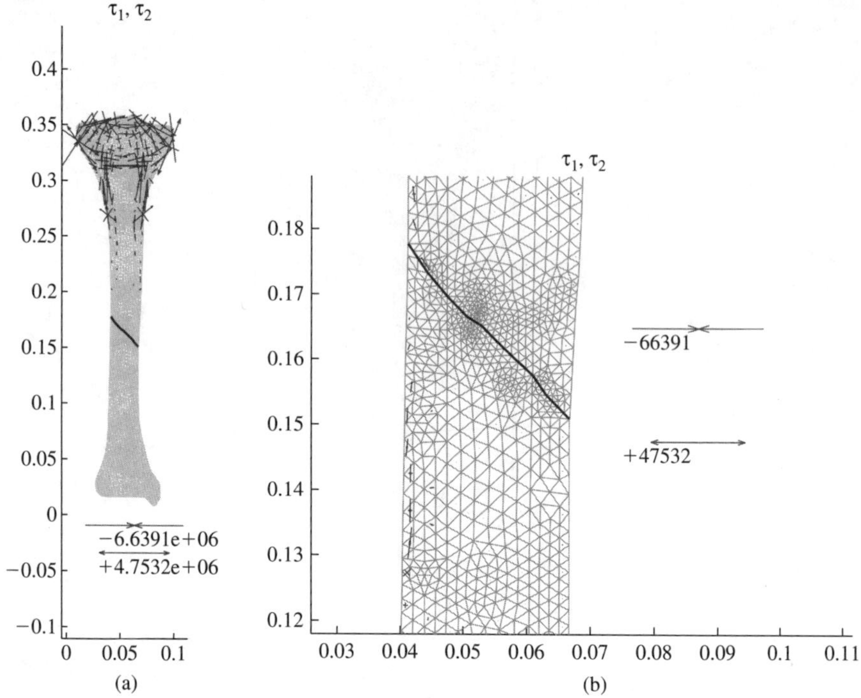

FIG. 7.9 Model of a surgical technique based on the internal miniosteosynthesis combined with the external fixation—the principal stresses in the tibia and in the detailed area near the fracture.

the head of the hip joint is loaded nonuniformly. A certain part is loaded by a pressure, that is, the **weight-bearing surface** (WBS), the remaining part of the articular surface of the head is less loaded, in some cases after overloading tensile stresses can occur. Disposition of bone lamellae is represented by the loading relation of the proximal femur. They are distinguished on the pressure systems, describing courses of more frequently applied compression loading, and on traction systems, originating in areas loaded by traction forces. This trabecular system is spread in a space, with the primary trabecular system behaving like a spiral system. The height of the inner corticalis in the area of the small trochanter is called the **Adams arch**.

The inner architecture of the acetabulum is given by genetic factors as well as by the functional influence. Flat bones of the pelvis represent a system composed from materials of different mechanical properties, which are very advantageous for this system as they demand functional loading. Combinations of rigid and elastic corticalis with porous spongious bone closed between them correspond to the well-known sandwich construction. On the X-ray radiograph we find areas of higher bone density corresponding to thickened bone beams, which lay in places of great stress streams. With respect to their distributions we then consider concentration of stresses on partial bone tissues (Fig. 7.10).

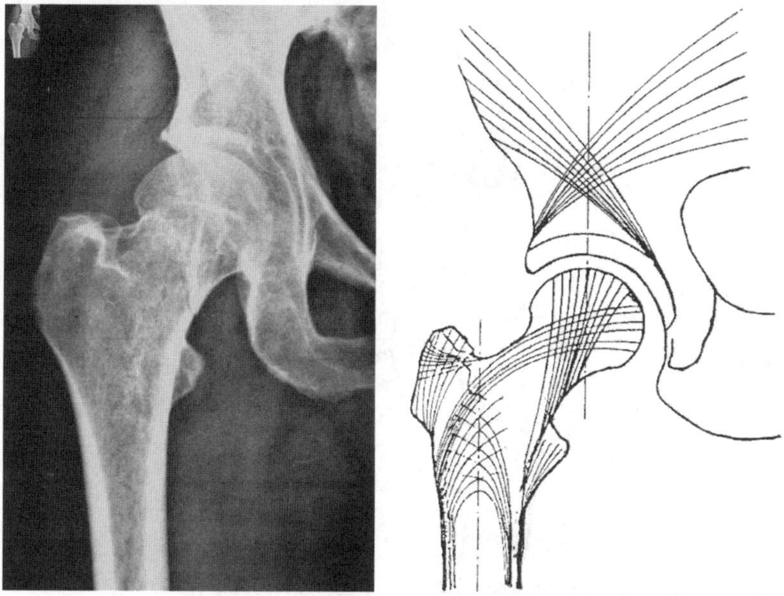

FIG. 7.10 (a) X-ray radiograph of the hip joint and the pelvis and (b) trabecular system in the proximal femur. [Modified after Bartoš (1998) and Nedoma et al. (2006).]

In the case of advanced stages of osteoarthritis in human joints, the occurrence of cysts of different sizes is observed. Their sizes are between 1 and 25 mm diameters and they are of different shapes: lentil-shaped, pear-shaped, heart-shaped, wedge-shaped, or sphere-shaped cysts. Small cysts are observed in the joints of the lower limbs, which are the most loaded joints in the human body. The cysts are observed on their heads, but they are also observed in the knee patella or in joints of the upper limbs.

Arthritic cysts are one type of cysts. They are observed in areas of a joint surface of the joint head, into which the system of a trajectory in the proximal part of the femoral bone is aimed but never out of it. Cysts can be often observed in a pair, the **coupled or kissing cysts**. Their situation is observed on both surfaces of the joint gap, one on the head of the joint and the second one in the opposite place in the acetabulum, nevertheless some cysts are in mutual contact between them. We observe the system of coupled cysts on both sides of the maximal loading of the head joint (see Fig. 7.11). In the case of osteoarthritic cysts their occurrence on many members can be also observed, and their number increases with advanced arthritis of the joint and with disappearance of its cartilage. Their shape is mostly irregular. In the single occurrence the cysts are pear shaped. Their size depends on the magnitude of loading forces and on the size of the joint. The cyst tissue depends on its origin, a connection between the cavity of the cyst and the joint cavity was evidenced. The cyst is created by the following:

1. The gelatinous tissue containing ligamental cells with fibers creating the mesh of different density (and thickness)

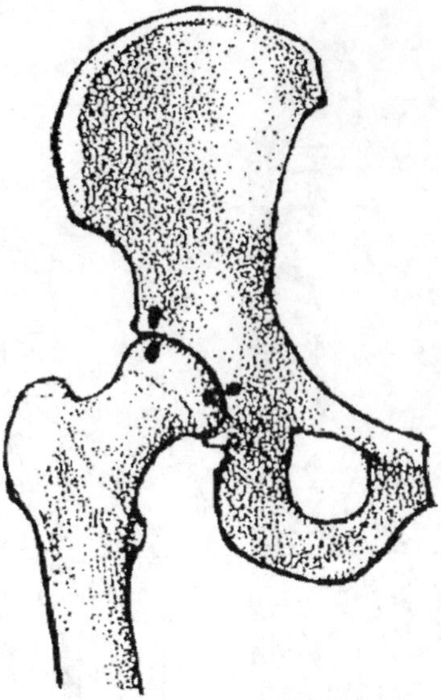

FIG. 7.11 System of coupled cysts. [After Stehlík and Nedoma (1989), Nedoma (1993b), and Nedoma et al. (2006).]

2. Rigid ligamental tissues
3. The cells of a fibrocartilaginous type
4. Liquid containing mononuclear cells

The tissue of cysts is often made of fragments of bone trabecula (often gradually extinct) and fragments of the hyaline cartilage and the synovial liquid. The cyst wall is created by the layer of the condensed spongious tissue. Explanations of the origins of the cysts abound; see, for example, Collins (1949), Landells (1957), Guest and Barnes (1959), Bombelli (1983), Petrtýl et al. (1985), Stehlík and Nedoma (1989), and Nedoma (1991, 1993b). The oldest theories of the origins of arthritic cysts assume a primal degeneration of bone tissues. But with the explanation of the evolution of osteoarthritis these theories are no longer accepted. By using photoelastometric analyses Petrtýl et al. (1985) demonstrated that the main reason for cysts is local stress concentration, which is defined as the damaged (rough) surface of the joint head, and therefore the cysts are a consequence of the arthritic process. Their assertions are based on experimental photoelasticometric analyses of a differently loaded plate with arbitrary stress concentrations and with the hyaline cartilage simulated by a layer of India rubber (see Fig. 7.12). These stress concentrations are seen with polarized light by different shapes of isochromates (i.e., by concentric colored streaks in the polarized light in the loaded plate), which then were compared with the cyst shapes of works by Collins (1949), Trueta (1957), and Guess and Barnes (1959). Landells (1957) thinks that the origin of the osteoarthritic cyst can be cleared up by overloading certain local places on the joint surface in the course of which traumatic destructions

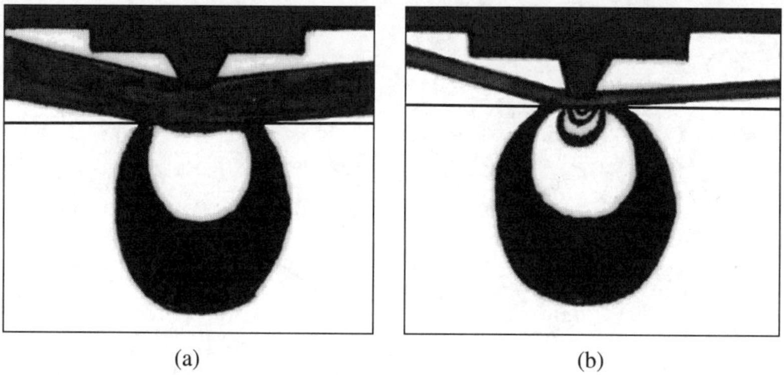

FIG. 7.12 Photoelastic measurins of cysts. [Modified after Petrtýl et al. (1985) and Nedoma et al. (2006).]

of the upper layer of the bone lamellae occur. By placing the fissures under high pressure, the synovial liquid is pressed into the spongy bone. The crust size is given by the shape and size of the cavity and by the length of time needed to fill it with synovial liquid.

Another explanation for the origin of the cyst is based on the theory of contact problems presented by Stehlík and Nedoma (1989), Nedoma (1991, 1993b), and Nedoma et al. (2006). Their explanations consider the analysis of contact conditions on the boundary between the head of the joint and the acetabulum from analyses of numerical simulations and numerical modeling of an overloaded human joint and from comparison of numerical results with the experimental photoelasticometric results of Petrtýl et al. (1985) (see Fig. 7.12).

Another explanation of the origins of the cyst and their evolution used the stress–strain analyses based on the theory of contact problems with friction in thermoelasticity. The mathematical analysis, based on the theory of contact problems discussed in Chapter 6, gives a new look at the cyst's origin [see also in Nedoma (1983, 1987, 1993b, 1994b), Nedoma and Stehlík (1995), Stehlík and Nedoma (1989), Stehlík et al. (1997), and Nedoma et al. (2006)]. The model problem used is a very simple one that corresponds to the photoelasticometric model, but it must simulate the mutual contact between the head of the joint and the acetabulum (Fig. 7.13). In contradiction to the photoelastometric model, the model used was simulated by an intertrochanteric part of the femur in the shape of the bevelled plate with a unique thickness, where the bevelled plate simulates the contact surface between the joint head and the acetabulum. For simplicity the acetabulum was assumed to be absolutely rigid.

From the analysis of the condition acting on the contact boundary between both main components, physically analyzed in previous chapters, we saw that in the course of deformation of both joint components, the opposite points are displaced in a way general, but always in such a way that one component of the joint cannot penetrate the second one. For contact forces, acting on the contact boundary between both components of the joint, the principle of action and reaction shows that the normal

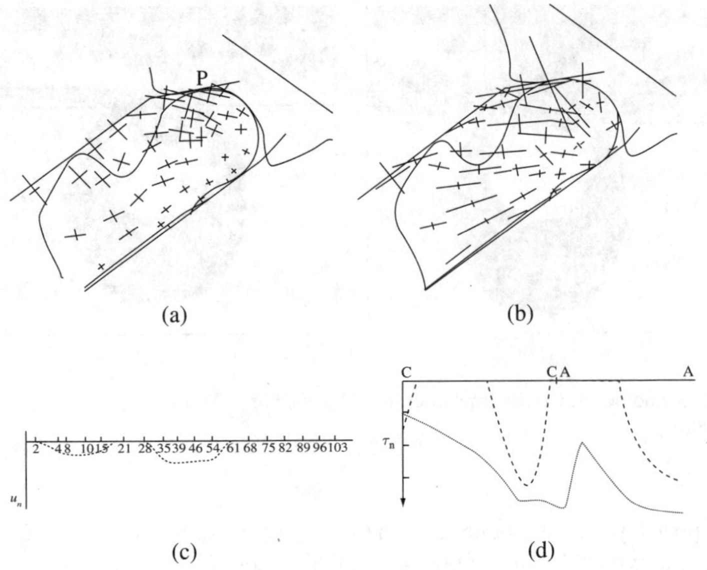

FIG. 7.13 Principal stresses τ_1, τ_2 (a) in the standard loaded hip joint, (b) in the overburdened hip joint, (c) normal (u_n) displacement component in the standard loaded and in the overburdened hip joint, and (d) normal (τ_n) stress component in the standard loaded and in the overburdened hip joint.

and tangential forces in the absolute value are equal and act in opposite directions, whereby the positive direction is taken to the surface of the head of the joint. It was shown that the contact forces cannot be tensile forces.

In the course of deformation of both particular components, these are or are not in a contact. If both particular components are in a contact, then nonzero contact forces exist, which keep both joint components in contact. If both joint components are not in a contact, then the contact forces are zero. The space between both joint components is filled with synovial liquid, which due to its excellent viscous and other properties, reduces the friction acting at the contact surface of both parts of the joint, and, moreover, it transfers the temperature and the heat flow and it is a nutrient to the articular tissue as well as the whole joint. The viscosity of the synovial fluid changes when the joint is burdened.

The mathematical analysis, based on the theory of contact problems discussed in Chapter 6, gives a new possibility to the cyst's origin (Stehlík and Nedoma, 1989; Nedoma, 1993b; Nedoma et al., 2006; Stehlík et al., 1997). In our analysis the simple model of the hip was studied, where the acetabulum was taken to be rigid and the shape of the upper part of the femur was taken only in its very simple form (Fig. 7.13). Nevertheless, the results are very important and sufficiently accurate to represent real situations in the hip joint under the normal loading and under overburden of the hip joint. In Fig. 7.13(d) the normal components of stress for the normal loading and overburdened joints are shown. The maximal pressure occurred at point P of the head of the femur. The neighboring area of point P corresponds to the spherical area as in the classical biomechanics but from the generalized point of view. At a certain distance from point P with maximal compressive stresses, there exist areas where the normal component of stress is very small and practically equal to zero. There are areas where only the condition $u_n^k - u_n^l < 0$ and $\tau_n^{kl} = 0$ on the contact surface of the joint are fulfilled. Below these surface areas the tensile stresses originate. Since cartilage and bone have different properties, according to the conditions $u_n^k - u_n^l < 0$ and $\tau_n^{kl} = 0$ the cartilage is more bulged then the bone tissue. This situation is shown in Fig. 7.13(c) where the normal component of displacement is given. The area of the bulged cartilage tissue is seen in the cyst zones, and this describes precisely the mechanism of the origin of cysts. These bulged areas are filled with synovial fluid. Tensile stresses also create microcracks, which after filling up with synovial fluid can represent another origin for the cysts. This idea attempts to explain the origin of two symmetrically situated cysts (see Fig. 7.11). After more burdening the bulged cartilage layer can be broken and thus create cysts with cartilage and bone fragments inside the cyst. It is evident that the direction of the acting forces also influences for the size and shape of the cyst. Since around the cyst, that is, around the bulged area, the acting forces increase, the areas with the relative strengthened beams of new types are originated. Petrtýl et al. (1985) present their experimental studies where cartilage was simulated with India rubber and the bone tissue of the femur by a sheet of Plexiglass. They explain the origin of the cysts with the idea that the overburdened areas are pathologically destroyed as a consequence of incongruence. Their experimental photoelastometric result is shown in Fig. 7.12(a) and (b). They suggest the shape of isochromates corresponds with the shape of the cysts. But their experimental photoelastometric results, given in Fig. 7.12(a) and (b), indicate that the cartilage in reality is swollen—the situation also given in Figs. 81a,b of Petrtýl et al. (1985). But the authors think that the results of Petrtýl et al. (1985), discussed above, support our idea of the origin, of the cyst that is, the bulged India rubber "cartilage" in Fig. 7.12(a) and (b) describes the origin based on the contact theory of bodies.

The characteristic property of the cysts when they occur as one pair of cysts, that is, one cyst in the head of the femur and the second one in the acetabulum, or when they occur in the double pair system, created one pair of the previous system (i.e., one cyst in the head of the femur and the second one in the acetabulum), occurring on the right-hand side from the acting force and the second pair occurring on the left-hand side from the acting forces, schematically given in Fig. 7.11. The mechanism for such a system of cysts is a consequence of unilateral contact conditions.

Human joints are formed by osseous and connective tissues capable of regeneration. Thus, microcracks caused by local overburdening do not increase because during the regeneration process microcracks are healed. Microcracks originate in places where tension stresses occur. Many authors predicate that osteophytes are found in places where tension stresses occurred. In the skeleton the thermal field is also associated with deformation. A change in heat content in an element produces a state of deformation and a state of stress. Conversely, the state of deformation evoked by burdening loads produces a thermal field in the bone. Then a part of the mechanical energy produced by the deformation of the body changes into heat. In Fig. 7.14 simulation of heat generation on the contact boundary between the head of the hip joint

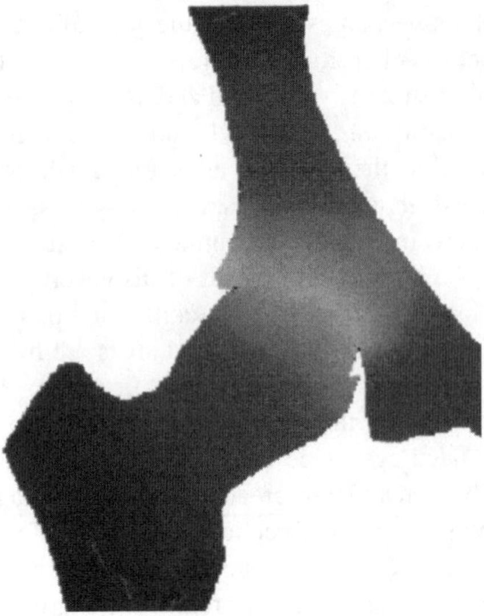

FIG. 7.14 Heat generation on the hip joint contact boundary evoked by deformation of the hip joint system.

and the acetabulum evoked by deformation is shown. Such a model problem is based on the coupled unilateral contact problem in thermoelasticity or thermoviscoelasticity [see Nedoma and Dvořák (1995) and Krejčí et al. (1997) based on a coupled unilateral contact problem in thermoelasticity]. Some heat is transported by the blood into places distant from the joint. Thermomechanical dissipation of energy takes place in the skeleton, which probably influences also the origin of osteophytes. But the main influence on osteophytes is tension forces on the joint.

In the biomechanical model of the human joint studied from the classical biomechanics point of view acting forces are approximated by a point force, and, therefore, their resultant is related to one point only. In the real human skeleton acting forces act at every point at the common boundary of both joint components, and in both parts of the human joint a stress–strain state is evoked. Further, in order to better understanding of the problem, we will speak about the situation in the hip joint only. However, generalizations can be made about other types of human joints. In our studies the simple approximation of the joint was used where the acetabulum was rigid and the femoral head is evident, as shown in Figs. 7.13(a) and (b), which represent the principal stresses in the femoral head. But this simplification of the femoral head is sufficient to prove the main idea of the generalized biomechanics of the dynamics of human joints in comparison with the results of Petrtýl et al. (1985). If the hip joint is burdened even a little, then compressive stresses in the hip joint are seen, which is illustrated in Fig. 7.13(a). More compressive stresses accumulate in the area around point P, where the joint is intensely loaded, which corresponds to the spherical sector (zone) in classical biomechanics. Figure 7.13(b) corresponds to this case, where the hip joint is overburdened. From the numerical results one can analyze the area of

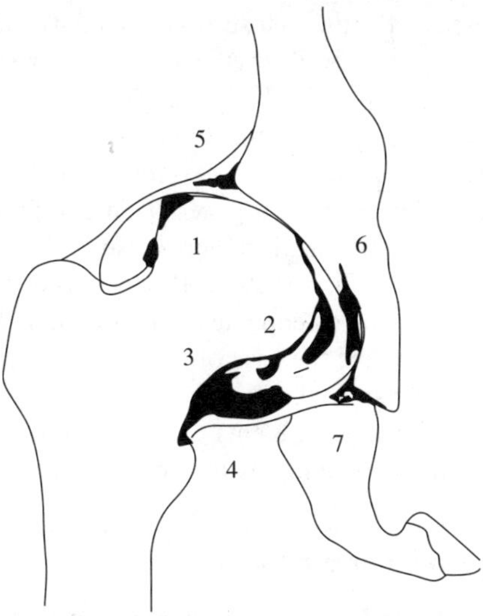

FIG. 7.15 Main types of osteophytes. [After Nedoma, et al. (2006)].

great compressive stresses, which occur above all in places where the femoral head is overburdened (in Fig. 7.13 denoted by $\longrightarrow\longleftarrow$ in the neighborhood of point P) and the area where the tensions originate (in Fig. 7.13 denoted by $\longleftarrow\longrightarrow$). These tensions originate in two fundamental zones, which in classical biomechanics Bombelli (1983) called positive and negative tensions.

In a healthy joint the friction between both joint components (the femoral head and the acetabulum in the case of the hip joint) is very small. That is ensured by the excellent viscoelastic properties of synovial fluid. In an impaired joint the friction on the contact surface between both joint components is greater than in the healthy joint. For arthritic joints the friction increases as the cartilage erodes and more bone tissue touch the contact surface between both joint components. Thus, the friction coefficient (in the Coulombian sense) is very high, and we then speak about **dry friction.** Such cases occur in late-state osteoarthritis.

In an osteoarthritic hip joint we can detect seven main osteophytes (see Fig. 7.15). Three are located on the femoral head, that is, the superior cervical osteophyte (1); second, the capital drop (2), comprising the fovea osteophyte, the inferior marginal osteophyte, and the cuplike osteophyte; and, third, the inferior cervical osteophyte (3), which sometimes, when hypertrophic, takes on an elephant trunk shape (4). The next three are located in and around the socket of the femoral head, that is, the roof osteophyte (5), the curtain osteophyte (6), and the floor osteophyte (7). The comparison of the numerical analyses with the observed osteophytes from an X ray gives good results. From the physiological point of view the formation of the osteophyses is also influenced by heat. The temperature of the human body varies between 35–37°C max. and 42°C in some cases of an illness. In a human joint heat is also associated with deformation of both joint parts. Change in the heat value produces a state of deformation and stresses—the thermal stress—and conversely the deformation of the joint evokes the burdening that produces the heat. Thus, one part of the mechanical energy produced by deformation of the body changes into heat. Thermomechanical dissipation of energy then takes place in the joint, which may influence the origin of the osteophytes. But it is evident that some heat is transported away from the hip joint. Finally, one can say that this hypothesis is generalized also for other human joints as well as for the spine. A similar mechanism works also in the case of deformation of the spine as a whole as well as of the vertebrae. Projections on the vertebrae originate from the same mechanisms as the osteophytes. This situation will be discussed later in this chapter. This model studies the case of the loaded thoracolumbar spine with osteophytes that press into the spinal cord in the spinal channel.

In the end we can study these problems as (visco)elastic biomaterials, but, in reality, the bone is a live mass operating by the laws of nature.

7.3.2 Biomechanics of Osteotomy

Introduction Intertrochanteric or subtrochanteric osteotomies of the femur, the knee, as well as other joints, and the osteotomy of the pelvis are therapeutic techniques that have been used for decades as solutions for the deformity and degenerative changes in human joints. Furthermore, we restrict ourselves to the hip joint. During the development of an osteoarthritic hip joint, osteophytes, cysts, as well arthrosis originate on the contact boundary between the femoral head and the acetabulum in highly loaded human joints. The purpose of proximal femur osteotomy is to bring the femoral head into a new position inside the acetabulum. The corrective options of intertrochanteric osteotomy most frequently used in orthopedic practice are valgization, varization, displacements, oblique displacements, lateralization, extension rotation, or shortening of the shaft (see Fig. 7.16).

Osteophytes and cysts on the contact boundary between the head of the joint and the acetabulum originate in highly loaded joints. Osteophytes originate from osseous metaplasia of the synovial membrane and joint capsule in which all stages of both phylogenesis and ontogenesis of bone formation may be reproduced. In a completely developed osteoarthritic hip joint, three principal osteophytes on the femoral head as well as three principal osteophytes on the acetabulum are observed (see Fig. 7.15). The purpose of proximal femur osteotomy is to position the femoral head into a new position inside the acetabulum. The most frequently used corrective options of intertrochanteric osteotomy are shown in Fig. 7.16 [modified after McMurray (1939) and Schneider (1993)], and the model of intertrochanteric osteotomy (valgization, varization, displacements, etc.) is shown in Fig. 7.17(a) and (b). The problems of osteotomy are discussed in Pauwels (1973), Schneider (1979), and Schatzker (1986).

The present widespread tomographic measurement methods, that is, computed tomography (CT) and magnetic resonance imaging (MRI), and automatic three-dimensional reconstructions also involve numerous new points of view for planning femoral osteotomy (Schneider, 1979; Bombelli, 1983; Schatzker, 1986; Nedoma, 1993b). Earlier methods were based on descriptive geometry methods, on the triangulation method [Walkin, and Klaue, 1988) and on the basic kinematic technique

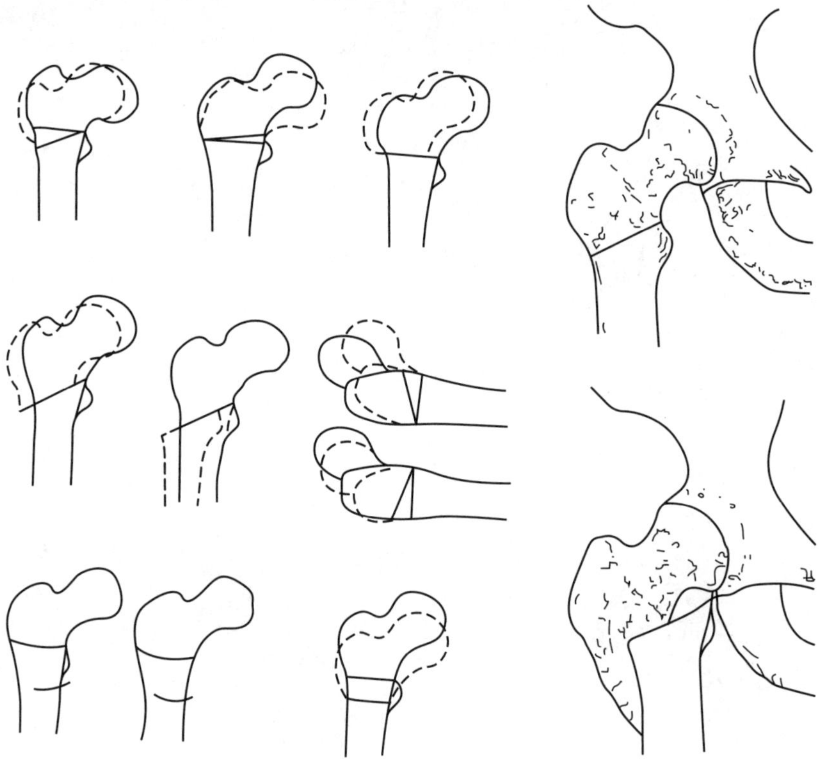

(a) (b)

FIG. 7.16 Intertrochanteric osteotomy: [(a) Modified after Schneider (1979), and (b) modified after McMurray (1939)].

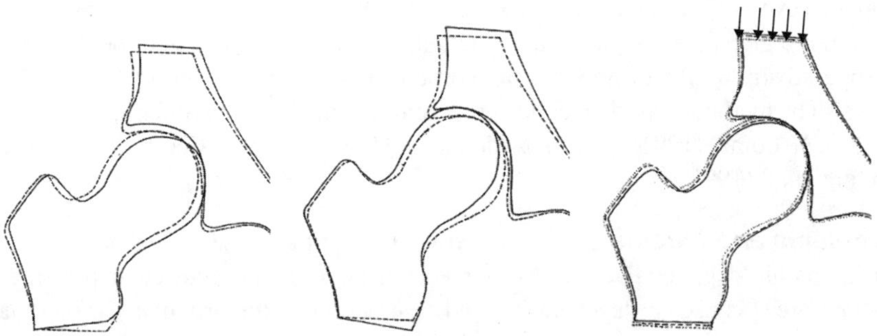

(a) (b) (c)

FIG. 7.17 Subtrochanteric osteotomy—numerical simulation: (a) varization, (b) valgization, and (c) displacement.

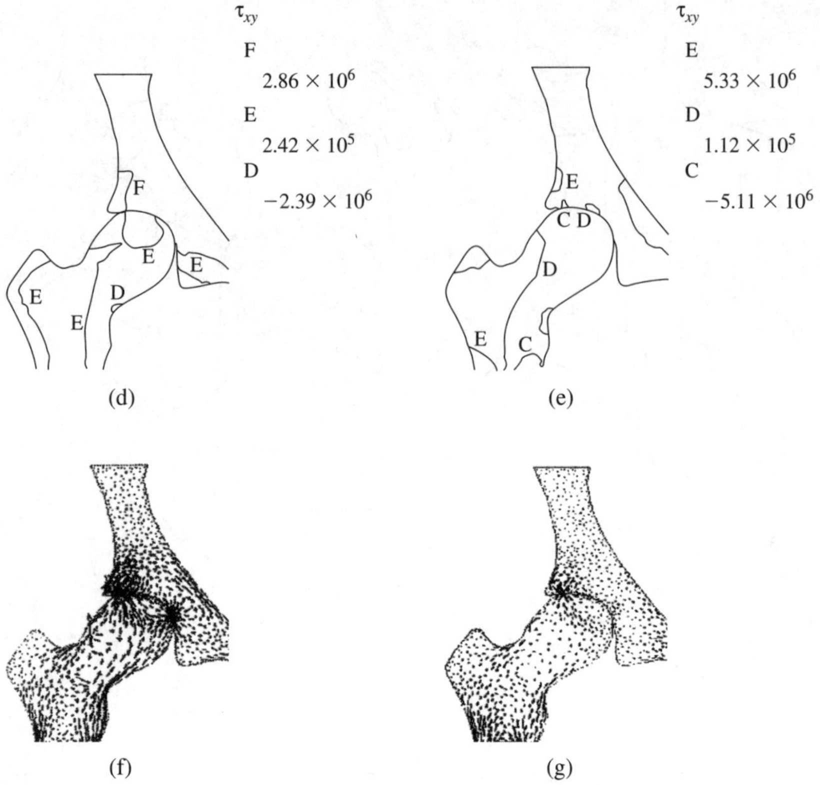

FIG. 7.17 *Continued*: Subtrochanteric osteotomy: (d) shear stress component τ_{xy}, (e) vertical stress component τ_y, (f) principal stressses in the case of varization osteotomy—the initial state, and (g) principal stressses—the state after osteotomy.

(Sangeorzan et al., 1989). In 1981, Charit elaborated a new method that has been applied several times since that time (Charit, 1982). He calculated the cross section from three given angles (flexion, abduction, rotation), which have to be determined preoperatively by the orthopedic surgeon under a fluoscopic control (Charit, 1982, 1988). The methods based on contact problems are discussed in Stehlík, and Nedoma (1989), Nedoma, (1991, 1993b), Nedoma, and Stehlík (1995), Eck et al. (1990), and Stehlík et al. (1997).

Problems and Models of Osteotomy One of the causes of osteoarthritis and an idiopathic degenerative disease is mechanical. High intra-articular pressure on contact areas causes a degeneration and destruction of the articular cartilage, and later a deterioration of the joint. Surgical treatment is based on the restoration of the joint (osteotomy) or on the replacement of the joint by arthroplasty.

Femoral intertrochanteric osteotomy was introduced as a routine surgical procedure after the introduction of biomechanical principles. The first discussion was presented by Pauwels in 1936 (Pauwels, 1973), where two main mechanical principles are involved. One of them improves the convergence of the contact surfaces of the femoral head and the acetabulum, which is followed by a reduction in the intra-articular pressure because of an enlargement of the contact area. Lower pressure leads to the restoration of the cartilage covering the contact surfaces. The second function of osteotomy is to change the load of the joint because of the change of geometry of the hip joint.

Besides these mechanical principles, there is a very important biological effect of osteotomy, the "McMurray effect" (McMurray, 1939). It is based on the biological stimulation of the articular structures by healing the bone after osteotomy.

For our models we use the theory of contact problems, assuming linear elasticity and the finite element technique. Mathematically, types of osteotomy are simulated by boundary conditions, for example, varization, valgization, and displacement by the Dirichlet type of the boundary condition, in which we cut off a part of the femoral bone to simulate a requested type of osteotomy, as well as in the other types of osteotomy. The models concern mainly the human hip joint, but the theory presented in Chapter 6 can also be applied to other types of human joints such as the shoulder and knee. The method presented here facilitates the simulation of surgical osteotomy problems and stress–strain analyses of the pelvis and of both hip joints. Then, on the basis of the stress–strain analyses of those model problems, the surgeon can determine which type of surgical techniques can be used depending on the condition of the hip joint.

Mathematical Model and Numerical Experiments For our models we use the results of the contact problem theory assuming linear elasticity. For the numerical solution of the corresponding mathematical problems we use the well-known finite element technique. The models deal mainly with the human hip joint, but the theory can also be applied to other types of human joints. The method presented in Chapter 6 facilitates the simulation of surgical osteotomy problems and the stress–strain analysis of the pelvis and both hip joints. Then, on the basis of the stress–strain analyses of those model problems, the surgeon can determine which type of surgical techniques can be used in depending on the present condition of the hip joint, as we mentioned above.

The geometry of the femur and the pelvis can be taken from an X-ray radiograph, a CT scan, on from MRI images. The present results were obtained using an X-ray radiograph, that is, from 2D or 3D cuts across the femur and the pelvis. The domain occupied by this part of the femur and pelvis is then triangulated by regular triangulation (2D or 3D). The models start with equilibrium equations, the well-known Hooke law as well as boundary and contact conditions [see, e.g., Nedoma, and Stehlík (1995), Stehlík et al. (1997), or Chapter 6]. Taking into account the boundary conditions, we simulate the function of the investigated human skeleton. The muscular tissue causes forces in the tendons, which can be transmitted by tendons to an osseous tissue. In addition to these forces there are other loading forces due to the weight of the subject that are transmitted to the skeleton. Based on the knowledge of the physiological distribution of the tendon tissues and skeletal sites across which the loading forces are transmitted due to body weight and the direction and magnitude of these two types of acting forces, we will find the condition for the loading, which has to be prescribed on the part of the skeleton. In the case of osteotomy, we define where the surgical cut across the femur (for an intertrochanteric or subtrochanteric osteotomy) will be made. We then simulate the osteotomy technique using the boundary conditions. In many cases the direction of loading is taken in the vertical direction. When the loaded area deviates from the horizontal plane, the biomechanical equilibrium is impaired and, depending on the inclination, migration of the head in a lateral or medial direction occurs. Some authors tried to improve the models of different joints regarding the action of burdening forces, which take into account the geometrical conditions in the area of the hip joint, the inertial forces, and the influence of the main muscle groups of abductors and adductors, respectively. The situation is, however, more complicated, in particular as regards the action of forces and the transmission of loading forces from the acetabulum to the head of the joint, that is, the action of forces in the contact with the surface between the acetabulum and the head of the hip joint. These conditions are called the contact conditions. Let us consider these conditions acting on the contact surfaces between the acetabulum and the head of femur. Thus, the first condition represents the condition of nonpenetration of both components of the joint, that is, the head of the femur and the acetabulum. The second condition follows from the principle of action and reaction and from the fact that the contact forces cannot be tensile (this condition describes the situation of acting forces in the contact area of the joint). The third one follows from the following deliberation: During the deformation of both joint components, the contact either occurs or not. If they are not in contact, then the contact forces are equal to zero, but if the joint components are in contact, then there exist nonzero contact forces. This third condition then describes the above deliberation mathematically.

Moreover, if both joint components are in contact, then friction forces act on the contact surface between both joint components in the Coulombian sense. The absolute value of these forces is proportional to the absolute value of acting normal forces. The coefficient of proportionality is the coefficient of the Coulombian friction. We do not know its magnitude, but it can be estimated using an upper bound. Due to the acting and friction forces we have the following cases: (i) If the absolute value of the tangential forces acting on the contact boundary is less than the frictional forces, then the frictional forces preclude mutual shifts of both joint components, and the deformation energy cumulates in both components of the joint. (ii) If the tangential forces are equal in their absolute value to the frictional forces, then both joint components shift mutually, and, at the same time, the opposite points of the contact boundary change their position in the direction opposite to the tangential forces. The space between both joint components is filled with synovial liquid, which reduces the friction on the joint and feeds the gristle of the joint. In the normal healthy joint the Coulombian friction coefficient is very small, while in the arthritic joint or the joint with cysts, the Coulombian friction coefficient increases.

The algorithm used is based, first, on the formulation of the displacements (the primary formulation) and, second, on the formulation in the stresses (the dual formulation). Both algorithms follow from the variational formulation of the problem studied and from the finite element method, which are used for the numerical approximation of the initial conditions problem. The first part leads to minimizing the functional of potential energy over the set of admissible displacements and the second part to finding the saddle point of the Lagrangian functional based on the dual variational formulation. The 2D and 3D numerical analyses can be used for loading the normal hip joint and for the joints after the varization, valgization, and displacement osteotomies. It is evident that the presented method allows one to analyze the distributions of stress and strain fields of the lower limb after the osteotomy surgical treatment.

For illustration we will present models of varization, valgization, and displacement osteotomies. The 2D mathematical models correspond to a real patient. The geometry of the models was obtained from an X-ray body scanner in the Clinic of Radiology, 3rd Medical Faculty, Charles University, Prague. The skeleton of the patient was loaded by a vertical force of 1000 N. In Figs. 7.17(a)–(c) the varization, valgization, and displacement osteotomies, based on the X-ray body scanner pictures, are simulated and the resulting state after the surgical treatments are presented. Figures 7.17(d)–(g) show the subtrochanteric osteotomy in the case where the cut is made below the small trochanter. The distribution of horizontal, shear, vertical, as well as principal stresses are presented. Moreover, the state before and after osteotomy applications are presented in detail. The coefficient of the Coulombian friction is assumed to be equal to zero. The dashed lines in Figs. 7.17(a) and (b) represent the initial state of the pelvis and the femur, while the dotted lines represent the situation after the surgical treatment. In these models we assume that the pelvis is fixed in the pubis as "symphysis ossium pubis," and that the pelvis is loaded by 1000 N in its upper part.

The obtained results concerning stress–strain data can be used for analyses of distributions of pressure and traction stresses, which have an influence on the course of the healing, the regeneration stimulation in the surgical joint and, therefore, also have an influence on the overall results of the clinical treatment. These results can be used for the optimization of surgical interventions in the course of their presurgical planning.

7.4 BIOMECHANICAL ANALYSIS OF THE LOOSENED TOTAL HIP ARTHROPLASTY (THA)

The object of our study will be a patient after implantation of an artificial hip joint replacement. In his case the partial loosening of the stem of the total hip joint replacement was indicated. The loosening of the acetabulum was not observed. The used model of the replaced hip joint was derived from the frontal-posterior X-ray body scanner image. In the X-ray image one-third of the loosening, beginning from the collar of the endoprosthesis, was indicated.

Our goal in the study was an investigation of the loosening of the endoprosthesis. For this reason three types of models were derived: (i) the model of unloosening total hip joint replacement of the patient; (ii) the model corresponding to the present state of the patient, with loosening approximately one-third of the stem length, and (iii) the model corresponding to the probably future loosening of the THA approximately to two-thirds of the stem length. The model was based on the quasi-static linear elasticity under the assumption that the hip joint was loaded uniformly, which means that the inertia forces can be omitted. This model was used for the analysis of three schematic situations corresponding to the three stages of evolution of the loosening total hip joint replacement. The skeleton was loaded by the vertical force, which is transmitted over the spine and the os sacrum onto the pelvis and the hip joint. The pelvis was fixed in the area of symphysis and loaded in the supra-acetabular part. The femur is fixed below the subtrochanteric area in a certain, but sufficiently great, distance below the end of the stem. All that was said about the model and the method as well as the algorithms used in Section 7.3 and in Chapter 6 is also valid in this case of loosened THA of the hip joint, and, therefore, these considerations cannot be repeated.

The detailed analysis of the numerical results then makes it possible to characterize the situation on the contact boundary between the head of the THA and the arbitrary acetabulum and on the contact boundary between the loosened stem and the femur, as well as the distribution and magnitude of stress components and the principal stresses in the pelvis, in the artificial acetabulum, in the stem of THA and the femur. The elastic coefficients are changed in the interval 0.28×10^{10} to 0.38×10^{12} Pa for the Young's modulus of elasticity and between 0.23 and 0.33 for the Poisson's constant. In Figs. 7.18(a)–(b) the horizontal, shear, and vertical stress components, and the principal stresses in the pelvis, femur, artificial acetabulum, and the stem for the unloosened THA are presented. Numerical results show the distribution of

τ_x		τ_{xy}	
A		A	
	-4.80×10^5		-3.23×10^5
B		B	
	-3.95×10^5		-2.51×10^5
C		C	
	-3.11×10^5		-1.80×10^5
D		D	
	-2.26×10^5		-1.08×10^5
E		E	
	-1.41×10^5		-3.72×10^4
F		F	
	-5.74×10^4		3.42×10^4
G		G	
	2.71×10^4		1.05×10^5
H		H	
	1.11×10^5		1.77×10^5
I		I	
	1.96×10^5		2.48×10^5
J		J	
	2.80×10^5		3.20×10^5
K		K	
	3.65×10^5		3.92×10^5
L		L	
	4.49×10^5		4.63×10^5

(a)   (b)

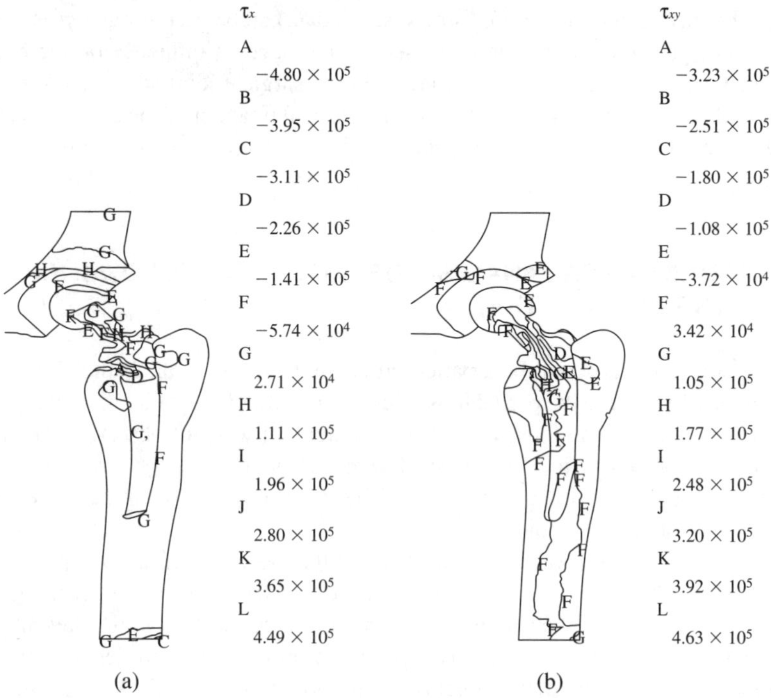

FIG. 7.18 Horizontal and shear stress components for the unloosened total hip joint replacement.

$$\tau_y$$

A
   -1.12×10^6
B
   -9.99×10^5
C
   -8.71×10^5
D
   -7.43×10^5
E
   -6.15×10^5
F
   -4.87×10^4
G
   -3.58×10^5
H
   -2.30×10^5
I
   -1.02×10^5
J
   2.54×10^5
K
   1.53×10^5
L
   2.81×10^5

(c)   (d)

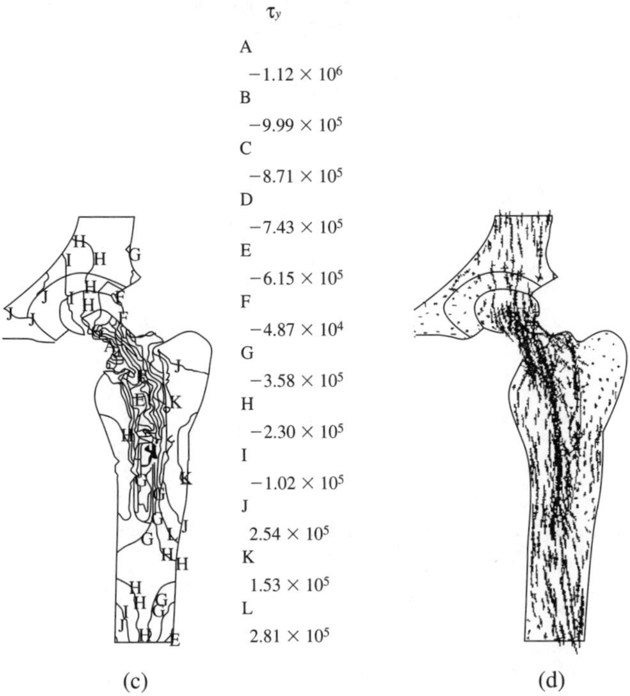

FIG. 7.18 *Continued*: Vertical stress component and the principal stresses for the unloosened total hip joint replacement.

stresses in the pelvis, femur, and the total hip joint replacement. From this one can deduce that for the THA application the great trochanter is rather more loaded than in the case of the healthy joint because the greater part of the stress is transmitted over the stem into the lower parts of the femur.

The isostresses between values for I and J show the boundary between compression and extension. In Fig. 7.18d we can indicate great extension (i.e., extension is denoted by the symbol ⟵ ⟶) in the trochanteric part of the femur near the collar of the THA and along the Adams arch. In the inner area of the stem we observe the compression (i.e., compression is denoted by the symbol ⟶ ⟵). However, we cannot consider the effect of loading by the muscle forces acting on the femur in the area of the great trochanter as well as in the area of the pelvis. In Fig. 7.19 the model of loosened hip joint replacement is given. Figures 7.20 to 7.22 analyze stresses in the loosened THA. The isostresses G and H in the Fig. 7.20 in the case of the partly loosened stem of the THA, respectively, and isostresses H and I in Fig. 7.21, which study the loosened stem of THA up to two-thirds of its length, define the area where we observe a compression and where we observe an extension. From both sets of figures we see that the great trochanter and the femur in the area of loosened THA are practically unloading, and thus the bone thins out and becomes gradually extinct, which has also negative influence on the fixation of the THA stem in the femur.

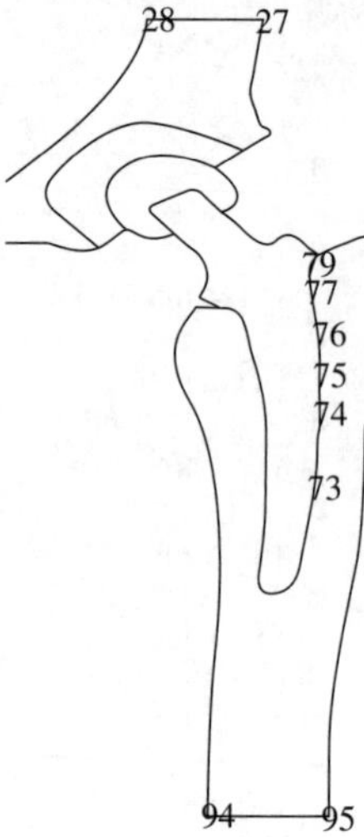

FIG. 7.19 Model of loosened hip joint replacement.

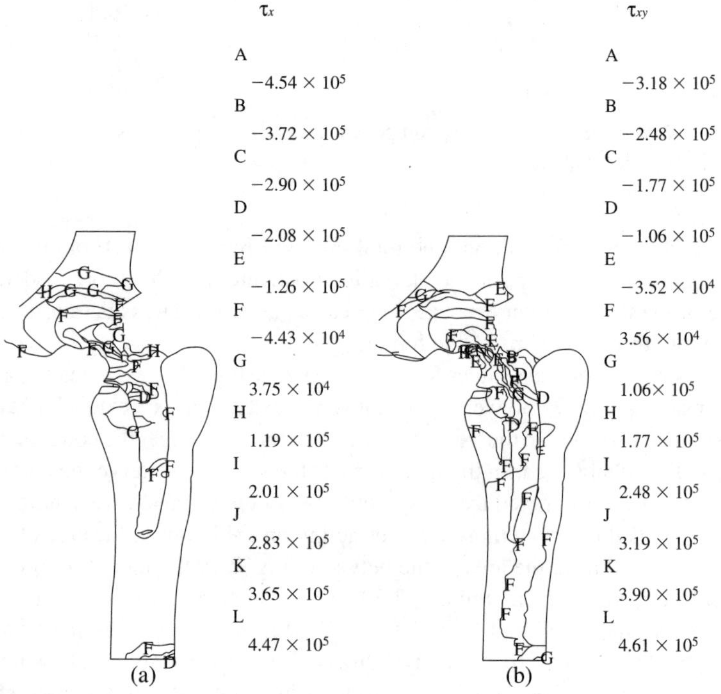

FIG. 7.20 Horizontal and shear stress components in partially ($\sim \frac{1}{3}$ of the stem length) loosened hip joint replacement.

τ_{xy}

A
 -1.34×10^6
B
 -1.15×10^6
C
 -9.57×10^5
D
 -7.63×10^5
E
 -5.69×10^5
F
 -3.74×10^5
G
 -1.80×10^5
H
 1.35×10^4
I
 2.07×10^5
J
 4.01×10^5
K
 5.96×10^5
L
 7.90×10^5

(c)   (d)   (e)

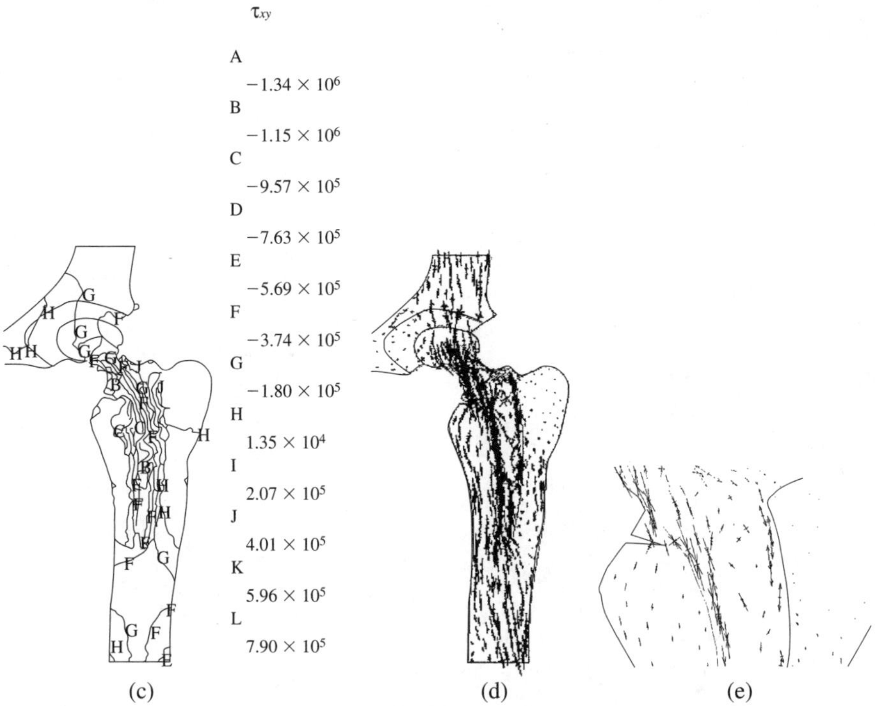

FIG. 7.20 *Continued*: Vertical stress components and the principal stresses in partially ($\sim \frac{1}{3}$ of the stem length) loosened hip joint replacement as well as the corresponding zoom in the upper area of the stem.

From the analysis of the principal stresses we see that the THA stem is more and more bent, that is manifested by the enlargement of the zone in the THA stem to the end of the stem. Figures 7.21 and 7.22 characterize further the evolution of the loosened THA and then indicates an extensive zone to the left of the femur in the case of $\sim$ two-thirds of the loosened THA stem, which means that in this state the loosened THA stem of the femur is also buckled above the loosened THA stem. Distribution of the principal stresses show the stress reduction in the upper part of the femur, including the great trochanter, which is practically without loading. At this point the cortical and spongious bones thin out and then both become gradually extinct. Remember that due to the effect of wear the THA stem is loosened. This effect was discussed in Chapters 4 and 6.

Knowledge about behavior of the loosened stem in the bone of the femur are very substantial for specialists in orthopedics. Such information can give normal and tangential components of displacement and stress vectors. The normal displacement component describes magnitudes of stem looseness, while the tangential displacement component tell something about stem movements in the femoral bone. The normal and tangential stress components give information about tensional relationships on the contact between the stem and the bone. In Fig. 7.22 the normal and tangential

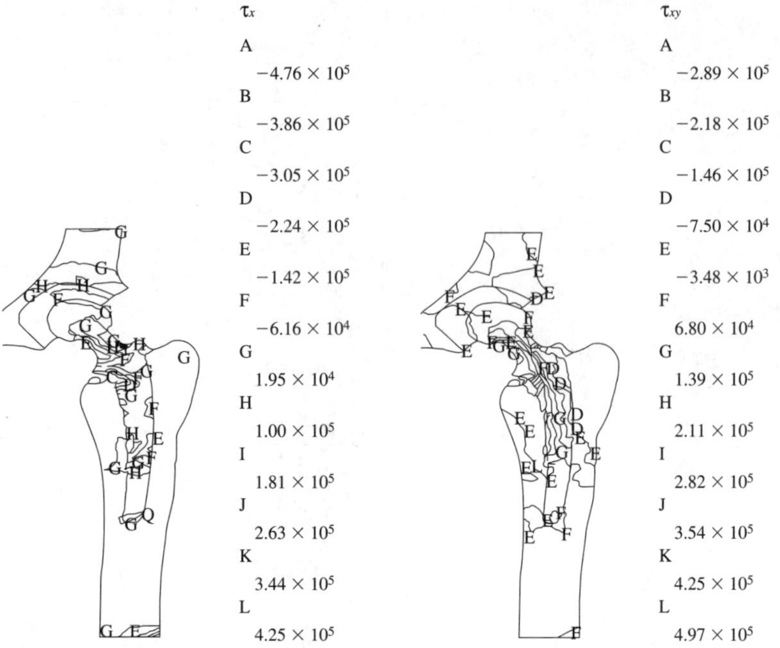

(a) (b)

FIG. 7.21 Horizontal and shear stress components in partially ($\sim\frac{2}{3}$ of the stem length) loosened hip joint replacement.

components of displacement and stress vectors are presented. In Fig. 7.22 we observe between the mesh points of the contact on the loosening stem (in Fig. 7.19 denoted by points 73–79) values of the loosened stem in interval 0 (at point 73) and $7 \times 10^{-6}\, m$ (at point 79). For complete information we may remark that in Fig. 7.22 the values on only one side of the contact arc are depicted. It is evident that on the opposite side the values have also nonzero values. For specialists in applied mathematics the useful information about used algorithm represents the rate of its convergence. Figure 7.23 illustrates the convergence of the method based on the conjugate gradient method with constraints and the comparison of the conjugate gradient method (CGM) with constraints and the preconditioned CGM with constraints (a) diagonal, (b) SOR, and (c) ILL are given.

7.5 BIOMECHANICAL ANALYSIS OF THE HIP JOINT AFTER THA IMPLANTING AND SUBTROCHANTERIC OSTEOTOMY HEALING

The clinical implantation of THA under alterable anatomical conditions is very topical because there are frequent cases. As a matter of fact that the total joint replacements where subjects have anatomical deflections from the normal hip joint. The most frequent causes are shape changes that originate as a consequence of inborn defects

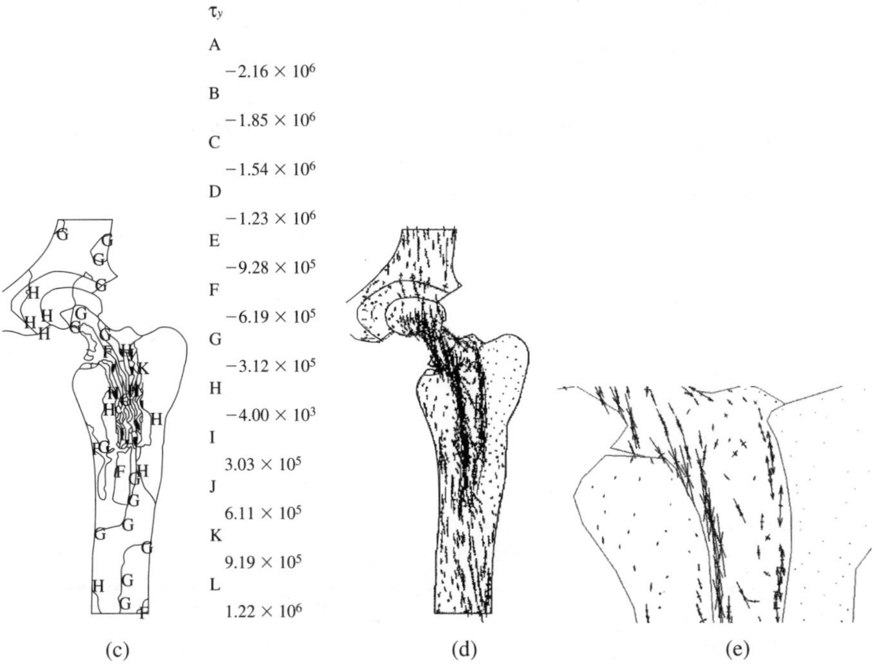

FIG. 7.21 *Continued*: Vertical stress components and the principal stresses in partially ($\sim\frac{2}{3}$ of the stem length) loosened hip joint replacement as well as the corresponding zoom in the upper area of the stem.

striking the proximal part of the femur as well as the acetabulum of the hip joint. Biomechanical problems of implantation of THA in these cases are discussed in the literature [see, e.g., Vasu et al. (1982), Huiskes and Chao (1983), Huiskes (1987), and Huiskes and Sloof (1986)]. Only in sporadic cases is the shape of the defects due to accidents or from surgical treatments. These clinical problems are very interesting and current because the cause of different secondary accidental arthritis of hip joints with the destruction of the joint with alterations in the function of the joint are very frequent and only one satisfactory solution represents an artificial total joint replacement.

From the surgeon's point of view this set of patients represents very complicated problems, where it is necessary to elect specifically individual solutions with regard to a surgical technique as well as with regard to a choice of useful implants. Such cases represent very complicated reconstructive performances on the skeleton, and the resulting morphology is very different from the anatomical norms. This considers high demands not only on the surgical technique but also on postoperative diagnostics, which determine also suitable progress.

At present demanding graphical procedures such as CT scans, NMRI imaging, and the 3D reconstruction image oriented on the shape characterization are frequently applied. On the contrary we are far from conditions that create positive function of the implant system. It is because of a traditional mindset that a surgeon selects a

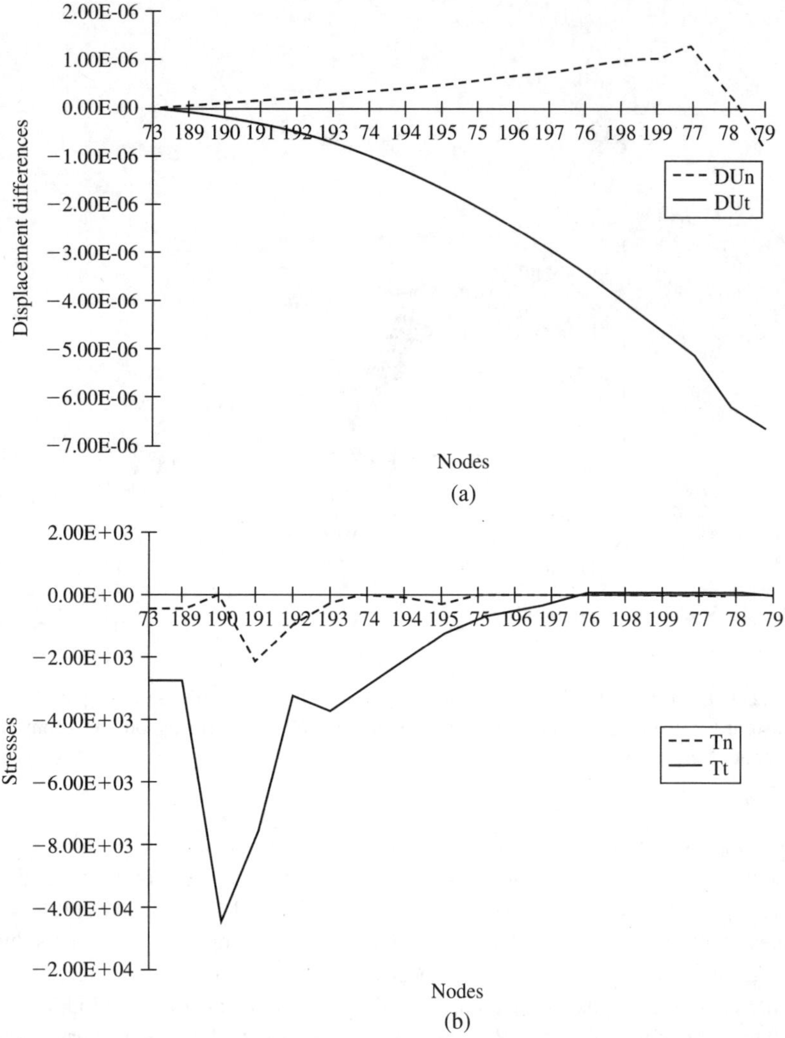

FIG. 7.22 Normal and tangential components of (a) displacement and (b) stress vectors on the contact surface between the head of femur and the acetabulum, the case of partially loosened hip stem ($\sim\frac{2}{3}$ of the stem length).

certain shape for reconstruction which he believes it will be a therapeutic success. However, in some stuations this is not the case. Practical experience shows that for a reliable result an implant system should not only apply an anatomical shape but should also find a shape modification of the skeleton.

It is necessary to obtain needed informations about the shape conditions of the skeleton or of its parts, as well as information about the biomechanical conditions that will determine the long-time function of the joint replacement system as it was demonstrated experimentally as well as by mathematical modeling and also

(a)

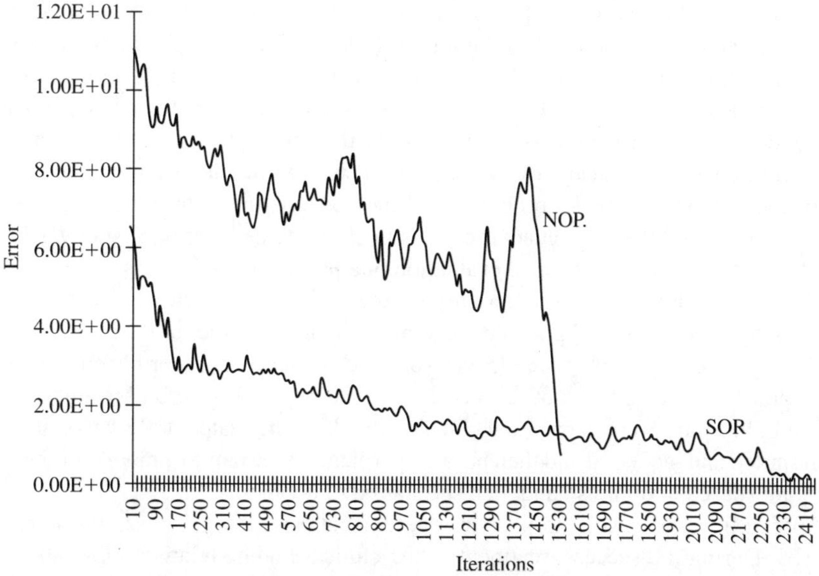

(b)

FIG. 7.23 Convergence of the conjugate gradient method with constraints and comparison of the CGM with constraints without preconditioning (NOP.—represents no preconditioning) and with preconditioning: (a) diagonal, (b) SOR, and (c) ILL.

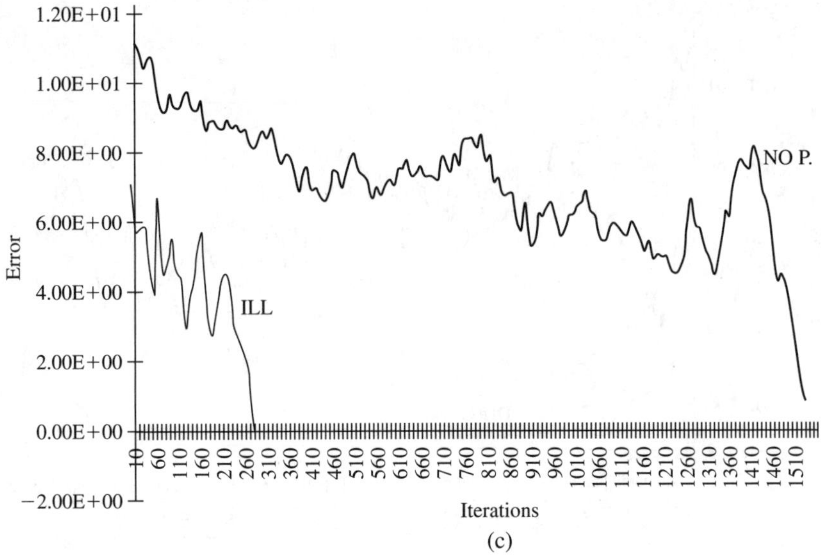

FIG. 7.23 *Continued*

as verified by clinical studies [see Brekelmans et al. (1973), Dunn (1976), Harris et al. (1977), Carter et al. (1982), Semlitsch (1983), Murphy et al. (1987), Huiskes (1987), Stehlík, and Nedoma (1989), Rubin et al. (1990), Nedoma (1993b), Harkess (1998), and Nedoma et al. (1999a,b)]. This information can be obtained on a universal model, but for specialists—surgeons—it will be better to obtain these requisite data from concrete clinical cases. Therefore, here we present special methods and algorithms as well as model problems and methods of stress–strain analyses based on unilateral and bilateral contact problems in elasticity and thermoelasticity, which can be applied for biomechanical evaluation of a particular clinical case.

This knowledge can be used not only for the evaluation of a clinical patient case, but also for instruction and guidance of similar situations in the future.

An object of our study was the 45-year-old woman treated in the orthopedic department since she was 8 years of age, when surgical treatments for an birth defect of the right hip joint were carried out. First, difficulties originated at 32 years of age, pain started and increased and then became permanent; the range of motion of the left hip joint started to diminish and then she started to hobble.

Difficulties worsened, the pain was permanent, and the gait was considerably altered. During the success treatment in the clinical finding relatively limitation in her movement was found. The right lower limb was in flexion and contracted 10° with movements to 80° of flexion, and rotation and duction were limited. The right lower limb was shortened by 1.5 cm. On an X-ray radiograph (Fig. 7.24) of the right hip joint the deformation of the proximal femur, corresponding to the situation after the valgized subtrochanteristic osteotomy, was evident, and the head was in a luxate position in the neocotyle above the level of the initial acetabulum; moreover, heavy degenerative arthritic changes also presented.

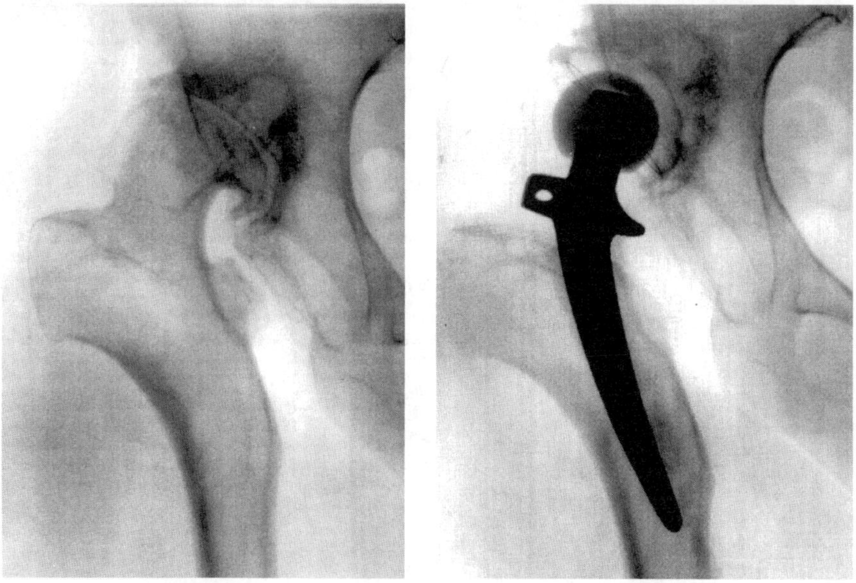

FIG. 7.24 X-ray radiograph of the degenerate right hip joint (a) before (the initial state) and (b) after the valgized subtrochanteric osteotomy and application of the THA (i.e., after the surgical treatment).

Total hip joint replacement was suggested. The surgical treatment as well as the postoperative process, including physiotherapy, were without complications. Later the patient expressed she felt no pain in the operated joint, and, in spite of some mobility limitations, she evaluated the joint as favorable; she walked without problems. For longer walks she used a cane.

Methods and Results The model of the hip joint used for the female patient (Fig. 7.25(a)) was derived from the X-ray radiograph of the postsurgery right hip joint. On this X-ray the atypical shape of the proximal part of the femur is observed, which is in the pronounced valgization position. As a consequence of this shape change the implanted femoral component of THA is also in an atypical position, higher and in a very sheer position. The massif of the great trochanter comes into prominence laterally. The whole joint is shifted against the normal cranial position because the acetabulum is implanted approximately at a "neocotyl" level. The condition was solved using the plane contact model by the finite element method discussed in Chapter 6. See also Nedoma (1987, 1993b), Nedoma et al. (2006), and Nečas, and Hlaváček (1981).

In Fig. 7.25(b) the horizontal stress component is presented. In Figs. 7.25(c) and (d) the shear and vertical components of stresses are presented; their importance is fundamental for the condition discussed. It is evident that from the cause of the isostresses the spreading of stresses in the area of the acetabulum is similar to the normal anatomical configuration and that the stress concentration in the cranial part of

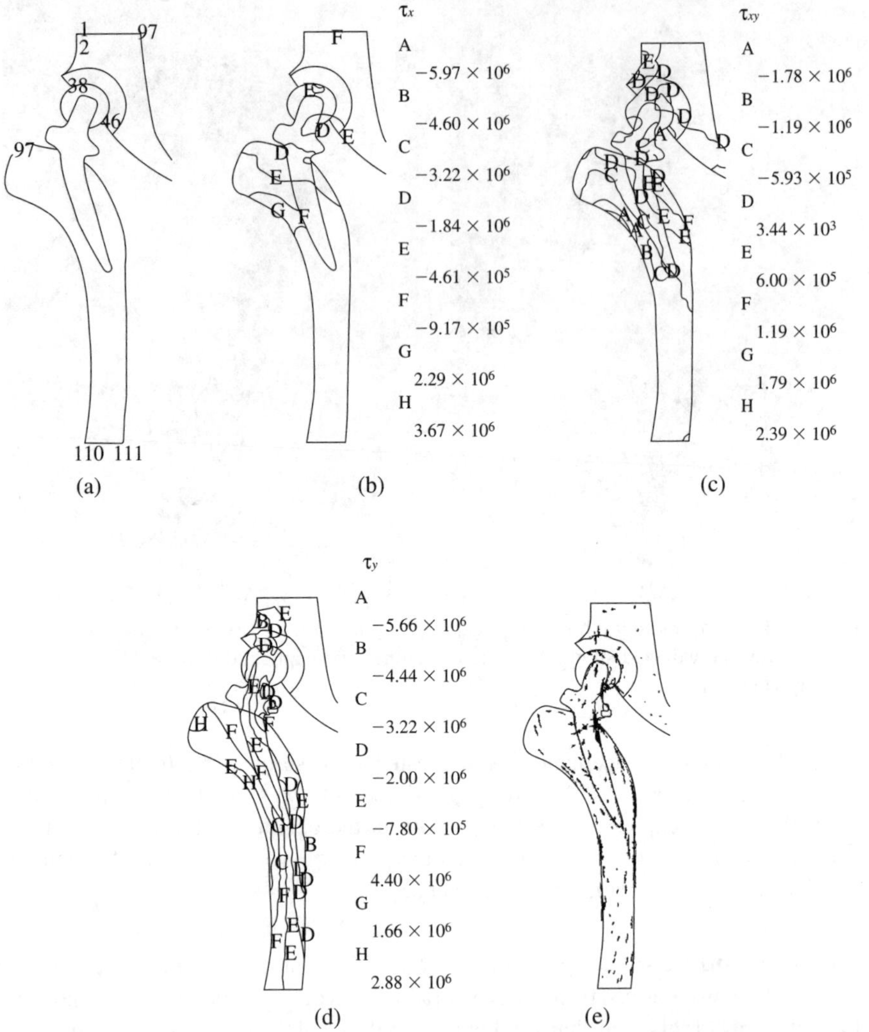

FIG. 7.25 Degenerate hip joint after the surgical treatment: (a) the model, (b) horizontal, and (c) the shear stress components, (d) vertical stress component and (e) principal stresses.

the artificial acetabulum corresponds to the conventional situation. The cranial shift of the whole hip joint in principle does not influence the course of the stress. Another situation is in the area of the proximal femur, in our case the concentration of stresses in the arc of the medial part of the femur below the collar of the endoprosthesis is not observed because the prosthesis is in an extremely sheer and valgization position. Moreover, the bending moment does not exist. This fact is also supported by the course of the principal stresses; see Fig. 7.25(e). Paradoxically, in this case of nonanatomical shape of the proximal femur, the strained prosthesis, as well as a part of the skeleton of the femur from the determinate point of view is even more favorable than in normal anatomical conditions. In our studies, the influence of the muscle displacement as it affects the muscle insertions was not considered.

Discussion Implantation of the THA of the hip joint is a generally accepted solution after fractures in the areas of the proximal femur with ensuing arthritis or "nonunion" of the femoral neck. It is evident that at present it is the best healing method. The deformation of the proximal femur after an injury creates technical difficulties during the surgical treatment, surgical techniques which are necessary and in some cases special types of implants are also needed. (Dunn, 1976), as well as characteristic changes of biomechanics of the hip joint. Clinical experiences show that these changes can have very unfavorable effects in the form of premature arthritis or even a failure of the THA. These effects will have biomechanical consequences on the shape changes. Then an evaluation of these biomechanical aspects give us information and ideas with respect to perspectives of the particular joint system. In this case there is a good correlation between a relatively positive biomechanical situation with slow the clinical devolopment of arthritis. We believe a very good prognosis for the implant system of the total hip joint replacement can be expected. Further clinical inquiry may confirm this.

7.6 ANALYSIS OF LOADED TUBULAR LONG BONE FILLED WITH MARROW TISSUE

7.6.1 Introduction

The long bone is represented as a tubular structure. The central part of the long bone is called the diaphysis; the ends of the long bone are known as the epiphysis, which is made up of the cancellous bone with a thin layer of the compact bone. The diaphysis is characterized by the medullar cavity, which is filled in with marrow. The medullar cavity is situated inside the tubular bone, which is also made up of the cancellous bone with a thin layer of the cortical bone. The cancellous bone is a trabecular structure. The ends of the epiphysis in the contact areas of the joints are covered with articular cartilage. Articular cartilage is structurally heterogeneous, and its primary functions include transmissions and distributions of loads across joints and provision of a smooth surface for relative gliding of joint surfaces. The tibia is one of the long bones, the second longest bone in the human body, and it will be the focus of our model study.

Bone marrow is a soft tissue and it is situated in the hollow interior of bones, namely in long bones, and we speak about the medullar cavity. These represent the places where most new blood cells are produced. There are two bone marrow types—red and yellow marrow. For more details see, for example, the Wiley Encyclopedia of Biomedical Engineering (2006).

The success of the artificial knee joint arthroplasty (TKA) depends on many factors. The mechanical factor is one of the important ones. The idea of a prosthesis being a device that transfers the knee joint loads to the diaphysis describes the mechanical factor in terms of the load transfer mechanism. A complex relation exists between this mechanism, the magnitude, and direction of the loads, the geometry of the TKA configuration, the material properties of the TKA, the bone and the marrow, which fill the medullar cavity of the tibia, and the physical connections at the different (bio-)material connections.

The aim of this section is to determine the distribution of stresses in the long bone and, namely, inside the medullar cavity filled with the yellow marrow, before and then after the application of the artificial joint replacements. In our study we are interested in the healthy long bone only, in our case in the tibia. In connection with the TKA application it will be necessary to evaluate the effect of axial angle changes on the weight-bearing total knee replacements and distributions of transferred stresses across the TKA stem into the lower parts of the tibia bone and into the marrow.

Many surgical treatments on pathological human joints are based on the improvement of bad anatomical and mechanical conditions, resulting from modification or reconstruction of its geometry or its function in connection with applications of their artificial total replacements. In order to study biomechanics of the pathological human joints, the computational models of the physiological human joints and of bones must be first created. The geometry of such models are created using the data from an MRI or a CT scan and/or from an X-ray radiograph.

Recent trends in total joint replacements and clinical failures have provided a strong impetus for the development of accurate stress analyses in normal and prosthetic joints. Theoretical stress analysis methods based on the finite element techniques are frequently used to evaluate the mechanical behavior of loaded bones with (or without) stems of prostheses. Nevertheless, stress–strain analyses of loaded bones filled up with marrow tissue and analyses of stress propagation through marrow tissue were not investigated up to the present.

This section addresses modeling of loaded long bones, such as the femur and the tibia, with (or without) the stem of the prosthesis and filled up with marrow tissue. It is evident that theoretical models using contact problems with (or without) friction in the elastic and viscoplastic (Bingham) rheologies would be able to apply structural stress analyses of loaded bones that depend on a transfer of loading through a marrow tissue.

Next, the model of the weight-bearing long bone (in our study the tibia) filled with marrow will be investigated. The model of the long bone will be investigated in two different rheologies—the first corresponds to the physiological bone with the marrow in elastic rheology, and the second one in viscoplastic rheology. Both models are based on the contact theory in elasticity and viscoplasticity (viscoplastic Bingham rheology). For the geometry of the long bone model the X-ray radiograph of the tibia was used. The stress–strain analyses were performed on computational models by means of finite element methods in 2D.

The model in elasticity is formulated as the primary coercive unilateral contact problem with given friction. For the numerical solution the nonoverlapping domain decomposition method is used, presented in Chapter 6. In the model the contact boundary is represented between the tissue of the bone and the marrow.

7.6.2 The Models

Elastic Model We will assume that the investigated long bone filled with marrow occupies the domain $\Omega = \Omega^1 \cup \Omega^2$, where Ω^1 is occupied by the cortical bone tissue and Ω^2 is occupied by the marrow tissue. About the boundary $\partial\Omega$ we assume to be sufficiently smooth and that it consists of three disjoint parts such that $\partial\Omega = \bar{\Gamma}_u \cup \bar{\Gamma}_\tau \cup \bar{\Gamma}_c$, where by Γ_u we denote the part of the boundary where the bone is fixed; by $\Gamma_\tau = {}^1\Gamma_\tau \cup {}^2\Gamma_\tau$ we denote the part of the boundary where the bone is loaded, that is, ${}^1\Gamma_\tau$, and the part of the boundary where the bone is unloaded, that is, ${}^2\Gamma_\tau$; and by Γ_c the contact boundary between the bone tissue and the marrow tissue (Fig. 7.26). The unilateral and bilateral contact boundaries are assumed to be sufficiently smooth. The bone is loaded by the weight of the human body, and the direction of the acting loading $\mathbf{P}$ corresponds to a real situation.

Let

$$\tau_{ij} = \tau_{ij}(\mathbf{u}) = c_{ijkl}(\mathbf{x})e_{kl}(\mathbf{u}), \qquad i,j,k,l, = 1,\ldots,N,$$

$$e_{ij}(\mathbf{u}) = \frac{1}{2}\left(\frac{\partial u_i}{\partial x_j} + \frac{\partial u_j}{\partial x_i}\right), \qquad i,j = 1,\ldots,N, \tag{7.6}$$

where $c_{ijkl}(\mathbf{x})$ are elastic coefficients, $e_{ij}(\mathbf{u})$ are components of the small strain tensor, and N is the space dimension. About the elastic coefficients $c^\iota_{ijkl} = c^\iota_{ijkl}(\mathbf{x})$ is assumed that they satisfy the conditions of symmetry

$$c_{ijkl} = c_{jikl} = c_{klij} = c_{ijlk},$$

and the condition (7.7)

$$c_{ijkl}e_{ij}e_{kl} \geq c_0 e_{ij}e_{ij} \quad \forall e_{ij}, e_{ij} = e_{ji} \quad \text{for a.e. } \mathbf{x} \in \Omega, c_0 > 0,$$

where $c_0 => 0$ independent of $\mathbf{x} \in \Omega$. A repeated index implies summation from 1 to N.

Then the problem to be solved has the following classical formulation:

Problem $\mathcal{P}$ Let $N = 2, 3$. Find a vector function $\mathbf{u} : \bar{\Omega} \to \mathbb{R}$, satisfying

$$\frac{\partial \tau^\iota_{ij}(\mathbf{u}^\iota, \mathbf{u}^{\iota'})}{\partial x_j} + F^\iota_i = 0, \quad i,j = 1,\ldots,N, \ \iota = 1,2, \quad \text{in } \Omega^\iota, \tag{7.8}$$

$$\tau^\iota_{ij} = \tau^\iota_{ij}(\mathbf{u}^\iota) = c^\iota_{ijkl}(\mathbf{x})e_{kl}(\mathbf{u}^\iota), \quad i,j,k,l = 1,\ldots,N, \ \iota = 1,2, \tag{7.9}$$

$$\tau_{ij}n_j = P_i, \ i,j = 1,\ldots,N, \quad \text{on } {}^1\Gamma_\tau, \tag{7.10}$$

$$\tau_{ij}n_j = 0, \ i,j = 1,\ldots,N, \quad \text{on } {}^2\Gamma_\tau, \tag{7.11}$$

$$u_i = 0, \quad i = 1,\ldots,N, \quad \text{on } \Gamma_u, \tag{7.12}$$

and the bilateral contact condition with given friction

$$
\begin{aligned}
& u_n^1 - u_n^2 = 0, \\
& |\tau_t^{12}| \le \mathcal{F}_c^{12} |\tau_n^{12}(\mathbf{u})| \equiv g_c, \quad \text{on } \Gamma_c, \\
& |\tau_t^{12}| < g_c \Rightarrow \mathbf{u}_t^1 - \mathbf{u}_t^2 = 0, \\
& |\tau_t^{12}| = g_c \Rightarrow \text{there exists } \vartheta \ge 0 \text{ such that } \mathbf{u}_t^1 - \mathbf{u}_t^2 = -\vartheta \tau_t^{12}(\mathbf{u}),
\end{aligned}
\tag{7.13}
$$

where the normal and tangential components of the displacement vector $\mathbf{u} = (u_i)$, $i = 1, \ldots, N$ and the stress vector $\tau = (\tau_i)$, $i = 1, 2$, are defined as $u_n = u_i n_i$, $\mathbf{u}_t = \mathbf{u} - u_n \mathbf{n}$, $\tau_n = \tau_{ij} n_j n_i$, and $\tau_t = \tau - \tau_n \mathbf{n}$, where $\mathbf{n}$ denotes the outward unit normal to the boundary $\partial\Omega$, and on Γ_c we assume that the positive direction of the normal $\mathbf{n}$ is related to ${}^1\Omega$. Moreover, $\mathcal{F}_c^{sm}$ is the coefficient of the Coulombian friction.

Let $W = \sqcap_{\iota=1}^{2} H^{1,N}(\Omega^\iota)$ be the Sobolev space in the usual sense, and let $\|\mathbf{v}\|_W = (\sum_\iota \sum_i \|v_i\|_{1,\Omega^\iota}^2)^{1/2}$ be the corresponding norm. Let us introduce the sets of virtual and admissible displacements

$$
V = \{\mathbf{v} \in W \mid \mathbf{v} = \mathbf{0} \quad \text{on } \Gamma_u, \ v_n^1 - v_n^2 = 0 \text{ on } \Gamma_c\},
$$

and let V_h be the finite element approximation of V.

Let $\mathbf{F}^\iota \in L^{2,N}(\Omega^\iota)$, $\mathbf{P} \in L^{2,N}({}^1\Gamma_\tau)$, $c_{ijkl}^\iota \in L^\infty(\Omega^\iota)$, $\iota = 1, 2$. Then the model problem leads to the following variational problem:

Problem $(\mathcal{P})_\mathbf{v}$ Find a function $\mathbf{u} \in V$ such that

$$
a(\mathbf{u}, \mathbf{v} - \mathbf{u}) + j(\mathbf{v}) - j(\mathbf{u}) \ge L(\mathbf{v} - \mathbf{u}) \quad \forall \mathbf{v} \in V,
\tag{7.14}
$$

where

$$
\begin{aligned}
a(\mathbf{u}, \mathbf{v}) &= \sum_{\iota=1}^{2} \int_{\Omega^\iota} c_{ijkl}^\iota(\mathbf{x}) e_{ij}(\mathbf{u}^\iota) e_{kl}(\mathbf{v}^\iota)\, d\mathbf{x}, \\
j(\mathbf{v}) &= \int_{\Gamma_c} g_c |\mathbf{v}_t^1 - \mathbf{v}_t^2|\, ds, \\
L(\mathbf{v}) &= \sum_{\iota=1}^{2} \int_{\Omega^\iota} F_i^\iota v_i\, d\mathbf{x} + \sum_{\iota=1}^{2} \int_{{}^1\Gamma_\tau \cap \partial\Omega^\iota} P_i v_i\, ds.
\end{aligned}
\tag{7.15}
$$

It can be shown that the problem has a unique solution. The finite element approximation of (7.14) leads to the following problem:

Problem $(\mathcal{P})_h$ Find a function $\mathbf{u}_h \in \mathbf{V}_k$ such that

$$
a(\mathbf{u}_h, \mathbf{v}_h - \mathbf{u}_h) + j(\mathbf{v}_h) - j(\mathbf{u}_h) \ge L(\mathbf{v}_h - \mathbf{u}_h) \quad \forall \mathbf{v}_h \in V_h.
\tag{7.16}
$$

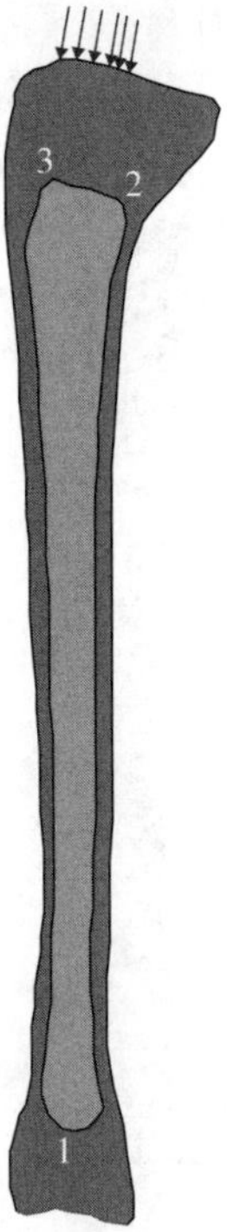

FIG. 7.26 Long bone (tibia) filled in with the marrow tissue: the model.

For the numerical solution of (7.16) the nonoverlapping domain decomposition method will be used [for details see Daněk et al. (2004, 2005a,b); see also Chapter 6].

Numerical Results The obtained results are based on the bilateral contact problem with or without friction in elasticity and the finite element method. The model problem is formulated as a primary contact problem (i.e., in displacements), where the geometry of the weight-bearing tibia is based on its cross section in the sagittal plane. The geometry of the model of the long bone (see Fig. 7.26) is based on the X-ray radiograph. In this section the long bone of the 2D model of the tibia is used. The bone tissue is approximated by the cortical bone tissue only. The biomaterial constants used in the computation are as follows: for the bone tissue $E = 1.71 \times 10^{10}$ Pa, $\nu = 0.25$ and for the marrow $E = 2.0 \times 10^6$ Pa, $\nu = 0.49$. The tibia is fixed on the boundary Γ_u, where the Dirichlet condition $\mathbf{u} = 0$ is described by the red part of the boundary in Fig. 7.26 [see the accompanying DVD disk]. On another part of the boundary the tibia is loaded on ${}^1\Gamma_\tau$ and unloaded on ${}^2\Gamma_\tau$, where the loading is modeled by the Neumann type conditions on ${}^1\Gamma_\tau \cup {}^2\Gamma_\tau$, that is, by the loading $\mathbf{P} = [-0.0035; -0.2] \times 10^7$ Pa on ${}^1\Gamma_\tau$ and by the unloading $\mathbf{P} = [0; 0]$ on ${}^2\Gamma_\tau$. The contact between the bone and the marrow tissues is modeled by the bilateral contact condition on Γ_c (contact boundary is given by the points 1–2–3).

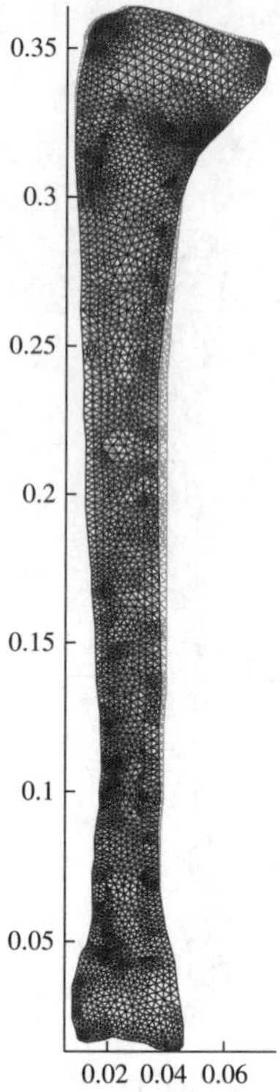

FIG. 7.27 Long bone (tibia) filled in with the marrow tissue: the deformation of the long bone.

Discussion of Numerical Results The numerical results are presented in Figs. 7.26–7.31. The geometry of the weight-bearing tibia is given in Fig. 7.26. The magnitude of the loading is 2.0×10^6 Pa. In Fig. 7.26 the bone tissue is colored in dark brown, while the marrow is colored in yellow [see the accompanying DVD disk]. The articular cartilage is not assumed in the investigated model. Deformation of the long bone, the tibia, is presented in Fig. 7.27 Small bending of the tibia in its middle part is observed. In Fig. 7.28 the horizontal and vertical components of the displacement vector are presented. The horizontal component of the displacement vector indicates the bending in the middle part of the tibia, while the vertical component of displacement vector indicates certain vertical displacement changes in the marrow, namely in the middle part of the tibia. Moreover, it indicates elementary shifts of the biomaterial points of the marrow tissue on the contact surface between the bone tissue and the marrow, as well as elementary shifts in the marrow tissue inside the tubular bone. Figure 7.29 represents distributions of the horizontal, vertical, and shear stresses. Numerical results show that greater changes of horizontal stresses are observed in the areas near points 2 and 3 and on both sides of point 1 in the epiphyses and smaller changes also in the bone tissue. The vertical stresses indicate changes in pressures on the outside right part of the bone tissue in the middle tibia and on the inside left part of the middle tibia tissue, while changes in tensile

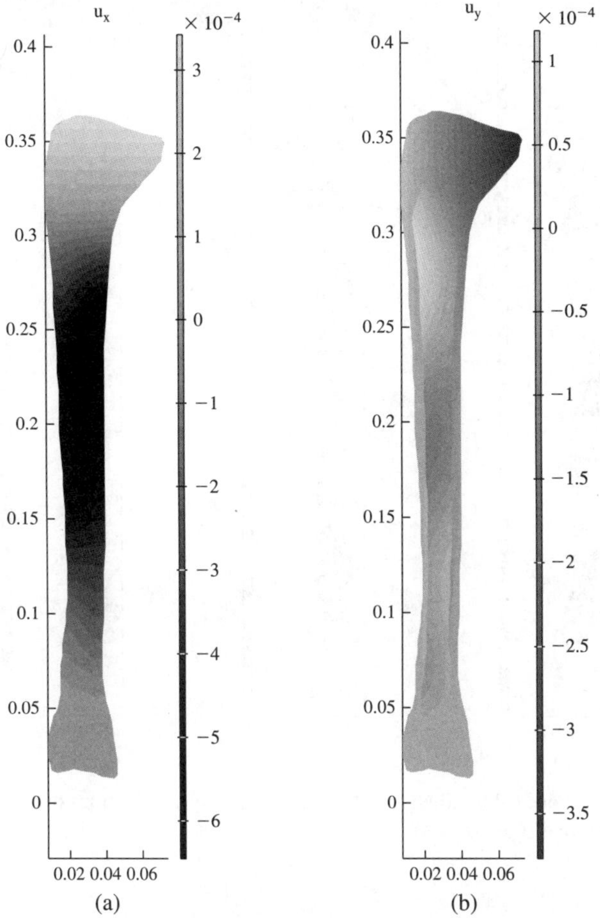

FIG. 7.28 Long bone (tibia) filled with the marrow tissue: distributions of (a) horizontal and (b) vertical displacement components.

stresses are observed in the inside right part of the bone tissue and in the outside left part of the middle tibia tissue. Moreover, it also indicates the probability of small changes in vertical stresses in the marrow area. Changes of shear stresses are observed in the middle part of the tibia and in the areas between the marrow and bone tissues in the epiphysis. Greater changes of shear stresses can be also observed in the epiphysis of the tibia. The elementary changes of tangential stresses on the contact surface between the marrow and the bone tissues are also indicated. The principal stresses in Fig. 7.30 show that the great value of loading is transported through the bone tissue as the compression predominantly on the right outer part of the compact bone and through the layer of small thickness near the left inside wall of the bone, while the outer left part of the bone tissue in the middle tibia is characterized by tensile stresses. Moreover, the compression is also observed in the cancellous bone of both

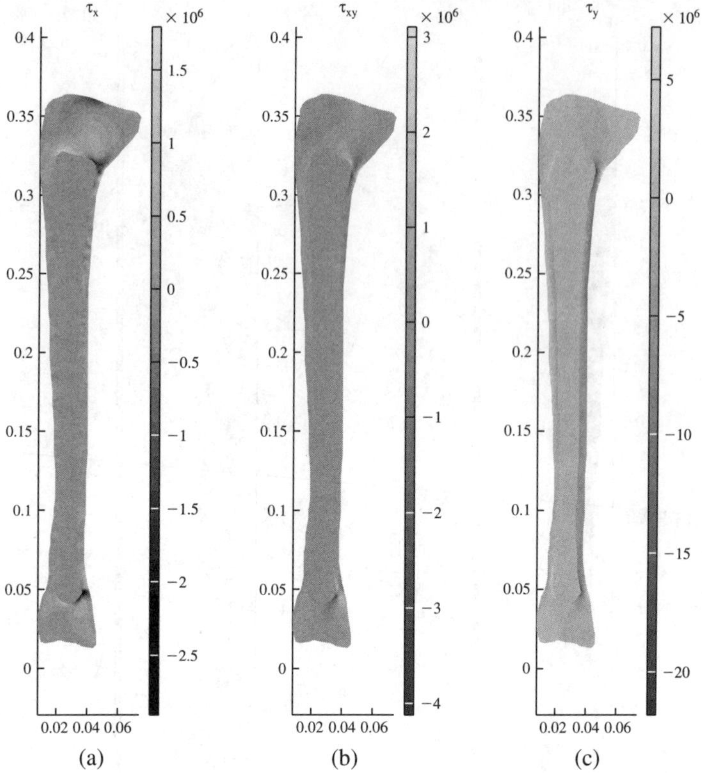

FIG. 7.29 Long bone (tibia) filled with the marrow tissue: distributions of (a) horizontal, (b) shear, and (c) vertical stresses.

epiphyses. The principal stresses presented in Fig. 7.30 show that the great loading is transported through the right side of the tubular tibia. On the contact boundary between the bone and the marrow tissues small tangential shifts can be expected. All these observed effects are consequences of bending in the middle part of the tibia. Furthermore, we see that Figs. 7.29 and 7.30 show that a great value of loading is transported through the bone tissue. In Fig. 7.30(b) the distribution of principal stresses in the medullar cavity filled in with the yellow marrow is presented in detail. We see that the yellow marrow in the medullar cavity in its upper and lower parts are compressed, and that the magnitude of stresses in both areas are 1000 × lower than in the corticalis of the tubular tibia, while the distribution of the stress field in the yellow marrow in the middle part of the tubular tibia as a consequence of bending is characterized by a greater area of tension, the magnitude of which is about several orders lower than tensions or compressions in the bone tissue and of one order lower than compressions in the marrow of the upper and lower parts of the medullar cavity.In Figs. 7.30a and 7.30b the blue arrows refer to tensile stresses while the red arrows refer to the pressures [see the accompanying DVD disk]. In Fig. 7.31 the normal

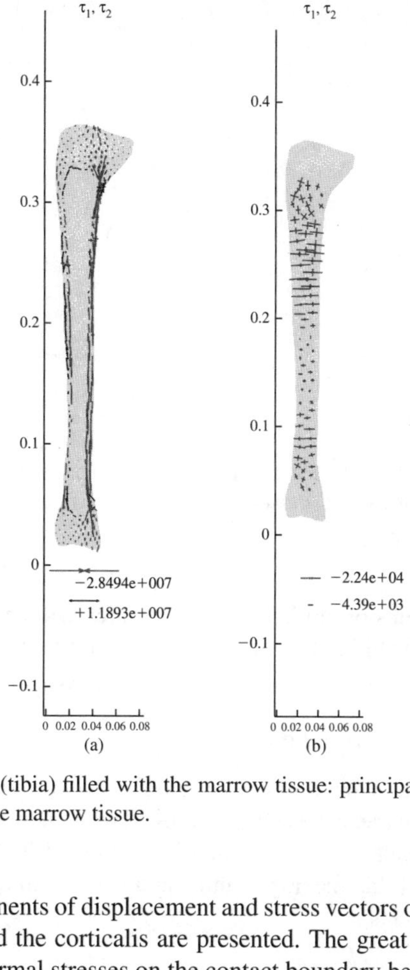

FIG. 7.30 Long bone (tibia) filled with the marrow tissue: principal stresses in (a) the tibia bone tissue and in (b) the marrow tissue.

and tangential components of displacement and stress vectors on the surface between the marrow tissue and the corticalis are presented. The great changes in tangential displacements and normal stresses on the contact boundary between the marrow tissue and the corticalis are connected with bending of the loaded bone. In each case it is a very interesting result.

Conclusion From the above-mentioned analyses of the weight-bearing tubular tibia, we see that a great value of loading is transported through the bone tissue, and, moreover, that small bending of the tibia in its middle part is observed. Numerical results indicate that the yellow marrow in the medullar cavity in its upper and lower parts are compressed and that the magnitude of stresses in both areas are 1000× lower than in the corticalis of the tubular tibia. Moreover, the distribution of the stress field in the yellow marrow in the middle part of the tubular tibia, as a consequence of bending, is characterized by a greater area of tension with the magnitude that is of several orders lower than tension or compressions in the bone tissue and of one order lower than compressions in the marrow of the upper and lower parts of the medullar cavity. Convincing biomechanical analyses can give the analyses of the TKA (and/or

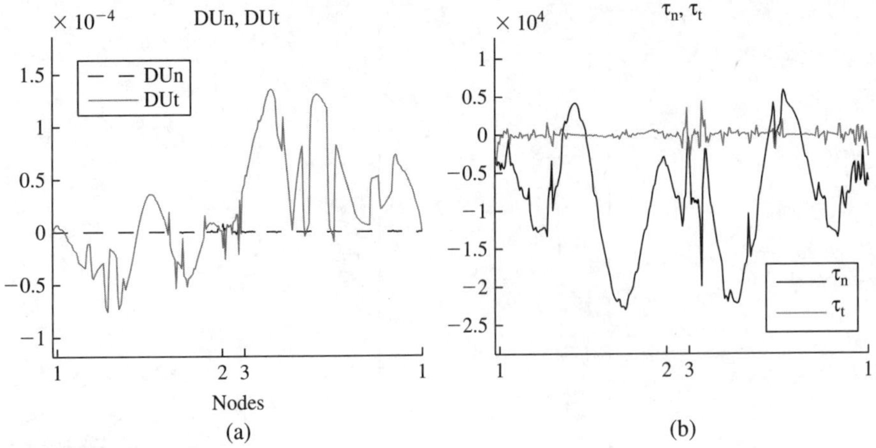

FIG. 7.31 Long bone (tibia) filled in with the marrow tissue: normal and tangential components of (a) displacements and (b) stresses on the contact surface between the marrow tissue and the corticalis bone.

the artificial replacements of hip joint) models, which will analyze distributions of stresses in areas near the tibial stem of the artificial knee joint.

7.6.3 Stationary Viscoplastic Model

Since the bone as well as marrow tissues are biomaterials of viscoplastic characteristics, then the problem of the loaded long bone filled with marrow tissue can be solved also as a contact problem with viscoplastic friction in the Bingham viscoplastic rheology. It is evident that this rheology can be applied also to other model problems discussed in biomechanics.

Formulation of the Problem Let $\mathcal{S}_N$ be the space of second-order symmetrical tensors on $\mathbb{R}^N$, that is, $\mathcal{S}_N = \mathbb{R}^{N\times N}$. Let $\mathbf{u}$ be the material velocity, $D = D(\mathbf{u}) = \frac{1}{2}(\nabla\mathbf{u} + \nabla^{\mathrm{T}}\mathbf{u})$ be the rate deformation tensor, $D^D = D - \frac{1}{3}(\mathrm{tr}\ D)I_N$ its deviator, I_N the identity map on $\mathbb{R}^N$. Let τ be the Cauchy stress tensor, $\tau^D = \tau + pI_N$ its deviator, where $-p$ represents the spherical part of the stress tensor and has a meaning as pressure.

Next, we will solve the following problem:

Problem $\mathcal{P}$ Find the velocity $\mathbf{u} : \Omega \to \mathbb{R}^N$, $N = 2, 3$, and a stress tensor $\tau : \Omega \to \mathbb{R}^{N\times N}$ satisfying

$$\mathrm{div}\,\tau + \mathbf{F} = 0 \quad \text{in } \Omega; \tag{7.17}$$

$$\mathrm{tr}\, D = \mathrm{div}\,\mathbf{u} = 0 \quad \text{in } \Omega, \tag{7.18}$$

$$\begin{aligned} \tau^D &= \hat{g}D|D|^{-1} + 2\hat{\mu}D \quad \text{if} \quad |D| \neq 0, \\ |\tau^D| &\leq \hat{g} \quad \text{if} \quad |D| = 0, \end{aligned} \tag{7.19}$$

with the boundary value conditions

$$\tau_{ij}n_j = P_i \text{ on } {}^1\Gamma_\tau, \tag{7.20}$$

$$\tau_{ij}n_j = 0 \text{ on } {}^2\Gamma_\tau, \tag{7.21}$$

$$\mathbf{u} = \mathbf{u}_1 \ (= 0 \text{ or } \neq 0) \quad \text{on} \quad \Gamma_u, \tag{7.22}$$

and with the bilateral contact condition with the local friction law on Γ_c^{12} :

$$\begin{aligned} &u_n^1 - u_n^2 = 0 \quad \text{and} \quad |\tau_t^{12}| \leq \mathcal{F}_c^{12}|\tau^{D12}|, \\ &\text{if } |\tau_t^{12}| < \mathcal{F}_c^{12}|\tau^{D12}|, \quad \text{then } \mathbf{u}_t^1 - \mathbf{u}_t^2 = 0, \\ &\text{if } |\tau_t^{12}| = \mathcal{F}_c^{12}|\tau^{D12}|, \quad \text{then there exists } \lambda \geq 0 \\ &\qquad\qquad\qquad\qquad\qquad \text{such that } \mathbf{u}_t^1 - \mathbf{u}_t^2 = -\lambda\tau_t^{12}, \end{aligned} \tag{7.23}$$

where $\mathcal{F}_c^{12}$ is a coefficient of friction, $u_n^1 = u_i^1 n_i^1$, $u_n^2 = u_i^2 n_i^1 = -u_i^2 n_i^2$, $\tau_n^{12} = \tau_n^1 = \tau_n^2$, $\tau_t^{12} = \tau_t^1 = -\tau_t^2$.

If we determine $|\tau^D|$ from (7.19), we obtain $|\tau^D| = \hat{g} + 2\hat{\mu}|D(\mathbf{u})|$. Then the contact condition (7.23) depends on the solution of the investigated problem.

Variational (Weak) Solution of the Problem We introduce the Sobolev spaces as the spaces of all vector functions having generalized derivatives of the (possibly fractional) order s of the type $[H^s(\Omega)]^k \equiv H^{s,k}(\Omega)$, where $H^s(\Omega) \equiv W_2^s(\Omega)$. The norm will be denoted by $\|\cdot\|_{s,k}$ and the scalar product by $(\cdot,\cdot)_s$ (for each integer k). We set $H^{0,k}(\Omega) \equiv L^{2,k}(\Omega)$.

We introduce the set of kinematic admissible fields defined by

$$K = \{\mathbf{v} \in H^{1,N}(\Omega) |\ \operatorname{div} D = 0 \text{ in } \Omega,\ \mathbf{v} = \mathbf{u}_1 \text{ on } \Gamma_u, v_n^1 - v_n^2 = 0 \text{ on } \Gamma_c\}.$$

Let $\mathbf{P} \in L^{2,N}({}^1\Gamma_\tau)$, $\mathbf{F} \in L^{2,N}(\Omega)$, $\hat{g}$, $\hat{\mu}$ be a piecewise constant and positive and let $\mathcal{F}_c^{12} \in L^\infty(\Gamma_c^{12})$, $\mathcal{F}_c^{12} \geq 0$ a.e. on Γ_c^{12} and $\mathbf{u}_1 \equiv 0$.

For $\mathbf{u}, \mathbf{v} \in H^{1,N}(\Omega)$ we put

$$a(\mathbf{u},\mathbf{v}) = \sum_{\iota=1}^{2} a^\iota(\mathbf{u}^\iota,\mathbf{v}^\iota) = 2\int_\Omega \hat{\mu}D(\mathbf{u})D(\mathbf{v})\,d\mathbf{x},$$

$$S(\mathbf{v}) = \sum_{\iota=1}^{2} S^\iota(\mathbf{v}^\iota) = \int_\Omega f_i v_i\,d\mathbf{x} + \int_{\Gamma_\tau} P_i v_i\,ds \equiv (\mathbf{F},\mathbf{v}),$$

$$j(\mathbf{v}) = \sum_{\iota=1}^{2} j^{\iota}(\mathbf{v}^{\iota}) = \int_{\Omega} \hat{g}|D(\mathbf{v})|\, d\mathbf{x},$$

$$j_g(\mathbf{v}) = \sum_{\iota=1}^{2} j_g^{\iota}(\mathbf{v}^{\iota}) = \int_{\Gamma_c^{12}} \mathcal{F}_c^{12}|\tau^{D12}|\, |\mathbf{v}_t^1 - \mathbf{v}_t^2|\, ds.$$

Let us multiply (7.17) with (7.19) by $\mathbf{v}-\mathbf{u}$, integrate over Ω, and apply the Green theorem satisfying the boundary conditions. Then after some modification, we obtain the following variational (weak) formulation:

Problem $(\mathcal{P})_v$ Find a function $\mathbf{u} \in K$ such that

$$a(\mathbf{u}, \mathbf{v}-\mathbf{u}) + j(\mathbf{v}) - j(\mathbf{u}) + j_g(\mathbf{v}) - j_g(\mathbf{u}) \geq S(\mathbf{v}-\mathbf{u}) \quad \forall \mathbf{v} \in K. \tag{7.24}$$

The bilinear form $a(\mathbf{u}, \mathbf{v})$ is symmetric, that is, $a(\mathbf{u}, \mathbf{v}) = a(\mathbf{v}, \mathbf{u})$. Moreover, for $\mathbf{u} \in V$ there exists constant $c_B > 0$ such that $a(\mathbf{u}, \mathbf{u}) \geq c_B \|\mathbf{u}\|_{1,N}^2$ for all $\mathbf{u} \in V$.

The analysis of the problem will be based on the penalization, regularization, and finite element techniques. The algorithm is the modification of Ionescu and Sofonea (1993). We shall assume that $\mathbf{u}_1 \neq 0$. Then by means of a translation, we will obtain the homogeneous boundary condition and $V = K - \mathbf{u}_1$, where V is a closed subspace of $H^{1,N}(\Omega)$.

Problem $\mathcal{P}_v$ is equivalent to the following problem:

Problem $(\mathcal{P}_{sf})_v$ Under the above-mentioned hypothesis if $(\mathbf{u}; \tau)$ is a smooth solution for Problem $\mathcal{P}$, then $\mathbf{u}$ is a minimum point of the functional $\mathcal{J}$ on K, that is,

$$\mathbf{u} \in K, \quad \mathcal{J}(\mathbf{u}) \leq \mathcal{J}(\mathbf{v}) \quad \forall \mathbf{v} \in K,$$

where (7.25)

$$\mathcal{J}(\mathbf{v}) = \tfrac{1}{2} a(\mathbf{v}, \mathbf{v}) + j(\mathbf{v}) + j_g(\mathbf{v}) - S(\mathbf{v}).$$

Numerical Solution Let us introduce the space $\mathcal{W}$, a closed subspace of $H^{1,N}(\Omega)$, by

$$\mathcal{W} = \{\mathbf{v} \mid \mathbf{v} \in H^{1,N}(\Omega), \mathbf{v}|_{\Gamma_u} = 0, v_n^1 - v_n^2 = 0 \text{ on } \Gamma_c^{12}\}, \tag{7.26}$$

in which the incompressibility condition $\operatorname{div} \mathbf{u} = 0$ is not introduced.

Since the linear space $V = K - \mathbf{u}_1, \mathbf{u}_1 \in V$ on which the variational problem is formulated contains the condition of incompressibility representing certain cumbersome elements for a numerical solution, therefore we apply a penalty technique, similarly as in the case of the incompressible Newtonian fluid [see Temam (1979)]. The penalty term will be introduced by

$$P(\mathbf{u}_\varepsilon) = \frac{1}{\varepsilon} c(\mathbf{u}_\varepsilon, \mathbf{u}_\varepsilon), c > 0,$$

where

$$c(\mathbf{u}, \mathbf{v}) = \int_\Omega (\text{div } \mathbf{u})(\text{div } \mathbf{v})\, d\mathbf{x} \quad \forall \mathbf{u}, \mathbf{v} \in H^{1,N}(\Omega).$$

It can be shown that for each $\varepsilon > 0$ the corresponding penalized variational inequality has a unique solution and that its corresponding solution converges strongly in $H^{1,N}(\Omega)$ to the solution of the initial problem when $\varepsilon \to 0$.

To solve the penalized problem numerically, we set $\bar{\mathbf{u}} = \mathbf{u} - \mathbf{u}_1$ and then the finite element technique will be used. Let $\mathcal{W}_h \subset \mathcal{W}$ be a family of finite element subspaces with the following property:

$$\forall \mathbf{v} \in \mathcal{W} \text{ there exists } \mathbf{v}_h \in \mathcal{W}_h \text{ such that } \mathbf{v}_h \to \mathbf{v} \text{ in } H^{1,N}(\Omega) \text{ for } h \to 0. \tag{7.27}$$

Then we will solve the following problem:

Problem $(\mathcal{P}_{sf})_h$ Find $\bar{\mathbf{u}}_{\varepsilon h} \in \mathcal{W}_h$ satisfying the variational inequality

$$\begin{aligned} &a_h(\bar{\mathbf{u}}_{\varepsilon h}, \mathbf{v}_h - \bar{\mathbf{u}}_{\varepsilon h}) + j_h(\mathbf{v}_h) - j_h(\bar{\mathbf{u}}_{\varepsilon h}) + j_{gh}(\mathbf{v}_h) \\ &\quad - j_{gh}(\bar{\mathbf{u}}_{\varepsilon h}) + \frac{1}{\varepsilon} c_h(\bar{\mathbf{u}}_{\varepsilon h}, \mathbf{v}_h - \bar{\mathbf{u}}_{\varepsilon h}) \geq S_h(\mathbf{v}_h - \bar{\mathbf{u}}_{\varepsilon h}) \; \forall \mathbf{v}_h \in \mathcal{W}_h. \end{aligned} \tag{7.28}$$

Lemma 7.1 Let $\bar{\mathbf{u}}_{\varepsilon h}$ be a solution of (7.28) for each $h > 0$ and let $\bar{\mathbf{u}}_\varepsilon$ be the solution of the penalized problem for a fixed $\varepsilon > 0$. Then

$$\bar{\mathbf{u}}_{\varepsilon h} \to \bar{\mathbf{u}}_\varepsilon \text{ strongly in } H^{1,N}(\Omega) \text{ when } h \to 0.$$

The proof follows from Lemma A4.2 of Ionescu and Sofonea (1993).

Since the functionals $j(\mathbf{v})$ and $j_g(\mathbf{v})$ are not differentiable in the Gâteaux sense, they can be regularized. Therefore, we introduce the function $\psi_\gamma : \mathbb{R} \to \mathbb{R}$ defined by

$$\psi_\gamma(x) = (x^2 + \gamma^2)^{1/2} - \gamma, \tag{7.29}$$

which regularizes the function $x \to |x|$. Then ψ_γ is differentiable and the following inequality

$$||x| - \psi_\gamma(|x|)| < \gamma \quad \forall x \in \mathbb{R}, \qquad \gamma \geq 0 \tag{7.30}$$

holds.

Then the functionals $j(\mathbf{v})$ and $j_g(\mathbf{v})$ will be regularized by their regularizations $j_\gamma(\mathbf{v})$ and $j_{g\gamma}(\mathbf{v})$, defined by

$$j_\gamma(\mathbf{v}) = \int_\Omega \hat{g}\psi_\gamma(|D(\mathbf{v}+\mathbf{u}_1)|)\,d\mathbf{x},$$

$$j_{g\gamma}(\mathbf{v}) = \int_{\Gamma_c^{12}} \mathcal{F}_c^{12}[\hat{g} + 2\hat{\mu}\psi_\gamma(D|\mathbf{v}+\mathbf{u}_1|)\psi_\gamma(|\mathbf{v}^1 - \mathbf{v}^2 + (\mathbf{u}_1^1 - \mathbf{u}_1^2)|)\,ds.$$

Then we will solve the penalized-regularized problem: Find $\bar{\mathbf{u}}_{\varepsilon h\gamma} \in \mathcal{W}_h$ satisfying

$$\begin{aligned} a_h(\bar{\mathbf{u}}_{\varepsilon\gamma h}, \mathbf{v}_h - \bar{\mathbf{u}}_{\varepsilon\gamma h}) + j_{\gamma h}(\mathbf{v}) - j_{\gamma h}(\bar{\mathbf{u}}_{\varepsilon\gamma h}) + j_{g\gamma h}(\mathbf{v}_h) - j_{g\gamma h}(\bar{\mathbf{u}}_{\varepsilon h}) \\ + \frac{1}{\varepsilon} c_h(\bar{\mathbf{u}}_{\varepsilon\gamma h}, \mathbf{v}_h - \bar{\mathbf{u}}_{\varepsilon\gamma h}) \geq S_h(\mathbf{v}_h - \bar{\mathbf{u}}_{\varepsilon\gamma h}) \quad \forall \mathbf{v}_h \in \mathcal{W}_h. \end{aligned} \tag{7.31}$$

It can be shown that the functionals $j_{\gamma h}(\mathbf{v})$ and $j_{g\gamma h}(\mathbf{v})$ are convex and continuous, and, therefore, problem (7.31) has a unique solution $\bar{\mathbf{u}}_{\varepsilon\gamma h} \in \mathcal{W}_h$. As a result we have the following theorem:

Theorem 7.1 Let $\mathbf{u}_\varepsilon = \bar{\mathbf{u}}_\varepsilon + \mathbf{u}_1$, where $\bar{\mathbf{u}}_\varepsilon$ is a solution of the penalized problem with the homogenous condition on Γ_u, $\mathbf{u}_{\varepsilon h} = \bar{\mathbf{u}}_{\varepsilon h} + \mathbf{u}_1$, $\mathbf{u}_{\varepsilon\gamma h} = \bar{\mathbf{u}}_{\varepsilon\gamma h} + \mathbf{u}_1$ for all $\varepsilon, \gamma, h > 0$. Let $\mathbf{u}$ be the solution of Problem $(\mathcal{P}_{sf})_v$. Then

$$\begin{aligned} &\text{(i)} \quad \mathbf{u}_\varepsilon \to \mathbf{u} \ \text{ strongly in } H^{1,N}(\Omega) \text{ when } \varepsilon \to 0, \\ &\text{(ii)} \quad \mathbf{u}_{\varepsilon h} \to \mathbf{u}_\varepsilon \text{ strongly in } H^{1,N}(\Omega) \text{ when } h \to 0, \\ &\text{(iii)} \quad \mathbf{u}_{\varepsilon\gamma h} \to \mathbf{u}_{\varepsilon h} \text{ strongly in } H^{1,N}(\Omega) \text{ when } \gamma \to 0. \end{aligned} \tag{7.32}$$

Numerically, the problem leads to solving the nonlinear algebraic system, which can be solved by, for example, the Newton iterative method and/or the PDAS method.

For the dynamically loaded long bone or the other parts of the human skeleton see Nedoma (2004) and generally Duvaut, and Lions (1976), Glowinski (2003), and Nedoma (2003, 2004, 2006, 2010).

7.6.4 Conclusion

Recent trends in total joint replacements and clinical failures have provided a strong impetus for the development of accurate stress analyses in normal and prosthetic joints. Theoretical stress analysis methods based on the finite element techniques are frequently used to evaluate the mechanical behavior of loaded bones with (or without) stems of prostheses. Nevertheless, stress–strain analyses of loaded bones filled up with a marrow tissue and analyses of stress propagation through marrow tissue have not been investigated up to the present. Our study demonstrates that the marrow tissue inside the corticalis of the long bone, in our study modeled by the tibia, was deformed during the loading and that stresses also in the marrow tissue were observed. Similar situations can be also observed in the case of total joint replacements and, namely, in the areas near the stems of the prostheses.

7.7 NUMERICAL ANALYSIS OF THE WEIGHT-BEARING TOTAL KNEE REPLACEMENT; ANALYSIS OF EFFECT OF AXIAL ANGLE CHANGES ON WEIGHT-BEARING TOTAL KNEE ARTHROPLASTY

7.7.1 Analysis of Effect of Axial Angle Changes on Weight-Bearing Total Knee Arthroplasty

Introduction and Clinical Aspect of the Problem The aim of this section is to compare the biomechanical influences of different grades of valgus deformity after the timely dependent total knee arthroplasty (TKA). Investigation of axial angle changes on the weight-bearing total knee arthroplasty has been studied by many authors (Ritter et al., 1994; Hungerford, 1995; Sparmann et al., 2003).

From the perspective of orthopedics, division load acting on the tibial component after implantation of the total knee arthroplasty is of primary importance. The pressure ratios in the knee after the TKA are decided according the soft tissue tension (capsule, ligaments, muscular insertions) in the vicinity of the replacement and the resulting axial position of the whole limb. Both these basic factors are influenced especially by the technique of implantation, and, in a decisive way, it determines the survival time of the implant. From many reports it follows unambiguously that nonobservance of the balance of both compartments (medial and lateral) or possible overloading of the posterior part of the tibia plate leads to quick wear out with considerable abrasion of the polyethylene insert. No implant tolerates mistakes in surgical techniques, older simple types of prostheses fail just like the more sophisticated and much more expensive modern implants do.

Due to asymmetrical overloading, the premature abrasion of a plastic insertion made from polyethylene elements, initiates a complicated reaction leading to a loosening of metallic components of the replacement from the bone.

Soft Tissue Tension Balancing the tension in soft tissue in the frontal plane (lateromedial) of the external and internal parts of the joint is very difficult, and surgeons often resort to compromises in cases with very severe deformity. A similar problem occurs in the sagittal plane. The contracture of the posterior cruciate ligament (PCL) unfavorably influences the load in the posterior part of the joint. It is imperative to loosen partly the ligament very often or to resect it completely. In such cases it is necessary to use a variant of the implant with mechanical posterior stabilization. Evaluation of the joint balance is, to a large extent, subjective and measurable with difficulty and, by this time, even the most up-to-date instrumental systems with a computer-assisted instrumentation do not solve the problem entirely. The question that measures the precision of the soft tissue balance and of the axial deviations still is essential for long-term survival of the implant.

Axial Position of the Limb Affecting the axial position of a limb there are many factors that must be considered and evaluated clinically and radiologically in detail before the surgery. Changes on the skeleton are: (i) acquired or congenital changes of the form of the femur or the proximal tibia and (ii) deviation in the axial knee joint axis under the influence of the adjoining joints.

Changes in the form of both bones that make up the knee joint can cause incorrect positioning of the component. In the case of the distal femur it is very often a gradually developing deformity, rarely hypoplasia of the lateral condyle causing a valgus deformity of the knee. To a certain extent the rotation of the vertical axis is a component of the valgus deformity.

A different situation arises on the posttraumatic axial deformities of the femur when gradual damage of the overloaded components of the femur and development of the joint deformity come to light. That means that the valgus deformity of the femoral diaphysis is caused by overloading the external side of the joint, of its gradual damage, and, in the final phase, to the valgus deformity of the knee. The varus knee position varies when the diaphysis is healing. As for the proximal tibia, the greatest difficulties exist in defects in individual compartments caused during implantation. In case of a defect on the medial side of the tibial plate varus deformity is present, and on the lateral side valgus deformity of the knee joint is present. Large defects on the medial side often extend dorsally and support formation of flection contracture. After tibial osteotomy we see with patients with a specific deformity in the external rotation of the metaphysis.

Axial deviations of the lower limb can be caused by the adjoining joints, that is, as a compensatory deformity in the area of the knee joint.

The Upper End of the Femur When deformities are acquired, the varus position of the femoral neck results in lateralization of diaphysis, and valgus position results in its medialization. Thereafter, compensatory mirror deformities emerge in the area of the knee joint.

The tibia: There is a noticeable varus position, which cannot be revised within the resection of the proximal tibia. In order to make it possible to walk with the plantigrade position of the foot, the compensatory position of the ankle joints emerge in the mentioned situations. Alternatively, rigid deformities of the leg and the talocrural (ankle) joint can unfavorably influence the position of the once healthy knee joint by the formation of a compensatory deformity.

Before the TKA operation of the knee joint, we evaluate the real axis of the lower limb from the X-ray radiograph, which must be taken not only of the knee joint but also the head of the femur and ankle joint. The conjunction of the femur center with the head of the femur in the center of the ankle joint determines the mechanical axis of the lower limb. After the adjustment of the mechanical axis, the angle with the anatomical femur axis determines the degree of physiological valgus, in which the resection of the distal end of the femur should be performed. If we maintain the correct technique of implantation, restore the mechanical axis of the lower limb and the collateral soft tissue balance, and prevent the flection gap of the knee from being too tight, we will create the basic conditions for the correct biomechanical function of the total knee arthroplasty. The results of the mathematical simulation of the load by the finite element method with numerical deviations in various degrees of resection of the distal

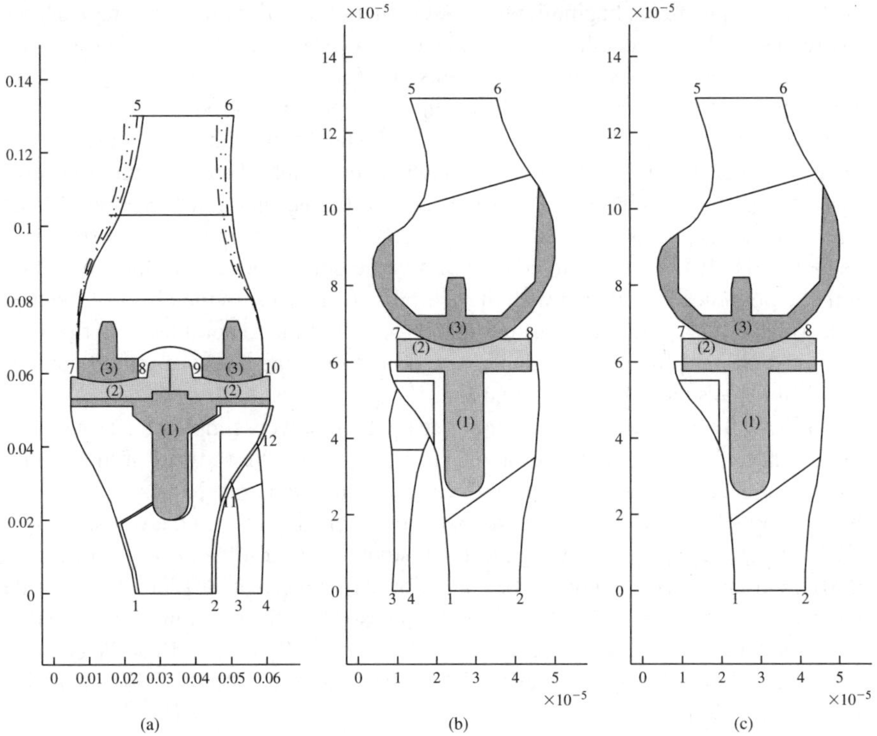

FIG. 7.32 Models of axial changes of the weight-bearing total knee replacements: the models of knee joint replacement in (a) frontal and (b) and (c) sagittal planes.

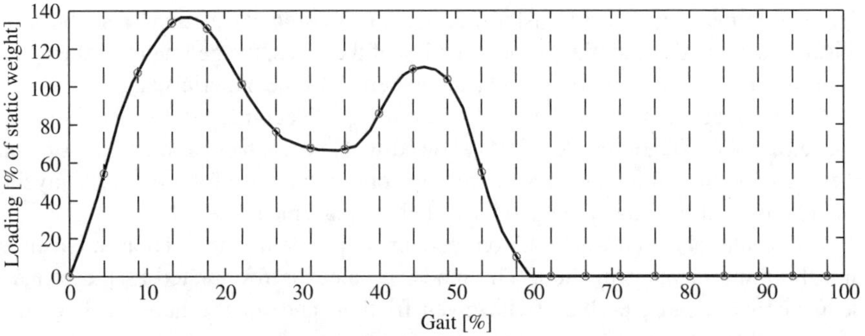

FIG. 7.33 Function of loading of the knee joint during gait. [Modified after Anderson and Pandy (2001), Fregly et al. (2003, 2005, 2008, and Fernandez and Pandy (2006).]

end of the femur are shown by individual models (Fig. 7.32) as well as the function of loading the knee joint during walking (Fig. 7.33).

The fundamental analyses of a healthy normal human gait were carried out by Fischer at the end of the nineteenth century [see Fischer (1907)]. He divided one complete stride into 31 positions, beginning with a stationary bipedal support, where positions 1–4 represent the monopedal support; positions 5–10 represent the monopedal support of a normal gait; position 11 represents the bipedal support; positions 12–23 the right monopedal support; position 24 the bipedal support; positions 25–30 are the left monopedal support; and position 31 is that of the final stop, and it is not included in the normal gait. During walking the hip and the knee joints change positions according to the particular phase of the stride. During the normal gait the femoral head constantly shifts position in relation to the sourcil, introduced by Pauwels (in 1963) [see Pauwels (1973)], as a curved area of a dense bone. Because of this the section of the femoral head in contact with the sourcil also changes. In the phase of the heel strike (phase 12) it is the superoanteromedial aspect of the femoral head, in the foot-to-ground phase (phase 16) the superior aspect, and in the phase of toe off (phases 22–23) it is the supero-posteromedial aspect.

During walking the situation, for the knee joint is relatively simplier than for the hip joint because the movements during the stride, during the loaded part of the leg, are small. In this case the loading acts in the direction of the biomechanical axis, during the unloading part of the strike and has a relatively small effect on the whole loading effect during these phases of the gait. In our study we used some of results of Pandy (2001), Fernandez and Pandy (2006), and Former-Cordero et al. (2006). The stride was divided into 24 steps. The loading function used for the knee joint is discussed in Anderson and Pandy (2001), Fregly et al. (2003, 2005, 2008), Herr and Wilkenfeld (2003), Bei et al. (2004), Geyer et al. (2006), Fernandez and Pandy (2006), and also in Kim et al. (2009).

The Model The success of the artificial replacements of the knee joint depends on many factors. The mechanical factor is an important one. The idea of a prosthesis being a device that transfers the knee joint loads to the bone explains the mechanical factor in terms of the load transfer mechanism. A complex relation exists between this mechanism, the magnitude and direction of the loads, the geometry of the bone–joint prosthesis configuration, the elastic properties of the materials and the physical connections at the material connections. The authors (Nedoma, 1993b; Nedoma et al., 2003a,b, 2006; Daněk et al., 2004) showed that the contact problems in suitable rheology and their finite element approximations are very useful tools for analyzing the relations for the knee joint and its artificial replacement.

The results obtained are based on the contact problem with friction in elasticity and the finite element method. The model problem is formulated as the primary semicoercive contact problem with given friction, and for the numerical solution of the problem the nonoverlapping domain decomposition method is discussed in Chapter 6. Next, two types of problems will be analyzed—the axial angle changes of the weight-bearing total knee replacement in the frontal plane and in the sagittal plane in static and dynamic loading.

Nine models of the total replacement of the left knee joint deal with linkage on the axial deviation, that is, MODEL I, MODEL II, and MODEL III. The models are considered in the frontal cut (MODEL I) where it is possible to analyze the influence of the axial deviation. The sagittal cuts (MODEL II—across the outer condyle, MODEL

III—across the inner condyle) tell little about the influence of the axial deviation but tell more about the overload of the posterior part of the tibial plate in the sagittal (antero-posterior) direction. For the loading function used during walking for the knee joint, see Anderson and Pandy (2001), Fregly et al. (2003, 2005, 2008), Herr and Wilkenfeld (2003), Bei et al. (2004), Geyer et al. (2006), Fernandez and Pandy (2006), and Kim et al. (2009).

The Frontal Plane—the Static Case MODEL I-k, where $k = 3, 5, 7, 9$, is connected with the angle of the valgus in the resection of the lower end of the femur k degrees, with MODEL II-k deg. and MODEL III-k deg. corresponding to models (b) and (c) in the sagittal plane (Fig. 7.32). In all models the following material parameters are used:

Material		Young's Modulus of Elasticity E (Pa)	Poisson Constant ν
(i)	Bone	1.71×10^{10}	0.25
(ii)	Ti6A14V	1.15×10^{11}	0.3
(iii)	UHMWPE	3.4×10^{8}	0.4
(iv)	CoCrMo	2.08×10^{11}	0.3

The examined femorotibial area of the knee joint occupies the region, we denote it by Ω and its boundary by $\partial\Omega = \Gamma_u \cup \Gamma_\tau \cup \Gamma_c$. The boundary $\partial\Omega$ is created by parts 1–2 and 3–4 (Γ_u), where the fibula and tibia are fixed, by parts 7–8 and 9–10 (Γ_c), which are contact boundaries between both parts of the femorotibial joint, by parts 11–12 (Γ_c), where the fibula is in bilateral contact with the tibia, and by parts 5–6 (${}^1\Gamma_\tau$), where the load is prescribed; on the remaining parts of the boundary (${}^2\Gamma_\tau$) the femoral joint is unloaded.

Boundary conditions prescribed on parts of the boundary are denoted by 1–2, 3–4, 5–6, 7–8, 9–10, and 11–12. Zero displacement is prescribed between points 1 and 2 (fixed tibia) and 3 and 4 (fixed fibula), between points 5 and 6 the femur is loaded by a loading 0.215×10^7 Pa, the unilateral contact conditions are between points 7 and 8 and 9 and 10, and the bilateral contact condition is between points 11 and 12.

The Sagittal Plane—The Static Case For the numerical analysis the finite element method (FEM) and the nonoverlapping domain decomposition method as in the previous case were used. Two corresponding models of the total knee replacements in linkage on the axial deviation 3, 5, 7, 9 deg. across both condyles were investigated, where MODEL II represents the cut across the outer condyle [the model in Fig. 7.32(b)] and MODEL III the cut across the inner condyle [the model in Fig. 7.32(c)]. The sagittal cut tells nothing about the influence of the axial deviation; it tells something about the overload of the posterior part of the tibial plate in the sagittal (antero-posterior) direction. Therefore, it informs us about a relevant strong wear of the polyethylene insert. In both models the same material parameters were considered.

The investigated femorotibial area of the knee joint in the sagittal plane occupies the region; we denote it also by Ω and its boundary also by $\partial\Omega = \Gamma_u \cup \Gamma_\tau \cup \Gamma_c$. The boundary $\partial\Omega$ is created by parts 1–2 and 3–4 (Γ_u), where the fibula and the tibia are fixed, by the part between points 5 and 6 (${}^1\Gamma_\tau$), where the loading is prescribed, by the part between points 7 and 8 (Γ_c), which is the contact boundary between both parts of the femorotibial joint, and by the part 9–10 (Γ_c), where the fibula through the ligament is joined with the tibia; on the remaining parts of the boundary (${}^2\Gamma_\tau$) the femorotibial joint is unloaded.

The boundary and contact conditions in the case of MODEL II were prescribed on parts of the boundary denoted by 1–2, 3–4, 5–6, 7–8, 9–10. Zero displacement is prescribed between points 1 and 2 (fixed tibia) and 3 and 4 (fixed fibula), between points 5 and 6 the femur is loaded by a loading 1.39×10^6 to 1.61×10^6 Pa, that is, 1.39×10^6Pa in the case of MODEL II-3, 1.46×10^6 Pa in the case of MODEL II-5, 1.52×10^6 Pa in the case of MODEL II-7, and 1.61×10^6 Pa in the case of MODEL II-9 are prescribed, the unilateral contact condition is between points 7 and 8 and the bilateral contact condition is between points 9 and 10.

The boundary and contact conditions in the case of MODEL III are prescribed on parts 1–2, 5–6, and 7–8. The zero displacement vector is prescribed on parts 1–2, on part 5–6 a loading 1.4×10^6 Pa in the case of MODEL III-3, 1.2×10^6 Pa in the case of MODEL III-5, 1.0×10^6 Pa in the case of MODEL III-7, and 0.86×10^6 Pa in the case of MODEL III-9. The unilateral contact condition is between points 7 and 8 and the remaining part of the boundary is unloaded.

Evaluation of Numerical Results in the Frontal Plane—The Static Case

Figures 7.34–7.142 show numerical models and numerical results for various axial angle changes (3°, 5°, 7°, and 9°): the static case. (see the accompanying DVD disk.) The results are presented in Figs. 7.35–7.124 for axial deviation of 3°, 5°, 7°, and 9°; Figs. 7.35–7.43 for the axial deviation of 3°, Figs. 7.62–7.70 for 5°, Figs. for 7°, and Figs. 7.116–7.124 for 9°. Deformations of the femorotibial part of the knee joint after the load with the given factor of enlargement are shown in Figs. 7.35, 7.62, 7.89, and 7.116. Horizontal and vertical components of the displacement vector (see Figs. 7.36–7.37, 7.63–7.64, 7.90–7.91, and 7.111–7.112) characterize the internal shift of the material points before and after the deformation. Spreading of the stress field in the investigated part of the skeleton is characterized by horizontal, vertical, and shear components of the stress tensor and the principal stresses in the considered femorotibial part of the lower limb (see Figs. 7.38–7.41, 7.65–7.68, 7.92–7.95, and 7.119–7.122). Normal and tangential components of the displacement and of the stress vectors on the contact boundary between the two components, femoral and tibial, of the knee joint in the area of both condyles can be seen in Figs. 7.42–7.43, 7.69–7.70, 7.96–7.97, and 7.123–7.124. Owing to a great number of data all these results can be found on the accompanying DVD disk.

Numerical results show the horizontal stress components balance the relatively small tensile stresses in the area around the external lateral and internal medial contact parts and in the area of the incisura intercondylica. Otherwise pressure stresses are predominantly indicated in the femur, especially for the contact boundary and also in the tibia. In the numerical results we can see that in the knee joint the stress gradients are equal in the epiphysis and metaphysis, and the diaphysis is already strained evenly. For the vertical components of stress, predominantly the pressure stresses are observed in the whole area, and the incisura intercondylica and the medial margin are reduced, and the stress grows further in the lateral direction, as observed also in the tibia. As shown by the numerical results, pressure stresses are transmitted over the lateral and medial parts of the contact area. Location of the maximal stresses in the direction of the x axis is nearly identically situated with the area of the minimal stresses in the direction of the y axis. In the course of a vertical component of stress, shear stresses, and the principal stresses, we can see that pressure stresses are more concentrated in the area of the external condyles of the femur and the tibia and less across the internal condyles. Tensile stresses are located in the incisura intercondylica area. The directions of the power stream agree with the beam structure in the epiphysis. The normal and tangential components of the displacement and stress vectors on the contact boundaries of both condyles can be tested so as as to analyze the total replacement of the knee joint on the axial deviation. The normal component DU_n characterizes both parts of the joint in the load of the knee joint and its subsequent deformation. From the analysis of the normal displacement on the overload of the knee joint, it results that when loading both parts of the joint in a certain phase of weighting, for example, during walking with a load, the contact boundaries retreat from one to the other at some points, even though the distance to the opposite points is relatively small. In our case, both components of the knee joint are constantly in close contact, and they are not separated during loading because the knee joint is not overloaded. The tangential component of the displacement vector DU_t characterizes a mutual shift of the contralateral points of both components of the knee joint in the tangential direction. The analysis of the tangential component of the displacement vector indicates relatively small shifts in both condylar components of the joint. The character of this process is a little different for the axial deviation studied. From the values of both components of the displacement vector we get a real mutual shift of the contralateral points of the contact in the space. The normal and tangential component of the stress vector on the contact in both condylar parts of the knee joint (see Figs. 7.42–7.43 and similarly in other cases) characterizes loading relations on both condylar parts of the knee joint. We see that on contact between both femorotibial parts of the knee joint the pressure is observed in both condyles. Figures 7.62–7.63 represent deformations of two condylar parts of the contact boundary between nodes 7 and 8 and 9 and 10 after the loading of the knee joint.

Results of Computations for 3° Valgus Construction of the model satisfies the requirement of maintaining the mechanical axis with symmetrical division of the load on the whole area. In Fig. 7.35 deformation of the knee joint is presented. In Figs. 7.36 and 7.37 horizontal and vertical components of the displacement vector are given. In the horizontal displacements we observe shifts in the area of the UHMWPE polyethylene insert, while the vertical displacement illustrates sequels of vertical loading of the knee joint. Distribution of the horizontal, vertical, and shear stresses are in Figs. 7.38–7.40, while the principal stresses are in Fig. 7.41. Maximal horizontal stresses appear below the insert in the tibial plato. Distribution of the vertical stresses and the principal stresses (Figs. 7.39 and 7.41) reveal the pressure areas, namely around the femoral replacement component and in the external condyle, as well as below it in the insert and the tibial plato and at the bottom part of the tibia. The tensile stresses follow from the distribution of the shear stresses (Fig. 7.40) and the principal stresses (Fig. 7.41) in the area above the incisura intercondylica and in the left part of the tibial plato and the conical part of the stem. The normal and tangential displacement and stress components are in Figs. 7.42 and 7.43. We see that both components are in contact and that the mutual movement on the internal condyle grows toward the center, and in the external condyle the tangential movement grows away from the incisura intercondylica up to the maximal movement behind the condyle center and then falls to the external margin of the knee joint. Normal stresses indicate a similar course on both condyles with two local maximal normal stresses that is, pressures. Tangential stresses are practically equal to zero.

Results of Computations for 5° Valgus A model was constructed so as to maintain the mechanical axis with symmetrical division of the load on the whole area. Figure 7.62 shows the deformity of the knee joint. Deformity of the knee joint reveals itself mainly in the femoral part of the joint. Figures 7.63 and 7.64 represent the horizontal and vertical components of the displacement vector. The horizontal component of the displacement vector indicates shifts in the area of the polyethylene insert of the external condyle in its medial part. The vertical component of the displacement reveals itself especially in the area of the femur and minimally in the area of the fibula. Figures 7.65–7.68 show the division of the horizontal and vertical stress tensor as well as the shear and the principal stresses in the area of the artificial replacement of the knee joint. The greatest changes in the horizontal stress component are indicated in the area of the tibial plate. From the vertical component of the stress tensor (Fig. 7.66) and the principal stresses (Fig. 7.68) it follows that the stress in the diaphysis is spread eventually; in the area of metaphysis it begins to be separated by pressure, in the area of epiphysis the pressure is transmitted over the fixed elements of the femoral component of the artificial replacement, and it is the external condyle that is more burdened and further transmitted over the insert on the tibial plate and over the fixed element of the tibial plato onto the tibia. The shear stresses (Fig. 7.67) and the principal stresses (Fig. 7.68) indicate the areas characterized by tensile stresses in the area above the incisura intercondylica and in the area of the tibial plato. From the normal and tangential components of the displacement vector on the contact, it follows that both components are in tight contact and that the movement on the internal condyle goes toward the center and in the external condyle the tangential component goes from the incisura intercondylica up to the maximum movement, which is practically behind the center of the condyle, and then falls markedly to the external rim (Figs. 7.69–7.70). The normal component of the stress vector has, on both condyles, a similar effect given the symmetry of the condyle and the greater load on the external condyle. An analysis of the normal contact stresses indicates only the pressure stresses. The tangential contact stresses are near zero.

Results of Computations for 7° Valgus The principle of maintaining the mechanical axis with symmetrical division of the whole surface of the joint has been preserved also in this case. The symmetrical division of forces for both knee compartments is documented also by the saturation of the coloration in the pictures based on the numerical results, especially the transmission of the pressure forces over both condyles in the case of the principal stresses (Fig. 7.95). The maximum transfer of the load in the area of the fixed femoral elements and of the stem of the tibial component is indicated in Figs. 7.92–7.95. The horizontal component of the displacement vector (Fig. 7.90) indicates the straightening of the deformity in the femorotibial part of the knee joint. The vertical component of the stress tensor (Fig. 7.93) indicates a straighten spread of stresses in comparison with the preceding case, where even shear stresses were indicated (Fig. 7.94). A balanced transmission of the load going symmetrically through both parts of the joint is represented in the case of the principal stresses in Fig. 7.95 by the course of the pressure load. Lowering the tensile stress is indicated in the area of the incisura intercondylica. The character of the normal components of the displacement vector does not change, that is, both knee joint components stay in tight contact, only the character of the course of the tangential displacement component on the internal condyle changes (Fig. 7.96). The values of this component change as well. Also a partial straightening of the pressure load on both condyles of the knee joint is indicated. The tangential component of the stress vector changes only a little (Fig. 7.97). The deformity of the knee joint is shown in Fig. 7.89.

Results of Computations for 9° Valgus In this model there was an effort to preserve the basic condition relevant to the mechanical axis. Horizontal and vertical components of the displacement vector (Figs. 7.117–7.118) indicate greater changes of deformity in both areas of the UHMWPE insert. The components of the stress tensor and the principal stress (Figs. 7.119–7.122) indicate signs of knee overload and increase of transmission of the load by an external compartment. The increased load passing through the external compartment is, in addition, emphasized in Fig. 7.122 by the course of the principal stresses. Numerical results show that also in this case both components of the replacement knee joint are in tight contact, the movement in tangential direction increases, which indicates greater deformity of the joint replacement (Figs. 7.123–7.124). Normal components of the stress vector on contact between both femorotibial parts of the knee joint (Fig. 7.124) indicate greater values of loading (overloading) the external condyle. A deformity of the knee joint is represented in Fig. 7.116.

Overloading of the posterior part of the tibial plate in the anteroposterior direction in the stable soft posterior structures, possibly the incorrect inclination of the resection of the proximal tibia, is studied in Figs. 7.125–7.126. The results display a certain overloading of the posterior part of the tibial plate as was the case of a 7° deviation, which indicates the possibility of wear of the polyethylene insert TKA. Figures 7.143(a + b) show the shift of the material points on the contact surface before and after the deformation for the medial and lateral condyle for 5°, 7°, and 9°.

Evaluation of Numerical Results in the Sagittal Plane—The Static Case

The same computations as in the previous case in the frontal plane were made. Numerical results show that the largest values have, on one hand, a horizontal component of the displacement vector u_x and, on the other hand, numerical results of normal and tangential components of the displacement vector DU_n, DU_t and of the stress vector τ_n and τ_t at the contact between both components of the knee joint. The contact between both parts of the joint is at boundary between points 7 and 8. Point 7 corresponds with the posterior part of the tibial insert. The horizontal component of the displacement vector u_x is shown in Figs. 7.45, 7.54, 7.72, and 7.81. The figures show the shifts within the bounds of approximately -1.598×10^{-5} m (-1.605×10^{-5} m), -1.667×10^{-5} m (-1.381×10^{-5} m), -1.798×10^{-5} m (-1.151×10^{-5} m), and -1.852×10^{-5} m (-9.203×10^{-6} m) in the case of the outer condyle for the axial deviations of 3°, 5°, 7°, and 9° in the area of the tibial plato, where the data in the round brackets represent data on the inner condyle. These results indicate that the polyethylene insert is pressed into the posterior part of the tibial plato, which is greater than in the case of the internal condyle, and, therefore, that the polyethylene insert is deformed and worn out.

Figures 7.51, 7.78, 7.105, and 7.133 suggest that joint components are in contact in the course of a loading for all investigated axial deviations of 3°, 5°, 7°, and 9°, and that the tangential displacements on the contact, which are within the bounds approximately -1.0×10^{-5} m, -1.05×10^{-5} m, -1.1×10^{-5} m, and -1.2×10^{-5} m in the case of the outer condyle and within the bounds approximately -1.08×10^{-5} m, -0.9×10^{-5} m, -0.75×10^{-5} m, and -0.6×10^{-5} m in the case of the internal condyle, show on the shifts of joint components in the course of loading the joint system for axial deviations 3°, 5°, 7°, and 9°.

Figures 7.52, 7.61, 7.79, 7.88, 7.106, 7.115, 7.133, and 7.142 present the normal and tangential components τ_n and τ_t on the contact boundary between both parts of the knee joint. While the tangential stress components are approximately equal to zero, the greatest values have normal stresses on the contact boundary. On the outer condyle the normal stress values within the bounds are approximately -0.75×10^5 Pa, -9.5×10^5 Pa, -10×10^5 Pa, and -10.5×10^5 Pa, while on the internal condyle within the bounds they are approximately -0.7×10^5 Pa, -6.5×10^5 Pa, -6×10^5 Pa, and -5×10^5 Pa, in the vicinity of the posterior part of the UHMWPE insert. Subsequently, the pressure increases and the next minimum increase occurs at the front part of the tibial plato, as in the previous case. The results demonstrate a certain overloading of the posterior part of tibial plato, which indicates the possibility of its being worn out and resulting in wear of the insert of the TKA. Numerical results show that the 3D models should be analyzed. Such models then clarify certain conclusions of the 2D models in the posterior area of the tibial plato and at certain rotations of the polyethylene insert toward the tibial plato.

Conclusions—The Static Case. From the intial numerical results we deduce that optimal distribution forces operated on TKA in the anteroposterior direction and well-balanced transition of forces in anteroposterior direction will correspond to the 7° case, which indicated also the distribution of the principal stresses. Optimal transfer of forces in the anteroposterior direction with the maximum amplification of the posterior corticalis of the femur, the fixation elements of femoral, and/or of the stem of the tibial component are also observed in practice. On the contrary the overloading of the posterior part of the tibial plato in the anteroposterior direction with soft posterior structures or incorrect inclination of the resection of the proximal tibia suggest the numerical results for the 9° case of valgus. Analyses of results of cuts across the condyles in the sagittal plane document certain overloading of the posterior part of the tibial plato, which suggests the possibility wear of the polyethylene insert of the TKA. Numerical results indicate the necessity to analyze the 2D and 3D models of the knee joint system with joint capsule and ligaments. Convincing biomechanical analyses can analyze 3D models of knee joint systems only.

7.7.2 Evaluation of Numerical Results in Frontal and Sagittal Planes—The Dynamic Case

Walking is an important activity that helps maintain good health. The same is true for patients after the knee joint replacements. Division of loads acting on the tibiofemoral components after implantation of the total knee replacement during walking is of a primary importance in orthopedic surgery. Pressure ratios in the knee after TKA of the soft tissue tension (i.e., ligaments, muscular insertions, capsules of joints) in the vicinity of the replacement, and the resulting axial position of the whole limb must be also determined in the case of the timely loaded replacement knee joint because these factors are influenced by the implantation technique used. The compliance and stability required in the human knee joint for optimal daily functions are provided by articulation and several other components such as ligaments, cartilage, and muscle forces that allow complex mechanical responses to different types of physiological loads. Due to the relative incongruence of the articular surfaces, ligaments play a key role in providing passive stability to the joint throughout its whole range of motion.

The proper understanding of knee joint biomechanics is, therefore, essential in the prevention and treatment of its disorders and injuries. Despite the many studies conducted, the exact mechanical behavior of the knee joint and the causes of many of its injures and its artificial replacements are not completely known yet. To ensure the optimal function and long-time survival of the artificial replacement, the simulation of loaded artificial knee joint during walking is desirable. A number of studies used direct measurements of joint contact forces obtained from implants to evaluate model predictions of muscle and joint loading during movements. These model predictions were made during last decade (Pandy, 2001; Anderson and Pandy, 2001; Zajac et al., 2002, 2003; Heller et al., 2003; Bei and Fregly, 2004; Bei et al., 2004; Taylor et al., 2006; Fregly et al., 2008; Xiao and Higginson, 2008; Kim et al., 2009).

In the previous section we studied and discussed axial angle changes on weight-bearing total knee replacements for the statically and axially loaded knee joint system. In this section we will study the same knee joint system, this time loaded by timely dependent loads corresponding with loading during movement. The loading function was derived from the data of Fernandez and Pandy (2006), Anderson and Pandy (2001), Bei et al. (2004), and Geyer et al. (2006). The model in this dynamic case is the same as in the previous case, only the loads change in time in dependence on the course of the load during the movement given by the acting vertical force versus the gait cycle in percent. Numerical results adhere to the same character of stresses and deformations as in the static case, but their values change from minimal values if the leg is in the phase of disburden and to maximal values if the leg is in the loading phase of the strike.

Numerical results are arranged as follows: First, the timely dependent fields of (i) deformation and displacements (files Movie_deformation.avi, Movie_u_x.avi, Movie_u_y.avi) and (ii) stresses (files Movie_pr_stress.avi, Movie_pr_stressloc.avi, Movie_tau_x_stress.avi, Movie_tau_xy_stress.avi, Movie_tau_y_stress.avi) are presented on the accompanying DVD disk in directory /Chapter_7/Section_7.7/Dynamic_case/Movies for the axial angle of 5° and for the frontal plane. The single figures of principal stresses with information about maximal values of pressures and tensions and information about local values of pressures and tensions (in the directory they are identified by the name with the indicator "loc") are introduced. By the same process the single figures of τ_x, τ_{xy}, τ_y, u_x, and u_y and deformation are presented. Moreover, the normal and tangential components of stress and displacement vectors and the shift of the material points on the contact boundary before and after deformation for the medial and lateral condyles are presented. All these results are included on the accompanying DVD disk, so that readers can find individual figures of interest.

Figures 7.144—7.304 show numerical models and numerical results for axial angle of 5° and for the frontal plane: the dynamic case. (See the accompanying DVD disk). From these results a new type of revolving plastic insert was developed. This problem will be discussed in the next section.

It is evident that other problems in the biomechanics of hip, knee, and other human joints can be studied in two or three dimensions. For example, Bei and Fregly (2004) studied the artificial knee contact model with linear material properties and loads of 750, 1500, 2250, and 3000 N and with flexion angles of 0°, 30°, 60°, and 90°.

7.8 TOTAL KNEE REPLACEMENT WITH ROTATIONAL POLYETHYLENE INSERT

7.8.1 Evaluation of Numerical Results and Observations in Practice

Because of the advanced destruction of the contact surfaces in the human knee joint, the total knee joint arthroplasty (TKA) is a favorable option. Todays TKA implants offer great variety and facilitate treating both surfaces of the femur and the tibia, as well as the surface of the patella and to compensate the cruciate ligaments. Optimal achievements that is, symmetrical distribution of loads onto the tibial component of the TKA is a fundamental demand. As significant as the distribution of forces in the frontal plane is the well-balanced distribution of loads in a sagittal plane. The inclination of the resection of the proximal tibia, the contracture of a posterior cruciate ligament, the dorsal osteophytes, as well as the increase of tension on posterior capsule decide the pressure relationship in the knee joint in the sagittal direction after the total replacement. These factors allow for a permanent increase of pressure in the posterior part of the joint. Relevant unbalancing in both planes are factors that in a crucial way influence survival time of the implant, which depends on the implant technique used. Overloading the posterior part of the tibial plate by an imbalance in both compartments—medial and lateral—leads to their quick wear and considerable abrasion of the polyethylene insert of the TKA. Not adhereing to the surgical technique and incorrect loading of the TKA then leads to the destruction of the implant.

During follow up of the primary TKA, surgeons identify sizable abrasion of a posteromedial part of the tibial polyethylene insert, caused by overloading after wrong inclination of resecting or imbalance of pressure on both parts of the joint. Considerable wear causes the release of polyethylene particles, which causes a wear granuloma, which then causes bone destruction and a loosening of the metal part of the TKA from the bone bed.

Lack of the adherence to the principle of symmetrical balancing on stresses on soft tissues in the anteroposterior direction of the joint is a great mistake: Increasing pressure in flexion increases pressure in the posterior part of the tibial plato.

A relevant statement for biomechanical analyses of the total replacement of the knee joint depends on axial deviation, which above all has horizontal and vertical components of the displacement vector, characterizing inner shifts of mass points before and after deformation and stress components (horizontal, shear, and vertical). In addition the principal stresses occur in the femorotibial part of the lower limb, and, moreover, the normal and tangential components of displacement and the stress vector on the contact surface between both components of the artificial joint as well as between the UHMWPE insert and the tibial (Ti) plato. Analyses of our numerical results in the previous section show that in the case of the frontal plane the optimal angle of the valgus position at resection of the lower end of the femur is approximately 7°. The cut in the sagittal plane for axial deviations has zero information, but it has a weighty information about overloading the posterior part of the tibial plato in the anteroposterior direction. These analyses of numerical results deduce that optimal forces operated on the TKA in the anteroposterior direction and a well-balanced transition of forces in the anteroposterior direction correspond to the 7° case, which was declared also by the distribution of the principal stresses. Optimal transfer of forces in the anteroposterior direction with maximum amplification of the posterior corticalis of the femur, femoral fixation elements, and tibial stem are also observed in practice. On the contrary the overloading of the posterior part of the tibial plato in the anteroposterior direction with unloosened soft posterior structures or incorrect inclination of the resection of the proximal tibia suggest our numerical results for the 9° case of valgus.

7.8.2 TKA with Rotating UHMWPE Insert

Balancing stresses or pressure forces, respectively, in the sagittal plane ensures balancing of the artificial joint in the anteroposterior, that is, the sagittal, plane. Normally, with the mediolateral balancing, it shares the survival time of the implant. Our analyses of the loaded knee joint and its artificial replacement in the frontal and sagittal planes and their common comparison indicate a mutual rotation of the separate components of TKA. This fact can be fully used for the construction of new types of TKA with a rotating polyethylene tibial insert, which fully recompenses the function of the posterior cruciate ligament. Moreover, it respects important rotating demands of the femur against the tibia. The construction of the TKA with a rotational insert can be made by partially rotating about 15°–20° or by rotating the TKA with 360°. These types were presented in Stehlik and Nedoma (2006). The more advantageous and profitable construction type is the TKA with a rotating UHMWPE insert of about 360°. This type of the TKA fully substitutes the function of the posterior cruciate ligament (PCL). Femoral components can be constructed as asymmetric (right and left), and they are manufactured from a CoCrMo (ISO 5832-4) alloy. Moreover, they can be manufactured for an application with or without bone cement. Tibial components can be manufactured as a modular type in the right and left realizations. The polyethylene insert will be made from UHMWPE (ISO 5834-2) and the tibial plato from the titanium alloy Ti6Al4V ELI (ISO 5832-3). The tibial component can be made from the Ti alloy with the stem with the cylindrical hole for the cylindrical pivot and with a detachable movable polyethylene insert, which rotates around the metal pivot. The anchored part of the tibial component is asymmetric, corresponding to the different shapes of the medial and lateral condyles of the tibia. This irregular space is filled with the Ti6Al4V alloy, which creates one tibial plato.

Next numerical results present the application of the TKA with a rotating polyethylene insert in the case of statically and dynamically loaded TKA that depend on an axial deviation of the TKA application. It can be expected that the wear of the polyethylene tibial plato will be substantially low because during loading the polyethylene insert does not turn much owing to acting tangential contact forces on the surfaces between femoral and tibial TKA components.

7.8.3 Evaluation of Numerical Experiments

As in the case of the weight-bearing total knee arthroplasty (TKA) discussed in Section 7.6, before the TKA operation of the knee joint, we evaluate the real axis of the lower limb and determine the mechanical axis. After the adjustment of the mechanical axis, the angle with the anatomical femur axis determines the degree of physiological valgus in which the resection of the distal end of the femur should be made. In this section we will limit ourselves to one angle of restriction of the distal end of the femur only, say a 7° valgus, under the static loadings. This type of prosthesis is a rotating device that transfers the knee joint loads to the bone. At the same time because of the possibility that the insert rotation also reduces the load on the posterior part of the UHMWPE insert, this explains the mechanical factor (which is one of the important factors for the successful use of the TKA) in terms of the load transfer and rotating mechanisms.

We see that a complex relation exists between this mechanism, the magnitude and direction of the loads, inclusive in the posterior part of the polyethylene insert, the geometry of the bone joint prosthesis configuration, the elastic properties of the materials, and the physical connections at the material connections. It is evident [see Stehlík and Nedoma (2006)] that the theory of contact problems in suitable rheology and their finite element approximations are, as in the previous case, also useful tools for analyzing these relations for the knee joint and its TKA, where the contact surfaces, instead of the condyle contact surfaces are also the contact surfaces between the rotating insert and the tibial plato and the contact surfaces of the pivot with its neighborhood.

The numerical results are based on the numerical analysis of the contact problem with small friction (due to the properties of the synovial liquid) in elasticity and the finite element method. The model problem leads to solving the static or dynamic primary semicoercive contact problems with given friction (see Chapter 6) for the axial angle changes of the weight-bearing total knee arthroplasty in the frontal and sagittal planes. The TKA models in the sagittal planes across the internal and external condyles inform us about the overload of the posterior part of the tibial plate in the anteroposterior directions, while in the frontal plane these models inform us about the effects of the axial deviations. These models are two-dimensional models. It is evident that three-dimensional models will give better approximations of real situations.

The knee joint and/or the TKA occupy the region, which we denote by Ω and its boundary as $\partial\Omega = \Gamma_u \cup \Gamma_\tau \cup \Gamma_c$. The boundary Γ_u is created by two parts (denoted by the red color in Fig. 7.305, see the accompanying disk), where the fibula and the tibia are fixed. The contact boundary, which we denote by Γ_c, is created by several parts, which are denoted by the yellow color (see Fig. 7.305). The boundary Γ_τ is created by two parts, where a load of 1.6×10^6 Pa is prescribed on ${}^1\Gamma_\tau$ (in Fig. 7.305 denoted by the green arrows) and on the remaining part of Γ_τ, denoted by ${}^2\Gamma_\tau$, is unloaded.

Discretization statistics are given for the model in the frontal plane: The number of mesh points is 1328, the number of elements is 2312, and for the model in the sagittal plane the number of mesh points is 884 and the number of elements is 1390. The numerical results are obtained only for the static loading.

The numerical results for (a) the frontal cross-section case and for (b) the sagittal cross-section case are presented in Figs. 7.306–7.313. Figures 7.306–7.313 show numerical results for the TKA with the rotating insert (see the accompanying DVD disk).

7.9 COMPUTER-ASSISTED SURGERY IN ORTHOPEDICS: A PERSPECTIVE

Computer-assisted orthopedic surgery (surgical navigation or computer-aided orthopedic surgery—CAOS) is a cutting edge technology that has improved the safety and accuracy of orthopedic procedures. Computer-assisted surgery was developed to provide a precision guidance for neurosurgical procedures. In orthopedics the computer-assisted system was developed to reduce errors in component alignment during total knee arthroplasty (Delp et al., 1998). Computer-assisted surgery combines and is driven by technological developments in the areas of imaging, tracking, image processing, sensor technology, and robotics.[1] Computer-assisted orthopedic

[1] In 1921 the Czech writer Karel Čapek introduced the notation "robot" in his book *Rossom's Universal Robot—RUR* for automatic machines with a very high intelligence.

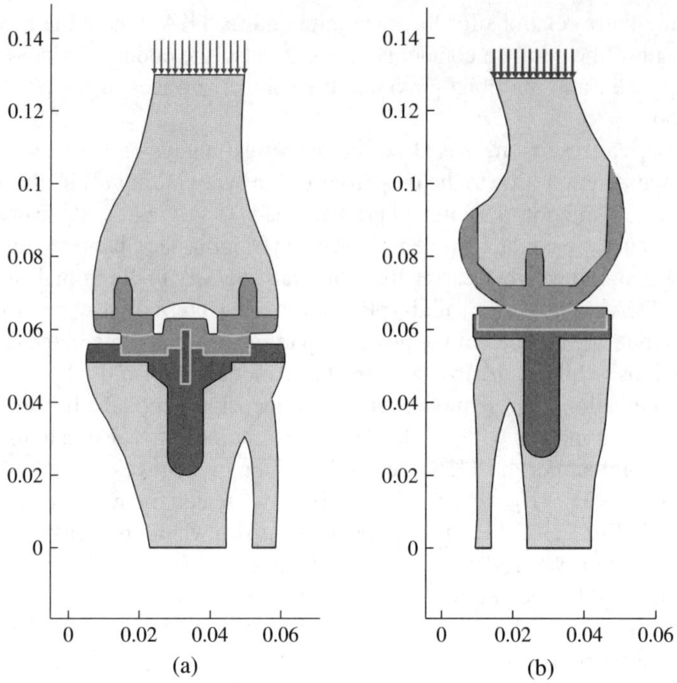

FIG. 7.305 Total knee replacement with rotating polyethylene insert: the model in (a) frontal plane and (b) sagittal plane.

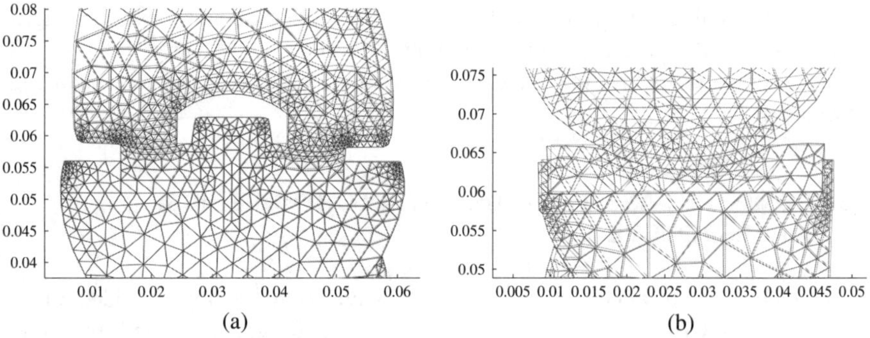

FIG. 7.307 Details of deformations for the TKA with the rotating polyethylene insert: the model in (a) the frontal plane (enlarge factor = 20) and (b) the sagittal plane (enlarge factor = 50).

surgery is getting popular and accepted in the various orthopedic disciplines. CAOS-based technologies are the surgical toolbox of the future and represent a spectrum of devices, represented by three-dimensional image guides and non-image-based navigation systems, robotic assistive tools, as well as intraoperative visualization devices. All orthopedic navigation systems can be divided into the following groups in which

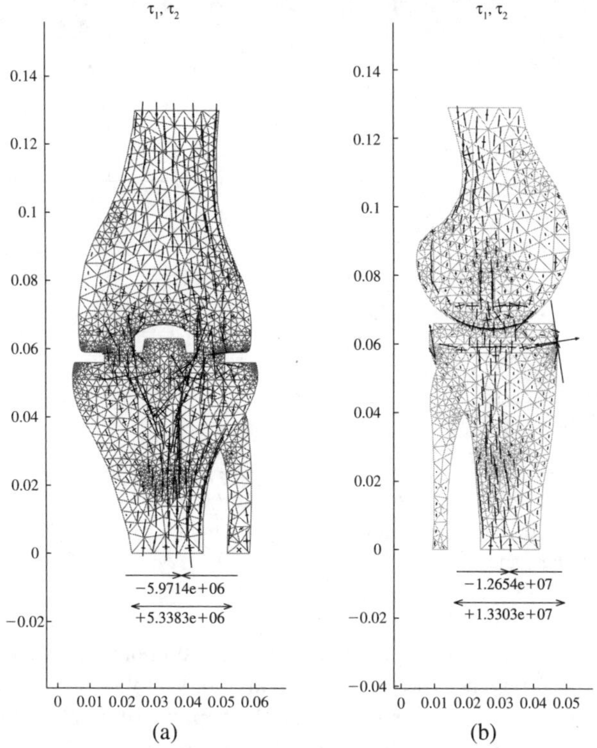

FIG. 7.313 Principal stresses for the TKA with the rotating polyethylene insert: the model in (a) frontal plane and (b) sagittal plane.

(i) the information is collected preoperatively (MRI or CT) scans (spine, knee, hip surgeries), (ii) that use intraoperative imaging (fracture surgery), and (iii) image-free systems. A typical computer-assisted surgery system consists of three major components: (a) control computer, (b) software for image processing, including registration, and display, (c) and localizer for tracking instruments and as well as the patient (Cleary et al., 2006) (see Fig. 7.314).

The patient's anatomy is directly input by the surgeon "mapping" specific landmarks and performing kinematic movements of specific joints from which the computer software builds up a working model on which the surgery will be based. Image-free systems are most frequently used during joint replacement surgery. If the system needs more information for a construction of the patient's working model, then it asks the surgeon for new information. As surgical knowledge advances, new information is quickly incorporated into the navigation software for the surgeon to use.

The application of the CAOS in orthopedics is now widespread in large areas ranging from hip and knee arthroplasty, other joints replacements, spinal surgery, up to fracture surgery.

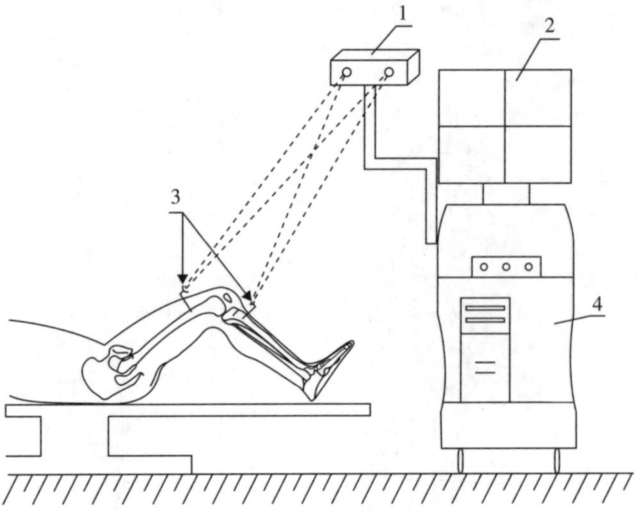

FIG. 7.314 Image-free navigation system: (1) optical tracking system, (2) computer monitor, (3) references frames, and (4) control computer.

The aim of this brief study is to explain how the computer-assisted surgical navigation technique may help to improve total joints replacements and show new perspectives in the CAOS techniques. A description of total knee arthroplasty will be presented. We will analyze only those steps that have a connection with the mathematical simulations during surgery.

Navigation systems are well established in orthopedic surgical procedures. Modular systems with implant-adopted or universal software applications are available, respecting the individual demands of surgeons and patients. Standard applications with optical tracking systems allow the routine use in total knee arthroplasty, total hip arthroplasty, and spinal surgery. The success of the total knee arthroplasty depends on many factors, that is, patient selection, design of the prosthesis used, the preoperative condition of the joint,the surgical technique, as well as postoperative rehabilitation. Division load acting on the tibial component after implantation of the TKA is of primary importance. The pressure ratios in the knee after the TKA application are decided according to the soft tissue tension (ligaments, capsule, muscular insertions) in the vicinity of the arthroplasty and the resulting axial position of the whole limb. These factors are influenced by the technique used for implantation. Stulberg et al. (2002) mentioned that the most common cause of revision TKA was errors in the surgical techniques used, as small changes in component positioning can lead to substantial changes in postoperative performance. Therefore, the computer-aided navigation systems have been developed to help to reduce errors in component alignment during total knee arthroplasty (Delp et al., 1998; Siston et al., 2007). These systems allow intraoperative recording of joint range of motion and kinematics, providing the capability to study the mechanics of knees with advanced point diseases. These systems have been developed to help surgeons to install implants more accurately and reproducibly. Moreover, they also record quantitative information such as joint range of motion, laxity, and kinematics intraoperatively.

The computer, the corresponding software, and the localizer that tracks instruments, as well as the patient are the heart of the computer-assisted orthopedic surgery system (Fig. 7.314). A typical sequence of steps in applying a computer-aided surgical navigation system is the following (Cleary et al., 2006):

1. Preoperative MRI and/or CT scans are obtained.
2. The data from MRI or CT images are transported and inserted into the computer memory in the DICOM format. The DICOM (Digital Imaging and Communications in Medicine) format is the standard medical image file format [see the DICOM reference guide (2001)].
3. A reference target is attacked to the body to elliminate undesirable motion of the patient's limbs or the camera.
4. Registration maps, the image data set of physical locations on the patient's body.
5. Tracking of surgical instruments, including probes or pointers, and display the anatomy beneath these instruments. On the visual display the axial, sagittal, coronal and 3D views are presented. These last steps—registration and tracking—are necessary to take that information and use it during surgery.
6. Oblique reformats—multiplanar reconstructions—can be provided at any angle to help carry out the procedure.

Registration of orthopedic procedures informs the system where the joint is situated in space. Registration is a tracking marker on the bone and joint that can be recognized by the computer's tracking system. Several methods of registration for navigation are available. One is fluoroscopy, which is the form of registration based on an X ray to register the patient's joint system or his bone. Optical tracking in which a light-emitting diode (LED) is either attached to a camera or actively attached to the patient, represents a good method. The other method is based on electromagnetic tracking, which uses a passive, excitable, electromagnetic marker that can be exicited in a magnetic field to create an image of the patient's anatomy. There exists also the other diagnostic method based on an ultrasound, which is in development.

In the fluoroscopic navigation, the registration process is automatic. Registration using an imageless system is a matter of identifying the centers of rotations in the joints though kinematic testing. By this technique a surgeon visually concentrates on landmark points. Tracking is very important because we need sensors and measurement devices that can provide feedback during surgery on the orientation and relative position of the tools to the bones. Imageless or CT-free systems use biomechanical models to assist, that is, a total knee replacement and anterior cruciate ligament reconstruction. One must determine the axes of motion and then to replicate this motion with the implanted components. It means that it is necessary to identify the axes from CT or MRI scans of the leg; however, it can be done by identifying the physiological axes of motion using a pure kinematic approach.

As mentioned in Clearly et al. (2006), first the trackers are implanted in the bones of a limb, and then the mechanical centers of the hip, knee, and ankle joints are determined. Then some anatomical landmarks are identified, and one can define the axes as well as the corresponding human anatomy. The navigation capability is then used to control orientation and position of a saw guide to execute the cuts in the femur and tibia and under satisfied conditions (i.e., when soft-tissue balancing satisfy all demands) to size the implant (TKA).

The main part of a computer-aided orthopedic surgery (CAOS) system is the localizer. It tracks the instruments and the patient in the 3D space. The localizer has the similar function as a GPS (global positioning system) unit. Surgical tracking systems are based on optical, infrared, electromagnetic, and acoustic signals; the optical tracking systems are predominently used for their high quality—that is, high accuracy and reliability. For more details see Delp et al. (2000), DiGioria III et al. (2004a,b), Cleary et al. (2006), Siston and Delp (2006), and Siston et al. (2005a,b, 2007).

Perspectives More of these computer-aided orthopedic surgery systems are based on the kinematic ideas. In real orthopedics the biomechanical reality-based models would be able to model tissues by determining the mechanical properties experimentally and by deriving mathematical models based on these determined biomaterials as well as biomechanical properties, and the corresponding discrete models based on the finite element methods, meshless methods, and so on [see Nedoma (1993b), Nedoma et al. (1999a,b, 2006), De et al. (2001), and also the methods presented in Chapter 6 of this book]. The progress of perspective CAOS will be based on several ideas and goals. The first goal is to develop a patient surgical plan based on the patient's internal anatomy such as in previous classical approaches. The second goal is to implant the plan in surgery and to obtain real-time feedback on the further realization of the plan. The third goal of CAOS is to provide a measurement tool for patient outcome. These data are the basic information for computer-assisted information management systems in the operating room, and make the CAOS system in interactive planning, measurement, and intraoperative management tool with the ability to successfully, with high security and accuracy, solve and realize surgical procedures. Realizations of surgical procedures will be based on verifying individual steps by biomechanical-mathematical (physical-mathematical) models. Therefore, after registration the CAOS system acquires the patient's joint system anatomy and limb alignment. At this step, the CAOS plans the best match of the patient's anatomy with the size and type of the implant. At this stage the CAOS verifies the optimal positions and distributions of stresses and deformation of the components of the patient's joint system of the surgeon interest, based on the biomechanical-mathematical model of axial angle changes and its discrete analysis discussed in Chapter 6 and earlier in this chapter.

After the numerically verifying operative plan, which is generated for the patient, the next step involves a process of implantation. The accuracy of the bone cuts, determined and verified by the discrete model, can be measured intraoperatively with a quantitative device, giving a feedback value that demonstrates deviations from the numerically verified plan. This verification can, first, be based on the statically loaded knee joint system and, secondly, on the dynamically loaded knee joint system after the application of total knee replacement, based on the analyses of dynamically loaded TKA as discussed in Chapter 6. Having the feasibility to accurately adjust the performance of the desired step intraoperatively potentially eliminates the accumulation of errors that occurs in a procedure with multiple parts, and also allows quantitative assessment of close surgical feedback. The continued step involves determination of the component-to-component position and soft tissue balance, which will be verified by the statically and/or dynamically loaded TKA system during the surgical treatment. Then the limb alignment and implant position will be quantitatively analyzed, recorded, and compared to patient outcome, corresponding radiographic and clinical outcomes, and resulting implant success or failure. But the main goal of this approach is to eliminate all failure cases because the stress–strain analyses in step-by-step verified surgical procedures cannot allow negative results. Therefore, the negative result corresponds only to the surgeon's mistakes during surgical procedures. This new approach in the CAOS technique needs optimal methods based on the theory of static and dynamic contact problems and their discretization by the finite element methods and very quick algorithms. Such methods and algorithms are given and discussed in Chapter 6.

Finally, we present some numerical analyses of a statically and dynamically loaded knee joint replacement system and the numerical analyses of axial angle changes during the planning process and when verifying stage-by-stage surgical processes of the surgical treatment. On the four-quadrant display (axial, sagittal, coronal, and 3D) the surgeon, due to his or her choice, can select the requested information about the present state of the knee joint replacement, that is, information about distributions of principal stresses, horizontal, vertical, and shear stresses, horizontal and vertical displacements in the whole area of interest during statically or dynamically (e.g., corresponding to loading during gait) loaded knee joint replacement system, as well as situations on the contact surfaces on both condyles, as follows from the possibilities of methods solving the static and dynamic contact problems discussed in Chapter 6.

The recommended process for the perspective CAOS can be the following:

1. Preoperative CT and/or MRI scans.
2. The obtained data of the CT/MRI scans transport and insert into the CAOS memory in the DICOM format.
3. On the basis of the preoperative CT/MRI scans (i) a preoperative simulation of the axial angle changes on the weight-bearing TKA and the ensuing analyses based on the simulations and methods discussed in Section 7.7; and (ii) after stress–strain analyses find an optimal situation for the future surgical treatment procedures.
4. After these preparatory stress–strain analyses, based on the data from the MRI and/or CT scans, transport and insert the data of stress–strain analyses into the computer memory.

5. In the course of the surgical treatment continue in the same manner as in the previous case, that is, apply procedures of the CAOS given above.
6. However, in the course of the surgical treatment verify the stress–strain state at the operational step and compare with the results of the preliminary simulation of TKA, carried out before the surgical treatment, so that (i) in every step of the operation procedures verify the current stress–strain state of the TKA system; (ii) according to circumstances during the operation that depend on the numerical data obtained at every step of the operational procedure so that the surgeon will use an optimal surgical procedure to obtain optimal distributions of stress–strain fields in the knee joint replacement area of interest, because only the optimally distributed stress–strain fields in the TKA system have a possibility to ensure an optimal result of the surgical treatment; (iii) the optimal algorithms used for verifying a current state of the TKA will be based on the algorithms for the static and dynamic contact problems and their further modifications presented in Chapter 6.

As a consequence of these ideas we briefly present the detailed model of the surgical procedure of the varus knee deformity (Fig. 7.315). The course of computer-aided orthopedic surgery will be divided into two separate procedures: (a) before the surgical treatment and (b) in the course of the surgical treatment.

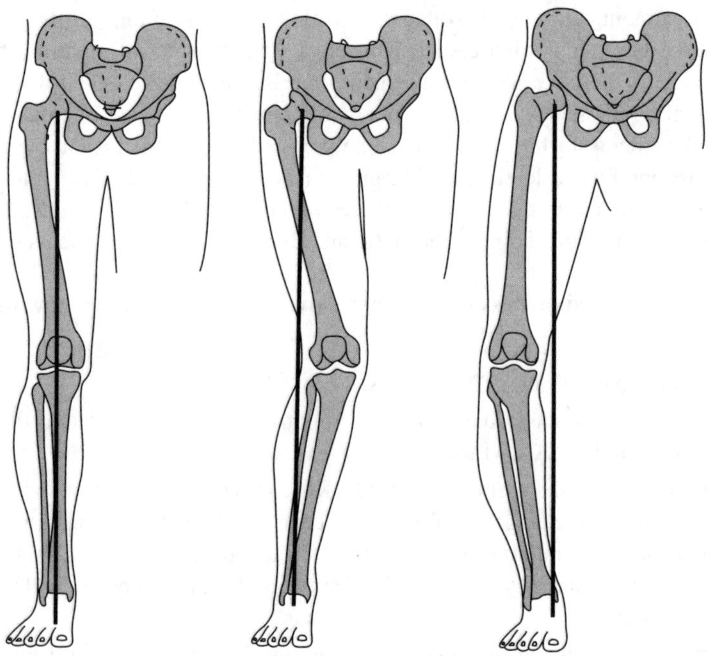

FIG. 7.315 Lower limb: (a) the physiological state, (b) the valgus knee deformity before the surgical treatment, and (c) the varus knee deformity before the surgical treatment.

A. The preoperative process is divided into the following steps:

a. Preoperative CT and/or MRI scans are obtained and results inserted into the CAOS memory in the DICOM format.
b. On the basis of the preoperative CT/MRI scans a surgeon simulates the deformed lower limb before the surgical treatment on the computer during his/her preparation for the treatment on the patient. He or she analyzes the initial situation (see Fig. 7.316). From the obtained 2D and 3D numerical results, that is, horizontal and vertical displacements, horizontal, vertical, shear, von Mises, and principal stresses, as well as the normal and tangential displacements (velocities) and stresses on the contact surfaces in the hip and knee (and the ankle) joints, he/she evaluates the initial situation of the patient.
c. On the basis of these CT/MRI data (if possible) the surgeon constructs the patient's healthy knee joint before its deformation and analyzes the situation of the loaded knee joint, for example, without any rotation (0°) or with a small rotation (~0.9°), which can occur during the real situation over the gait. He/she analyzes horizontal and vertical displacements, horizontal, vertical, shear, von Mises, and principal stresses, as well as the normal and tangential displacements

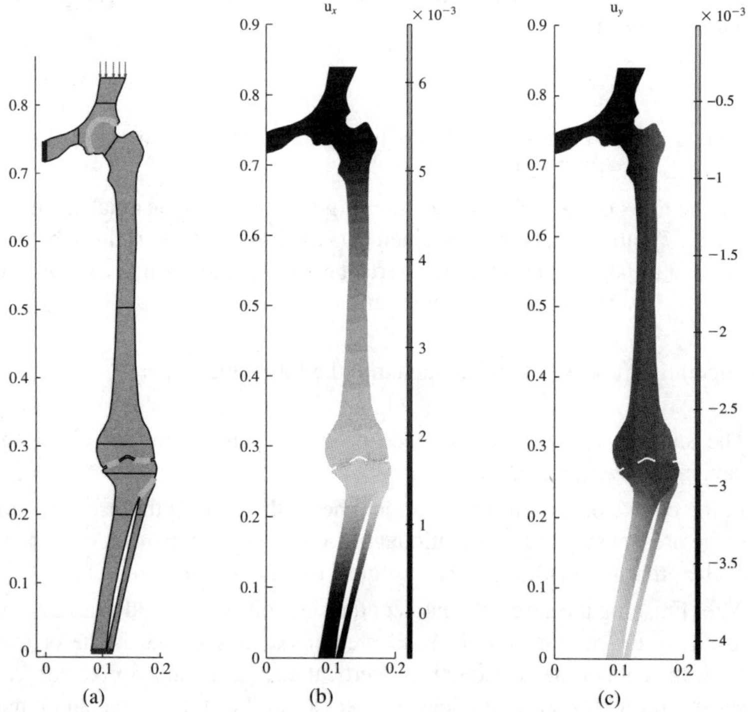

FIG. 7.316 Deformed lower limb before the surgical treatment: (a) the model, (b) the horizontal displacement component, and (c) the vertical displacement component.

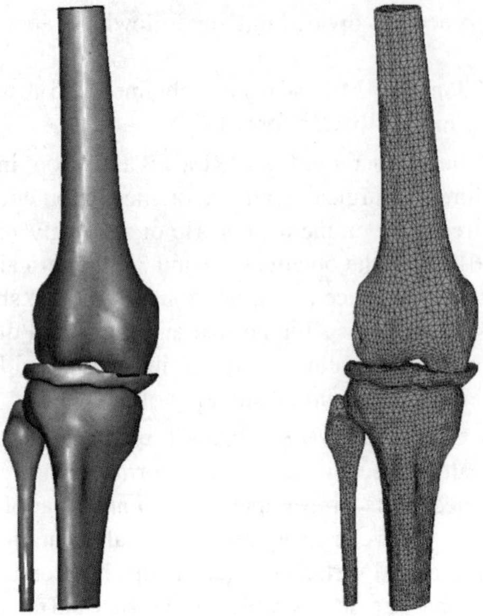

(a) (b)

FIG. 7.317 3D model of a healthy knee joint before surgical treatment: (a) the geometry, and (b) the finite element mesh.

(velocities) and stresses on the contact surfaces of the joints of the lower limb (see Figs. 7.317–7.321).

d. On the basis of the CT/MRI data the surgeon simulates the axial changes at 0°, 3°, 5°, 7°, and 9°, analyzes obtained data as above and determines the probable optimal tibial cut. The surgeon inserts the data for the optimal case of resection into the CAOS computer for future use.

B. The operative process can be divided into the following steps:

a. The surgeon applies the sequence of steps of the computer-aided surgical navigation system as above.
b. In the course of the surgical treatment he or she verifies the stress–strain state at the present step of the operational procedure and compares it with the results of the proposed studies before a surgical treatment (see point A).
c. After finishing the surgical treatment the surgeon verifies and analyzes the final result by the approach used in all previous analyses, that is, he or she analyzes the data of the horizontal and vertical displacements, horizontal, vertical, shear, von Mises, and principal stresses, as well as the normal and tangential displacements (velocities) and stresses on the contact surfaces on both condyles of the knee joint.

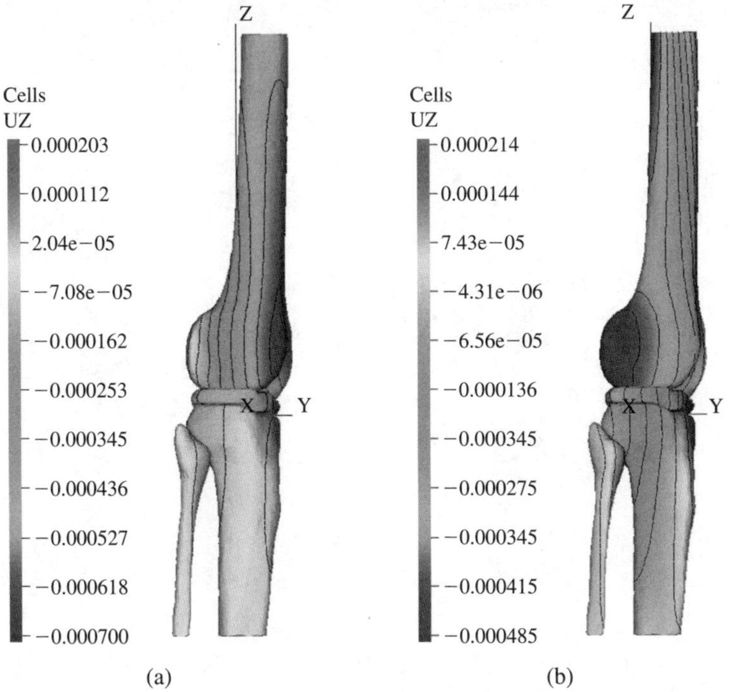

FIG. 7.318 Healthy knee joint before surgical treatment: vertical displacements: (a) rotation about 0° and (b) rotation about 0.9°.

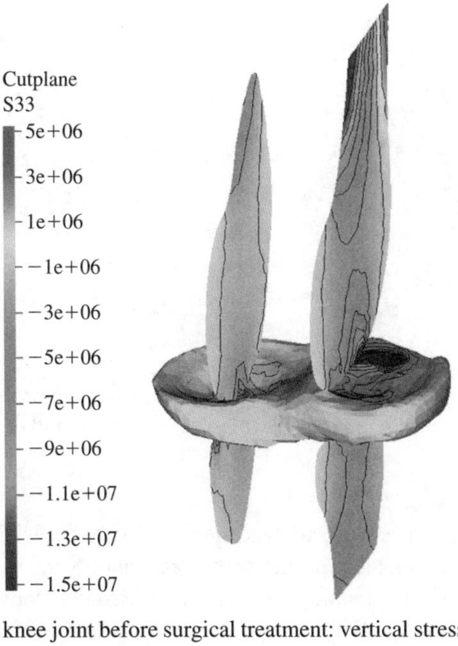

FIG. 7.319 Healthy knee joint before surgical treatment: vertical stresses $\tau_{zz} \equiv s_{33}$ in sagittal planes situated across both condyles.

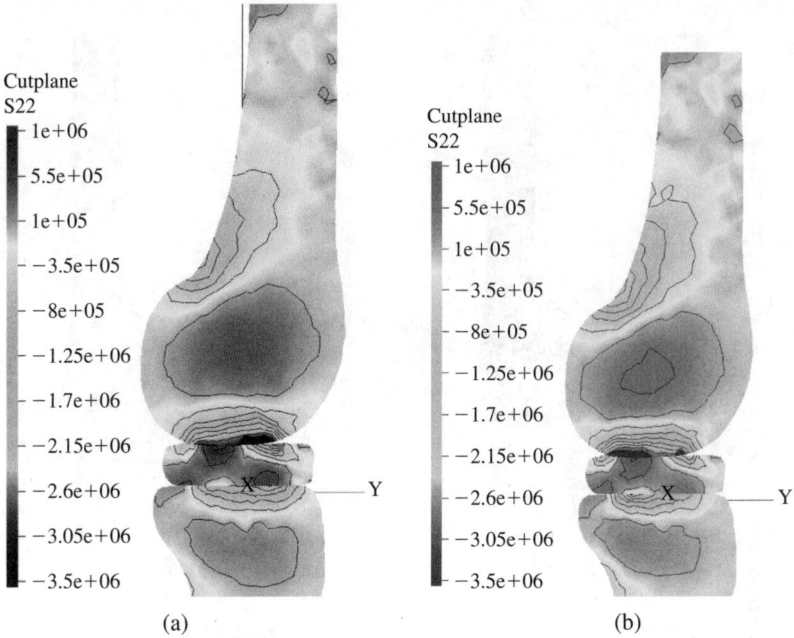

FIG. 7.320 Healthy knee joint before surgical treatment: a stress component τ_{yy}: (a) rotation about 0° and (b) rotation about 0.9°.

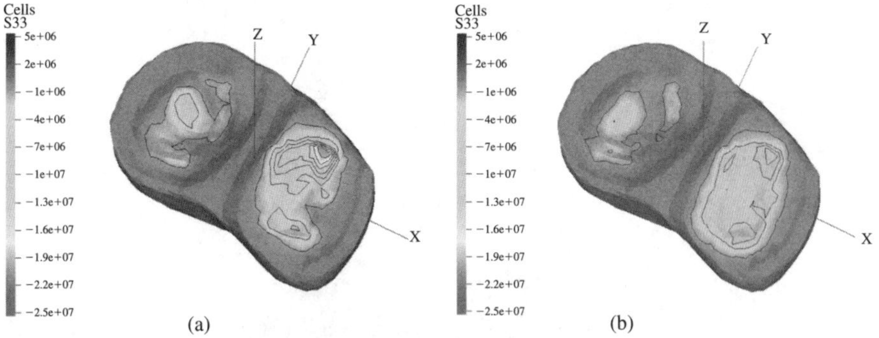

FIG. 7.321 Healthy knee joint before surgical treatment: distribution of a vertical component of stresses on the meniscus contact surface: (a) rotation about 0° and (b) rotation about 0.9°.

Moreover, the surgeon can also analyze the situation on the UHMWPE insert under the different loading in time; for simplicity, we present only the vertical component of stresses τ_{zz} for the case if the loading forces act in the frontal plane under the direction for example, of −4°, 0°, +4° with regard to the vertical axis z (see Figs. 7.322). Similarly, other analyses can be made according to the surgeon's demands.

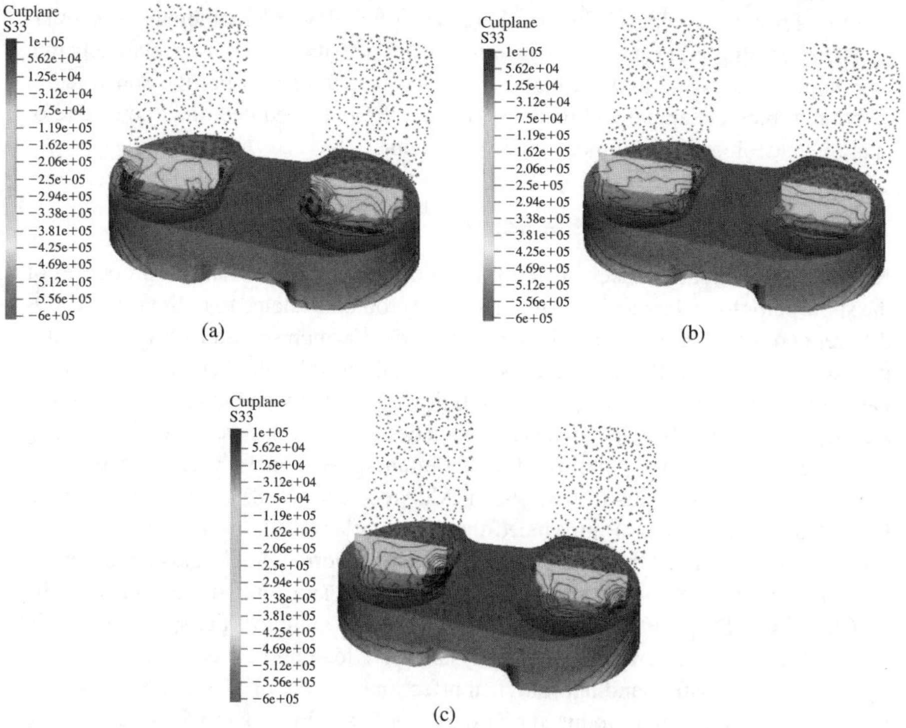

FIG. 7.322 Knee joint after implantation TKA: a situation on the UHMWPE insert: a vertical component of stresses $\tau_{zz} \equiv s_{33}$; loading forces act in the frontal plane under the direction (a) −4°, (b) 0°, and (c) 4° with regard to the vertical axis z.

The CAOS approach can be applied also in hip joint surgery, in other great joint surgeries, in spinal surgery, as well as in dental surgery, for example, in the case of the temporomandibular joint and its artificial replacements. Finally it should be mentioned that as a suitable rheology the viscoelasticity with short memory, due to viscoelastic properties of a bone and soft tissues, appears to be optimal for use in computer-assisted orthopedic surgery.

7.10 BIOMECHANICAL AND MATHEMATICAL MODELS OF THE THORACOLUMBAL SPINE

7.10.1 Introduction

Present results in medicine, biomechanics, and technological disciplines contributed to a considerable extent to the many processes in our body. The spine is a very important part of the human skeleton because it lends support for the head and the whole body. Therefore, it has been studied in detail during last several decades in order to understand and scientifically explain traumatic and degenerative processes of the spine column. From this point of view mathematical modeling and mathematical simulations are very important and advantageous because with them we can simulate separate pathological states and find optimal methods for new therapies. For the construction of optimal biomechanical and mathematical models we need detailed knowledge of anatomy, histology, and biomechanics of the spine.

7.10.2 Biomechanics of the Spine

Knowledge of spinal biomechanics and functions as well as mechanical properties of the spine helps to understand all pathogenetical problems related to it. Since the 1930s different concepts of the biomechanics of the spine, which started to generate therapeutic directives as well as the direction for optimal surgical treatment, have appeared. Determination of spinal stability and instability has always been a challenge. Different biomechanical models of the spine have been designed to improve the understanding of spinal pathology. Two- and the three-column spine models were introduced by Holdsworth (1963), Denis (1983, 1984), McAfee et al. (1983), and Louis (1985). Later Magerl et al. (1994) introduced the present classification of spinal fractures. A two-column model of the spine was given by Holdsworth (1963), and later by White and Panjabi (in 1978), who introduced spine stability and studied the spine as a 3D structure in a 3D space. They defined stability as the ability of the spine to contain the displacements of vertebrae under physiological loading in such dimensions that significant deformation and neurological affections did not originated. Similar definitions of stability were introduced by Louis (1985) and Frymoyer with Krag (1986). The model introduced by Louis was a three-column biomechanical spine model. His model is created from three columns, where the vertebral body builds one column and the intervertebrae joints build two separable columns. In 1983 Denis [see Denis (1983, 1984)] defined a three-column model, further refining the principle of spinal stability. He divided the vertebral body into two parts, the frontal column is created from the frontal part of the vertebral body and the frontal longitudinal ligament, the middle column from the back part of the vertebral body and the posterior longitudinal ligament, and the posterior column is the same as in the Holdsworth's model. Magerl et al. (1994) modeled a two-column spinal model. The anterior column is represented by the vertebral body and the longitudinal ligaments and the intervertebral disk, the posterior column by the lamina, the intervertebral joints, and the posterior ligamentous complex. It consists of three main types A, B, and C. The present wide use of CT and MRI makes it possible to create automatic three-dimensional reconstructions, which involve numerous new points of view for the reconstruction of new spine models. Computer techniques have produced powerful methods for modern biomechanical modeling, the promise being the ability to assess the stability of a construct prior to implantation. The application of spinal biomechanical knowledge spans many industries and supports improved medical diagnoses and more effective treatments.

The human spine is a complex structure, created from 33 vertebrae, whose principal functions are to protect the spinal cord and to transfer loads from the head and trunk to the pelvis and the lower limbs, and it is characterized by the human spinal column.

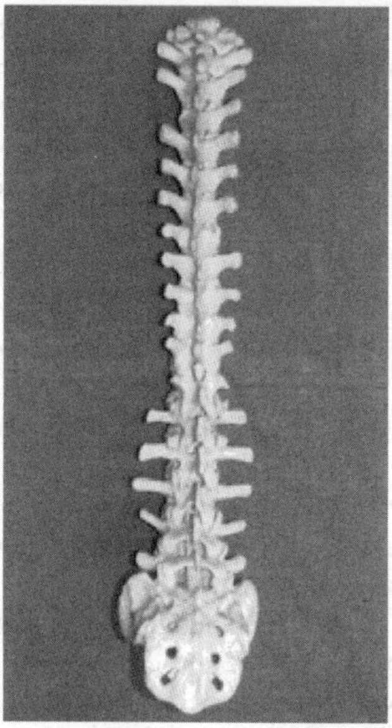

FIG. 7.323 Cervical, thoracic, and lumbar spine with the sacrum.

The principal functions of the final column are to protect the spinal cord and to permit a variety of movements. The 33 vertebrae of the spinal column are divided into five parts: cervical (7), thoracic (12), lumbar (5), sacral (5), and coccygeal (4) regions (Fig. 7.323). The two cranial vertebrae C1, the atlas, and C2, the axis, are atypical and play a role in the articulation between the head and the cervical spine. The articulation between the vertebrae is carried out by amphiarthrodial joints. Between two neighboring vertebrae a fibrocartilaginous disk is situated. Its primary functions are to sustain loads transmitted from segments lying above, to absorb shocks, to eliminate a bone-to-bone contact, and to reduce the effect of impact forces. The articulations of each vertebra with a neighboring vertebra permit movements in three planes so that the structure of the spine permits movements including flexion–extension, lateral flexion, and rotation. The inner part of the disk—the nucleus pulposus—is created from a gelatinous material, and it is surrounded by a tough outer covering, the annulus fibrosus, composed of fibrocartilage. During human activities the disk is loaded with a combination of compression, bending, and torsion. Flexion, extension, and lateral flexion of the spine produce tensile and compressive stresses in the disk, while rotations produce shear stresses. The spinal column is extremely vulnerable and spinal fractures occur often. Similarly, as in the case of human joints, in the spine the osteophytes are also observed. In this section two models will be presented—the fracture of the Chance type and the model of spinal cord pressed by the osteophytes inside the spinal cord channel during static and dynamic loading.

Anatomy of the Thoracic Spine The thoracic spine consists of 12 vertebral bodies and intervertebral disks. Their size increases with distal location. The vertebrae consist of the vertebral body and the vertebral arch with transverse and spinous processes. The vertebral body is connected with the vertebral arch through pedicles. The posterior part of the vertebral body, both the pedicles and the vertebral arch, form the osseous spinal canal—the foramen vertebrale. On both sides of the upper part of the vertebral bodies, the articular surfaces are present to accommodate rib heads. From the second to the tenth vertebral body, there are two articular surfaces on each side, the superior and the inferior one; the last three bodies usually have only one articular surface. The transverse processes are massive and oriented dorsolaterally. The spinous processes aim dorsocaudally and overlap one another like tiles on a roof. The articular processes lie in the frontal plane, the articular surfaces of the upper processes are oriented dorsally and the lower processes ventrally.

Anatomy of the Lumbar Spine The lumbar vertebrae are the largest vertebrae in the body. There are usually five (or six) vertebral bodies in the lumbar spine. This is caused by lumbalization of the proximal part of the sacrum. The vertebral bodies are concave on the sides, the anulus fibrosus originates at the circumference of the upper and lower end plates, which are concave. The shape of the body is basically round with slight straightening next to the spinal canal, which in comparison with the thoracic spine is more spacious. It has the shape of an irregular triangle. The transverse processes originate laterally at the pedicle-lamina connection, and they are located anterior to the articular processes. At the base of the transverse processes are the processus accessorii, which are rudimentary transverse processes. The articular processes are located posterior to the transverse processes, and their orientation is almost vertical. The articular surfaces are in the sagittal plane, the upper is facing medially, the lower laterally. The processus mamillaris is located on the dorsolateral edge of the upper articular process. The spinous processes are large and strong and aim straight dorsally.

During human activities the vertebrae as well as disks are loaded by static and dynamic loads. The loads are maximal around toe-off and increase linearly with walking speed. Cappozzo (in 1984) shows that the compressive loads at the L3–L4 motion segment are between 0.2 and 2.5 times the body weight.

The body of vertebrae (corpus vertebrae) is created from spongy and cortical bones and it is bordered by a plane plateau, on which the discus intervebralis of the practically same form as the plane plateau is jointed. The number of disks is 23 or 24 in the case of L6. Each disk has the marginal layer from the hyaline cartilage, which is jointed with the bone of both vertebrae with the Young's modulus $E = 24.3$ MPa and Poisson's constant $\sigma = 0.45$. The periphery of the disk is created from a ligamental tissue, the annulus fibrosus is created from the colagenial lamellae (fibrocartiloges). Inside every disk the nucleus pulposus is situated, which is created from the aqueous material containing $\sim$ 88% water (H_2O) in youth and 67% water in old age ($\sim$77 years). The nucleus pulposus is a supporting element that creates some deposit, around which the vertebrae rotate. But the direct mathematical model is an open problem up to the present, and it depends on new anatomical research. The load on the intervertebral disk in the region of the lumbar spine, that is, between L3 and L4, is $\sim$1375 N in the sitting position, $\sim$960 N in the basic position, and $\sim$250 N in the at rest position (bed-ridden position). Under a load of more than 5 kg the motion of stretching arms forward place a load on the intervertebral disk of about 1990 N.

A spine fracture can originate from an axial compression, axial distraction, axial rotation, or shear stresses. The mechanism of the spine injury is understood by Magarel's classification. Two fundamental injury mechanisms with their combination create flexion in the sagittal plane and translation, more precisely the shear stresses and rotation in the transversal plane, can occur. Each operational approach (medical, treatment operation) applied to the spine takes into account: (i) reposition, representing a restoration of normal relations between single vertebrae in the case of luxation or translation fractures, respectively, or a shape restoration of the vertebral body; (ii) stabilization, having a crucial meaning for the maintenance of a reposition and also of a decompression; and (iii) in the case of compression also decompression with the main goal to release the nervous tissue.

For the Th/L spine fracture the operational approach as well as the implants used represent a problem. The operational approaches are anterior and posterior (Dick, 1984). In the case of the anterior approach we can carry out the spinal channel, replacement of the vertebral body by its "tìp" (implant), and stabilization. The implant ensures the position of the mutual vertebral situation. In case the posterior approach is applied, a surgeon has more possibilities; the fundamental significance is stability and the reposition abilities of the implant.

7.10.3 The Spinal Model

Let $0, x_1, \ldots, x_N$ be the orthogonal Cartesian coordinate system, where N is the space dimension and let $\mathbf{x} = (x_1, \ldots, x_N)$ be a point in this Cartesian system. Let $I = (0, t_p)$ be the time interval.

Let the thorax and/or lumbar spines together with ligaments and the sacrum occupy the region $\Omega = \bigcup_{\iota=1}^{s} \Omega^\iota \subset \mathbb{R}^N, N = 2, 3$, where the regions Ω^ι, $\iota = 1, \ldots, s$, are represented by the vertebrae, ligaments, and the sacrum. Let Ω^ι have Lipschitz boundaries $\partial\Omega^\iota$ and let us assume that $\partial\Omega = \Gamma_\tau \cup \Gamma_u \cup \Gamma_c$, where $\Gamma_\tau = {}^1\Gamma_\tau \cup {}^2\Gamma_\tau$ represents loaded and unloaded parts of the boundary, where ${}^1\Gamma_\tau$ is a part of the boundary, where loads are prescribed, for example, where muscle forces operate on the spine column in places of crosswise spinal and joint protuberance or where the weight of a body or an outer burden act and Γ_u is a part of the boundary, where the spinal system is fixed, and $\Gamma_c = \bigcup_{k,l} \Gamma_c^{kl}$, $\Gamma_c^{kl} = \partial\Omega^k \cap \partial\Omega^l$, $k \neq l$, $k, l \in \{1, \ldots, s\}$ is a part of the boundary, where the vertebrae are in mutual contacts, for example, in areas of intervebral joints or in areas of spinal fractures, respectively. Let $Q = I \times \Omega$ denote the time–space domain and let $\Gamma_\tau(t) = \Gamma_\tau \times I$, $\Gamma_u(t) = \Gamma_u \times I$, $\Gamma_c(t) = \Gamma_c \times I$ denote the parts of its boundary $\partial Q = \partial\Omega \times I$ (Fig. 7.324).

Furthermore, let $\mathbf{n}$ denote the outer normal vector of the boundary, $u_n = u_i n_i$, $\mathbf{u}_t = \mathbf{u} - u_n \mathbf{n}$, $\tau_n = \tau_{ij} n_j n_i$, and $\boldsymbol{\tau}_t = \boldsymbol{\tau} - \tau_n \mathbf{n}$ be normal and tangential components of displacement and stress vectors $\mathbf{u} = (u_i)$, $\boldsymbol{\tau} = (\tau_i)$, $\tau_i = \tau_{ij} n_j$, $i, j = 1, \ldots, N$. Let $\mathbf{F}, \mathbf{P}$ be the body and surface forces, ρ the density. Let us denote by $\mathbf{u}' = (u'_k)$ the velocity vector. Let on Γ_c^{kl} the positive direction of the outer normal vector $\mathbf{n}$ be assumed with respect to Ω^k. Let us denote by $[u_n]^{kl} = u_n^k - u_n^l$ and $[u'_n]^{kl} = u_n'^k - u_n'^l$ the jump of

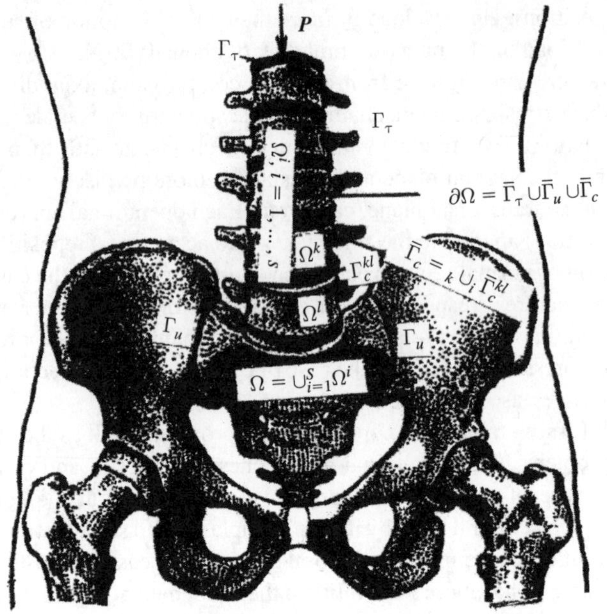

FIG. 7.324 Mathematical model of the lumbar spine with the sacrum.

the normal displacement and normal velocity, respectively, across the contact zone between neighboring spinal parts Ω^k and Ω^l. Similarly, we denote $[\mathbf{u}_t]^{kl} \equiv \mathbf{u}_t^k - \mathbf{u}_t^l$ and $[\mathbf{u}_t']^{kl} \equiv \mathbf{u}_t'^k - \mathbf{u}_t'^l$. If one of the colliding bodies is absolutely rigid, say Ω^l, then $[u_n']^{kl} \equiv u_n'^k \equiv u_n'$ and $[\mathbf{u}_t']^{kl} \equiv \mathbf{u}_t'^k \equiv \mathbf{u}_t'$ and similarly for the normal and tangential displacements.

Such models can be described as follows:

A. Statically Loaded Spinal System

Problem $\mathcal{P}_{\text{stat}}$ Distribution of the displacement field $\mathbf{u}$ and the stress field τ in the studied spinal system satisfy in the case of the statically loaded spine the equilibrium equations for every domain Ω^ι of Ω read as follows:

$$\frac{\partial \tau_{ij}(\mathbf{u}^\iota)}{\partial x_j} + F_i^\iota = 0, \qquad i,j = 1,\ldots,N,\ \iota = 1,\ldots,s,\ \text{in } \Omega^\iota, \tag{7.33}$$

where $\mathbf{F}^\iota$ are body forces. A repeated index implies the summation from 1 to N. The relation between the displacement vector $\mathbf{u} = (u_i)$, $i = 1,\ldots,N$, and the small strain tensor e_{ij} is defined by

$$e_{ij} = e_{ij}(\mathbf{u}) = \frac{1}{2}\left(\frac{\partial u_i}{\partial x_j} + \frac{\partial u_j}{\partial x_i}\right), \qquad i,j = 1,\ldots,N.$$

The relation between the stress and strain tensors is defined by the generalized Hooke law

$$\tau_{ij}^\iota = c_{ijkl}^\iota e_{kl}(\mathbf{u}^\iota), \qquad i,j,k,l = 1,\ldots,N,\ \iota = 1,\ldots,s,$$

in the anisotropic case, whereas in the isotropic case

$$c^{\iota}_{ijkl} = \mu^{\iota}(\delta_{ik}\delta_{jl} + \delta_{il}\delta_{jk}) + \lambda^{\iota}\delta_{ij}\delta_{kl}, \ \beta^{\iota}_{ij} = \gamma^{\iota}\delta_{ij},$$

where λ^{ι}, μ^{ι} represent the Lamé coefficients. The coefficients c^{ι}_{ijkl} form a matrix of the type $(N \times N \times N \times N)$ and satisfy the symmetry and Lipschitz conditions

$$c^{\iota}_{ijkl} = c^{\iota}_{jikl} = c^{\iota}_{klij} = c^{\iota}_{ijlk},$$

$$0 < a^{\iota}_0 \leq c^{\iota}_{ijkl}(\mathbf{x})\xi_{ij}\xi_{kl}|\xi|^{-2} \leq A^{\iota}_0 < +\infty, \xi \in \mathbb{R}^{N^2}, \qquad \xi_{ij} = \xi_{ji}, \text{for a.e. } \mathbf{x} \in \Omega^{\iota},$$

where a^{ι}_0, A^{ι}_0 are constants independent of $\mathbf{x} \in \Omega^{\iota}$. For the isotropic case $a^{\iota}_0 = 2\min\{\mu^{\iota}(\mathbf{x}); \mathbf{x} \in \Omega^{\iota}\}$ and $A^{\iota}_0 = \max\{2\mu^{\iota}(\mathbf{x}) + 3\lambda^{\iota}(\mathbf{x}); \mathbf{x} \in \Omega^{\iota}\}$.

We will consider the following conditions: Let us denote the outward unit normal to the boundary $\partial\Omega$ by $\mathbf{n} = (n_i)$. Let $\Gamma_\tau = {}^1\Gamma_\tau \cup {}^2\Gamma_\tau$ and let us assume that a portion of the investigated spinal system is loaded at a certain boundary, which will be denoted by ${}^1\Gamma_\tau$ and it is unloaded at ${}^2\Gamma_\tau$. We thus have, for example, the conditions

$$\begin{aligned} \tau_{ij}n_j &= P_i \quad \text{on } {}^1\Gamma_\tau, \\ \tau_{ij}n_j &= 0 \quad \text{on } {}^2\Gamma_\tau. \end{aligned} \tag{7.34}$$

Let the investigated spinal system be fixed on the boundary Γ_u, that is,

$$u_i = 0 \quad \text{on } \Gamma_u. \tag{7.35}$$

Movement between bodies of vertebrae is facilitated by disks and by intervertebral joints. The disks facilitate flexion in all directions, small movements in the sense of rotation and translation, which are limited by the construction of the disk. The motion in the intervertebral joints is determined by the shapes of the joint surfaces. These surfaces are contact surfaces. The unilateral contact conditions with or without friction is sufficient to describe the function of the intervertebral joints, and, therefore, together with the ligament apparatus they facilitate both mobility of the spine as well as stability. Hence, we can use the following unilateral contact conditions: Denote by $\mathbf{u}^k$, $\mathbf{u}^l$ (indices k, l correspond to the neighboring bodies in contact) the displacements and the temperatures in the neighboring bodies. All these quantities are functions of spatial coordinates. Then on the contact boundaries Γ^{kl}_c the condition of nonpenetration

$$u^k_n(\mathbf{x}) - u^l_n(\mathbf{x}) \leq 0 \quad \text{on } \Gamma^{kl}_c \tag{7.36}$$

holds, where $u^l_n = u^l_i n^k_i$. For the contact forces, due to the law of action and reaction, we find

$$\tau^k_n(\mathbf{x}) = \tau^l_n(\mathbf{x}) \equiv \tau^{kl}_n(\mathbf{x}), \qquad \boldsymbol{\tau}^k_t(\mathbf{x}) = -\boldsymbol{\tau}^l_t(\mathbf{x}) \equiv \boldsymbol{\tau}^{kl}_t(\mathbf{x}). \tag{7.37}$$

Since the normal components of contact forces cannot be positive, that is, cannot be tensile forces,

$$\tau^k_n(\mathbf{x}) = \tau^l_n(\mathbf{x}) \equiv \tau^{kl}_n(\mathbf{x}) \leq 0 \quad \text{on } \Gamma^{kl}_c. \tag{7.38}$$

During the deformation of the bodies they are in contact or they are not in contact. If they are not in contact, then $u_n^k - u_n^l < 0$, and the contact forces are equal to zero, that is, $\tau_n^k = \tau_n^l \equiv \tau_n^{kl} = 0$. If the bodies are in contact, that is, $u_n^k - u_n^l = 0$, then there may exist nonzero contact forces $\tau_n^k = \tau_n^l \equiv \tau_n^{kl} \leq 0$. These cases are included in the following condition:

$$(u_n^k(\mathbf{x}) - u_n^l(\mathbf{x}))\tau_n^{kl}(\mathbf{x}) = 0 \quad \text{on } \Gamma_c^{kl}. \tag{7.39}$$

Further, if both bodies are in contact, then on the contact boundary the Coulombian type of friction acts. The frictional forces g_c^{kl} acting on the contact boundary Γ_c^{kl} are, in their absolute value, proportional to the normal stress component, where the coefficient of proportionality is the coefficient of the Coulombian friction $\mathcal{F}_c^{kl}(\mathbf{x})$, that is,

$$g_c^{kl}(\mathbf{x}) = \mathcal{F}_c^{kl}(\mathbf{x})|\tau_n^{kl}(\mathbf{x})|. \tag{7.40}$$

Due to the acting and frictional forces we have the following cases:

If the absolute value of tangential forces $\boldsymbol{\tau}_t^{kl}(\mathbf{x})$ is less than the frictional forces g_c^{kl}, then the frictional forces preclude the mutual shifts of both bodies being in contact. If the tangential forces $\boldsymbol{\tau}_t^{kl}$ are equal in their absolute value to the frictional forces, so that there are no forces that can preclude the mutual motion of both elastic spine parts (elastic bodies). Thus the contact points change their position in the direction opposite to that in which the tangential stress component acts. These conditions are described as follows:

1. If $u_n^k - u_n^l = 0$, then $|\boldsymbol{\tau}_t^{kl}(\mathbf{x})| \leq g_c^{kl}(\mathbf{x})$.
2. If $|\boldsymbol{\tau}_t^{kl}(\mathbf{x})| < g_c^{kl}(\mathbf{x})$, then $\mathbf{u}_t^k(\mathbf{x}) - \mathbf{u}_t^l(\mathbf{x}) = 0$, which means that the friction forces are sufficient to preclude the mutual shifting between the fractured vertebrae.
3. If $|\boldsymbol{\tau}_t^{kl}(\mathbf{x})| = g_c^{kl}$, then there exists a function $\vartheta \geq 0$ such that $\mathbf{u}_t^k(\mathbf{x}) - \mathbf{u}_t^l(\mathbf{x}) = -\vartheta\boldsymbol{\tau}_t^{kl}(\mathbf{x})$, which means that the friction forces are not sufficient to preclude the mutual shifting of both parts of vertebrae. This shift acts in an opposite direction to the acting tangential forces. Note that here we have $\mathbf{u}_t^l = \mathbf{u}^l - u_n^l\mathbf{n}^k$ and $u_n^l = \mathbf{u}^l.\mathbf{n}^k \equiv u_i^l n_i^k$.

These conditions are valid also for spinal joints. In some problems the bilateral conditions with or without friction can be used, for example, for the case without friction

$$u_n = 0, \tau_{tj} = 0, \qquad j = 1, \ldots, N. \tag{7.42}$$

These conditions describe the conditions on the axis Γ_0 (or plane) of symmetry.

B. Dynamically Loaded Spinal System In the study we will investigate the dynamic multibody contact problem with the Coulomb friction in linear viscoelasticity with short memory in N dimensions formulated in velocities.

The stress–strain relation will be defined by Hooke's law:

$$\begin{aligned} \tau_{ij} &= \tau_{ij}(\mathbf{u}, \mathbf{u}') = c^{(0)}_{ijkl}(\mathbf{x}) e_{kl}(\mathbf{u}) + c^{(1)}_{ijkl}(\mathbf{x}) e_{kl}(\mathbf{u}'), \\ e_{ij}(\mathbf{u}) &= \frac{1}{2}\left(\frac{\partial u_i}{\partial x_j} + \frac{\partial u_j}{\partial x_i}\right), \qquad i,j,k,l = 1,\ldots,N, \end{aligned} \tag{7.43}$$

where $c^{(n)}_{ijkl}(\mathbf{x})$, $n=0,1$, are elastic and viscous coefficients and $e_{ij}(\mathbf{u})$ are components of the small strain tensor, and N is the space dimension. For the tensors $c^{(n)}_{ijkl}(\mathbf{x}), n=0,1$, we assume

$$\begin{aligned} &c^{(n)}_{ijkl} \in L^{\infty}(\Omega), \qquad n = 0,1, \qquad c^{(n)}_{ijkl} = c^{(n)}_{jikl} = c^{(n)}_{klij} = c^{(n)}_{ijlk}, \\ &c^{(n)}_{ijkl} e_{ij} e_{kl} \geq c^{(n)}_0 e_{ij} e_{ij} \quad \forall e_{ij}, e_{ij} = e_{ji} \quad \text{and a.e. } \mathbf{x} \in \Omega, c^{(n)}_0 > 0. \end{aligned} \tag{7.44}$$

A repeated index implies the summation from 1 to N.

The unilateral conditions will be formulated in velocities, that is, we will assume the following contact conditions:

$$[u'_n]^{kl} \leq 0, \qquad \tau^k_n = \tau^l_n \equiv \tau^{kl}_n \leq 0, \qquad [u'_n]^{kl} \tau^{kl}_n = 0 \quad \text{on } \cup \Gamma^{kl}_c. \tag{7.45}$$

If one of the colliding bodies is absolutely rigid, say Ω^l, then $[u'_n]^{kl} \equiv u'^k_n \equiv u'_n$ and the unilateral conditions are the following:

$$u'_n \leq 0, \qquad \tau_n \leq 0,\ u'_n \tau_n = 0 \quad \text{on } \Gamma^k_c.$$

Then we have the following problem:

Problem $(\mathcal{P}_{\mathbf{dyn}})$ Let $N=2,3$, $s \geq 2$. Find a vector function $\mathbf{u}: \overline{\Omega} \times I \to \mathbb{R}$, satisfying

$$\rho^{\iota} \frac{\partial^2 u^{\iota}_i}{\partial t^2} = \frac{\partial \tau^{\iota}_{ij}(\mathbf{u}^{\iota}, \mathbf{u}^{\iota\prime})}{\partial x_j} + F^{\iota}_i, \qquad i,j = 1,\ldots,N,\ \iota = 1,\ldots,s,\ (t,\mathbf{x}) \in Q^{\iota} = I \times \Omega^{\iota}, \tag{7.46}$$

$$\begin{aligned} \tau_{ij} n_j &= P_i, \qquad i,j = 1,\ldots,N,\ (t,\mathbf{x}) \in {}^1\Gamma_\tau(t) = I \times \cup^s_{\iota=1} ({}^1\Gamma_\tau \cap \partial\Omega^{\iota}), \\ \tau_{ij} n_j &= 0, \qquad i,j = 1,\ldots,N,\ (t,\mathbf{x}) \in {}^2\Gamma_\tau(t) = I \times \cup^s_{\iota=1} ({}^2\Gamma_\tau \cap \partial\Omega^{\iota}), \end{aligned} \tag{7.47}$$

$$u_i = 0, \qquad i = 1,\ldots,N, \quad \text{on } \Gamma_u(t) = I \times \cup^s_{\iota=1} (\Gamma_u \cap \partial\Omega^{\iota}), \tag{7.48}$$

the unilateral contact conditions in velocities

$$[u'_n]^{kl} \leq 0, \qquad \tau^k_n = \tau^l_n \equiv \tau^{kl}_n \leq 0, \qquad [u'_n]^{kl}\, \tau^{kl}_n = 0, \tag{7.49}$$

and the Coulombian law of friction,

$$[\mathbf{u}_t']^{kl} = 0 \Longrightarrow |\boldsymbol{\tau}_t^{kl}| \leq \mathcal{F}_c^{kl}(0)|\tau_n^{kl}|,$$
$$[\mathbf{u}_t']^{kl} \neq 0 \Longrightarrow \boldsymbol{\tau}_t^{kl} = -\mathcal{F}_c^{kl}([\mathbf{u}_t']^{kl})|\tau_n^{kl}|\frac{[\mathbf{u}_t']^{kl}}{|[\mathbf{u}_t']^{kl}|}, \tag{7.50}$$

for $(t,\mathbf{x}) \in \Gamma_c(t) = I \times \cup_{k,l}\Gamma_c^{kl}$, and the initial conditions

$$\mathbf{u}(\mathbf{x},0) = \mathbf{u}_0(\mathbf{x}), \qquad \mathbf{u}'(\mathbf{x},0) = \mathbf{u}_1(\mathbf{x}), \;\; \mathbf{x} \in \Omega, \tag{7.51}$$

where $\mathbf{u}_0$, $\mathbf{u}_1$, $\mathbf{u}_2$ are the given functions, $\mathbf{u}_2$ has the time derivative $\mathbf{u}_2'$, and $\mathbf{u}_0, \mathbf{u}_1$ satisfy the static linear contact problems in elasticity with or without the Coulombian friction $\mathcal{F}_c^{kl}$ [see Nedoma (1983, 1987)] or they can be equal to zero. The coefficient of friction $\mathcal{F}_c^{kl} \equiv \mathcal{F}_c^{kl}(\mathbf{x},\mathbf{u}')$ is globally bounded, nonnegative, and it satisfies the Carathéodory property, that is, $\mathcal{F}_c^{kl}(\cdot,\mathbf{v})$ is measurable, $\mathbf{v} \in \mathbb{R}$, and $\mathcal{F}_c^{kl}(\mathbf{x},\cdot)$ is continuous for a.e. $\mathbf{x} \in \Gamma_c^{kl}$.

Numerical Experiments Numerical experiments are based on the statically loaded spine system with a fracture of the vertebrae (Chance type of fracture) and on the statically loaded spinal system where the lumbar spine with the spinal cord channel is pressed by the internal osteophytes situated in the spinal cord channel.

As a first model problem the case of the **Chance fracture** was investigated, which is a relatively simple model for the mathematical spine modeling (Fig. 7.325). In

FIG. 7.325 Chance fracture—schematic view.

Nedoma et al. (2000, 2003a, 2006) the real Chance fracture model was discussed. The model was derived from the 2D CT image of the lumbar spine and from the analysis of the force stream between L3 and L4 transmitted by the spine in the prone position, which is assumed to a load of $\sim 250\,N$. The value of loading was recalculated into the axis of the L1 vertebrae. The spongy bone was assumed to be homogeneous and isotropic with Young's modulus $E = 75$ MPa and Poisson's constant $\sigma = 0.25$. The compact part of the vertebrae was assumed to be homogeneous and also isotropic with Young's modulus $E = 16100$ MPa and Poisson's constant $\sigma = 0.25$. The average value for the vertebral body was found to be $E = 3500$ MPa and $\sigma = 0.25$. The ligamental ring can be modeled as a homogeneous orthotropic material corresponding to an average value of lamellae. In the model the average value together with a hyaline cartilage was used. The hyaline cartilage was modeled by a homogeneous isotropic material with $E = 24.3$ MPa and $\sigma = 0.45$. For the disk together with the cartilage the average values of Young's modulus and Poisson's constant, $E = 104$ MPa, $\sigma = 0.45$, were used. The nucleus pulposus was modeled as a noncompressed material with Young's modulus $E = 1000$ MPa and Poisson's constant $\sigma = 0.4$. Our numerical tests were based on the anterior and posterior approaches of the Chance fracture with or without rotation (see Fig. 7.326). In the anterior approach the fixator was fixed at L1 and L3 in the first case, at L1 and L3 in the second, as well as at L2 at some distance above the Chance fracture line. In the case of the posterior approach the fixator is fixed at L1 and L3. The load on the spine is assumed to be 196 N in the direction of the axis of L1.

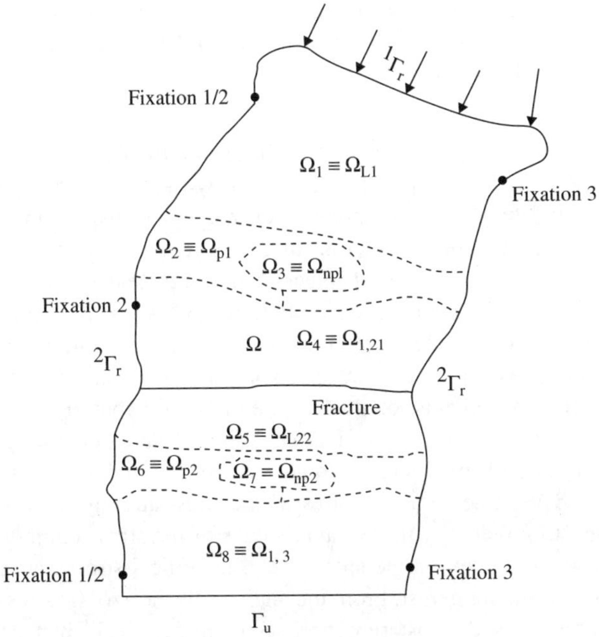

FIG. 7.326 Mathematical model of the Chance fracture with and without rotation after the reposition: the anterior and posterior approaches (• denotes the point of fixation).

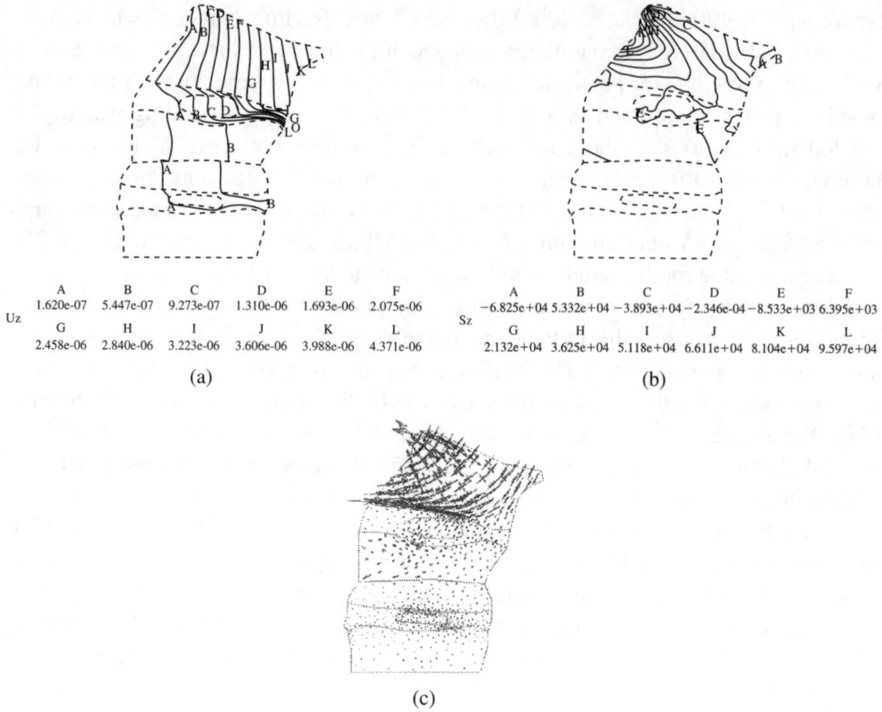

FIG. 7.327 Chance fracture with rotation after reposition with fixed L1 and L3—the anterior approach: (a) the vertical displacement component u_z, (b) the vertical stress component τ_{zz}, and (c) the principal stresses τ_1 and τ_2.

The results of our analyses, based on linear elastic rheology, are represented by stress components $\tau_{11} \equiv \tau_{xx}$, $\tau_{12} \equiv \tau_{xz}$, and $\tau_{22} \equiv \tau_{zz} \equiv S_z$ and principal stresses τ_1 and τ_2, normal and tangential components of stress and displacement vectors on the contact zone of the Chance fracture of L2 vertebrae (Figs 7.327–7.329). For illustration we present the vertical displacement component u_z and stress component τ_{zz} and in Fig. 7.329 also the shear stresses $\tau_{12} \equiv \tau_{xz} \equiv S_{xz}$, and the principal stresses τ_1 and τ_2. Figs 7.327a, 7.238a and 7.329a represent the vertical displacement component u_z. From the courses of displacement u_z isolines we can observe the small shifts (i.e. movements) on the fracture lines. Distribution and magnitude of the stress component and of the principal stresses facilitate comparing the burdening of particular parts of the fractured spine and their mutual stress–strain relations after reposition. The same values of stresses are illustrated by isostresses, that is, by curves connecting places with the same values of the stress component in MPa. Since the load was assumed on the spine axis, the vertical stress and displacement components are our main interest. From the analysis the horizontal stress component and the shear stresses for the posterior approach in the classical Chance fracture case we can see that both stress components in the area of a fracture move arround the same value, that is, in the interval -2.83×10^4 MPa, 4.6×10^4 MPa for τ_{11} and in

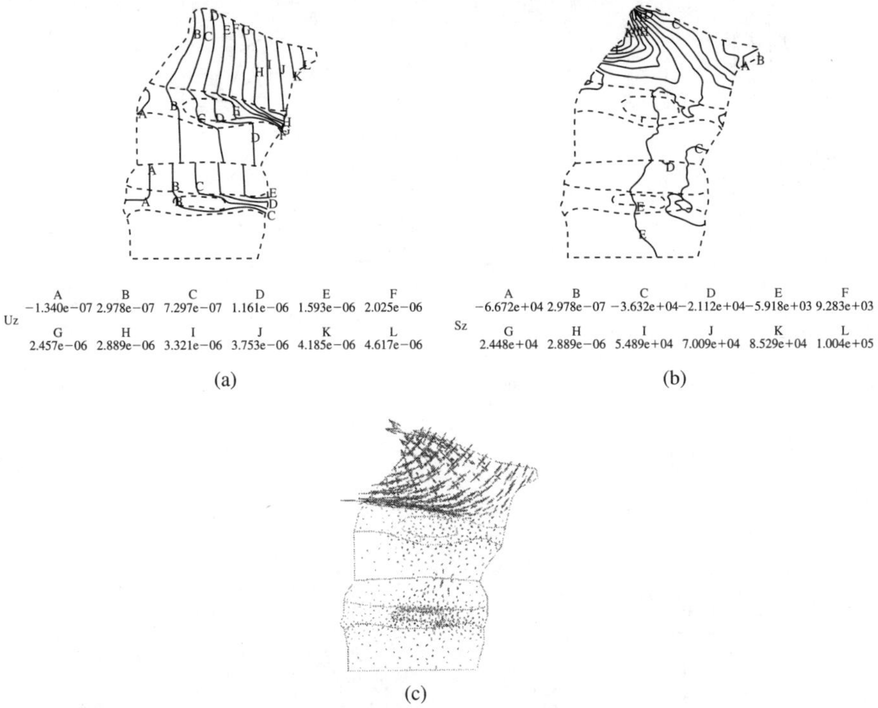

FIG. 7.328 Chance fracture with rotation after reposition with fixed L1, L2, and L3—the anterior approach: (a) the vertical displacement component u_z, (b) the vertical stress component τ_{zz}, and (c) the principal stresses τ_1 and τ_2.

the neighborhood of 5.33×10^3 MPa for the shear stresses. These values indicate a qualitative evaluation. Greater informative values represent principal stresses and their distributions in the vertebral bodies, disks and nucleus pulposus. The analysis of the vertical stress component indicates that in spite of a vertebral body fixation microscopic movements along the fracture plane are (i) of the order of 1.62×10^{-4} mm up to 5.45×10^{-4} mm for the fixation, and at L1 and L3 was observed and with the maximal vertical displacement in the spinal cord zone at L1 with a value of $\sim 4.37 \times 10^{-3}$ mm was observed; (ii) of the order of -1.34×10^{-4} mm on the inner part of the spine up to 1.59×10^{-3} mm in the area of the spinal cord channel, with the maximal value 4.6×10^{-3} mm in the area of the spinal cord channel at L1 vertebrae; (iii) of the order 7.475×10^{-3} mm to 1.07×10^{-2} mm on the inner side of the spine in the direction to the center of the fracture at the L1 vertebrae, with the maximal value of $\sim 3.33 \times 10^{-2}$ mm on the inner side of a lower part at the L1 vertebrae.

From the comparison of both surgical approaches we can see that the displacements in the anterior approach are of about 0.5 order lower than in the case of the posterior approach. The displacements are very small so that in all the cases studied they have only positive effects on the healing of the fracture because the microscopic pressure movements accelerate a bone coalescence.

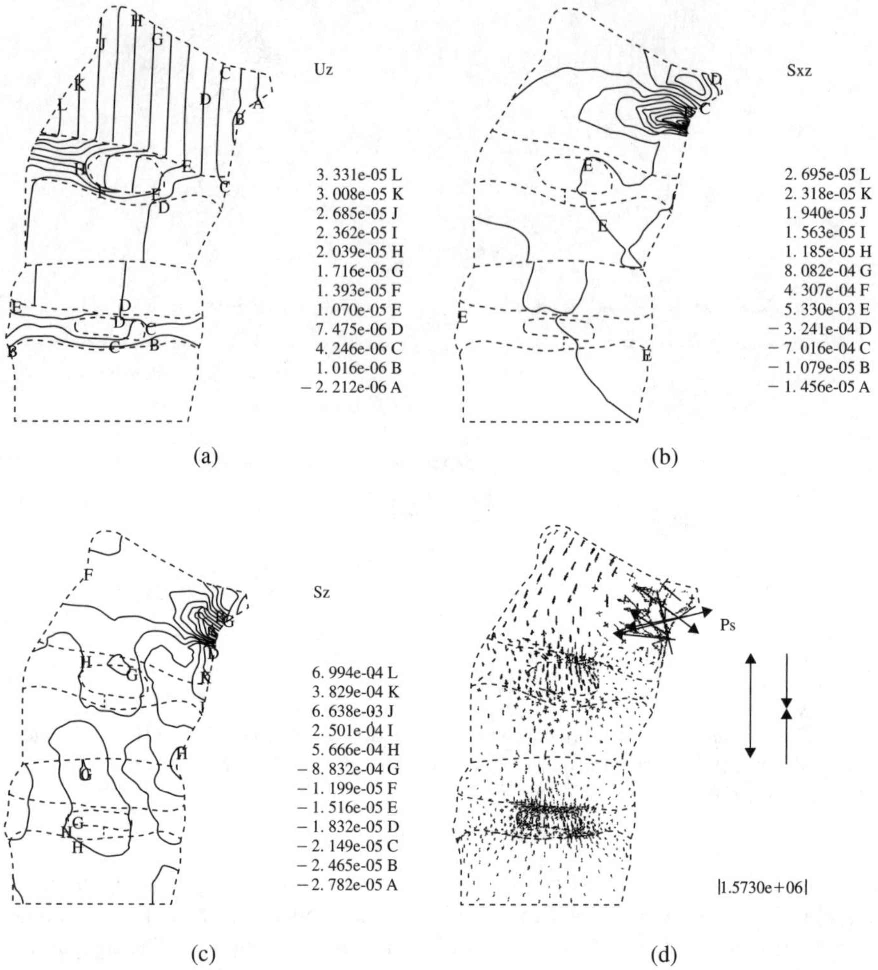

FIG. 7.329 Chance fracture with rotation after reposition with fixed L1 and L3—the posterior approach: (a) the vertical displacement component u_z, (b) the shear stress component τ_{xz}, (c) the vertical stress component τ_{zz}, and (d) the principal stresses τ_1 and τ_2.

The analyses of principal stresses show that in all investigated cases the maximal concentration of stresses, that is, extension (denoted by $\longleftarrow\longrightarrow$), accumulate in places of fixation, representing a very good situation for the fractured spine. Compression is denoted as $\longrightarrow\longleftarrow$. The analyses show that in the case of the anterior approach the stresses accumulate at the L1 vertebral body, while in the posterior approach stresses accumulate in the neighborhood of the fixation and into the area of intervertebral disks near the nucleus pulposus, but its values are lower than in the area of fixation in the L1 vertebral body. In the case of fixation L1, L2, and L3 in the area of the fracture, compression is indicated near the spinal cord channel. Its value is lower than in the area of fixation but greater than on the nucleus pulposus. In the case of the posterior approach the stresses are indicated in the area of intervertebral disks near the nucleus pulposus and in the upper part of the L1 vertebrae. The values are substantially lower than in the area of fixation.

The second model represents the **model of the loaded lumbar spine with the spinal cord channel** in which the spinal cord is pressed by the deformed loaded inner osteophytes in the spinal cord channel.

The discussion of kinetics covers the thoracic and lumbar spine. The discussed model involves only the lumbar spine because it is subjected to significantly greater loads than in the rest of the spine. During daily activities, loads on the disks are complex, and they are usually subjected to a combination of compression, bending, and torsion. It was observed that flexion, extension, and lateral flexion of the spine produce mainly tensile and compressive stresses on the disks, whereas rotations produce above all shear stresses. When motion segments are transected vertically, the nucleus pulposus of the disks protrudes, which means that the nuclei pulposi are under pressure. In the lumbar spine the tensile stress in the posterior part of the annulus fibrosus has been estimated to be $\sim 5 \times$ the applied axial compressive load. Degeneration of disks reduces their proteoglycan content and thus their hydrophylic capacities. Since disks become less hydrated, their elasticity and their ability to store energy and distribute loads gradually decrease. Thus these changes make the disks more vulnerable to stresses.

However, when large parts of the lumbar spine are fixed with the help of osteophytes, the stress–strain relations in the spine are radically changed. In the normal state the vertebrae have 6 degrees of freedom, that is, rotation about and translation along transverse, sagittal, and longitudinal axes. Then the motion, which is produced during flexion, extension, lateral flection, and axial rotation of the spine, is a complex of combined motion, which is a result of simultaneous rotation and translation.

In the normal state the intervertebral disks exhibit viscoelastic properties, that is, creep and relaxations, and hysteresis. Creep occurs more slowly in normal healthy disks than in degenerated disks. All properties of the spine vertebrae and ligaments change when the spine column is harden and fixed by osteophytes and their further coalescence into the rigid rod. When the spinal cord channel starts to narrow, then the spinal cord is compressed, and, due to the changes in load, spinal cord is more and more compressed, causing pain and at times a paralysis of lower limbs.

The model is similar to the previous case of the Chance fracture. The model is different from previous case only in the sense that no fracture is present. The ligaments and other soften tissues are assumed. Thus, this model is similar to the model problem presented above, only material properties of the lumbar spine are different.

The model of the thoracolumbar spine was derived from the MRI of a 75-year-old patient (see Fig. 7.330) and for our numerical computation it was simplified (see Fig. 7.331) and shows Th11/12-L5. It is well established that the failure of the spinal column can occur through various combinations of bending moments as well as from compressive and shear forces (Langrana et al., 2002; Cusick and Yoganandan, 2002). In our model information about the force stream between L3 and L4 transported by the spine in the prone position, which for the load is 250 N, and in the vertical position it is the largest in the sitting position at approximately 1375 N, in the vertical position it

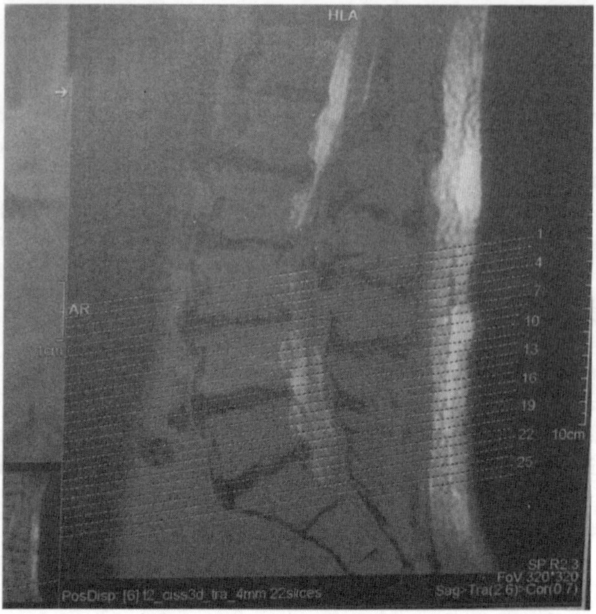

FIG. 7.330 Magnetic resonance imaging (MRI) picture of the thoracolumbar spine.

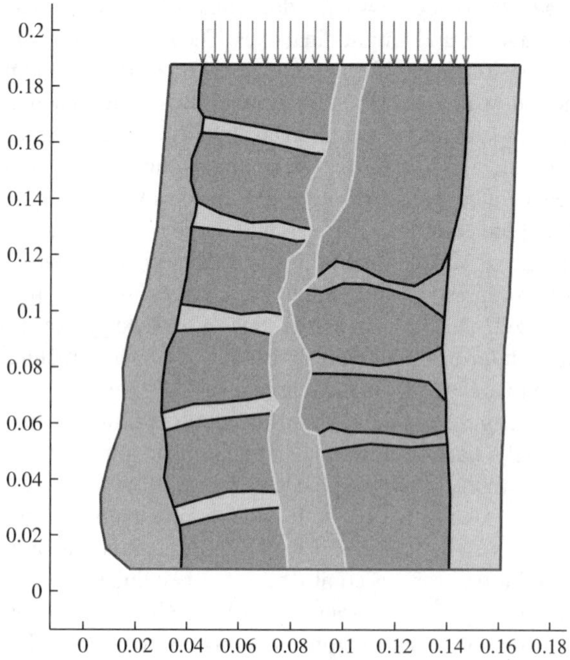

FIG. 7.331 Model of the thoracolumbar spine based on the MRI analysis.

is approximately 960 N and in the basic position with a load of 5 kg with arms stretched forward it is 1990 N. The spongy bone was modeled by a homogeneous and isotropic biomaterial with a Young's modulus of $E = 75$ MPa and Poisson's ratio of $\sigma = 0.25$, the compact bone with $E = 17{,}100$ MPa and Poisson's ratio $\sigma = 0.25$, and the average value for the vertebral body with $E \cong 3500$ MPa, $\sigma = 0.25$. The hyaline cartilage was modeled by a homogeneous isotropic biomaterial with $E = 24.3$ MPa, $\sigma = 0.45$, and the ligament tissues with $E = 0.7$ MPa and $\sigma = 0.49$. The observations shown in the MRI (see Fig. 7.330) indicate the stiffened areas of the osteoligamental complex by the system of osteophytes, so that the back column transports loads from the upper parts of the human body into the lower parts of the human body. Then the relatively large loads are transmitted also through this stiffened osteoligamental complex.

Results for the deformation of the thoracolumbar spine, the distributions of stress components $\tau_{11}, \tau_{12}, \tau_{22}$, and the principal stresses τ_1 and τ_2 as well as the displacement components in this part of the spine were obtained. The values have the horizontal displacement component, the vertical and shear stresses, as well as the principal stresses (global and local). All results show that the stresses are transported through two spinal columns into the lower parts of the human body. The cases studied had loads of 250, 960, 1375, and 1990 N. The results show that the critically strained places of the thoracolumbar spine are the same for all loads, and they are found at L2/L3 of the narrow spinal cord and below the stiffened area of the back column. Our numerical results indicate that the three column theories of the spine (Denis, 1983) explain better the distribution of stresses in the thoracolumbar spine, and thus the stability of the spine, than the two-column theories of the spine (Holdsworth, 1970; Whitesides 1977; White and Panjabi, 1978).

Numerical computations indicate the following:

1. Maximal global pressures at the narrow part of spinal cord at L2 and L3 vertebrae have the following values: under a load of 250 N the pressure has a value of -1.0559×10^5 Pa, under a load of 960 N a value of -4.065×10^5 Pa, under a load of 1375 N a value -5.6862×10^5 Pa, and under the load 1990 N a value of -8.392×10^5 Pa.
2. Since the back column is created from the osteoligamental complex, which is stiffened by the osteophytes, then large tensions are observed at L3/L4 in the back column with the following values: under a load of 250 N the tension has a value of 1.9406×10^5 Pa, under a load of 960 N a value of 7.4663×10^5 Pa, under a load of 1375 N a value of 1.0503×10^6 Pa, and under a load of 1990 N a value of 1.543×10^6 Pa.

The numerical analyses of the spinal cord indicate local pressures and tensions at the L1 and L2 vertebra as follows: with maximal pressures of -680.29 Pa under a load of 250 N and tensions at 268.43 Pa on both sides of the maximal pressure: and moreover, a pressure of -4780.4 Pa and tension of 3834.7 Pa under a load of 960 N; a pressure of -3973 Pa, which in this area is doubled, and tension at 33,288 Pa; a pressure of $-8653, 5$ Pa, where the second value of the doubled pressure is $\sim$ one-half the value of the first one, while the maximal tension increases up to a value of 7351.4 Pa below the maximal pressures, and the tension upward of the maximal pressures increases little against the previous loading (i.e., 1375 N). Numerical analyses show that the lower part of the spinal cord functions as a stress guide, but with relatively small values of stresses. It is evident that from the obtained numerical results distributions of displacements, stresses, and deformations of the vertebrae in the overloaded spine as wells as the normal and tangential components of the displacement and stress vectors between the spinal cord and the vertebral column and on the contact surfaces of the intervertebral joints can be determined and analyzed.

Figures 7.322–7.363 show numerical results for the model of thoracolumbar spine—cases with loading of 250, 960, 1375, and 1990 N (see the accompanying DVD disk).

7.11 BIOMECHANICAL AND MATHEMATICAL MODELS OF JOINTS OF THE UPPER LIMBS

7.11.1 Introduction

Should the upper limbs of the human body lose their ability to function, the very complicated movements needed for work are not possible. Therefore, these limbs are provided with the ability for versatile mobility. Anatomically, it is reflected in the different arrangements of the joints. Their closures have a gripping function. Muscles of the upper limb are adapted to the function of gripping, and they are created from spindle-shaped muscles. Owing to their great mobility these muscles are more numerous and of different types.

The upper limbs are created from the scapula, the clavicula, the humerus, the radius, the ulna, the ossa carpi, the ossa metacarpi, and the phalanges (see Fig. 7.364). The motion of the shoulder girdle runs in the glenohumeral acromioclavicular, the sternoclavicular joints, and in the space between the scapula and the wall of the thorax. These muscles produce the most dynamic and mobile joints in the human body, so that we can say that the shoulder forms the base for all upper extremity movements.

7.11.2 Anatomy and Biomechanics of Shoulder Joints

The complex of the shoulder is created from two main structures—the shoulder joints and the shoulder girdle. The shoulder joint, known as the glenohumeral articulation, is a ball-and-socket joint between the nearly hemispherical humeral head and the shallow concave glenoid fossa of the scapola. The shallowness of the glenoid fossa allows for substantial freedom of movements of the humeral head on the joint surface of the glenoid. The movements are (i) flexion, representing the movements of the humerus to the front, that is, forward upward movements; (ii) extension, representing returning from flexion; (iii) abduction, representing a horizontal upward movement of the humerous to the side; and (iv) adduction, representing a return from abduction and outward rotation representing movements of the humerous around its long axis of the lateral side as well as inward rotation representing a return from

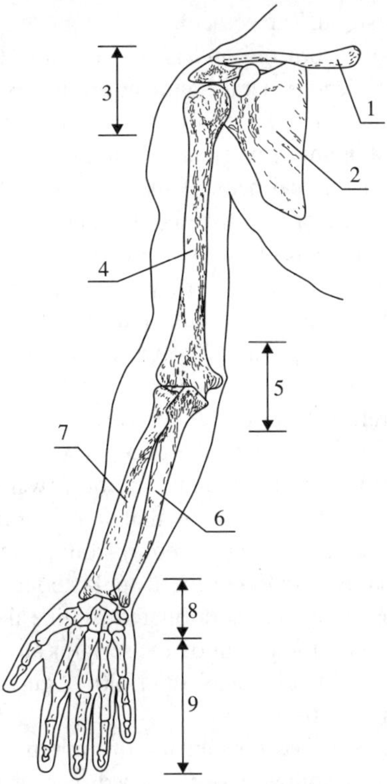

FIG. 7.364 Upper limb — the shoulder joint, the elbow joint, and the hand [modified after Zrzavý (1977), Jazravi et al. (2001), Baar and Bear-Lehman (2001)]: (1) clavicle, (2) scapula, (3) shoulder joint, (4) humerus, (5) elbow joint, (6) radius, (7) ulna, (8) wrist, and (9) hand.

outward rotation. The stability of the joint is provided by the glenohumeral and coracohumeral ligaments and the muscles crossing the joint. Numerous muscles act on the various components of the shoulder system to provide mobility as well as dynamic stability. Dynamic stability is ensured by passive muscle tension, via a barrier effect of the contracted muscle, compressive forces brought about by muscular contraction, joint motion that induces tightening of the passive or ligamentous constraints, and a redirection of the joint force toward the center of the glenoid. The muscles of the shoulder joint are as follows: the deltoideus, the superspinatus, the pectoralis major, the coracobrachialis, the latissimus dorsi, the teres major, the teres minor, the infraspinates, and the subscapularis.

The glenohumeral joint enables the body to make major exertions. The three glenohumeral ligaments—the superior, the middle, and inferior—are discrete extensions of the anterior glenohumeral joint capsule, and they are critical to shoulder stability and function. The superior glenohumeral ligament originates from the anterosuperior labrum, just anterior to the long head of biceps, and it inserts pressure onto the lesser tuberosity, and it acts as the main restraint to inferior translation with the arm in the resting or adducted position. The coracohumeral ligament originates from the lateral side of the base of the coracoid and is inserted into the anatomical neck of the humerus. The middle glenohumeral ligament originates inferior to the superior glenohumeral ligament and it is inserted further laterally into the lesser tuberosity. The inferior glenohumeral ligament originates from the inferior aspect of the labrum and is inserted into the anatomical neck of the humerous. Synovial liquid acts by cohesion and adhesion to further stabilize the glenohumeral joint. The bony structure of the shoulder girdle is represented by the clavicle and the scapula. The small acromioclavicular synovial joint is situated between the lateral end of the clavicle and the acromion of the scapula. The acromioclavicular and the sternoclavicular joints have menisci, layers of cartilage, situated between the bony surfaces. The scapular movements are (i) elevation, that is, movement of the scapula in the frontal plane; (ii) depression, that is, return from elevation; (iii) upward rotation, that is, turning the glenoid fossa upward and with the exception of the lower medial border of the scapula away from the spinal column, and downward rotation, representing return from upward rotation; (iv) protraction, that is, movements of the distal end of the clavicle forward; (v) retraction, that is, return from protection; and (vi) forward and backward rotation, that is, rotation of the scapula about the shaft of the clavicle. These movements are controlled and coordinated by like the trapesius, the levator scapulae, the rhomboid, the pectoralis minor, the serratus auterior, and the subclavius muscles. For more details see Morray and An (1990), Agur et al. (1991), Valle et al. (2001), and Nedoma et al. (2006).

The shoulder is the spherical free joint and it is the most movable joint in the human body. The shape of the joint hollow is oval shaped or pear shaped, and it is much smaller (five times) than the surface of the humer head. Therefore, to a great extent the head is in a contact with the joint. The joint is widespread, relatively thin, and on the frontal side is heighten by the ligament, which is the suspension ligament of a free limb. The joint head in the hollow is secured by a muscular tonus surrounding the joint from all sides. The roof above the head of humer creates any secondary hollows of the shoulder and then it limits the movements in this joint. The shoulder has 4 (from 6) degrees of freedom, that is, the rotations about the x, y, and z axes and one shift movement, which occurs in the luxation case only.

7.11.3 Anatomy and Biomechanics of the Elbow

The elbow joint (articulatio cubiti) is a complicated pulley-type joint connecting the arm with the forearm, and it is composed from three separate joints. From the anatomical point of view it is created from three joints with a common joint. Anatomically, it connect the ulna with the humerus (articulatio humeroulnaris), the radius with the humerus (articulatio humeroradialis), and the radius with the ulna (articulatio radioulnaris proximalis). One of them is a pulley type, between the humerus and the ulna; the second one is spherical, between the humerus and the radius; and the third one is cylindrical, between the ulna and the radius. In the first and second joints the joint bending takes place and in the third one rotation takes place. The elbow joint is relatively important because it enables the gripping function of the upper limb.

The rotational movements of the forearm are limited by the contact surfaces of the humerus. The center of rotation is the same as the center of bending. The humeroulnar joint is a hinge joint formed by the articulation between the spool-shaped trochlea of the distal humerus and the concave trochanter fossa of the proximal ulna.

The humeroradial joint is also a hinge joint formed between the capitulum of the distal humerus and the head of the radius. The proximal radioulnar joint is a pivot formed by the head of the radius and the radial notch of the proximal ulna. The muscles coordinating and controlling the movement of the elbow joint are the biceps, the brachioradialis, the brachialis, the pronator teres, the triceps brachii, the anconeus, and the supinator. The biceps brachii muscle is the most powerful flexor of the elbow joint, which on the distal side is attached to the tuberosity of the radius, and on the proximal side it has attachments at the top of the corocoid process and the upper lip of the glenoid fossa. Another important flexor is the brachialis muscle, which is attached at the lower half of the anterior portion of the humerus and the coronoid process of the ulna. The triceps brachii muscle is the most important muscle controlling the extension movement of the elbow. The pronator teres and the supinator muscles perform the pronation and supination movements of the forearm. For more details see Chao and Murray (1978), An et al. (1981), Morray (1993, 1994), and Jazrawi et al. (2001).

7.11.4 Biomechanics of the Shoulder and Elbow Joint Replacements

Shoulder joint replacement was first applied in Paris by Péan in 1892 [see Péan (1973), Lugli (1978)]. The replacement was made by using platinum and India rubber. This replacement was unsuccessful because the patient suffered from tuberculosis of the joints. The second replacement of the elbow joint was made by Judet in 1950, and the experience from the total hip joint replacements was used and Plexiglass was applied. Further experiments were made with metal replacements by Krüger in 1951 and by Neer in 1955 [see Krüger (1951), Neer (1955)]. With the new knowledge in aloarthroplasty of large human joints new types of shoulder and elbow joint replacements were developed. In the case of the shoulder joint the following replacements can be made:

1. Partial replacement of the shoulder joint, which replaces the independent head or the head with partial replacement of the upper part of the humerus
2. Total shoulder joint replacement as a hanging system, which replaced the head as well as the hollow
3. Total shoulder joint replacement as a nonhanging system

Elbow replacements are done very in frequently. For biomechamical specialists and orthopedic surgeons seem to be more demanding than other surgical treatment techniques which can be applied with greater success.

Behavior of both types of joints, the shoulder and elbow joints, under the dynamic conditions is defined by the constitutional properties of separate components of the shoulder or elbow joint systems, respectively. Elastic and viscoelastic properties of both joints determine the conditions in the joint. These conditions, as in the case of larger human joints, are influenced by the structure and power of the surrounding ligament and the muscular system, which does not have a resting function but a gripping function. By application of total replacements of these joints under the ill joints the inner-joint relations start to be realised. Therefore, the total replacement must be constructed in such a way that it ensures good functioning of the shoulder or the elbow joints as well as good fixation in the joint sector of the skeleton. Moreover, the contact surfaces of joint components in both cases are determined for joint replacements that function well from the point of view of its function the distribution of stresses in the skeleton. Also, as in the case of all large human joints, the construction of the contact surfaces must change the function of the cartilage and the synovial liquid in the healthy shoulder or the elbow joints, respectively, and to ensure transportation of dynamic loads that originate as a consequence of dynamical operation of muscle forces, above all in the shoulder joint, and at the same moment to ensure absorbtion of forces originated at the change into a quiet state.

7.11.5 Model of the Main Joints of the Upper Limb

It is evident that methods introduced in Chapter 6 can be applied in the case of both shoulder and elbow joints and their total replacements. Thus under similar considerations as in the case of large human joints and their total replacements, we will solve the problem that will help us study the stress–strain analysis of the area of the shoulder or elbow joints, that is, the following mathematical model problem:

Let $I=(0,t_p)$ be the time interval and let the shoulder or the elbow joint together with ligaments occupy the region $\Omega = \cup_{\iota=1}^{s} \Omega^{\iota} \subset \mathbb{R}^N$, $N=2,3$, where the regions Ω^{ι}, $\iota=1,\ldots,s$, are represented by the separate parts of the shoulder or elbow joint system and the ligaments. Let Ω^{ι} have Lipschitz boundaries $\partial\Omega^{\iota}$ and let us assume that $\partial\Omega=\Gamma_\tau\cup\Gamma_u\cup\Gamma_c\cup\Gamma_0$, where $\Gamma_\tau={}^1\Gamma_\tau\cup{}^2\Gamma_\tau$ represents loaded and unloaded parts of the boundary, where ${}^1\Gamma_\tau$ is a part of the boundary, where loads are prescribed, for example, where muscle forces operate on the different parts of the shoulder or elbow joint systems or where the weight of the body or an outer burden acts and $\Gamma_u={}^1\Gamma_u\cup{}^2\Gamma_u$ is a part of the boundary where the blade bone (shoulder blade) is assumed to be fixed on ${}^1\Gamma_u$ and on ${}^2\Gamma_u$, where displacements corresponding to movements of the limb (e.g., the blade bone), $\Gamma_c=\cup_{k,l}\Gamma_c^{kl}$, $\Gamma_c^{kl}=\partial\Omega^k\cap\partial\Omega^l$, $k\neq l$, $k,l\in\{1,\ldots,s\}$ is a part of the boundary, where the different parts of the assumed systems are in mutual contact, for example, in areas of joints or of artificial replacements of joints, respectively. Let $Q=I\times\Omega$ denote the time–space domain and let $\Gamma_\tau(t)=\Gamma_\tau\times I$, $\Gamma_u(t)=\Gamma_u\times I$, $\Gamma_c(t)=\Gamma_c\times I$ denote the parts of its boundary $\partial Q=\partial\Omega\times I$.

Furthermore, let $\mathbf{n}$ denote the outer normal vector of the boundary, $u_n=u_in_i$, $\mathbf{u}_t=\mathbf{u}-u_n\mathbf{n}$, $\tau_n=\tau_{ij}n_jn_i$, $\tau_t=\tau-\tau_n\mathbf{n}$ be normal and tangential components of displacement and stress vectors $\mathbf{u}=(u_i)$, $\tau=(\tau_i)$, $\tau_i=\tau_{ij}n_j$, $i,j=1,\ldots,N$. Let $\mathbf{F},\mathbf{P}$ be the body and surface forces and ρ the density. Let us denote by $\mathbf{u}'=(u'_k)$ the velocity vector. Let on Γ_c^{kl} the positive direction of the outer normal vector $\mathbf{n}$ be assumed with respect to Ω^k. Let us denote by $[u_n]^{kl}=u_n^k-u_n^l$ and $[u'_n]^{kl}=u_n'^k-u_n'^l$ the jump of the normal displacement and velocity, respectively, across the contact zone between neighboring assumed joint system parts Ω^k and Ω^l. Similarly, we denote $[\mathbf{u}_t]^{kl} \equiv \mathbf{u}_t^k - \mathbf{u}_t^l$ and $[\mathbf{u}_t']^{kl} \equiv \mathbf{u}_t'^k - \mathbf{u}_t'^l$. If one of the colliding bodies is absolutely rigid, say Ω^l, then $[u_n']^{kl} \equiv u_n'^k \equiv u_n'$ and $[\mathbf{u}_t']^{kl} \equiv \mathbf{u}_t'^k \equiv \mathbf{u}_t'$ and similarly for the normal and tangential displacements.

Such models can be described as follows:

A. Statically Loaded Shoulder and Elbow Joint Systems Distribution of the displacement field $\mathbf{u}$ and stress field τ in shoulder or elbow joint systems satisfy the equilibrium equations for every domain Ω^ι:

$$\frac{\partial \tau_{ij}(\mathbf{u}^\iota)}{\partial x_j} + F_i^\iota = 0, \qquad i,j = 1,\ldots,N, \qquad \iota = 1,\ldots,s, \text{ in } \Omega^\iota, \tag{7.52}$$

where $\mathbf{F}^\iota$ are body forces. A repeated index implies the summation from 1 to N. The relation between the displacement vector $\mathbf{u} = (u_i)$, $i = 1,\ldots,N$, and the small strain tensor e_{ij} is defined by

$$e_{ij} = e_{ij}(\mathbf{u}) = \frac{1}{2}\left(\frac{\partial u_i}{\partial x_j} + \frac{\partial u_j}{\partial x_i}\right), \qquad i,j = 1,\ldots,N.$$

The relation between the stress and strain tensors is defined by the generalized Hooke law

$$\tau_{ij}^\iota = c_{ijkl}^\iota e_{kl}(\mathbf{u}^\iota), \quad i,j,k,l = 1,\ldots,N, \quad \iota = 1,\ldots,s,$$

in the anisotropic case, whereas in the isotropic case

$$c_{ijkl}^\iota = \mu^\iota(\delta_{ik}\delta_{jl} + \delta_{il}\delta_{jk}) + \lambda^\iota \delta_{ij}\delta_{kl}, \qquad \beta_{ij}^\iota = \gamma^\iota \delta_{ij},$$

where λ^ι, μ^ι represent the Lamé coefficients. The coefficients c_{ijkl}^ι form a matrix of the type $(N \times N \times N \times N)$ and satisfy the symmetry and Lipschitz conditions

$$c_{ijkl}^\iota = c_{jikl}^\iota = c_{klij}^\iota = c_{ijlk}^\iota$$

$$0 < a_0^\iota \le c_{ijkl}^\iota(\mathbf{x})\xi_{ij}\xi_{kl}|\xi|^{-2} \le A_0^\iota < +\infty, \xi \in \mathbb{R}^{N^2}, \qquad \xi_{ij} = \xi_{ji}, \text{ for a.e. } \mathbf{x} \in \Omega^\iota,$$

where a_0^ι, A_0^ι are constants independent of $\mathbf{x} \in \Omega^\iota$. For the isotropic case $a_0^\iota = 2\min\{\mu^\iota(\mathbf{x}); \mathbf{x} \in \Omega^\iota\}$ and $A_0^\iota = \max\{2\mu^\iota(\mathbf{x}) + 3\lambda^\iota(\mathbf{x}); \mathbf{x} \in \Omega^\iota$.

We will consider the following conditions: Let us denote the outward unit normal to the boundary $\partial\Omega$ by $\mathbf{n} = (n_i)$. Let $\Gamma_\tau = {}^1\Gamma_\tau \cup {}^2\Gamma_\tau$ and let us assume that a portion of the investigated shoulder or elbow joint system is loaded at a certain boundary, which will be denoted by ${}^1\Gamma_\tau$ and it is unloaded at ${}^2\Gamma_\tau$. We thus have, for example, the conditions

$$\begin{aligned} \tau_{ij}n_j &= P_i \quad \text{on } {}^1\Gamma_\tau, \\ \tau_{ij}n_j &= 0 \quad \text{on } {}^2\Gamma_\tau. \end{aligned} \tag{7.53}$$

Let the boundary $\Gamma_u = {}^1\Gamma_u \cup {}^2\Gamma_u$ of the investigated system be fixed on ${}^1\Gamma_u$ and/or on ${}^2\Gamma_u$, where displacements are prescribed, characterizing movements of the limb, for example, the movement of the blade bone:

$$\begin{aligned} u_i &= 0 \quad \text{on } {}^1\Gamma_u, \\ u_i &= u_{0i} \quad \text{on } {}^2\Gamma_u. \end{aligned} \tag{7.54}$$

The movements between bodies of different assumed joint systems are facilitated by the shoulder or elbow joints, respectively. These joints facilitate flexion in all directions, small movements in the sense of rotation, and small translation. These surfaces are similar to contact surfaces. The unilateral contact conditions with or without friction are sufficient to describe the function of these joints of the upper limb, and, therefore, together with the ligament apparatus and muscles they facilitate both mobility of the limb as well as their stability during movement. Hence, we can use the following unilateral contact conditions: Denote by $\mathbf{u}^k$, $\mathbf{u}^l$ (indices k, l correspond with the neighboring bodies in contact) the displacement and the temperature of neighboring bodies. All these quantities are functions of spatial coordinates. Then on the contact boundaries Γ_c^{kl} the condition of nonpenetration

$$u_n^k(\mathbf{x}) - u_n^l(\mathbf{x}) \leq 0 \quad \text{on } \Gamma_c^{kl} \tag{7.55}$$

holds, where $u_n^l = u_i^l n_i^k$. For the contact forces, due to the law of action and reaction, we find

$$\tau_n^k(\mathbf{x}) = \tau_n^l(\mathbf{x}) \equiv \tau_n^{kl}(\mathbf{x}), \qquad \tau_t^k(\mathbf{x}) = -\tau_t^l(\mathbf{x}) \equiv \tau_t^{kl}(\mathbf{x}). \tag{7.56}$$

Since the normal components of contact forces cannot be tensile forces,

$$\tau_n^k(\mathbf{x}) = \tau_n^l(\mathbf{x}) \equiv \tau_n^{kl}(\mathbf{x}) \leq 0 \quad \text{on } \Gamma_c^{kl}. \tag{7.57}$$

During the deformation of the bodies, they are in contact or they are not in contact. If they are not in contact, then $u_n^k - u_n^l < 0$, and the contact forces are equal to zero, that is, $\tau_n^k = \tau_n^l \equiv \tau_n^{kl} = 0$. If the bodies are in contact, that is, $u_n^k - u_n^l = 0$, then there may exist nonzero contact forces $\tau_n^k = \tau_n^l \equiv \tau_n^{kl} \leq 0$. These cases are included in the following condition:

$$(u_n^k(\mathbf{x}) - u_n^l(\mathbf{x}))\tau_n^{kl}(\mathbf{x}) = 0 \quad \text{on } \Gamma_c^{kl}. \tag{7.58}$$

Further, if both parts of the joint (shoulder and elbow) systems are in contact, then the Coulombian type of friction acts on the contact boundary. Similarly, as in the other large human joints, the frictional forces g_c^{kl} acting on the contact boundary Γ_c^{kl} are, in their absolute value, proportional to the normal stress component, where the coefficient of proportionality is the coefficient of the Coulombian friction $\mathcal{F}_c^{kl}(\mathbf{x})$, that is,

$$g_c^{kl}(\mathbf{x}) = \mathcal{F}_c^{kl}(\mathbf{x})\,|\tau_n^{kl}(\mathbf{x})|. \tag{7.59}$$

Due to acting and frictional forces we have the following cases:

If the absolute value of tangential forces $\tau_t^{kl}(\mathbf{x})$ is less than the frictional forces g_c^{kl}, then the frictional forces preclude the mutual shifts of both bodies being in contact. If the tangential forces τ_t^{kl} are equal in their absolute value to the frictional forces, so that there are no forces that can preclude the mutual motion of both parts of joint system (elastic bodies) of the studied joint system. Thus the contact points change their position in the direction opposite to which the tangential stress component acts.

These conditions are described by the following conditions:

1. If $u_n^k - u_n^l = 0$, then $|\tau_t^{kl}(\mathbf{x})| \le g_c^{kl}(\mathbf{x})$.
2. If $|\tau_t^{kl}(\mathbf{x})| < g_c^{kl}(\mathbf{x})$, then $\mathbf{u}_t^k(\mathbf{x}) - \mathbf{u}_t^l(\mathbf{x}) = 0$, which means that the friction forces are sufficient to prelude the mutual shifting between both parts of the assumed joint system. (7.60)
3. If $|\tau_t^{kl}(\mathbf{x})| = g_c^{kl}$, then there exists a function $\vartheta \ge 0$ such that $\mathbf{u}_t^k(\mathbf{x}) - \mathbf{u}_t^l(\mathbf{x}) = -\vartheta\tau_t^{kl}(\mathbf{x})$, which means that the friction forces are not sufficient to preclude the mutual shifting of both parts of assumed joint systems. This shift acts in an opposite direction to the tangential forces. Note that here we have $\mathbf{u}_t^l = \mathbf{u}^l - u_n^l\mathbf{n}^k$ and $u_n^l = \mathbf{u}^l.\mathbf{n}^k \equiv u_i^l n_i^k$.

In some problems the bilateral conditions with or without friction can be used, for example, for the case without friction

$$u_n = 0, \tau_{tj} = 0, \qquad j = 1, \ldots, N. \tag{7.61}$$

These conditions with or without friction can be used to describe the conditions, for example, on the scapula or the clavicula.

B. Dynamically Loaded Shoulder and Elbow Joint Systems In the study of the shoulder and elbow joint systems we will investigate the dynamic multibody contact problem with the Coulomb friction in linear viscoelasticity with short memory in N dimensions formulated in velocities.

The stress–strain relation will be defined by Hooke's law

$$\begin{aligned} \tau_{ij} &= \tau_{ij}(\mathbf{u}, \mathbf{u}') = c_{ijkl}^{(0)}(\mathbf{x})e_{kl}(\mathbf{u}) + c_{ijkl}^{(1)}(\mathbf{x})e_{kl}(\mathbf{u}'), \\ e_{ij}(\mathbf{u}) &= \frac{1}{2}\left(\frac{\partial u_i}{\partial x_j} + \frac{\partial u_j}{\partial x_i}\right), \quad i,j,k,l = 1, \ldots, N, \end{aligned} \tag{7.62}$$

where $c_{ijkl}^{(n)}(\mathbf{x})$, $n = 0, 1$, are elastic and viscous coefficients and $e_{ij}(\mathbf{u})$ are components of the small strain tensor, and N is the space dimension. For the tensors $c_{ijkl}^{(n)}(\mathbf{x})$, $n = 0, 1$, we assume

$$\begin{aligned} &c_{ijkl}^{(n)} \in L^\infty(\Omega), \qquad n = 0, 1, \qquad c_{ijkl}^{(n)} = c_{jikl}^{(n)} = c_{klij}^{(n)} = c_{ijlk}^{(n)}, \\ &c_{ijkl}^{(n)} e_{ij} e_{kl} \ge c_0^{(n)} e_{ij} e_{ij} \quad \forall e_{ij}, e_{ij} = e_{ji} \quad \text{and} \quad \text{a.e. } \mathbf{x} \in \Omega, \qquad c_0^{(n)} > 0. \end{aligned} \tag{7.63}$$

A repeated index implies the summation from 1 to N.

The unilateral conditions will be formulated in velocities, that is, we will assume the following contact conditions:

$$[u_n']^{kl} \leq 0, \quad \tau_n^k = \tau_n^l \equiv \tau_n^{kl} \leq 0, \qquad [u_n']^{kl}\tau_n^{kl} = 0 \quad \text{on } \cup\Gamma_c^{kl}. \tag{7.64}$$

If one of the colliding bodies is absolutely rigid, say Ω^l, then $[u_n']^{kl} \equiv u_n'^k \equiv u_n'$ and the unilateral conditions are the following:

$$u_n' \leq 0, \qquad \tau_n \leq 0, \qquad u_n'\tau_n = 0 \quad \text{on } \Gamma_c^k. \tag{7.65}$$

Then we have the following problem:

Problem $\mathcal{P}_{\mathbf{dyn}}$ Let $N = 2, 3$, $s \geq 2$. Find a vector function $\mathbf{u} : \overline{\Omega} \times I \to \mathbb{R}$, satisfying

$$\begin{aligned} \rho^\iota \frac{\partial^2 u_i^\iota}{\partial t^2} &= \frac{\partial \tau_{ij}^\iota(\mathbf{u}^\iota, \mathbf{u}^{\iota\prime})}{\partial x_j} + F_i^\iota, \qquad i,j = 1,\ldots,N, \qquad \iota = 1,\ldots,s, \\ (t,\mathbf{x}) \in Q^\iota &= I \times \Omega^\iota, \end{aligned} \tag{7.66}$$

$$\begin{aligned} \tau_{ij}n_j &= P_i, \qquad i,j = 1,\ldots,N, \qquad (t,\mathbf{x}) \in {}^1\Gamma_\tau(t) = I \times \cup_{\iota=1}^s ({}^1\Gamma_\tau \cap \partial\Omega^\iota), \\ \tau_{ij}n_j &= 0, \qquad i,j = 1,\ldots,N, \qquad (t,\mathbf{x}) \in {}^2\Gamma_\tau(t) = I \times \cup_{\iota=1}^s ({}^2\Gamma_\tau \cap \partial\Omega^\iota), \end{aligned} \tag{7.67}$$

$$\begin{aligned} u_i &= 0, \qquad i = 1,\ldots,N, \qquad \text{on } {}^1\Gamma_u(t) = I \times \cup_{\iota=1}^s ({}^1\Gamma_u \cap \partial\Omega^\iota), \\ u_i &= u_{2i}, \qquad i = 1,\ldots,N, \qquad \text{on } {}^2\Gamma_u(t) = I \times \cup_{\iota=1}^s ({}^2\Gamma_u \cap \partial\Omega^\iota), \end{aligned} \tag{7.68}$$

$$\begin{aligned} &\text{the unilateral contact conditions in velocities} \\ &[u_n']^{kl} \leq 0, \qquad \tau_n^k = \tau_n^l \equiv \tau_n^{kl} \leq 0, \qquad [u_n']^{kl}\tau_n^{kl} = 0, \\ &(t,\mathbf{x}) \in \Gamma_c(t) = I \times \cup_{k,l}\Gamma_c^{kl}, \\ &\text{and the Coulombian law of friction} \\ &[\mathbf{u}_t']^{kl} = 0 \Rightarrow |\tau_t^{kl}| \leq \mathcal{F}_c^{kl}(0)|\tau_n^{kl}|, \\ &[\mathbf{u}_t']^{kl} \neq 0 \Rightarrow \tau_t^{kl} = -\mathcal{F}_c^{kl}([\mathbf{u}_t']^{kl})|\tau_n^{kl}| \frac{[\mathbf{u}_t']^{kl}}{|[\mathbf{u}_t']^{kl}|}, \end{aligned} \tag{7.69}$$

and the initial conditions

$$\mathbf{u}(\mathbf{x},0) = \mathbf{u}_0(\mathbf{x}), \qquad \mathbf{u}'(\mathbf{x},0) = \mathbf{u}_1(\mathbf{x}), \qquad \mathbf{x} \in \Omega, \tag{7.70}$$

where $\mathbf{u}_0$, $\mathbf{u}_1$, $\mathbf{u}_2$ are the given functions, $\mathbf{u}_2$ has the time derivative $\mathbf{u}_2'$, and $\mathbf{u}_0$, $\mathbf{u}_1$ satisfy the static linear contact problems in elasticity with or without the Coulombian friction $\mathcal{F}_c^{kl}$ [see Nedoma (1983, 1987)]. The coefficient of friction $\mathcal{F}_c^{kl} \equiv \mathcal{F}_c^{kl}(\mathbf{x}, \mathbf{u}')$ is globally bounded, nonnegative, and it satisfies the Carathéodory property, that is, $\mathcal{F}_c^{kl}(\cdot, \mathbf{v})$ is measurable, $\mathbf{v} \in \mathbb{R}$, and $\mathcal{F}_c^{kl}(\mathbf{x}, \cdot)$ is continuous for a.e. $\mathbf{x} \in \Gamma_c^{kl}$.

7.11.6 Biomechanical and Mathematical Model of the Wrist and Hand

Introduction The wrist (carpus) is a compound joint, and it is a collection of bones, as well as soft tissues that connect the hand with the forearm. This joint system is capable of a substantial arc of motion that enlarges the function of the hand and the fingers. Although the function of joints of the upper limb is to position the hand so as to allow activities of the daily living, the wrist represents the key function of the hand. Stability of the wrist is essential in order for the flexor and extensor muscles to function, and the wrist position affects the ability of fingers to flex and extend maximally and to grip perfectly during prehension. The hand is a complex and multifunctional mobile organ. The hand is the final part of the limb. The mobility and stability of the shoulder, the elbow, and the wrist, which operate in different planes, allow the hand to move within a large space. The hand's extraordinary functional adaptability is based on the arrangement and mobility of the 19 bones and 14 joints of the hand (Stuchin 1992; Steiberg and Plancher, 1995; Ruby, 1995; Barr and Bear-Lehman, 2001; Nedoma et al., 2006).

Anatomy of the Wrist and Hand The compound joint of the wrist is created from the ulna and by the eight carpal bones with the distal radius, the structures within the ulnocarpal space, and the metacarpals. The eight carpal bones, which are sesamoid bones, are situated into two rows, that is, into the proximal and distal rows. The bones of the distal row are the trapezium, trapezoid, capitate, and hamate. The distal carpal row forms a relatively immobile transverse unit that articulates with the metacarpals and forms the carpometacarpal joints. These four bones in the distal row fit closely into each other and are held together by interosseous ligaments. The proximal row is created from the os scaphoideum, os lunatum, os triquetrum, and os pisiforme (which is an additional sesamoid bone in the tendon of the musculus (musculi) flexor carpi ulnaris). This row articulates the distal radius, forming the radiocarpal joint, that is, the scaphoid fossa of radius (~46), lunate fossa of radius (~43), and ulnar soft tissue structures (~11). The os scaphoideum spans both rows anatomically and functionally and it articulates with the radius. The lunate bone articulates in part with the ulnar soft tissue structures. Between the proximal and distal rows of carpal bones is the midcarpal joint, and between adjacent bones of these rows are the intercarpal joints. From the frontal view the joint slot has a goose-necked shape. The palmar surface of the carpus is concave. The distal radius, os lunate, and os triquetrum articulate with the distal ulna through a ligamentous and cartilaginous structure, the ulnocarpal complex (Barr, Bear-Lehman, 2001, Nedoma et al., 2006) (see Fig. 7.364).

The distal end of the radium and ulna, the eight wrist bones, and the first and fifth metacorpus evoke a compound joint, which can be divided into four units, that is, the radiocarpal joint, distal radioulnar joint, mediocarpal joint, and carpometacarpal joint, where a joint is created from the metacarpal bones and the surfaces of bones of the distal wrist row. From a functional point of view this joint has only a small sense. For more details see Ruby (1995), Stuchin (1992), Barr and Bear-Lehman (2001), Steiberg and Plancher (1995), and Nedoma et al. (2006).

The main componets of the hand are the fingers and the thumb. The digital rays of fingers are numbered from the radial to the ulnar side, that is, I, thumb; II, index finger; III, middle finger; IV, ring finger, V, little finger. Each finger ray articulates proximally with a particular carpal bone in a carpometacarpal joint. The next joint in each ray, the metacarpophalangeal joint (MCP joint), links the metacarpal bone to the proximal phalanx. Between the phalanges (distal, middle, proximal) of the fingers there exists a proximal (PIP) and a distal (DIP) interphalangeal joint. The thumb has only one interphalangeal joint (IP joint). The bones of the hand are situated in three arches—two transverse (proximal transverse arch and distal transverse arch) and one longitudinal (longitudinal arch). Two transverse arches are connected by a portion of the longitudinal arch, composed of four digital rays and the proximal carpus. The second and third metacarpal bones create the central pillar of this arch. The longitudinal arch is created from the individual digital rays, and the mobility of the thumb and fourth and fifth finger rays around the second and third fingers allow the palm to flatten or cup itself to gather objects of various sizes and shapes (Strickland, 1987; Barr and Bear-Lehman, 2001).

The entrinsic flexor and extensor muscles are largely responsible for the change in the shape of the hand when in motion. The intrinsic muscles of the hand are primarily responsible for maintaining the configuration of the three arches. The flexor muscles in the wrists are the flexor carpi ulnaris (which allows for flexion of the wrist and ulnar deviation of the hand), the flexor carpi radialis (which allows for flexion of the wrist and radial deviation of the hand), and the palmaris longus (which allows for tension of the palmar fascia). The extensors are the extensor carpi radialis longus and brevis (which allows for extension of the wrist and radial deviation of the hand) and the extensor carpi ulnaris and brevis (which allow for extension of the wrist and ulnar deviation of the hand). The pronators–supinators are the pronator teres (which allows for forearm pronation), the pronator quadrans (which allows for forearm pronation), the supinator quadrans (which allows for forearm supination), and the brachioradialis (which allows for pronation or supination, which depends on the position of the forearm). The muscles of the hand with the corresponding muscle actions are the following: (i) the extrinsic muscles: the flexors—the flexor digitorum superficialis (flexion of the PIP and MCP joints), the flexor digitorum profundus (flexion of the DIP, PIP, and MCP joints), the flexor policis longus (flexion of the IP and MCP joints of the thumb); the extensors—the extensor policis longus (extension of the IP and MCP joints of the thumb and secondary adduction of the thumb), the abductor policis longus (abduction of the thumb), the extensor indicis proprius (extension of the index finger), the extensor digitorum communis (extension of fingers), and the extensor digiti quinti proprius (extension of V finger); (ii) the intrinsic muscles: the interossei (extension of the PIP and DIP joints and flexion of the MCP joints), the dorsal interossei (spread of the index and ring fingers away from the long finger), the palmar interossei (adduction of the index, ring, and little fingers toward the long finger), the abductor pollicis brevis (abduction of the thumb), the flexor pollicis brevis (flexion and rotation of the thumb), the opponens pollicis (rotation of the first metacarpal toward the palm), the abductor digiti quinti (abduction of the little finger; extension of the PIP and DIP joints), the flexor digiti quinti brevis (flexion of the proximal phalanx of the little finger and forward rotation of the fifth metacarpal), and the adductor pollicis (abduction of the thumb).

The shape of the joint surfaces and ligaments connecting individual bones of the wrist are very important for movements in the wrist and the hand with containment of relative stabilities (Taleisnik, 1985; Strickland, 1987; Barr and Bear-Lehman, 2001). The function of the wrist and hand ligaments is to restrict joint motion and appose joint surfaces.

The wrist ligaments are capable of inducing bony displacements and of transmitting loads originating in the proximal or distal segments. The wrist ligaments are the palmar and dorsal ligaments. The palmar ligaments are thick and strong, while the dorsal ligaments are much thinner and fewer in number. The wrist ligaments are divided into extrinsic (running from radius to carpus and from carpus to metacarpals) and intrinsic (which originate and insert into the carpus) components. The palmar extrinsic ligaments are created from the radial collateral ligament, the palmar radiocarpal ligaments, and components of the triangular fibrocartilage complex (TFCC), while the radial collateral ligament is more palmar than lateral and is viewed as the most lateral of all the palmar radiocarpal fasciles rather than as a collateral ligament. The palmar wrist ligaments are the radioscaphocapite ligament, the radial collateral ligament, the radiolunate ligament, the radioscapholunate ligament, the ulnolunate ligament, the radiotriquetral ligament (meniscus homologue), the ulnar collateral ligament, the superficial palmar radiocarpal ligament, the triangular fibrocartilage, which are the entrinsic ligaments, and the intrinsic ligaments, that is, the scapholunate ligament, lunotriquetral ligament, and the palmar intercarpal ligament as well as the short palmar intrinsiocs. The triangular fibrocartilage complex is created from the radiotriquetral ligament, the triangular fibrocartilage (articular disk), the ulnolunate liganent, the ulnar collateral ligament, and the poorly distinguishable dorsal and palmar radioulnar ligaments. The dorsal wrist ligaments are the radiotriquetral ligament, the radiolunate ligament, and the radioscapoind fascicles of the dorsal radiocarpal ligament, which are the extrinsic ligaments and the intrinsic ligaments, that is, the dorsal intercarpal ligament, the trapeziotrapeziod ligament, the trapeziocapitate ligament, the capitohamate fascicles of the short intrinsic ligaments and the scaphotrapeyium ligament [for more details see Taleisnik (1985)].

The hand has an intricate retinacular system that encloses, compartmentalizes, and restrains the joint and tendons, as well as the skin, nerves, and blood vessels. This system surrounds each digit to create balanced forces of the entrinsic and intrinsic musculature and stability and control of the hand [for more details see Strickland (1987) and Barr and Bear-Lehman (2001)].

Owing to the biomechanics of wrist loads a detailed description of the blood and nervous systems is not necessary therefore, we will not discuss these systems. We will only mention that the wrist and the hand are innervated for motor and sensory functions by three peripherical nerves descending from the brachial plexus—radial, median, and ulnar and that they are supplied by the radialis, ulnaris, interossea anterior arteries with a great number of capillaries.

Biomechanics and the Model In this section we will discuss the wrist only. The wrist is the compound joint with 3 degrees of freedom. Owing to the forearm, the hand can be in a dorsal or palmar flexion, in an ulnar or radial duction, in pronation, or in supination, respectively. This complicated joint allows for the rotation of the forearm of up to 150°, and the distal end of the radius with the hand rotates around the relatively stable head of the elbow bone. The flexion–extension and duction movements take place in the radiocarpal and mediocarpal joints, which together create the unit (Stuchin, 1992; Ruby, 1995). The flexion–extension movement is approximately 120° with the range between 84° and 169°. Approximately 60% of extension takes place in the radiocarpal joint and 40% in the mediocarpal joint. In the case of flexion 40% of the movement takes place in the radiocarpal joint and 60% in the mediocarpal joint. The range of duction movements is approximately 50°, from which 15°–20° correspond to the radial duction and 35° to the ulnar duction. Approximately 40% of this movement takes place in the radiocarpal joint and 60% in the mediocarpal joint. The center of these rotational movements is the os capitanum. The multiplicity of wrist articulations and the complexity of carpal motion make it difficult to determine the exact center of motion for the primary axes of flexion–extension or radial–ulnar deviation. The actual center of rotation is assumed to be in the head of the os capitanum, the flexion–extension axis is oriented from the radial to the ulnar styloid process, and the radial–ulnar deviation axis is oriented orthogonal to the flexion–extension axis.

The hand is an extremely mobile part of the human body that can coordinate an infinite variety of movements. The blending of hand and wrist movements enables the hand to change its shape so as to palpate or graspe many different objects. The mobility of the hand is the result of the position of the bones in relation to one another, the articular countours, and of muscle actions.

The mathematical wrist model was constructed on the basis of an MRI (see Fig. 7.365), and it is created from the end of the radium and the ulna and by the system of eight carpal bones and by metacarpal bones (I–V). The bones of both rows are connected by stiff ligaments with Young's modulus E, which is of several orders lower than in the case of the carpal bones, the radius and the ulna, which have characteristics of cortico-spongy bones. The proximal row of the wrist bones is created from the os scaphoideum (Sc), the os lunatum (L), and the os triquetrum (Tq) as well as the os pisiforne. These bones are mutually connected by the ligaments, so that from the point of view of the biomechanics of the whole wrist system these bones create one body with different physical properties and parameters. A similar situation is found in the distal row created from the os trapezium (Tz), the os capitatum (C), and the os hamatum (H), which are mutually connected by strong ligaments, and, therefore, they create one biomechanical system—in one body with different physical properties and parameters. This system, the distal system, is then connected by strong ligaments with the basis of I–V metacarpal bones.

The following joints provide mobility to the wrist: (i) the radiocarpal joint created from the joint surface of radium and of the proximal rows of the bones Sc and L and (ii) the mediocarpal joint created from the systems of proximal and distal carpal bones, which are mutually connected by ligaments into one biomechanical body.

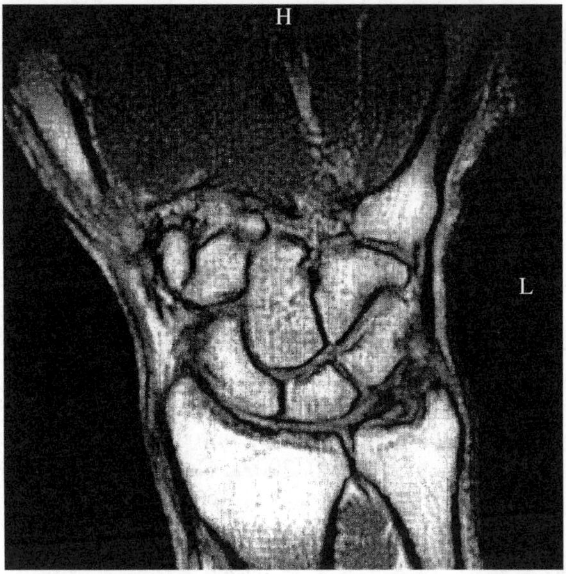

FIG. 7.365 MRI of the wrist.

In constructing the mathematical wrist model we need to mathematically and accurately simulate the function of the wrist ligaments. Relatively simply the interosseal ligaments (the interbones ligaments), that is, the ligaments that in single rows mutually connect individual wrist bones, can be simulated. The capsular ligaments, that is, the ligaments, that strengthen the capsule, can be simulated as the effect of tendono-ligamental forces acting from external onto the individual exterior sides of both systems, proximal and distal, and the ulna and the radius. The geometry of the wrist system can be obtained from an MRI of the ligaments and then simulated as viscoelastic zones (strips) fixed under certain stresses between individual bones, which, moreover, strengthened individual biomechanical bodies between themselves. The flexion–extensional movement and the ductial movements take place in the radiocarpal and mediocarpal joints, which are characterized by the contact surfaces and then by conditions of similar types as in the case of other human joints. In addition to the mutual movements of both rows—distal and proximal—the position of bones of proximal row changes. Both biomechanical bodies, starting from the basis of wrist bones, are fixed by strong ligaments fixed on the elbow bone and the radius.

The mathematical model will be based on the theory of contact problems and nonlinear rheology (Nedoma and Hlaváček, 2002; see also Chapter 6).

Construction of the Mathematical Model Let $I = (0, t_p)$ be the time interval. Let $\Omega = \cup_{\iota=1}^{s} \Omega^{\iota} \subset \mathbb{R}^N$, $N = 2, 3$, be a region occupied by a part of the human hand created from the wrist bones together with the ligaments and with the ulna and the radius, where the regions Ω^{ι}, $\iota = 1, \ldots, s$, are represented by the single parts of the wrist system, the ligaments, and the ulna and the radius. Let Ω^ι have Lipschitz boundaries $\partial\Omega^\iota$ and let us assume that $\partial\Omega = \overline{\Gamma}_\tau \cup \overline{\Gamma}_u \cup \overline{\Gamma}_c$, where Γ_τ, Γ_u and Γ_c are disjunct, and where $\Gamma_\tau = {}^1\Gamma_\tau \cup {}^2\Gamma_\tau$ represents loaded and unloaded parts of the boundary, where ${}^1\Gamma_\tau$ is a part of the boundary, where loads $\mathbf{P} \neq \mathbf{0}$ are prescribed, for example, where muscle forces operate on the different part of the wrist systems or where an outer burden acts, and where on ${}^2\Gamma_\tau$ the load $\mathbf{P} = \mathbf{0}$ and Γ_u is a part of the boundary, where the wrist system is assumed to be fixed, $\Gamma_c = \cup_{k,l} \Gamma_c^{kl}$, $\Gamma_c^{kl} = \partial\Omega^k \cap \partial\Omega^l$, $k \neq l$, $k, l \in \{1, \ldots, s\}$ is a part of the boundary, where the different parts of the wrist systems studied are in mutual contacts, for example, in areas of radiocarpal and mediocarpal joints or of artificial replacements of some of wrist and/or hand joints, respectively. Let $Q = I \times \Omega$ denote the time–space domain and let $\Gamma_\tau(t) = \Gamma_\tau \times I$, $\Gamma_u(t) = \Gamma_u \times I$, $\Gamma_c(t) = \Gamma_c \times I$ denote the parts of its boundary $\partial Q = \partial\Omega \times I$.

The relation between stresses and strains and the positive definite function of density of the deformation energy W will be introduced (Belytschko et al., 1974; Kulak et al., 1976, Nedoma et al., 2000, 2003a, Nedoma and Hlaváček, 2002). Let the function of the density of deformation energy W be defined by

$$W = A^\lambda(e_{ij}), \tag{7.71}$$

where A is a function of the strain tensor of a small deformation and λ is a nonnegative parameter. Let the function A be defined by

$$A = c_{ijkl}(\mathbf{x})e_{ij}(\mathbf{u})e_{kl}(\mathbf{u}), \qquad e_{ij} = \frac{1}{2}\left(\frac{\partial u_i}{\partial x_j} + \frac{\partial u_j}{\partial x_i}\right), \qquad i,j,k,l = 1,\ldots,N, \tag{7.72}$$

where c_{ijkl} are elastic coefficients and e_{ij} are components of the small strain tensor and where

$$\begin{aligned} &c_{ijkl} \in L^\infty(\Omega), \qquad c_{ijkl} = c_{klij} = c_{jikl} = c_{ijlk}, \\ &c_{ijkl}(\mathbf{x})e_{ij}(\mathbf{u})e_{kl}(\mathbf{u}) \geq c_0 e_{ij}(\mathbf{u})e_{ij}(\mathbf{u}) \end{aligned} \tag{7.73}$$

is valid for all symmetric matrices e_{ij} and for almost all $\mathbf{x} \in \cup_{\iota=1}^s \Omega^\iota$, where a repeated index implies the summation from 1 to N.

From the theory of continuum mechanics it is known that the relation between stresses and strains is given by the relation

$$\tau_{ij} = \frac{\partial W}{\partial e_{ij}}. \tag{7.74}$$

Then from (7.74), (7.71), and (7.72) we obtain

$$\begin{aligned} \tau_{ij} &= \lambda[A(e_{ij})]^{\lambda-1}\frac{\partial A(e_{ij})}{\partial e_{ij}} = c^*_{ijkl}(\mathbf{u})e_{kl}, \\ &\text{where } c^*_{ijkl}(\mathbf{u}) = 2\lambda[A(e_{ij})]^{\lambda-1}c_{ijkl}(\mathbf{x}). \end{aligned} \tag{7.75}$$

The scalar coefficient $2\lambda[A(e_{ij})]^{\lambda-1}$ depends on the state of deformation, and, therefore, it simulates properties of materials creating bones, ligaments, and tendons, that is, softness and hardness. The coefficients $c^*_{ijkl}(\mathbf{u})$ are functions of a state of deformation in the bone, ligaments, and tendons, and they represent nonlinear coefficients of elasticity. The parameter λ determines an order of nonlinearity of elastic coefficients $c^*_{ijkl}(\mathbf{u})$. If $\lambda < 1$, then parameter λ is from the interval $0 < \lambda < 1$, and it represents the effect of material softness; in the case if $\lambda = 1$, the elastic coefficients do not depend on the state of deformation, while for $\lambda > 1$ this parameter characterizes the hardness of the material of bones, ligaments, and tendons.

Let $\mathbf{n}$ denote the outer normal vector of the boundary, $u_n = u_i n_i$, $\mathbf{u}_t = \mathbf{u} - u_n \mathbf{n}$, $\tau_n = \tau_{ij} n_j n_i$, $\tau_t = \tau - \tau_n \mathbf{n}$ be normal and tangential components of displacement and stress vectors $\mathbf{u} = (u_i)$, $\tau = (\tau_i)$, $\tau_i = \tau_{ij} n_j$, $i, j = 1, \ldots, N$. Let $\mathbf{F}$, $\mathbf{P}$ be the body and surface forces, ρ the density. Let us denote by $\mathbf{u}' = (u'_i)$ the velocity vector. Let on Γ^{kl}_c the positive direction of the outer normal vector $\mathbf{n}$ be assumed with respect to Ω^k. Let us denote by $[u_n]^{kl} = u^k_n - u^l_n$ and $[u'_n]^{kl} = u'^k_n - u'^l_n$ the jump of the normal displacement and velocity, respectively, across the contact zone between neighboring joint system parts Ω^k and Ω^l. Similarly, we denote $[\mathbf{u}_t]^{kl} \equiv \mathbf{u}^k_t - \mathbf{u}^l_t$ and $[\mathbf{u}'_t]^{kl} \equiv \mathbf{u}'^k_t - \mathbf{u}'^l_t$.

Let the body forces be $\mathbf{F} \in L^{2,N}(\Omega)$, the surface forces (loads) be $\mathbf{P} \in L^{2,N}(\Gamma_\tau)$, and the frictional forces be $g^{kl}_c \in L^2(\Gamma^{kl}_c)$. Then the model problem represents static or dynamic nonlinear contact problems in nonlinear elastic rheology, respectively.

Statically Loaded Wrist and/or Hand We have the following problem:

Problem $\mathcal{P}^{\mathbf{wrist}}_{\mathbf{stat}}$ Let $N = 2, 3$, $s \geq 2$. Find a vector function $\mathbf{u} : \overline{\Omega} \to \mathbb{R}$, satisfying

$$\frac{\partial \tau^\iota_{ij}(\mathbf{u}^\iota)}{\partial x_j} + F^\iota_i = 0, \qquad i, j = 1, \ldots, N, \qquad \iota = 1, \ldots, s, \qquad \mathbf{x} \in \Omega^\iota, \tag{7.76}$$

$$\begin{aligned} \tau_{ij} n_j &= P_i, \quad i, j = 1, \ldots, N, \qquad \mathbf{x} \in \cup^s_{\iota=1} ({}^1\Gamma_\tau \cap \partial\Omega^\iota), \\ \tau_{ij} n_j &= 0, \quad i, j = 1, \ldots, N, \qquad \mathbf{x} \in \cup^s_{\iota=1} ({}^2\Gamma_\tau \cap \partial\Omega^\iota), \end{aligned} \tag{7.77}$$

$$u_i = 0, \qquad i = 1, \ldots, N, \quad \text{on } \cup^s_{\iota=1} (\Gamma_u \cap \partial\Omega^\iota), \tag{7.78}$$

the unilateral contact condition

$$[u_n]^{kl} \leq 0, \qquad \tau^{kl}_n \leq 0, \qquad [u_n]^{kl} \tau^{kl}_n = 0,$$

and the Coulombian law of friction:

1. If $[u_n]^{kl} = 0$, then $|\tau^{kl}_t(\mathbf{x})| \leq g^{kl}_c(\mathbf{x})$.
2. If $|\tau^{kl}_t(\mathbf{x})| < g^{kl}_c(\mathbf{x})$, then $[\mathbf{u}_t]^{kl} = 0$, which means that the friction forces are sufficient to preclude the mutual shifting between both parts of the joint system.
3. If $|\tau^{kl}_t(\mathbf{x})| = g^{kl}_c$, then there exists a function $\vartheta \geq 0$ such that $[\mathbf{u}_t]^{kl} = -\vartheta \tau^{kl}_t(\mathbf{x})$, which means that the friction forces are not sufficient to preclude the mutual

shifting of both parts of the joint systems. This shift acts in an opposite direction to the acting tangential forces. Note that here we have $\mathbf{u}_t^l = \mathbf{u}^l - u_n^l \mathbf{n}^k$ and $u_n^l = \mathbf{u}^l \mathbf{n}^k \equiv u_i^l n_i^k$.

Dynamically Loaded Wrist and Hand As in the case of the shoulder joint in the previous section the dynamic model problem will be as follows:

Problem $\mathcal{P}_{\mathbf{dyn}}^{\mathbf{wrist}}$ Let $N = 2, 3,\ s \geq 2$. Find a vector function $\mathbf{u} : \overline{\Omega} \times I \to \mathbb{R}$, satisfying

$$\rho^\iota \frac{\partial^2 u_i^\iota}{\partial t^2} = \frac{\partial \tau_{ij}^\iota(\mathbf{u}^\iota, \mathbf{u}^{\iota\prime})}{\partial x_j} + F_i^\iota, \quad i,j = 1,\ldots,N, \quad \iota = 1,\ldots,s,\ (t,\mathbf{x}) \in Q^\iota = I \times \Omega^\iota, \tag{7.79}$$

$$\begin{aligned} \tau_{ij} n_j &= P_i, \qquad i,j = 1,\ldots,N, \qquad (t,\mathbf{x}) \in {}^1\Gamma_\tau(t) = I \times \cup_{\iota=1}^s ({}^1\Gamma_\tau \cap \partial\Omega^\iota), \\ \tau_{ij} n_j &= 0, \qquad i,j = 1,\ldots,N, \qquad (t,\mathbf{x}) \in {}^2\Gamma_\tau(t) = I \times \cup_{\iota=1}^s ({}^2\Gamma_\tau \cap \partial\Omega^\iota), \end{aligned} \tag{7.80}$$

$$u_i = 0, \qquad i = 1,\ldots,N, \qquad \text{on } \Gamma_u(t) = I \times \cup_{\iota=1}^s (\Gamma_u \cap \partial\Omega^\iota), \tag{7.81}$$

the unilateral contact conditions in velocities

$$\begin{aligned} &[u_n']^{kl} \leq 0, \qquad \tau_n^k = \tau_n^l \equiv \tau_n^{kl} \leq 0, \qquad [u_n']^{kl} \tau_n^{kl} = 0, \\ &(t,\mathbf{x}) \in \Gamma_c(t) = I \times \cup_{k,l} \Gamma_c^{kl}, \end{aligned} \tag{7.82}$$

and the Coulombian law of friction

$$\begin{aligned} &[\mathbf{u}_t']^{kl} = 0 \Rightarrow |\tau_t^{kl}| \leq \mathcal{F}_c^{kl}(0) |\tau_n^{kl}|, \\ &[\mathbf{u}_t']^{kl} \neq 0 \Rightarrow \tau_t^{kl} = -\mathcal{F}_c^{kl}([\mathbf{u}_t']^{kl}) |\tau_n^{kl}| \frac{[\mathbf{u}_t']^{kl}}{|[\mathbf{u}_t']^{kl}|}, \end{aligned}$$

and the initial conditions

$$\mathbf{u}(\mathbf{x},0) = \mathbf{u}_0(\mathbf{x}), \qquad \mathbf{u}'(\mathbf{x},0) = \mathbf{u}_1(\mathbf{x}), \qquad \mathbf{x} \in \Omega, \tag{7.83}$$

where $\mathbf{u}_0$, $\mathbf{u}_1$, $\mathbf{u}_2$ are the given functions and $\mathbf{u}_2$ has the time derivative $\mathbf{u}_2'$.

For simplicity we can assume that the Coulombian friction can be neglected because the friction coefficient (dry friction) is approximately 0.0025, then we can put $\tau_t^{kl} \simeq 0$.

Remark 7.1 The model can be simplified in such a way that the unilateral contact conditions can be replaced by the bilateral contact conditions.

Biomechanical Analysis of the Loaded Wrist Computer graphics on the basis of concrete data, instead of reconstructions and analyses of geometry of the wrist, render possible some simulation of new situations as well as construction of new formations such as joint replacements (Murphy et al., 1985, 1986, 1987). Such 2D and 3D simulations bring new views for the evaluation of reconstructive performance (surgical treatment) in the area of the wrist. From the surgical point of view these surgical treatments are very complicated and difficult.

The approach used was the finite element method for the contact problem in nonlinear elasticity. The model was derived from the 2D magnetic resonance image scan of a real patient. The metacarpal bones of the wrist were exposed to a load of ~350 N, corresponding to about one-fourth of the patients weight and acting on the boundary denoted by Γ_τ^M. For numerical realization the model was simplified, that is, the biomaterials of the bone, ligaments, and tendons were assumed to be homogeneous and isotropic (parameter $\lambda = 1$), and the ulna and radius were fixed at a sufficient distance from the wrist. On the other parts of the upper limb no loads were assumed. The contact boundaries were assumed between the ulna, the radius, and the system of carpal bones and between both rows of carpal bones. With greater loads an enlargement of the contact areas in the wrist may occur. The model used takes into account all of these realities.

For the cancellous compact bones the following values of Young's modulus $E = 1.89 \times 10^4$ MPa and Poisson's constant $\nu = 0.28$ for the radius and $E = 1.88 \times 10^4$ MPa, $\nu = 0.28$ for the ulna (Yamada, 1970) were used. For individual joints of the wrist and the metarcarpal bones the following values of Young's modulus $E = 1.88 \times 10^4$ MPa and Poisson's constant $\nu = 0.28$ and for the ligaments $E = 34.18$ MPa, $\nu = 0.1$, were assumed. The hyaline cartilage was simulated by a homogeneous isotropic material with $E = 24.3$ MPa, $\nu = 0.45$.

The model of the wrist is shown in Fig. 7.366 The wrist occupies the region $\Omega = \Omega_1 \cup \Omega_2 \cup \Omega_3 \cup \Omega_4 \cup \Omega_5 \cup \Omega_6 \cup \Omega_M$, where Ω_1 and Ω_2 are regions occupying the radium and ulna, $\Omega_3 = \Omega_{\mathrm{Sc}} \cup \Omega_{\mathrm{L}} \cup \Omega_{\mathrm{Tq}} \cup \Omega_{vp}$, where all are as defined above, and Ω_{vp} denotes the interligaments, $\Omega_4 = \Omega_H \cup \Omega_C \cup \Omega_{\mathrm{Td}} \cup \Omega_{\mathrm{Tz}} \cup \Omega_M \cup \Omega_{vz}$, where all are as defined above, and Ω_M denotes the metacarpal bones with the intercarpal ligaments Ω_{vz}. The wrist load affects the metacarpal bones and ligaments, which are simulated in the model by the continuous load layer denoted as Ω_M. The load forces act on the upper boundary Γ_τ^M. On the other part of the boundary $\Gamma_\tau = \partial\Omega \backslash \Gamma_\tau^M \backslash \Gamma_u$ part of the upper limb is unloaded, where by Γ_u we denote the boundary where the radius and ulna are fixed. The contact boundaries between Ω_1 and Ω_3, that is, between the radius and the proximal row of metacarpal bones, we denote as Γ_c^R and between Ω_3 and Ω_4, that is, between proximal and distal rows of metacarpal bones, we denote as Γ_c^{HC}. In the figure of the model in the os scaphoideum (i.e., in Ω_{Sc}) the fracture line, denoted by a dashed line Γ_c^f, is delineated.

In Fig. 7.367 the deformation of the wrist is presented. Figures 7.368 and 7.369 show the distribution of the horizontal and vertical displacement components in the wrist. We can see that the values of both components are in the interval between 5.0×10^{-7} m and 5.0×10^{-6} m. Figures 7.370–7.374 show the distribution of the components of stresses ($\tau_{11}, \tau_{22}, \tau_{12}$) and of the principal stresses (τ_1, τ_2). In Fig. 7.370 the horizontal stresses $\tau_{11} \equiv s_x$ (in MPa) are presented. In the figure the scale illustrates the corresponding values of isostresses. In Figs. 7.371 and 7.372

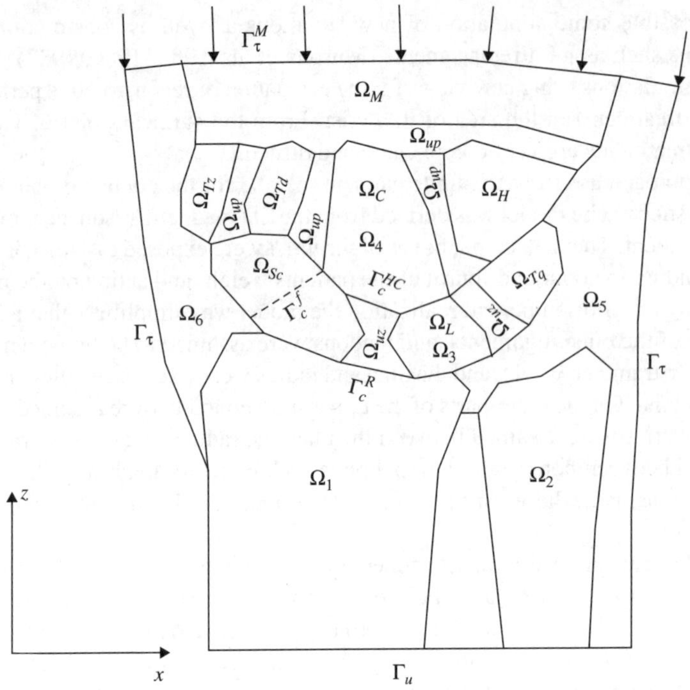

FIG. 7.366 Model of the wrist.

the vertical and shear stresses are presented. In Fig. 7.373 the principal stresses are given and in Fig. 7.374 the reduced (von Mises) stresses are presented. The reduced stresses as well as the components of stresses illustrate areas of both rows, proximal and distal, which are practically without stresses. The reduced stresses show that when the wrist is loaded by relatively small loads, a certain anomalous zone, situated probably in a fracture zone in the os scaphoideum, can be observed. Comparing these obtained stress values with the determined properties of bones, we can see that the values of stresses do not exceed the strength limit, and, therefore, in the os scaphoideum no fracture can originate. Figures 7.375 and 7.376 illustrate values of normal and tangential components of displacement and stress vectors on the contact boundaries. The normal component of the displacement vector u_n illustrates mutual separation of individual bones in the contact zones in the wrist and the tangential component of the displacement vector $\mathbf{u}_t$ describes their mutual movement. Normal and tangential components of the stress vector (τ_n and τ_t) describe a force relation on the contact boundaries in the wrist. Force and movement relations on the contact boundaries in the wrist represent new information for an analysis in the case of a damage of the function of the hand. In the 3D approximation this methodology renders it possible to simulate and then analyze and cure pathological cases of damaged wrists. Figures 7.367–7.376 show numerical results for the model of the loaded wrist. (See the accompanying DVD disk).

Comments The submitted model is constructed on the basis of real geometry of individual parts of the wrist, and, therefore, it renders it possible to write the function of the real wrist in the normal and pathological state of a concrete patient relatively well. The fact that at given loading it is possible to find distribution of stress–strain field and deformation in the individual parts of the wrist as well as in the wrist as a whole, and because it renders it possible to find a magnitude of normal and tangential component of stresses on the contact boundaries between individual biomechanical bodies, then from the analyses of numerical results, which are in graphical as well as in the numerical forms, then the surgeon can evaluate a more optimal therapy for a specific patient. Moreover, the surgeon, before the surgical treatment, can simulate a whole treatment operation and its result. Without fail it is necessary to probe the patient using an MRI, which helps to simulate the bone as well as soft (ligamental) structures and which renders it possible to create 2D or 3D models and 2D and 3D meshes by using the finite element method. This approach enables the surgeon to persuade a patient to undergo the proposed therapy.

7.12 MATHEMATICAL AND BIOMECHANICAL ANALYSES OF THE TEMPOROMANDIBULAR JOINT

7.12.1 Introduction

The temporomandibular joint (TMJ) is a bilateral composed joint connecting the mandible to the temporal bone. Both joints function bilaterally and any change on one side influences the contra lateral joint. TMJ has two articulating bone components—mandibular condyle (caput mandibulae) and glenoid fossa (fovea articularis) of the temporal bone. The condyle has the shape of an ellipsoid with a mediolateral diameter of 20 mm and an antero-posterior diameter of 10 mm. The articular surface is covered with fibrocartilage. The disk (discus articularis) of a biconcave shape fills the space between the articular surfaces, thus compensating their incongruities. The disk is changed by degenerative processes and may be perforated in the middle, cracked, or torn apart.

Therefore, detailed knowledge of the function and morphology of the temporomandibular joint is necessary for clinical applications and for analyses of the function of a TMJ prosthesis. Movements of the TMJ are biomechanically sophisticated and they are still not clear. Therefore, mathematical modeling of movements of the TMJ can be used for a better understanding of TMJ biomechanical aspects, its morphology, and functions (Röhrle and Pullan, 2007; Leader et al., 2003; Ichim et al., 2006; Peréz del Palomar and Dobraré, 2006; Beek et al., 2000; 2001a,b, 2003; Gal et al., 2004; Reona et al., 2007; Donzelli et al., 2004; Hannam and Langenbach, 1995; Korioth and Versluis, 1997; Peck et al., 1997; Jeffrey, 1996; Langenbach and Hannam, 1999; Koolstra and van Eijden 2004, 2005; Koolstra et al., 2001).

Mathematical modeling is an invaluable method to help understand the biomechanics of the anatomically and functionally complex masticatory system. There are practical ways to study variables such as muscle tensions and joint loads in humans, causes and effects, and how to predict the likely outcomes of interventions. Several finite element simulations of the TMJ can be found in the literature. Three-dimensional FE models were developed by Nagahara et al. (1999) who analyzed the biomechanical reactions in the mandible and in the TMJ during clenching under various restraint conditions. Beek et al. (2000, 2001a) developed a 3D FE model of the articular disk under clenching and in an opening movement for different condylar positions. Hu et al. (2003) studied the influence of the cartilage that covers the joint surfaces. Tanaka et al. (2001, 2002, 2004) investigated the stress distribution in the TMJ during jaw opening, analyzed the differences in the stress distribution of the disk between subjects with and without internal dearrangement as well as the biomechanical response of the retrodiscal tissue in the TMJ under compression. Röhrle and Pullan (2007) introduced the 3D finite element model to calculate and investigate the directions and magnitude of muscle forces generated by the left and right masseter muscle during one chewing cycle. Winter et al. (2004) studied a time-dependent healing function for immediate loaded implants. Their analyses are based on the mechanics of the damage.

Three-dimensional reconstructions of the TMJ images usually use two medical paging modalities: MRI and CT scan. Contrary to CT scans MRI enables simultaneous visualization of all tissues, and it is the most preferred imaging technique in many institutions for the study of the TMJ soft tissue pathology (Brooks et al., 1997). MRI is also a medical modality that has been used routinely in the evaluation of TMJ abnormalities since 1985 (Hannan and Langenbach, 1995; Korioth and Versluis, 1997; Peck et al., 1997).

7.12.2 The Temporomandibular Joint—Anatomy and Physiology

The temporomandibular joint is one of the most complicated joints of the human skeleton. The TMJ contains the temporal bone, part of the cranium; the mandible, the movable part; the condyle, the form of an oblate spheroid; the condylar axis runs between the tubercles, angle 150°–160°; articulating surfaces are covered with a thick layer of fibrocartilage; articular disk fills the space between the condyle and the temporal bone; shock absorber; muscles; and ligaments.

Jaw movements are complicated and biomechanically not totally clarified. The masticatory system is an anatomical and functional complex with multiple muscle groups that interact to produce forces on a mandible constrained by irregularly shaped joints, multiple dental contacts, and soft tissues. There are two types of movements: **rotational movement** (opening and closing movements), the disk is compressed; **and translational movement** (sliding the lower jaw forward or side to side), the disk is protracted and abbreviates. The majority of the joint movements are **combined**. Movements are mostly combined together resulting in jaw movements.

In humans, the movement of the mouth requires a complex combination of rotation in the lower TMJ compartment (condyle–disk) and of translation in the upper compartment (glenoid fossa–disk) (Gallo et al., 1997). In the three-dimensional global view, mandibular motion can be modeled using the pathways of a helical axis (Gallo et al., 1997; Sadat-Khonsari et al., 2003a,b). Current technology allows the non-invasive detection and recording of free human movements in a free dimensional space, and provides data for the quantitative assessment and biomechanical modeling of most body joints (Ferrario et al., 2002). Data collected from these observations, together with suitable mathematical and biomechanical models, allowed a more complete description of mouth movement, formulating hypotheses even about the motion of the condyle–disk complex within the joint (Naeije, 2003). A detailed knowledge of normal movements is necessary for a better understanding of TMJ disorders and for their treatment (Naeije, 2003).

It is difficult to measure directly many of the physical and physiological variables involved, human and animal experimental findings are limited, and they offer incomplete information on jaw biomechanics. Models clarify relationships between structures and functions, and they have been used successfully in other musculoskeletal regions. Models of the human jaw have demonstrated plausible musculoskeletal mechanics.

7.12.3 Biomechanics of TMJ and Its Function

The TMJ functions symmetrically, and this harmony allows biting, chewing, and speaking. Movements of the TMJ are biomechanically sophisticated and still not well understood [see Koolstra and van Eijden (1997)]. There are two types of movement: (i) rotary movement and (ii) sliding movement. The disk is compressed during the rotary movement and stretched and shortened during the sliding movement. Movements are mostly combined together, resulting in these jaw movements:

1. Depression of the mandible is performed by suprahyoid muscles (digastricus muscle, mylohyoideus muscle, geniohyoideus muscle). The condyles rotate around the condyle axis at the beginning of the opening, the next step is sliding of condyles and disks forward, then with the hinge terminal position they reach the articular eminence. The axis of rotation gets outside the glenoid fossa in the terminal part of the rotary movement.
2. Elevation of the mandible (the closing movement) is regulated by chewing muscles. The disk is sliding backward together with the condyle. The disk returns to the basic position after relaxation of the pterygoid muscles.
3. Propulsion (the forward movement). Condyles are sliding forward and below during propulsion. The bilateral contraction of medial pterygoid muscles together with a superficial portion of the masseteric muscle and lateral pterygoid muscle creates the propulsion movement.
4. Lateropulsion (side movement). The mandible slides from the physiological position to both sides with chewing. Lateropulsion is composed of protraction on one side and retraction on the other side. The mandible motion is mediated mainly by chewing muscles. The TMJ participates in the position of the mandible only during extreme movements.

One of the most relevant characteristics of the TMJ is that it is composed of two joints (Norman and Bramley, 1990; Gray, 1995). Since the TMJ is composed of two joints, the movement of one will consequently move the other. Thus it is important to analyze the different behavior of each joint during jaw movements, and above all, during nonsymmetrical movements such as lateral excursions of the mandible (Peréz del Palomar and Dobharé, 2006).

Positions of the mandible from the static point of view:

1. Physiological rest position is a balanced state of chewing muscles. It is affected by static and dynamical factors, that is, the vertical dimension of a rest distance between alveolar processes and a tension or tiredness of muscles, position and posture of the head, and the like.
2. Centric position of the mandible—the position of the mandible after swallowing.
3. Habitual position is an intermaxillary relation at the end of chewing or in maximal intercuspidation.

Mandibular kinematics have been used to study condylar movements, speech, mastication, muscle movements, as well as stomato-surgery. The experiences in practice addressed the need to standardize representation of mandibular kinamatics (Hannam, 1994; Wu and Cavanagh, 1995).

The stress and strain distributions within the articular disk under physiological loads are not well understood. Beek et al. (2001b) discussed experimental tests on removed disks of fresh cadavers and showed that these tests have been conducted in order to determine their mechanical properties. However, to the best of our knowledge, no experimental or theoretical studies of the articular disk in vivo have been carried out. Therefore, it is necessary to combine realistic constructive models, geometries, and loads in order to predict the biomechanical response of the different components of the joint both for symmetric and nonsymmetric movements.

From the dynamical point of view the TMJ is the part of the chewing apparatus with the mastication function. Mastication is a sequence of movements starting from a physiological rest or a habitual position, continuing with depression and elevation with retrusion and lateropulsion, and leading back to maximal intercuspidation. During mastication the chewing muscles generate the masticatory force. This force may be up to 4000 N (maximal pressure). This force is transferred from the mandible to the cranium basis through TMJ with all its structures and through the dentoalveolar way—from teeth and alveolar processes to the temporal bone. The disk of TMJ transfers a considerable part of the masticatory force and functions as an absorber of a muscular force. It has been said that the shape of the disk is irregular. It is obvious that the dispersion of the pressure is irregular as well [see Koolstra and van Eijden (1999)].

Models make clear relationships between structure and function, and they should show optimal musculoskeletal mechanics. Modeling uses CTs and MRIs, with their high-resolution ability (contrast resolution, serial tomography together with radiography).

Mathematical model allow the application of mechanical and biomechanical aspects of TMJ on prosthesis of the TMJ (TMJP). The construction and application of a joint prosthesis must respect these parameters. Only then will it function faultlessly and for a long time, even if the prosthesis is loaded maximally. The natural joint presents a system in equilibrium, where the shape is best suitable for its function.

An imbalance will result in a failure of the function and integrity of the TMJ. This applies equally to the prosthesis. Therefore, the aim of the mathematical model of TMJ and TMJ prosthesis functions is to establish conditions for preventing any imbalance of the harmony and/or potential destruction of TMJ and TMJP. TMJ is strained by traction and pressure. Contact surfaces tend to separate (traction) or press together. The effect of traction power is contained by ligaments, the joint functions as a mobile connection between the mandible and other parts of the skeleton, within certain limits. In the mathematical representation of our models this is described as a "gap." Contact surfaces are pushed together during pressure, so the movements between articulating structures happen within a tight connection of sliding articular surfaces.

The synovial fluid is essential for the TMJ function. The surface tension of the synovial fluid allows the fluid to spread over the articular surfaces as a capillary film that permits the lubrication of the joint during the condylar movements. The synovial fluid contains the glycoprotein lubricin, which serves to lubricate and minimize the friction between the surfaces. The friction between contact surfaces causes wear of the structure. And it is the same with the surface of the prostheses. The intensity of friction is influenced by the state of articulating surfaces. The friction (according to the Coulomb law) depends on the normal component of an acting force, thus with an increase of the thrust force (normal force), the friction, deformations, and wear also increase. The mathematical models have to respect all these facts, enabling us to analyze situations on articulating surfaces.

The models used represent a simple approximation of a TMJ function. The aim of this discussion is to simulate boundary conditions for which the obtained results correspond to the real situation in the TMJ and according to the knowledge and experiences of specialists in stomatology.

7.12.4 Formulation of the Contact Problem with Given Friction

In this section we will deal with a generalized semicoercive contact problem in linear elasticity describing the global model of the TMJ joint. The problem is formulated as the primary variational inequality problem, that is, in terms of displacements [see Hlaváček and Nedoma (2002a), Nedoma (1987, 1998a), Nedoma et al. (2006), and Chapter 6]. We will assume the generalized case of bodies of arbitrary shapes that are in mutual contact. On one part of the boundary the bodies are loaded and on the second one they are fixed. Therefore, as a result, some of the bodies can shift and rotate.

Let $I = (0, t_p)$, $t_p > 0$, be the time interval. Let $0, x_1, \ldots, x_N$ be the orthogonal Cartesian coordinate system, where N is the space dimension, and let $\mathbf{x} = (x_1, \ldots, x_N)$ be a point in this Cartesian system. Let the body, being in an initial stress–strain state and created by a system of elastic anisotropic or isotropic bodies, occupy a region $\Omega \subset \mathbb{R}^N, N = 2, 3$, consists of s bodies $\Omega^\iota, \iota = 1, \ldots, s$, so that $\Omega = \cup_{\iota=1}^s \Omega^\iota$, and let several neighboring bodies, say Ω^k and Ω^l, be in mutual contact. Denote the common contact boundary between both bodies Ω^k and Ω^l before deformation by Γ_c^{kl}. Let the boundary $\partial\Omega$ be divided into disjoint parts $\Gamma_\tau = \cup_{\iota=1}^s ({}^1\Gamma_\tau^\iota \cup {}^2\Gamma_\tau^\iota)$, $\Gamma_u = \cup_{\iota=1}^s ({}^1\Gamma_u^\iota \cup {}^2\Gamma_u^\iota)$, $\Gamma_c = \cup_{k,l} \Gamma_c^{kl}$ and Γ_0 such that $\partial\Omega = \Gamma_\tau \cup \Gamma_u \cup \Gamma_c \cup \Gamma_0 \cup \mathcal{R}$, where the surface measure of $\mathcal{R}$ is zero and the parts Γ_τ, Γ_u, Γ_c^{kl}, and Γ_0 are open sets in $\partial\Omega$. Let $\mathbf{n} = (n_i)$ be the outward normal vector to $\partial\Omega$, resp. to $\partial\Omega^k$ for the boundary Γ_c^{kl}. Let $Q \equiv \Omega(t) = I \times \Omega$ denote the time–space region, and let its boundary be $\partial Q = \partial\Omega \times I$, where $\Gamma_\tau(t) = \Gamma_\tau \times I$, $\Gamma_u(t) = \Gamma_u \times I$, $\Gamma_c(t) = \Gamma_c \times I$, $\Gamma_0(t) = \Gamma_0 \times I$ are the parts of its boundary $\partial Q = \partial\Omega \times I$.

Assume that Lamé coefficients $\lambda^{(0)}$ and $\mu^{(0)}$ as well as anisotropic elastic coefficients $c_{ijkl}^{(0)}$ are bounded functions. Then we will assume that $\lambda^{(0)\iota}, \mu^{(0)\iota}, c_{ijkl}^{(0)\iota} \in C^1(\overline{\Omega}^\iota)$. A repeated index implies summation from 1 to N.

The distribution of stress–strain fields in the area of the TMJ system is described (i) by the equation of equilibrium in the statically loaded TMJ

$$\frac{\partial \tau_{ij}^\iota(\mathbf{u}^\iota)}{\partial x_j} + F_i^\iota = 0, \qquad i,j = 1, \ldots, N,\ \iota = 1, \ldots, s, \ \text{in } \Omega^\iota, \tag{7.84}$$

and (ii) by the equation of motion

$$\frac{\partial \tau_{ij}^\iota(\mathbf{u}^\iota)}{\partial x_j} + F_i^\iota = \rho^\iota \frac{\partial^2 u_i}{\partial t^2}, \qquad i,j = 1, \ldots, N,\ \iota = 1, \ldots, s, \ \text{in } \Omega^\iota. \tag{7.85}$$

The relation between the displacement vector $\mathbf{u} = (u_i)$, $i = 1, \ldots, N$, and the small strain tensor e_{ij} is defined by

$$e_{ij} = e_{ij}(\mathbf{u}) = \frac{1}{2}\left(\frac{\partial u_i}{\partial x_j} + \frac{\partial u_j}{\partial x_i}\right), \qquad i,j = 1, \ldots, N.$$

The relation between the stress and strain tensors is defined by the generalized Hooke law for the elastic case and by the stress–strain law for the viscoelastic case

$$\begin{aligned} \tau_{ij}^\iota &= c_{ijkl}^{(0)\iota} e_{kl}(\mathbf{u}^\iota), \\ \tau_{ij}^\iota &= c_{ijkl}^{(0)\iota} e_{kl}(\mathbf{u}^\iota) + c_{ijkl}^{(1)\iota} e_{kl}(\mathbf{u}'^\iota), \end{aligned} \qquad i,j,k,l = 1, \ldots, N,\ \iota = 1, \ldots, s, \tag{7.86}$$

in the anisotropic case, whereas in the isotropic case

$$c_{ijkl}^{(n)\iota} = \lambda^{(n)\iota}\delta_{ij}\delta_{kl} + \mu^{(n)\iota}(\delta_{ik}\delta_{jl} + \delta_{il}\delta_{jk}),\ n = 0, 1, \tag{7.87}$$

where δ_{ij} is the Kronecker symbol. The coefficients $c_{ijkl}^{(n)\iota}$ form a matrix of the type $(N \times N \times N \times N)$ and satisfy the symmetry and the Lipschitz conditions

$$c_{ijkl}^{(n)\iota} = c_{jikl}^{(n)\iota} = c_{klij}^{(n)\iota} = c_{ijlk}^{(n)\iota} \quad \text{on}\Omega^{\iota}, \qquad n = 0, 1, i, j = 1, \ldots, N, \iota = 1, \ldots, s, \text{and}$$
$$0 < a_0^{(n)\iota} \leq c_{ijkl}^{(n)\iota}(\mathbf{x})\xi_{ij}\xi_{kl}|\xi|^{-2} \leq A_0^{(n)\iota} < +\infty \text{ for a.e. } \mathbf{x} \in \Omega^{\iota}, \xi \in \mathbb{R}^{N^2}, \xi_{ij} = \xi_{ji}, \tag{7.88}$$

where $a_0^{(n)\iota}$, $A_0^{(n)\iota}$ are constants independent on $\mathbf{x} \in \Omega^{\iota}$ and for the isotropic case $a_0^{(n)\iota} = 2\min\{\mu^{(n)\iota}(\mathbf{x});\ \mathbf{x} \in \Omega^{\iota}\}$ and $A_0^{(n)\iota} = \max\{2\mu^{(n)\iota}(\mathbf{x}) + 3\lambda^{(n)\iota}(\mathbf{x});\ \mathbf{x} \in \Omega^{\iota}$. A repeated index implies summation from 1 to N.

Furthermore, let us denote by $\tau_i = \tau_{ij}n_j$ components of the stress vector, normal and tangential components of the stress vector by $\tau_n = \tau_i n_i = \tau_{ij}n_j n_i$, $\boldsymbol{\tau}_t = \boldsymbol{\tau} - \tau_n\mathbf{n}$, and the normal and tangential components of the displacement vector by $u_n = u_i n_i$ and $\mathbf{u}_t = \mathbf{u} - u_n\mathbf{n}$. Let us denote by $\mathbf{u}^k$, $\mathbf{u}^l$ (indices k, l correspond with the neighboring bodies being in contact) the displacements in the neighboring bodies. Let us denote by $\mathbf{u}' = (u_i')$ the velocity vector. Let on Γ_c^{kl} the positive direction of the outer normal vector $\mathbf{n}$ be assumed with respect to Ω^k. Let us denote $[u_n]^{kl} = u_n^k - u_n^l$ and $[u_n']^{kl} = u_n'^k - u_n'^l$ the jump of the normal displacement and normal velocity across the contact zone between neighboring bodies of Ω^k and Ω^l. Similarly, we denote by $[\mathbf{u}_t]^{kl} = \mathbf{u}_t^k - \mathbf{u}_t^l$ and $[\mathbf{u}_t']^{kl} = \mathbf{u}_t'^k - \mathbf{u}_t'^l$ the tangential displacement and tangential velocities on Γ_c^{kl}. All these quantities are functions of spatial coordinates in statically loaded TMJ, and they are functions of spatial coordinates and time in dynamically loaded TMJ. On the contact boundaries Γ_c^{kl} the condition of nonpenetration

$$\begin{aligned} &u_n^k(\mathbf{x}) - u_n^l(\mathbf{x}) \leq d^{kl}(\mathbf{x}) && \text{in a statically loaded TMJ on } \Gamma_c^{kl}, \\ &u_n^k(\mathbf{x}, t) - u_n^l(\mathbf{x}, t) \leq d^{kl}(\mathbf{x}, t) && \text{in a dynamically loaded TMJ on} \Gamma_c^{kl}(t) \end{aligned} \tag{7.89}$$

hold, where d^{kl} is a gap. For the contact forces, due to the law of action and reaction, we find

$$\tau_n^k(\mathbf{u}) = \tau_n^l(\mathbf{u}) \equiv \tau_n^{kl}(\mathbf{u}), \qquad \boldsymbol{\tau}_t^k(\mathbf{u}) = -\boldsymbol{\tau}_t^l(\mathbf{u}) \equiv \boldsymbol{\tau}_t^{kl}(\mathbf{u}). \tag{7.90}$$

Since the normal components of contact forces cannot be positive, that is, cannot be tensile forces, then

$$\begin{aligned} &\tau_n^k(\mathbf{u}(\mathbf{x})) = \tau_n^l(\mathbf{u}(\mathbf{x})) \equiv \tau_n^{kl}(\mathbf{u}(\mathbf{x})) \leq 0 \\ &\qquad \text{in a statically loaded TMJ on } \Gamma_c^{kl}, \\ &\tau_n^k(\mathbf{u}(\mathbf{x}, t)) = \tau_n^l(\mathbf{u}(\mathbf{x}, t)) \equiv \tau_n^{kl}(\mathbf{u}(\mathbf{x}, t)) \leq 0 \\ &\qquad \text{in a dynamically loaded TMJ on } \Gamma_c^{kl}(t). \end{aligned} \tag{7.91}$$

During the deformation of the bodies they are in contact or they are not in contact. If they are not in contact, then $u_n^k - u_n^l < d^{kl}$, and the contact forces are equal to zero, that is, $\tau_n^k = \tau_n^l \equiv \tau_n^{kl} = 0$. If the bodies are in contact, that is, $u_n^k - u_n^l = d^{kl}$, then there may exist nonzero contact forces $\tau_n^k = \tau_n^l \equiv \tau_n^{kl} < 0$. These cases are included in the following condition:

$$\begin{aligned}
&(u_n^k(\mathbf{x}) - u_n^l(\mathbf{x}) - d^{kl}(\mathbf{x}))\tau_n^{kl}(\mathbf{u}(\mathbf{x})) = 0 \qquad \text{on } \Gamma_c^{kl} \\
&\qquad\qquad \text{in a statically loaded TMJ, and} \\
&(u_n^k(\mathbf{x},t) - u_n^l(\mathbf{x},t) - d^{kl}(\mathbf{x},t))\tau_n^{kl}(\mathbf{u}(\mathbf{x},t)) = 0 \qquad \text{on } \Gamma_c^{kl}(t) \\
&\qquad\qquad \text{in a dynamically loaded TMJ.}
\end{aligned} \tag{7.92}$$

Further, if both bodies are in contact, then on the contact boundary the Coulombian type of friction acts. The frictional forces g_c^{kl} acting on the contact boundary Γ_c^{kl} are, in their absolute value, proportional to the normal stress component, where the coefficient of proportionality is the coefficient of the Coulombian friction $\mathcal{F}_c^{kl}(\mathbf{x})$, that is,

$$\begin{aligned}
&g_c^{kl}(\mathbf{x}) = \mathcal{F}_c^{kl}(\mathbf{x})|\tau_n^{kl}(\mathbf{u})| && \text{in the statically loaded TMJ and} \\
&g_c^{kl}(\mathbf{x},t) = \mathcal{F}_c^{kl}(\mathbf{x},t)|\tau_n^{kl}(\mathbf{u}(\mathbf{x},t))| && \text{in the dynamically loaded TMJ.}
\end{aligned}$$

Due to the acting and frictional forces we have the following cases:

If the absolute value of tangential forces $\boldsymbol{\tau}_t^{kl}(\mathbf{u})$ is less than the slip limits g_c^{kl}, then the frictional forces preclude the mutual shifts of both bodies being in contact. If the tangential forces $\boldsymbol{\tau}_t^{kl}$ are equal in their absolute value to the frictional forces, so that there are no forces that can preclude the mutual, that is, bilateral, motion of both elastic bodies. Thus the contact points change their position in the direction opposite to that in which the tangential stress component acts. These conditions are described in a statically loaded TMJ by the following conditions:

$$\left.\begin{aligned}
&\text{If } u_n^k(\mathbf{x}) - u_n^l(\mathbf{x}) = d^{kl}(\mathbf{x}), \text{ then } |\boldsymbol{\tau}_t^{kl}(\mathbf{x})| \le g_c^{kl}(\mathbf{x}) \text{ and} \\
&\text{(a) if } |\boldsymbol{\tau}_t^{kl}(\mathbf{u}(\mathbf{x}))| < g_c^{kl}(\mathbf{x}), \text{ then } \mathbf{u}_t^k(\mathbf{x}) - \mathbf{u}_t^l(\mathbf{x}) = 0, \\
&\text{(b) if } |\boldsymbol{\tau}_t^{kl}(\mathbf{u}(\mathbf{x}))| = g_c^{kl}, \text{ then there exists a function } \vartheta \ge 0 \\
&\qquad \text{such that } \mathbf{u}_t^k(\mathbf{x}) - \mathbf{u}_t^l(\mathbf{x}) = -\vartheta\boldsymbol{\tau}_t^{kl}(\mathbf{u}(\mathbf{x})),
\end{aligned}\right\} \text{ on } \Gamma_c^{kl}, \tag{7.93}$$

where (a) means that the friction forces are sufficient to preclude the mutual shifting between the assumed parts of joint and (b) means that the friction forces are not sufficient to preclude the mutual bilateral shifting of both assumed parts of the joint. This shift acts in an opposite direction to the acting tangential forces.

In a dynamically loaded TMJ we have the following conditions:

$$\left.\begin{aligned}
&\text{If } u_n^k(\mathbf{x},t) - u_n^l(\mathbf{x},t) = d^{kl}(\mathbf{x},t), \text{ then } |\boldsymbol{\tau}_t^{kl}(\mathbf{u}(\mathbf{x},t))| \le g_c^{kl}(\mathbf{x},t) \text{ and} \\
&\text{(a) if } |\boldsymbol{\tau}_t^{kl}(\mathbf{u}(\mathbf{x},t))| < g_c^{kl}(\mathbf{x},t), \text{ then } \mathbf{u}_t^k(\mathbf{x},t) - \mathbf{u}_t^l(\mathbf{x},t) = 0, \\
&\text{(b) if } |\boldsymbol{\tau}_t^{kl}(\mathbf{u}(\mathbf{x},t))| = g_c^{kl}(\mathbf{x},t), \text{ then there exists a function } \vartheta \ge 0 \\
&\qquad \text{such that } \mathbf{u}_t^k(\mathbf{x},t) - \mathbf{u}_t^l(\mathbf{x},t) = -\vartheta\boldsymbol{\tau}_t^{kl}(\mathbf{u}(\mathbf{x},t)),
\end{aligned}\right\} \text{ on } \Gamma_c^{kl}(t). \tag{7.94}$$

Moreover, on the boundary $\cup_{\iota=1}^{s}({}^1\Gamma_u \cap \partial\Omega^\iota)$ the following conditions

$$\begin{aligned} &\mathbf{u}(\mathbf{x}) = \mathbf{u}_2(\mathbf{x}) \quad \text{on } \cup_{\iota=1}^{s}({}^1\Gamma_u \cap \partial\Omega^\iota) \text{ in a statically loaded TMJ,} \\ &\mathbf{u}(\mathbf{x},t) = \mathbf{u}_2(\mathbf{x},t) \quad \text{on } \cup_{\iota=1}^{s}({}^1\Gamma_u \cap \partial\Omega^\iota) \times I \text{ in a dynamically loaded TMJ,} \end{aligned} \tag{7.95}$$

will be given; on the boundary $\cup_{\iota=1}^{s}({}^2\Gamma_u \cap \partial\Omega^\iota)$ the following conditions

$$\begin{aligned} &\mathbf{u}(\mathbf{x}) = \mathbf{0} \quad \text{on } \cup_{\iota=1}^{s}({}^2\Gamma_u \cap \partial\Omega^\iota) \text{ in a statically loaded TMJ,} \\ &\mathbf{u}(\mathbf{x},t) = \mathbf{0} \quad \text{on } \cup_{\iota=1}^{s}({}^2\Gamma_u \cap \partial\Omega^\iota) \times I \text{ in a dynamically loaded TMJ,} \end{aligned} \tag{7.96}$$

will be given; on the boundary $\cup_{\iota=1}^{s}({}^1\Gamma_\tau \cap \partial\Omega^\iota)$ the following conditions

$$\begin{aligned} &\tau_{ij}(\mathbf{u}(\mathbf{x}))n_j = P_i(\mathbf{x}) \quad \text{on } \cup_{\iota=1}^{s}({}^1\Gamma_\tau \cap \partial\Omega^\iota) \text{ in a statically loaded TMJ,} \\ &\tau_{ij}(\mathbf{u}(\mathbf{x},t)) = P_i(\mathbf{x},t) \quad \text{on } \cup_{\iota=1}^{s}({}^1\Gamma_\tau \cap \partial\Omega^\iota) \times I \text{ in a dynamically loaded TMJ,} \end{aligned} \tag{7.97}$$

will be given; on the boundary $\cup_{\iota=1}^{s}({}^2\Gamma_\tau \cap \partial\Omega^\iota)$ the following conditions

$$\begin{aligned} &\tau_{ij}(\mathbf{u}(\mathbf{x}))n_j = 0 \quad \text{on } \cup_{\iota=1}^{s}({}^2\Gamma_\tau \cap \partial\Omega^\iota) \text{ in a statically loaded TMJ,} \\ &\tau_{ij}(\mathbf{u}(\mathbf{x},t))n_j = 0 \quad \text{on } \cup_{\iota=1}^{s}({}^2\Gamma_\tau \cap \partial\Omega^\iota) \times I \text{ in a dynamically loaded TMJ,} \end{aligned} \tag{7.98}$$

will be given; on the boundary $\cup_{\iota=1}^{s}({}^1\Gamma_0 \cap \partial\Omega^\iota)$ the bilateral contact condition

$$\begin{aligned} &u_n(\mathbf{x}) = 0,\ \tau_{tj}(\mathbf{u}(\mathbf{x})) = 0, j = 1,\ldots,N, \quad \text{in a statically loaded TMJ and} \\ &u_n(\mathbf{x},t) = 0,\ \tau_{tj}(\mathbf{u}(\mathbf{x},t)) = 0, j = 1,\ldots,N, \quad \text{in a dynamically loaded TMJ} \end{aligned} \tag{7.99}$$

will be given.

In the case of the dynamically loaded TMJ the initial conditions

$$\mathbf{u}(\mathbf{x},0) = \mathbf{u}_0(\mathbf{x}), \qquad \mathbf{u}'(\mathbf{x},0) = \mathbf{u}_1(\mathbf{x}), \qquad \mathbf{x} \in \Omega, \tag{7.100}$$

where $\mathbf{u}_0$, $\mathbf{u}_1$, $\mathbf{u}_2$ are the given functions, we assume that $\mathbf{u}_2$ has a time derivative $\mathbf{u}_2'$.

In what follows we will deal with the following model problem of the statically loaded temporomandibular joint:

Let us assume that the TMJ occupies the region; we denote it as Ω, and let $\Omega = \cup_{\iota=1}^{s} \Omega^\iota \subset \mathbb{R}^N, N = 2,3$, is a union of domains with Lipschitz boundaries $\partial\Omega^\iota$, occupied by the bodies about which we will assume to be elastic. Let several neighboring bodies, say Ω^k and Ω^l, be in a mutual contact. Let us denote the common contact boundary between both bodies Ω^k and Ω^l before deformation by Γ_c^{kl}. Let the boundary $\partial\Omega$ be divided into disjoint parts Γ_τ, Γ_u, $\Gamma_c = \cup_{k,l}\Gamma_c^{kl}$ and Γ_0 such that $\partial\Omega = \Gamma_\tau \cup \Gamma_u \cup \Gamma_c \cup \Gamma_0 \cup \mathcal{R}$, where the surface measure of $\mathcal{R}$ is zero and the parts $\Gamma_\tau = \cup_{\iota=1}^{s}({}^1\Gamma_\tau^\iota \cup {}^2\Gamma_\tau^\iota)$, $\Gamma_u = \cup_{\iota=1}^{s}({}^1\Gamma_u^\iota \cup {}^2\Gamma_u^\iota)$, Γ_c^{kl} and Γ_0 are open sets in $\partial\Omega$. Let $\mathbf{n} = (n_i)$ be the outward normal vector to $\partial\Omega$, resp. to $\partial\Omega^k$ for the boundary Γ_c^{kl}.

Let us assume that body forces $\mathbf{F} \in L^{2,N}(\Omega)$, surface tractions $\mathbf{P} \in L^{2,N}(\Gamma_\tau)$, boundary displacements $\mathbf{u}_0 \in H^{\frac{1}{2},N}(\Gamma_u)$, and slip limits $g_c^{kl} \in L^2(\Gamma_c^{kl})$ are given.

Problem $\mathcal{P}$ Find a vector function $\mathbf{u}$ satisfying

$$\frac{\partial \tau_{ij}^{\iota}(\mathbf{u}^{\iota})}{\partial x_j} + F_i^{\iota} = 0, \qquad i,j = 1,\ldots,N,\ \iota = 1,\ldots,s,\ \text{ in } \Omega^{\iota}, \tag{7.101}$$

$$\mathbf{u}^{\iota} = \mathbf{0} \quad \text{on } {}^1\Gamma_u^{\iota},\ \ \iota = 1,\ldots,s, \tag{7.102}$$

$$\mathbf{u}^{\iota} = \mathbf{u}_2 \quad \text{on } {}^2\Gamma_u^{\iota},\ \ \iota = 1,\ldots,s, \tag{7.103}$$

$$\tau_{ij}^{\iota} n_j^{\iota} = P_i^{\iota} \quad \text{on } {}^1\Gamma_u^{\iota},\ \ \iota = 1,\ldots,s, \tag{7.104}$$

$$\tau_{ij}^{\iota} n_j^{\iota} = 0 \quad \text{on } {}^2\Gamma_u^{\iota},\ \ \iota = 1,\ldots,s, \tag{7.105}$$

$$u_n^{\iota} = 0, \qquad \tau_{tj}^{\iota} = 0 \quad \text{on } \Gamma_0^{\iota},\ j = 1,\ldots,N,\ \iota = 1,\ldots,s, \tag{7.106}$$

$$\left.\begin{aligned} & u_n^k - u_n^l \le d^{kl},\ \tau_n^{kl} \le 0, \qquad (u_n^k - u_n^l)\tau_n^{kl}(\mathbf{u}) = 0 \text{ and} \\ & \text{if } u_n^k - u_n^l = d^{kl}, \text{ then } |\boldsymbol{\tau}_t^{kl}(\mathbf{u})| \le g_c^{kl} \text{ and} \\ & \text{(a) if } |\boldsymbol{\tau}_t^{kl}(\mathbf{u})| < g_c^{kl}, \text{ then } \mathbf{u}_t^k - \mathbf{u}_t^l = 0, \\ & \text{(b) if } |\boldsymbol{\tau}_t^{kl}(\mathbf{u})| = g_c^{kl}, \text{ then there exists a function } \vartheta \ge 0 \\ & \qquad\qquad \text{such that } \mathbf{u}_t^k - \mathbf{u}_t^l = -\vartheta \tau_t^{kl}(\mathbf{u}), \end{aligned}\right\} \quad \text{on } \Gamma_c^{kl}. \tag{7.107}$$

Further, we will deal with the following variational problem:

Let us look for the displacement vector $\mathbf{u} = (u_i) \in W \equiv [H^1(\Omega^1)]^N \times \ldots \times [H^1(\Omega^s)]^N$, $N = 2, 3$, where $H^1(\Omega^{\iota})$ is the Sobolev space in the usual sense. We will assume that (we adopt the same notation) $F_i^{\iota} \in L^2(\Omega^{\iota})$, $P_i \in L^2(\Gamma_\tau)$, $u_{2i} \in H^{\frac{1}{2}}(\Gamma_u)$. Let us denote by $(\cdot,\cdot)$ the scalar product in $[L^2(\Omega)]^N$, by $\langle \cdot,\cdot \rangle$ the scalar product in $[L^2(\Gamma_c)]^N$,

$$V_0 = \{\mathbf{v} \mid \mathbf{v} \in W, \mathbf{v} = \mathbf{0} \text{ on } {}^1\Gamma_u \cup {}^2\Gamma_u, v_n = 0 \text{ on } \Gamma_0\},$$

$$V = \{\mathbf{v} \mid \mathbf{v} \in W, \mathbf{v} = \mathbf{0} \text{ on } {}^1\Gamma_u,\ \mathbf{v} = \mathbf{u}_2 \text{ on } {}^2\Gamma_u,\ v_n = 0 \text{ on } \Gamma_0\}$$

the sets of virtual displacements and by

$$K = \{\mathbf{v} \mid \mathbf{v} \in V,\ v_n^k - v_n^l \le d^{kl} \text{ on } \cup_{k,l} \Gamma_c^{kl}\}$$

the set of all admissible displacements, which for $\mathbf{u}_0 = \mathbf{0}$ and $d^{kl} = 0$ is a convex cone with the vertex at the origin.

Multiplying (7.84) by $\mathbf{v} - \mathbf{u}$, integrating over Ω^ι using the boundary conditions, then we obtain the following variational problem:

Problem $(\mathcal{P})_v$ Find a function $\mathbf{u} \in K$, such that

$$a(\mathbf{u}, \mathbf{v} - \mathbf{u}) + \langle g_c^{kl}, |\mathbf{v}_t^k - \mathbf{v}_t^l| - |\mathbf{u}_t^k - \mathbf{u}_t^l| \rangle \geq S(\mathbf{v} - \mathbf{u}) \quad \forall \mathbf{v} \in K, \tag{7.108}$$

where for $\mathbf{u}, \mathbf{v} \in K$ we have

$$\begin{aligned}
a(\mathbf{u}, \mathbf{v}) &= \sum_{\iota=1}^{s} a(\mathbf{u}, \mathbf{v}) = \int_\Omega c_{ijkl} e_{ij}(\mathbf{u}) e_{kl}(\mathbf{v}) \, d\mathbf{x}, \\
S(\mathbf{v}) &= \sum_{\iota=1}^{s} S(\mathbf{v}^\iota) = \int_\Omega F_i v_i \, d\mathbf{x} + \int_{\Gamma_\tau} P_i v_i \, ds, \\
j_{gn}(v) &= \int_{\bigcup_{k,l} \Gamma_c^{kl}} g_c^{kl} |\mathbf{v}_t^k - \mathbf{v}_t^l| \, ds = \langle g_c^{kl}, |\mathbf{v}_t^k - \mathbf{v}_t^l| \rangle.
\end{aligned}$$

This problem can be rewritten in the equivalent variational formulation:

Problem $(\mathcal{P})_{ve}$ Find a function $\mathbf{u} \in K$, such that

$$L(\mathbf{u}) \leq L(\mathbf{v}) \quad \forall \mathbf{v} \in K, \tag{7.109}$$

where $L(\mathbf{v})$ is defined by

$$L(\mathbf{v}) = L_0(\mathbf{v}) + j_{gn}(v), \quad L_0(\mathbf{v}) = \tfrac{1}{2} a(\mathbf{v}, \mathbf{v}) - S(\mathbf{v}). \tag{7.110}$$

It can be proved by the following results:

Theorem 7.2 There exists a unique weak solution of Problem $(\mathcal{P})_v$, provided ${}^2\Gamma_u^\iota$ is not empty for all $\iota \leq s$, for instance.

For the proof see Hlaváček et al., (1988), and Hlaváček and Nedoma (2002a).

It can be shown that any classical solution of Problem $\mathcal{P}$ is a weak solution, and on the other hand, if the weak solution is smooth enough, it is a classical solution.

7.12.5 Finite Element Solution of the Problem

Let the domain $\Omega \subset \mathbb{R}^N$ be a bounded domain, and let it be approximated by a polygonal (for $N=2$) or polyhedral (for $N=3$) domain Ω_h. Let the domain Ω_h be "triangulated," that is, the domain $\overline{\Omega}_h = \Omega_h \cup \partial\Omega_h$ is divided into a system of m triangles T_{h_i} in the 2D case and into a system of m tetrahedra T_{h_i} in the 3D case, generating a triangulation $\mathcal{T}_h$ such that $\overline{\Omega}_h = \cup_{i=1}^m T_{h_i}$ and such that two neighboring triangles have only a vertex or an entire side common in the 2D case, and that two neighboring tertrahedra have only a vertex or an entire edge or an entire face common in the 3D case. Denote by $h = \max_{1 \leq i \leq m} (\text{diam } T_{h_i})$ the maximal side of the triangle T_{h_i} in the 2D case and/or the maximal edge of the tetrahedron in the 3D case in $\mathcal{T}_h$. Let ρ_{T_i} denote the radius of the maximal circle (for 2D case) or maximal ball (for 3D case), inscribed in the simplex T_{h_i}. A family of triangulation $\{\mathcal{T}_h\}$, $0 < h \le h_0 < \infty$, is said to be regular if there exists a constant $\vartheta_0 > 0$ independent of h and such that $h/\rho_{T_i} \le \vartheta_0$ for all $h \in (0, h_0)$. We will assume that the sets $\overline{\Gamma}_u \cap \overline{\Gamma}_\tau$, $\overline{\Gamma}_u \cap \overline{\Gamma}_c$, $\overline{\Gamma}_u \cap \overline{\Gamma}_0$, $\overline{\Gamma}_c \cap \overline{\Gamma}_\tau$, $\overline{\Gamma}_c \cap \overline{\Gamma}_0$, $\overline{\Gamma}_\tau \cap \overline{\Gamma}_0$ coincide with the vertices or edges of T_{h_i}.

Let $R_i \in \Omega_h$ be an arbitrary interior vertex of the triangulation $\mathcal{T}_h$. Generally, the basis function w_h^i (where w_h^i is a scalar or vector function) is defined to be a function linear on each element $T_{h_i} \in \mathcal{T}_h$ and taking the values $w_h^i(R_j) = \delta_{ij}$ at the vertices of the triangulation, where δ_{ij} is the Kronecker symbol. The function w_h^i represents a pyramid of height 1 with its vertex above the point R_i and with its support (supp w_h^i) consisting of those triangles or tetrahedra that have the vertex R_i in common. The basis function has a small support since diam(supp w_h^i) $\le 2h$ and the parameter $h \to 0$.

Let V_h be the set of linear finite elements, that is, the set of continuous vector functions in $\overline{\Omega}_h$, piecewise linear over $\mathcal{T}_h$, that is,

$$V_h = \{\mathbf{v} \in [C(\overline{\Omega}_h^1)]^N \times \cdots \times [C(\overline{\Omega}_h^s)]^N \cap V \mid \mathbf{v}|_{T_{h_i}} \in [P_1(T_{h_i})]^N \ \forall T_{h_i} \in \mathcal{T}_h\}$$

and

$$K_h = \{\mathbf{v}_h \in V_h \mid v_{hn}^k - v_{hn}^l \le d^{kl} \text{ on } \cup \Gamma_c^{kl}\} = K \cap V_h.$$

Numerically, the problem is solved by the finite element technique. The problem then leads to the following:

Problem $(\mathcal{P})_h$ Find a vector function $\mathbf{u}_h \in K_h$ such that

$$L(\mathbf{u}_h) \le L(\mathbf{v}_h) \quad \forall \mathbf{v}_h \in K_h. \tag{7.111}$$

As in Hlaváček and Nedoma (2002a) it can be shown that the seminorm $|\mathbf{u} - \mathbf{u}_h| = O(h)$, where $\mathbf{u}$ is the solution of the problem $(\mathcal{P})_v$ and $\mathbf{u}_h$ is the solution of the problem $(\mathcal{P})_h$.

7.12.6 Dynamically Loaded Temporomandibular Joint

Mathematical modeling of movements of the TMJ will be based on the dynamically formulated contact problem (7.85) – (7.100). The derivation of the problem and its analysis will be parallel with the dynamic contact problem formulated and discussed in Sections 6.5 and 6.6. The algorithm can be based on the explicit or semi-implicit schemes in time, the mortar finite element method, and the primal dual active set strategy method, discussed in detail in Chapter 6.

7.12.7 Numerical Results

The purpose of the section is to characterize processes in TMJ during its function. All models are counted according to MRI images.

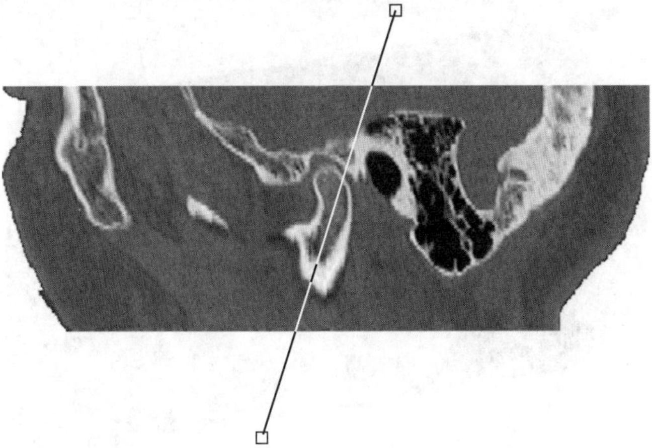

FIG. 7.377 MRI scan of the TMJ in the lateral view.

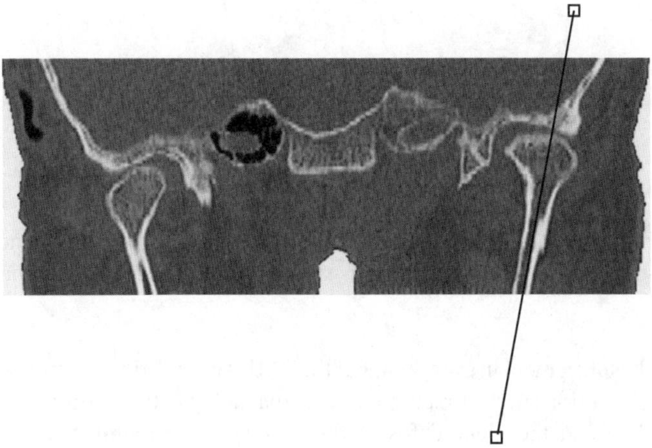

FIG. 7.378 MRI scan of the TMJ in the frontal view.

Models in 2D The geometry of the two-dimensional models was obtained from appropriately chosen cut. Figures 7.377 and 7.378 show MRI scans in lateral and frontal views. There is a line segment for better orientation. Slices contain this line segment, which was found by rotating a plane of the slice around this line segment (see Fig. 7.379).

Three two-dimensional models were created in this study. The geometry and the material parameters (Young's modulus E and Poisson's ratio ν) for all of them are the same. The used values are $E = 1.71 \times 10^{10}$ Pa, $\nu = 0.25$ for the bone tissue and $E = 0.492 \times 10^9$ Pa, $\nu = 0.1$ for the cartilage. The TMJ is loaded by loading 1×10^4 Pa on the top side of the mandible and by loading 1.3×10^4 Pa on the bottom and the right-hand side of the mandible (see green arrows in Figs. 7.380–7.382). We assume slip limits $g_c^{kl} = 0$ for all models. The temporal bone is fixed on two parts

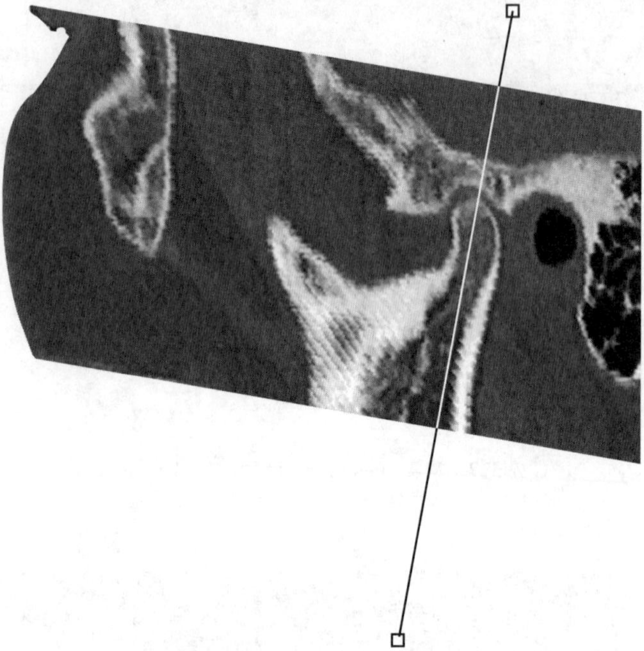

FIG. 7.379 MRI scan of the TMJ—the cut used for the 2D model.

of the boundary [in Figs. 7.380(a), 7.381(a), and 7.382(a) they are denoted by red lines]. The differences among the models studied are in specified additional boundary conditions.

Model 1 Displacement in every point of the TMJ is bounded by ± 1 mm in horizontal and vertical directions in the first model. Maximal deformation is observed in the area of the condyle and the glenoid fossa; the disk is also strongly loaded. The main strain goes vertically with a peak through the processus coronoideus [Fig. 7.380(b)]. The main pressure area runs through the condyle and the glenoid fossa, partly through the processus muscularis. The projection of the traction is in incisura mandibulae [Fig. 7.380(c)].

Model 2 In Fig. 7.381(a) a blue line characterizes the possibility of the jaw mobility, which is realized by the bilateral contact condition. The main deformation area concentrates on the glenoid fossa and the disk. The deformation appears on the ventral side of the processus coronoideus [Fig. 7.381(b)]. The peak pressure is in the condyle and the fossa glenoid. The traction component almost does not appear [Fig. 7.381(c)].

Model 3 The movement in the third model simulates typical depression (opening), elevation (closing), and lateropulsion (side movements) of the mandible. In this model the displacement (-1 mm, 1 mm) for a red point [see Fig. 7.382 (a)] is prescribed. The highest pressure is observed in the area of the condyle and the distal

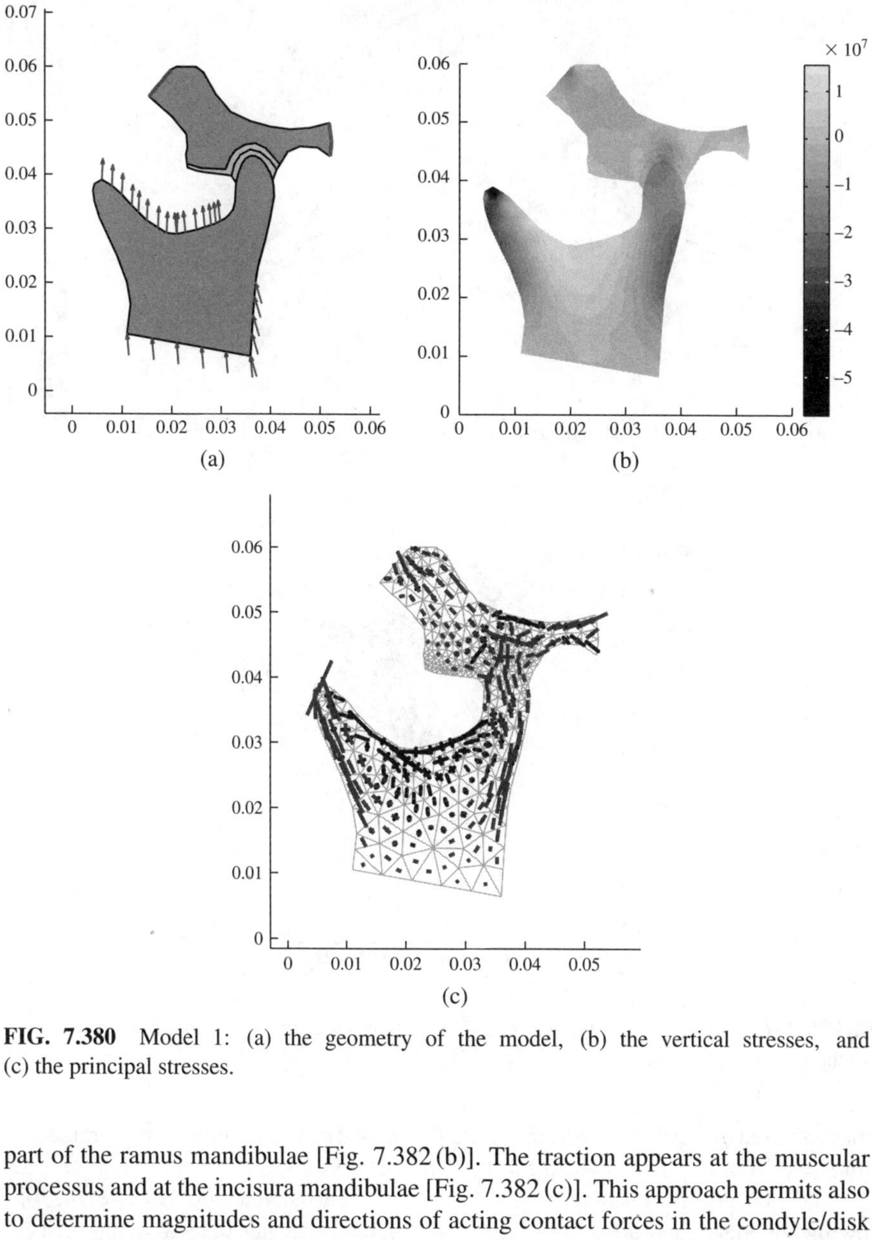

FIG. 7.380 Model 1: (a) the geometry of the model, (b) the vertical stresses, and (c) the principal stresses.

part of the ramus mandibulae [Fig. 7.382 (b)]. The traction appears at the muscular processus and at the incisura mandibulae [Fig. 7.382 (c)]. This approach permits also to determine magnitudes and directions of acting contact forces in the condyle/disk areas and their deformations.

The 3D TMJ model A three-dimensional model of the mandible was created using the data set of axial MRI of a male patient that was obtained from the Visible Human Project. The finite element mesh (see Fig. 7.383) was prepared using the Open Inventor 3D toolkit by the Mercury Visualization Sciences Group. Discretization statistics are

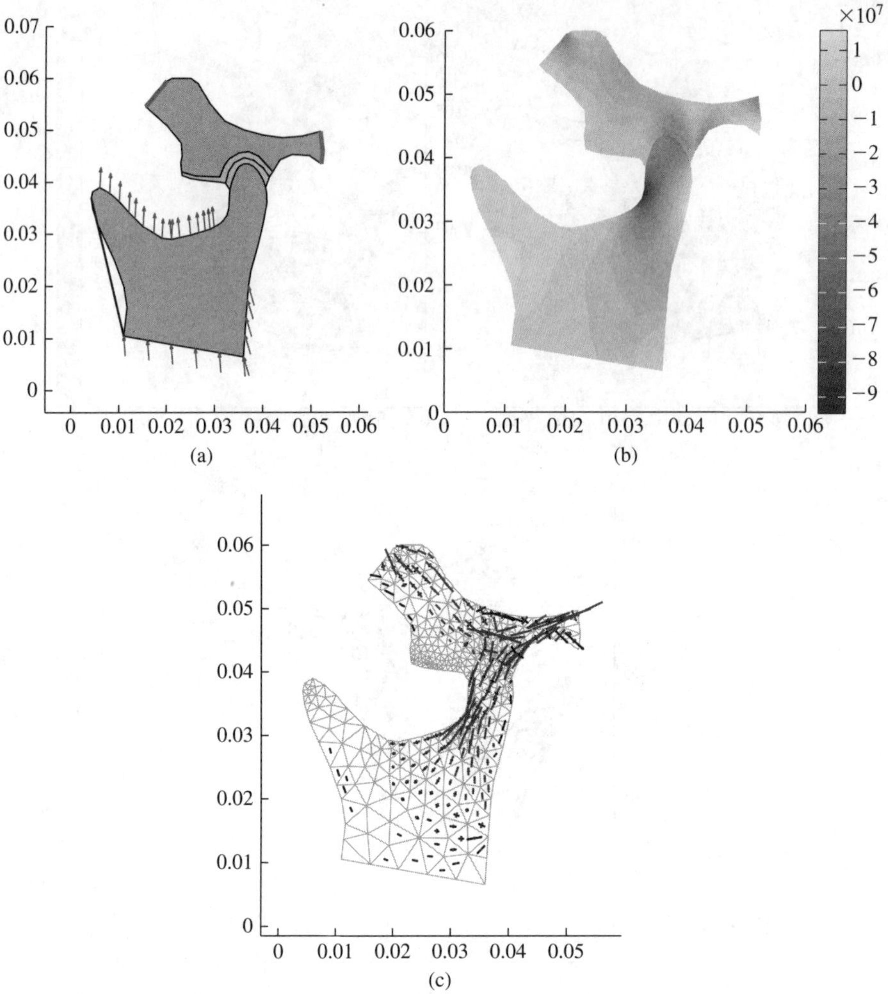

FIG. 7.381 Model 2: (a) the geometry of the model, (b) the vertical stresses, and (c) the principal stresses.

characterized by 7295 tetrahedrons and 2043 nodes. The values of material parameters are $E = 1.71 \times 10^{10}$ Pa, $\nu = 0.25$ for the bone tissue.

On the basis of the results obtained for the two-dimensional models we set the following boundary conditions (see Fig. 7.384). For the upper side of teeth we prescribed the vertical displacement of 1 mm. Working masticatory muscles on the right-hand side of the mandible are characterized by the loading $(-0.6; -0.6; 2.8) \times 10^6$ N/mm^2 for the musculus pterygoideus lateralis, $(-0.6; 0; 2.8) \times 10^6$ N/mm^2 for the musculus masseter, and $(0; 0.5; 1.5) \times 10^6$ N/mm^2 for the musculus pterygoideus medialis (on the left-hand side we consider symmetric loading). We assume the model case in which the heads (condyles) of the TMJ and the acetabulae are in mutual contact,

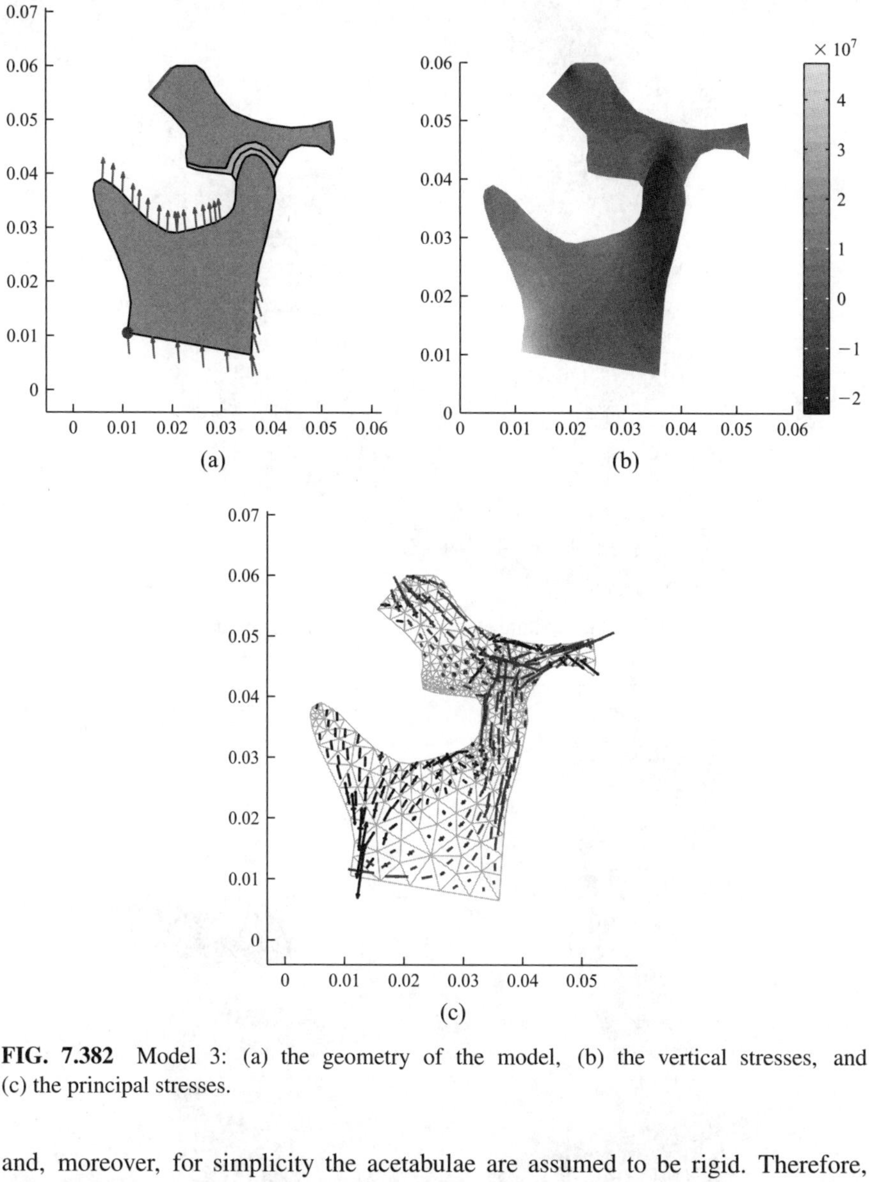

FIG. 7.382 Model 3: (a) the geometry of the model, (b) the vertical stresses, and (c) the principal stresses.

and, moreover, for simplicity the acetabulae are assumed to be rigid. Therefore, the unilateral contact condition on the contact boundary (in Fig. 7.384 denoted by yellow color) can be replaced by the bilateral contact condition (7.106). For the 3D model the principal stresses are displayed in Fig. 7.385 and the von Mises stresses in Fig. 7.386 in the same cut, which was assumed in the two-dimensional model. On the basis of the TMJ analysis the artificial TMJ replacement will be designed and numerically analyzed. In the first step, presented in this section, the healthy TMJ is investigated and analyzed. The numerical analysis of the principal and von Mises stresses of the healthy TMJ shows that the highest pressures are observed in the areas

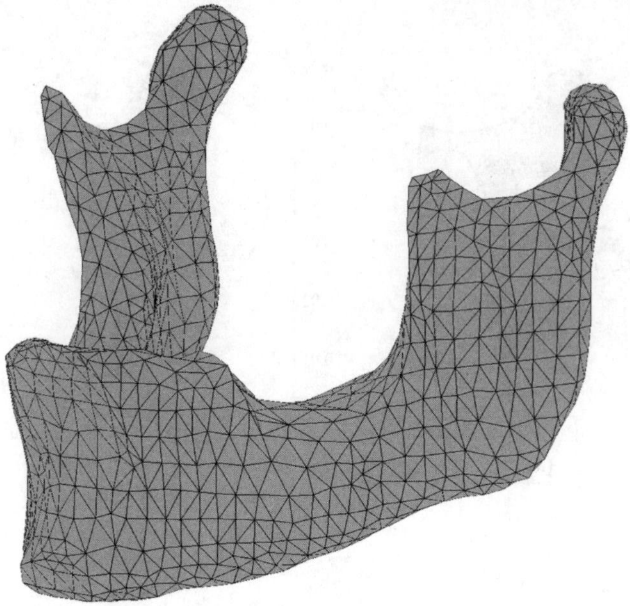

FIG. 7.383 Finite element mesh for the 3D model.

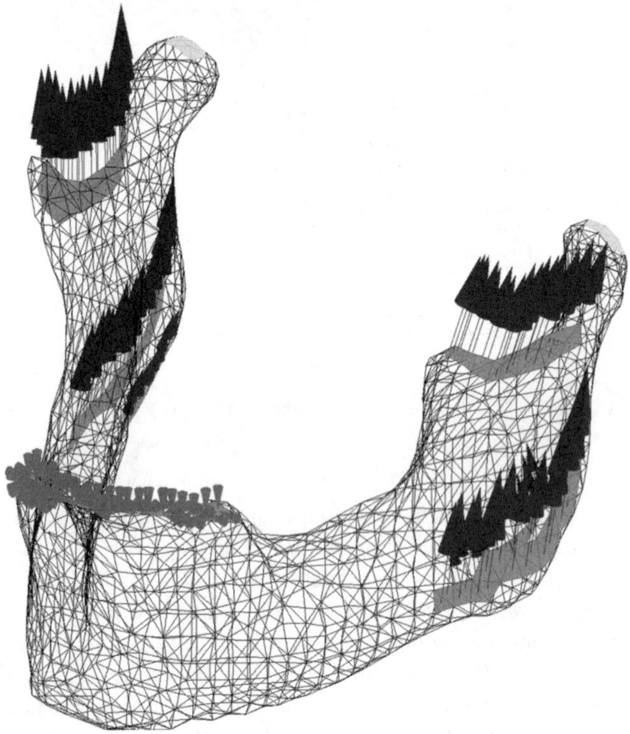

FIG. 7.384 Boundary conditions for the 3D model.

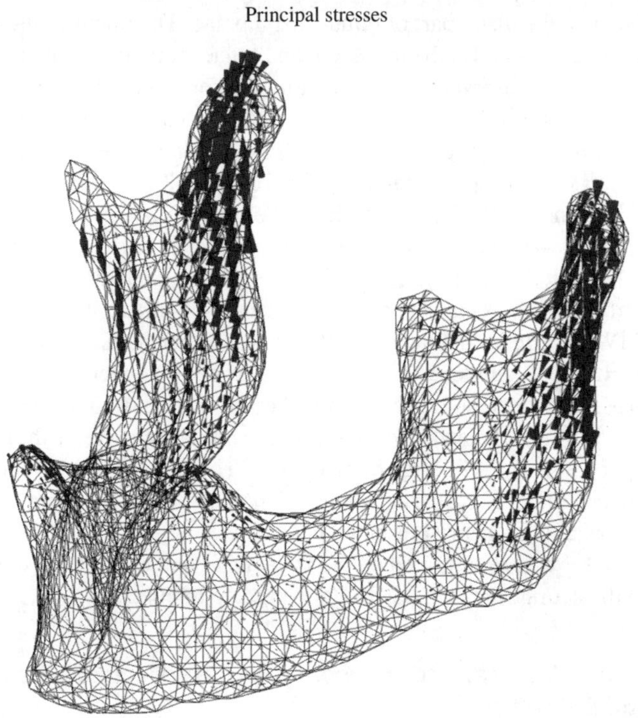

FIG. 7.385 Distribution of the principal stresses in the 3D model.

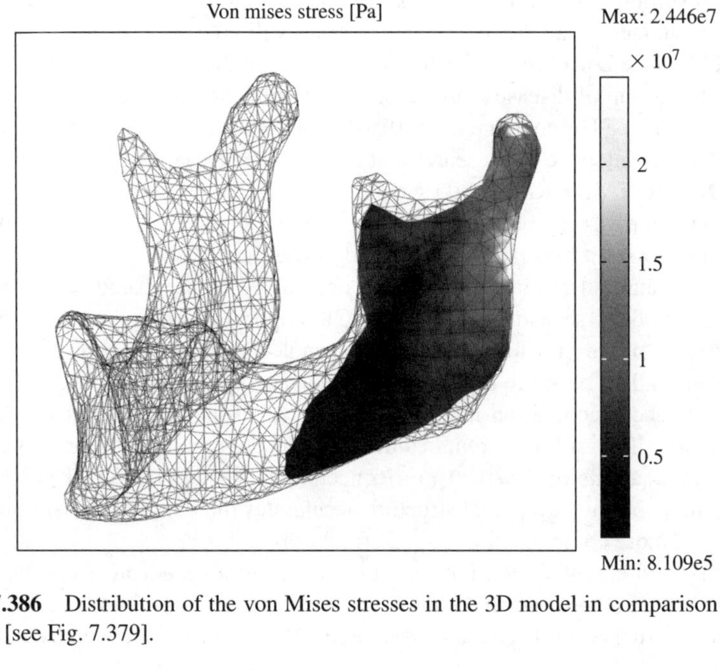

FIG. 7.386 Distribution of the von Mises stresses in the 3D model in comparison with the 2D cut [see Fig. 7.379].

of the condyle and the distal part of ramus mandibulae. The tractions are observed at the muscular processus and at the incisura mandibulae, both in agreement with the 2D model results and the observations in stomatological practice. Maximal deformations are observed in the area of condyles and the glenoid focca. From the obtained numerical data the maximal pressure is 2.13×10^7 Pa in the contact area and the magnitudes of principal stresses are approximately 1.87×10^7 Pa. All obtained numerical results are in good agreement with the stomatological observations.

Similar models will be also developed for artificial TMJ replacements. These models can be developed on the theories presented in Chapter 6. After the application of artificial TMJ replacement, the artificial compartment of the TMJ, based on the UHMWPE, can be damaged similarly as in other human joints. Ultra-high-molecular-weight polyethylene (UHMWPE) wear debris generated at the articular surfaces of TMJ replacements enters periprosthetic tissues and stimulates bone resorption, which may ultimately lead to loosening of the implant and create the necessity for revision surgery as in cases of large human joints. Therefore, the theoretical analyses are needed also in the TMJ case. Mathematical models will be based on the numerical wear simulation and the theory presented in Chapter 6 [see also Winter et al. (2004)]. Polyethylene wear will be enforced on the nodal basis and estimated using Archard's law because the acting forces in the TMJ area are up to 4000 N.

7.12.8 Role of Medical Informatics in Dentist's Decision-Making TMJ Disorders

The unclear etiology of TMD (temporomandibular joint disorders), the same clinical findings resulting from various causes and the proven relation between TMD and psychological factors are the main reason there is still no consensus in the classification of TMD [see Dworkin and LeResche (1992)]. All these facts make TMD a very complicated group of diseases for creating a suitable and compact electronic health record system [see Hippmann et al. (2009)]. One of the most commonly used diagnostic schemes intended for research purposes is the Research Diagnostic Criteria for TMD (RDC/TMD). RDC/TMD standardizes the clinical examination of patients with TMD, improves reproducibility among clinicians, and facilitates comparison of results among researchers [see Dworkin and LeResche (1992)].

Temporomandibular joint disorders are considered to be a subgroup of musculoskeletal disorders [see Goldstein (1999)]. Our application was prepared to produce a user-friendly program, which will unify the whole masticatory system and its problems because all its parts are directly connected. The system was enhanced with the automatic speech recognition module [see Fig. 7.387 and Zvárová et al. (2008)]. The structure allows a transparent health record on the whole dentition and accomplished examinations of a patient in a concentrated form. The dental information recorded in a common graphical structure accelerates the dentist's decision making and brings a more complex view on gathered information.

The first and essential step for decision making in all medical fields, including dentistry, is to structure data and knowledge and to remove or minimize information stored in a textual form. If data are structured, then they can be coded by means of

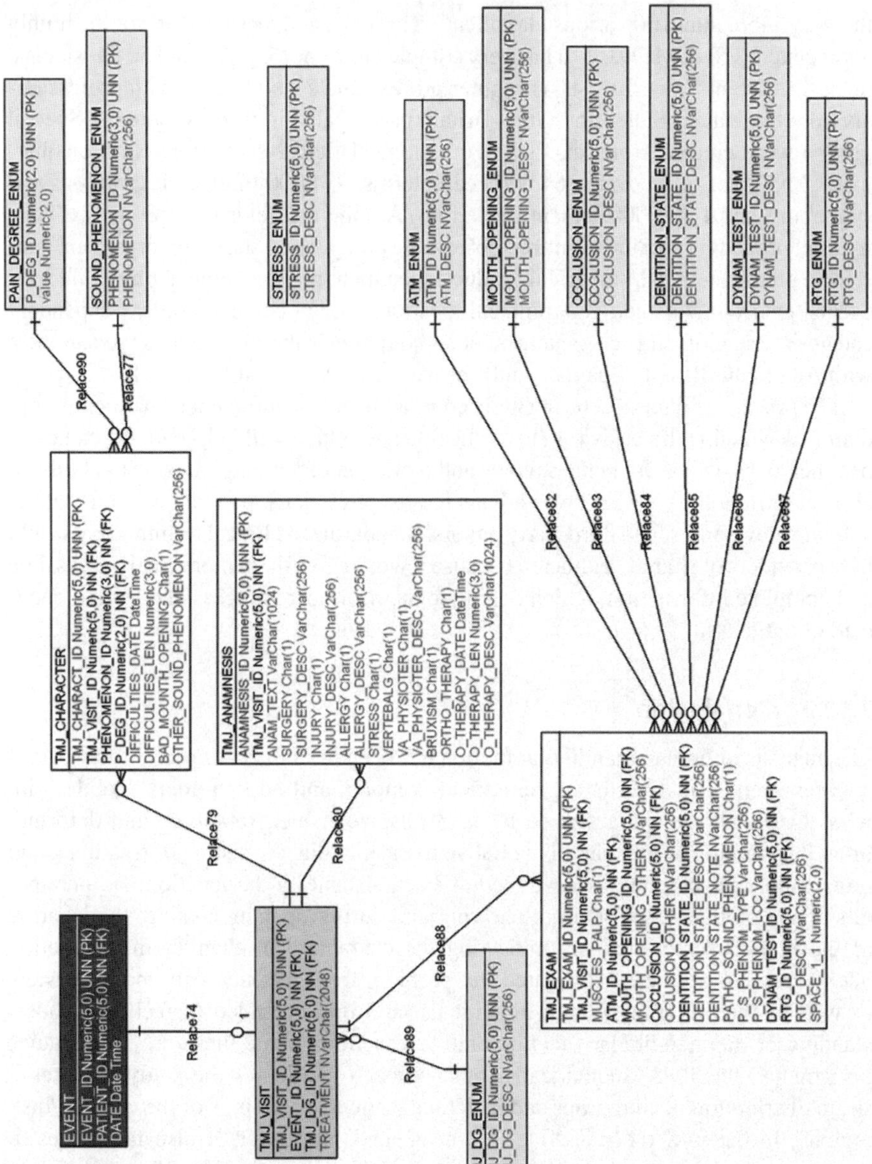

FIG. 7.387 Automatic speech recognition module. [After Daněk et al. (2010)].

international classification systems. Usage of these systems is a crucial step toward interoperability of heterogeneous healthcare records [see Nagy et al. (2008)]. The formation of classification systems has been motivated mostly by their practical usability in registration, sorting, and statistical processing of medical information. The first interest has been to register incidence of diseases and causes of deaths. Nowadays, there are more than 100 various classification systems and thesauri. One of the highly appreciated is SNOMED CT. It has very complex terminology. Around 50 physicians, nurses, assistants, pharmacists, computer professionals, and other health professionals from the United States and Great Britain participated in its development. Special groups were created from specific terminological fields, such as nursing or pharmacy. SNOMED CT covers 364,400 health terms, 984,000 English descriptions and synonyms, and 1,450,000 semantic relations. Among the fields in SNOMED CT are finding, procedure and intervention, observable entity, body structure, organism, substance, pharmaceutical/biological product, specimen, physical object, physical force, events, environments and geographical locations, social context, context-dependent categories, staging and scales, attribute, and qualifier value. Nowadays we can meet with American, British, Spanish, and German versions of SNOMED CT.

The process of data structuring and coding will lead to sufficient semantic interoperability, which is the basis for shared healthcare, which will help with efficiency in healthcare, as well as financial savings and reduction of patients' stress. Standardized clinical terminology would bring advantages to physicians, patients, administrators, software developers, and third-party payors. Standardized clinical terminology would help providers of shared healthcare because it would give them more easily accessible and complete information, which belong to the healthcare process and lead to better care of patients.

7.12.9 Conclusion

The analysis of human mandibular motion has been the subject of present-day extensive research. The TMJ is a geometrical complex and an extremely mobile joint whose motions are characterized by large displacements, rotations, and deformations. The motion of the mandible relative to the maxilla, consisting of rotational and translational movements, is described at every instant by the location and parameters of the motion screw. During chewing and during opening/closing, the location of the motion screw changes, reflecting the characteristic changes in the motion itself. Previous studies investigated the problem from the kinematic point of view as well as the variation among individuals, and they intended to facilitate understanding of the mandibular function and dysfunction. While these previous studies investigated the TMJ kinematically, our approach facillitates the study the stress–strain distribution in all regions and in details in the surroundings of the condyle/disk regions, in the jaw, the mandible, the head, and, moreover, it also facillitates to determine the acting contact forces in the condyle/disk areas. Therefore, the future main goal of our investigations leads to the development of optimal algorithms for the dynamic contact model problems based on the 3D dynamic primal dual active set strategy approach in the mathematical case. Mathematical modeling is essential to the subscribed biomechanical process of the TMJ. Very important for diagnostics and treatments is knowledge of forces acting in the TMJ. This information is necessary for the determination of locations of the highest loadings. Mathematical modeling completes our information about actions in the TMJ, and it can help to improve the treatment of TMJ disorders or to produce a replacement of the TMJ in the future.

7.12.10 Temporomandibular Joint Reconstruction—Clinical Case Reports

The temporomandibular disorder (TMD) is a collective term embracing a number of clinical problems that involve the masticatory musculature, the temporomandibular joint (TMJ), and associated structures. A typical TMD patient suffers from a pain/discomfort in the jaw, mainly in the region of the TMJ and/or muscles of mastication, and limitation of a mandibular function.

Many of these patients can be managed with nonsurgical therapies, but some end-stage TMJ patients require a surgical TMJ repair or reconstruction. The patient can develop the end-stage TMJ as a result of trauma, osteoarthritis, reactive arthritis, ankylosis, idiopathic condylar resorption, connective tissue/autoimmune diseases (e.g., rheumatoid arthritis, psoriatic arthritis, lupus, scleroderma, Sjogren's syndrome, ankylosing spondylitis) or other TMJ pathologies.

The joint structures can be destroyed following the use of alloplastic implants. A joint replacement is a surgical procedure in which the severely damaged part of the TMJ is removed and replaced with a prosthetic device. If either a condyle or a fossa component of the TMJ is replaced, the surgery is called a partial joint replacement (Fig. 7.388). In total joint replacement, the condyle and fossa are both replaced (see Figs 7.389 and 7.390).

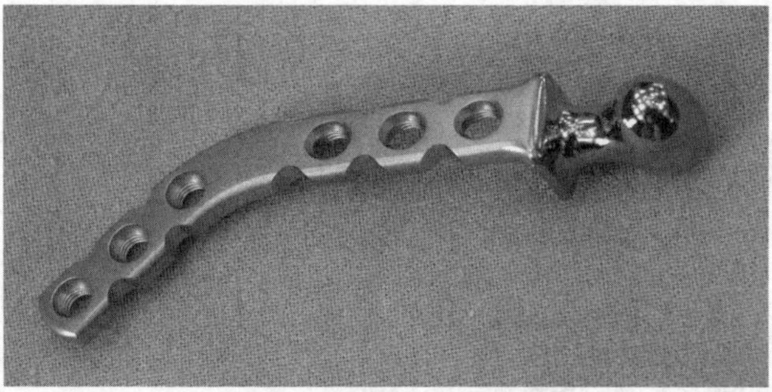

FIG. 7.388 Partial TMJ prosthesis.

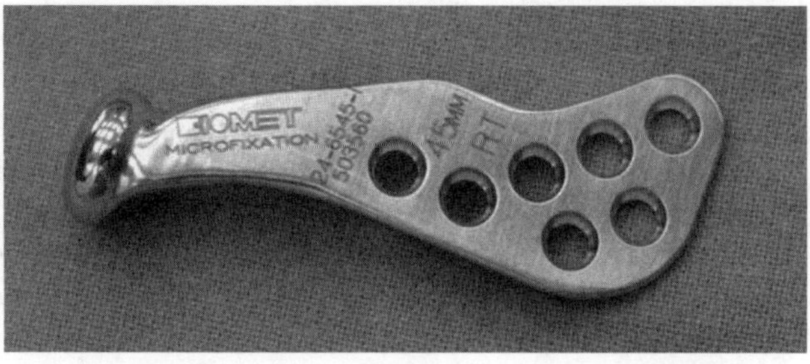

FIG. 7.389 Total TMJ prosthesis—condylar prostheses.

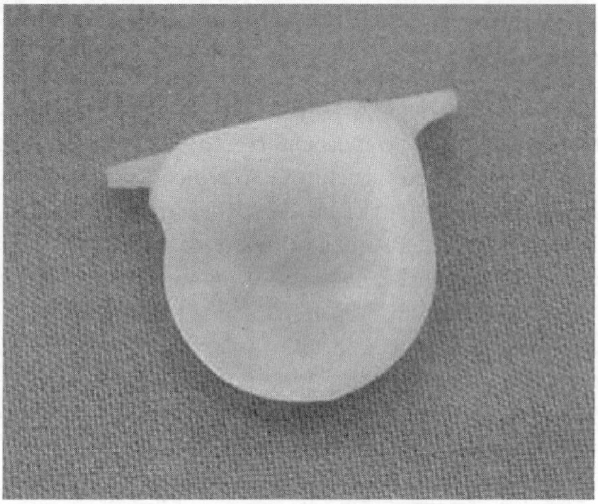

FIG. 7.390 Total TMJ prosthesis—fossa eminence.

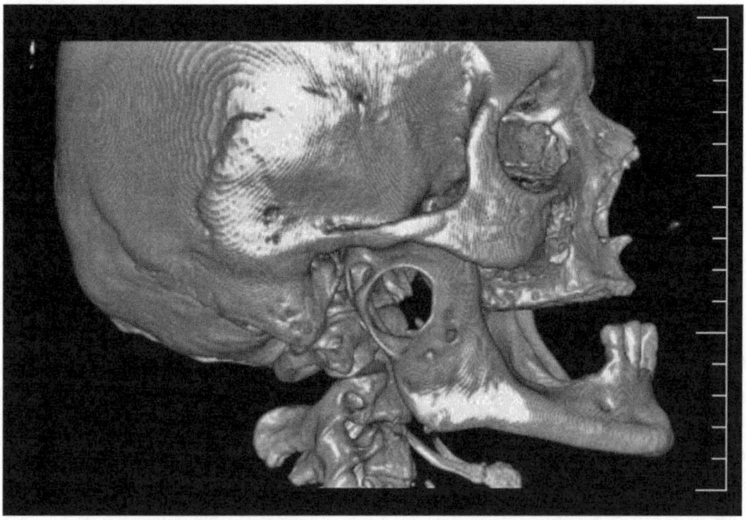

FIG. 7.391 Ramus mandibulae with a residual cyst.

The joint replacement can be demonstrated in two case reports. The first one was a 66-year-old woman with a residual cyst in the right ramus mandibulae (Fig. 7.391). After ramus mandibulae resection the partial TMJ prosthesis was inserted (Fig. 7.392). A 19-year-old woman had osteomyelitis in both temporomandibular joints after newborn sepsis (Fig. 7.393). Left and right total TMJ prostheses were used during the therapy (Figs. 7.394 and 7.395).

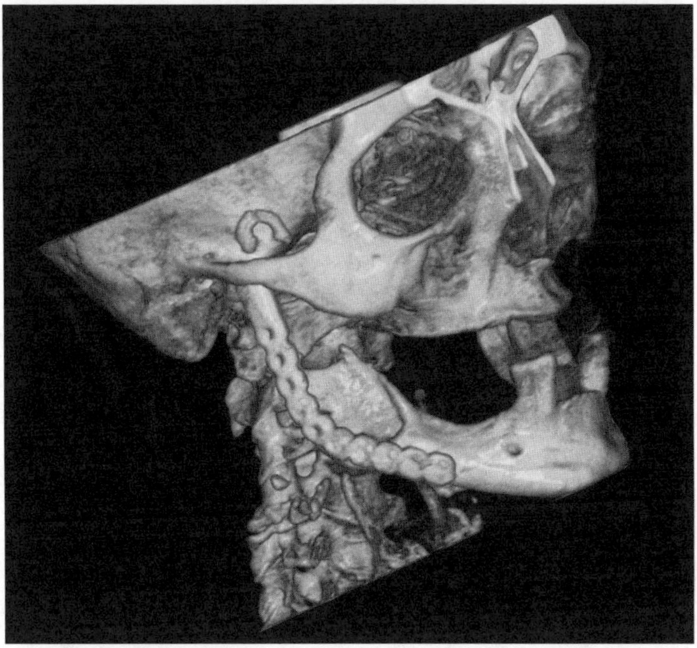

FIG. 7.392 Partial TMJ prosthesis 1 year after treatment.

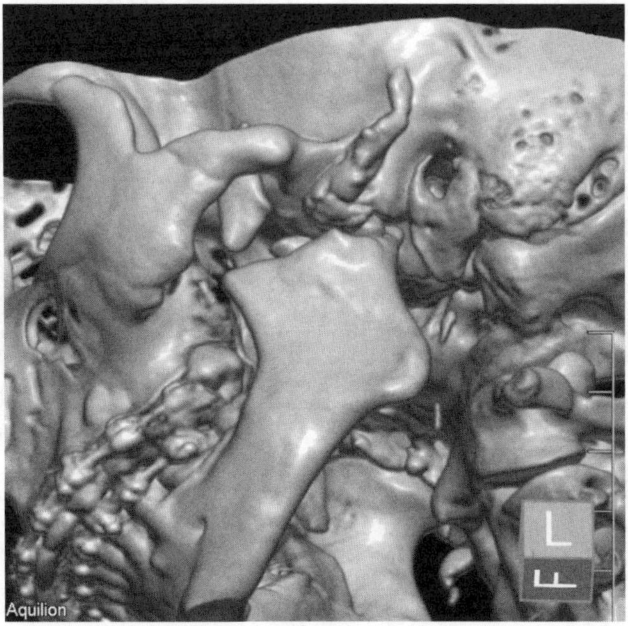

FIG. 7.393 Ankylosis of left TMJ before the therapy—CT (computerized tomography) reconstruction.

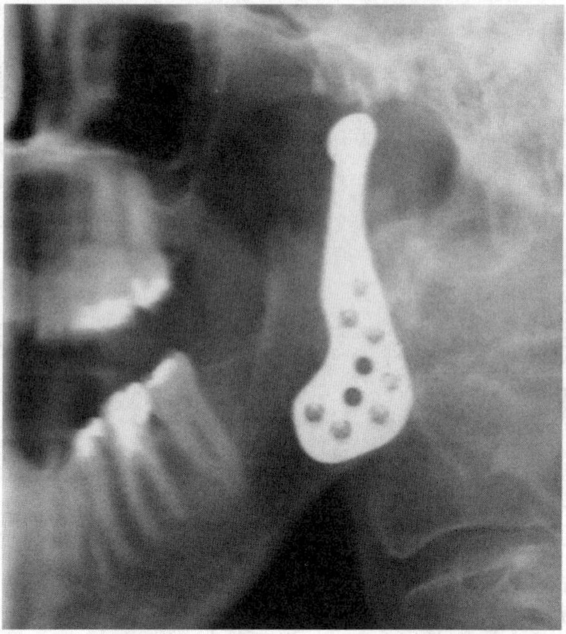

FIG. 7.394 Total TMJ prosthesis after treatment.

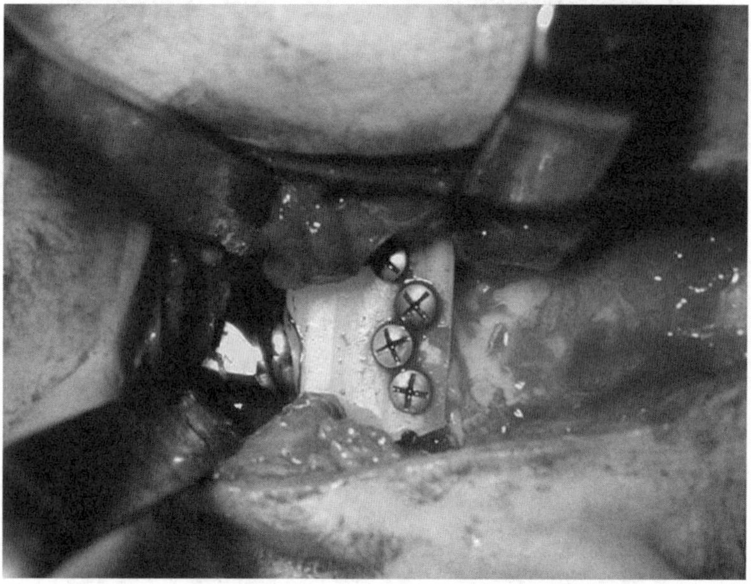

FIG. 7.395 Total TMJ prosthesis—state after surgical treatment.

APPENDIX

A.1 LIST OF NOTATIONS

(a,b)	open interval, i.e., the set of numbers $a < x < b$
$[a,b]$	closed interval, i.e., the set of numbers $a \le x \le b$
$I = [t_0, t_p]$	time interval, $0 \le t_0 < t \le t_p$
$\forall$	for all
$\exists$	there exists
a.e.	almost everywhere
$\Rightarrow$	implication, i.e., $A \Rightarrow B$ means that A implies B
$\Leftrightarrow$	equivalence, $A \Leftrightarrow B$ means that A is equivalent to B
$\prod_{s=1}^{m} (1 - X(s)) = (1 - X(1)) \times \cdots \times (1 - X(m))$,	where $X(s)$ is a non-negative sequence
N	positive integer, in applications having its value in $\{1, 2, 3\}$
c	generic positive constant
h	finite element mesh size
k	time step size
$r_+ = \max\{r, 0\}$	positive part of r
$A \cup B$	union of sets A and B
$A \cap B$	intersection of sets A and B
$A \backslash B$	difference of sets A and B
$A \times B$	Cartesian product of sets A and B
$\mathbb{R}^N$	N-dimensional vector space over the field of real numbers (Euclidean space)
$\mathbb{S}^N$	space of second-order symmetric tensors on $\mathbb{R}^N$
$\mathbb{N}$	set of positive integers
$\mathbf{x} = (x_1, \ldots, x_n)$	point in $\mathbb{R}^n$
$\lvert\mathbf{x} - \mathbf{y}\rvert = (\sum_{i=1}^{n} (x_i - y_i)^2)^{1/2}$	distance between two points $\mathbf{x}, \mathbf{y} \in \mathbb{R}^N$
$\mathbf{u} \in V$ or K	vector $\mathbf{u}$ belongs to space V or to set K

$(\mathbf{x}, \mathbf{y})$	inner product of vectors $\mathbf{x} \in \mathbb{R}^n$ and $\mathbf{y} \in \mathbb{R}^n$, $(\mathbf{x}, \mathbf{y}) = \sum_{i=1}^{n} x_i y_i$
Ω	domain in $\mathbb{R}^N$, i.e., an open, bounded, and connected set in $\mathbb{R}^N$ with the Lipschitz boundary
$\partial\Omega, \Gamma_i$	boundary of domain Ω and its parts, $i = 0, u, \tau, c$
Γ_c	contact boundary
$\overline{\Omega} = \Omega \cup \partial\Omega$	closure of set (domain) Ω
diam Ω	diameter of set Ω
supp f	support of function f, i.e., the closure of the set of those points $\mathbf{x}$ at which $f(\mathbf{x}) \neq 0$
$\text{meas}_N\, M$	N-dimensional Lebesgue measure of set M
$f : x \mapsto f(x)$	function correspondence, i.e., image of element x is element $f(x)$
$f : \Omega \to \mathbb{R}$	mapping f from the set (domain) Ω into set $\mathbb{R}$
$\mathrm{x}_n \to x$	strong convergence of the sequence $\{\mathrm{x}_n\}$
$\mathrm{x}_n \rightharpoonup x$	weak convergence of the sequence $\{\mathrm{x}_n\}$
$\mathbf{n} = (n_i)$	unit outward normal vector on boundary $\partial\Omega$
$\mathbf{t} = (t_i)$	unit tangential vector on boundary $\partial\Omega$
$\mathbf{u} = (u_i), u_n, \mathbf{u}_t$	displacement vector and its normal and tangential components
$\boldsymbol{\varepsilon}$	linearized or small deformations tensor, i.e., $\boldsymbol{\varepsilon}(\mathbf{u}) = (\varepsilon_{ij}(\mathbf{u}))$
τ_{ij}	stress tensor
$\boldsymbol{\tau} = (\tau_i),\ \tau_i = \tau_{ij} n_j,$	stress vector and its Cartesian components
$\tau_n = \tau_i n_i,\ \boldsymbol{\tau}_t = \boldsymbol{\tau} - \tau_n \mathbf{n}$	normal and tangential components of the stress vector
∇	gradient operator
Δ	Laplace operator
$\mathbf{I}$	identity operator
$L^p(\Omega)$	Lebesgue space of p-integrable functions
$H^k(\Omega) \equiv W^{k,2}(\Omega)$	Sobolev space of functions whose weak derivatives of orders less than or equal to k are p integrable in Ω
$H^{1/2}(\Gamma)$	Sobolev space on Γ ($\Gamma \subset \partial\Omega$), defined as the range of the trace operator on $H^1(\Omega)$
$H^{-1/2}(\Gamma)$	dual of $H^{1/2}(\Gamma)$
$\prod_{\iota=1}^{s} H^1(\Omega^\iota) = H^1(\Omega^1) \times H^1(\Omega^2) \times \cdots \times H^1(\Omega^s).$	Sobolev space of functions defined on the domain $\Omega = \cup_{\iota=1}^{s} \Omega^\iota$
$\langle \cdot, \cdot \rangle_\Gamma$	duality pairing between $H^{1/2}(\Gamma)$ and $H^{-1/2}(\Gamma)$
$L^p(I; X)$	$\{v : I \to X \text{ measurable: } \|v\|_{L^p(I;X)} < \infty\}$
$H^k(I; X) \equiv W^{k,2}(I; X)$	$\{v \in L^2(I; X) \ :\ D^\alpha v \in L^2(\Omega)$ for each nonnegative multi-index α such that $|\alpha| \leq k\}$
V	$\{\mathbf{v} \in [H^1\Omega]^N; \mathbf{v} = 0 \text{ on } \Gamma_u\}$
V'	dual space of V
V_h	finite element space approximating V

$K = \{\mathbf{v} \in V \mid v_n \leq 0$ on $\Gamma_c\}$, K_h	finite element approximation of K
δ_{ij}	Kronecker delta
$O(h)$	of order h, i.e., there exists a constant $c > 0$, independent of h, such that $\lvert O(h)\rvert \leq ch$
$\delta u_n = \frac{1}{k}(u_n - u_{n-1})$	backward divided difference
ψ_K	indicator function of set K
$d\mathbf{x}$	N-dimensional volume element
ds	$(N-1)$-dimensional surface area element

A.2 CARTESIAN TENSORS

Orthogonal Transformation of Coordinates

Let us assume two- or three-dimensional Euclidean spaces in $\mathbb{R}^N$, $N = 2, 3$, with two Cartesian coordinate systems. Let an orthogonal transformation from $\mathbb{R}^N \to \mathbb{R}^N$ be given. Let coordinates of an arbitrary point P in the first system be denoted as x_i, $i = 1, \ldots, N$, and in the second one by x_i', $i = 1, \ldots, N$. The relation between these two sets of components is

$$x_i' = a_{ij}x_j + c_i', \tag{A.1}$$

where the matrix (a_{ij}) is orthogonal, that is,

$$a_{ki}a_{kj} = \sum_{k=1}^{N} a_{ki}a_{kj} = \delta_{ij}, \qquad i,j = 1, \ldots, N, \tag{A.2}$$

where δ_{ij} is the so-called Kronecker delta, $\delta_{ij} = 1$ for $i = j$ and $\delta_{ij} = 0$ for $i \neq j$, and where the Einstein summation convention (about the repeated indices) is used and c_i' are coordinates of the origin O. For the inverse transformation of (A.1) we find

$$x_k = a_{ik}x_i' + c_k, \tag{A.3}$$

where $c_k = -a_{ik}c_i'$. The relation (A.1) represents the transition from one Cartesian system to another.

Vectors and Tensors

An ordered N-tuple $A = (A_i)$, satisfying transformation (A.1) the relations $A_i' = a_{ij}A_j$, is called a tensor of the first order or the vector. A zero-ordered tensor is a scalar. A matrix $\mathbf{T} = (T_{ij})$, $i, j = 1, \ldots, N$, we call a tensor of order 2, if under transformation (A.1) the relations

$$T_{ij}' = a_{ik}a_{jl}T_{kl} \tag{A.4}$$

are satisfied. A tensor of order 2 is symmetric if $T_{ij} = T_{ji}$ and is skew-symmetric if $T_{ij} = -T_{ji}$. A tensor of order 2 can be divided into its symmetric and skew-symmetric parts, that is, $T_{ij} = \frac{1}{2}(T_{ij} + T_{ji}) + \frac{1}{2}(T_{ij} - T_{ji})$ holds. The symmetric properties under the transformation are not changed. The tensor of the second order δ_{ij}, for which it holds $\delta_{ij} = 1$ if $i = j$ and $\delta_{ij} = 0$ if $i \neq j$ is the so-called Kronecker delta.

A tensor of order n ($n = 1, 2, \ldots$) is a mapping from the set of all rectangular Cartesian coordinate systems into the set of n-dimensional matrices $\mathbf{T} = (T_{i_1 \ldots i_n})$, $i_k = 1, \ldots, N$; $\forall k = 1, \ldots, n$, which under transformation (A.1) satisfy the relations

$$T'_{i_1 \ldots i_n} = a_{i_1 j_1} \cdots a_{i_n j_n} T_{j_1 \ldots j_n}. \tag{A.5}$$

Such tensors can be also symmetric or skew-symmetric. For example, the Levi–Cività tensor ε_{ijk} is a tensor of order 3 in $\mathbb{R}^3$, and it is the skew-symmetric tensor such that transformations between coordinate systems are restricted to proper rotations, that is,

$$\varepsilon'_{ijk} = a_{jl} a_{jm} a_{kn} \varepsilon_{lmn},$$

and, moreover, $\varepsilon_{ijk} = 0$ if two of the integers i, j, k are equal or if all three are equal, $\varepsilon_{ijk} = -1$ if (ijk) is an even permutation of (123) and $\varepsilon_{ijk} = -1$ if (ijk) is an odd permutation of (123).

The determinant $|A|$ of a three-by-three matrix $A = (A_{ij})$ is given by $|A| = \varepsilon_{ijk} A_{1i} A_{2j} A_{3k}$. For tensors the algebraic sum operation is valid for the case if tensors are of the same orders, and moreover, the algebraic operations of multiplying and contraction are also valid. A contraction of an nth order tensor, $n \geq 2$, is formed by setting two of its indices equal and performing the resulting sum. Generally, the order of tensors is determined by the number of free indices. For more details about tensors see Rektorys et al. (1995).

Principal Axes and Invariants of a Symmetric Tensor of the Second Order

Let T_{ij} be a symmetric tensor of the second order in $\mathbb{R}^3$ describing the property of a certain physical quantity. For the tensor T_{ij} we can define a quadratic surface, the so-called quadric. Let us assume that it is represented by a simple surface. Then we can find the coordinate system such that the coordinate axes are identical with the principal axes of the quadric, that is, we can find a coordinate system such that in the quadric only quadratic terms of coordinates can occur. Then we can write

$$T_{ij} x_i x_j = \pm 1, \tag{A.6}$$

or

$$F(x_1, x_2, x_3) = T_{ij} x_i x_j \mp 1 = 0. \tag{A.7}$$

From differential geometry it follows that the components of the normal vector to the surface, defined by $F(x_1, x_2, x_3) = 0$, are proportional to the partial derivatives $\partial F / \partial x_i$, $i = 1, \ldots, 3$. We find a direction of the normal $\mathbf{n} = (n_i)$, $i - 1, 2, 3$, in such a way that the tangential component will be equal to zero, that is, $T_{ij}n_j = Tn_i$, $i,j = 1,2,3$. The necessary condition for the existence of the nonzero solution is

$$\det(T_{ij} - T\delta_{ij}) = 0, \tag{A.8}$$

or equivalently

$$-T^3 + \theta_1 T^2 - \theta_2 T + \theta_3 = 0. \tag{A.9}$$

This equation is the so-called characteristic equation with three real solutions T_i, $i = 1, \ldots, 3$, independent of the coordinate system, where the coefficients θ_i, $i = 1, \ldots, 3$, satisfy

$$\theta_1 = T_1 + T_2 + T_3 = T_{11} + T_{22} + T_{33} = T_{ii}, \tag{A.10}$$

$$\begin{aligned} \theta_2 &= T_1T_2 + T_2T_3 + T_3T_1 \\ &= \begin{vmatrix} T_{11} & T_{12} \\ T_{12} & T_{22} \end{vmatrix} + \begin{vmatrix} T_{22} & T_{23} \\ T_{23} & T_{33} \end{vmatrix} + \begin{vmatrix} T_{11} & T_{13} \\ T_{13} & T_{33} \end{vmatrix}, \end{aligned} \tag{A.11}$$

$$\theta_3 = T_1T_2T_3 = \begin{vmatrix} T_{11} & T_{12} & T_{13} \\ T_{12} & T_{22} & T_{23} \\ T_{13} & T_{23} & T_{33} \end{vmatrix}. \tag{A.12}$$

The values of coefficients θ_i, $i = 1, \ldots, 3$, are independent of the choice of the coordinate system, and, therefore, they are invariants of the symmetric tensor T_{ij}.

A.3 SOME FUNDAMENTAL THEOREMS

Theorem A.1 (Green) Let Ω be a bounded domain with the Lipschitz boundary $\partial\Omega$ and let $f, g \in H^1(\overline{\Omega})$. Then

$$\int_\Omega \frac{\partial f}{\partial x_i} g \, d\mathbf{x} = \int_{\partial\Omega} fgn_i \, ds - \int_\Omega f \frac{\partial g}{\partial x_i} d\mathbf{x}, \tag{A.13}$$

where n_i is the ith component of the unit outward normal to the boundary $\partial\Omega$.

Lemma A.1 (Fatou's Lemma) Let $\{f_i\}_{i=1}^{\infty}$ be a sequence of measurable functions that are nonnegative almost everywhere on Ω. Then $\liminf_{i\to\infty} f_i(x)$ is integrable and

$$\int_\Omega \liminf_{i\to\infty} f_i(x) \, dx \le \liminf_{i\to\infty} \int_\Omega f_i(x) \, dx.$$

Theorem A.2 (Fubini's Theorem) Let $\Omega_i \subset \mathbb{R}^N$, $i = 1,2$, be measurable and put $\Omega = \Omega_1 \cup \Omega_2$. Let $f(x, y)$ be integrable over Ω. Then for almost all $x \in \Omega_1$ and $y \in \Omega_2$ the integrals

$$\int_{\Omega_1} f(x, y) \, dx \quad \text{and} \quad \int_{\Omega_2} f(x, y) \, dy$$

exist. Moreover,

$$\int_{\Omega} f(x,y)\,dx\,dy = \int_{\Omega_1}\left(\int_{\Omega_2} f(x,y)\,dy\right)dx = \int_{\Omega_2}\left(\int_{\Omega_1} f(x,y)\,dx\right)dy.$$

Theorem A.3 (Trace Theorem) There exists a continuous linear mapping $\mathcal{Z}: H^1(\Omega) \to L^2(\partial\Omega)$, such that $\mathcal{Z}u = u$ for $u \in C^1(\overline{\Omega})$. $\mathcal{Z}u$ is the so-called trace of the function u.

For the proof see Nečas (1967) and Nečas and Hlaváček (1981).

Theorem A.4 (Reynolds Transport Theorem) For any sufficiently smooth function f and the material region $G(t)$

$$\frac{d}{dt}\int_{G(t)} f\,d\mathbf{x} = \int_{G(t)} \frac{\partial f}{\partial t}d\mathbf{x} + \int_{\partial G(t)} f\mathbf{v}\cdot\mathbf{n}\,ds \tag{A.14}$$

holds, where $\mathbf{n}$ is the unit exterior normal to G, $d\mathbf{x}$ is a volume element, $\mathbf{v}$ is a velocity in spatial coordinates, ds is a surface element, and ∂G denotes the boundary of G.

Lemma A.2 (Dubois–Raymond) Suppose that

$$\int_{G} f(\mathbf{x})\,d\mathbf{x} = 0 \tag{A.15}$$

for every region G contained in a domain Ω. If $f(\mathbf{x})$ is continuous for $\mathbf{x}$ in Ω, then $f(\mathbf{x}) \equiv 0$ for $\mathbf{x}$ in Ω.

A.4 ELEMENTARY INEQUALITIES

In some proofs the following inequalities are used.

(A1) Let $a, b \in \mathbb{R}^1$, $\varepsilon > 0$. Then $ab \le \frac{1}{2}\varepsilon a^2 + (2\varepsilon)^{-1}b^2$.

(A2) Let $a, b \ge 0$, and $x^2 \le ax + b$. Then $x^2 \le a^2 + 2b$.

(A3) *Young's inequality:* Let f be a continuous real, strictly increasing function defined on $[0,\infty)$ such that

$$\lim_{u\to\infty} f(u) = \infty \quad \text{and} \quad f(0) = 0.$$

Let $g = f^{-1}$. For all $x \in [0,\infty)$, define $\varphi(x) = \int_0^x f(u)\,du$, $\psi(x) = \int_0^x g(v)\,dv$. Then for all $a, b \in [0,\infty)$ we have $ab \le \varphi(a) + \psi(b)$ and the equality occurs if and only if $b = \varphi(a)$.

Corollary 1 Given $p > 1$ and nonnegative real numbers a and b, we have

$$ab \leq \frac{a^p}{p} + \frac{b^{p'}}{p'}, \quad \text{where } \frac{1}{p} + \frac{1}{p'} = 1, \quad \text{i.e., } p' = \frac{p}{p-1}.$$

The equality holds if and only if $a^p = b^{p'}$.

(A4) *Hölder's inequality:* Let $f \in L^p(\Omega)$ and $g \in L^{p'}(\Omega)$, $p > 1$. Then $f \cdot g \in L^1(\Omega)$ and

$$\left| \int_\Omega f(x)g(x)\,dx \right| \leq \int_\Omega |f(x)g(x)\,dx| \leq \|f\|_p \|g\|_p.$$

(A5) *Gronwall lemma*

(i) Continuous version: Let $f \in L^1(t_0, t_1)$ be a nonnegative function, and g and φ be continuous functions on $\langle t_0, t_1 \rangle$. If φ satisfies

$$\varphi(t) \leq g(t) + \int_{t_0}^t f(\tau)\varphi(\tau)\,d\tau \quad \forall t \in \langle t_0, t_1 \rangle,$$

then

$$\varphi(t) \leq g(t) + \int_{t_0}^t f(s)g(s) \exp\left(\int_s^t f(\tau)\,d\tau \right) ds \quad \forall t \in \langle t_0, t_1 \rangle.$$

Moreover if g is nondecreasing, then

$$\varphi(t) \leq g(t) \exp\left(\int_{t_0}^t f(\tau)\,d\tau \right) \quad \forall t \in \langle t_0, t_1 \rangle.$$

(ii) Discrete version: Assume that k_n is a nonnegative sequence, and that the sequence φ_n satisfies

$$\varphi_0 \leq g_0,$$
$$\varphi_n \leq g_0 + \sum_{i=0}^{n-1} p_i + \sum_{i=0}^{n-1} k_i \varphi_i, \quad n \geq 1.$$

Then φ_n satisfies

$$\varphi_1 \leq g_0(1 + k_0) + p_0,$$
$$\varphi_n \leq g_0 \prod_{i=0}^{n-1}(1 + k_i) + \sum_{i=0}^{n-2} p_i \prod_{j=i+1}^{n-1}(1 + k_j) + p_{n-1}, \quad n \geq 2.$$

Moreover, if $g_0 \geq 0$ and $p_n \geq 0$ for $n \geq 0$, then it follows that

$$\varphi_n \leq \left(g_0 + \sum_{i=0}^{n-1} p_i\right) \exp\left(\sum_{i=0}^{n-1} k_i\right), \quad n \geq 1.$$

[For the proof see Quarteroni and Valli (1994) and Lions and Magenes (1972).]

A.5 FINITE ELEMENT METHOD

Variational Formulation of the Problem and Its Solution

Let V be a real Hilbert space with a scalar product $(\cdot, \cdot)$ and a norm $\|\cdot\| = (\cdot, \cdot)^{1/2}$ and let $\mathcal{U} \subseteq V$ be a nonempty, convex, and bounded subset. Let V' be a dual space to the space V, that is, the space of continuous linear functionals on V. Let $a : \mathcal{U} \times \mathcal{U} \to \mathbb{R}$ be a bilinear form and let $f \in V'$, then the norm of the functional f is defined by

$$\|f\|_* = \sup_{v \neq 0, v \in V} \frac{|\langle f, v \rangle|}{\|v\|},$$

where by $\langle f, v \rangle$ we denote the value of f at point v.

Definition A.1 A bilinear form $a : \mathcal{U} \times \mathcal{U} \to \mathbb{R}$ is symmetric on $\mathcal{U}$ if

$$a(u, v) = a(v, u) \quad \forall u, v \in \mathcal{U}.$$

A bilinear form $a : \mathcal{U} \times \mathcal{U} \to \mathbb{R}$ is bounded on $\mathcal{U}$ if there exists a constant $c > 0$ such that

$$|a(u, v)| \leq c\|u\| \, \|v\| \quad \forall u, v \in \mathcal{U}.$$

A bilinear form $a : \mathcal{U} \times \mathcal{U} \to \mathbb{R}$ is V elliptic on $\mathcal{U}$ if there exists a constant $c_0 > 0$ such that

$$a(v, v) \geq c_0 \|v\|^2 \quad \forall v \in \mathcal{U}.$$

Definition A.2 Let $\mathcal{U} \subseteq V$ be a nonempty, convex, and bounded subset, let $a : \mathcal{U} \times \mathcal{U} \to \mathbb{R}$ be a V-elliptic bilinear form, and $f \in V'$. Then $u \in \mathcal{U}$ is a solution of elliptic variational inequality if

$$a(u, v - u) \geq \langle f, v - u \rangle \quad \forall v \in \mathcal{U}. \tag{A.16}$$

If $\mathcal{U} = V$, then (A.16) leads to the following form:

$$a(u, v) = \langle f, v \rangle \quad \forall v \in V, \tag{A.17}$$

and we speak about a variational equation.

The existence and uniqueness of a variational equation follows from the Lax–Milgram theorem.

Theorem A.5 (Lax–Milgram) Let $\mathcal{U} = V$ and let a bilinear form $a : V \times V \to \mathbb{R}$ be bounded and V elliptic on V and $f \in V'$. Then problem (A.17) has a unique u and

$$\|u\| \leq c_0^{-1} \|f\|_*. \tag{A.18}$$

For the proof see Ciarlet (1978).

Let us assume that a bilinear form $a : V \times V \to \mathbb{R}$ is symmetric on V, then problem (A.17) can be rewritten in the following form:

$$\text{Find } u \in V \text{ such that } J(u) \leq J(v) \quad \forall v \in V, \tag{A.19}$$

where $J : V \to \mathbb{R}$ is a quadratic functional defined by

$$J(v) = \tfrac{1}{2} a(v, v) - \langle f, v \rangle. \tag{A.20}$$

Theorem A.6 Let a bilinear form $a : V \times V \to \mathbb{R}$ be symmetric, bounded, and V-elliptic on V. Then there exists a unique solution $u^* \in V$ of problem (A.19). Moreover, $u^* \in V$ solves (A.19) if and only if it solves problem (A.17).

Proof Let $u^* \in V$ be a solution of (A.19). Let us put $\varphi(t) = J(u^* + tv)$. In order to attain the function $\varphi(t)$ for $t = 0$ its minimum and then its derivative at point 0 must be equal to zero, that is, $\varphi'(0) = 0$. Then

$$\varphi'(0) = \lim_{t \to 0} \frac{\varphi(t) - \varphi(0)}{t} = a(u^*, v) - \langle f, v \rangle,$$

and thus u^* solves (A.17).

On the contrary let us assume that $u^* \in V$ is a solution of problem (A.17). Then

$$\begin{aligned} J(v) - J(u^*) &= \tfrac{1}{2} a(v, v) - \langle f, v \rangle - \tfrac{1}{2} a(u^*, u^*) + \langle f, u^* \rangle \\ &= a(u^*, v - u^*) - \langle f, v - u^* \rangle + \tfrac{1}{2} a(v - u^*, v - u^*) \\ &= \tfrac{1}{2} a(v - u^*, v - u^*) \geq \tfrac{1}{2} c \|v - u^*\|^2 \geq 0. \end{aligned}$$

Then the functional J attains on V its minimum at point $u^* \in V$ and thus u^* is a solution of problem (A.19). The existence and uniqueness follow from the equivalence of problems (A.17) and (A.19) and the Lax–Milgram theorem. □

The next theorem gives information about the solution of the problem if function f changes.

Theorem A.7 Let $f_1, f_2 \in V'$ and $u_1, u_2 \in V$ be two solutions of problem (A.17), that is,

$$a(u_i, v) = \langle f_i, v \rangle \quad \forall v \in V, \quad i = 1, 2. \tag{A.21}$$

Then

$$\|u_1 - u_2\| \leq c_0^{-1} \|f_1 - f_2\|_*. \tag{A.22}$$

Proof The proof follows from (A.18) setting $u = u_1 - u_2$ and $f = f_1 - f_2$. □

We see that the solution u depends continuously on the right-hand side, that is, if f in the right-hand side changes a little, then solution u changes also a little.

The above-defined problems can be also formulated as dual problems, which means that we find, for instance, the gradient of u if the second-order elliptic problems are considered; see, for example, Glowinski, Lions, and Trémolières (1976, 1981) and Hlaváček et al. (1988).

Finite Element Approximation

Let us assume that the set S is a subset of V and assume that S is a nonempty and finite-dimensional subspace. Let $\mathbf{u}$ and $\mathbf{u}_S$ be unique solutions of the problems

$$\text{Find } u \in V \text{ such that } a(u, v) = \langle f, v \rangle \quad \forall v \in V, \tag{A.23}$$

$$\text{Find } u_S \in S \text{ such that } a(u_S, v) = \langle f, v_s \rangle \quad \forall v \in S. \tag{A.24}$$

Let us study solutions of both problems (A.23) and (A.24), that is, a behavior of the norm $\|u - u_S\|$.

Theorem A.8 Let a bilinear form $a : V \times V \to \mathbb{R}$ be V-elliptic and bounded on the space V. Then

$$\|u - u_S\| \leq \frac{c}{c_0} \inf_{v \in S} \|u - v\|. \tag{A.25}$$

For the proof see Axelsson and Baker (1984).

Let $\{e_i\}, i = 1, \ldots, n$, be a basis of the subspace S. Then a finding solution u_S can be written as a linear combination of a basis of the space V, that is,

$$u_S = \sum_{i=1}^{n} d_i e_i, \quad d_i \in R, \tag{A.26}$$

where d_i are unknown coefficients. Then substituting (A.26) into (A.24) we obtain the system of algebraic equations of the form

$$\sum_{j=1}^{n} d_j a(e_j, e_i) = \langle f, e_i \rangle, \quad i = 1, \ldots, n, \tag{A.27}$$

or of the form

$$B\mathbf{d} = \mathbf{F}, \tag{A.28}$$

where $B = (b_{ij})$, $i, j = 1, \ldots, n$, $b_{ij} = a(e_j, e_i)$, $\mathbf{d} = (d_i)$, $\mathbf{F} = (F_i)$, $F_i = \langle f, e_i \rangle$. Since the bilinear form $a : V \times V \to R$ is$V$–elliptic, the matrix B is regular. Moreover, if the bilinear form a is symmetric, the matrix of the system B is also symmetric. The solution of the problem then leads to a solution of the system of linear algebraic equations for unknown coefficients of the linear combination (A.26).

The finite element method is based on the above-defined idea. Let $0 < h < 1$. For every h we adjoin a finite-dimensional subspace S_h of a dimension $\dim S_h = n(h)$, $n(h) \to \infty$ if $h \to 0_+$.

Let $u_h \in S_h$ be a solution of the problem:

$$\text{Find } u_h \in S_h \text{ such that } a(u_h, v) = \langle f, v \rangle \quad \forall v \in S_h. \tag{A.29}$$

Then a problem of this type leads to a solution of a system of linear algebraic equations. This method is known as the **Galerkin** or **finite element method**. If the bilinear form $a : V \times V \to \mathbb{R}$ is symmetric, problem (A.29) is equivalent to

$$\text{Find } u_h \in S_h \text{ such that } J(u_h) \leq J(v) \quad \forall v \in S_h. \tag{A.30}$$

We see that the minimum of the functional J, instead on the whole space V, we find only on its subspace S_h. This method is known as the **Ritz method**. Generally, we speak about the **Ritz–Galerkin method.**

If $\|u - u_h\| \to 0$ as $h \to 0_+$, then we say that the finite element method converges [see Ciarlet (1978) and Axelsson and Barker (1984)].

If the Lax–Milgram theorem is applicable and the bilinear form a is symmetric, then the Ritz–Galerkin method leads to minimize the energy functional J over an n-dimensional subspace of V. This leads directly to the problem of minimizing a quadratic functional with a positive definite Hessian over the vector space $\mathbb{R}^n$. For such problems the direct and iterative methods can be used. Construction of the primitive stiffness matrix, the matrix of the system of linear algebraic equations B, will depend on the choice of the basis functions $\{e_i\}$, $i = 1, \ldots, n(h)$. By a suitable choice of basis functions, the support of each individual basis function is small, the bilinear form satisfies $a(e_i, e_j) = 0$ if the indices i and j of enumeration refer to nodes that are enough mutually far. More precisely $a(e_i, e_j) = 0$ if $\text{meas}_N(\text{ supp } e_i \cap \text{ supp } e_j) = 0$ for indices i and j of the used triangulation and enumeration. Then we obtain a very sparse matrix B, that is, the matrix with a great number of zero elements and, moreover, with the strip and the block diagonal structures.

The finite element system (A.28) can be solved by the direct methods such as the Gausian elimination method and its modifications as well as by the iterative methods such as the Jacobi, Gauss–Seidel, and relaxation (successive overrelaxation (SOR) method and symmetric successive overrelaxation (SSOR) method) methods as well as the conjugate gradient method without and with preconditioning. For more details see Varga (1962), Young (1971), Axelsson and Baker (1984), Golub and Van Loan (1989), Axelsson (1994), Quarteroni and Valli (1994), and Nedoma (1998b).

REFERENCES

The list of presented references obtains the used references in the book as well as the references close connected with the problems discussed in the book.

Adams, R.A. (1975). *Sobolev Spaces*. Academic Press, New York.

Adams, R.A., and Fournier, J.J.F. (2003). *Sobolev Spaces*, 2nd ed. Academic Press/Elsevier, Amsterdam.

Adams, M.A., Bogduk, N., Burton, K., and Dolan, P. (2002). *The Biomechanics of Back Pain*. Churchill Livingstone, Edinburgh, London, New York.

Adomian, G. (1983). *Stochastic Systems*. Academic Press.

Agur, A.M.R. (1991). *Grant's Atlas of Anatomy*, 9th ed. Williams & Wilkins, Baltimore.

Agur, A.M.R., Lee, M., and Grand, J.C. (1991). *Atlas of Anatomy*, 9th ed. Lippincott Williams & Wilkins, Philadelphia.

Aliabadi, M.H. (2002). *The Boundary Element Method. Applications in Solids and Structures*, Vol. 2. Wiley, West Sussex, England.

Aliabadi, M.H., and Brebbia, C.A. (1993). *Contact Mechanics, Computational Techniques*. Computational Mechanics Publications, Southampton, Boston.

Allard, P., Cappozzo, A., Lundberg, A., and Vanghan, C. (Eds) (1998). *Three-Dimensional Analysis of Human Locomotion*. Wiley, London.

Allard, A., Stokes, I.A.F., and Blanchi, J.P. (Eds) (1995). *Three-Dimensional Analysis of Human Movement*. Human Kinetics. Champaign, IL.

Amontons, G. (1699). On the Resistence Originating in Machines (in French). *Mem. Acad. Roy.*, 296–222.

An, Y.H., and Draughn, R.A. (Eds) (2000). *Mechanical Testing of Bone and the Bone-Implant Interface*. CRC Press, Boca Raton, FL.

An, K.N., Huy, F.C., Morray, B.F., Linscheid, R. L., and Chao, E. Y. (1981). Muscles Across the Elbow Joint: A Biomechanical Analysis. *J. Biomech.*, 14(10), 659–669.

Anderson, F.C., and Pandy, M.G. (2001). Dynamic Optimization of Human Walking. *J. Biomech. Eng.*, 123, 381–390.

Anderson, F.C., Ziegler, J.M., Pandy, M.G., and Whalen, R.T. (1995). Application of High-Performance Computing to Numerical Simulation of Human Movement. *J. Biomech. Eng.*, 117, 155–157.

Andersson, E. (2000). Existence Results for Quasi-Static Contact Problem with Coulomb Friction. *Appl. Math. Optim.*, 42, 169–202.

Archard, J.F. (1953). Contact and Rubbing of Flat Surfaces. *J. Appl. Phys.*, 24, 981.

Archad, J.F., and Hirst, W. (1957). An Examination of a Mild Wear Process. *Proc. Roy. Soc. London*, 515–528.

Atkinson K., and Han, W. (2001). *Theoretical Numerical Analysis. A Functional Analysis Framework*. Springer, New York.

Axelsson, O. (1994). *Iterative Solution Methods*. Cambridge University Press, Cambridge, New York, NY.

Axelsson, O., and Barker, V.A. (1984). *Finite Element Solution of Boundary Value Problems*. Academic, Orlando, FL.

Babuška, I., and Reinboldt, W.C. (1980). Reliable Error Estimation and Mesh Adaptation for the Finite Element Method. In: Oden, J.T. (Ed.). *Computational Methods in Nonlinear Mechanics*. North-Holland, Amsterdam.

Baillet, L., and Sassi, T. (2002). Méthodes d'eléments finis avec hybridisation frontière pour les problèmes de contact avec frottement. *C.R. Acad. Sci. Paris, Ser.I*, 334, 917–922.

Baillet, L., and Sassi, T. (2006). Mixed Finite Element Methods for the Signorini Problem with Friction. *Numer. Meth. Part. Diff. Eqs.*, vol.22, Issue 6, 1489–1508.

Baiocchi, C., and Brezzi, F. (1983). Optimal Error Estimates for Linear Parabolic Problems under Minimal Regularity Assumptions. *Calcolo*, 20, 143–176.

Baker, G.A., Bramble, J.H., and Thomeé, V. (1977). Single Step Galerkin Approximations for Parabolic Problems. *Math. Comput.*, 31, 818–847.

Banerjee, P.K., and Butterfield, R. (1981). *Boundary Element Method in Engineering Science*, McGraw-Hill, New York.

Barr, A.E., and Bear-Lehman, J. (2001). Biomechanics of the Wrist and Hand. In: Nordin, M., and Frankel, V.H. (Eds.). *Basic Biomechanics of Musculoskeletal System*. Lippincott Williams & Wilkins, Philadelphia.

Bartel, D.L., Davy, D.T., and Keaveny, T.K. (2006). Orthopaedic Biomechanics: Mechanics and Design. In: *Musculoskeletal System*. Pearson-Prentice Hall, Upper Saddle River, NJ.

Bartoš, M. (1998). *Employment of the Mathematical Model for Judgement of Some Arthoplastic Hip Operation*. Habilitation, UVN, Prague.

Bathe, K.J. (1982). *Finite Element Procedures in Engineering Analysis*. Prentice-Hall, Englewood Cliffs, NJ.

Bathe, K.J. (1996). *Finite Element Procedures*. Prentice-Hall, Englewood Cliffs, NJ.

Beek, M., Koolstra, J., van Ruijven, L., and van Eijden, T. (2000). Three-Dimensional Finite Element Analysis of the Human Temporomandibular Joint Disc. *J. Biomech.*, 33, 307–316.

Beek, M., Koolstra, J., van Ruijven, L., and van Eijden, T. (2001a). Three-Dimensional Finite Element Analysis of the Cartilaginous Structures in the Human Temporomandibular Joint. *J. Dental Res.*, 80, 1913–1918.

Beek, M., Aarnts, J.K., Feilzer, A., and van Eijden, T. (2001b). Dynamical Properties of the Human Temporomandibular Joint Disc. *J. Dental Res.*, 80, 876–880.

Beek, M., Koolstra, J., and van Eijden, T. (2003). Human Temporomandibular Joint Disc Cartilage as a Poroelastic Material. *Clin. Biomech.*, 18, 69–76.

Beer, G., Smith, J., and Duenser, Ch. (2008). *The Boundary Element Method with Programming. For Engineers and Scientists*. Springer, Wien.

Bei, Y., and Fregly, B.J. (2004). Multibody Dynamic Simulation of Knee Contact Mechanics. *Med. Eng. Phys.*, 26(9), 777–789.

Bei, Y., Fregly, B.J., Sawyer, W.G., Banks, S.A., and Kim, N.H. (2004). The Relationship between Contact Pressure, Insert Thickness, and Mild Wear in Total Knee Replacements. *Comput. Model. Eng. Sci.*, 6(2), 145–152.

Bell, A.L., Brand, R.A., and Pedersen, D.R. (1989). Prediction of Hip Joint Centre Location from External Landmarks. *Human Move. Sci.*, 8, 3–16.

Bell, A.L., Pedersen, D.R., and Brand, R.A. (1990). A Comparison of the Accuracy of Several Hip Center Location Prediction Methods. *J. Biomech.*, 23, 617–621.

Belytschko, T., Liu, W.K., and Moran, B. (2000). *Nonlinear Finite Elements for Continua and Structures*. Wiley, Chichester.

Belytschko, T., Kulak, R.F., Schultz, A.B., and Galante, J.O. (1974). Finite Element Stress Analysis of an Intervertebral Disc. *J. Biomech.*, 7, 277–285.

Bergman, G., Deuretzbacher, G., Heller, M., Graichen, F., Rohlmann, A., Straus, J., and Duda, G.N. (2001). Hip Contact Forces and Gait Patterns from Routine Activities. *J. Biomech.*, 34(7), 859–871.

Besier, T.F., Beaupre, G.S., and Delp, S.L. (2004). Subject Specific Modeling to Estimate Patello-Femoral Joint Contact Stress. *Med. Sci. Sports Exert.*, 36(5) (Suppl.), S1–S2.

Besov, O.V., Il'in, V.P., and Nikol'skii, S.M. (1979). *Integral Representation of Functions and Embedding Theorems*. English Translation, Nauka, Moscow (in Russian).

Besov, O.V., Il'in, V.P., and Nikol'skii, S.M. (1979). English Translation, Vols I, II, Winston, Washington, DC; Wiley, New York.

Bevill, S.L., Bevill, G.R., Penmetsa, J.R., Petrella, A.J., and Rullkoetter, P.J. (2005). Finite Element Simulation of Early Creep and Wear on Total Hip Arthroplasty. *J. Biomech.*, 38, 2365–2375.

Beznoska, S., Čech, O., and Löbl, K. (1987). *Artificial Replacement of Human Joints*. SNTL, Prague (in Czech).

Birolli, M. (1977). G-Convergence for Elliptic Equations, Variational Inequalities and Quasi-variational Inequalities. *Rend. Sem. Mat. Fis., Milano*, 47, 269–328.

Black, J. (1988). *Orthopaedic: Biomaterials in Research and Practice*. Churchil Livingstone, New York.

Blanc, P., Gasser, L., and Rappay, J. (1995). Existence of a Stationary Model of Binary Alloy Solidification. *M^2AM, Model Math. Anal.*, Numer. 29.

Blankenvoort, L., Kuiper, J.H., Huiskes, R., and Grootenboer, H.J. (1991). Articular Contact in a Three-Dimensional Model of the Knee. *J. Biomech.*, 24, 1019–1031.

Bombelli, R. (1983). *Osteoarthritis of the Hip*, 2nd ed. Springer, Berlin.

Bonet, J., and Wood, R.D. (1997). *Nonlinear Continuum. Mechanics for Finite Element Analysis*. Cambridge University Press, New York.

Bonfield, W., and Datta, P.K. (1974). Young's Modulus of Compact Bone. *J. Biomech.*, 7, 147–149.

Boone, D.C., and Azen, S.P. (1979). Normal Range of Motion of Joints in Male Subjects. *J. Bone Jt. Surg.*, 61, 756–759.

Bottema, O., and Roth, B. (1979). *Theoretical Kinematics*. North-Holland, Amsterdam.

Bouchala, J., Dostál, Z., and Sadovská, M. (2008). Theoretically Supported Scalable BETI Method for Variational Inequalities. *Computing*, 82, 53–75.

Boutin, P. (1971). L'alumine et sou utilisation en chirurgie de la hande (tude experimentale). *Presse Med.*, 79, 14–19.

Bramble, J.H., and Thomeé, V. (1974). Discrete Time Galerkin Methods for a Parabolic Boundary Value Problem. *Ann. Mat. Pura Appl.*, 101, 115–152.

Bramble, J.H., and Zlámal, M. (1970). Triangular Elements in the Finite Method. *Math. Comput.*, 24, 809–820.

Brauer, G.M., Steiberger, D.R., and Stansbury, J.W. (1986). Dependence of Curing Time Peak Temperature and Mechanical Properties on the Composition of Bone Cement. *J. Biomed. Mater. Res.*, 20(6), 839–852.

Brdička, M. (1959). *Continuum Mechanics*, CSAV Publ. House, Prague.

Brebbia, C.A. (1978a). *The Boundary Element Method for Engineers*. Prentech Press, London.

Brebbia, C.A. (Ed.) (1978b). *Recent Advances in Boundary Element Methods*. McGraw-Hill, Southampton.

Brebbia, C.A. (Ed.) (1988). *Boundary Element X. Vol. 2, Stress Analysis*. CMP & Springer, Berlin, Heidelberg.

Brebbia, C.A., and Dominguez, J. (1989). *Boundary Elements—A Course for Engineers*. Computational Mechanics, McGraw-Hill, Southampton, NY.

Brebbia, C.A., Telles, J.C.F., and Wrobel, L.C. (1984). *Boundary Element Techniques. Theory and Applications in Engineering*. Springer, Berlin.

Brebbia, C.A., Wendland, W.L., and Kuhn, G. (Eds) (1978). *Boundary Elements IX. Vol. 2, Stress Analysis*. Springer, Berlin, Heidelberg.

Brekelmans, W.A.M., Poort, H.W. and Slooff, T.J.J.H. (1972). A New Method to Analyze the Mechanical Behaviour of Skeletal Parts. *Acta Orthop. Scand.*, 43, 301–317.

Breme, H.J., Barbosa, M.A., and Rocha, L.A. (1998). Adhesion to Ceramics. In: Helsen, J.A., and Brene, H.J. (Eds.). *Metals as Biomaterials*, pp. 219–264, Wiley, Chichester.

Brezzi, F., Hager, W.W., and Raviart, P.A. (1977). Error Estimates for the Finite Element Solution of Variational Inequalities. Part I. *Primal Theory. Numer. Math.*, 28, 431–443.

Britton, N.F. (2003). *Essential Mathematical Biology* (Springer Undergraduate Mathematics Series). Springer, Berlin.

Brooks, S.L., Brand, J.W., Gibbs, J., Hollender, L., Lurie, A.G., Omnell, K-A., Westesson, P-L., White, S.C. (1997). Imaging of the Temporomandibular Joint. A Position Paper of the American Academy of Oral and Maxillofacial Radiology. *Oral Surg Oral Med. Oral. Pathol. Oral. Radiol. Endod.*, 83(5), 609–18.

Brouse, P. (1988). *Optimization in Mechanics: Problems and Methods*. North-Holland, Amsterdam.

Brown, T.D., and Ferguson, A.B. (1980). Mechanical Property Distribution in the Cancellous Bone of the Human Proximal Femur. *Acta Orthop. Scand.*, 429–437.

Brunsson, S., Schmid, F., Schäafer, M., and Wohlmuth, B. (2007). A Fast and Robust Iterative Solver for Nonlinear Contact Problems Using a Primal-Dual Active Set Strategy and Algebraic Multigrid. *Int. J. Numer. Math. Eng.*, 69, 524–543.

Burnstein, A.H., and Wright, T.M. (1994). Fundamentals of Orthopaedic Biomechanics. Williams & Wilkins, Philadelphia.

Burwell, J.T. (1958). Survey of Possible Wear Mechanisms. *Wear*, 1, 119–141.

Cappozo, A. (1984). Gait Analysis Methodology. *Human Movement Science*, 3, 27–50.

Carter, D.R., Vasu, R., and Harris, W.H. (1982). Stress Distributions in the Acetabular Region-II. Effects of Cement Thickness and Metal Backing of the Hip Acetabular Component. *J. Biomech.*, 15(3), 165–170.

Céa, J. (1971). *Optimisation, théorie et algorithms*. Dunod, Paris.

Chan, T. (2001a,b). A Review of Analytical Techniques for Gait Data. Part I. Fuzzy, Statistical and Fractal Methods. Part II. Neural Network and Wavelet Methods. *Gait Posture*, Part I: 13(1), 49–66; Part II: 13(2), 102–120.

Chan, S.H., and Tuba, I.S. (1971). A Finite Element Method for Contact Problems in Solid Bodies. *Int. J. Mech. Sci.*, 13, 615–639.

Chao, E.Y., and Morrey, B.F. (1978). Three Dimensional Rotation of the Elbow. *J. Biomech.*, 11, 57–73.

Chao, E.Y.S., Kasman, R.A., and Au, K.N. (1982). Rigidity and Stress Analyses of External Fracture Fixation Device. A Theoretical Approach. *J. Biomech.*, 15, 971–983.

Charit, Y. (1982). The Geometry of the Oblique Osteotomy. Research Report 87, Univ. of Witwatersrand, Johannesburg, South Africa.

Charit, Y. (1988). Applications of a 3D Geometrical Analysis to a Case of Orthopaedic Surgery. Research Report 88, Univ. of Witwatersrand, Johannesburg, South Africa.

Charnley, J. (1960). Anchorage of the Femoral Head Prosthesis of the Shaft of the Femur. *J. Bone Joint Surg. Br.*, 42-B, 28–30.

Charnley, J. (1961). Arthroplasty of the Hip: A New Operation. *Lancet*, 1129–1132.

Charnley, J. (1970). *Acrylic Cement in Orthopedic Surgery*. Livingstone, Edinburgh, London.

Charnley, J. (1979). *Low Friction Arthroplasty*. Springer, New York.

Charnley, J., and Halley, D.K. (1975). Rate Wear in Total Hip Replacement. Clin. Orthop., 112, 170–179.

Charnley, J., Kamangar, A., and Longfield, M.D. (1969). The Optimum Size of Prosthetic Heads in Relation to the Wear of Plastic Sockets in Total Replacement of the Hip. *Med. Biol. Eng.*, 7, 31–39.

Chau, O., Fernández, J.R., Han, W., and Sofonea, M. (2003a). Variational and Numerical Analysis of a Dynamic Frictionless Contact Problem with Adhesion. *J. Comput. Appl. Math.*, 156, 127–157.

Chau, O., Fernández, J.R., Shillor, M., and Sofonea, M. (2003b). Variational and Numerical Analysis of a Quadistatic Viscoelastic Contact Problem with Adhesion. *J. Comput. Appl. Math.*, 159, 431–465.

Chau, O., Shillor, M., and Sofonea, M. (2004). Dynamic Frictionless Contact with Adhesion. *J. Appl. Math. Phys.* (ZAMP), 55, 32–47.

Ciarlet, P.G. (1978). *The Finite Element Method for Elliptic Problems*. North-Holland, Amsterdam.

Ciarlet, P.G, (1988). *Mathematical Elasticity*. North-Holland, Amsterdam.

Ciarlet, P.G., and Nečas, J. (1985). Unilateral Problems in Non-Linear Three-Dimensional Elasticity. *Arch. Ration. Mech. Anal.*, 87, 319–338.

Cleary, K., Buersmayer, T., Wood, B.J., and Glossop, N. (2006). Computer Aided Surgery. *Wiley Encyclopedia of Biomedical Engineering*. Wiley, Chichester.

Cocu, M. (1984). Existence of Solutions of Signorini Problem with Friction. *Int. J. Eng.*, 22, 567–575.

Cocu, M., and Rocca, R. (2000). Existence of Results for Unilateral Contact Problems with Friction and Adhesion. *Math. Model. Numer. Anal.*, 34, 981–1001.

Cocu, M., and Scarella, G. (2006). Analysis of a Class of Dynamic Unilateral Contact Problems with Friction for Viscoelastic Bodies, In: Wriggers, P., and Nackenhorst, V. (Eds). *Analysis and Simulation of Contact Problems. Lecture Notes in Applied and Computational Mechanics*, Vol. 27, pp. 137–144. Springer, Berlin, Heidelberg.

Cocu, M., Cangemi, L., and Raous, M. (1999). Approximation Results for a Class of Quasistatic Contact Problems Including Adhesion and Friction. In: Argoul, P., Frémond, M., and Ngueyn Quoc, S. (Eds). *Proc. of IUTAM Symposium on the Variations of Domains and Free Boundary Problems in Solid Mechanics*. pp. 211–218. Kluwer, Dordrecht, The Netherland, Boston.

Cocu, M., Pratt, E., and Raous, M. (1995). Analysis of an Incremental Formulation for Frictional Contact Problems. In: Raous, M., Jean, M., and Moreau, J.J. (Eds). *Contact Mechanics*. pp. 13–20. Plenum, New York.

Cocu, M., Pratt, E., and Raous, M. (1996). Formulation and Approximation of Quasi-Static Frictional Contact. *Int. J. Eng. Sci.*, 34(t), 783–798.

Collins, D.H. (1949). *The Pathology of Articular and Spinal Diseases*. Arnold, London.

Cottrell, J.M., Townsend, E., Lipman, J., Sculco, T.P., and Wright, T.M. (2007). Bearing Surface Design Changes Affect Contact Patterns in Total Knee Arthroplasty. *Clin. Orthop. Relat. Res.*, 464, 127–131.

Covin, S.C. (Ed.) (2001). *Bone Mechanics Handbook*. CRC Press, Boca Raton, FL.

Crisfield, M.A. (1991). *Non-Linear Finite Element Analysis of Solids and Structures*. Vol. 1. Wiley, New York.

Cusick, J.F., and Yoganandan, N. (2002). Biomechanics of the Cervical Spine 4: Major Injuries. *Clin. Biomech.* 17(1), 1–20.

Čech, O., Stryhal, F., Sosna, A., Beznoska, S. (1982). *Stable Osteosynthesis in Traumatology and Orthopaedics*. Avicenum, Prague (in Czech).

Daněk, J. (2002). Domain Decomposition Algorithm for Solving Contact Elastic Bodies. In: Sloot, P.M.A. et al. (Eds). *ICCS'2002, Lecture Notes in Computer Science, 2331*, pp. 820–829. Springer, Berlin, Heidelberg.

Daněk, J., Denk, F., Hlaváček, I., Nedoma, J., Stehlík, J., and Vavřík, P. (2004). On the Stress-Strain Analysis of the Knee Replacement. In: *Lecture Notes in Computer Science*, 3044, pp. 456–466. Springer, Berlin, Heidelberg.

Daněk, J., Hliňáková, P., Přečková, P., Dostálová, T., Nedoma, J., and Nagy, M. (2010). *Modelling of the Temporomandibular Joints and the Role of Medical Informatics in Stomatology*. ICCSA'2010, Part IV, LNCS 6019, pp. 62–71. Springer, Berlin, Heidelberg.

Daněk, J., Hlaváček, I., and Nedoma, J. (2005a). Domain Decomposition for Generalized Unilateral Semi-Coercive Contact Problem with Friction in Elasticity. *Math. Comput. Simul.*, 68(3), 271–300.

Daněk, J., Stehlík, J., Vavřík, P., Nedoma, J., and Hlaváček, J. (2005b). On the Effect of Axial Changes on the Weight-Bearing Total Knee Replacements. In: *Proc. of the 17th IMACS Congress, Session on Mathematical Biomechanics*, Paris, 2005.

Dunford, N., and Schwartz, J.T. (1958). *Linear Operators. 1. General Theory*. Wiley–Interscience, New York, London.

Daniel, J.W. (1971). *The Approximate Minimization of Functionals*. Prentice-Hall, Englewood Cliffs, NJ.

De, S., Kim, J., and Srinisavan, (2001). A Meshless Numerical Technique for Physically Based Real-Time Medical Simulations. In: *Medicine Meets Virtual Reality*. ISO Press, Newport Beach, CA.

Debruner, H.V. (1975). Studien zur Biomechanik des Hüftgelenkes, I. *Z. Orthop.*, 377–388.

Delp, S.L., DiGioia III, A.M., and Jaramaz, B. (2000). *MIVVAI 2000: Third International Conference*, Springer, Pittsburgh, PA.

Delp, S.L., Stulberg, S.D., Davies, B., Picard, F., and Leitner, F. (1998). Computer Assisted Knee Replacement. *Clin. Orthop.*, 49–56.

Denis, F. (1983). The Three Column Spine and Its Significance in the Classification of Acute Thoracolumbar Spinal Injuries. *Spine*, 8, 817–831.

Denis, F. (1984). Spinal Instability as Defined by the Three-Column Spine Concept in Acute Spinal Trauma. *Clin. Orthop.*, 189(Oct.), 65–76.

Desai, J.P., Kresh, J.Y., Wechsler, A.S., Castellanos, A.E., and Mayers, W.C. (2006). Robotic Surgery. *Wiley Encyclopedia of Biomedical Engineering*. Wiley, Hoboken, NJ.

Dick, W. (1984). *Internal Fixation of Thoracic and Lumbar Spine Fractures*. Hans Huber, Toronto.

DICOM Reference Guide. (2001). Health Devices, 30(1–2), 5–30.

DiGioia, III, A.M., Blendea, S., and Jaramaz, B. (2004a). Computer-Assisted Orthopedic Surgery: Minimally Invasive Hip and Knee Reconstruction. *Orthop. Clin. North Am.*, 35(2), 183–189.

DiGioia III, A.M., Jaramaz, B., Picard, F., and Nolte, L.P. (Eds) (2004b). *Computer and Robotic Assisted Knee and Hip Surgery*. Oxford University Press, New York.

Dingwell, J.B., and Cusumano, J.P. (2000). Nonlinear Time Series Analysis of Normal and Pathological Human Walking. *Chaos*, 10(4), 848–863.

Dobryns, J.H. (1992). *Carpal Instability. A Review: Wrist Disorders*, pp. 239–246. Springer, Tokyo.

Donzelli, P.S., Gallo, L.M., Spilker, R.L., and Palla, S. (2004). Biphasic Finite Element Simulation of the TMJ Disc from in Vivo Kinematic and Geometric Measurements. *J. Biomech.*, 37, 1787–1791.

Douglas Jr. J., and Dupont, T. (1970). Galerkin Methods for Parabolic Equations. *SIAM J. Numer. Anal.*, 7, 575–626.

Dowson, D. (1979). *History of Tribology*. Longman, New York.

Dowson, D., and Wright, V. (1981). *An Introduction to the Biomechanics of Joints and Joint Replacement*. Mechanical Engineering, London.

D'Souza, N., Stachiewicz, J.W., Miller, J., Ahmed, A.M., and Burke, D.L. (1977). Thermal Analysis of Axisymmetric Bone Implants Using Self-Curing Polymethylmethacrylate. Paper No. 57, 3rd Am. Meeting of the Soc. for Biomaterials, New Orleans, Louisiana.

Dunn, H.K. (1976). Total Hip Reconstruction in Chronically Dislocated Hip. *J. Bone Jt. Surg.*, 58-A(6), 838–845.

Dunne, N.J., and Orr, J.F. (1998). Flow Characteristics of Curing Polymethalmethacrylate Bone Cement. *J. Eng. Med.*, 212, 199–207.

Dupont, T. (1982). Mesh Modification for Evolution Equations. *Math. Comput.*, 39, 85–107.

Duvaut, G. (1980). Equilibre d'un solide elastique avec contact unilateral et frottement de Coulomb. *C.R. Acad. Sci. Paris, Serie A*, 290, 263–265.

Duvaut, G., and Lions, J.L. (1976). *Inequalities in Mechanics and Physics*. Springer, and Berlin, Heidelberg, New York.

Dworkin, S.F., and LeResche, L. (1992). Research Diagnostic Criteria for Temporomandibular Disorders: Review, Examinations and Specifications Criteria. *J. Craniomand. Disord. Facial Oral Pain*, 6, 301–355.

Eck, M., Hoschek, L., and Weber, V. (1990). 3D Determination of an Oblique Osteotomy in the Hip Mathematical Optimization Fulfilling Some Anatomical Demands. *J. Biomech.*, 23, 1061–1067.

Eck, Ch., Steibach, D., and Wendland, W.L. (1999). A Symmetric Boundary Element Method for Contact Problems with Friction. *Math. Comput. Simul.*, 50, 43–61.

Eck, Ch., Jarušek, J., and Krbec, M. (2005). *Unilateral Contact Problems: Variational Methods and Existence Theorems*. Marcel Dekker, New York.

Ekeland, I., and Temam, R. (1976). *Convex Analysis and Variational Problems*. North-Holland, Amstedam.

Elliott, C.M. (1981). On the Finite Element Approximation of an Elliptic Variational Inequality Arising from an Implicit Time Discretization of the Stefan Problem. *IMA J. Numer. Anal.*, 1, 115–125.

Elliott, C.M., and Ockendon, J. (1982). *Weak and Variational Methods for Moving Boundary Problems*. Pitman, Boston.

Emshoff, R., Brandlmaier, I., Bosch, R., Gerhard, S., Rudisch, A., and Bertram, S. (2002). Validation of the Clinical Diagnostic Criteria for Temporomandibular Disorders for the Diagnostic Subgroup—Disc Derangement with Reduction. *J. Oral Rehabil.*, 29(12), 1139–1145.

Falk, R.S. (1974). Error Estimates for the Approximations of a Class of Variational Inequalities. *Math. Comp.*, 28, 963–971.

Feistauer, M., and Sobotíková, V. (1990). Finite Element Approximation of Non-linear Elliptic Problems with Discontinuous Coefficients. *M^2AN Math. Model. Numer. Anal.*, 24(4), 457–500.

Feldmann, D.G. (2005). The Use of Ceramic Materials for Fluid Power Components. *Mechanika*, 54(4), 60–64.

Fernandez, D.L. (1992). *Scaphoid Non-Union: Current Approach to Management. Wrist Disorder*, pp. 153–161. Springer, Berlin.

Fernandez, J.W., and Hunter, P.J. (2005). An Anatomically Based Patient-Specific Finite Element Model of Patella Articulation: Towards a Diagnostic Tool. *Biomech. Model. Mechanobiol.*, 4, 28–38.

Fernandez, J.W., and Pandy, M.G. (2006). Integrating Modelling and Experiments to Assess Dynamic Musculoskeletal Function in Humans. *Exp. Physiol.*, 91(2), 971–382.

Ferrario, V.F., Sforza, C., Serrao, G., Grassi, G.P., and Mossi, E. (2002). Active Range of Motion of the Head and Cervical Spine: A Threedimensional Investigation in Healthy Young Adults. *J. Orthop. Res.*, 20, 122–129.

Fischer, O. (1907). *Kinematik Organischer Gelenke*. Vierweg, Braumsweig.

Fleute, M., and Lavallée, S. (1998). Building a Complete Surface Model from Sparse Data Using Statistical Shape Models: Application to Computer Assisted Knee Surgery. In: Wells,

W.M., Colehester, A., and Delp, S. (Eds). *MICCAI'98*, LNCS 1496, pp. 879–887. Springer, Berlin.

Forner-Cordero, A., Koopman, H.J.F.M., and van der Helm, F.C.T. (2006). Describing Gait as a Sequence of States. *J. Biomech.*, 39, 948–957.

Franců, J. (1982). Homogenization of Linear Elasticity Equations. *Apl. Mat.*, 27, 96–117.

Fregly, B.J., Bei, Y., and Sylvestr, M.E. (2003). Experimental Evaluation of an Elastic Foundation Model to Predict Contact Pressures in Knee Replacements. *J. Biomech.*, 36(11), 1659–1668.

Fregly, B.J., Sawyer, W.G., Harman, M.K., and Banks, S.A. (2005). Computational Wear Prediction of a Total Knee Replacement from in Vivo Kinematics. *J. Biomech.*, 38, 305–314.

Fregly, B.J., Reinbolt, J.A., and Chmielewski, T.L. (2008). Evaluation of a Patient—Specific Cost Function to Predict the Influence of Foot Path on the Knee Adduction Torque During Gait. *Comput. Meth. Biomech. Biomed. Eng.*, 11(1), 63–71.

Frémond, M. (1982). Equilibre des structures qui adhèrant à lleur support. *C.R. Acad. Sci. Paris 295*, Série II, 913–916.

Frémond, M. (1987). Adhérence des solides. *J. Mécan. Théor. Appl.*, 6(3), 383–407.

Frémond, M. (1988a). Contact with Adhesion. In: Moreau, J.J., Panagiotopoulos, P.J., and Strang, G. (Eds). *Topics in Nonsmoothh Mechanics*. Birkhäuser, Bassel.

Frémond, M. (1988b). Contact with Adhesion. In: Moreau, J.J., Panagiotopoulos, P.D., and Strang, G. (Eds.). *Topics in Nonsmooth Mechanics*, pp. 177–221. Birkhäuser, Bassel.

Frémond, M. (2002). *Non-Smooth Thermomechanics*. Springer, Berlin.

Friedel, J. (1965). *High-Strength Materials*. Wiley, New York.

Friedman, A. (1982). *Variational Principles and Free Boundary Problems*. Wiley-Interscience, New York.

Fučík, S., and Kufner, A. (1980). *Nonlinear Differential Equations*. Elsevier, Amsterdam.

Fung, Y.C. (1965). *Generalized Hooke´s Law. Foundations of Solid Mechanics*. Prentice-Hall, Englewood Cliffs, NJ.

Fung, Y.C. (1968). Biomechanics (Its Scope, History and Some Problems of Continuum Mechanics in Physiology). *Appl. Mech. Rev.*, 21(1), 1–20.

Fung, Y.C. (1977). *A First Course in Continuum Mechanics*. Prentice-Hall, Englewood Cliffs, NJ.

Fung, Y.C. (1981). *Biomechanics: Mechanical Properties of Living Tissues*. Springer, New York.

Fung, Y.C. (1990). *Biomechanics: Motion, Flow, Stress and Growth*. Springer, New York.

Fung, Y.C. (1993). *Biomechanics: Mechanical Properties of Living Tissues*. Springer, New York.

Gal, J.A., Gallo, L.M., Palla, S., Murray, G., and Klineberg, I. (2004). *J. Biomech.*, 37, 1405–1412.

Gallo L.M., Airoldi G.B., Airoldi R.L., and Palla S. (1997). Description of Mandibular Finite Helical Axis Pathways in Asymptomatic Subjects. *J. Dent Res.*, 76, 704–713.

Garner, E., Lakes, R. Lee, T., Swan, C., and Brand, R. (2000). Viscoelastic Dissipation in Compact Bone: Implications for Stress-Induced Fluid Flow in Bone. *Trans. ASME*, 122, 166–172.

Gasser, L. (1995). Existence Analysis and Numerical Schemes for Models of Binary Alloy Solidification. EPPL Thesis, Lausanne.

Gatica, G.N., and Sayas, F.-J. (2006). An a Priori Error Analysis for the Coupling of Local Discontinuous Galerkin and Boundary Element Methods. *Math. Comp.*, 75, 1675–1696.

Geyer, H, Seyforth, A., and Blickhan, R. (2006). Compliant Leg Behaviour Explains Basic Dynamics of Walking and Running. *Proc. R. Soc. B.* DOI: 10.1098/rspb, 3637.

Giddings, V.L., Kurtz, S.M., and Edidin, A.A. (2001). Total Knee Replacement Polyethylene Stresses during Loading in a Knee Simulator. *J. Tribol.*, 123, 842–847.

Glowinski, R. (1979). Finite Element and Variational Inequalities. In: Whiteman, J.R. (Ed.) *The Mathematics of Finite Elements and Applications III*, pp. 135–171. Academic Press, New York.

Glowinski, R. (1984). *Numerical Methods for Nonlinear Variational Problems*. Springer, New York.

Glowinski, R. (2003). Finite Element Methods in Incompressible Viscous Flow. In: Ciarlet, P.G., and Lions, J.L. (Eds.). *Handbook of Numerical Analysis, Numerical Methods for Fluids* (Part 3), Vol. IX. Elsevier, Amsterdam.

Glowinski, R., Lions, J.-L., and Trémolières, R. (1976). *Analyse numérique des inéquations variationnelles. Parts I and II*. Dunod, Paris.

Glowinski, R., Lions, J.-L., and Trémoliéres (1981). *Numerical Analysis of Variational Inequalities*. North-Holland, Amsterdam.

Godest, A.C., Meaugonin, M., Hang, E., Taylor, M., and Gregson, P.J. (2002). Simulation of a Knee Joint Replacement During a Gait Cycle Using Explicit Finite Element Analysis. *J. Biomech.*, 35, 267–276.

Goldsmith, A.A.J., Dowson, D., Issac, G.H., and Lancaster, J.G. (2000). A Comparative Joint Simulator Study of the Wear of Metal-on-Metal and Alternative Material Combinations in Hip Replacements. Proc. Inst. Mech. Engrs, Part H, *J. Eng. Med.*, 214, H1, 39–47.

Goldstein, B. (1999). Temporomandibular Disorders: A Review of Current Understanding. *Oral. Surg. Oral Med. Oral Pathol. Oral Radiol. Endod.*, 88, 379–385.

Golub, G.H., and Van Loan, C.F. (1989). *Matrix Computations*, 2nd ed. Johns Hopkins University Press, Baltimore, MD.

Gopferich, A. (1996). Mechanism of Polymer Degradation and Erosion. *Biomaterials*, 16(1), 103–114.

Gray, H. (1995). The Muscles and Fascie. In: *Gray's Anatomy*, 38th ed. Churchill Livingstone, New York.

Guest, K.E., and Barnes, R. (1959). Epiphysical Retardation by Stapling. *J. Bone Jt. Surg.*, 41-B, 215.

Gunzburg, R., Parkinson, R., Moore, R. Cantraine, F., Hutton, W., Vernon–Roberts, B., and Fraser, R. (1992). A Cadaveric Study Comparing Discography, Magnetic Resonance Imaging, Histology, and Mechanical Behavior of the Human Lumbar Disc. *Spine*, 17(4), 417–428.

Hackbush, W. (1981). Multi-Grid Convergence Theory. In: Hackbush, W., and Trottenberg, U. (Eds.). *Multi-Grid Methods*, pp. 177–219. Springer, Berlin.

Halloran, J.P., Petrella, and A.J., and Rullkoetter, P.J. (2005). Explicit Finite Element Modeling of Total Knee Replacement Mechanics. *J. Biomech.*, 38, 323–331.

Hamill, J., and Knutzen, K.M. (2008). *Biomechanical Basis of Human Movement*. Lippincott Williams & Wilkins, Philadelphia.

Han, W., and Reddy, B.D. (1999). *Plasticity: Mathematical Theory and Numerical Analysis*. Springer, New York.

Han, W., and Sofonea, M. (2002). *Quasistatic Contact Problems in Viscoelasticity and Viscoplasticity*. AMS, Intern. Press, Providence, RI.

Han, W., Jensen, S., and Shimansky, I. (1997). The Kačanov Method for Some Nonlinear Problems. *Appl. Numer. Math.*, 24, 57–79.

Hannam, A.G. (1994). Musculoskeletal Biomechanics in the Human Jaw. In: Zarb, G.A., Carlsson, G.E., Sessle, B.L., and Mohl, N.D. (Eds.). *Temporomandibular Joint and Masticatory Muscle Disorders*, Chapter 4. Mosby, Munksgaard.

Hannam, A.G., and Langenbach, G.E.J. (1995). Modelling the Masticatory System during Function. In: Morimoto, T. Matsuya, T., and Takada, K. (Eds.). *Brain and Oral Functions. Oral Motor Function and Dysfunction*, pp. 217–226. Elsevier Science, Amsterdam.

Harkess, J.W. (1998). Arthroplasty of Hip. In: Canale, S.T. (Ed.). *Cambell's Operative Orthopaedic*, (9th ed.), Mosby, St. Louis.

Harris, W., Crothers, O., and Oh, I. (1977). Total Hip Replacement and Femoral–Head–Bone—Graphing for Severe Acetabular Deficiency in Adults. *J. Bone Jt. Surg.*, 59-A(6), 752–759.

Haslinger, J. (1991). *Shape Optimization in Unilateral Boundary Value Problems*. ISNM, Vol. 101, pp. 51–55. Birkhäuser, Basel.

Haslinger, J., and Hlaváček, I. (1980,1981). Contact between Elastic Bodies. *Apl. Mat.*, 25 (1980), 324–348; 26 (1981) 263–290, 321–344.

Haslinger, J., and Hlaváček, I. (1982). Approximation of the Signorini Problem with Friction by a Mixed Finite Element Method. *J. Math. Anal. Appl.*, 86, 99–122.

Haslinger, J., and Neittaanmäki, P. (1996). *Finite Element Approximation for Optimal Shape, Material and Topology Design*, 2nd ed. Wiley, Chichester.

Haslinger, J., and Tvrdý, M. (1983). Approximation and Numerical Realization of Contact Problems with Friction. *Apl. Mat.*, 28, 55–71.

Haslinger, J., Hlaváček, I., and Nečas, J. (1996). Numerical Methods for Unilateral Problems in Solid Mechanics. In: *Handbook of Numerical Analysis*, Vol. IV, pp. 313–486. Elsevier Science, Amsterdam.

Hatton, A., Ingham, E., Fischer, J., and Nevelos, A. (2000). Histological Analysis of Tissue from Revised Uncemented Ceramic-on-Ceramic Total Hip Arthroplastics. In: Proceedings of the 6th World Biomaterials Congress, Hawai, p. 1188.

Hausdorff, J.M., Mitchell, S.L., Firtion, R., Peng, C.K., Cuokowicz, M.E., Wei, J.Y., and Goldberger, A.L. (1997). Altered Fractal Dynamics of Gait: Reduced Stride-Interval Correlations with Aging and Huntington's Disease. *J. Appl. Physiol.*, 82(1), 262–289.

Hausdorff, J.M., Peng, C.K., Ladin, Z., Wei, J.Y., and Goldberger, A.L. (1995). Is Walking a Random Walk? Evidence for Long-Range Correlations in Stride Interval of Human Gait. *J. Appl. Physiol.*, 78(1), 349–358.

Head, W.W., Bank, D.J., and Emerson, R.H. (1995). Titanium as the Material of Choice for Cementless Femoral Components in Total Hip Arthroplasty. *Clin. Orthop. Rel. Res.*, 311, 85–90.

Heller, M.O., Taylor, W.R., Perka, C., and Duda, G.N. (2003). The Influence of Alignment on the Musculo-Skeletal Loading Conditions of the Knee. *Langenbecks Arch. Surg.*, 388, 291–297.

Helsen, J.A., and Breme, H. (Eds.) (1998). *Metals as Biomaterials*. Willey, Chichester.

Herr, H., and Willenfeld, A. (2003). User-Adaptive Control of Magneto-Rheological Prosthetic Knee. *Industrial Robot: Int. J.*, 30(1), 41–55.

Hertz, H. (1881). Uber die Beruhrung fester elastischer Korper. *J.F.Math.*, 92, 156–171.

Hintermüller, M., Ito, K., and Kunish, K. (2002). The Primal-Dual Active Set Strategy as a Semismooth Newton Method. *SIAM J. Optim.*, 13, 865–888.

Hintermüller, M, Ito, K., and Kunish, K. (2003). The Primal-Dual Active Set Strategy as Semi-Smooth Newton Method. *SIAM J. Optim.*, 13, 865–888.

Hintermüller, M, Kovtunenko, V.A., and Kunish, K. (2004). The Primal-Dual Active Set Method for a Crack Problem with Non-Penetration, *IMA J. Appl. Math.*, 69, 1–26.

Hintermüller, M, Kovtunenko, V.A., and Kunish, K. (2005). Generalized Newton Methods for Crack Problems with Nonpenetration Condition. *Numer. Meth. Part. Diff. Eqs.*, 21, 586–610.

Hippmann, R., Dostálová, T., Zvárová, J., Nagy, M., Seydlová, M., Hanzlíček, P., Kříž, P., Šmídl, L., and Trmal, J. (2009). Voice-Supported Electronic Health Record for Temporomandibular Joint Disorders. *Methods Inf. Med.*, 49(2), 168–172.

Hlaváček, I. (1986). Shape Optimization of Elasto-Plastic Bodies Obeying Hencky´s Law. *Appl. Math.*, 31, 486–499.

Hlaváček, I. (1989). Korn´s Inequality Uniform with Respect to a Class of Axisymmetric Bodies. *Appl. Math.*, 34, 146–154.

Hlaváček, I. (1994). Shape Optimization by Means of the Penalty Method with Extrapolation. *Appl. Math.*, 39, 449–477.

Hlaváček, I. (2003). Unilateral Contact with Coulomb Friction and Uncertain Input Data. *Funct. Anal. Optimiz.*, 24, 509–530.

Hlaváček, I. (2004). Mixed Finite Element Analysis of Semi-Coercive Unilateral Contact Problem with Given Friction. Tech. Report No. 921, ICS AS CR, Prague.

Hlaváček, I. (2007). Mixed Finite Element Analysis of Semi-Coercive Unilateral Contact Problems with Given Friction. *Appl. Math.*, 52, 25–58.

Hlaváček, I., and Lovíšek, J. (1977). A Finite Element Analysis for the Signorini's Problem in Plane Elastostatics. *Appl. Math.*, 22(3), 215–228.

Hlaváček, I., and Lovíšek, J. (1980). Finite Element Analysis of the Signorini Problem in Semi-Coercive Cases. *Appl. Math.*, 25, 274–285.

Hlaváček, I., and Nečas, J. (1970). On Inequalities of Korn's Type. *Arch. Rational. Mech. Anal.*, 36, 305–334.

Hlaváček, I., and Nedoma, J. (2002a). On a Solution of a Generalized Semi-Coecive Contact Problem in Thermo-Elasticity. *Math. Comput. Simul.*, 60, 1–17.

Hlaváček, I., and Nedoma, J. (2002b). Reliable Solution of a Unilateral Frictionless Contact Problem in Quasi-Coupled Thermoelasticity with Uncertain Input Data. In: Sloot, P.M.A., et al. (Eds.). Proc. of the Conference ICCS'2002, the Workshop on Numerical Models in Geomechanics, Amsterdam, April 2002. *Springer Lecture Notes in Computer Sciences LNCS*, 2331, 820–829.

Hlaváček, I., and Nedoma, J. (2004). Reliable Solution of a Unilateral Contact Problem with Friction and Uncertain Data in Thermo-Elasticity. *Math. Comput. Simul.*, 64, 559–580.

Hlaváček, I., Chleboun, J., and Babuška, I. (2004). *Uncertain Input Data Problems and the Worst Scenario Method.* Elsevier, Amsterdam.

Hlaváček, I., Haslinger, J., Nečas, J., and Lovíšek, J. (1988). *Solution of Variational Inequalities in Mechanics.* Springer, New York.

Hlaváček, I., Nedoma, J. and Daněk J. (2006). Worst Scenario and Domain Decomposition Methods in Geomechanics, Future Generation Computer Systems 22, 468–483.

Holdsworth, F. (1963). Fractures, Dislocations and Fracture-Dislocation of the Spine. *J. Bone Jt. Surg.*, 45-B, 6–20.

Holdsworth, F.W. (1970). Fractures, Dislocations and Fracture-Dislocation of the Spine. *J. Bone Jt. Surg.*, 52-A, 1534–1551.

Holm, R. (1946). *Electric Contacts*. Almquist and Wiksells. Stockholm.

Hsiao, G.C., and Wendland, W.L. (1977). A Finite Element Method for Some Integral Equation of the First Kind. *J. Math. Anal. Appl.*, 18, 449–481.

Hu, K., Qiguo, R., Fang, J., and Mao, J. (2003). Effects of Condylar Fibrocartilage on the Biomechanical Loading of the Human Temporomandibular Joint in a Three-Dimensional, Nonlinear Finite Element Model. *Med. Eng. Phys.*, 25, 107–113.

Hüeber, S., and Wolhmuth, B.I. (2005a). A Primal-Dual Active Set Strategy for Non-Linear Multibody Contact Problems. *Comput. Meth. Appl. Mech. Eng.*, 194, 3147–3166.

Hüeber, S., and Wolhmuth, B.I. (2005b). An Optimal a Priori Estimate for Non-Linear Multibody Contact Problems, *SIAM J. Numer. Anal.*, 43, 156–173.

Hüeber, S., Mair, M., and Wolhmuth, B.I. (2005). A Priori Estimates and an Inexact Primal-Dual Active Set Strategy for Linear and Quadratic Finite Elements Applied to Multibody Contact Problems. *Appl. Numer. Math.*, 54, 555–576.

Huggler, A.H., and Schreiber, A. (1978). *Alloarthroplastik des Hüftgelenkes*. Georg Thieme, Stuttgart.

Hughes, T.R. (1987). *The Finite Element Method: Linear Static and Dynamic Finite Element Analysis*. Prentice-Hall, Englewood Cliffs, NJ.

Huiskes, R. (1979). Thermal Tissue Damage by Acrylic Cements. 25th Meeting Ortop. Res. Soc., San Francisco, California.

Huiskes, R. (1980). Some Fundamental Aspects of Human Joint Replacement. *Acta Ortop. Scand. Suppl.*, No. 185, Munksgaard, Copenhagen.

Huiskes, R. (1987). Finite Element Analysis of Acetabular Reconstruction. *Acta Orthop. Scand.*, 58, 620–625.

Huiskes, R., and Chao, E.Y.S. (1983). A Survey of Finite Element Analysis in Orthopaedic Biomechanics. The First Decade. *J. Biomech.*, 6, 385–409.

Huiskes, R., and Sloof, T.J. (1987). Stress Transfer across the Hip Joint in Reconstructed Acetabuli. In: Bergmann, G., Kôlbel, R., and Rohlmann, A. (Eds.). Biomechanics. Basic and Applied Research, pp. 333–340, Martins Nijhoff, Dordrecht.

Hungerford, D.S. (1995). Alignment in Total Knee Replacement. *Instr. Course Lect.*, 44, 455–468.

Hvid, I., Christensen, P., Sandergaard, F., Christensen, P.B., Larsen, Ch.G. (1983). Compressive Strength of Tibial Cancellous Bone. *Acta Ortop. Scand.*, 54, 819–825.

Ichim, I., Swain, M.V., and Kieser, J.A. (2006). Mandibular Stiffness in Humans: Numerical Predictions. *J. Biomech.*, 39, 1903–1913.

Inman, V.T. (1976). *The Joints of the Ankle*. Williams & Wilkins, Baltimore, MD.

Ionescu, I.R., and Sofonea, M. (1993). *Functional and Numerical Methods in Viscoplasticity*. Oxford University Press, Oxford.

Jarm, T., Kramor, P., and Županič, A. (Eds.) (2007). *Medison 2007, IFMBE Proceedings 16* Springer, Berlin, Heidelberg.

Jarušek, J. (1983). Contact Problems with Bounded Friction. *Czech. Mat. J.* (1983, 1984), 33(1983), 2, 237–261, 34(1984) 109, 619–629.

Jarušek, J. (1996). Dynamic Contact Problems with Given Friction for Viscoelastic Bodies. *Czech. Math. J.*, 46(121), 475–487.

Jay, G.D., Torres, J.R., Rhee, D.K., Helminen, H.Y., Hytinnen, M.M., Cha, Ch.-J., Elsaid, K., Kim, K.-S., Cui, Y., and Warman, M.L. (2007). Association between Friction and Wear in Diarthrodial Joints Lacking Lubricin. *Arthritis Rheum.*, 56(11), 3662–3669.

Jazrawi, L.M., Rokito, A.S., Birdzell, M.G., and Zuckerman, J.D. (2001). Biomechanics of Elbow. In: Nordin, M., and Frankel, V.H. (Eds.). *Basic Biomechanics of Musculoskeletal System*. Lippicott Williams & Wilkins, Philadelphia.

Jefferis, C.D., Lee, A.J.C., and Ling, R.S.M. (1975). Thermal Aspects of Self-Curing Polymethylmathacrylate. *J. Bone Jt. Surg.*, 57B(4), 511–518.

Jeffrey, A. (1996). *Numerical Analysis. Mathematics for Engineers and Scientist*, 5th ed., pp. 744–811. Chapman and Hall, London.

Jenny, J.Y., Boeri, C. (2001). Computer-Assisted Implantation of Total Knee Prosthesis: A Case-Control Comparative Study with Classical Instrumentation. *Comput. Aided Surg.*, 6, 217–220.

Jensen, J.S. (1978). A Photoelastic Study of a Model of the Proximal Femur. A Biomechanical Study of Unstable Trochanteric Fractures. I. *Acta Orthop. Scand.*, 49, 54–59.

Jerome, J.W. (1977). Nonlinear Equations of Evolution and a Generalized Stefan Problem. *J. Diff. Eqs.*, 26, 240–261.

Jerome, J.W. (1983). *Approximation of Non-Linear Evolution Systems*. Academic Press, New York.

Jerome, J.W., and Rose, M.E. (1982). Error Estimates for Multi-Dimensional Two-Phase Stefan Problem. *Math. Comp.*, 39, 377–414.

Jinno, T., Goldberg, V.M., Davy, D., and Stavenson, S. (1998). Osseointegration of Surface-Blasted Implants Made of Titanium Alloy and Cobalt-Chromium Alloy in a Rabbit Intramedullary Model. *J. Biomed. Mater. Res.*, 42, 20–29.

John, P.S. (2005). Navigating the Future—Surgery Using Computer. *Indian J. Orthop.*, 39, 1–3.

Johnson, R.C. (1979). Reconstruction of the Hip. A Mathematical Approach to Determine Optimum Geometric Relationships. *J. Bone Jt. Surg.*, 61-A(5), 639–652.

Johnson, R.C., and Larson, C.B. (1964). Biomechanics of Cup Arthroplasty. *Clin. Orthop.*, 66, 56–69.

Johnson, R.C., Wenger, D.R., Roberts, J.M., Burke, S.W., and Reach, J.W. (1986). Acetabular Coverage: Three-Dimensional Anatomy and Radiographic Evaluation. *J. Ped. Orthop.*, 6, 548.

Jupiter, J.B. (1991). Fractures of the Distal End of the Radius. Current Concepts Review. *J. Bone Jt. Surg.*, 73-A, 461–469.

Kaasschietes, E.F., Frijus, A.J.H., and Huyghe, J.M. (2003). Mixed Finite Element Modelling of Cartilaginous Tissue. *Math. Comput. Simul.*, 61, 549–560.

Kačur, J. (1985). *Method of Rothe in Evolution Equations, Vol. 80, Teubner Texte zur Mathematik*, Teubner, Leipzig.

Karas, V., Komárek, P., and Straus, J. (1985). The Magnitude and Direction of Reaction Forces Operating in the Hip Joint Walking. *Proc. Intern. Congress of Biomechanics*. Umeá University, Umeá.

Karas, V., Straus, J., and Otáhal, S. (1986). An Influence of the Geometrical Relationship of the Pelvis, the Hip Region and the Gait Speed on the Loading of the Hip Joint. Technical Report FTVS UK, Prague (in Czech).

Katsikadelis, J.T. (2002). *Boundary Elements. Theory and Applications*. Elsevier Science, Amsterdam.

Kauer, J.M.G. (1980). Functional Anatomy of the Wrist. *Clin Orthop.*, 149, 9–20.

Kestřánek, Z. (1995). Numerical Analysis of the Contact Problem. Comparison of Methods Finding the Approximate Solution. Technical Report No V-648, ICS AS CR, Prague.

Kestřánek, Z. (1999). Numerical Analysis of 3D Contact Problem of Signorini Type with Friction in Thermo-Elasticity, h-Version of Finite Element Approximation. PhD Thesis, FJFI CVUT, Prague (in Czech).

Kestřánek, Z., and Nedoma, J. (1996). The Conjugate Projected Gradient Method. Numerical Tests and Results, Tech. Report No 677, ICS AS CR, Prague.

Kestřánek, Z., and Nedoma, J. (1998). FEC—A Code for Contact Problems in Thermo-elasticity with Friction. Technical Report V-740, ICS AS CR, Prague.

Khan, A.S., and Huang, S. (1995). *Continuum Theory of Plasticity*. Wiley, New York.

Kikuchi, N., and Oden, J.T. (1988). *Contact Problems in Elasticity: A Study of Variational Inequalities and Finite Element Methods*. SIAM, Philadelphia, PA.

Kim, H.J., Fernandez, J.W., Akbarshahi, M., Walter, J.P., Fregly, B.J., and Pandy, M.G. (2009). *J. Orthop. Res.*, 106, Publ. on line in Wiley Inter Science, DOI: 10.1002/jor.20876.

Kirkwood, R.N., Culham, E.G., and Costigan, P. (1999). Radiographic and Non-Invasive Determination of the Hip Joint Center Location: Effect on Hip Joint Moments. *Clin. Biomech.*, 14, 227–234.

Kleiber, M., and Hien, T.D. (1992). *The Stochastic Finite Element Method*. Wiley, Chichester.

Kolář, V., Kratochvil, J., Leitner, F., and Ženíšek, A. (1979). *Calculation of Surface and Space Construction by Finite Element Method*. SNTL, Prague (in Czech).

Koolstra, J.H., and van Eijden, T.M. (1997). The Jaw Open-Close Movements Predicted by Biomechanical Modelling. *J. Biomech.*, 30, 943–950.

Koolstra, J.H., and van Eijden, T.M. (1999). Three-Dimensional Dynamical Capabilities of the Human Masticatory Muscles. *J. Biomech.*, 32, 145–152.

Koolstra, J.H., and van Eijden, T.M. (2004). Functional Signifucance of the Coupling between Head and Jaw Movements. *J. Biomech.*, 37, 1387–1392.

Koolstra, J.H., and van Eijden, T.M. (2005). Combined Finite-Element and Rigid-Body Analysis of Human Jaw Joint Dynamics. *J. Biomech.*, 38(12), 2431–2439.

Koolstra, J.H., Naeije, M., and van Eijden, T.M. (2001). The Three-Dimensional Active Envelope of Jaw Border Movement and Its Determinants, *J. Dent. Res.*, 80(10), 1908–1912.

Korioth, W., and Versluis, A. (1997). Modeling the Mechanical Behavior of the Jaws and Their Related Structures by Finite Element Analysis. *Crit. Rev. Oral Biol. Med.*, 8(1), 90–104.

Krejčí, R., Bartoš, M., Dvořák, J., Nedoma, J., and Stehlík, J. (1997). 2D and 3D Finite Element Pre- and Post-Processing in Orthopaedy. *Int. J. Med. Inform.*, 45, 83–89.

Krüger, F.J. (1951). A Vitalium Replica Artroplasty on the Shoulder; a Case Report of Aseptic Necrosis of the Proximal end of the Humerus. *Surgery*, 30, 1005–1011.

Křížek, M., and Naittaanmäki, P. (1990). *Finite Element Approximation of Variational Problems and Applications*. Longman, Harlow.

Kuhn, G. (1988). Boundary Element Technique in Elastostatics and Linear Fracture Mechanics. Theory and Engineering Applications, In: *CISM Courses Lectures*, Vol. 301, pp. 109–169. Springer, Vienna.

Kühn, K.D. (2000). *Bone Cements: Up-to-Date Comparison of Physical and Chemical Properties of Commercial Materials*. Springer, Berlin.

Kuiper, J.H. (1993). Numerical Optimization of Artificial Hip Joint Design. Ph.D. Thesis, University of Nihmegen, Nijmegen.

Kulak, R.F., Belytschko, T., and Schultz, A.B. (1976). Nonlinear Behaviour of the Human Intervertebral Disc under Axial Load. *J. Biomech.*, 9, 377–386.

Lambert, J.D. (1991). *Numerical Methods for Ordinary Differential Systems: The Initial Value Problem*. Wiley, Chichester.

Landells, R.F. (1957). *J. Bone Jt. Surg.*, 35-B, 643–649.

Langenbach, G.E.J., and Hannam, A.G. (1999). The Role of Passive Muscle Tensions in a Three-Dimensional Dynamic Model of the Human Jaw. *Arch. Oral Biol.*, 44(7), 557–573.

Langrana, N.A., Harten, R.D., Lin, D.C., Reiter, M.F., and Lee, C.K. (2002). Acute Thoracolumbar Burst Fractures: A New View of Loading Mechanisms. *Spine*, 27(5), 498–508.

Laursen, T.A. (2003). *Computational Contact and Impact Mechanics*. Springer, Berlin, Heidelberg.

Leader, J.K., Boston, J.R., Debski, R.E., and Rudy, T.E. (2003). Mandibular Kinematics Represented by a Non-Orthogonal Floating Axis Joint Coordinate System. *J. Biomech.*, 36, 275–281.

Leardini, A., O'Connor, J.J., Catani, F., and Giannini, S. (2000). The Role of the Passive Structures in the Mobility and Stability of the Human Ankle Joint: A Literature Review. *Food Anable Int.*, 21(7), 602–615.

Léon, E.P., Aliabadi, M.H., and Dominguez, O. (2008). Boundary Element Analysis for Primary and Secondary Creep Problem. *Rev. Mexi. Física*, 54(5), 341–348.

Le Tallec, P. (1994). Domain Decomposition Methods in Computational Mechanics. *Comput. Mech. Adv.*, 1, 121–220.

Lewis, G. (1998). Contact Stress at Articular Surfaces in Total Joint Replacements. Part I. Experimental Methods, *Bio-Med. Mat. Eng.*, 8, 91–110; Part II. *Anal. Numer. Meth.*, 259–278.

Lim, S.C., and Ashby, M.F. (1987). Wear Mechanism Maps. *Acta Metall.*, 35(1), 1–24.

Linz, P. (1985). *Analytical and Numerical Methods for Voltera Equations*. SIAM, Philadelphia.

Lions, J.L. (1969). *Quelques méthods de résolution des problémes aux limites non-linéaires*. Dunod, Paris.

Lions, J.L., and Magenes, R. (1972). Non-Homogenous Boundary Value Problems and Applications. Springer, Berlin, Heidelberg, New York [Vols. I–II (1972), Vol. III (1973)].

Louis, R. (1985). Spinal Stability as Defined by the Three-Column Spine Concepts. *Anat. Clin.*, 7, 33–42.

Luenberger, D.G. (1984). *Introduction to Linear and Non-Linear Programming*. Addison-Wesley, Reading, MA.

Lugli, T. (1978). Artificial Shoulder Joint by Péan (1893). *Clinical Orthopaedics & Related Research*, vol. 133, 215–218.

MacDonald, N., Cannings, C., Hoppensteadt, F.C., and Segel, L.A. (2008). *Delays in Biological Systems: Linear Stability Theory (Cambridge Studies in Mathematical Biology)*. Cambridge University Press, Cambridge, MA.

Magerl, F. (1984a). External Spinal Skeletal Fixation. In: Weber, B.G., and Magerl, F. (Eds.). *External Skeletal Fixation*. Springer, Berlin.

Magerl, F. (1984b). Stabilization of the Lower Thoracic and Lumbar Spine with External Skeletal Fixation. *Clin. Orthop.*, 189, 125–141.

Magerl, F., Aebi, S.D., Gertzbein, S.D., Harms, S.J., and Nazarian, S.A. (1994). Comprehensive Classification of Thoracic and Lumbar Inhuries. *Eur. Spine J.*, 3, 184–201.

Mangis, D. (1982). Adherence of Solids. In: Georges, J. (Ed.). *Microscopic Aspects of Adhesion and Lubrication*, pp. 221–252. Elsevier, Amsterdam.

Mann, R.A. (1993). Biomechanics of the Foot and Ankle. In: *Surgery of the Foot and Ankle*. pp. 3–43. Mosby, St. Louis.

Marsden, A.C., and Hughes, T.J.R. (1983). *Mathematical Foundations of Elasticity*. Prentice-Hall, Englewood Cliffs, N.J.

—— (1994). Dover Publ. Inc., New York.

Marsden, J.E., and Hughes, T.J.R. (1983, 1994). *Mathematical Foundations of Elasticity*. Prentice-Hall, Englewood Cliffs, NJ.

Martin, A.C., and Carrey, G.F. (1969). *Introduction to Finite Element Analysis*. McGraw-Hill, New York.

Martins, J.A.C., and Oden, J.T. (1985). Models and Computational Methods for Dynamical Friction Phenomena. *Comp. Math. Appl. Mech. Eng.*, 52, 527–631.

Maugis, D. (1990). Fracture Mechanics and Solid Adhesion. In: Charmet, J.C., Roux, S., and Guyon, E. (Eds.). *Disorder and Fracture*, pp. 187–218. Plenum Press, New York.

Mazzullo, S., Paolini, M., and Verdi, C. (1991). Numerical Simulation of Thermal Bone Necrosis during Cementation of Femoral Prosthesis. *J. Math. Biol.*, 29, 475–494.

McAfee, P.C., Yuan, H.A., Frederickson, B.E., and Lubicky, J.P. (1983). The Value of Computed Tomography in Thoracolumbar Fractures; An Analysis of One Hundred Consecutive Cases and a New Classification. *J. Bone Jt. Surg.*, 65-A, 461–473.

McMurray, T.P. (1939). Osteoarthritis of the Hip Joint. *J. Bone It. Surg.*, 21, 1–11.

Merloz, P. Tonetti, J., Pittet, L., Coulomb, M., Lavallée S., Troacqz, Y., Cinquin, P., and Sautot, P. (1998). Computer-Assisted Spine Surgery. *Comput. Aided Surg.*, 3(6), 297–305.

Middleton, J.C., and Tipton, A.J. (2000). Synthetic Biodegradable Polymers as Orthopedic Devices. *Biomaterials*, 21(23), 2335–2346.

Mikhlin, S.G. (1957). *Integral Equations*. Pergamon Press, Oxford, London.

Milka, Z. (1989). Numerical Solution of the Model Describing Geodynamic Processes in Medium with Periodic Structure. Thesis. Faculty of Mathematics and Physics of the Charles University, Prague (in Czech).

Moore, K.L. (1999). *Clinically Oriented Anatomy*. Lippicott Wiliams & Wilkins, Philadelphia.

Morasso, P.G., and Sanguineti, V. (2006). Human Motion Analysis. In: *Wiley Encyclopedia of Biomedical Engineering*. Wiley, Chichester.

Moreau, J.J. (1974). On Unilateral Constraints, Friction and Plasticity. In: Capriz, G., and Stampachia, G. (Eds.). *New Variational Techniques in Mathematical Physics*. Edizione Cremonese, Roma.

Morrey, B.F. (1993). *The Elbow and Its Disorders*, 2nd ed. W.B. Saunders, Philadelphia

Morrey, B.F. (1994). Biomechanics of the Elbow and Forearm. In: Delee, J.C., and Drez, D. (Eds.). *Orthopedic Sports Medicine*, Chapter 7, W.B. Saunders, Philadelphia.

Morrey, B.F., and An, K.N. (1990). Biomechanics of the Shoulder. In: Rockwood, C.A., and Matsen III, F.A. (Eds.). *The Shoulder*. W.B. Saunders, Philadelphia.

Morrey, B.F., Askew, L.J., An, L.J., and Chao, E.Y. (1981). A Biomechanical Study of Functional Elbow Motion. *J. Bone Jt. Surg.*, 63A, 872.

Morrison, J.B. (1970). The Mechanics of the Knee Joint in Relation to Normal Walking. *J. Biomech.*, 3, 51–61.

Mosco, V., and Strang, G. (1974). One-Sided Approximation and Variational Inequalities. *Bull. Am. Math. Soc.*, 80, 308–312.

Mow, Van C., and Hayes, W.C. (1997). *Basic Orthopaedic Biomechanics*, 2nd ed., Liippincott-Raven, Philadelphia.

Mow, Van C., and Huiskes, R. (2004). *Basic Orthopaedic Biomechanics and Mechanobiology*. Lippicott Williams & Wilkins, Baltimore, MD.

Murphy, S.B., Kijewski, P.K., and Scott, R.D. (1985). *Computer-Assisted Radiology*. Springer, Berlin.

Murphy, S.B., Kijewski, P.K., Simon, S.R., Griffin, P.P., Chandle, H.P., Reilly, D.T., Penenberg, B.L., and Lauoly, M.L. (1986). Computer-Aided Analysis, Simulation and Design in Orthopaedic Surgery. *Orthop. Clin. North. Am.*, 17, 637–649.

Murphy, S.B., Kijewski, P.K., Simon, S.R., and Millis, M.B. (1987). Simulation of Orthopaedic Reconstructive Surgery. In: Lemke, H.V. (Ed.), *Computer-Assisted Radiology*. pp. 411–415. Springer, Berlin.

Murray, W.M., Buchanann, T.S., and Delp, S.L. (2002). Scaling of Peak Moment Arms of Elbow Muscles with Upper Extremity Bone Dimensions. *J. Biomech.*, 35, 19–26.

Müller, M.E., Nazarin, S., Koch, P., and Scatzker, J. (1990). *The Compressive Classification of Fractures of Long Bones*. Springer, Bern.

Naeije, M. (2003). Measurement of Condylar Motion: A Plea for the Use of the Condylar Kinematic Centre. *J. Oral Rehabil.*, 30, 225–230.

Nagahara, K., Murata, S., Nakamura, S., and Tsuchiya, T. (1999). Displacement and Stress Distribution in the Temporomandibular Joint During Clenching. *Angle Orthodont.*, 69, 372.

Nagy, M., Hanzlíček, P., Přečková, P., Kolesa, P., Mišúr, J., Dioszegi, M., and Zvárová, J. (2008). Building Semantically Interoperable EHR Systems Using International Nomenclatures and Enterprise Programming Technique. In eHealth: *Combining Health Telematics, Telemedicine, Biomedical Engineering and Bioinformatics to the Edge*, pp. 105–110. IOS Press, Amsterdam.

Nečas, J. (1967). *Les méthodes directes en théorie des équations elliptiques*. Academia, Prague.

Nečas, J. (1983). *Introduction to the Theory of Non-Linear Elliptic Equations. Teubner-Texte zur Mathematik*, Vol. 52. Teubner, Leipzig.

Nečas, J., and Hlaváček, I. (1981). *Mathematical Theory of Elastic and Elasto-Plastic Bodies: An Introduction*. Elsevier, Amsterdam.

Nečas, J., and Hlaváček, I. (1982). Optimization of the Domain in Elliptic Unilateral Boundary Value Problems by Finite Element Method. *RAIRO Anal. Numér.*, 16, 351–373.

Nečas, J., and Hlaváček, I. (1983). Solution of Signorini's Contact Problems in Deformation Theory of Plasticity by Secant Modules Methods. *Appl. Math.*, 28(3), 199–214.

Nečas, J., Jarušek, J., and Haslinger, J. (1980). On the Solution of the Variational Inequality to the Signorini Problem with Small Friction. *Boll. Un. Mat. Ital.*, 17-B(5), 796–811.

Nedoma, J. (1976). *Finite Element Technique in Geophysics*. Geophys. Inst. CSAS, Prague.

Nedoma, J. (1983). On One Type of Signorini's Problem without Friction in Linear Thermo-Elasticity. *Appl. Math.*, 28(6), 393–407.

Nedoma, J. (1987). On the Signorini Problem with Friction in Linear Thermo-Elasticity: The Quasi-Coupled 2D-Case, *Appl. Math.*, 32(3), 186–199.

Nedoma, J. (1991). Static Stress-Strain Analysis of Human Joint and Their Artificial Substitutes. Research Report V-516, Institute of Computer Science AS CR, Prague.

Nedoma, J. (1993a). Finite Element Analysis in Nuclear Safety. Technical Report V-550. Institute of Computer Science AS CR.

Nedoma, J. (1993b). Mathematical Modelling in Biomechanics. Bone- and Vascular-Implant Systems. Habilit. Thesis. Inst. of Comput. Sci. AS CR, Prague.

Nedoma, J. (1994a). Finite Element Analysis of Contact Problem in Thermo-Elasticity. The Semi-Coercive Case. *J. Comput. Appl. Math.*, 50, 411–423.

Nedoma, J. (1994b). FEM Analysis of Artificial Substitutes of Human Hip Joints and Their Optimal Design. In: Whiteman J.R. (Ed.). *The Mathematics of Finite Elements and Applications*. Wiley, Chichester.

Nedoma, J. (1997a). Finite Element Approximation of a Coupled Contact Stefan-Like Problem Arising from the Time Discretization in Deformation Theory of Thermo-Plasticity. *J. Comput. Appl. Math.*, 82, 313–334.

Nedoma, J. (1997b). On a Coupled Stefan-Like Problem in Thermo-Viscoplastic Rheology. *J. Comput. Appl. Math.*, 84, 45–80.

Nedoma, J. (1998a). *Numerical Modelling in Applied Geodynamics*. Wiley, Chichester, New York.

Nedoma, J. (1998b). Contact Problem in Biomechanics and Iterative Solution Methods for Constrained Optimization. Theory. Technical Report No. V-756, Inst. of Computer Sci, AS CR, Prague.

Nedoma, J. (2000a). Dynamic Contact Problems with Friction in Elasticity and Thermoelasticity. In: Feistauer, M., Rannacher, R., and Kozel, K. (Eds.). *Proc. 4th Conference on Numerical Modelling in Continuum Mechanics*, pp. 256–266. Matfyzpress, Prague.

Nedoma, J. (2000b). Finite Element Analysis of Elastic (Seismic) Wave Propagation in Region with Unilateral Contacts, In: Arabnia, H.R. (Ed.), Proc. Conference PDPTA'2000, Las Vegas, NV.

Nedoma, J. (2003). Numerical Solution of a Stefan-Like Problem in Bingham Rheology. *Math. Comput Simul.*, 61, 271–281.

Nedoma, J. (2004). On the Solution of Contact Problems with Visco-Plastic Friction in the Bingham Rheology: An Application in Biomechanics. ICCSA'2004, *Lecture Notes in Computer Science* (3044), Springer, Berlin, Heidelberg.

Nedoma, J. (2005). Mathematical Models of Artificial Total Replacements of Joints. I. Dynamical Loading of THA and TKR, Mathematical 2D and 3D Models, TR 950, ICS AS CR, (in Czech).

Nedoma, J. (2006). On a Solvability of Contact Problems with Visco-plastic Friction in the Thermo-visco-plastic Bingham Rheology. *Future Gen. Comput. Syst.*, 22, 484–499.

Nedoma, J. (2010). Special Problems in Landslide Modelling. Mathematical and Computational Methods. In: Werner, E.D., and Friedmann, H.P. (Eds.). *Landslides: Causes, Types and Effects*. Nova Sci. Publ., New York, NY.

Nedoma, J., and Dvořák, J. (1995). On the FEM Solution of a Coupled Contact Two-Phase Stefan Problem in Thermo-Elasticity. Coercive Case. *J. Comput. Appl. Math.*, 63, 411–420.

Nedoma, J., and Hlaváček, I. (2002). Solution of a Semi-Coercive Contact Problem in a Non-Linear Thermo-Elastic Rheology. *Math. Comput. Simul.*, 60, 119–127.

Nedoma, J., and Stehlík, J. (1995). Mathematical Simulation of Osteotomy. Numerical Analysis and Results. *J. Comput. Appl. Math.*, 63, 421–438.

Nedoma, J., Stehlík, J., Bartoš, M., Denk, F., Džupa, V., Fousek, J., Hlaváček, I., Klézl, Z., and Květ, I. (2006). *Biomechanics of Human Skeleton and Artificial Replacements of Its Parts*. EUROMISE – Charles University, Karolinum Press, Prague (in Czech).

Nedoma, J., Bartoš, M., Hornátová, H., Kestřánek, Z., and Stehlík, J. (1999a). Numerical Analysis of the Loosened Total Hip Replacements THR. *Math. Comput. Simul.*, 50, 285–304.

Nedoma, J., Bartoš, M., Kestřánek, Z. Sen., Kestřánek, Z., Jr., and Stehlík, J. (1999b). Numerical Methods for Constrained Optimization in 2D and 3D Biomechanics. *Numer. Linear Algebra Appl.*, 6, 577–586.

Nedoma, J., Klezl, Z., Bartoš, M., Kestřánek, Z. Sen., and Kestřánek, Z., Jr. (2000). Numerical Modelling in Spinal Biomechanics. In: Proc. 16th IMACS World Congress, Lausanne, Switzerland.

Nedoma, J., Klezl, Z., Fousek, J., Kestřánek, Z., and Stehlík, J. (2003a). Numerical Simulation of Some Biomechanical Problems. *Math. Comput. Simul.*, 61, 283–295.

Nedoma, J., Hlaváček, I., Daněk, J., Vavřík, P., Stehlík, J., and Denk, F. (2003b). *Some Recent Result on a Domain Decomposition Method in Biomechanics of Human Joints*. LNCS I-III, Vols. 2667–2669., Springer, Berlin.

Neer, C.S. (1955). Artificial Replacement for the Humeral Head. *J. Bone Joint Surg. (Am)*, 37, 215–228.

Neer, C.S., Watson, K.C., and Stanton, F.J. (1982). Recent Experience on Total Shoulder Replacement. *J. Bone Joint Surg. (Am)*, 64, 319–337.

Neumann D.A., and Wong D.L. (2002). *Kinesiology of the Musculoskeletal System: Foundations for Physical Rehabilitation*. Elsevier Science, Mosby, St. Louis.

Niinouri, M. (1998). Mechanical Properties of Biomedical Titanium Alloys. *Mater. Sci. Eng.*, A243, 231–236.

Nofrini, L. Slomczykowski, M., Iacono, F., and Marcacci, M. (2004). Evaluation of Accuracy in Ankle Center Location for Tibial Mechanical Axis Identification. *J. Investig. Surg.*, 17, 23–29.

Nordin, M., and Frankel, V.H. (2001). Basic Biomechanics of the Musculoskeletal System. Lippicott Wiliams & Wilkins, Baltimore, MD.

Norman, J.E., and Bramley, P. (1990). *Textbook and Colour Atlas of Temporo-Mandibular Joint: Diseases, Disorders, Surgery*. Mosby, St. Louis, MO.

Oden, J.T., and Reddy, J.N. (1976). *An Introduction to the Mathematical Theory of Finite Elements*. Wiley, New York.

Onsager, L. (1931). Reciprocal Relations in Irreversible Processes. *Phys. Rev.*, 37, 405–427.

Ortega, J.M., and Rheinboldt, W.C. (1970). *Iterative Solution of Non-Linear Equations in Several Variables*. Academic Press, New York.

Özkaya, N., and Nordin, M. (1999). *Fundamentals of Biomechanics. Equilibrium, Motion and Deformation*. Springer, New York.

Palmar, A.K., and Werner, F.W. (1981). The Triangular Fibrocartilage Complex of the Wrist—Anatomy and Function. *J. Hand Surg.*, 6, 153.

Palmar, A.K., and Werner, F.W. (1984). Biomechanics of the Distal Radioulnar Joint. *Clin. Orthop.*, 187, 26.

Panagiotopoulos, P.D. (1975). A Nonlinear Programming Approach to the Unilateral Contact and Friction Boundary Value Problem. *Ingenieur-Archiv.*, 44, 421–432.

Panagiotopoulos, P.D. (1985). *Inequality Problems in Mechanics and Applications. Convex and Non-Convex Energy Functions*. Birkhäuser, Boston.

Panagiotopoulos, P.D. (1993). *Hemivariational Inequalities. Applications in Mechanics and Engineering*, Springer, Berlin.

Pandy, M.G. (2001). Computer Modelling and Simulation of Human Movements. *Ann. Rev. Biomech. Eng.*, 3, 245–273.

Parsons, J.R., and Ruff, A.W. (1973). Survey on Metallic Implant Materials. NTIS. NBSIR 73-420, 55pp., Washington, D.C.

Pauwels, F. (1973). *Atlas für Biomechanik der Gesurden und Kranken Hefte, Principen, Technik und Resultäte einer Kausalen Terapie*. Springer, Berlin.

Pauwels, F. (1976). *Biomechanics of the Normal and Diseased Hip: Theoretical Foundation, Technique and Results*. Springer, Berlin.

Péan, J.E. (1973). On Prosthetic Methods to Repair Bone Fragments. *Clinical Orthopaedics & Related Research*, vol. 94, 4–7.

Peck, Ch.C., and Hannam, A.G. (2006). Human Jaw and Muscle Modelling. *Arch. Oral Biol.*, 52, 300–304.

Peck, C.C., Murray, G.M., Johnson, C.W.L., and Klineberg, I.J. (1997). The Variability of Condylar Point Pathways in Open-Close Jaw Movements. *J. Prosthet. Dent.*, 77, 394–404.

Penrose, J.M.T., Holt, G.M., Blaugonin, M., Hose, D.R. (2002). Development of an Accurate Three-Dimensional Finite Element Knee Model. *Comput. Meth. Biomech. Biomed. Eng.*, 4(4), 291–300.

Périé, D., Hobartho, M.C. (1998). In vivo determination of contact areas and pressure of the femorotibial joint using non-linear finite element analysis. *Classical Biomechanics* 13, 394–402.

Pérez del Palomar, A., and Doblaré, M. (2006). Finite Element Analysis of the Temporomandibular Joint During Lateral Excursions of the Mandible. *J. Biomech.*, 39, 2153–2163.

Petrtýl, M., Ondrouch, A., and Milbauer, M. (1985). *Experimental Biomechanics of the Solid Phase of the Human Skeleton*. Akademia, Prague (in Czech).

Piazza, S.J., Okita, N., and Cavanagh, P.R. (2001). Accuracy of the Functional Method of Hip Joint Center Location: Effects of Limited Motion and Varied Implementation. *J. Biomech.*, 34, 967–973.

Picard, F., DiGioia III, A.M., Moody, J., and Jamaraz, B. (2003). Probe and Associated System and Method for Facilitating Planar Osteotomy during Arthroplasty. U.S. Patent. Application Number US20011000776497, Carregie Mellon University, Pittsburgh, PA.

Poitout, D.G. (Ed.) (2004). *Biomechanics and Biomematerials in Orthopedics*. Springer, New York.

Possart, W. (1998). Adhesion of Polymers. In: Helsen, J.A., and Brene, H.J. (Eds). *Metals as Biomaterials*, pp. 219–264. Wiley, Chichester.

Proctor, P., and Paul, J.P. (1982). Ankle Joint Biomechanics. *J. Biomech.*, 15, 627.

Pschenichny B.N., and Danilin, Yu. M. (1978). *Numerical Methods in Extremal Problems*. Mir, Moscow.

Quarteroni, A., and Valli, A. (1994). *Numerical Approximation of Partial Differential Equations*. Springer, Berlin.

Rabinowicz E. (1995). Friction and Wear of Materials. Wiley-Interscience, New York.

Rankin, J.S. (1926). The Elastic Range of Friction. *Phil. Mag.*, 7th Series, 806–816.

Raous, M. (1999). Quasistatic Signorini Problem with Coulomb Friction and Coupling Adhesion. In: Wriggers, P., and Panagiotopoulos, P. (Eds.). *CISM Courses and Lectures*, 384, pp. 101–178, Springer, Wien.

Raous, M. (2009). Personal communication.

Raous, M., and Monerie, Y. (2002). Unilateral Contact, Friction and Adhesion in Composite Materials: 3D Cracks in Composite Materials. In: Martins, J.A.C., and Monteiro Morques, M.D.P. (Eds.). *Collection Solid Mechanism and Its Application*, pp. 333–346. Kluwer, Dordrecht, The Netherlands.

Raous, M., Cangémi, L., and Cocu, M. (1999). A Consistent Model Coupling Adhesion, Friction, and Unilateral Contact. *Comput. Meth. Appl. Mech. Eng.*, 177, 383–399.

Raous, M., Jean, M., and Moreau, J.J. (Eds.). (1995). *Contact Mechanics*. Plenum Press, New York.

Raviart, P.A., and Thomas, J.M. (1983). *Introduction à l´analyse numérique des équations aux dérivées partielles*. MASSON, Paris.

Reggiani, B., Leardini, A., Corazza, F., and Taylor, M. (2006). Finite Element Analysis of a Total Ankle Replacement During the Stance Phase of Gait. *J. Biomech.*, 39, 1435–1443.

Reilly, D.T., and Burstein, A.H. (1975). The Elastic and Ultimate Properties of Compact Bone Tissue. *J. Biomech.*, 8, 393–405.

Reilly, D.T., Burstein, A.H., and Frankel, V.H. (1974). The Elastic Modulus for Bone. *J. Biomech.*, 7, 271–275.

Rektorys, K. (1983). *The Method of Discretization in Time and Partial Differential Equations*. Reidel, Dordrecht and Boston.

Rektorys, K. (Ed.) (1994). *Survey to Applied Mathematics*. Vol. 1 and 2, 2nd Revised Edition, Kluwer, Dordrecht, The Netherlands.

Reona, J.M., Garzía-Aznar, J.M., Domínguez, J., and Doblaré, M. (2007). Numerical Estimation of Bone Density and Elastic Constants Distribution in a Human Mandible. *J. Biomech.*, 40, 828–836.

Richtmayer, R.D., and Morton, K.W. (1967). *Difference Methods for Initial Value Problems*. Interscience, New York.

Ritter, M.A., Farris, P.M., Keating, E.M., and Meding, J.B. (1994). Post Operative Alignment of Total Knee Replacement, Its Effect on Survival. *Clin. Orthop.*, 299, 153–156.

Rodin, E.L. (1980). Biomechanics of the Human Hip. *Clin. Orthop.*, 152, 30–34.

Röhrle, O., and Pullan, A.J. (2007). Three-Dimensional Finite Element Modelling of Muscle Forces During Mastication. *J. Biomech.*, 40, 3363–3372.

Rojek, J., and Telega, J.J. (2001a,b). Contact Problems with Friction, Adhesion and Wear in Orthopaedic Biomechanics I: General Developments. II. (together with S. Stupkiewicz.). Numerical Implementation and Application to Implanted Knee Joints. *J. Theor. Appl. Mech.*, 39, (I) 655–677, (II) 679–706.

Rojek, J., Telega, J.J., and Bednarz, P. (1999). Adhesion in Hip and Knee Implantes: Modelling and Numerical Analysis. *Acta Bioeng. Biomech.*, 1(Suppl 1), 377–380.

Rojek, J., Telega, J.J., and Stupkiewicz, S. (2001). Contact Problems with Friction, Adhesion and Wear in Orthopaedic Biomechanics. Part II. Numerical Implementation and Application to Implanted Knee Joints. *J. Theor. Appl. Mech.*, 3(39), 679–706.

Rubin, C.T., Kenneth, J., and Steven, D.B. (1990). Functional Strains and Cortical Bone Adaptation: Epigenetic Assurance of Skeletal Integrity. *J. Biomech.*, 23, 1, 43–49.

Ruby, K.L. (1995). Carpal Instability. *J. Bone Jt. Surg.*, 77-A(3), 476–482.

Rudin, W. (1991). *Functional Analysis*, 2nd ed. McGraw-Hill, New York.

Rybka, V., and Vavřík, P. (Eds.) (1993). *Aloplastics of Knee Joint.* Arcadia, Prague (in Czech).

Rydell, N.W. (1966). Forces Acting on the Femoral Head-Prosthesis. *Acta Ortop. Scand. Suppl.*, 88, 1–132.

Sadat-Khonsari, R., Fenske, C., Kahl-Nieke, B., Kirsch, I., and Jude, H.D. (2003a). The Helical Axis of the Mandible during the Opening and Closing Movement of the Mouth. *J. Orofac. Orthop.*, 64, 178–185.

Sadat-Khonsari, R., Fenske, C., Kahl-Nieke, B., Kirsch, I., and Jude, H.D. (2003b). Mandibular Instantaneous Centers of Rotation in Patients with and without Temporomandibular Dysfunction. *J. Orofac. Orthop.*, 64, 256–264.

Salsich, G.B., and Perman, W.H. (2007). Patellofemoral Joint Contact Area Is Influenced by Tibiofemoral Rotation Alignment in Individuals Who Have Patellofemoral Pain. I. *Orthop. Sports Phys. Ther.*, 37(9), 521–528.

Sammarco, G.J., and Hockenburg, R.T. (2001). Biomechanics of the Foot and Ankle. In: Nordin, M., and Frankel, V.H. (Eds.). *Basic Biomechanics of the Musculoskeletal System.* Lippicott Wiliams & Wilkins, Philadelphia.

Sangeorzan, B.P., Judd, R.P., Sangeorzan, R.J. (1989). Mathematical Analysis of Single-Cut Osteotomy for Complex Long Bone Deformity. *J. Biomech.*, 22, 1271–1278.

Santavirta, S., Gristina, A., and Konttinen, Y.T. (1992). *Acta Ortop. Scand.*, 63, 225–232.

Schaldach, M., Hohmann, D. (Eds.) (1976). *Advances in Artificial Hip and Knee Joint Technology*. Springer, Heidelberg, New York.

Schatz, A.H., Thomeé, V., and Wahlbin, L.B. (1980). Maximum Norm Stability and Error Estimates in Parabolic Finite Element Equations. *Comm. Pure Appl. Math.*, 33, 265–304.

Schatzker, J. (1986). *The Intertrochanteric Osteotomy*. Springer, Berlin.

Schneider, R. (1979). *Die intertrochantere Osteotomie bei Coxarthrose*. Springer, Berlin.

Segal, L.A. (1977). An Introduction to Continuum Theory. In: DiPrima, R.C. (Ed.). *Modern Modelling of Continuum Phenomena*. AMS, Providence, RI.

Seidel, G.K., Marchinda, D.M., Dijkers, M., and Soutas-Little, R.W. (1995). Hip Joint Center Location from Palpable Bony Landmarks—A Cadaver Study. *J. Biomech.*, 28, 995–998.

Semlitsch, M. (1983). Metallic Implant Materials for Hip Joint Endoprotheness Designed for Cemented and Cementless Fixation. *MEP*, 12(4), 1–21.

Semlitsch, M. (1986). Klassische und neue Titanlegitrungen zur Herstellunh Künstlicher Hüftgelenke. Intern. Conference of Titanium Products and Applications. San Francisco.

Serbetci, K., and Hasirci, N. (2004). Recent Developments in Bone Cement. In: Yaszemski, et al. (Eds.). *Biomaterials in Orthopedics*. Marcel Dekker, New York.

Sfantos, G.K., and Aliabadi, M.H. (2006). Wear Simulation Using an Incremental Sliding Boundary Element Method. *Wear*, 260, 1119–1128.

Sfantos, G.K., and Aliabadi, M.H. (2007a). A Boundary Element Formulation for Three-Dimensional Sliding Wear Simulation. *Wear*, 262, 672–683.

Sfantos, G.K., and Aliabadi, M.H. (2007b). Total Hip Arthroplasty Wear Simulation Using the Boundary Element Method. *J. Biomech.*, 40, 378–389.

Sharma, A., Komistek, R.D., Scuderi, G.R., and Cates, Jr., H.E. (2007). High-Flexion TKA Designs: What Are Their in Vivo Contact Mechanics? *Clin. Ortop. Relat. Res.*, 464, 117–126.

Shaw, S., and Whiteman, J.R. (1997). Towards Robust Adaptive FEMs for Partial Differential Volterra Equation Problem Arising in Viscoelasticity Theory. In Whiteman, J.R. (Ed.). *The Mathematics of Finite Elements and Applications*, pp. 55–80. MAFELAP 1996, Wiley, Chichester.

Shaw, S., Warby, M.K., Whiteman, J.R., Daeoson, C., and Wheeler, M.F. (1994a). Numerical Techniques for Treatment of Quasi-Static Viscoelastic Stress Problems in Linear Isotropic Solids. *Comput. Methods Appl. Mech. Eng.*, 118, 211–237.

Shaw, S., Warby, M.K., and Whiteman, J.R., (1994b). An Error Bound via the Ritz-Volterra Projection for a Fully Discrete Approximation to a Hyperbolic Integrodifferential Equation. Technical Report, 94/3, BICON, Brunel University, Urbridge.

Shaw, S., Warby, M.K., and Whiteman, J.R., (1997). Error Estimates with Sharp Constants for a Fading Memory Volterra Problem in Linear Solid Viscoelasticity. *SIAM J. Numer. Anal.*, 34, 1237–1254.

Shi, P., and Shillor, M. (1992). Existence of a Solution to the *n*-Dimensional Problem of Thermo-Elastic Contact, *Comm. PDE*, 17, 1597–1618.

Shi, J.F., Wang, C.J., Laoui, T., Hart, W., and Hall, R. (2007). A Dynamic Model of Simulating Stress Distribution in the Distal Femur after Total Knee Replacement. *Proc. Inst. Mech. Eng. [H]*, 221(8), 903–912.

Shillor, M. (1998). Recent Advances in Contact Mechanics. Special Issue of *Math. Comput. Modelling*, 28, 4–8.

Shillor, M., Sofonea, M., and Telega, J.J. (2004). Models and Analysis of Quasi-Static Contact. *Lecture Notes in Physics*, 655, Springer, Berlin.

Showalter, R.E. (1997). *Monotone Operators in Banach Spaces and Nonlinear Partial Differential Equations*. American Mathematical Society, Providence, RI.

Signorini, A. (1933). *Sopra alcume questione di elastostatica*. Alli della Società Italiana per il Progresso della Scienze.

Simo, J.C., and Hughes, J.J.R. (1997). *Elastoplasticity and Viscoplasticity. Computational Aspects*. Springer, Berlin.

Siston, R.A., and Delp, S.L. (2006). Evaluation of a New Algorithm to Determine the Hip Joint Center. *J. Biomech.*, 39, 125–130.

Siston, R.A., Daub, A.C., Giori, N.J., Goodman, S.B., and Delp, S.L. (2005a). Evaluation of Methods That Locate the Center of the Ankle for Computer-Assisted Total Knee Arthroplasty. *Clin. Ortho. Related Res.*, 439, 129–135.

Signorini, A. (1933). Sopra alcume questioni di statica dei sistemi continui. Ann. Scuola Normale Pisa 2(2), 231–257.

Siston, R.A., Patel, J.J., Goodman, S.B., Delp, S.L., and Giori, N.J. (2005b). The Variability of Femoral Rotational Alignment in Total Knee Arthroplasty. *J. Bone Jt. Surg. — Am.*, 87, 2276–2280.

Siston, R.A., Giori, N.J., Goodman, S.B., and Delp, S.L. (2007). Surgical Navigation for Total Knee Arthroplasty: A Perspective. *J. Biomech.*, 40, 728–735.

Sofonea, M., Han, W., and Shillor, M. (2006). *Analysis and Approximation of Contact Problems with Adhesion or Damage*. Chapman & Hall/CRC, Boca Raton, FL.

Soong, T.T. (1973). *Random Differential Equations in Science and Engineering*. Academic Press, New York.

Sparmann, M., Wolke, B., Czupalla, H., Banzer, D., and Zin K.A. (2003). Positioning of Total Knee Arthroplasty with and without Navigation Support. A Prospective Randomised Study. *J. Bone Jt. Surg.*, 85B, 830–835.

Stańczyk, M., and Reitbergen, B. (2004). Thermal Analysis of Bone Cement Polymerization at the Cement-Bone Interface. *J. Biomech.*, 37, 1803–1810.

Stehlík, J. (1995). *Treatment of Fractures by External Fixation Poldi-7*. Scientia Medica, Prague (in Czech).

Stehlík, J., Bartoš, M., Kestřánek, Z., Nedoma, J., and Novický, M. (1997). Application of Numerical Modelling of Osteotomy to Orthopaedic Practice. *Inter. J. Med. Inform.*, 45, 75–82.

Stehlík, J., Tvrdek, M., Bartoniček, Y. (1992). The Technique of the Osteosynthesis in the Replantation of the Upper Limb. *Acta Chir. plast.*, 34, 241–248.

Stehlík, J., and Nedoma, J. (1989). Mathematical Simulation of the Function of Large Human Joints and Optimal Design of Their Artificial Replacements, I, II. Technical Report No. 406, 407, ICS AS CR, Prague (in Czech).

Stehlík, J., and Nedoma, J. (2006). The Proposition of the Total Knee Replacement with Rotary Polyethylene Inlay. Technical Report No. 957, ICS AS CR, Prague (in Czech).

Stehlík, J., Vavřík, P., Daněk, J., Nedoma, J., Hlaváček, I., and Denk, F. (2006). Analysis of Axial Angle Changes on the Weight-Bearing Total Knee Replacements. Technical Report No. 959, ICS AS CR, Prague.

Steinbach, O., and Wendland, W.L. (1998). The Construction of Some Efficient Preconditioners in the Boundary Element Methods. *Adv. Comput. Math.*, 9, 191–216.

Steinberg, B.D., and Plancher, K.D. (1995). Clinical Anatomy of the Wrist and Elbow. *Clin. Sports Med.*, 14(2), 299–313.

Stevens, R. (1986). *Zirconia and Zirconia Ceramics*. Magnesium Electron, Twickenham, UK.

Stiehl, J.B. , Konermann, W.H., Haaker, R.G., DiGioia III, A.M., Langloty, F., Zheng, G., and Nolte, L.P. (2007). *Navigation and MIS in Orthopedic Surgery*. Springer, Berlin, Heidelberg.

Strickland, J,W. (1987). Anatomy and Kinesiology of the Hand. In: Fess, E.E., and Philips, C.A. (Eds.). *Hand Spliting: Principles and Methods*, 2nd ed., pp. 3–41. Mosby, St. Louis.

Strömberg, N. (1997). Thermomechanical Modelling of Tribological Systems. Ph.D. Thesis, p. 497, Linköping University, Sweden.

Strömberg, N., Johansson, L., and Klarbring, A. (1995). Generalized Standard Model for Contact Friction and Wear. In: Raous, M., Jean, M., and Moreau, J.J. (Eds.), *Contact Mechanics*, Plenum Press, New York.

Strömberg, N., Johansson, L., and Klarbring, A. (1996). Derivation and Analysis of a Generalized Standard Model for Contact, Friction and Wear. *Int. J. Solids Struct.*, 33(13), 1817–1836.

Stuchin, S.A. (1992). Wrist Anatomy. *Hand Clinic* 8(4), 603–609.

Stulberg, S.D., Loan, P., and Sarin, V. (2002). Computer-Assisted Navigation in Total Knee Replacement: Results of an Initial Experience in Thirty-Five Patients. *J. Bone Jt. Surg.—Am.*, 84-A(Suppl 2), 90–98.

Suh, N.P. (1973). The Delamination Theory of Wear. *Wear*, 25, 111–124.

Suh, N.P. (1982). Surface Interactions. In: Senholzi, P. (Ed.). *Tribological Technology*, Proc. NATO Study Institute on Tribological Technology, Maratea, Italy, Mirtinus Nijhoff, Dordrecht, Hague, The Netherlands, Vol. 1, pp. 34–208.

Sulzer Medica Product Information. Computer Assisted Surgery System. Available at: http://www.sulzer.orthopedics.cz/asp/article58.asp.

Swanson, A.B. (1972). Flexible Implant Resection Arthroplasty. *Hand*, 4, 119–132.

Swanson, A., and de Groot, G. (1985). Flexible Implant Arthroplasty in the Upper Extremity. In: Tubiana, A.M. (Ed.). *The Hand*, Vol. 2, Saunders, Philadelphia.

Swenson, Jr., L.W., Schurman, D.J., and Piziali, R. (1976). Thermal Analysis of Total Hip Replacements Using PMMA Bone Cement. 22nd Ann. Meeting Ortop. Res. Soc., February, San Francisco.

Šolín, P. (2006). *Partial Differential Equations and the Finite Element Method*. Wiley-Interscience, Hoboken, NJ.

Šolín, P., Segeth, K., and Doležel, I. (2004). *Higher-Order Finite Element Methods*. Chapman & Hall/CRC, Boca Raton, FL.

Tabor, D. (1951). *The Hardness of Metals*. Oxford University Press, Clarendon Press, Oxford.

Taleisnik, J. (1985). *The Wrist*. Churchill Livingstone, New York.

Tanaka, E., del Pozo, R., Sugiyama, M., and Tanne, K. (2002). Biomechanical Response of Retrodiscal Tissue in the Temporomandibular Joint under Compression. *J. Oral Maxillofacial Surg.*, 60, 546–551.

Tanaka, E., del Pozo, R., Tanaka, M., Asai, D., Hirose, M., Iwabe, T., and Tanne, K. (2004). Three Dimensional Finite Element Analysis of Human Temporomandibular Joint with and without Disc Displacement During Jaw Opening. *Med. Eng. Phys.*, 26, 503–511.

Tanaka, E., Rodrigo, P., Tanaka, M., Hawaguchi, A., Shibazaji, T., and Tanne, K. (2001). Stress Analysis in the TMJ During Jaw Opening by Use of a Three Dimensional Finite Element Model Based on Magnetic Resonance Images. *Int. J. Oral Maxillofacial Surg.*, 30, 421–430.

Tashiro, Y., Miura, H., Matsuda, S., Okazaki, K., and Iwamoto, Y. (2007). Minimally Invasive versus Standard Approach in Total Knee Arthoroplasty. *Clin. Orthop. Relat. Res.*, 463, 144–150.

Tau, C.L. (1979). Three-Dimensional Boundary Integral Equation Stress Analysis of Cracked Components. Ph.D. Thesis, University of London.

Taylor, A.E. (1967). *Introduction to Functional Analysis*. Wiley, New York.

Taylor, W. Heller, M., Bergmann, G., Duda, G. (2006). Tibio-Femoral Loading During Human Gait and Stair Climbine. *J. of Orthopaedic Research*, 22, 625–632.

Temam, R, (1979). *Navier-Stokes Equations. Theory and Numerical Methods*. North-Holland, Amsterdam.

Tipper, J.L., Ingham, E., Hailey, J.L., Besong, A.A., Wroblewski, B.M., Stone, M.H., and Fisher, J. (2000). Quantitative Analysis of Polyethylene Wear Debris, Wear Rate, and Head Damage in Retrived Charnlay Hip Prostheses. *J. Mat. Sci: Mats Med.*, 11, 117–127.

Tipper, J.L., Mathews, J.B., Ingham, E., Stewart, T.D., Fisher, J., and Stone, M.H. (2003). Wear and Functional Biological Activity of Wear Debris Generated from UHMWPE-on-Zirconia Ceramic, Metal-on-Metal and Alumina Ceramic-on-Ceramic Hip Prostheses During Hip Simulator Testing. In: Hutchings, I.M. (Ed.). *Friction, Lubrication and Wear of Artificial Joints*. Professional Engineering Publishing, Bury St. Edmunds and London.

Thomée, V. (1984). *Galerkin Finite Element Methods for Parabolic Problems*. Springer, Berlin.

Tortora, G.J., and Anagnostakos, N.P. (1984). *Principles of Anatomy and Physiology*, 4th ed. Harper & Row, New York.

Trommsdorf, E. (1963). Polymerisate der Acrylsaüre, ihrer Holomoge und Derivate. In: Houwelink, R., and Staverman, A.J. (Eds.). *Chemie und Technologie der Kunststoffe*, Vol. II/1, pp. 541–599. Akad. Verlagsges, Leipzig.

Trueta, J. (1957). Consideration on the Pathology of Osteoarthritis of the Hip. Proc. of the VII Congress Intern. Orthop. and Traumat. Society, pp. 857–874.

Tscherne, H., and Gotzen, L. (Eds.). (1984). *Fractures with Soft Tissue Injuries*. Springer, Berlin.

Valenta, J. (Ed.) (1985). *Biomechanics*. Academia, Prague (in Czech).

Valenta, J. (Ed.) (1993). *Biomechanics*, Academia, Prague (English translation).

Valle, C.J.D., Rokito, A.S., Birdzell, M.G., and Zuckerman, J.D. (2001). Biomechanics of Shoulder. In: Nordin, M., and Frankel, V.H. (Eds.). *Basic Biomechanics of Musculoskeletal System*. Lippicott Williams & Wilkins, Philadelphia.

Varga, R.S. (1962). *Matrix Iterative Analysis*. Prentice-Hall, Englewood Cliffs, NJ.

Vasu, R., Carter, D.R., and Harris, W.H. (1982). Stress Distributions in Acetabular Region I. Before and after Total Joint Replacement. *J. Biomech.*, 15(3), 155–164.

Viceconti, M., Zannoni, C., Testi, D., and Cappello, A. (1999). CT Data Sets Surface Extraction for Biomechanical Modeling of Long Bones. *Comput. Meth. Prog. Biomed.*, 59, 159–166.

Visintin, A. (1996). *Models of Phase Transitions*. Birkhäuser, Boston.

Walkin, A., and Klaue, K. (1988). 3D and in Vivo Modelling and Evaluation of Hip Coverage. In: Bemann, J.V., and Herron, R.E. (Eds.). Proc. Biostereometrics 88, 5th Inter. Conf. Mtg., Basel.

Walker, P.S., Blunn, G.W., Broome, D.R., Perry, J., Watkins, A., Sathasivam, S., Dewar, M.E., and Paul, J.P. (1997). A Knee Simulating Machine for Performance Evaluation of Total Knee Replacements. *J. Biomech.*, 30, 83–89.

Washizu, K. (1968). Variational Methods in Elasticity and Plasticity. Pergamon Press (2nd ed. in 1975). Oxford-New York.

Waterhouse, R.B. (1984). Fretting Wear. *Wear*, 100, 111–124.

Weber, B.G., and Čech, O. (1973). *Pseudoarthrosen*. Huber, Bern.

Weber, B.G., and Magerl, F. (1984, 1985). *The External Fixator*. Springer, Berlin, New York.

Weinans, H. (1991) Mechanically Induced Bone Adaptations Arround Orthopaedic Implants. Thesis, University of Nijmegen, Nijmegen.

Wheeler, M.F. (1973). A Priori L_2-Error Estimates for Galerkin Approximations to Parabolic Partial Differential Equations. *SIAM J. Numer. Anal.*, 10, 723–759.

White, R.E. (1982a). An Enthalpy Formulation of the Stefan Problem. *SIAM J. Numer.*, 19, 1129–1157.

White, R.E. (1982b). A Numerical Solution of the Enthalpy Formulation of the Stefan Problem. *SIAM J. Numer. Anal.*, 19, 1158–1172.

White, A.A., and Panjabi, M.M. (1978). *Clinical Biomechanics of the Spine*. Lippicott, Philadelphia.

Whitesides, T.E. (1977). Traumatic Kyphosis of the Thoracolumbar Spine. *Clin. Ortop.*, 128, 78–92.

Wiley Encyclopedia of Biomechanical Engineering (2006). John Wiley & Sons, Chichester, New York.

Wilkens, K.J., Duong, L.V., McGarry, M.H., Kim, W.C., and Lee, T.Q. (2007). Biomechanical Effects of Kneeling after Total Knee Arthroplasty. *J. Bone Jt. Surg. Am.*, 89(12), 2745–2751.

Winter, D.A., Yack, H.J. (1987). EMG Profiles During Normal Human Walking: Stride to Stride and Inter-Subject Variability. *EEG and Clinical Neurophysiology*, 67, 402–411.

Winter, W., Heckmann, S.M., and Weber, H.P. (2004). A Time-Dependent Healing Function for Immediate Loaded Implants. *J. Biomech.*, 37, 1861–1867.

Witten, M. (1987). *Mathematical Models in Medicine: Diseases and Epidemics*, Vol. 1. Pergamon Press, New York.

Wolhmuth, B.I. (2000). A Mortar Finite Element Method Using Dual Spaces for the Lagrange Multiplier. *SIAM J. Numer. Anal.*, 38, 989–1012.

Wolhmuth, B.I., and Krause, R. (2003). Monotone Multigrid Methods on Nonmatching Grids for Non-Linear Multibody Contact Problems, *SIAM J. Sci. Comput.*, 25, 1, 324–347.

Wood, W.L. (1990). *Practical Time-Stepping Schemes*. Clarendon Press, Oxford.

Wriggers, P. (2002). *Computational Contact Mechanics*. Wiley, Chichester.

Wrobel, L.C., and Aliabadi, M.H. (2002). *The Boundary Element Method*. Wiley, New Jersey.

Wu, W.-L. (2004). Boundary Element Formulations for Fracture Mechanics Problem. Ph.D. Thesis, School of Mathematics and Applied Statistics, University of Wollongong, Australia. Available at: http://ro.now.edu.au/thesis/253.

Wu, G., and Cavanagh, P.R. (1995). ISB Recommendation for Standardization in the Reporting of Kinematic Data. *J. Biomech.*, 28, 1257–1261.

Wu, J.Z., Herzog, W., and Epstein, M. (1998). Effects of Inserting a Pressensor Film into Articular Joints on the Actual Contact Mechanics. *J. Biomech. Eng.*, 120, 655–659.

Xiao, M., and Higginson, J.S. (2008). Muscle Function May Depend on Model Selection in Forward Simulation of Normal Walking. *J. Biomech.*, 41(15), 3236–3242.

Yamada, H. (1970). *Strength of Biological Materials*. Wiliams & Wilkins, Baltimore, MD.

Yamazaki, F., Schinozuka, M., and Dasgupta, G. (1985). Neumann Expansion for Stochastic Finite Element Analysis. Technical Report, Dept. Civ. Eng., Columbia University.

Yamazaki, F., Schinozuka, M., and Dasgupta, G. (1988). Neumann Expansion for Stochastic Finite Element Analysis. J. Eng. Mech., 144(8), 1335–1354.

Yang, Y., Ong, J.L., and Bessho, K. (2004). Plasma-Sprayed Hydroxyapatite-Coated and Plasma-Sprayed Titanium-Coated Implants. In: Yaszemski, M.L., Trantolo, D.J., Lewandrowski, K.U., Hasirci, V., Altobelli, D.E., and Wise, H.L. (Eds.). *Biomaterials in Orthopedics*. Marcel Dekker, New York, Basel.

Yaszemski, M.J., Trantolo, D.J., Lewandrowski, K.U., Hasirci, V., Altobelli, D.E., and Wise, D.L. (Eds.) (2004). *Biomaterials in Orthopedics*. Marcel Deker, New York, Basel.

Yoon, T.R., Rowe, S.M., Jung, S.T., Seon, K.J., and Malonay, W.J. (1998). Osteolysis in Association with a Total Hip Arthoplasty with Ceramic Bearing Surfaces. *J. Bone Jt. Surg.*, 80-A, 1459–1468.

Yosida, K. (1974). *Functional Analysis*, Springer, New York, Heidelberg.

Young, D.M. (1971). *Iterative Solution of Large Linear Systems*. Academic Press, New York.

Yuehuei, H., and Draughn, R.A. (Eds.) (2000). *Mechanical Testing of Bone and the Bone-Implant Interface*. CRC Press, Boca Raton, FL.

Zajac, F.E., Neptune, R.R., and Kautz, S.A. (2002, 2003). Biomechanics and Muscle Coordination of Human Walking. Part I. Introduction to Concepts, Power Transfer, Dynamics and Simulations. *Gait Posture*, 2002, 16, 215–232; Part II. Lessons from Dynamical Simulations and Clinical Implications. *Gait Posture*, 2003, 17, 1–17.

Zarb, G.A., Carlsson, G.E., Sessle, J.B., and Mohl, N.D. (1996). Temporomandibular Joint and Mastication Muscular Disorders. *J. Oral Maxillofacial Surg.*, 54, 1201–1211.

Zarb, G.A., Carlsson, G.E. (1997). Temporomandibular joint function and dysfunction Jn: Boeva, A.D., Duskin, J.F. Heely, J.D., Helkimo, M.I. (Eds.). Functional disturbance of the temporo-manolibular joint. C.V. Mosby Company, St. Louis, Missouri, P.204.

Zeidler, E. (1990a). *Applied Functional Analysis: Applications to Mathematical Physics*, Vol. 108 of Appl. Math. Sci., Springer, New York.

Zeidler, E. (1990b). *Applied Functional Analysis: Main Principles and Their Applications*, Vol. 109 of Appl. Math. Sci., Springer, New York.

Ženíšek, A. (1990). *Non-Linear Elliptic and Evolution Problems and Their Finite Element Approximations*. Academic Press, London.

Zhong Z.-H. (1993). *Finite Element Procedures for Contact-Impact Problems*. Oxford University Press, Oxford.

Ziegler, H. (1958). An Ettempt to Generalize Onsager's Principle, and Its Significance for Rheological Problems. *ZAMP*, 9, 748–763.

Ziegler, H. (1963). Some Extremum Principle in Irreversible Thermodynamics with Application to Continuum Mechanics. *Prog. Solid Mech.*, 4, 93–193.

Zienkiewicz, O.C., and Taylor, R.L. (1989, 1991). *The Finite Element Method*, 4th ed. McGraw-Hill, New York and (2000) (5th ed.), Butterworth-Heinemann, Oxford.

Zienkiewicz, D.C., and Taylor, R.L. (2000). *The Finite Element Method*, 5th ed. Butterworth, Heinemann, Oxford.

Zhao, D., Sawyer, W.G., and Fregly, B.J. (2006). Computational Wear Prediction of UHMWPE in Knee Replacements, *J. ASTM Int.*, 3, 45–50.

Zlámal, M. (1974). Finite Element Methods for Parabolic Equation. *Comput. Math.*, 28, 393–404.

Zlámal, M. (1977). Finite Element Methods for Nonlinear Parabolic Equations. *RAIRO Anal. Numer.*, 11, 93–107.

Zlámal M. (1980). A Finite Element Solution of the Non-Linear Heat Equation. *RAIRO Anal. Numer.*, 14, 203–216.

Zmitrowicz, A. (1987). Thermodynamical Model of Contact, Friction and Wear. Part I. Governing Equations, II. Constitutive Equations for Materials and Linearized Theories,

III. Constitutive Equations for Friction, Wear and Frictional Heat. *Wear*, 114, 135–221.

Zrzavý, J. (1977). *Anatomy*. Avicenum, Prague (in Czech).

Zvárová, J., Dostálová, T., Hanzlíček, P., Teuberová, Z., Nagy, M., Seydlová, M., Eliášová, H., and Šimková, H. (2008). Electronic Health Record for Forensic Dentistry. *Methods Inf. Med.*, 47(1), 8–13.

Information from www pages of the following companies: Ticona, JJOsly, SCI, Johnson & Johnson–De Puy, Sulzer, Walter, Beznoska, M.I.L.

INDEX